JOHN S. KARLING

Chytridiomycetarum Iconographia

CHYTRIDIOMYCETARUM ICONOGRAPHIA

An Illustrated and Brief Descriptive Guide to the Chytridiomycetous Genera with a Supplement of the Hyphochytriomycetes

by

JOHN S. KARLING

Purdue University

1977 · J. CRAMER

In der A.R. Gantner Verlag Kommanditgesellschaft

FL-9490 VADUZ

MONTICELLO, N.Y. - LUBRECHT & CRAMER

Printed in Germany
by Strauss & Cramer GmbH, 6945 Hirschberg II
ISBN 3-7682-1111-8

TO THE MEMORY OF MY LATE DAUGHTER, SAYRE

PREFACE

This iconograph was begun several years ago primarily as a teaching aid to give students a pictorial view of the Chytridiomycetes and other zoosporic fungi and a better understanding of their morphology, development, and life cycles. The plates of drawings were the source of lantern slides for lecture purposes, and at the same time the original plates were hung on the walls of the laboratory to supplement the microscopic observations in progress. As additional data, drawings, and descriptions were published, the plates were altered and brought up to date. This, then was the original purpose of this iconograph. However, as new information on the Chytridiomycetes accumulated in the literature and general systematic accounts of these fungi were published, it became evident, to me at least, that such accounts were not adequately illustrated. I observed from teaching experiences that beginning students in mycology often had great difficulty in visualizing what a fungus looked like from the diagnoses and descriptions in such systematic accounts. Accordingly, a second purpose for publishing this iconograph became evident — to supplement existing general systematic accounts of the Chytridiomycetes and make the plates of drawings available to other mycologists as aids in the identification of genera and in teaching.

Since the chytridiomycetous species are so numerous, this iconograph is limited to the genera, but enough species of each genus are illustrated to give a fairly comprehensive picture of each group. Also, drawings of a single genus and species from several authors are included to indicate differences in observations and interpretations. All copies of their drawings have been made by me, and I alone am responsible for any inaccuracies.

In many cases, for lack of space, only a part of an author's drawing is included, and I apologize for this brevity. Unfortunately, in some cases a plate includes drawings of genera which appear to be unrelated, as in the case of *Olpidiomorpha, Physorhizophydium,* and *Sporophlyctidium*, of which there are only a few drawings in the literature. It is to be particularly noted that the dimensions of the planospores, zoosporangia, and resting spores given in the respective descriptions relate to the variations in sizes which they exhibit instead of to those of a single organ or planospore. A fairly full description is given of each drawing, but inasmuch as this may not give a comprehensive view of the genera, a brief descriptive account of each genus precedes the illustrations of it. A list of references or bibliography is given after each family. It is necessarily brief and includes only references relative to the drawings and a few others which concern debatable points of observation, identification, and classification.

ACKNOWLEDGMENTS

A large number of mycologists have contributed to the preparation of this iconograph, particularly by their permission to copy their drawings, and I am grateful to them for this courtesy. Acknowledgment is expressed to them in the descriptions of the respective plates of drawings. Special thanks go to Professor Donald P. Rogers for nomenclatural suggestions and to Professor W. J. Koch for use of drawings of *Dipolium* prior to their publication. Also, I am grateful to the following journals for the use of drawings and references:

Adv. Frontiers Pl. Sci.; Akad. Nauk. U.S.S.R.; Amer. J. Bot.; Amer. Naturalist; Ann. Bot.; Ann. Inst. Oceanogr.; Ann. Mycologici; Ann. Parasitol.; Ann. Protistol.; Ann. Sci. Nat. Bot.; Ann. Soc. Belge Micro.; Ann. Univ. Stellenbosh; Arch. Bot. Nord. France; Arch. Mikrobiol.; Arch. Protistenk.; Arch. Zool. Exp. et Gen.; A. V. Leeuwenhoek; Ber. deut. Bot. Gesel.; Ber. Schweiz Bot. Gesel.; Beitr. Biol. Pflanz.; Biochem.; Bot. Centralbl.; Bot. Gaz.; Bot. Mag. Tokyo; Bot. Tidsskr.; Bot. Zeit.; Bull. Acad. Roy. Belg.; Bull. Coll. Agric. Imp. Univ. Tokyo; Bull. Intern. Acad. Sci. Bohême; Bull. Nat. Sci. Mus. (Tokyo); Bull. Soc. Bot. France; Bull. Soc. Bot. Suisse; Bull. Soc. Mycol. France; Bull. Torrey Bot. Club; Canad. J. Bot.; Canad. J. Microbiol.; Canad. J. Res.; Canad. J. Zool.; Cell Biol.; Centralbl. Bakt. Parasitk. Inf.; Cesk. Parasitol.; Comp. Rendu Acad. Sci. Paris; C. R. Soc. Biol.; Dansk. Bot. Ark.; Experimenta; Farlowia; Flora; Folio Cryptogam.; Forsch. Chem. Org. Naturst.; Hedwigia; Indian Phytopath.; Jap. J. Bot.; Jahb. wiss. Bot.; J. Agric. Res.; J. Amer. Chem. Soc.; J. Bact.; J. Biophys. Biochem. Cytology; J. Cell Biol.; J. de Botanique; J. Elisha Mitchell Sci. Soc.; J. Exp. Bot.; J. Gen. Mikrobiol.; J. Linn. Soc. Bot. London; J. Morph.; J. Roy. Micro. Soc.; J. Wash. Acad. Sci.; Kulturpflanze; Kgl. Danske Videnskap. Selskabs. Forh.; Kgl. Vetern. O. Landboh.; Le Botaniste; Lloydia; Mem. l Herb. Boissier; Mem. Torrey Bot. Club; Michigan Acad. Sci. Arts, Letters; Mikol. Fitopatol.; Mycologia; Mycopath. et Mycol. Appl.; Nagaoa; Nature; New Phytol.; Nova Hedwigia; Oesterr. Bot. Zeitschr.; Parasitologica; Philos. Trans. Roy. Soc. London; Physiol. Plantarum; Phytopathology; Preslia; Proc. Amer. Philos. Soc.; Proc. Amer. Soc. Protozool.; Proc. Indiana Acad. Sci.; Proc. Imp. Acad. (Tokyo); Proc. Linn. Soc. NSW; Proc. Roy. Soc. Victoria; Protoplasma; Quart. Circ. Ceylon Rubber Res. Inst.; Quart. Rev. Biol.; Rev. Cytol. et Biol.; Rev. Soc. Cubana Bot.; Science; Sci. Repts. Nat. Tsing Univ.; Sci. Repts. Tokyo; Bunrika Daigaku; Scripta Bot. Horti, Univ. Imp. Petro.; Studia Bot. Cechica; Sydowia; Symb. Bot. Upsaliensis; The Naturalist; Trans. Brit. Mycol. Soc.; Trans. Illinois Acad. Sci.; Trans. Mycol. Soc. Japan; Torreya; Univ. S. Calif. Publ.; Zeitschr. f. Bot.; .Zeitschr. f. Hydrologie; Zeitschr. f. Vererbungsl.; and Zeitschr. f. Zellforsch.

The drawings and references from the Canad. J. Bot.; Canad. J. Zool.; Canad. J. Microbiol., and Canad. J. Res. are published by permission of the National Research Council of Canada.

TABLE OF CONTENTS

PREFACE .... p. vii.
ACKNOWLEDGMENTS .... p. viii.
INTRODUCTION .... p. 1.
The Planospore .... p. 1, pls. 1-2.
Systematics .... p. 3.

Chapter I

Family Olpidiaceae .... p. 13.
Olpidium .... p. 14, pls. 3-6.
Rozella .... p. 18, pl. 7.
Dictyomorpha .... p. 20, pl. 8.
Plasmophagus .... p. 22, pl. 9, figs. 1-5.
Dipolium .... p. 22, pl. 9, figs. 6-17.
Sphaerita .... p. 24, pl. 10, figs. 1-19.
Olpidiomorpha .... p. 26, pl. 10, figs. 20-23.
Genera of doubtful affinity .... p. 28.
Reesia .... p. 28, pl. 11.
Nucleophaga .... p. 32, pl. 12, figs. 1-24.
Blastulidium .... p. 32, pl. 12, figs. 25-36.
Morella .... p. 33, pl. 13, figs. 1-9.
Morella?, *Sphaerita*? .... p. 33, pl. 13, figs. 10-34.
Coelomycidium .... p. 33, pl. 14, figs. 1-20.
Johnkarlingia .... p. 36.
Endoblastidium .... p. 36, pl. 14, figs. 21-27.
Dermatocystidium .... p. 33.
Chytridiopsis .... p. 38, pl. 14, figs. 28-29.
Chytridioides .... p. 38, pl. 14, figs. 30-31.
Sagittospora .... p. 38.
Chytridhaema .... p. 38.
Literature references to the Olpidiaceae and genera of doubtful affinity .... p. 38.

Chapter II

Family Synchytriaceae .... p. 43.
Endodesmidium .... p. 46, pl. 15, figs. 1-5.
Micromycopsis .... p. 46, pl. 15, figs. 6-33.
Micromyces .... p. 48, pl. 15.
Synchytrium .... p. 50.
Subgenus *Microsynchytrium* .... p. 52, pl. 17.
Subgenus *Mesochytrium* .... p. 54, pl. 18, figs. 1-10.
Subgenus *Synchytrium* .... p. 54, pl. 18, figs. 11-42.
Subgenus *Exosynchytrium* .... p. 56, pl. 19.
Subgenus *Pycnochytrium* .... p. 56, pl. 20.
Subgenus *Woroninella* .... p. 56, pl. 21.
Literature references to the Synchytriaceae .... p. 57.

Chapter III

Family Achlygetonaceae .... p. 59.
Achlygeton .... p. 60, pl. 22, figs. 1-11.

Septolpidium . . . . . p. 60, pl. 22, figs. 12–16.
Bicricium . . . . . p. 62, pl. 22, figs. 17–19.
Myiophagus . . . . . p. 62, pl. 23.
Literature references to the Achlyogetonaceae . . . . . p. 62.

Chapter IV

Family Rhizideaceae . . . . . p. 63.
Subfamily Rhizidioideae . . . . . p. 63.
Rhizophydium . . . . . p. 64, pls. 24–29.
Septosperma . . . . . p. 78, pl. 30.
Loborhiza . . . . . p. 78, pl. 31.
Physorhizophidium . . . . . p. 80, pl. 32, figs. 1–10.
Podochytrium . . . . . p. 82, pl. 33.
Phlyctochytrium . . . . . p. 84, pls. 34–37.
Polyphlyctis . . . . . p. 92, pl. 38.
Dangeardia . . . . . p. 96, pls. 39–40, figs. 1–7.
Dangeardiana . . . . . p. 100, pl. 40, figs. 8–23.
Sporophlyctidium . . . . . p. 104, pl. 32, figs. 11–16.
Rhizidium . . . . . p. 104, pls. 41, 42.
Obelidium . . . . . p. 108, pl. 43.
Solutoparies . . . . . p. 108, pl. 44.
Nowakowskia . . . . . p. 108, pl. 45.
Rhizoclosmatium . . . . . p. 112, pl. 46.
Siphonaria . . . . . p. 114, pl. 47.
Asterophlyctis . . . . . p. 118, pl. 48.
Zygorhizidium . . . . . p. 122, pls. 49–51.
Chytriomyces . . . . . p. 128, pls. 52–54, figs. 1–12.
Sparrowia . . . . . p. 130, pl. 55.
Pseudopileum . . . . . p. 132, pl. 54, figs. 13–23.
Amphicypellus . . . . . p. 134, pl. 56.
Rhopalophlyctis . . . . . p. 136, pl. 57.
Rhizophlyctis . . . . . p. 140, pls. 58–61.
Karlingia . . . . . p. 148, pls. 62–64.
Allochytridium . . . . . p. 156, pl. 65.
Subfamily Chytridioideae . . . . . p. 160.
Chytridium . . . . . p. 160, pls. 66, 67.
Diplochytridium . . . . . p. 163, pls. 68, 69.
Blyttiomyces . . . . . p. 164, pls. 70, 71.
Catenochytridium . . . . . p. 166, pls. 72, 73.
Subfamily Polyphagoideae . . . . . p. 168.
Polyphagus . . . . . p. 170, pls. 74–76, figs. 43–49.
Arnaudovia . . . . . p. 172, pl. 76, figs. 50–57.
Sporophlyctis . . . . . p. 174, pl. 77.
Endocoenobium . . . . . p. 176, pl. 78.
Saccomyces . . . . . p. 176, pl. 79.
Literature references to the Rhizidiaceae and genera of doubtful affinity . . . . . pp. 178–187.

Chapter V

Family Entophlyctaceae . . . . . p. 189.
Subfamily Entophlyctoideae . . . . . p. 190.
Entophlyctis . . . . . p. 192, pls. 80–81.
Endochytrium . . . . . p. 196, pls. 82–84.

Cylindochytridium .... p. 198, pls. 85-86.
Phlyctorhiza .... p. 200, pl. 87.
Truittella .... p. 202, pl. 88.
Scherffeliomyces .... p. 204, pl. 89, figs. 1-19.
Coralliochytrium .... p. 204, pl. 89, figs. 20-38.
Scherffeliomycopsis .... p. 206, pl. 90.
Mitochytridium .... p 208, pl. 91.
Subfamily Diplophlyctoideae .... p. 208.
Diplophlyctis .... p. 210, pls. 92, 93.
Nephrochytrium .... p. 214, pls. 94, 95.
Rhizosiphon .... p. 218, pl. 96.

Chapter VI

Monocentric eucarpic genera of doubtful affinity .... p. 225.
Aphanistis .... p. 225, pl. 97, figs. 1-7.
Haplocystis .... p. 225, pl. 97, figs. 8-15.
Achlyella .... p. 226, pl. 97, figs. 16-18.
Mastigochytrium .... p. 226, pl. 97, figs. 19-28.
Macrochytrium .... p. 228, pl. 98.
Caulochytrium .... p. 230, pl. 99.
Canteria .... p. 232, pl. 100.
Literature references to the Entophlyctaceae .... pp. 222-223.

Chapter VII

Cladochytriaceae .... p. 237.
Cladochytrium .... p. 238, pls. 101-104.
Physocladia .... p. 240, pl. 104, figs. 102-108.
Nowakowskiella .... p. 242, pls. 105-108.
Amoebochytrium .... p. 245, pl. 109.
Polychytrium .... p. 245, pl. 110.
Septochytrium .... p. 250, pls. 111-112.
Megachytrium .... p. 252, pl. 113.
Coenomyces .... p. 254, pl. 114.
Literature references to the Cladochytriaceae .... p. 254.

Chapter VIII

Family Physodermataceae .... p. 261.
Physoderma .... p. 262, pls. 115-120.
Literature references to the Physodermataceae .... p. 274.
Polycentric genera of doubtful affinity .... p. 275.
Zygochytrium .... p. 276, pl. 121, figs. 1-20.
Tetrachytrium .... p. 278, pl. 121, figs. 21-40.
Saccopodium .... p. 278, pl. 122, figs. 1-3.
Chytrid sp. .... p. 278, pl. 122, figs. 4-24.
Literature references to genera of doubtful affinity .... p. 281.
Imperfectly known, doubtful, and excluded genera .... p. 280.
Rhizidiocystis .... p. 280.
Myceliochytrium .... p. 280.
Nephromyces .... p. 280.

Chapter IX

HARPOCHYTRIALES . . . . p. 283.
Family Harpochytriaceae . . . . p. 282.
Harpochytrium . . . . p. 284, pl. 123.
Oedogoniomyces . . . . p. 285, pl. 124.
Literature references to the Harpochytriales . . . . p. 286.

Chapter X

BLASTOCLADIALES . . . . p. 289.
Family Catenariaceae . . . . p. 290.
Catenophlyctis . . . . p. 292, pls. 125, 126.
Catenomyces . . . . p. 294, pl. 127.
Catenaria . . . . p. 298, pls. 128–130.
Family Coelomomycetaceae . . . . p. 303.
Coelomomyces . . . . p. 304, pls. 131, 132, 132A.
Family Blastocladiaceae . . . . p.312.
Blastocladiella . . . . p. 314, pls. 133–140.
Subgenus *Blastocladiella* . . . . p. 320, pls. 133–137.
Subgenus *Cystocladiella* . . . . p. 322, pl. 138.
Subgenus *Eucladiella* . . . . p. 322, pls. 139–140.
Allomyces . . . . p. 324, pls. 141–150.
Subgenus *Allomyces* . . . . p. 324, pls. 141–145.
Subgenus *Cytogenes* . . . . p. 338, pls. 146–148.
Subgenus *Brachyallomyces* . . . . p. 342, pls. 149–150, figs. 1–7.
Microallomyces . . . . p. 344, pl. 150, figs. 8–15.
Blastocladiopsis . . . . p. 344, pl. 151.
Blastocladia . . . . p. 346, pls. 152–155.
Imperfectly known and doubtful genera . . . . p. 352.
Ramocladia . . . . p. 350, pl. 156, figs. 1–5.
Allocladia . . . . p. 352, pl. 156, fig. 6.
Brevicladia . . . . p. 352, pl. 156, fig. 8.
Leptocladia . . . . p. 352, pl. 156, fig. 7.
Callimastix . . . . p. 352.
Literature references to the Blastocladiales . . . . p. 354.

Chapter XI

MONOBLEPHARIDALES . . . . p. 363.
Family Gonopodyaceae . . . . p. 364.
Gonopodya . . . . p. 366, pl. 157.
Monoblepharella . . . . p. 370, pls. 158, 159.
Family Monoblepharidaceae . . . . p. 376.
Monoblepharis . . . . p. 378, pls. 160–164.
Literature references to the Monoblepharidales . . . . p. 381.

Chapter XII

HYPHOCHYTRIOMYCETES . . . . p. 383.
Family Anisolpidiaceae . . . . p. 383.
Anisolpidium . . . . p. 384, pls. 166, 167.
Family Rhizidiomycetaceae . . . . p. 390.
Rhizidiomyces . . . . p. 392, pls. 168–171.

Latrostium . . . . p. 398, pl. 172.
Family Hyphochytriaceae . . . . p. 400.
Hyphochytrium . . . . p. 400, pls. 173–175.
Imperfectly known, doubtful, and excluded species . . . . . p. 402.
Reesia . . . . p. 402.
Chytridium mesocarpii . . . . p. 402.
Cystochytrium . . . . p. 404.
Catenariopsis . . . . p. 404.
References to the Hyphochytriomycetes . . . . p. 404.
AUTHOR INDEX . . . . p. 406.
SUBJECT INDEX . . . . p. 412.

Introduction

THE PLANOSPORE

Plates 1, 2

Inasmuch as the comparative structure of the planospore body and the number, site of insertion, relative lengths, and structure of the flagella are commonly interpreted as indicative of relationships among the zoosporic fungi, it is essential that special attention be devoted to such cells at the outset. Much emphasis is now being placed on the ultrastructure of the planospores, and although limited, such studies have revealed a fairly common pattern of structure, with some variations, throughout the Chytridiomycetes and Hyphochytriomycetes. The similarities and differences shown by light and electron microscopy will be presented first for the Chytridiales and followed by those of the other orders.

CHYTRIDIALES

Although the flagellum is posteriorly inserted and directed in the majority of the chytrids, it is to be noted at the outset that the early studies of Dangeard (1886), pl. 1, fig. 1, and later ones by Scherffel (1926), pl. 1, fig. 6, Ingold (1952), pl. 1, fig. 5, Rieth (1962), pl. 51, fig. 80; Canter (1963) pl. 1, fig. 2; Umphlett and Olson (1967), and Dogma (1969), pl. 84, fig. 52, showed that in some species the flagellum is anteriorly, subapically, or laterally, attached. In some species the flagellum may appear to be anterior when planospores elongate and swim on a thin agar film, in a mixture of 1% "Carrageen," "methocel," or a gum arabic solution (figs. 23, 24, 26), according to Gaertner (1964). He reported that they swim by an anteriorly inserted flagellum which bends around the body, becomes posteriorly directed, and drives the spore ahead. The present writer interprets such behavior to be a backward movement or swimming whereby the flagellum appears to be anteriorly inserted. When such elongate planospores are transferred back to water they resume a spherical shape and swim by the beating of a posteriorly attached flagellum. Koch (1968), also, showed that in a few species the flagellar attachment may vary from lateral to sub-basal and basal when planospores are swimming in water, "methocel," or on the surface of a thin agar plate.

Moreover, Zopf (1884), pl. 1, figs. 9–11; Serbinov (1907), figs. 7, 8; Sparrow (1938), and Geitler (1962), figs. 12–14, found that in some typically chytrid-like species endospores, or aplanospores, and amoebospores without a flagellum are formed. Also in all but a few species of *Sphaerita*, and in all identified species of *Nucleophaga* only non-flagellate endospores are produced. However, the author regards the species of the last genus as questionable chytrids on the basis of present knowledge. Thus, to characterize the Chytridiales as producing only planospores with a flagellum inserted at the posterior end is not altogether true, according to the reports in the literature. Also it is to be noted at this point that the chytrid planospore has been described as being monoplanetic, but in *Achlygeton* and *Achlyella* they encyst in clusters of cystospores at the exit orifice and later germinate to produce secondary swarmers, leaving the empty cysts behind. The same thing may occur with dispersed and encysted planospores of *Caulochytrium*.

As to shape, the planospores of most chytrids are spherical (figs. 15–17) to subspherical or ovoid with a large or small highly refractive globule which is usually colorless but may be red, orange, golden, or yellow in some species. In other species it may be lacking or is replaced by numerous minute refractive granules (figs. 16, 17). Other planospores may be oblong (fig. 19) or elongate, or ellipsoidal (fig. 18), and in 4 genera minute secondary planospores are produced from an encysted primary planospore (figs. 20–22). The type of movement of the flagellate planospores is quite characteristic in most species, and anyone who has studied chytrids can readily recognize their presence by this movement when they occur in a mixture of aquatic debris and other fungi. It is usually a free swimming erratic, or gliding, or darting movement with abrupt pauses and a change of direction. During the pauses the body of the planospore may elongate and become amoeboid, vacuolate, and creep about, after which it may round up and dart away again. In a few species the flagellate planospores never become actively motile but creep about in the surrounding medium.

So far relatively few of the numerous chytrids and harpochytrids have been studied with the electron microscope. Up to the end of 1974 these include *Olpidium brassicae* (Manton *et al*, 1952; Timmink and Campbell, 1969; Lesemann and Fuchs, 1970a,b), *Rozella allomycis* (Held, 1973), *Chytridium* sp., *Rhizophydium sphaerotheca* and *Rhizophlyctis* sp. (Koch, 1956), *Karlingia* (*Rhizophlyctis*) *rosea* (Chambers and Willoughby, 1964), *Rhizophydium sphaerotheca* (Fuller, 1966), *Nowakowskiella profusa* (Chambers *et al*, 1967), *Phlyctochytrium dichotomum* (Umphlett and Olson, 1967), *P. kniepii* and *P. punctatum* (Olson and Fuller, 1968), *P. irregulare* (McNitt, 1973, 1974), *Chytridium* sp. (Schnepf *et al*, 1971), *Phlyctochytrium* sp. (Kazama, 1972), *P. arcticum* (Chong and Barr, 1973), *Entophlyctis confervae-glomeratae* and *Rhizo-*

phydium patellarium (Chong and Barr, 1974), *Entophlyctis* sp. (Powell, 1974), *Harpochytrium hedenii* (Travland and Whisler, 1971; Whisler and Travland, 1973), and *Oedogoniomyces* (Reichle, 1972). These ultrastructural studies show that the planospores body and flagellum are very complex structures. The latter is composed of a sheath, 2 central and 9 peripheral fibrils (Fig. 34) and an attenuated tail piece or whiplash (figs. 29–31. This attenuation results from a reduction in the number of fibrils toward the end (fig. 31) according to Koch (1956). The flagellum is anchored to the spore body by the kinetosome, or "functional blepharoplast" (figs. 32, 33). The latter structure is a very complicated organelle by itself, according to Olson and Fuller (1968), and is composed of 9 sets of 3 microtubules, terminated by a plate, and joined with the cell surface by a system of props (figs. 34, 35). In addition a second, vestigial centriole, or "kinetosome" or "vestigial nonfunctional blepharoplast" has been reported in most species which have been studied ultrastructurally. It consists of a system of 9 doublets at one end and 9 triplets at the other end of its length. Also, a relatively long fibrillar rhizoplast (pl. 2, figs. 41–48) with a terminal grid-like enlargement connects the functional blepharoplast with the nucleus (Koch, 1956). The planospore body contains, in addition, a nucleus with an interrupted membrane, aggregations of ribosomes, mitochondria, lipoidal bodies, endoplasmic reticulum, dictyosomes, and microtubules. A membrane-bound nuclear cap appears to be lacking in most species studied so far, although Koch's diagrams (figs. 41–44, 46, 47, 49) would indicate its presence in several species. However, in *Phlyctochytrium articum* (Chong and Barr, 1973) *P. dichotomum* (Olson and Fuller, 1967), *Rhizophydium sphaerotheca* (Fuller, 1966), *Olpidium brassicae* (Temmink and Campbell, 1969), and *Entophlyctis confervae-glomeratae* (Chong and Barr, 1974), the ribosomes are scattered in the cytoplasm, but in *Phlyctochytrium* sp. (Kazama, 1972), *P. irregulare* (McNitt, 1973), and *Rhizophydium patellarium* (Chong and Barr, 1974) the ribosomes are aggregated in a partially membrane-bound area between the nucleus and mitochondria. Chambers *et al* deduced from studies of mature zoosporangia that a membrane-enveloped cap, invaginated at intervals by the mitochondria, is present (fig. 37) in *N. profusa*. They, also, found a "fibrous body" (fig. 37 fb) below the presumptive refractive globule which they believed might function as a photoreceptor. As shown in longitudinal and cross sections (figs. 38, 39) it is strikingly similar to the "rumposome" (pl. 159, figs. 34–36) which Fuller (1966) and Fuller and Reichle (1967) described in the planospore of *Monoblepharella* and more recently (1971) by Travland and Whisler in *Harpochytrium hedenii* and Reichle (1972) in *Oedogoniomyces*.

On the basis of their internal structure and the appearance and arrangement of the organelles Koch (1958, 1961) recognized 6 major types of planospores among 18 chytrid species which he studied by light microscopy. These types are shown in figs. 41–49 and vary by the presence (figs. 41–44, 46, 47, 49) or absence (fig. 45, 48) of a nuclear cap, thickness or massiveness and position of the cap relative to the nucleus, size of nucleus, the number and size of lipoidal bodies, angle of the rhizoplast relative to the nucleus, and the presence or absence of a vestigial or nonfunctional blepharoplast. However, these structural types have not been fully substantiated by subsequent ultrastructural studies, and it is quite likely that future electron microscope studies will reveal other variations.

HARPOCHYTRIALES

As observed by light microscopy, the planospores of this questionable order are spherical to obpyriform and ovoid in shape (pl. 123, fig. 40; pl. 124, figs. 10, 11, 20) with a posterior whiplash flagellum, and usually include several small globules and a nuclear cap. Ultrastructural studies of *Harpochytrium hedenii* Wille by Travland and Whisler (1971) show that the internal structure of the planospore body is generally similar to that of other chytrids but more so to that of the Monoblepharidales. It includes a functional and a vestigial kinetosome, numerous microtubules, several "spherical" mitochondria, dense granular bodies, a nucleus, a nuclear cap composed of ribosomes and bounded by a double membrane, endoplasmic reticulum, lipoidal bodies, and a rumposome— "all apparently integrated by striated rootlet material." The rumposome consists of a gridded plate which is connected with the striated rootlet by a short unbanded arm and defined by a membrane. Except for minor differences, Reichle (1972) found the same structures in planospores of *Oedogoniomyces*.

BLASTOCLADIALES AND MONOBLEPHARIDIALES

Both orders are characterized by posteriorly uniflagellate planospores borne in thin-walled zoosporangia, and there are no authentic reports in the literature of anteriorly, subapically, or laterally attached flagella as in the Chytridiales. As seen through the light microscope, the planospores of the Blastocladiales are basically similar in appearance, internal structure and type or motility. Some are spherical when they first emerge but usually become ovoid while motile and move in a relatively even manner while swimming. The cytoplasm may contain a few small anterior, hyaline or colored globules (pl. 136, figs. 87, 88), but in *Catenaria sphaerocarpa* (pl. 129, fig. 30) a large hyaline refractive globule is present as in many chytrids. Also, a dull-gleaming triangular or turbinate nuclear cap is visible near the center (Chong and Barr, 1974) and in some species a lateral lachrymose "side body" is present (pl. 133, figs. 1, 2). In addition a bright refractive granule or blepharoplast is usually visible at the point of insertion of the long whiplash flagellum.

The planospores of the Monoblepharidales are relatively large, predominantly obturbinate, and swim smoothly with a gliding motion. The small refractive globule or granules occupy the anterior portion (pl. 161, fig. 27) of the planospore, and may fuse into a single broadly cone-like body (pl. 161, fig. 27). A narrow strand connects the granules with the remainder of the planospore body. Immediately beneath the granules is an area devoid of granular material, and the remainder of the cytoplasm is finely granular and slightly refractive. Also, a highly refractive body is visible at the point of insertion of the flagellum.

Ultrastructurally, the planospores of these two orders have been studied more intensively and extensively, i.e., *Catenaria anguillulae, Coelomomyces psorophorae, C. punctatus, Blastocladiella emersonii, B. britannica, Blastocladia ramosa, Allomyces arbuscula, A. macrogynus, Allomyces* × *Javanicus, Allomyces* sp., and *Monoblepharella* sp., than those of the Chytridiales. Illustrations and descriptions of the organelles revealed by such studies will be presented in relation to the genera and species later, but mention is made here of the presence of 2 centrioles or kinetosomes (see Renaud and Swift, 1964) in the gametes of *A. arbuscula*, apparently similar to those described for chytrid and harpochytrid species, and the development of the flagellum in relation to the larger of these organelles (pl. 141, figs. 24–29). Also, it may be noted here that a single mitochondrion is present in *Coelomomyces psorophorae, C. punctatus, Blastocladiella emersonii* and *B. britannica*, while several are present in the other species studied. A membrane-bound nuclear cap is present in *C. psorophorae, C. punctatus, Blastocladiella emersonii, B. britannica, Blastocladia ramosa*, and *Allomyces* species (pls. 137, 143) but lacking in *Monoblepharella* sp. where the ribosomes lie free in the cytoplasm but are aggregated at the nucleus (pl. 159, fig. 34). Also, in the latter genus a conspicuous rumposome (pl. 159, figs. 34–36) is present.

HYPHOCHYTRIOMYCETES

The planospores of this group are anteriorly uniflagellate with lateral tinsels on the relatively short flagellum. As seen through the light microscope, they are predominantly ovoid to elongate in shape and contain a few to many refractive granules which gives them a greyish-granular appearance (pl. 166, figs. 12, 13). Their motion while swimming is quite characteristic and different from that of most chytrids and other zoosporic fungi. They seem to dart forward in a spiral fashion, stop, back up, and then go forward again, and under low magnifications planospores of these anisochytrids are readily recognized by their behavior. In *Rhizidiomyces apophysatus*, the only species studied so far by electron microscopy, two notable structures may be mentioned here, while the other organelles of the planospores will be illustrated and described later in relation to the species. A membrane-enveloped nuclear cap is lacking, and the ribosomes are aggregated in the cytoplasm at the nucleus. Also, 2 centrioles or kinetosomes are present (pl. 168, fig. 22) and connected by a fibril; one centriole becomes the basal body of the flagellum, and the other probably a vestigial blepharoplast.

The presence of the 2nd centriole or kinetosome in *R. apophysatus* as well as the "vestigial and nonfunctional blepharoplast" reported in *Allomyces arbuscula, A. macrogynus, Blastocladiella emersonii, B. britannica, Phlyctochytrium kniepii, P. punctatum, Chytridium* sp., *Rhizophydium sphaerotheca, Olpidium brassicae, Harpochytrium hedenii, Oedogoniomyces* and other species suggests, at least, that it might be a relic of a lost flagellum and that the uniflagellate planospore may have originated from an ancestor with biflagellate planospores. This concept was suggested much earlier by DeBary (1884) and Metz (1929) and more recently by Bessey (1942, 1950), Koch (1956), and Olson and Fuller (1968). Pertinent to this concept is the presence of 2 functional kinetosomes in the planospores of biflagellate species such as *Phytophthora parasitica*, etc., as shown by Reichle (1969) and other workers. Additional ultrastructural studies may reveal that a 2nd centriole or vestigial flagellum base is a common structure of the planospores and motile gametes of the Chytridiomycetes and Hyphochytriomycetes, although Martin (1971) failed to find one in *Coelomomyces punctatus*.

SYSTEMATICS

CHYTRIDIALES

This is the largest order of the Chytridiomycetes and as interpreted here includes 7 families, approximately 100 genera and nearly 1000 species, but many of the species are incompletely known and some of them will likely prove to be identical. At present, family distinctions are based primarily on thallus morphology, type of development, and organization. In these respects the families fall into broad categories of holocarpy, eucarpy, monocentricity and polycentricity. In the monocentric category some species are holocarpic without a vegetative absorbing system of rhizoids, while others are eucarpic with well-defined reproductive and vegetative portions. Among the holocarpic species, families are distinguished on the basis of whether the thallus develops into a single reproductive portion, or cleaves internally into a large number of them. Among the eucarpic monocentric species a familial distinction is usually made on whether the reproductive portion develops directly or indirectly from the planospore body, or from an enlargement in the germ tube. In the polycentric species several centers of reproduction occur along the length and at the tips of fine or coarse mycelioid filaments which may or may not bear rhizoids. In this category a familial separation is made on the absence or presence of an epi- endobiotic monocentric, eucarpic zoosporangial phase in addition to an endobiotic polycentric phase. These distinctions are indicated in the following key to the families and subfamilies.

Key to the Families and Subfamilies

A. Thallus monocentric, holocarpic, without rhizoids, or a vegetative absorbing system.
   1. Sexual reproduction, where known, by fusion of isoplanogametes, plasmogamy of thalli, or fusion of the contents of unequal gametangia.
      a. Initial cell enlarging to form one zoosporangium; zygote where known developing into a resting sporangium.
      Family **Olpidiaceae**
      b. Initial cell enlarging and eventually forming several zoosporangia.
         (1). Initial cell cleaving internally into several zoosporangia; sexual reproduction where known by fusion of isogametes; zygote where known developing into a resting sorus or prosorus.
         Family **Synchytriaceae**
         (2). Initial cell elongating and septating to form a linear series of zoosporangia; sexual reproduction unknown.
         Family **Achylogetonaceae**

B. Thallus monocentric, eucarpic; consisting of a zoosporangium or prosporangium, or resting spore with vegetative rhizoids or haustoria; zoosporangia operculate or inoperculate.
   1. Sexual reproduction where known by fusion of isogametes, fusion of the content of equal or unequal gametangia, or conjugation between male and female thalli.

Plate 1

Fig. 1. Four planospores of *Sphaerita dangeardii* Chatton and Broadsky with anteriorly inserted flaggelum. (Dangeard, 1933.)

Fig. 2. Two planospores of *Rhizophydium oblongum* Canter with anteriorly attached flagellum. (Canter, 1954.)

Fig. 3. Two anteriorly uniflagellate planospores of *Zygorhizidium vaucheriae* Rieth. (Rieth, 1967.)

Fig. 4. Two planospores (gametes?) of a *Nucleophaga*-like parasite in the nucleus of *Pseudospora volvocis* with a laterally attached flagellum. Possibly not a chytrid. (Robertson, 1905.)

Fig. 5. Two planospores of *Olpidium wildemanni* (Petersen) Karling with a laterally attached flagellum. (Ingold, 1952.)

Fig. 6. Two planospores of *Olpidiomorpha pseudosporae* with a laterally attached flagellum. (Scherffel, 1926.)

Fig. 7. Endogenous aflagellate planospores or aplanospores in a sporangium of *Sporophlyctis rostrata* Serbinov. (Serbinov, 1907.)

Fig. 8. Germination *in situ* of aplanospores. (Serbinov, 1907.)

Figs. 9–11. Aflagellate amoebospores of *Amoebochytrium rhizidioides* Zopf with a large plastic yellowish refractive globule which changes in shape as the spores move about. (Zopf, 1884.)

Figs. 12-14. Amoebospores of *Scherffeliomycopsis coleochaetes* Geitler with pseudopods and 2 contractile vacuoles. (Geitler, 1962.)

Fig. 15. Sketch of a spherical planospore with a refractive globule and posterior flagellum; representative of the majority of chytrids.

Figs. 16, 17. Spherical planospores of *Rhizophydium rarotonganensis* Karling and *R. condylosum* Karling, resp.; large refractive globule replaced by refractive granules. (Karling, 1968.)

Fig. 18. Ellipsoidal planospore of *Physoderma maculare* Wall. with a lateral refractive globule. (Clinton, 1902.)

Fig. 19. Oblong planospore of *Polyphagus euglenae* Nowakowski with a large basal refractive globule. (Nowakowski, 1876.)

Fig. 20. Primary planospore of *Micromycopsis fischeri* Scherffel. (Canter, 1949.)

Fig. 21. Encysted primary planospore of *M. fischeri* which will become a zoosporangium and form 4 secondary planospores. (Canter, 1949.)

Fig. 22. Minute secondary planospore (gamete?) of *M. fischeri.* (Canter, 1949.)

Figs. 23, 24, 26. Elongate planospores of *Phlyctochytrium* Sv. 43 Gaertner with an anteriorly (?) inserted flagellum swimming on a thin agar — 1% Carrageen film. (Gaertner, 1954.)

Fig. 25. Elongate planospore body of *Phlyctochytrium punctatum* "La 6b" Koch swimming in a liquid film on the surface of an agar plate culture; flagellum inserted at the anterior (?) end and posteriorly directed. (Koch, 1968.)

Fig. 27, 28. Planospores of *P. punctatum* swimming in 1% Methocel. (Koch, 1968.)

Figs. 29, 30. Planospores of *Rozella* sp. and *Rhizophydium carpophilum* resp., with tail pieces at the end of the flagellum. (Couch, 1941.)

Fig. 31. Sketch drawn from an electron micrograph of the tail piece or whiplash portion of the flagellum of *Chytridium* sp. showing a reduction in the number of fibrils toward the end. (Koch, 1956.)

Fig. 32. Sketch from an electron micrograph of *Chytridium* sp. showing vestigial blepharoplast (nfb), functional blepharoplast (fb), rhizoplast fiber (rf), and flagellum (ff). (Koch, 1956.)

Fig. 33. Sketch from an electron micrograph of a longitudinal section of the kinetosome (k) of *Phlyctochytrium punctatum* and the associated vestigial kinetosome (vk). (Olson and Fuller, 1968.)

Fig. 34. "A diagramic illustration of a longitudinal section of the chytrid kinetosome. Cross sections of various portions of the kinetosome and flagellum are illustrated to the sides of the kinetosome and flagellum." (Olson and Fuller, 1968.)

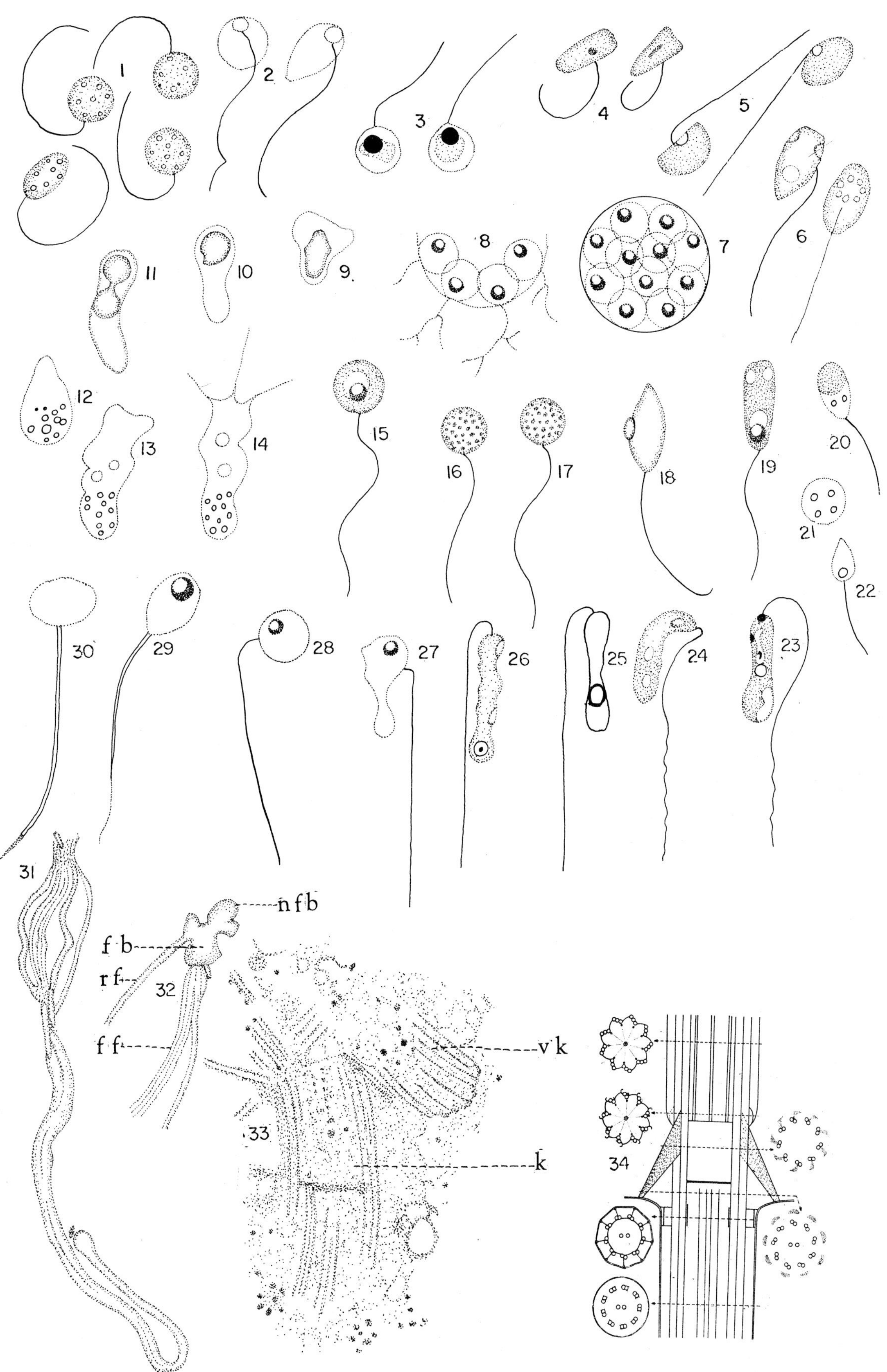

Planospores

a. Planospore cyst usually enlarging directly into a zoosporangium, or a prosporangium.
Family **Rhizidiaceae**

(1). Planospore cyst usually enlarging into zoosporangium or resting spore.
Subfamily **Rhizidioideae**

(2). Planospore cyst, or part of it, enlarging directly into a zoosporangium as in the Rhizidioideae, or development of it is delayed until the content of the absorbing system moves upward.
Subfamily **Chytridioideae**

(3). Planospore cyst enlarging into a prosporangium.
Subfamily **Polyphagoideae**

b. Planospore cyst not directly functional; zoorangium or prosporangium (apophysis) formed as an enlargement in the germ tube; sexual reproduction doubtful.
Family **Entophlyctaceae**

(1). Portion of germ tube enlarging into a zoosporangium.
Subfamily **Entophlyctoideae**

(2). Portion of germ tube enlarging into a subsporangial swelling or prosporangium (apophysis) which contributes to development of the zoosporangium.
Subfamily **Diplophlyctoideae**

C. Thallus wholly polycentric or polycentric in the dominant phase, with a monocentric eucarpic ephemeral phase.

1. Sexual reproduction unknown, or where known by fusion or isoplanogametes.

a. Thallus wholly polycentric, consisting of a rhizomycelium with tenuous filaments, rhizoids, intercalary swellings, zoosporangia and resting spores; zoosporangia operculate or inoperculate; sexual reproduction unknown.
Family **Cladochytriaceae**

b. Thallus monocentric and polycentric in different phases; life cycle consisting of a monocentric, epi-endobiotic eucarpic phase which alternates with an endobiotic polycentric resting sporangial phase; sexual reproduction by fusion of isoplanogametes.
Family **Physodermataceae**

The above key to the families and subfamilies follows that of Whiffen (1944) and Karling (1975) fairly closely and is artificial to a large degree because it involves mostly vegetative characters and types of development and organization. This is to be expected of any key at the present time because so many of the species are imperfectly known. It is difficult, if not almost impossible at present, to devise a workable key on the basis of type of sexual reproduction because it is unknown in most identified species. Furthermore, so far as is known there does not appear to be much correlation between type of sexual reproduction, morphology, and types of development and organization. Accordingly, species with different types of sexual reproduction are included or lumped together in various families listed above.

In this classification the families Phlyctidiaceae and Chytridiaceae are merged with the Rhizidiaceae, and the Megachytriaceae is merged with the Cladochytriaceae. The family name Phlyctidiaceae is illegitimate (Intern. Code, Edinb. ed., Art. 18, note 1) because it is based on the stem of the illegitimate genus *Phlyctidium*, which has not been conserved. Furthermore, the family has not been validly published by a Latin diagnosis, according to Cooke and

PLATE 2

Fig. 35. "A three-dimensional, diagramatic illustration of the kinetosome and the props which surround it." (Olson and Fuller, 1968.)

Fig. 36. Diagram of a longitudinal section of a planospore of *Olpidium brassicae* (Wor.) Dang. R = rhizoplast; A = axoneme; K = kinetosome; N = nucleus; AS = axoneme sheath; ER = endoplasmic reticulum; LB = lipoidal bodies; RV = rhizoplast vesicles; V = vacuole; M = mitochondria; ZE = ectoplast or membrane of planospore; MVB = multivesicular body. (Temmink and Campbell, 1969.) Copied from the Canad. J. Bot. 47, 1969, by permission of the Research Council of Canada.

Fig. 37. Diagram of the planospore body of *Nowakowskiella profusa* Karling. M = mitochondria partly embedded in the periphery of the nuclear cap, NC; GP = presumptive refractive globule; FB = fibrous body or rumposome; N = nucleus, G = dark granule. (Chambers et al., 1967.)

Fig. 38. Sketch from an electron micrograph of a longitudinal section of the fibrous body or rumposome. GP = presumptive refractive globule; FB = fibrous body.

Fig. 39. Diagram of a cross section of a portion of the fibrous body of *N. profusa*, based on an electron micrograph by Chambers *et al.*, 1967.

Fig. 40. Sketch from an electron micrograph of a cross section of the flagellum of *N. profusa*. (Chambers *et al.*, 1969.)

Figs. 41–49. Diagrams illustrating the internal structure of the 6 major types of chytrid planospores. (Koch, 1958, 1961.)

Fig. 41. *Nowakowskiella ramosa* Butler (Type I). LB = lipoidal body; NC = nuclear cap; N = nucleus; R = rhizoplast; Eg = extra-nuclear granule; FB = functional blepharoplast; F = flagellum; SG = side granule.

Fig. 42. *Catenochytridium carolineanum* Berdan (Type 1).

Fig. 43. *Septochytrium variabile* Berdan (Type 1). NFB = non-functional blepharoplast.

Fig. 44. *Rhizophydium sphaerotheca* (Type 1).

Fig. 45. *Entophlyctis* sp. (Type 2). Nuclear cap lacking.

Fig. 46. *Chytriomyces hyalinus* Karling (Type 3).

Fig. 47. *Phlyctochytrium irregulare* (Type 4).

Fig. 48. *Phlyctochytrium punctatum* (Type 5).

Fig. 49. *Rozella allomycis* (Type 6).

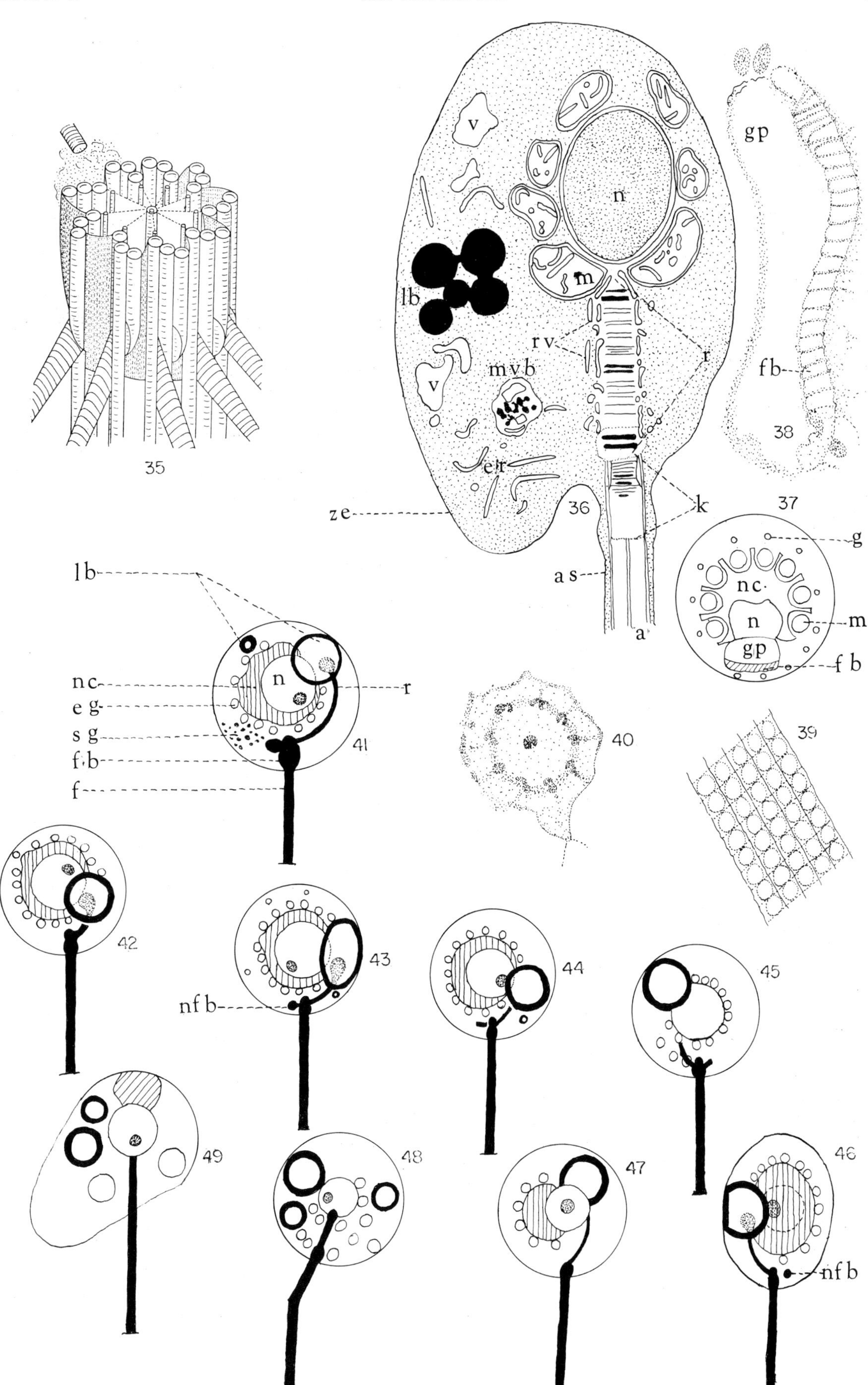

Planospores

Hawksworth (1970). Nonetheless, Sparrow (1974) still regarded it as a legitimate family name.

The classification of the chytrids into Inoperculatae and Operculatae and their separation into different series, families and subfamilies on the basis of the presence or absence of an operculum on the zoosporangium is largely artificial and places too much emphasis on this character above the generic level as pointed out by Whiffen (1944) and Karling (1966). It relegates the striking similarities in morphology, types of development, organization and sexual reproduction to a secondary level of distinction. Furthermore, the distinctiveness of the operculum as a criterion of classification has become questionable in light of the observation of Willoughby (1956), Koch (1957), pl. 36, fig. 72, Karling (1968), pl. 59, fig. 43–45, Umphlett and Koch (1969), and Roane and Paterson (1974) that in some inoperculate species an operculum-like structure or "quasi operculum" may be pushed up or off in the dehiscence of the zoosporangium and that dehiscence in operculate species may occasionally be inoperculate (Chambers, Markus, and Willoughby, 1967; Miller, 1968; and Johnson, 1973). While the separation of the chytrids into the Inoperculatae and Operculatae may be convenient in classification, it is largely arbitrary, superficial, and not indicative of natural relationships, and for these reasons it is not recognized here. Also, there is no concrete evidence that the inoperculate and operculate species have evolved along separate but parallel lines.

Inasmuch as this iconograph is primarily an illustration of genera, no key is presented for these taxa, and their assignment to particular subfamilies in the chytrids is in many cases arbitrary and only tentative. Furthermore, many of the genera are not clearly defined, and this becomes particularly evident as one looks at the illustrations and reads the descriptions of them. In many cases such taxa are hardly more than convenient form genera like those of the Fungi Imperfecti. The creation of most of the present genera was based primarily on characteristics exhibited by species in nature on so-called "natural" substrata such as plant and animal hosts, vegetable and animal debris in water and soil, etc., and classification among the chytrids at present is usually determined on size, shape, structure and flagellar attachment of the planospores; size, shape, and wall structure of the zoosporangia and resting spores; number, shape, and location of the discharge papillae or tubes, presence or absence and shape of an apophysis; number, diameter, position, branching and extensiveness of the rhizoids, type of thallus development, operculation, monocentricity, polycentricity, and substrate relations. However, in axenic cultures on synthetic media these characteristics may vary so greatly that they are of little value in classification. Under such conditions and even in nature, a genus and species may vary so much developmentally and morphologically that it may fit very well into more than one family or subfamily as the chytrids are presently classified.

Some of the newly erected genera do not fit well in any of the subfamilies, and new categories may have to be established for them. Our knowledge of the chytrids is very incomplete and many new genera will probably be discovered which will necessitate a marked revision of the present classification. This is clearly indicated by the recent discovery of *Caulochytrium*, which develops aerial zoosporangia from its zygote and does not belong in any of the known families. Also, recent studies of *Harpochytrium* and *Oedogoniomyces* and the erection of a new order, *Harpochytriales*, show how present concepts of classification are effected by new discoveries. This is further indicated by Dogma's pure culture studies on many of the long-known genera and species. Although the present classification of families is based on reported data and the writers' interpretation of them, it is obviously *only tentative* and offered as a convenience. Quite likely, as more and more species are grown in axenic culture and their overlapping variability becomes known, several genera and even families may be merged.

BLASTOCLADIALES AND MONOBLEPHARIDALES

These two orders combined are smaller in number of taxa than the Chytridiales, and at present include only 5 families, 12 or possibly 16 genera, approximately 95 species, and several varieties. Most of the genera appear more clearly delimited than many in the Chytridiales, and a larger number of species have been grown in axenic cultures on synthetic media and studied more intensively than any of the other Chytridiomycetes. Their larger size, relative ease of culture, and well-defined sexuality and alternation of gametophytic and sporophytic generations in many species have made them objects of great interest to mycologists. As to family distinctions, however, they are not always sharply defined. One family, the Catenariaceae as presently classified, appears to blend into or parallel the Rhizidiaceae and Cladochytriaceae in morphology, development and organization, and in light of present knowledge may perhaps, at least, be regarded as a transition family from the Chytridiales. Also, another family, the Coelomomycetaceae, stands apart from the others by the presence of wall-less thalli and the lack of thin-walled evanescent zoosporangia.

At present the Blastocladiales are classified into 3 families on the basis of type of sexual reproduction and thallus structure and organization as follows:

Blastocladiales

Sexual reproduction where known isogamous; thallus coenocytic, lacking walls; wall-less gametangia and resistant sporangia known.
Family **Coelomomycetaceae**

Sexual reproduction unknown or where known isogamous; thallus predominantly polycentric: tubular, catenulate, branched or unbranched with true or false septa, and bearing rhizoids along its length.

Family **Catenariaceae**

Sexual reproduction isogamous or anisogamous where known; thallus of varying complexity, monocentric and usually unbranched with or without true septa and rhizoids at the base, or polycentric (?), and branched with a central axis bearing rhizoids at the base and reproductive structures at the apex, with or without pseudosepta.

Family **Blastocladiaceae**

The Monoblepharidales include 2 families at present which are distinguishable primarily by their type of sexual reproduction, thallus structure, and organization, and are generally classified as follows:

Monoblepharidales

Sexual reproduction oogamous; female gametangium bearing one or more large gametes; male gametangium bearing numerous small male gametes; zygote motile and propelled by the flagellum of the male gamete; mycelium with or without constrictions and pseudosepta.

Family **Gonopodyaceae**

Sexual reproduction oogamous; oogonium usually bearing 1 egg; antheridium bearing numerous small antherozoids one of which is engulfed by the egg in fertilization; zygote non-flagellate, remaining endogenous or emerging to the orifice of the oogonium; mycelium without septa or pseudosepta.

Family **Monoblepharidaceae**

HYPHOCHYTRIOMYCETES

(Anisochytridiales)

Inasmuch as this group of species is generally regarded as a phycomycetous taxon, it is illustrated and described as an appendix to the Chytridiomycetes, but whether or not it constitutes a distinct class remains to be seen. At present it is classified as including 3 families, 4 genera and approximately 21 species. As in the Chytridiales, family distinctions are based on whether the thallus is holocarpic or eucarpic and monocentric, or mycelioid and polycentric, and on this basis 3 families are recognized at present. However, the distinctions are not always sharply-defined, and the thallus structure, development, and organization may vary markedly in axenic cultures on synthetic media.

Thallus olpidioid, monocentric, holocarpic, endobiotic; sexual reproduction where known isogamous; planospores delimited within or outside of the zoosporangium.

Family **Anisolpidiaceae**

Thallus eucarpic, predominantly monocentric; sexual reproduction unknown or doubtful; planospores delimited within or outside of the zoosporangium.

Family **Rhizidiomycetaceae**

Thallus eucarpic, predominantly polycentric and mycelioid; sexual reproduction unknown; planospores delimited within or outside of the zoosporangium.

Family **Hyphocytriaceae**

REFERENCES TO PLANOSPORES AND SYSTEMATICS

Bessey, E. A. 1942. Some problems in fungus phylogeny. Mycologia 34:355-379.

———. 1950. Morphology and taxonomy of fungi. Blakiston Co., Philadelphia.

Canter, H. M. 1949. Studies on British chytrids. VI. Aquatic Synchytriaceae. Trans. Brit. Mycol. Soc. 32:69-94, pls. 7-11, 13 text figs.

———. 1954. Fungal parasites of the phytoplankton. III. Ibid. 34:111-133, pls. 3-5, 9 text figs.

———. 1963. Studies of British chytrids. XXII. New species on chrysophycean algae. Ibid. 46: 305-320, 7 figs.

Chambers, T. C., and L. G. Willoughby. 1964. The fine structure of *Rhizophlyctis rosea*, a soil Phycomycete. J. Roy. Micro. Soc. 83:355-365, pls. 152-158.

———, K. Markus, and L. G. Willoughby. 1967. The fine structure of the mature zoosporangium of *Nowakowskiella profusa*. J. Gen. Microbiol. 46:135-141, pls. 1-5.

Chong, J., and D. J. S. Barr. 1973. Zoospore development and fine structures in *Phlyctochytrium arcticum* (Chytridiales). Can. J. Bot. 51:1411-1420.

———. 1974. Ultrastructure of the zoospores of *Entophlyctis confervae-glomeratae, Rhizophydium patellarium* and *Catenaria anguillulae*. Ibid. 52:1197-1204, 38 figs.

Clinton, G. P. 1902. *Cladochytrium alismatis*. Bot. Gaz. 33:49-61, pls. 1-6.

Cooke, W. B. and D. L. Hawksworth. 1970. A preliminary list of families proposed for fungi (including lichens). Mycol. Paper no. 111, Commonwealth, Mycol. Publ.

Couch, J. N. 1941. The structure and action of the cilia in some aquatic Phycomycetes. Amer. J. Bot. 28:704-713, 58 figs.

Dangeard, P. A. 1886. Sur les organismes inférieurs. Ann. Sci. Nat. Bot. 4:241-341, pls. 11-14.

———. 1933. Nouvelles observations sur les parasites des Eugléniens. Le Botan. 24: 1-48, pl. 4.

DeBary, A. 1884. Vergleichende Morphologie und Biologie der Pilze, Wm. Engelmann, Leipzig.

Dogma, I. J., Jr. 1969. Observations on some cellulosic chytridiaceous fungi. Arch. f. Mikrobiol. 66:208-219, 70 figs.

Fuller, M. S. 1966. Structure of the uniflagellate zospores of aquatic Phycomycetes. Colston Papers 18:67-84, 26 figs.

———, and L. W. Olsen. 1971. The zoospore of *Allomyces*. J. Gen. Microbiol. 66:171-183, 5 pls.

Gaertner, A. 1954. Beobachtungen über die Bewegungsweise von Chytridineenzoosporen. Arch. f. Mikrobiol. 20:423-426, 1 fig.

Geitler, L. 1962. Entwickelung und Beziehung zum Wirt der Chytridiale *Scherffeliomycopsis coleochaetis* n. gen., n. spec. Osterr. Bot. Zeitschr. 109:250-274, 8 figs.

Held, A. A. 1973. Encystment and germination of the parasitic chytrid *Rozella allomyces* on host hyphae. Canad. J. Bot. 51:1825-1835, 23 figs.

Ingold, C. T. 1952. *Funaria* rhizoids infected with *Pleotrachelus wildemanni*. Trans. Brit. Bryol. Soc. 2:53-54, 1 fig.

Johnson, T. W., Jr. 1973. Aquatic fungi of Iceland: some polycentric species. Mycologia 65:1337-1355, 42 figs.

Karling, J. S. 1966. The chytrids of India with a supplement of other zoosporic fungi. Beih. Sydowia. VI:1-125.

———. 1967. Some zoosporic fungi of New Zealand. IX. *Polyphlyctis* gen. nov., *Phlyctochytrium*, and *Rhizidium*. Sydowia 20:86-95, pls. 15, 16.

———. 1968. Zoosporic fungi of Oceania. III. Monocentric chytrids. Arch. f. Mikrobiol. 61: 112-127, 3 figs.

———. 1969. Zoosporic fungi of Oceania. VII. Fusions in *Rhizophlyctis*. Amer. J. Bot. 56:211-221, 108 figs.

———. 1975. The taxonomy of the Chytridiomycetes and Hyphochytridiomycetes. Proc. Intern. Symp. on Taxonomy of Fungi, Madras Univ., India.

Kazama, F. Y. 1972. Ultrastructure and phototaxis of the zoospores of *Phlyctochytrium* sp., an estuarine chytrid. J. Gen. Microbiol. 71:555-566.

Koch, W. J. 1956. Studies on the motile cells of chytrids. I. Electron microscope observations of the flagellum, blepharoplast and rhizoplast. Amer. J. Bot. 43:811-819, 26 figs.

———. 1957. Two new chytrids in pure culture, *Phlyctochytrium punctatum* and *Phlyctochytrium irregulare*. J. Elisha Mitchell Sci. Soc. 73:108-122, 24 figs.

———. 1958. Studies on the motile cells of chytrids. II. Internal structure of the body observed with light microscopy. Amer. J. Bot. 45:59-72, 143 figs.

———. 1961. Studies on the motile cells of chytrids. III. Major types. Ibid. 48:786-788, 8 figs.

———. 1968. Studies on the motile cells of chytrids. IV. Planonts in the experimental taxonomy of aquatic Phycomycetes. J. Elisha Mitchell Sci. Soc. 84:69-83, 14 figs.

———. 1969. Studies of the motile cells of chytrids. V. The Monoblepharidales and Blastocladiales types of posteriorly uniflagellate motile cell. Mycologia 61:422-426, 3 figs.

Lesemann, D. E., and W. H. Fuchs. 1970a. Elektronenmikroskopische Untersuchung über Vorbretung der Infektion in encystierten zoosporen von *Olpidium brassicae*. Arch. f. Mikrobiol. 71:9-17, 7 figs.

———. 1970b. Die Ultrastruktur des penetrationsvorganges von *Olpidium brassicae* an Kohlrabi-Wurzeln. Ibid. 71:9-17, 7 figs.

Litvinov, M. A. 1958. On the criteria of the species in Chytridiales. Akad. nauk USSR no. 1, pp. 68-84 (Russ. Trans.).

Manton, I., B. Clark and A. D. Greenwood. 1952. Further studies on the structure of plant cilia, by a combination of visual and electron microscopy. J. Exp. Bot. 3:204-215, 8 pls.

Martin, W. W. 1971. The ultrastructure of *Coelomomyces punctatus* zoospores. J. Elisha Mitchell Sci. Soc. 87:209-221, 20 figs.

McNitt, R. 1973. Mitosis in *Phlyctochytrium irregulare*. Canad. J. Bot. 2065-2074, 10 pls.

———. 1974a. Centriole ultrastructure and its possible role in microtubule formation in an aquatic fungus. Protoplasma 80:91-108, 27 figs.

———. 1974b. Zoosporogenesis in *Phlyctochytrium irregulare*. Cytobiol. 9:290-306, 24 figs.

———. 1974c. Ultrastructure of *Phlyctochytrium irregulare* zoospores. Ibid. 9:307-320, 29 figs.

Metz, C. 1929. Versuch einer Stammesgeschichte des Pilzreiches. Schrift. Konigsberg Gelehrt. Ges. Naturw. Klasse 6:1-58.

Miller, C. E. 1968. Observations concerning taxonomic characteristics in chytridiaceous fungi. J. Elisha Mitchell Sci. Soc. 84:100-107, 44 figs.

Nowakowski, L. 1876. Beitrag zur Kenntniss der Chytridiaceen. II. *Polyphagus euglenae*. *In* Cohn, Beitr. Biol. Pflanzen 2:201-219, pls. 8, 9.

Olson, L. W., and M. S. Fuller, 1968. Ultrastructural evidence for the biflagellate origin of the uniflagellate zoospore. Arch. f. Mikrobiol. 62: 237-250, 18 figs.

Powell, M. J. 1974. Fine structure of plasmodesmata in a chytrid. Mycologia 66:606-614, 7 figs.

———, and W. J. Koch. 1973. Mitosis in the aquatic fungus *Entophlyctis* sp. (Chytridiomycetes, Chytridiales, Phlyctidiaceae) Assoc. Southeast Biol. Bull. 20:76.

Reichle, R. E. 1969. Fine structure of *Phytophthora parasitica* zoospores. Mycologia 61: 30-51, 27 figs.

———. 1972. Fine structure of *Oedogoniomyces* zoospores, with comparative observations of *Monoblepharella* zoospores. Canad. J. Bot. 50: 819-824, pls. 1-7.

Rieth, A. 1962. Beitrag zur Kenntniss der Phycomyceten. IV. *Pleotrachelus wildemanni* Petersen neue für Deutschlands. Die Kulturpf. 10:93-105,

5 figs., pls. 1, 2.

Renaud, F. L., and H. Swift. 1964. The development of basal bodies and flagella in *Allomyces arbuscula*. J. Cell Biol. 23:240-265, 24 figs.

Roane, M. K., and R. A. Paterson. 1974. Some aspects of morphology and development in the Chytridiales. Mycologia 66:147-164, 9 figs.

Robertson, J. A. 1971. Phototaxis in a new *Allomyces*. Arch. f. Mikrobiol. 85:259-266, 5 figs.

Robertson, M. 1905. *Pseudospora volvocis* Cienkowski. Quart. J. Micro. Sci. 49:2183-230, pl. 12.

Scherffel, A. 1926. Beitrage fur Kenntnis der Chytridineen. Teil III. Arch. f. Protistenk. 54:510-528, pls. 9-11.

Schnepf, E., G. Diechraber, E. Hegewald, and C. J. Soeder. 1971. Elektronenmikroskopische Beobachtungen an Parasiten aus *Scenedesmus* Masskulturen. Arch. f. Mikrobiol. 75:230-245, 21 figs.

Serbinov, J. L. 1970. Kenntniss der Phycomyceten. Organization u. Entwickelungsgeschichte einiger Chytridineen Pilze (Chytridineae Schröter). Scripta Bot. Horti Univ. Imp. Petro. 24:1-173, pls. 1-6.

Sparrow, F. K., Jr. 1938. Some chytridiaceous fungi from North Africa and Borneo. Trans. Brit. Mycol. Soc. 21:145-151, 2 figs.

________. 1974. Chytridiomycetes, Hyphochytriomycetes. Chap. 6, in The Fungi, Vol. IV, B, edited by G. C. Ainsworth, F. K. Sparrow, and A. S. Sussman. Acad. Press, New York, N.Y.

Timmink, J. H. M., and R. N. Campbell. 1969. The ultrastructure of *Oplidium brassicae*. II. Zoospores. Canad. J. Bot. 47:227-231, 26 figs.

Travland, L. B., and H. C. Whisler. 1971. Ultrastructure of *Harpochytrium hedenii*. Mycologia 63: 767-789, 18 figs.

Umphlett, C. J. and L. W. Olson. 1967. Cytological and morphological studies of a new species of *Phlyctochytrium*. Mycologia 59:1085-1096, 41 figs.

________, and W. J. Koch. Two new dentigerate species of *Phlyctochytrium*. Mycologia 61: 1021-1030, 17 figs.

Whiffen, A. J. 1944. A discussion of taxonomic criteria in the Chytridiales. Farlowia 1:583-597, 2 figs.

Whisler, H. C., and L. B. Travland. 1973. Mitosis in *Harpochytrium*. Arch. f. Protistenk. 115: 69-74, pls. 2-6.

Willoughby, L. G. 1956. Studies on soil chytrids. I. *Rhizidium richmondense* and its parasites. Trans. Brit. Mycol. Soc. 39:125-141, 9 figs.

Zopf, W. 1884. Zur Kenntniss der Phycomyceten. I. Zur Morphologie und Biologie der Ancylisteen and Chytridiaceen, zugleich ein Beitrag zur Phytopathologie, Nova Acta Acad. Leop. Carol. 47:143-236, pls. 12-21.

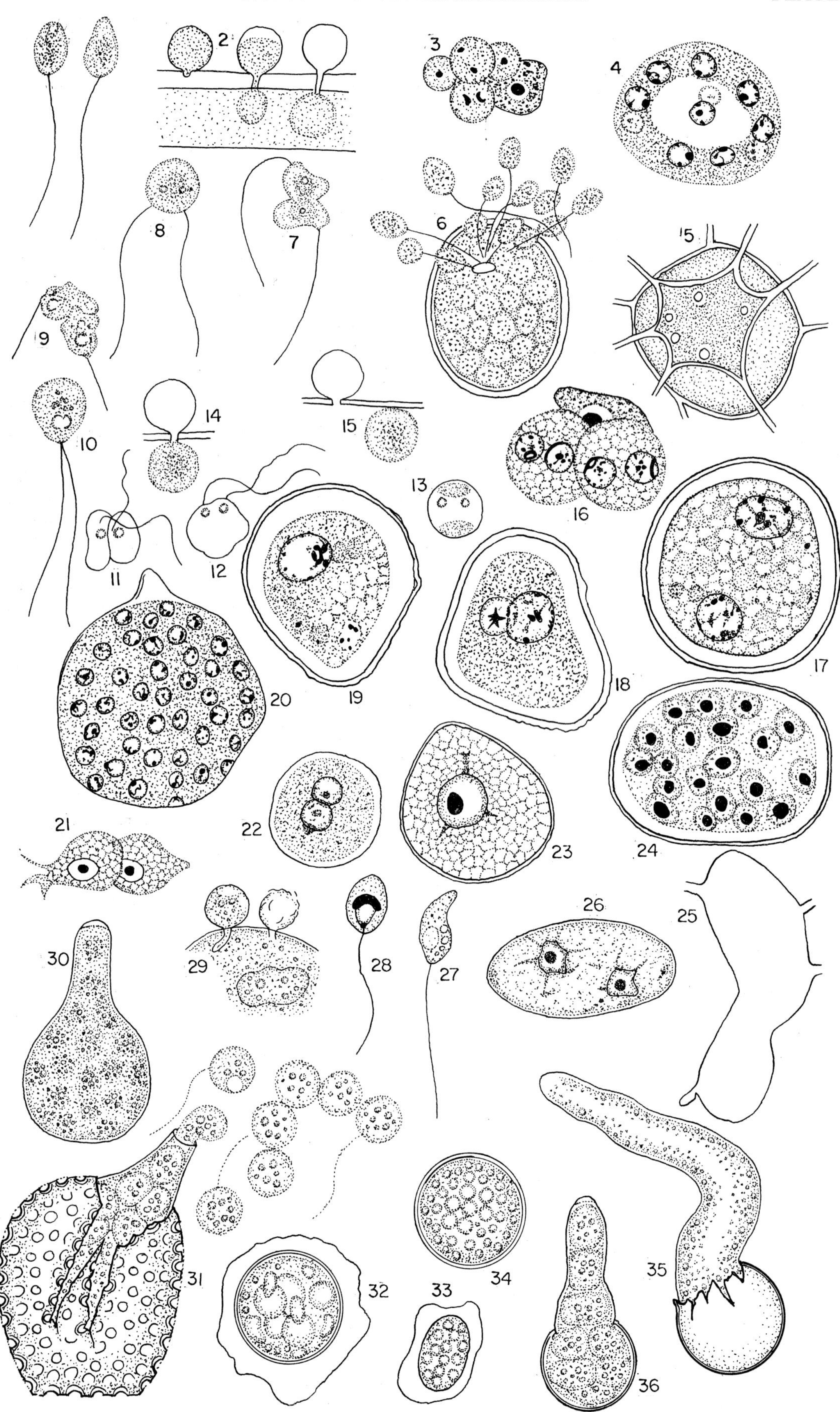
2
3
4
8
7
6
5
9
10
14
15
16
13
11
12
17
18
19
20
21
22
23
24
25
26
27
28
29
30
31
32
33
34
35
36

Olpidium

Chapter I

OLPIDIACEAE

This large family is reported to include 10 or more genera and more than 100 species which are intracellular parasites of freshwater algae, marine and freshwater Chytridiomycetes, Oomycetes, Zygomycetes, aquatic microscopic animals, mosses, pollen grains and flowering plants. In addition several other genera and species have been included in or said to have close affinities with this family, but their relationship with and status as olpidioid chytrids are questionable. Nevertheless, some of these genera and species are illustrated herewith and described briefly. As will be described and illustrated for the separate genera the thallus in this family, derived from an infecting planospore, is holocarpic without a specialized vegetative absorbing system, and becomes transformed at maturity into a zoosporangium or resting sporangium. In rare instances more than one resting sporangium is derived from a thallus. In species where sexuality is known to occur the resting sporangia develop as zygotes, and like the asexually developed ones they function directly as zoosporangia when they germinate.

Sexual reproduction has been reported in several genera and species, and it occurs by fusion of isoplanogametes, or plasmogamy of two protoplasts (isoplanogametes?) derived from separate gametes within the host cell, or the fusion of the contents of a larger female and a smaller male thallus through a short tube or papilla within the host cell. From the studies of monosporangial isolates and combinations of their planospores or gametes several species have been reported to be genotypically heterothallic, i.e., *O. brassicae*, or phenotypically homothallic, i.e., *O. virulentes* (Sahtiyanic) comb. nov. and *O. bornovanus* (Sahtiyanic) comb. nov.

The following order of presentation of the genera of this family is arbitrary and not indicative of relationships and evolution.

PLATE 3

Figs. 1-8, 14-20. *Olpidium viciae* Kusano. (Kusano, 1912, 1932.)

Fig. 1. Ovoid, 4.0-4.5 μ, planospores.

Fig. 2. Infection of *Vicia unijuga.*

Fig. 3. Four young parasites aggregated at the host nucleus.

Fig. 4. Multinucleate young thallus.

Fig. 5. Surface view of a mature zoosporangium with 5 exit pores.

Fig. 6. Discharge of planospores.

Fig. 7. Fusion of isoplanogametes.

Fig. 8. Biflagellate motile zygote.

Figs. 14, 15. Infection of the host by the zygote.

Fig. 16. Two young binucleate zygotes adjacent to the host nucleus.

Fig. 17. Mature binucleate resting sporangium.

Fig. 18. Karyogamy in the resting sporangium in preparation for germination.

Fig. 19. Diploid nucleus in synaptic (?) stage of meiosis.

Fig. 20. Multinucleate resting sporangium with an exit papilla shortly before cleavage; outer wall omitted.

Figs. 9, 10. *Olpidium trifolii* Kusano. (Kusano, 1929.)

Fig. 9. Fusion of isoplanogametes.

Fig. 10. Motile biflagellate zygote.

Figs. 11-13. *Olpidium bothriospermi* Sawada. (Sawada, 1922.)

Fig. 11. Pairing and fusion of isoplanogametes.

Fig. 12. Biflagellate zygote.

Fig. 13. Encysted zygote.

Figs. 21-24. *Olpidium radicale* Schwartz and Cook. (Schwartz and Cook, 1928.)

Fig. 21. Plasmogamy within the host cell.

Fig. 22. Binucleate mature resting sporangium with nuclei about to fuse.

Fig. 23. Resting sporangium with a diploid nucleus shortly before germination.

Fig. 24. Germinated resting sporangium with planospores.

Figs. 25, 26. *Olpidium agrostidis* Sampson. (Sampson, 1932.)

Fig. 25. Empty zoosporangium with 4 exit canals.

Fig. 26. Binucleate mature resting sporangium.

Figs. 27-36. *Olpidium allomycetos* Karling. (Karling, 1948.)

Figs. 27, 28. Living and fixed and stained planospores, 4 × 5.5 μ respectively.

Fig. 29. Infection of *Allomyces anomalus.*

Fig. 30. Mature zoosporangium shortly before dehiscence.

Fig. 31. Discharge of planospores from a zoosporangium parasitizing a resting spore of *A. anomalus.*

Fig. 32, 33. Large and small resting sporangia lying in a vesicle, probably formed by contraction of thallus content.

Fig. 34. Resting sporangium formed from the entire thallus without contraction.

Fig. 35, 36. Germination of resting sporangia.

OLPIDIUM (Braun) Rabenhorst

(sensu recent. Schroeter, Flora Europeae algarum 3:288, 1868.
Kryptoganen-Fl. Schlesiens 3:180, 1885)

Chytridium, subgen. *Olpidium* Braun, 1856. Abhandl. Konigl. Akad. Wiss. Berlin 1855:75.
Cyphidium Magnus, 1875. Wissensch. Meeresuntersuch. Abt. Kiel 2-3: 77.
Pleotrachelus Zopf, 1884. Nova Acta Acad. Leo-Carol. 47:173.
Olpidiella Lagerheim, 1888. J. de Botanique 2:438.
Asterocystis de Wildemann, 1893. Ann. Soc. Belge Micro. (Mem.) 17:21.
Endolpidium de Wildemann, 1894. Ann. Soc. Belge Micro. (Mem.) 18:153.
Olpidiaster Pascher, 1917. Beiheft. Bot. Centralbl. 35^2: 578.
Pseudolpidiella (pro parte) Cejp, 1959. Flora CSR. Oomycetes I, p. 460.

Plates 3-6

This genus is reported to include approximately 50 species, a large number of which are incompletely known or doubtful (see Litvinov, 1959). These species are world-wide in distribution and parasitize primarily freshwater algae, fungi, moss protonema, pollen grains, flowering plants and microscopic aquatic animals. One species, *O askaulos* (Bradley) comb. nov. has been described as an Eocene fossil (Bradley, 1967). In most known species the content of the encysted planospore enters the host (figs. 2, 29) through a short or long infection tube and develops into a holocarpic zoosporangium (figs. 4–6, 30, 31) or a thick-walled resting sporangium (figs. 32–34) which functions directly as a zoosporangium when it ger-

PLATE 4

Fig. 37, 37A. *Olpidium endogenum* (Braun) Schroeter.

Fig. 37. Empty zoosporangia in *Closterium lunula.* (Braun, 1856.)

Fig. 37A. Resting sporangia lying free in vesicles. (Cejp, 1933.)

Figs. 38-40. *Olpidium uredinis* (Lagerheim) Fischer. (Lagerheim, 1888.)

Fig. 38. Zoosporangium discharging planospores in a uredospore of *Puccinia airae.*

Fig. 39. Spherical, 3-4 μ diam., planospore with a hyaline refractive globule.

Fig. 40. Resting sporangium in the uredospore of *P. airae.*

Fig. 41. *Olpidium gregarium* (Nowak.) Schroeter; 2 resting sporangia and 3 zoosporangia in a rotifer egg. (Sparrow, 1936.)

Figs. 42, 43. *Olpidium saccatum* Sorokin. (Scherffel, 1926.)

Fig. 42. Two empty zoosporangia in *Cosmarium* sp.

Fig. 43. Resting sporangium formed in the end of a curved thallus or vesicle.

Figs. 44-46. *Olpidium longicollum* Uebelmesser. (Uebelmesser, 1956.)

Fig. 44. Zoosporangium with a long neck discharging planospores from a pollen grain.

Fig. 45. Large, spherical, 10 μ diam., planospores with 2 refractive globules; flagellum 80 μ (?) long.

Figs. 46, 46A. Large resting sporangia.

Figs. 47-49. *Olpidium utriculiforme* Scherffel. Fig. 47 after Scherffel, 1926; figs. 48, 49 after Canter, 1949.

Fig. 47. Empty zoosporangium in *Cosmarium botrytis.*

Fig. 48. Elongate, tubular, branched, empty zoosporangium in *Closterium lunula.*

Fig. 49. Spherical, 2.4-2.8 μ diam., planospores with a hyaline globule.

Figs. 50-52. *Olpidium granulatum* Karling. (Karling, 1946.)

Fig. 50. Mature zooaporangium in rotifer egg.

Fig. 51. Elongate planospore, 3.5 × 4.5 μ, with granular content; flagellum 38-45 μ long.

Fig. 52. Resting sporangium lying in a vesicle.

Figs. 53, 54. *Olpidium rotiferum* Karling. (Karling, 1946.)

Fig. 53. Oblong 3-3.5 × 6.5-7 μ, tapering planospore with nuclear cap (?) and 2 sides bodies.

Fig. 54. Resting sporangium lying in a zoosporangium-like vesicle.

Figs. 55-58. *Olpidium indum* Karling. (Karling, 1964.)

Fig. 55. Zoosporangium of *Rhizophlyctis fuscus* parasitized by 6 polyhedral zoosporangia in various stages of development.

Fig. 56. Two resting sporangia formed in 1 vesicle.

Fig. 57. One resting sporangium in a vesicle.

Fig. 58. Germination of resting sporangium.

Figs. 59-62. *Olpidium appendiculatum* Karling. (Karling, 1965.)

Fig. 59. Appendiculate zoosporangium discharging planospores from pollen grain of *Pinus sylvestris.*

Fig. 60. Spherical, 2.8-3.4 μ diam., planospore with a minute refractive globule.

Fig. 61. Development of thallus at the end of the germ tube.

Fig. 62. Resting sporangium formed in entirety from a thallus with the zoospore cyst and germ tube persistent as an appendage.

Figs. 63-66. *Olpidium hyalothecae* Scherffel. Figs. 63, 66 after Scherffel, 1926; Figs. 64, 65 after Canter, 1949.

Fig. 63. Spiny resting sporangium in *Hyalotheca dissiliens.*

Figs. 64, 65. Resting sporangium and mature zoosporangium, resp., in *H. mucosa.*

Fig. 66. Spherical, 3-4 μ diam., planospores with a minute hyaline refractive globule.

Fig. 67. *Olpidium nematodeae* Skvortzow. Empty zoosporangia with long necks protruding from a nematode. (Skvortzow, 1927.)

Fig. 68. *Olpidium protonemae* Skvortzow; zoosporangia in an enlarged moss protonema cell. (Skvortzow, 1927.)

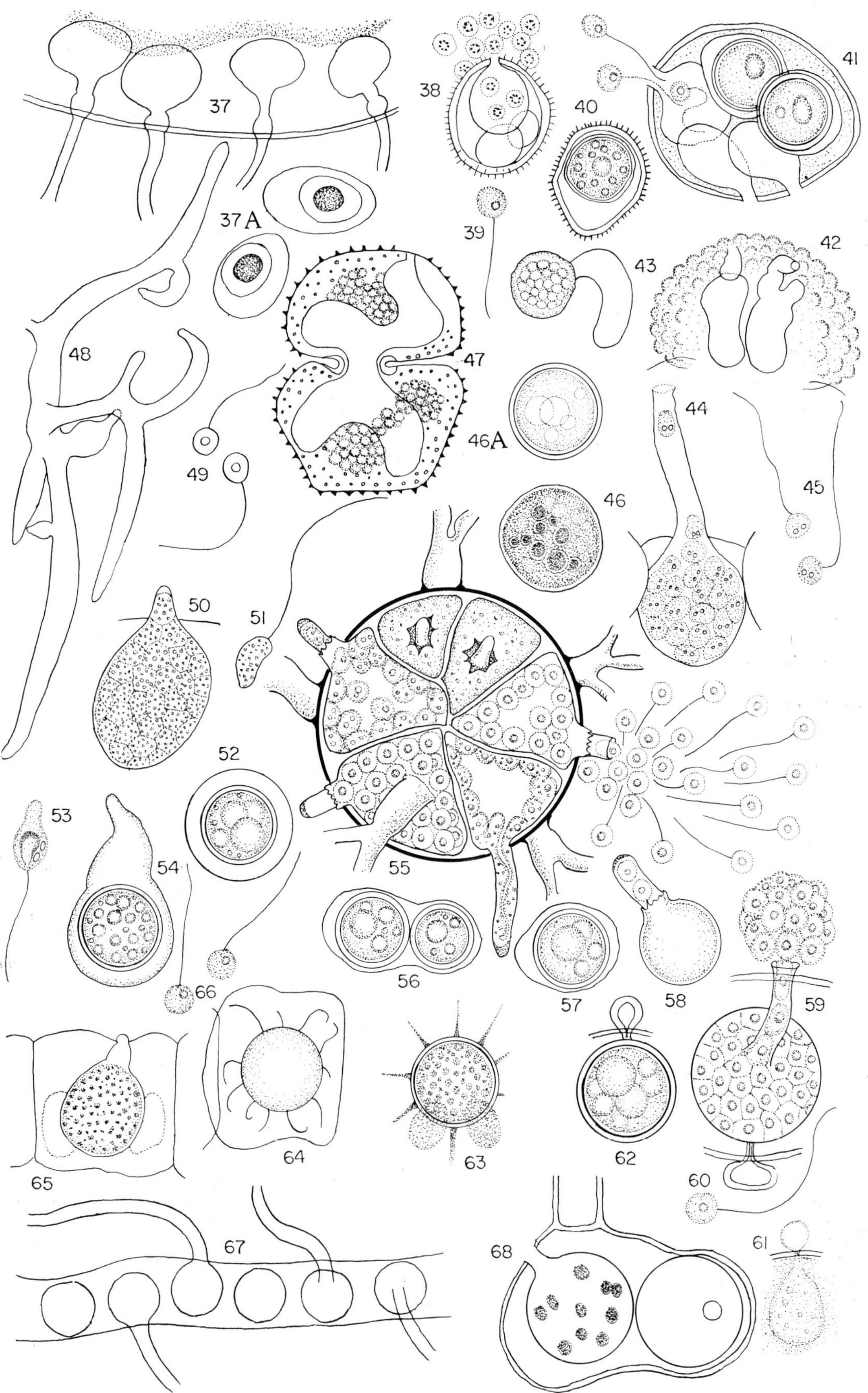
37
37A
38
39
40
41
42
43
44
45
46
46A
47
48
49
50
51
52
53
54
55
56
57
58
59
60
61
62
63
64
65
66
67
68

Olpidium

minates and produces planospores (figs. 35, 36, 76, 77). In some species the content of the incipient resting sporangium or thallus contracts and encysts so that the resting sporangium partly fills and lies in a vesicle (figs. 32, 33).

The thalli and zoosporangia vary from spherical to ovoid, irregular, tubular and elongate in shape in different species, and some of these variations are illustrated in plates 3–6 to emphasize such differences. At maturity they develop short exit tubes which end flush with the periphery of the host (figs. 65, 79) or elongate ones (figs. 37, 44, 67, 72) which extend for long distances. In *O. longicollum* Uebelmesser (fig. 44) the exit canal may be 60 to 80 μ long, but in *O. hantzschiae* Skvortzow (fig. 91) no papillae or tubes are formed. The zoosporangia are reported to open by a pore, but it is not improbable that Skvortzow's species may be a monad instead of a species of *Olpidium*. Also, in *O. euglenae* Dangeard no recognizeable papilla or tube is developed, but at maturity the zoosporangium forms a hernia or protrusion through the host wall which becomes globular and may attain a size as great as or greater than that of the zoosporangium proper (fig. 94). The refractive granular material coalesces to form the definitive yellowish-orange globules of the planospores, and the content of both structures cleaves into planospores (fig. 95). If Dangeard's account is confirmed, it is obvious that the zoosporangium functions both as a prosporangium and a zoosporangium, and, in that event, this species may be representative of another genus. However, Dangeard showed one incipient zoosporangium (fig. 96) in which the globules had been formed before the development of an epibiotic hernia or vesicle, and this suggests the possibility that the vesicle around the planospores in figure 95 is the periphery of a surrounding layer of matrix which was extruded with the planospores. In contrast to this and other species, some *Olpidium* members develop 2 (fig. 84) to several exit tubes (fig. 25).

For this reason and the fact that it is similar to *Olpidium* in morphology, development, organization, and sexuality, *Pleotrachelus* is merged with the former genus. It is very doubtful that the presence of several discharge tubes is a generic character be-

PLATE 5

Figs. 69, 70. *Olpidium rhizophlyctidis* Sparrow in *Rhizophlyctis* spp. Fig. 69 after Johnson, 1969; fig. 70 after Sparrow, 1948.

Fig. 69. Zoosporangium of host with 3 zoosporangia of the parasite, one of which is discharging planospores.

Fig. 70. Zoosporangium of host filled with incipient zoosporangia and resting spores of the parasite.

Figs. 71-77. *Olpidium synchytrii* Karling in the zoosporangia of *Synchytrium namae*. (Karling, 1958.)

Fig. 71. Zoosporangium of host with 5 thalli and zoosporangia of the parasite.

Fig. 72. Large zoosporangium of the parasite with a long curved neck outside of the host.

Fig. 73. Discharge of planospores.

Fig. 74. Subspherical, 1.8-2.2 μ diam., planospores with a minute refractive globule and a 7-9 μ long flagellum.

Fig. 75. Three spherical, hyaline, smooth resting spores in a zoosporangium of the host.

Figs. 76, 77. Stages in the germination of the resting spore.

Figs. 78, 79. *Olpidium pendulum* Zopf in pollen grains of *Pinus taeda*. (Johnson, 1969.)

Fig. 78. Zoosporangium with a short neck or papilla which does not extend beyond the host wall, discharging spherical, 4-5 μ diam., planospores.

Fig. 79. Germinated pollen grain with 6 zoosporangia in the pollen tube and 3 resting spores in the pollen grain proper.

Figs. 80-84. *Olpidium luxurians* (Tomaschek) Fischer in pollen grains. Figs. 80-82 after Tomaschek, 1879; fig. 83 after Peterson, 1910; fig. 84 after Johnson. 1969.

Fig. 80. Planospores, 2μ diam., with a rounded apex and a tapered posterior end.

Fig. 81. Ten zoosporangia in a pollen grain of *Pinus maritima*, 5 of which have long curved exit canals.

Fig. 82. Apparently a resting spore of *O. luxurians* which Tomaschek named *Diplochytrium* sp. because of its double contoured wall.

Fig. 83. Resting sporangium.

Fig. 84. Zoosporangium in dead pollen grain of *Pinus taeda* with 2 exit canals, discharging planospores.

Figs. 85, 86. *Olpidium vampyrellae* Scherffel in a zoocyst of *Vampyrella*; sporangia, and planospores, 2 × 3- 4 μ; showing lack of refractive globule, resp. (Scherffel, 1926.)

Fig. 87. *Olpidium pseudosporearum* Scherffel zoosporangium in a zoocyst of *Pseudospora*; planospores delimited and containing a hyaline refractive globule. (Scherffel, 1926.)

Fig. 88. *Olpidium leptophrydis* Scherffel in a zoocyst of *Leptophrys vorax*; 3 empty zoosporangia and 1 with large hyaline refractive globules. (Scherffel, 1926.)

Figs. 89-91. *Olpidium hantzschiae* Skvortzow in *Hantzchia amphioxys*. (Skvortzow, 1927.)

Figs. 89, 90. Planospores, 1.5-1.7 × 3-4 μ.

Fig. 91. Host filled with zoosporangia and resting spores; zoosporangia lacking exit papillae or necks but opening by a pore.

Figs. 92-96. *Olpidium euglenae* Dangeard in *Euglena* sp. (Dangeard, 1894-1895.)

Fig. 92. Spherical planospores with yellow-orange refractive globule.

Fig. 93. Full-grown zoosporangium.

Fig. 94. Zoosporangium forming a globular protuberance outside of host cell.

Fig. 95. Planospores formed in the endobiotic zoosporangium and the epibiotic protuberance.

Fib. 96. Formation of definitive refractive globules in a zoosporangium within the host.

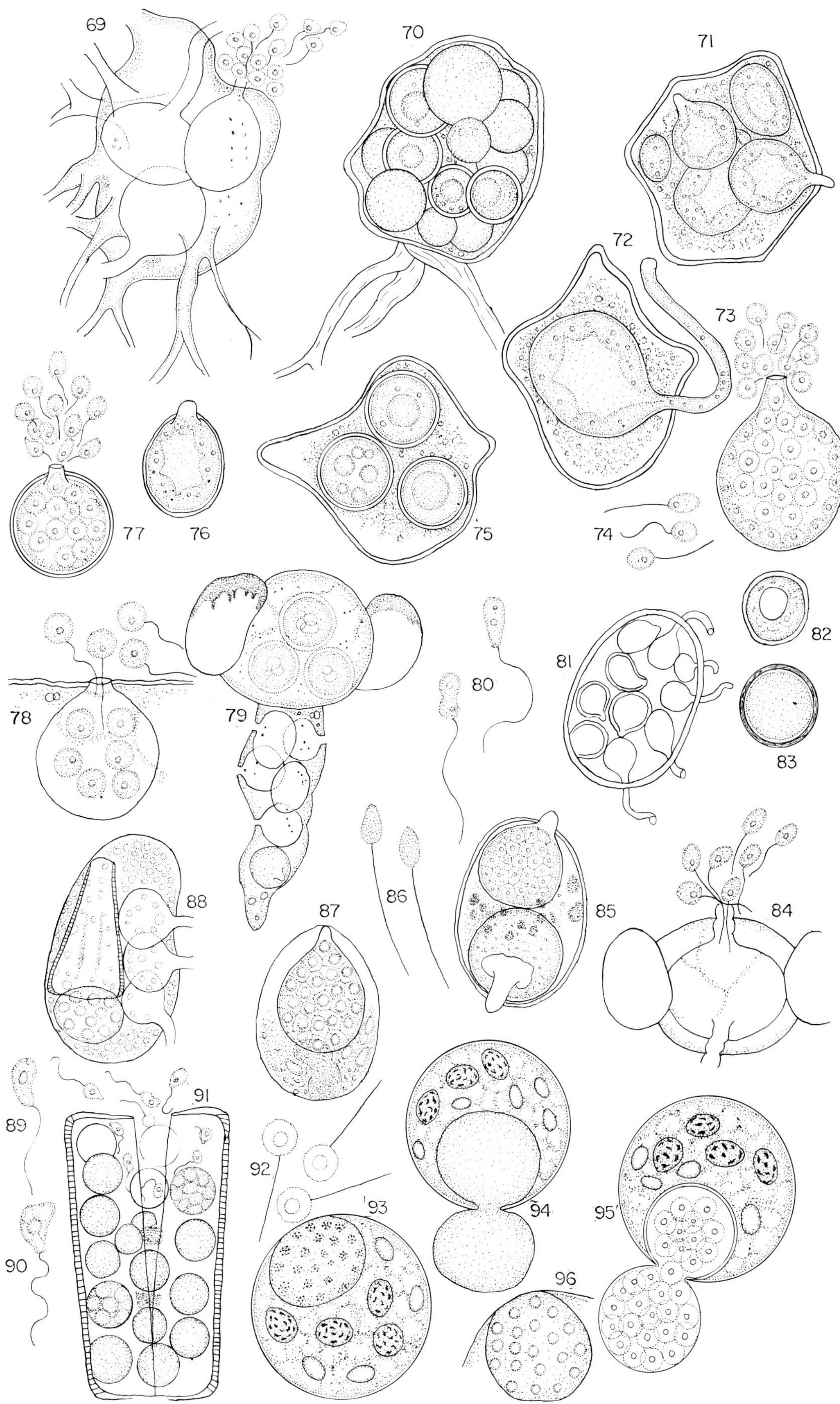
69
70
71
72
73
74
75
76
77
78
79
80
81
82
83
84
85
86
87
88
89
90
91
92
93
94
95
96

Olpidium

cause it is characteristic of some species of *Olpidium* also. Of more significance perhaps is the subapical or lateral insertion of the flagellum on the planospores of one species, *Olpidium wildemanni* (Petersen) comb. nov., but this is characteristic of the genus *Olpidiomorpha*, also. The former *Pleotrachelus* species are, nevertheless, illustrated separately in plate 6 to show their striking similarity to *Olpidium*.

In a few species, *O. viciae* Kusano, *O. trifolii* Schroeter, *O. bothriospermi* Sawada (figs. 11, 12) and *O. cucurbitacearum* Barr (1968) in which sexual reproduction has been described, the planospores are facultative. They may infect the host directly and give rise to sporangial thalli, or function as isoplanogametes and fuse (figs. 7, 9), producing biflagellate motile zygotes (figs. 8, 10) which infect the host (figs. 14, 15) and develop into binucleate resting sporangia (figs. 16, 17). In preparation for germination the two nuclei fuse (fig. 18), after which meiosis (fig. 19) presumably occurs. Following nuclear multiplication (fig. 20) and cleavage, planospores are produced and discharged. In other species, *O. radicale* Schwartz and Cook and possibly *O. agrostidis* Sampson, plasmogamy of two protoplasts derived from separate planospores occurs within the host (fig. 21) and results in the formation of binucleate resting sporangia (figs. 22, 26). Prior to germination the nuclei fuse (figs. 22, 23), and eventually the resting sporangia become multinucleate and produce planospores (fig. 24). In *O. viciae* Kusano (1912) fusions occur between gametes from the same sporangium, and from this description it appears that *O. viciae* is phenotypically homothallic. Later (1929) in more intensive studies Kusano found that very few gametes from the same sporangium copulated, and he concluded that *O. viciae* is properly heterothallic although it exhibits a tendency more or less towards homothallism. On the other hand, he found that *O. trifolli* is typically heterothallic when single gametangia are isolated and studied. So far no convincing evidence of sexuality has been found in the purely aquatic species of *Olpidium*, but it is quite probable that homothallic and heterothallic species will be demonstrated when monosporangial lines have been isolated and studied.

ROZELLA Cornu

Ann. Sci. Nat. Bot. V, 15:148, 1872.

Rozia Cornu, 1872. Bull. Soc. Bot. France 19:71 (non *Rozea* Becherelle, 1871-1872. Mem. Soc. Nationelle Sci. Nat. Cherbourg 16:241).

Pleolpidium Fischer, 1892. Rabenhorst Kryptogamen–Fl. 1(4); 43.

Plate 7

This genus includes approximately 22 identified and several unidentified species which parasitize

PLATE 6

Figs. 1-3. *Olpidium fulgens* (Zopf) comb. nov. (Zopf, 1892.)

Figs. 1, 2. Portions of infected hypertrophied mycelium, an enlarged portion with 3 parasitic zoosporangia which bear numerous exit canals.

Fig. 3. Ovoid planospore, 2.2-3 μ diam.

Figs. 4, 5. *Olpidium zopfianus* (Morini) comb. nov. (Morini, 1913.)

Figs. 4, 5. Resting spore and its germination, respectively.

Figs. 6-12. *Olpidium wildemanni* Petersen comb. nov. Fig. 6 after Ingold, 1952; figs. 7-12 after Rieth, 1962.

Figs. 6, 7. Ovoid planospores, 3×5 μ, with sublateral attachment of the flagellum.

Figs. 8, 9. Quiescent and amoeboid planospores, respectively.

Fig. 10. Enlarged quiescent planospore with 3 pulsating vacuoles.

Fig. 11. Hypertrophied protonema cells of *Funaria hygrometrica* parasitized by numerous zoosporangia bearing 1 exit canal.

Fig. 12. Enlarged protonema cell with 1 large empty zoosporangium which bears 7 exit canals.

Figs. 13-23. *Olpidium (Pleotrachelus) brassicae*. Fig. 13-17, 21-23 after Sahtiyanic, 1962; figs. 18-20 after Jacobsen, 1943.

Figs. 13, 14. Free swimming and amoeboid planospores, respectively.

Fig. 15. Infection of root hair by a short penetration papilla.

Fig. 16. Young thallus in root hair.

Fig. 17. Elongate zoosporangium with numerous long exit canals.

Figs. 18-20. Stages in the development of the resting sporangium.

Figs. 21, 22. Resting sporangia of genotypically heterothallic *O. brassicae*, formed by fusion of + and − isogametes.

Fig. 23. Germination of resting sporangium.

Figs. 24-33. *Olpidium virulentus*. (Sahtiyanic, 1962.)

Figs. 24-26. Quiescent, free swimming, and amoeboid planospores.

Fig. 27. Enlarged drawing of a planospore showing blepharoplast, nucleus, and 2 pulsating vacuoles to the right and left of the nucleus.

Figs. 28-30. Some variations in the shapes of the zoosporangia and number of exit canals.

Figs. 31, 32. Resting sporangia of phenotypically homothallic *O. virulentus*, formed by fusion of isogametes.

Fig. 33. Germination of resting sporangium.

Figs. 34-43. *Olpidium bornovanus*. (Sahtiyanic, 1962.)

Figs. 34, 35. Citriform and quiescent spherical planospores, respectively.

Fig. 36. Infection by a long penetration tube.

Fig. 37. Zoosporangium with 5 exit canals.

Fig. 38. Discharge of planospores from a zoosporangium with 1 exit canal.

Figs. 39-41. Resting sporangium of phenotypically homothallic *O. bornovanus*, formed by fusion of isogametes.

Figs. 42, 43. Germination of resting sporangia with 1 and 6 exit canals, respectively.

1
2
3
4
5
6
7
8
9
10
11
12
13
14
15
16
17
18
19
20
21
22
23
24
25
26
27
28
29
30
31
32
33
34
35
36
37
38
39
40
41
42
43

Olpidium

freshwater aquatic Chytridiomycetes and Oomycetes, and one species, *R. marina* (Sparrow) Johnson is marine and parasitic in *Chytridium*. Quite likely some of the identified species will prove to be identical when extensive host range studies have been made, and it is possible that biological races occur among some species. Characteristically, the zoosporangia of the parasite fill the host cells or segments of it so fully (figs. 4–6, 15–18) that the wall of the former seems to be continuous with that of the host. For convenience in identification Karling (1942) classified the species into two groups: (1) nonseptigenous monosporangiate (figs. 1–11), in which the parasites fail to cause septation of the host and develop only one zoosporangium or resting sporangium; and (2) septigenous polysporangiate which cause septation of the host and develop one (figs. 15–18) or usually more than one zoosporangium or resting sporangium within the host segment (figs. 21, 22). Cornu (1872) and Foust (1937) believed that thalli or primary incipient zoosporangia in the latter group may divide into a number of segments which become delimited by the development of host septa, mature in basipetal succession, and develop into zoosporangia or resting sporangia, but this has not been proven. Quite likely the presence of several parasites in a segment of the host is the result of multiple infection as in *Dictyomorpha*. Butler (1907) concluded that in some cases the protoplasts from several infecting planospores might fuse to form a plasmodium from which a single zoosporangium develops, but he also admitted the possibility that one of the protoplasts, in such instances, might become dominant and develop at the expense of the others.

The resting sporangia have been reported to develop asexually in all species except *R. allomycis* Foust. In this species Sörgel (1952) reported from a study of 15 stock cultures that resting sporangia are formed only when sexually compatible isolations are combined, and he concluded that this species is heterothallic. He did not observe fusion of isoplanogametes and concluded, therefore, that plasmogamy of sexually opposite and compatible protoplasts occurs within the host cell as in *Olpidium radicale*. Possibly heterothallism as well as homothallism will be demonstrated in other species of *Rozella* when they are studied intensively in single sporangial isolations and combinations.

DICTYOMORPHA Mullins

Amer. J. Bot. 48:378, 1961.

Pringsheimiella Couch, 1939. J. Elisha Mitchell Sci. Soc. 55:409.

Plate 8

The mode of development and morphology of *Dictyomorpha dioica* (Couch) Mullins, the type species of this monotypic genus, are so similar to those of *Rozella* members that it will probably be reduced to a species of the latter genus, as Mullins suggested. Because of this similarity it is not necessary to describe the developmental stages in detail. It is to be noted, however, that the zoosporangia of *D. dioica*

PLATE 7

Figs. 1, 2, 4, 5, 7-9. *Rozella chytriomycii* Karling. (Karling, 1946.)

Fig. 1. Motile elongate, 1.5 × 3 μ, and quiescent planospores.

Fig. 2. Infection of young *Chytriomyces hyalinus* zoosporangium; host nucleus a translucent body.

Fig. 4. Vacuolate parasite with 3 exit papillae filling host zoosporangium; host nucleus at right of vacuole.

Fig. 5. Simultaneous discharge of planospores from 3 orifices.

Fig. 7. Young stage of resting sporangium development; host nucleus in the center of the parasite.

Fig. 8. Later stage; host nucleus degenerating in vacuole of the parasite's incipient resting sporangium.

Fig. 9. Mature sparsely spiny resting sporangium.

Figs. 3, 10, 11. *Rozella cladochytrii* Karling. (Karling, 1942.)

Fig. 3. Young parasite in intercalary swelling of the rhizomycelium of *Nowakowskiella*, empty planospore case on the outside.

Fig. 10. Rare smooth-walled resting sporangium.

Fig. 11. Germination of a spiny resting sporangium.

Fig. 6. *Rozella apodyae-brachynematis* Cornu. Discharge of planospores from apical orifice of a zoosporangium in *Apodachlya*. (Cornu, 1872.)

Figs. 12-26. *Rozella allomycis* Foust. Figs. 12, 15-25 after Foust, 1937; fig. 13 after Koch, 1958; figs. 14, 26 after Sörgel, 1952.

Fig. 12. Motile planospores, 3-4 μ diam.

Fig. 13. Enlarged planospore showing nuclear cap, nucleus, lipoid body at right, and flagellum.

Fig. 14. Infection of *Allomyces*.

Fig. 15. Vacuolate parasite filling the host cell which has been delimited by a septum.

Fig. 16. Stage in formation of planospore initials.

Fig. 17. So-called disappearance stage of planospore initials.

Fig. 18. Mature zoosporangium of the parasite with an apical exit papilla and planospores; arrows indicate swirling of planospores immediately before dehiscence.

Figs. 19, 20. Deliquescence of papilla and discharge of planospores, resp.

Fig. 21. Portion of segmented host filament with empty sporangia at the apex and segments with resting sporangia below.

Fig. 22. Three terminal empty primary segments of the host divided into secondary segments, each of which contains a zoosporangium.

Fig. 23. Early stage of resting sporangium development in a host segment.

Fig. 24. Later stage, incipient resting sporangium surrounded by a hyaline zone.

Fig. 25. Later stage, spines developing in the hyaline zone.

Fig. 26. Germination of zygotic resting sporangium.

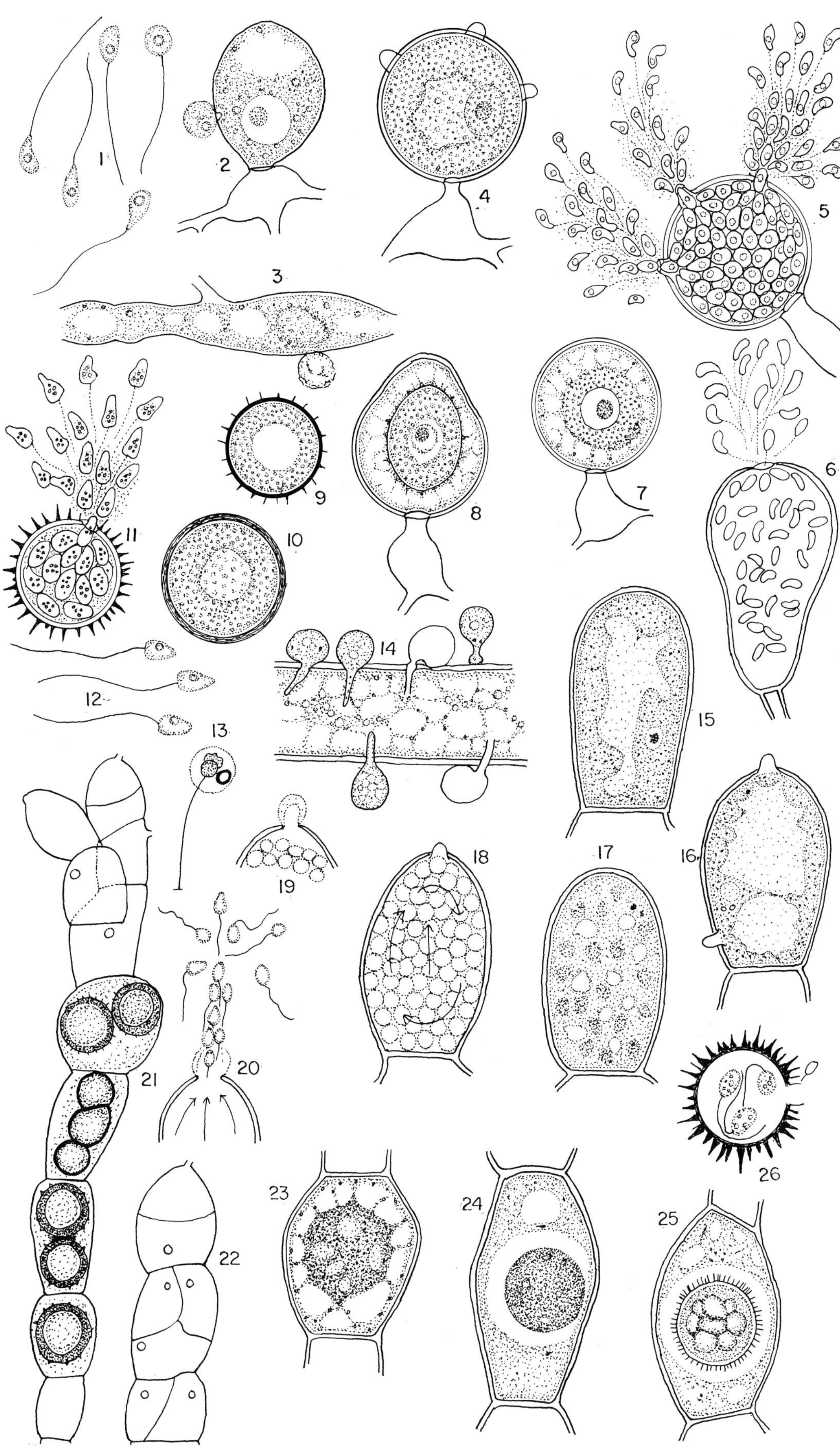

Rozella

do not fill the host cell completely as in *Rozella* but are distinguishable as separate entities. Multiple infections (fig. 7) result in the development of clusters of zoosporangia (fig. 1) which give the host hyphal tips or incipient sporangia an appearance similar to the sporangium of *Dictyuchus*. In zygote or resting sporangium development of the parasite the boundaries of the host spore initials and sporangium become thick-walled (figs. 32–34), and the transformed initials bear one or more zygotes or resting sporangia of the parasite.

In sexual reproduction compatible isoplanogametes fuse (figs. 20–24) to form a motile biflagellate zygote (fig. 25) which infects the host spore initials (figs. 26–29). From studies of compatible isolates and combinations of them Couch and Mullins demonstrated that *D. dioica* is heterothallic.

Dictyomorpha dioica has been found to parasitize 8 identified and 2 unidentified isolates of *Achlya*, and *Traustotheca clavata* by Mullins and Barksdale (1965).

PLASMOPHAGUS de Wildemann

Ann. Soc. Belge Micro. 19:219, 1895.

Plate 9, Figs. 1-5

This genus includes one species, *P. oedogoniorum* de Wildemann, which parasitizes filaments of *Oedogonium* and *Tribonema*. Its planospores (fig. 3) infect the algae and give rise to an irregular thallus which is difficult to distinguish at first from the host protoplasm. The presence of the parasite leads to enlargement and elongation of the host cells (figs. 1, 2, 4, 5), and prevents the formation of cross septa in *Oedogonium*. Division of the host nuclei is not inhibited at first, with the result that infected cells may become multinucleate. As the thallus matures its enveloping membrane or wall becomes visible (fig. 1), and the whole structure develops into a zoosporangium which completely or partly fills the host cell (figs. 1, 5). Its wall is usually closely pressed against that of the host, and may sometimes be indistinguishable from the latter. But it does not appear to be fused with the wall of the host as has been reported for *Rozella*. At maturity the zoosporangium develops an inconspicuous, non-projecting papilla, and its content is transformed into planospores. These escape in a slimy mass from the exit papilla (figs. 2, 5) and soon dart away. Sometimes several zoosporangia develop in an algal filament (fig. 4). So far, no resting spores have been observed in this genus.

Plasmophagus resembles *Olpidium* very closely, and further studies on its life cycle and development will probably show that it is identical with the latter genus. It, also, resembles *Rozella* in that it usually nearly fills the host cell, causes hypertrophy, and develops only an inconspicuous exit papilla. Although the wall of the sporangium may be closely pressed against that of the host, it is not fused (?) with the latter as has been described for *Rozella*.

DIPOLIUM Koch

(Unpublished)

Plate 9, Figs. 6-17

This monotypic genus parasitizes the gametangia of *Hyalotheca*, and is characterized by olpidioid zoosporangia with short or fairly long exit canals, and sexually formed spiny resting spores or zygotes. The development and dehiscence of the zoosporangia (figs. 7–9) are so similar to those of *Olpidium* that a description of them is unnecessary. Apparently,

PLATE 8

Figs. 1-37. *Dictyomorpha dioica* (Couch) Mullins. Figs. 1-5, 7-13, 15, 17, 19-29, 32-37 drawn from living material by Karling; figs. 6, 14 after Mullins. 1961; figs. 16, 18, 30, 31 drawn from parts of photographs by Mullins, 1961; fig. 34 drawn from photograph by Couch.

Fig. 1. Hyphal tips of *Achyla flagellata* with 10 sporangia in various stages of maturity.

Figs. 2-5. Various shapes of motile planospores.

Fig. 6. Enlarged view of stained planospore with nuclear cap, nucleus, rhizoplast, blepharoplast, and whiplash flagellum in a linear descending series, and a lipoid body on the right.

Fig. 7. Dense infection of hyphal tip by planospores.

Figs. 8-12. Stages of infection of host.

Fig. 13. Parasites in the spore initials of the host.

Fig. 14. Tetranucleate parasite in host spore initial with host nucleus adjacent.

Fig. 15. Vacuolate incipient zoosporangium.

Fig. 16. Similar stage in sectioned and stained material.

Fig. 17. Later stage resembling cleavage.

Fig. 18. Surface view of multinucleate parasite zoosporangium stained with aceto-orcein.

Fig. 19. Discharge of planospores.

Fig. 20. Pairing of compatible gametes.

Figs. 21-23. Changes in shape and position of gametes during fusion.

Fig. 24. Fusion of lipoid bodies.

Fig. 25. Motile biflagellate zygote.

Fig. 26-29. Infection of the host by the zygote.

Fig. 30. Three young zygotes enveloped by host protoplasm of hyphal tip.

Fig. 31. Young binucleate resting sporangium after treatment with ribonuclease and aceto-orcein.

Fig. 32. Incipient resting sporangia in spore initials of the host.

Fig. 33. "Containing" cell with 4 immature resting sporangia or zygotes.

Fig. 34. Two "containing" cells with one resting sporangium or zygote in each.

Fig. 35, 36. Mature resting sporangia or zygotes with reticulate and almost spiny outer walls.

Fig. 37. Germination of resting sporangium or zygote.

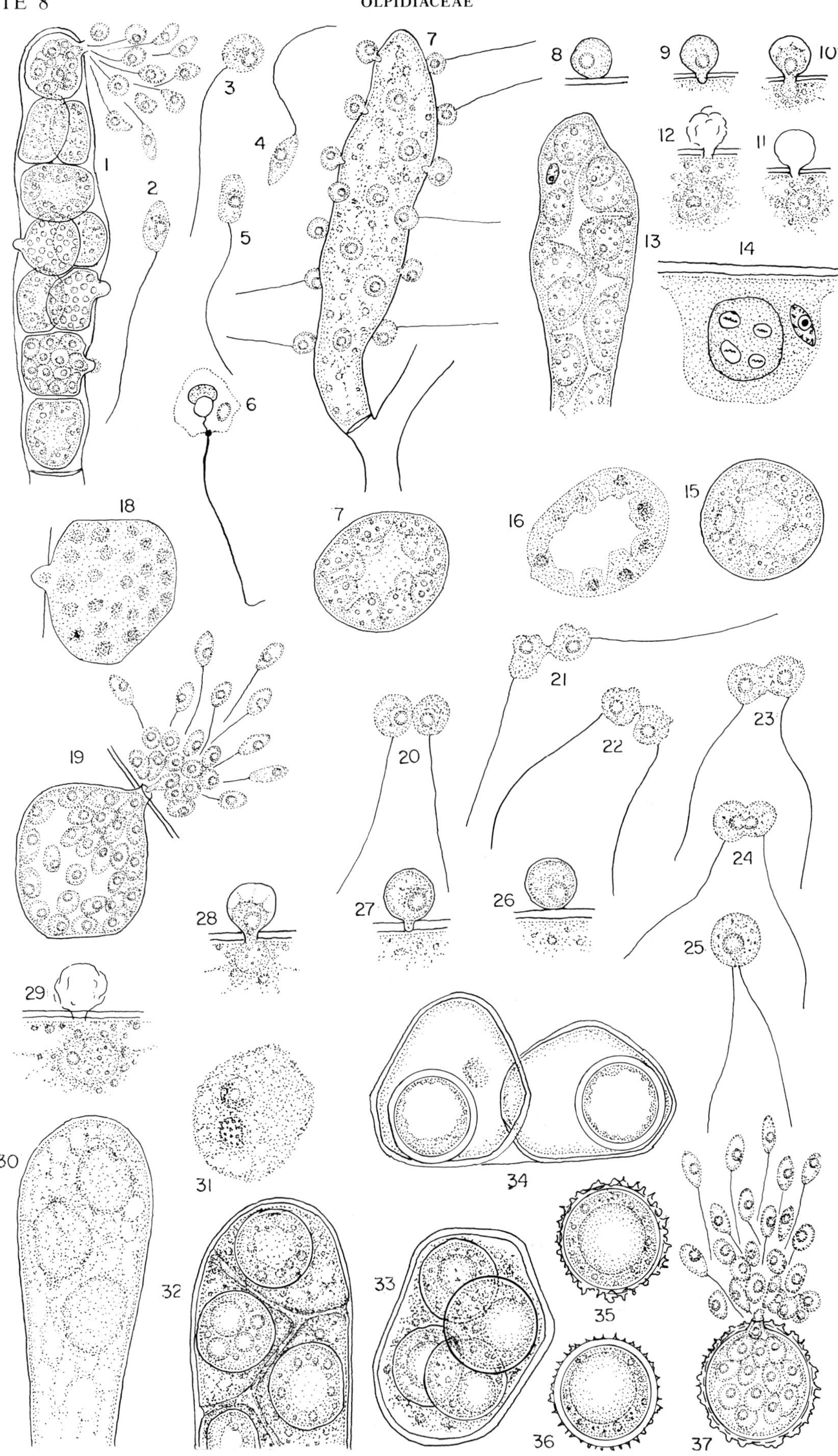

Dictyomorpha

the planospores (fig. 6) infect the host as in *Olpidium* species, and give rise to (figs. 7–9, 11) a holocarpic zoosporangium. In some instances multiple infection occurs (fig. 10).

Sexual reproduction in this genus is different in some respect from that of other species of the Olpidiaceae in that it is distinctly heterogamous. A small male thallus or gametangium becomes attached to a larger female gametangium and apparently develops a short conjugation papilla (figs. 12, 14), according to figures contributed by Koch prior to publication. Its content flows into the female gametangium, and after plasmogamy is completed the zygote develops spines on its surface (figs. 13–16). The empty male gametangium and conjugation tube (figs. 14, 15) become thick-walled and persist on the surface of the zygote or resting spore. Germination of the latter has not been observed.

SPHAERITA Dangeard

Bull. Soc. Bot. France 33:241, 1886a;
Ann. Sci. Nat. Bot. 4:277, 1886b

Plate 10, figs. 1-19

Since the time Dangeard created *Sphaerita* for a chytrid parasite in the cytoplasm of 2 rhizopods and *Euglena*, this genus has become a dumping ground for several other parasites of protozoa and some algae which develop nonmotile endospores in sporangia with a morula-like appearance (plate 13). Dangeard (1886a) named the parasite in *Nuclearia* and *Heterophrys* as *S. endogena* Dang. and described the movement of the uniflagellate planospores as jerky and very rapid or consisting of a single rotation because of the movement of a single, strongly curved, anteriorly attached flagellum (fig. 13). Later (1866b) he gave the same name to a parasite in *Euglena viridis*, illustrated the life cycle of *S. endogena*, and described it in greater detail in 1889. In 1900 Chatton and Broadsky maintained that the parasite they found in *Amoeba limax* is different from the one occurring in *Euglena*. Accordingly, they retained the name *S. endogena* for the parasite in *Amoeba* and renamed the one in *Euglena* as *S. dangeardii*. Later Dangeard (1933) suggested that because the latter species was better known than *S. endogena* he would designate it as the type of the genus if it did not violate the rules of nomenclature. However, his (1886a) binomial *S. endogena* was validly published (combined genus and species), and it must typify the genus as pointed out by Karling (1973). In the event *S. endogena* Chatton and Broadsky non Dang. proves to be a different species, it might be transferred to *Morella* and renamed as a new combination if this genus is validated, emended, and legitimatized.

As interpreted by Karling (1973) *Sphaerita* is a genus of olpidioid, monocentric, holocarpic endobiotic inoperculate parasites in which uniflagellate planospores and resting sporangia have been reported. As such it is limited to relatively few species for the reason that most investigators did not report the occurrence of flagellate planospores in the species they identified. *Sphaerita endogena* Dang., *S. dangeardii* Chatton and Broadsky and possibly *S. simplex* Dang. (1929) and the species described by Ivanic (1925) in *Amoeba jollos* belong in this genus. Dangeard did not illustrate flagella on the planospores of *S. simplex*, but Ivanic described the discharge of planospores *in vivo* in his species and their copulation without illustrating these stages. Possibly, his species is identical with *S. endogena* Dang. Resting sporangia have not been observed in the latter species as it occurs in rhizopods, and it is not unlikely that those illustrated by Dangeard (1886b) in *Euglena* relate instead to *Pseudosphaerita* because he stated that he could see a 2nd flagellum on some of the planospores from a germinated resting sporangium. The biflagellate planospores described by Dangeard (1889, 1895), Pumaly (1927), Cejp (1935), Réyes and Gomez (1958, 1961), Réyes (1963), and others probably relate to *Pseudosphaerita*, also.

PLATE 9

Figs. 1-5. *Plasmophagus oedogoniorum* de Wildemann. Figs. 1-4 after de Wildemann, 1895; fig. 5 after Sparrow, 1933.

Fig. 1. Thallus filling 2 cells and part of a 3rd cell of *Oedogonium* sp. and causing hypertrophy.

Fig. 2. Zoosporangium discharging planospores.

Fig. 3. Ovoid, 3 × 2 μ, planospores with a small eccentric hyaline refractive globule.

Fig. 4. Part of a host filament with 3 empty zoosporangia.

Fig. 5. Enlarged cell of *Tribonema bombycina* with a zoosporangium discharging planospores.

Figs. 6-17. *Dipolium philosexualis* Koch. After Koch (unpublished).

Fig. 6. Spherical planospores with a conspicuous lipid globule, nucleus, nucleolus, and possibly a nuclear cap.

Fig. 7. Zoosporangium inside of gametangium of *Hyalotheca*; resting spore with an attached male thallus beneath.

Fig. 8. Large zoosporangium with a short exit tube and delimited planospores shortly before dehiscence.

Fig. 9. Discharge of planospores.

Fig. 10. Host cell with 6 small zoosporangia.

Fig. 11. Partly empty zoosporangium with a fairly long exit canal.

Fig. 12. Female gametangium with small attached male cell; host cell omitted.

Fig. 13. Later stage; male cell or gametangium partly empty; zygote or resting spore developing spines.

Fig. 14. Young zygote with an empty attached male gametangium.

Figs. 15, 16. Mature zygotes with thick walls and spines, filled with refractive globules.

Fig. 17. Variations in the shapes of the spines on the resting zygotes.

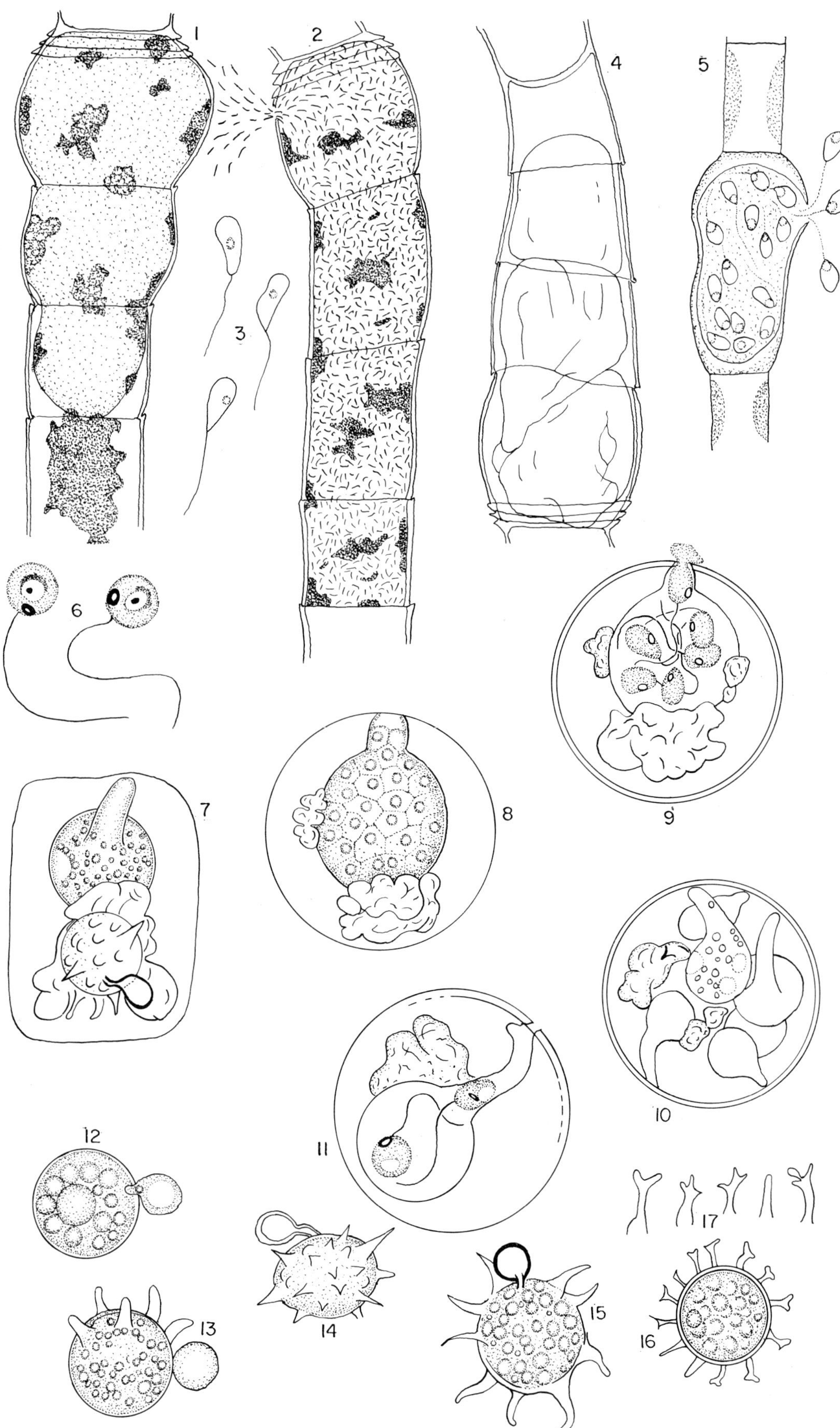

Plasmophagus, Dipolium

Infection of rhizopods apparently occurs by engulfment of the planospores of *S. endogena* Dang. (fig. 4), but the manner of infection by *S. dangeardii* is not fully known. Dangeard (1889) reported that the planospores fix themselves onto the host, but he did not describe their germination, infection and entry. Gojdics (1939) and Jahn (1939) suggested that the planospores enter through the cytopharynx. Within the host they develop into a multinucleate thallus which has been described as a plasmodium but is enveloped by a membrane (figs. 7, 8) as in species of *Olpidium*. At maturity it is converted holocarpically into a zoosporangium (figs. 10, 11) and may bear 2 oppositely directed exit papillae (fig. 9). In some cases papillae appear to be lacking, and the planospores are probably discharged through a rupture of the sporangium wall (figs. 1, 12). Later, spiny or smooth resting sporangia are reported to develop in the host (figs. 15–18), but it is not certain that these relate to *Sphaerita* inasmuch as Dangeard (1889, p. 49) described the planospores produced by them (fig. 18), as being biflagellate and heterocont. Karling (1973) suggested that they might relate to *Pseudosphaerita* instead.

The evidence of gametic fusions in *Sphaerita* is meager and not all conclusive. Dangeard (1889) illustrated pairing of planospores (fig. 13a) and noted that these simulated fusions of gametes which usually separated later. By 1933, however, he stated that such pairings probably represent an early stage of sexual reproduction similar to that of *Olpidium viciae* Kusano and results in the development of resting sporangia. But inasmuch as his fig. 8, pl. III (1889) includes both uni- and biflagellate planospores, it is not certain whether the biflagellate ones and pairs relate to *Sphaerita* or *Pseudosphaerita*. Ivanic stated that he had often observed discharge of flagellate planospores *in vivo* and their copulation in a *Sphaerita* in *Amoeba jollos*, but did not illustrate the fusions nor the number and point of insertion of the flagella. Perez Réyes and Salas Gomez (1961) and Perez Réyes (1963) believed that in what they call *S . dangeardii* the presence of zoosporangia and resting sporangia might represent alternating sexual and asexual generations.

Other parasites with morula-like sporangia which have been identified as *Sphaerita* species are shown in relation to *Morella endamoebae* (Becker) Réyes in plate 13.

OLPIDIOMORPHA Scherffel

Arch. f. Protistenk. 54:515, 1926.

Plate 10, Figs. 20-23

This monotypic genus has been observed but once in the zoocysts of *Pseudospora leptoderma*, living in filaments of *Vaucheria*. It resembles *Ol-*

PLATE 10

Figs. 1, 7-14, *Sphaerita dangeardii* Chatton and Broadsky; Figs. 2-5, *S. endogena* Dangeard; Figs. 15-18, *S. dangeardii*? (possibly *Pseudosphaerita euglenae*). Figs. 1-5 after Dangeard, 1886b; Figs. 6, 12, 15-18 after Dangeard, 1889; Figs. 7, 8, 10, 11, 13a after Dangeard, 1895; Figs. 3, 14 after Dangeard, 1933; Fig. 9 after Serbinov, 1907.

Fig. 1. Discharge of planospores from a zoosporangium in *Euglena viridis*.

Fig. 2. A zoosporangium with planospores in *Nuclearia simplex*.

Fig. 3. Discharge of spores from *N. simplex*; some spores are apparently uniflagellate.

Fig. 4. Ingestion of spores by *Nuclearia*.

Fig. 5. Discharge of spores from a zoosporangium in *Heterophrys dispersa*; some spores are apparently uniflagellate.

Fig. 6. Parasite in *Euglena sanguinea*.

Fig. 7. Young multinucleate thallus from an encysted *Euglena*.

Fig. 8. Mature multinucleate incipient zoosporangium shortly before cleavage.

Fig. 9. Broadly fusiform zoosporangium with exit papillae at opposite ends.

Fig. 10. Zoosporangium with cleavage segments.

Fig. 11. Zoosporangium with planospores shortly before dehiscence.

Fig. 12. Discharge of planospores.

Fig. 13. Enlarged view of planospores, 1.5–2 μ diam.; with a curved anteriorly attached but posteriorly directed flagellum; arrow indicates direction of motion.

Fig. 13a. Pairing (fusion?) of planospores.

Fig. 14. Encysted planospores.

Fig. 15. Young spiny resting sporangium in an encysted *Euglena*.

Fig. 16. Expulsion of spiny resting sporangium.

Fig. 17. Resting sporangium with planospores.

Fig. 18. Germinated resting sporangium discharging planospores.

Fig. 19. ?*Sphaerita trachelomonadis* Skvortzow. Spiny resting spore in *Trachelomanas teres* var. *glabra*. (Skvortzow, 1927.)

Figs. 20-23. *Olpidiomorpha pseudosporae* Scherffel. (Scherffel, 1926.)

Figs. 20, 21. Zoosporangia with long exit tubes in zoocysts of *Pseudospora leptoderma* which were living in dead *Vaucheria* filaments.

Fig. 22. Ovoid planospores, 2 $\times$ 3-4 μ, with a subapically attached flagellum; arrow indicated direction of motion.

Fig. 23. Planospores enlarged; inside and surface views showing basal vacuole, ring of globules at the apex, and subapically attached flagellum which is posteriorly directed.

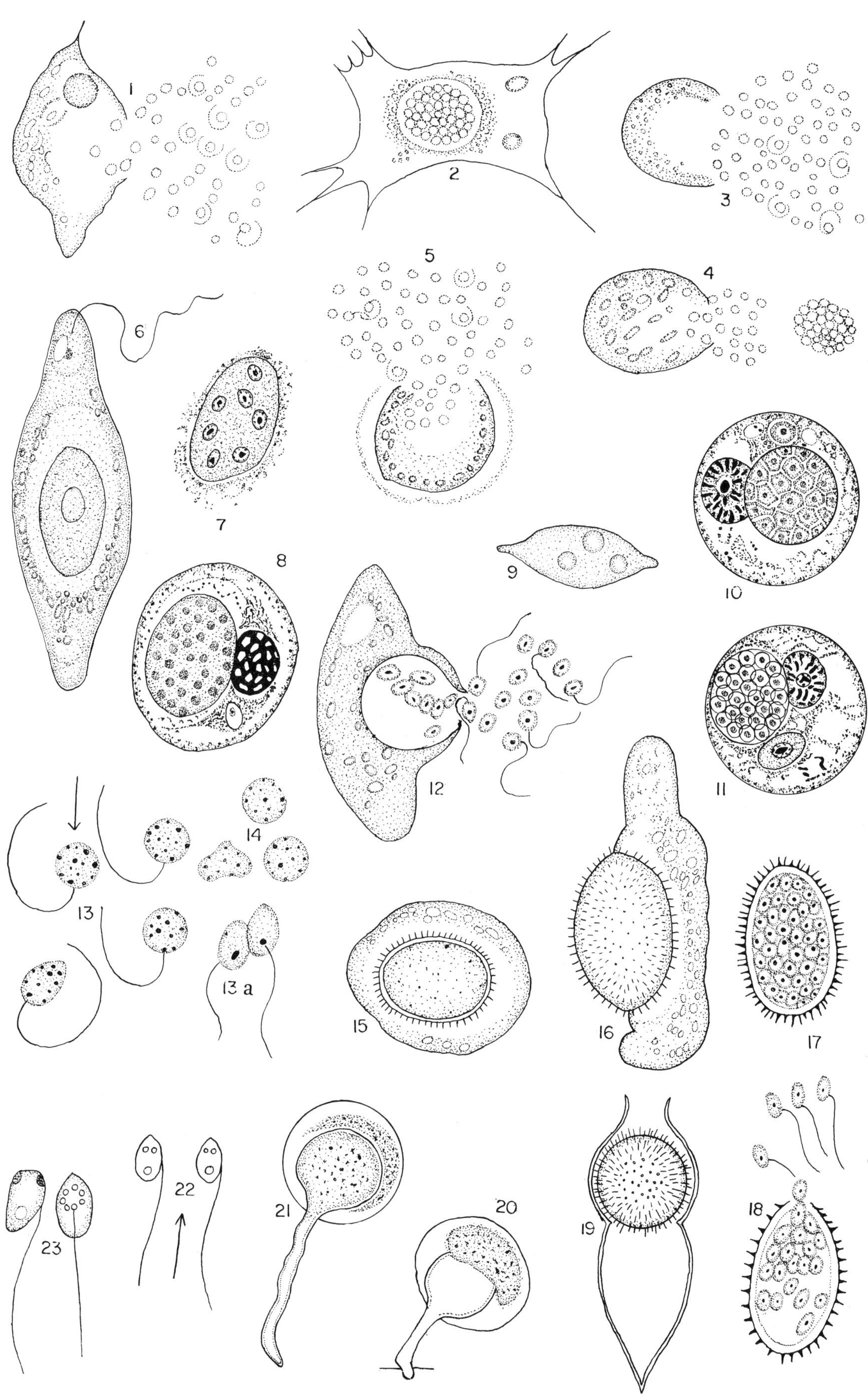

Sphaerita

pidium closely, but differs in that the content of the thallus lacks fat globules and is refractive with glistening irregular lumps, according to Scherffel. Discharge of planospores is through a single relatively long exit tube (figs. 20, 21), and the planospores bear a lateral of subapically attached trailing flagellum, contain a basal vacuole, and an anterior ring of refractive granules (figs. 22, 23). Resting sporangia or spores are unknown in this genus.

GENERA OF DOUBTFUL AFFINITY

Several genera and species parasitic in microorganisms have either been included in the Olpidiaceae or described as having affinities with members of this family, but their identity and relationships with the olpidioid holocarpic chytrids are not at all certain. Protozoologists and parasitologists established these genera at a time when the limits and definitions of the chytrids were not sharply defined, and affinities which seemed fairly close then appear much less so at the present time. Whether they are to be regarded as close or distant relatives of the Olpidiaceae is obviously open to question, but a few of them are illustrated and described herewith to emphasize their similarities to as well as differences from the true olpidioid chytrids. Planospores borne in zoosporangia with a single flagellum posteriorly, or axially, or subapically attached and directed backward during motility have been reported in *Reesia* Fisch, *Chytridhaema* Moniez, *Blastulidium* Perez, *Coelomycidium* Debaisieux, *Endoblastidium* Codreanu, and from a body in the nucleus of *Pseudospora volvocis* Cienkowski (Robertson, 1905) which Kirby (1941) interpreted to be a species of *Nucleophaga*.

REESIA Fisch

Sitzungsb. Phys.-Med. Soc. Erlangen 16:31, 1884.

Plate 11

The classification and relationships of this genus have become more confusing with Wagner's (1969) report that the planospores of *R. amoeboides* Fisch are posteriorly uniflagellate, although Fisch had described them as anteriorly uniflagellate. In view of Wagner's observations *Reesia* is described and illustrated here in relation to the doubtful genera of the Olpidiaceae without any implications that it is a taxon of true chytrids or anisochytrids.

On the basis of Fisch's observations Karling (1943) classified this genus in the family Anisolpidiaceae of the anisochytrids or Hyphochytriales. Sparrow (1960), also, included it as an imperfectly known genus of this order. Earlier mycologists such as Fischer (1892), Schroeter (1885, 1893–1897) and Minden (1911–1915) apparently did not regard flagellar insertion and position as a significant criterion of classification. Schroeter (1885) renamed Fisch's *R. lemnae* as *Olpidium lemnae*, but later (1893) he listed *Reesia* as the first and separate genus of the Olpidiaceae. Fischer regarded this genus as a doubtful member of the "Olpidien," and Minden listed it as a synonym of *Olpidium*.

Reesia is reported to include 3 species, *R. amoeboides* Fisch and *R. lemnae* Fisch parasitic in species of *Lemna*, and *R. cladophorae* Fisch in *Cladophora*. Fisch described the second species as *Chytridium lemnae*, but inasmuch as he reported that the planospores are anteriorly uniflagellate as in *R. amoeboides* Karling (1943) renamed it and placed it in *Reesia*. Very little is known about the third species which Fischer referred to as *R. cladophorae*, and it differs from the two previous species by the presence of a mycelial or rhizoidal thread on the thallus. The latter structure excludes it from *Reesia*, and Fischer and Minden believed that it might relate to *Olpidium entophytum*.

PLATE 11

Figs. 1-13, 17-26. *Reesia amoeboides* Fisch in *Lemna* species. Figs. 1-13 after Fisch, 1884; figs. 17-26 after Wagner, 1969.

Figs. 1, 2. Ovoid anteriorly uniflagellate planospores with a hyaline refractive globule.

Figs. 3, 4. Amoeboid thalli in the host cell.

Fig. 5. Rounded up spherical thallus with a distinct wall.

Fig. 6. Spherical zoosporangium with a long exit tube and planospores.

Fig. 7. Empty zoosporangium with a curved exit tube.

Figs. 8, 9. Pairing and fusion of isoplanogametes.

Figs. 10, 11. Biflagellate and quiescent zygotes.

Figs. 12, 13. Resting spores or mature zygotes (?) with a smooth light-brown exospore and a hyaline endospore and several large refractive globules.

Fig. 17. "Host cell with two large" posteriorly uniflagellate planospores.

Fig. 18. "Zoospore development" (?).

Fig. 19. Planospores with greatly elongate posterior flagella.

Fig. 20. Amoeboid flagellate thallus migrating through the host cell wall.

Fig. 21. Amoeboid thallus prior to resting spore development.

Figs. 22, 23. Immature and mature resting spores, resp.

Fig. 24. Empty germinated resting spores (?) with long and contorted exit tubes.

Fig. 25. "Amoeboid thallus with pseudopodia."

Fig. 26. "Solitary zoosporangium."

Figs. 14-16. *Reesia lemnae* (Fisch) Karling in *Lemna*. (Fisch, 1884.)

Fig. 14. Zoosporangium discharging anteriorly uniflagellate (?) planospores.

Figs. 15, 16. Two spherical resting spores with large refractive globules; exospore light-golden, endospore hyaline.

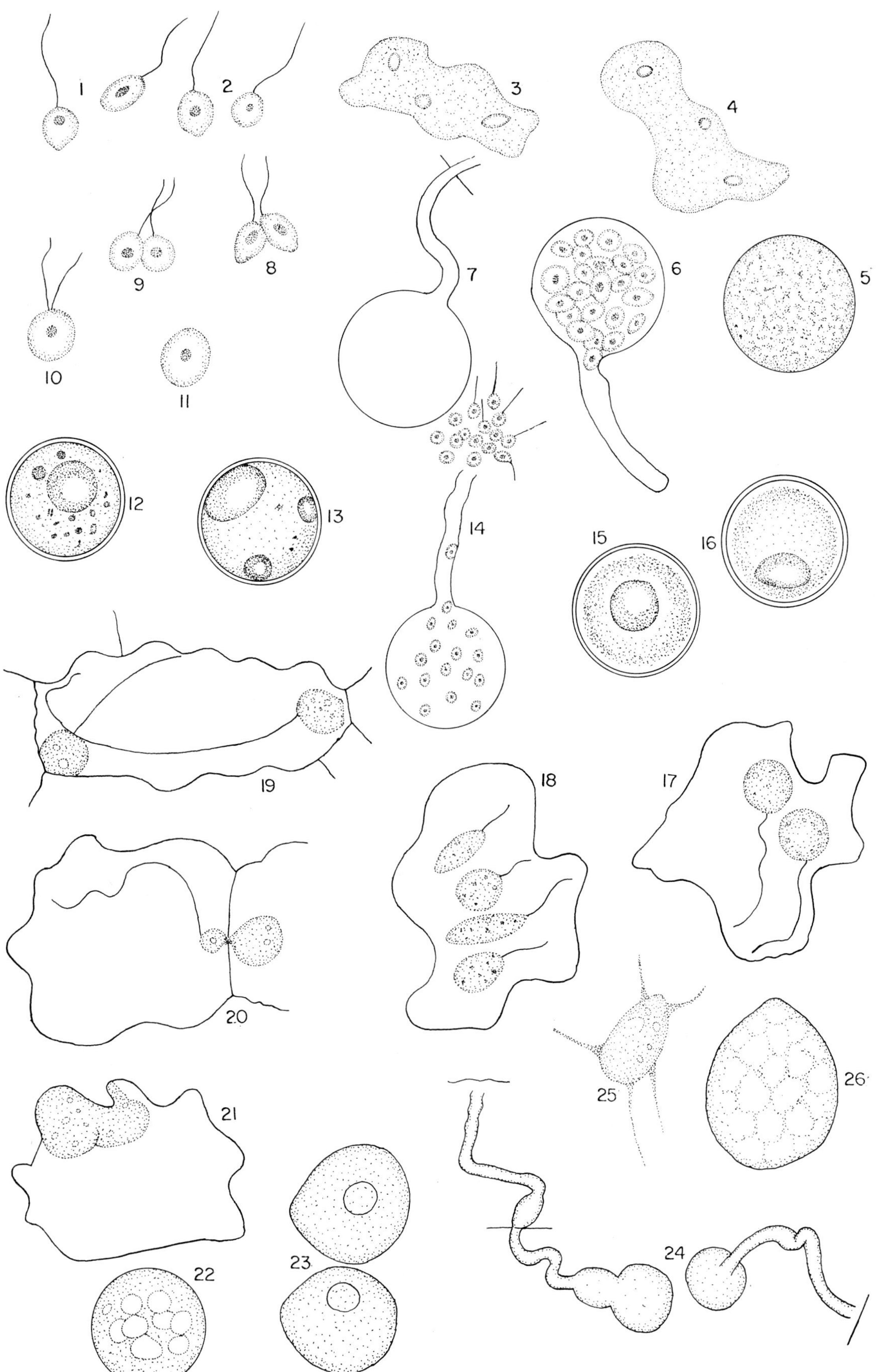

Reesia

Fisch, also, pointed out that *Reesia* is very similar to *Olpidium* except for its amoeboid thallus and anteriorly uniflagellate planospores. According to him, the latter are quite large in *R. amoeboides*, although he gave no measurements of their size. They are spherical to ovoid (figs. 1, 2) and hyaline with a conspicuous refractive globule. They penetrate the host by their flagellum (?) and develop into a thallus which is amoeboid (figs. 3, 4) with several pseudopodia in the early developmental stages. It is reported to be naked but immiscible with the host protoplasm at this stage, but eventually it rounds up and develops a distinct wall or membrane (fig. 5). Later, it is transformed into a zoosporangium with a long cylindrical straight or curved exit canal (figs. 6, 7) which may project considerably beyond the surface of the host cell. The discharged planospores may be quiescent in a globular mass for a few moments before dispersing or swim directly away after emerging.

Fisch's account indicates that the planospores may be facultative. They may either function as planospores and give rise to additional sporangial thalli, or as isoplanogametes and fuse in pairs (figs. 8, 9) to form motile biflagellate zygotes (fig. 10). These infect the host and develop into resting sporangia (figs. 12, 13) which have a yellowish to brown exospore, a hyaline endospore, and contain several refractive globules. The exospore is ruptured during germination by the expanding endospore which forms a protruding vesicle in which the planospores are delimited. Sexual reproduction is unknown in *R. lemnae* although light-golden resting sporangia (figs. 15, 16) occur and germinate as in *R. amoeboides*. The walls of these resting sporangia stain light-blue when tested with chloroiodide of zinc.

Compared with Fisch's above description of *R. amoeboides* Wagner's account of this species is somewhat confusing, and her illustrations do not clearly indicate the successive stages which she observed. She reported that small, ovoid to spherical, 1-1.5 × 2-3 μ, hyaline, posteriorly uniflagellate planospores are produced in zoosporangia (fig. 26) with a short discharge papilla. In addition she found large, 8 × 10 μ, planospores aggregated in the host epidermal cells (figs. 17, 18) which she presumed to be posteriorly uniflagellate. After a period of motility they "migrate through broken cells, or perhaps through very large wall pits." Initially they are ovoid (fig. 18), but as they enlarge and become spherical the flagellum elongates to a length about 6 times the greatest diameter of the planospores (fig. 19). Later, they develop into flagellate amoeboid cells which migrate for 24 to 48 hours before developing into resting sporangia (figs. 22, 23). No sexual fusions were observed in relation to their development. Also, germination was not observed, but Wagner illustrated several empty light- to yellowish-brown bodies with long irregular exit tubes which penetrated adjacent cell walls and extended to the outside of the host (fig. 24). However, her figures of these bodies suggest to this writer that they might relate to empty zoosporangia. Her account of the fate of the small planospores is incomplete and needs clarification. Do they reinfect the host and give rise to additional sporangial thalli, or enlarge in the host cell and become the large flagellate planospores?

In this writer's opinion the question of whether *Reesia* is a chytrid or an anisochytrid has not been answered conclusively. The answer turns on the insertion and direction of the flagellum, and if Wagner's

PLATE 12

Figs. 1-4. Invasion of *Endolimax williamsi* by two spores of *Nucleophaga intestinalis* Brug; parasite lying next to nucleus; entrance into nucleus; and parasite within nucleus, respectively. (Brug, 1926.)

Figs. 5, 6, 8-15. *Nucleophaga amoebae* Dang. (Dangeard, 1894-1895.)

Fig. 5. Uninucleate parasite in enlarged nucleolus.

Fig. 6. Multinucleate stage of parasite within the nucleus.

Fig. 8. Nucleus of parasite.

Fig. 9. Amoeba whose nucleus contains one sporangium filled with spores.

Fig. 10. Four parasites within the nucleolus.

Figs. 11, 12. Three and five incipient sporangia within host nucleus, resp.

Fig. 13. Nucleus with four sporangia filled with uninucleate spores.

Fig. 14. Two large sporangia filled with spores.

Fig. 15. Partly empty sporangia in nucleus of a dead amoeba.

Fig. 16. Cyst or resting spore of *N. peranemae* Hollande and de Balsac. (Hollande and de Balsac, 1942.)

Figs. 17-21. Uniflagellate gametes, fusion and development of an amoeboid zygote in a nuclear parasite of *Amoeba vespertilio*. (Doflein, 1907.)

Figs. 7, 22-24. After Robertson, 1905.

Fig. 7. Multinucleate body in enlarged nucleus of *Pseudospora volvocis*.

Fig. 22. Hypertropheid nucleus of *Pseudospora volvocis* filled with spherical bodies; nucleolus compressed to one side.

Fig. 25. Gametes from a hypertrophied nucleus of *Pseudospora volvocis* with a curved subapical flagellum.

Fig. 24. Fusion of gametes to form a biflagellate zygote.

Figs. 25-36. *Blastulidium paedophthorum* Perez. Figs. 25-28, 30-32, 35 after Perez, 1903, 1905; figs. 29, 33, 34, 36 after Jirovec, 1955.

Fig. 25. Ellipsoidal vacuolate multinucleate thallus.

Figs. 26, 27. Cleavage of protoplasm into uninucleate segments or spores.

Figs. 28, 29. Spores.

Figs. 30-34. Yeast-like budding, branching, and septate thalli.

Fig. 35. Enlarged view of a vacuolate segment.

Fig. 36. Content of an elongate segment cleaving into spores.

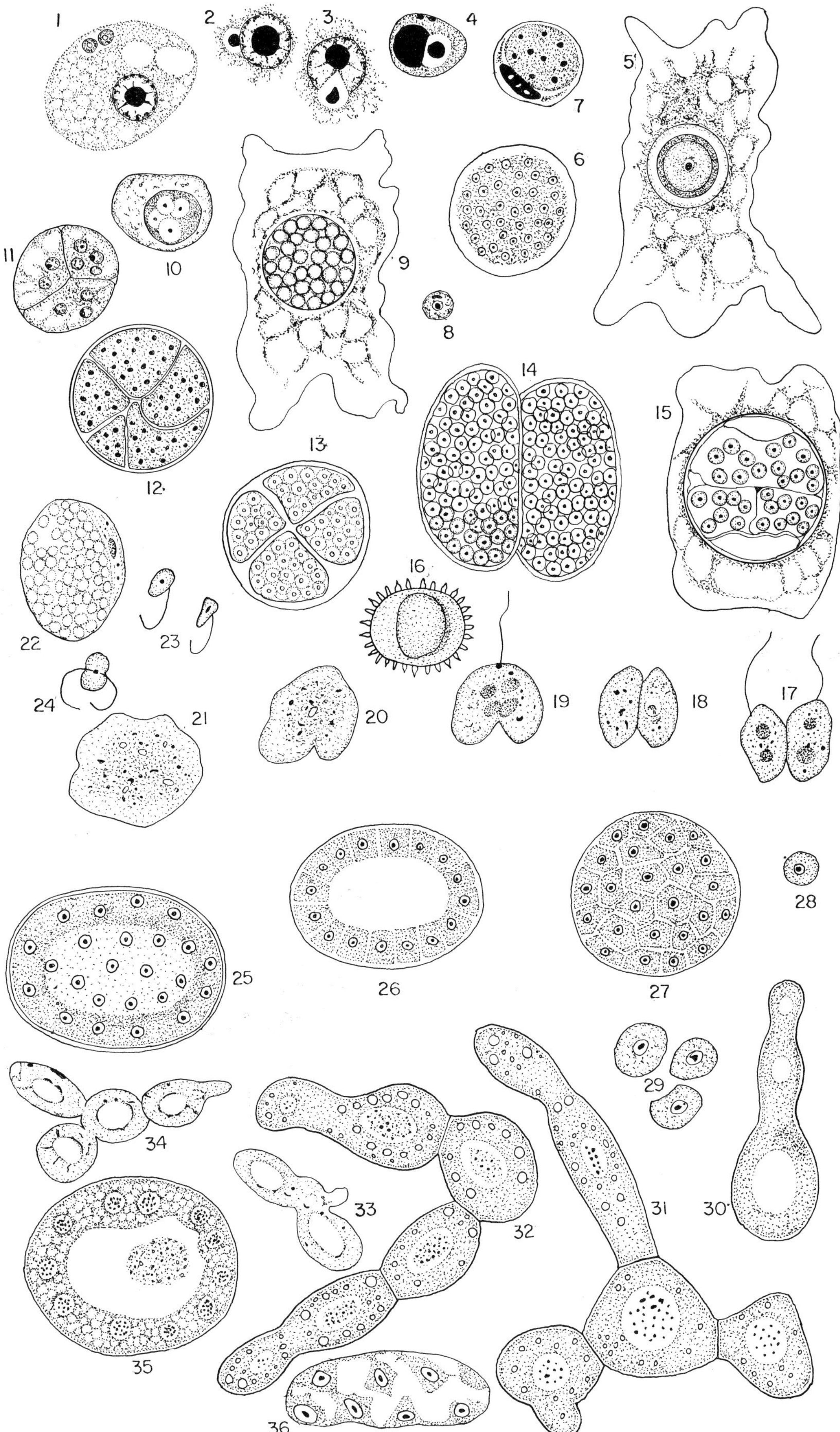

Nucleophaga, Blastulidium

report of posteriorly uniflagellate planospores is confirmed this genus should be removed from the anisochytrids. It should be noted here that Fisch's description of flagellar position is somewhat confusing, also. While he definitely stated that the flagellum is anterior, he, nevertheless, described the planospores as being exactly similar to those of previously known chytrids. Such contradictions may lead one to doubt the accuracy of his observations, and these discrepancies may have influenced Schroeter's, Fischer's, and Minden's interpretations. Nonetheless, Fisch's report of anteriorly uniflagellate planospores in *Olpidium*-like species has become more significant with the subsequent discovery of several olpidioid anisochytrids with similar planospores (see Karling, 1943, 1968; Canter, 1950). The inconsistencies between Fisch's and Wagner's accounts of the life history of *R. amoeboides* might be due to differences in species. Possibly, Wagner may have been studying a different organism.

NUCLEOPHAGA Dangeard

Le Botan. 4:214, 1895

Plate 12, Figs. 1-24

This genus includes several identified but poorly-known parasites of the nuclei of amoebae, flagellates, and some algae which resemble very closely species of *Morella* or others described as aplanosporic *Sphaerita* spp., and differ only by their presence in the nuclei of the hosts. The sporangia and spores of *Nucleophaga* were frequently mistaken by the early protozoologists for a reproductive phase of the hosts until Dangeard showed that they relate to a parasite. Apparently, liberated spores are engulfed by the host (fig. 1), migrate to the nucleus (fig. 2), and enter it (fig. 3). According to Dangeard, the spore enters and infects the nucleolus (fig. 5), but Brug (1926) reported that after entering the nucleus it develops between the nuclear membrane and nucleolus (fig. 4), compressing the latter to one side. As a result of infection the nucleolus and chromatic material are absorbed or destroyed by the parasite, and at the same time the host nucleus may enlarge 3 to more times its original diameter. Several parasites may infect the nucleus (fig. 10) and develop into sporangia (figs. 11–14). The thallus of the parasite becomes multinucleate (figs. 6, 11, 12) and is gradually transformed into a sporangium full of endospores (figs. 9, 13, 14). At this stage it is strikingly similar in appearance to the sporangial phase of species of *Morella*. The wall of the sporangium apparently ruptures, liberating the spores in the body of the dying host from which they eventually disperse into the surrounding medium.

No conclusive evidence of flagellate planospores has been found in *Nucleophaga*. Doflein (1907) described and illustrated motile cells from parasitized nuclei of *Amoeba* which conjugate in pairs (figs. 17–19) to form biflagellate and amoeboid zygotes (figs. 20, 21), but inasmuch as the flagellum is said to be anterior, it is not certain that the parasite is a *Nucleophaga* species. As noted earlier, Kirby (1941) believed it probable that the uniflagellate gametes (fig. 23) described by Robertson (1905) in *Pseudospora volvocis* Cienkowski might relate to a species of *Nucleophaga* instead of to a monad. Robertson reported that the nucleus of this organism becomes greatly enlarged and filled with a multinucleate globular body (fig. 7) which flattens the nucleolus against the inner periphery. Eventually, the nucleus becomes filled with small globular bodies (fig. 22) which pierce the nuclear membrane and escape as uniflagellate gametes (fig. 23). The flagellum is slightly curled and attached latterly or a short distance back of the anterior end, extends posteriorly, and propels the gamete forward. In these characteristics the gametes are somewhat similar to the planospores of *Sphaerita*. Shortly after becoming free, gametes from the same or different hypertrophied nuclei fuse in pairs to form biflagellate zygotes (fig. 24). These withdraw their flagella after a time, encyst as globular bodies, and later become amoeboid and infect *Volvox*. Robertson's figures of hypertrophied host nuclei with a globular body (fig. 7) which has pushed the nucleolus aside are very similar to those illustrated by subsequent workers of nuclei parasitized by *Nucleophaga*, and Kirby's interpretation may be correct. So far, resting spores or cysts have been found only in *N. peranemae* Hollande and de Balsac (1942) (fig. 16).

BLASTULIDIUM Perez

C. R. Soc. Biol. 55: 715. 1903

Plate 12, Figs. 25-36

Blastulidium includes but one species, *B. paedophthorum* Perez, which parasitizes the eggs and larvae of various species of *Daphnia, Sida, Simocephalus, Chydora, Lynceus, Coretha* and *Ceriodaphnia*. Perez regarded this species as a haplosporidian but Chatton (1908) considered it to be a chytrid, related to *Olpidium* and *Synchytrium*, after he had observed ovoid motile planospores with a single axially inserted flagellum. Such planospores were developed in zoosporangia and discharged through a short neck. Unfortunately, he did not illustrate any of the developmental stages and planospores, and it is not clear what he meant by the term "axial insertion." His observation of planospores has not been confirmed, and at the present time there is no illustrated evidence for including this genus among the chytrids.

The mature thallus exhibits a variety of forms. It usually consists of an ovoid, vacuolate multinucleate body (fig. 25) whose protoplasm cleaves into uninucleate segments (figs. 26, 27, 36). These round up as spores (fig. 28) and are released by the rupture of the thallus membrane. Perez (1903) also described some ellipsoidal bodies attached to the exterior of the hosts which he believed might represent resting stages of the parasite, but Chatton regarded these as single or conjugated planospores which had become attached to and infected the host. Chatton, also, observed thick-walled citriform bodies which he believed is a resting stage. These, he thought, might have been formed after the parasite had left the dead host by amoeboid movement, and encysted.

Later, Perez (1905) observed budding (fig. 30), branching, and septation (figs. 31, 32) of the thalli, and his observations have been confirmed by Jirovec (1955), who found typical yeast-like budding of thalli (figs. 33, 34).

MORELLA Perez Réyes

Rev. Soc. Mex. Hist. Nat. 24:5, 1963.

Plate 13, Figs. 1-9

Réyes created this genus without diagnosis for parasites of intestinal protozoa which produce non-motile endospores in morula-like sporangia and in which he could not find biflagellate planospores. He designated *M. endoamoebae* Réyes (*Sphaerita endoamoebae* Becker) as the type species. As such this genus is a questionable one, but the word *Morella* is, nevertheless, an excellent descriptive one for the general appearance of such parasites in their mature sporangial stage (figs. 8, 13, 14, 17). Karling (1973) suggested that should *Morella* prove to be valid it might be amended and extended to include all holocarpic, olpidioid species with sporangia filled with non-motile endospores, regardless of whether or not they parasitize intestinal protozoa. In this sense *Morella* might alleviate temporarily, at least, some of the taxonomic problems relative to the many morula-like species which have been classified in *Sphaerita*. Almost a score have been named as *Sphaerita* species and illustrated briefly in the literature without identity, principally by protozoologists who were studying the hosts and incidentally found the parasites in fixed and stained preparations. Kirby gave a full account of these parasites as well as those of *Sphaerita* and *Nucleophaga* which were known up to 1941, and since then Ball (1969), with some omissions, has brought the list of additional species up to date. A number of these parasites are illustrated in figs. 10–34 for comparison with *Morella endamoebae* (figs. 1–9).

Very little is known about infection and mode of entry by the endospores, but apparently they are engulfed by the host. As to method of development in the host Becker (1926) described the nuclei of uninucleate and multinucleate stages of *M. endoamoebae* (Becker) Réyes as dividing by constriction (figs. 4–7) and eventually being transformed directly into endospores (figs. 8, 9) without involvement and cleavage of any cytoplasm during sporogenesis.

Validating, emending and extending *Morella* to include all so-called *Sphaerita* species with mulberry-like sporangia does not seem prudent at present because these are so incompletely known, and in its present state the genus should probably be merged with *Nucleophaga*. So far as is known its general morphology, development, and life cycles seem to be very similar to those of *Nucleophaga*, and the two genera differ primarily in site of occurrence. Site of occurrence, whether in the cytoplasm or nucleus, is a doubtful generic distinction, particularly in view of the reported presence of *Sphaerita* (*Morella*?) *nucleophaga* Mattes (1924) in the nucleus and the rare occurrence of *Nucleophaga hypertrophica* Brumpt and Lavier (1935) in the cytoplasm. However, so little is known about species of both genera that no definite conclusions can be drawn about their identity at this time.

COELOMYCIDIUM Debaisieux

C. R. Soc. Biol. 82:1919; La Cellule 30:249, 1920.

Serumsporidium Nöller, 1920. Arch. Protistenk 41:183.

Plate 14, Figs. 1-20

This genus is reported to include 2 identified species, *C. simulii* Debaisieux and *C. ephemerae* Weiser, and an unidentified species collected by Shcherban and Golberg (1971), but only *C. simulii* is well known. The first 2 species parasitize the larvae of blackflies and ephemerids throughout most parts of the world while *Coelomycidium* sp. Shcherban and Golberg parasitizes the ova of mosquitos (*Culex modestus* and *C. pipiens*). According to the description and illustrations of Debaisieux (1920) the planospores of *C. simulii*, 6×8 μ, (fig. 1) with a 20 μ long flagellum, rhizoplast, blepharoplast, nucleus and nuclear cap are similar in appearance and structure to those of many chytridiomycetes, but those of *Coelomycidium* sp. are only 2×3.9 μ in size and shaped like a tear drop with a 17 μ long flagellum. Apparently *Serumsporidium melusinae* Nöller in larvae of *Simulium* (*Melusina*) *reptans* is identical with *C. simulii*.

After a period of motility the planospores of *C. simulii* become amoeboid (fig. 3), retract their flagellum, and probably infect the host larvae in this state if they have been released to the outside of the host. Within the larvae the young parasites increase in size and become multinucleate (figs. 4–7). Debaisieux described the thallus as being enveloped by a membrane. Shcherban and Golberg, on the other hand, described it as a sporangium when it is mature in their species. At this stage it is usually subspherical to spherical in shape (figs. 7, 8) and up to 100 μ or more in diameter, and in *Coelomycidium* sp. it may become 130 to 300 μ in diameter. Its protoplasm cleaves into uninucleate segments whose cytoplasm contain mitochondrial bodies (fig. 11) which gradually coalesce to (figs. 12, 13) form an apical body, according to Debaisieux. In addition a few granules or bodies may occur at the side of the nucleus (fig. 14). Debaisieux did not describe dehiscence of the holocarpic zoosporangium or membrane-bound plasmodium but only stated that the planospores are liberated in such quantities within the larvae that the latter burst spontaneously.

Weiser and Zizka's (1974a and b)* ultrastructural studies of *C. simulii* reveal some differences from the account of Debaisieux. They, too, describe the thallus as a plasmodium bounded by a 6–9 layered wall with lenticular vacuoles in the periphery, many nuclei, lipidic vacuoles, and ovoid mitochondria in the central portion contain a few cristae. During growth of the plasmodium the mitoses are intranuclear with tubular centrioles and tubular spindle fibers, and the nuclear membrane remains intact in the process. At maturity cleavage occurs by progressive radial furrowing from the center, eventually cutting out uninucleate segments connected by cytoplasmic bridges. Apparently the flagella are formed by this time, and their beating plays an important role in the final separation of the segments with the result that the segments become a spinning sponge-like mass in the zoosporangium.

Young planospores contain at least two mitochondria which coalesce to form a single, giant lobate one in mature spores, and it participates in the formation of the rumposome, according to Weiser and Zizka. In addition the mature spores contain other organelles, i.e., nucleus, nuclear cap, kinetosome, lomasomes, and a system of vacuoles usually found in most Chytridiomycetes. However, a "side body" or lipid sac is lacking. The whiplash flagellum with its sheath and 9 + 2 fibrils is inserted tangentially near the posterior end and anchored on the kinetosome, which in turn is "fixed in the membrane of the nuclear apparatus by a conical funnel of fibrils."

Cysts or resting spores develop in the fall and winter (figs. 15–19), and these contain several nuclei and numerous densely stained bodies or globules which may be so abundant as to obscure the nuclei, according to Debaisieux. The cysts may vary from 18 to 180 μ in diameter and are usually enveloped by a thin chitinous membrane or zone, but in others (fig. 18) the zone may be very thick. In addition to these, Debaisieux described polycystic bodies in which the small cysts are enveloped by a common capsule (fig. 19) secreted by the host. Such cysts probably originate from the development of planospores, according to Debaisieux, who did not find any evidence of gametic fusion. Weiser (personal communication) reported that the cysts are smaller than those described by Debaisieux and have several papillary caps.

Debaisieux classified *C. simulii* in the family Coelosporidiidae Caullery and Mesnil next to *Polycarum*, *Coelosporidium*, and *Blastulidium* and concluded that this family should be placed in the Chytridiales. Weiser

*In this connection see Loubes and Mannier, Protistologia 10:47-57, 1974.

PLATE 13

Figs. 1-9. *Morella endomoebae* Réyes in *Endoamoeba citelli*. (Becker, 1926.)

Fig. 1. Amoeba infected by uni- and binucleate parasites.

Fig. 2. Same host with an 8-nucleate parasite; each dumbbell shaped nucleus preparing for division.

Fig. 3. "Large sporangium with nuclei developing into spores."

Figs. 4-9. Developmental cycle of the parasite in the cytoplasm of the amoeba.

Figs. 10-13. *Sphaerita* (?) *endogena* Chatton and Broadsky non Dang. in *Amoeba limax*. (Chatton and Broadsky, 1909.)

Fig. 10. Amoeba with 5 uninucleate parasites.

Fig. 11. Similar amoeba with 6 uni-, bi-, and multinucleate parasites.

Fig. 12. Amoeba with 3 large multinucleate and several small uninucleate parasites.

Fig. 13. Later morella-like stage; 4 sporangia with endospores.

Fig. 14. *Sphaerita* (?) sp. sporangia in *Endolimax nana*. (Dobell and O'Connors, 1921.)

Fig. 15. *Sphaerita* (?) *plasmophaga* Mattes resting spore in *Amoeba sphaeronucleolus*. (Mattes, 1924.)

Fig. 16. *Sphaerita* (?) sp. fusion of endospores. (Penard, 1912.)

Figs. 17-21. *Sphaerita* (?) *trichomonadis* Crouch in *Trichomonas wenrichi*. (Crouch, 1933.)

Fig. 17. Trophozoite with a sporangium filled with endospores.

Figs. 18-21. Developmental stages of the parasite.

Figs. 22-27. *Sphaerita* (?) *hoari* Lubinsky in *Eremoplastron bovis*. (Lubinsky, 1955.)

Fig. 22. Elongate, somewhat V-shaped thallus in the body of the host.

Fig. 23. Ovoid thallus adjacent to macronucleus of the host; endoplasm of the parasite contracting.

Fig. 24. Later stage, rounding up of endoplasm in the center and surrounded by the homogeneous ectoplasm or envelope.

Figs. 25, 26. Multinucleate "microsporus" and "macrosporus" sporangia with broadly cone-shaped papillae; refractive globules next to nuclei in fig. 26.

Fig. 27. Empty sporangium with an exit papilla; outer envelope possibly the same as in fig. 24.

Figs. 28-34. *Sphaerita* (?) *dinobryoni* Canter in *Dinobryon sertularia*. (Canter, 1968.)

Fig. 28. Sporangium with elongate endospores in the base of the host.

Fig. 29. Curved endospores.

Fig. 30. Early stage of resting spore development.

Figs. 31-34. Smooth, hirsute, and spiny resting spores.

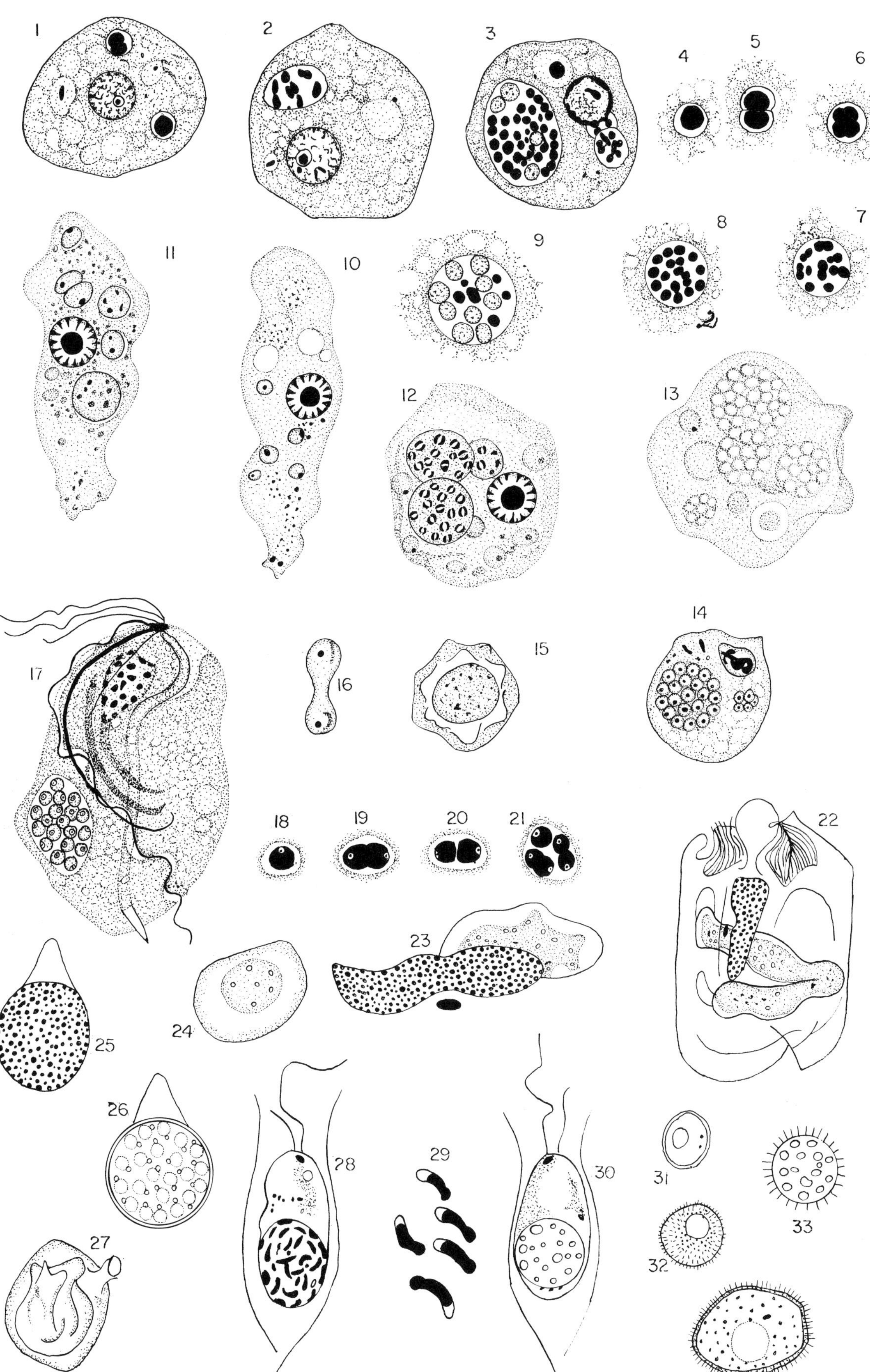

Sphaerita?, Morella

(1966) placed it temporarily among the chytrids, and later he and Zizka indicated more strongly that it is a chytrid. Maurand and Manier (1968) likewise indicated in their title that it may be a chytrid or a microsporid, but Rubcov (1969) was noncommital about its relationships with the chytrids. The appearance and ultrastructure of the planospores and the developmental stages of the thallus and zoosporangia are very similar to those of many Chytridiomycetes, but it is questionable, in this writer's opinion that they are indicative of relationships. So far as they are known, the cysts do not resemble closely the resting sporangia of known olpidioid chytrids. Nevertheless, the recent studies on this genus and future studies and discoveries of other similar organisms suggest, at least, that the creation of additional categories may become necessary among the simple fungi.

JOHNKARLINGIA Pavgi and Singh

Singh, Ph.D. Thesis, Banaras Hindu University Varanasi, India, 1975.

This genus of one species, *J. brassicae*, parasitizes the roots of crucifers and is characterized by minute posteriorly uniflagellate planospores and spiny resting spores. The planospores (gametes) fuse in pairs to form a zygote which infects root hairs and roots and develops into a resting spore. In germination it may function directly as a zoosporangium, or a sorus, or a prosorus, and in this process it combines characteristics of the Olpidiaceae and Synchytriaceae, according to Pavgi and Singh. Apparently, no asexual zoosporangia occur in its life cycle so far as is known. Pavgi and Singh regard this genus as a transition form between the families Olpidiaceae and Synchytriaceae.

ENDOBLASTIDIUM Codreanu

C. R. Acad. Sci. Paris 192:772, 1931.

Plate 14, Figs. 21-27

Endoblastidium was established by Codreanu for two species, *E. caulleryi* Codreanu and *E. legeri* Codreanu which parasitize *Baetis rhodani* Pict. and *Rhithrogena semicolorata* Curt. and have a predilection for the adipose tissues of the hosts. These parasites are characterized by plasmodia (figs. 21, 22) which are 12–18 μ in diameter and multinucleate in the vegetative or early developmental stages. At maturity they become enveloped by a distinct wall and are transformed into multinucleate zoosporangia (fig. 23). These are extruded to the outside of the host by rupture of the rectal wall. At this stage the zoosporangia are 30 to 50 μ in diameter and are usually ellipsoidal in shape. In less than 24 hours in contact with water dehiscence occurs by a median split and several lateral tears (fig. 24) through which the planospores escape. These are ovoid, 4–5 μ in diameter, with a compact lateral nucleus (fig. 25) surmounted by an apical capsule and a 10 μ long posteriorly attached flagellum. Resting spores or cysts are not known in this genus.

Codreanu regarded this genus as closely related

PLATE 14

Figs. 1-20. *Coelomycidium simuli* Debaisieux. Figs. 2, 7 after Strickland, 1913; figs. 1, 3-6, 8-28 after Debaisieux, 1920.

Figs. 1, 2. Planospores with an apical body ("accessory nucleus") or nuclear cap (?), nucleus, blepharoplast, rhizoplast, and posterior flagellum.

Fig. 3. Amoeboid planospore after losing its flagellum.

Figs. 4-6. Uni-, bi-, and multinucleate stages, resp.

Fig. 7. Mature multinucleate parasite.

Fig. 8. Same stage of living material.

Fig. 9. Cleavage into primary segments; nuclei in prophases of mitosis.

Fig. 10. Secondary cleavage segments or planospore initials; blepharoplasts and rhizoplasts present in some segments.

Figs. 11-13. Presence of and coalescence of mitochondrial chromatic bodies in the cytoplasm to form an apical body, respectively.

Fig. 14. Planospore with side bodies in the cytoplasm.

Fig. 15. Young uninucleate cyst or resting spore.

Fig. 16. Similar stage with numerous deeply stained bodies in the cytoplasm.

Fig. 17. Multinucleate cyst or resting spore surrounded by a hyaline zone.

Fig. 18. Mature living cyst filled with globules or bodies and surrounded by a thick hyaline zone or membrane.

Fig. 19. Cysts grouped in a common capsule secreted by the host.

Fig. 20. One cyst from such a capsule with several irregularly shaped and branched nuclei.

Fig. 21-25. *Endoblastidium caulleryi* Codreanu. Figs. 26, 27; *E. legeri* Codreanu. (Codreanu, 1931.)

Fig. 21. Small multinucleate plasmodium.

Fig. 22. Large plasmodium with nuclei in prophases of mitosis.

Fig. 23. Multinucleate sporangium with a sharply defined wall.

Fig. 24. Nearly empty sporangium with longitudinal branched split in the wall.

Fig. 25. Planospore.

Fig. 26. Small multinucleate plasmodium.

Fig. 27. Empty sporangium with a local invagination of the wall.

Fig. 28-30. *Chytridiopsis socius* Leger and Duboscq. (Leger and Duboscq, 1909.)

Fig. 28. Epithelial intestinal cell of *Blaps* with an apical sporangium of the parasite.

Fig. 29. Uninucleate amoeboid schizozoites from a sporangium.

Fig. 30. Resting cyst with spores.

Fig. 31. Resting cyst of *Chytridioides schizophylli* Tregouboff with spores. (Tregouboff, 1913.)

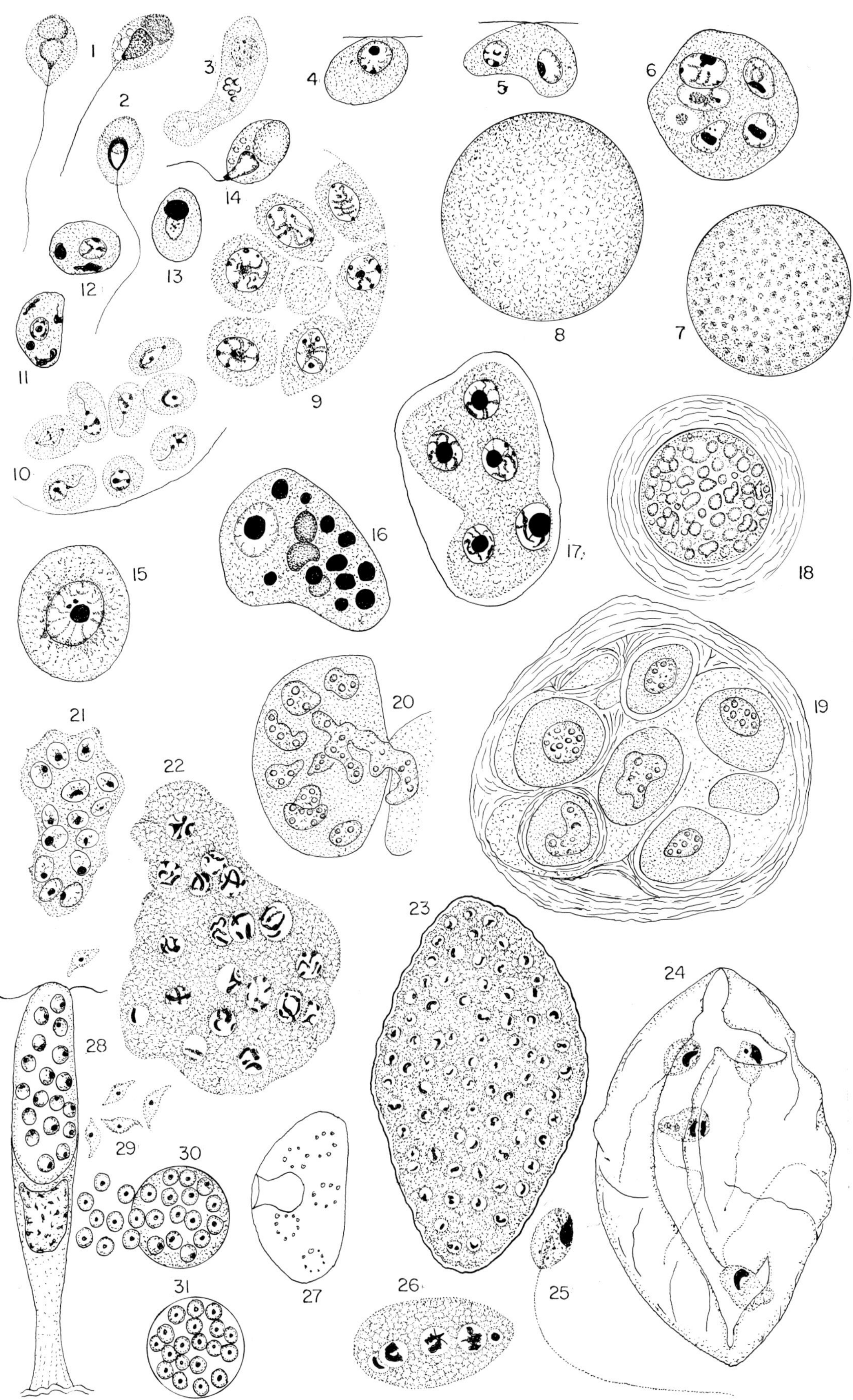

Coelomycidium, Endoblastidium, Chytridiopsis, Chytridioides

to *Coelomycidium* and believed that both genera, because of their holocarpic sporangia and posteriorly uniflagellate planospores, should be included in the Olpidiaceae. However, he concluded that they should constitute a distinct group within this family because of their cavitory parasitism and the lack of papillae or tubes for the discharge of the planospores. At the same time he stated that these genera exhibit more distinct affinities with the genera *Polycaryum*, which forms elongate aflagellate spores in sporangia with an exit tube (Stempell, 1903) and *Blastulidium*, which are temporarily annexed to the family Coelosporidiidae.

The question of whether or not *Dermocystidium pusula* Perez (1908), which causes large cutaneous pustules on the gills of newts and trout, is to be included among the olpidioid holocarpic chytrids has not been solved, but on the basis of present knowledge this is a doubtful genus and species. Most workers described the pustules as filled with nonmotile uninucleate bodies, but de Beauchamp (1914) reported that such bodies enlarge as the nuclei divide and eventually become transformed into ovoid, 10×13 μ sporangia without preformed openings or exit papillae. These produce spherical or slightly pyriform planospores which by the aid of a single long flagellum become motile, but de Beauchamp did not indicate the place of attachment of the flagellum. He regarded this parasite as a fungus and placed it among the chytrids.

Chytridiopsis Schneider (1884) and *Chytridioides* Treguoboff (1913) are additional genera which are reported to have affinities with the chytrids, but so far there is little or nothing in the life cycles that would warrant their inclusion in the Olpidiacae. *Chytridiopsis socius* Schneider (plate 14, figs. 28, 29) parasitizes the intestinal cells of *Blaps* and forms an oblong or ovoid, 7–20 μ diameter sporangium in which spherical (fig. 28) cells or incipient schizozites, 1–7 μ diam., develop. Subsequently, these become arched (fig. 29) in shape and amoeboid, and as the sporangium is expulsed from the epithelium they may divide once to several times. The durable cysts (fig. 30) are formed as a result of fusion of micro- and macrogametes, according to Leger and Duboscq (1909), and contain a number of spherical spores. These are expulsed into the intestine and to the outside with the excrement of the host. Leger and Dubosq pointed out the resemblance of this species to certain development aspects of Cienkowski's zoosporic monadineae and *Sphaerita*.

Chytridioides species parasitize the intestinal epithelia cells of *Schizophyllum mediterraneum* where they first appear as small spherical bodies, 1–5 μ diameter, which later attain a diameter of 15–20 μ and become multinucleate. Their protoplasm cleaves into numerous segments which round up as spores, and at this stage the parasite (fig. 31) is said to resemble the so-called morula stage of *Sphaerita*. The spores are released and become amoeboid schizozoites which infect other epithelial cells. Later fusion of gametes occurs, according to Tregouboff, which results in the formation of thick-walled durable cysts. Tregouboff believed that this genus had close affinity with the chytrids.

Sagittospora, a parasite of Ophryoscoleciae, was placed temporarily in the Olpidiaceae by Lubinsky (1955), but at present there isn't any sound reason for including it among the holocarpic chytrids. It develops into a plasmodial holocarpic thallus in which club-shaped aflagellate endospores are formed which apparently are released as the body of the host disintegrates.

CHYTRIDHAEMA Moniez

C. R. Acad. Sci., Paris 104:183, 1887.

This genus of one species, *C. cladocerarum* Moniez, was given the name *Chytridhaema* because its uniflagellate planospores occur in the blood of the host. It parasitizes the crustaceans *Sinocephalus retulus* and *Acroperus leucocephalus* and fills their body cavities with sac-like flattened, holocarpic sporangia which lack exit papillae or discharge tubes. Such sporangia produce numerous turbinate or top-shaped, 3 μ long, planospores whose content is dense and which have a refractive protuberance on the broad base. Moniez regarded this protuberance as an antheridium (?). The opposite end of the zoospore is greatly prolonged into a flagellum, but Moniez did not state whether it is posterior or anterior. He believed that this parasite resembles olpidiaceous and lagenidiaceous fungi, but it is difficult to classify it from his description. Unfortunately, he did not illustrate this species.

REFERENCES TO THE OLPIDIACEAE

Ball, G. H. 1969. Organisms living on and in protozoa. *In* Chen: Research in Protozoology 3: 565-718, 120 figs.

Barr, D. J. S. 1968. A new species of *Olpidium* parasitic on cucumber roots. Canad. J. Bot. 46:1087-1091, 22 figs.

Beauchamp, de. P. 1914. L'evolution et les affinities des protistes du genre *Dermocystidium*. Comp. Rendu. Acad. Sci., Paris 158:1359.

Becker, E. R. 1926. *Endamoeba citelli* sp. nov. from the striped ground squirrel *Citellus tridecem-lineatus*, and the life history of its parasite, *Sphaerita endamoebae* sp. nov. Biol. Bull. 5D:444-454, pl. 1.

Bradley, W. H. 1967. Two aquatic fungi (Chytridiales) of Eocene age from the Green River formation of Wyoming. Amer. J. Bot. 54:577-582, 9 figs.

Braun, A. 1856. Über *Chytridium* eine Gattung

Schmarotzergewäsche auf algen und Infusorien. Abhandl. Konigl. Akad, Wiss. Berlin 1855: 21-83, pls. 1-5.

Brug, S. E. 1926. *Nucleophaga intestinalis* n. sp., parasiet der kern van *Endolimax Williamsi* (Prow.) (*Jodamoeba butschlii* Pros.). Meded. dienst voksgesundheid. Nederlandsct.-Indie 4: 466-468, 17 figs.

Brumpt, E., and G. Lavier. 1935. Sur une *Nucleophaga* parasite d'*Endolimax nana*. Ann. Parasitol. 13:439-441, pl. 11.

Butler, E. J. 1907. An account of the genus *Pythium* and some Chytridiaceae. Mem. Dept. Agric. India, Bot. Ser. 1:1-160, pls. 1-10.

Canter, H. M. 1949. Studies on British chytrids V. *Olpidium hyalotheaceae* Scherffel and *Olpidium utriculiforme* Scherffel. Trans. Brit. Mycol. Soc. 32:22-29, pl. 5, 5 figs.

________. 1950. Studies on British chytrids. IX. *Anisolpidium stigeocloni* (de Wildemann) n. comb. Ibid. 33:335-344, 6 fig., pls. 24-26.

________. 1968. On an unusual fungoid organism, *Sphaerita dinobryoni* n. sp., living in species of *Dinobryon*. J. Elisha Mitchell Sci. Soc. 84: 56-61, 1 fig., pls. 1, 2.

Cejp, K. 1933. Further studies on the parasites of Conjugates in Bohemia. Bull. Intern. Acad. Sci. Boheme 42:1-11, 2 pls.

________. 1935. *Sphaerita*, parasite Paramecii Prespevek posnani houbovych parasitii Protozoi. Spisy. Prirodovedeckou Fakultow, Karlovy University, Praga (Publ. Fac. Sci. Univ. Charles) 141:3-7.

Chatton, E. 1908. Sur la reproduction et les affinities de *Blastulidium paedophthorum* Perez. C. R. biol. 64:34-36.

________, and A. Broadsky. 1909. Le parasitism d'une Chytridinee du genre *Sphaerita* Dangeard Chez *Amoeba limax* Dujard. Etude comparative. Arch f. Protistenk. 17:1-8, 6 figs.

Codreanu, R. 1931. Sur l'evolution des *Endoblastidium* nouveau genere de Protiste parasite coelomique des larves d'Ephermeres. C. R. Acad. Sci. Paris 192:772-775, 4 figs.

Cornu, M. 1872. Monographic des Saprolégniées; etude physologique et systématique. Ann. Sci. Nat. Bot. V. 15:1-198, pls. 1-7.

________. 1872b. Affinite des Myxomycetes et des Chytridinees. Bull. Soc. Bot. France 19:70-71.

Couch, J. N. 1939. Heterothallism in the Chytridiales. J. Elisha Mitchell Sci. Soc. 55:409-414, pl. 49.

Crouch, H. B. 1933. Four new species of *Trichomonas* from the wood-chuck (*Marmota Monax* Linn.) J. Parasit. 19:293-401, pl. 4.

Dangeard, P. A. 1886a. Sur une nouveau genre de Chytridinées parasite des Rhizopodes et des Flagellates. Bull. Soc. Bot. France 33:240-242.

________. 1886b. Recherches sur les organisms inférieurs. Ann. Sci. Nat. Bot. VII, 4:241-341, pls. 11-14.

________. 1889. Mémoire sur les Chytridinées. Le Botan. 1:39-74, pls. 2, 3.

________. 1895. Mémoire sur les parasites du noyau et du protoplasm. Ibid. 4:199-248, 10 figs.

________. 1933. Nouvelles observations sur les parasites des Eugléniens. Ibid. 25:3-56, pls. 1-4.

Debaisieux, P. 1919. Une Chytridinee nouvelle: *Coelomycidium simulii* nov. gen., nov. sp. Comp. Rendu Soc. Biol. 82:899-900.

________. 1920. *Coelomycidium simulii* nov. gen., nov. sp. et remarques sur l'*Amoebidium* des larves de *Simulium*. La Cellule 30:249-271, 2 pls.

Dobell, C., and F. W. O'Connors. 1921. The intestinal protozoa of man. New York.

Doflein, F. 1907. Fortpflanzungerscheinungen bei amöben and verwandten organismem. Sitz'b. München Gesel. Morph. Physiol. 23:10-181, figs. 1-3.

Fisch, C. 1884. Beitrage zur Kenntnis der Chytridiaceen. Sitzungsb. Phys.-Med. Soc. Erlangen 16:29-66., 1 pl.

Fischer, E. 1892. Die Pilze Deutschlands, Oesterreichs und der Schweiz. Rabenh. Kryptogamen.-Fl. (1), 4:1-490. Leipzig.

Foust, F. K. 1937. A new species of *Rozella* parasitic on *Allomyces*. J. Elisha Mitchell Sci. Soc. 53:197-204, pls. 22, 23.

Garrett, R. G., and J. A. Tomlinson. 1967. Isolate differences in *Olpidium brassicae*. Trans. Brit. Mycol. Soc. 50:429-435.

Gojdics, M. 1939. Some observations on *Euglena sanguinea* Ehrbg. Trans. Amer. Micro. Soc. 58:241-258.

Hollande, A., and H. H. DeBalsac. 1952. Parasitisme du *Peranema trichophorum* par une Chytridinee du genere *Nucleophaga*. Arch. Zool. Exp. et Gen. 82:37-46, 2 figs.

Ingold, C. T. 1952. *Funaria* rhizoids infected by *Pleotrachelus wildemanni*. Trans. Brit. Bryol. Soc. 2:53-54, 1 fig.

Ivanic, M. 1925. Zur Kenntnis der Agamogonieperiode einige Amoebenparasiten, Zool. Anz. 63:205-256.

Jacobsen, B. 1943. Studies on *Olpidium brassicae* (Wor.) Dang. Meddel. f. Plantepath Afd. D. Kgl. Vetern. o. Lanboh, Kobenhavn, No. 24: 5-51, pls. 1-7.

Jahn, T. L. 1933. On certain parasites of *Phacus* and *Euglena*; *Sphaerita phaci* sp. nov. Arch. f. Protistenk. 19:349-355, pls. 16, 17.

Jirovec, O. 1955. Parasites of our *Cladocera*. II. Cesk. Parasitol. 2:95-98, figs. 1-3A.

Johnson, T. W., Jr. 1969. Aquatic fungi of Iceland: *Olpidium* (Braun) Rabenhorst. Arch. f. Mikrobiol. 69:1-11, 34 figs.

Karling, J. S. 1942. Parasitism among chytrids. Amer.

J. Bot. 29:24-35, 47 figs.

———. 1943. The life history of *Anisolpidium ectocarpii* gen. nov. et sp. nov., and a synopsis and classification of other fungi with anteriorly uniflagellate zoospores. Ibid. 30:637-648, 21 figs.

———. 1946. Brazilian chytrids. VIII. Additional parasites of rotifers and nematodes. Lloydia 9:1-12, 53 figs.

———. 1948. An *Olpidium* parasite of *Allomyces*. Amer. J. Bot. 35:503-510, 27 figs.

———. 1958. *Olpidium synchytrii* sp. nov., a parasite of *Synchytrium namae*. Mycologia 50: 944-947, 8 figs.

———. 1964. Indian chytrids. II. *Olpidium indicum* sp. nov. Sydowia 17:302-307, 21 figs.

———. 1965. Some zoosporic fungi of New Zealand. I. Ibid. 19:213-266, pl. 46.

———. 1973. The present status of *Sphaerita*, *Pseudosphaerita*, *Morella* and *Nucleophaga*. Bull. Torry Bot. Club 99: 223-228.

Kirby, H., Jr. 1941. Organisms living on and in protozoa. Chap. 20:1009-1113, figs. 208-226., in G. N. Calkins and F. M. Summars, Protozoa in biological research. Columbia Univ. Press. New York.

Koch, W. J. 1958. Studies on the motile cells of chytrids, II. Internal structure of the body observed by light microscopy. Amer. J. Bot. 45: 59-72, 143 figs.

Kusano, S. 1912. On the life history and cytology of a new *Olpidium* with species reference to the copulation of motile isogametes. J. Coll. Agric. Imp. Univ. Tokyo 4:141-199, pls. 15-17.

———. 1929. Observations on *Olpidium trifolii* Schroet. Ibid. 10:83-99, 7 figs.

———. 1932. The host-parasite relationship in *Olpidium*. Ibid. 11:359-426, 10 figs.

Lagerheim, G. 1888. Sur un genre nouveau de Chytridiacees, parasite des uredospores certaines Uredinees. Jour. de Bot. 2:432-440, pl. 10.

Leger, L., and O. Duboscq. 1910. Sur les *Chytridiopsis* et leur evolution. Arch. Zool. Exp. Gen. 5 ser., 1:IX-XIII.

Litvinow, M. A. 1959. The taxonomy of the genus *Olpidium* (Russ. Trans.) Tr. Bot. Inst. Akad. Nauk USSR ser. 2, 12:188-212.

Lubinsky, G. 1955. On some parasites of parasitic protozoa. Canad. J. Microbiol. 1:440-450, 7 figs. Ibid. 1:675-684, 12 figs.

Macfarlane, I. 1968. Problems in the systematics of the Olpidiaceae. In Marine Mykologie. Veröffen. Inst. Meeresforsch. Bremerhaven, Sonderband 3:39-57, 11 figs.

Magnus, P. 1875. Die botanischen Ergebnisse der Nordseefahrt von 21 Juli bis 9 Sept. 1872. Meeresunters. Abt. Kiel 2-3:59-80, pls. 1, 2.

Mattes, O. 1924. Über Chytridineen im Plasma und Kern von *Amoeba sphaeronucleolus* und *Amoeba terricola*. Arch. f. Protistenk. 47:413-430, pls. 19, 20.

Maurand, J., and J. F. Manier. 1968. Actions histopathologiques comparies de parasites coelomiques, der larves de Simulies (Chytridiales, Microsporedies). Ann. Parasitol. Hum. et Comp. 43:79-85, 6 figs.

Minden, M. von. 1911-1915. Chytridineae, Ancylistineae, Monoblepharidineae, Saprolegiineae. Kryptogamenfl. Mark Brandenburg. 5:193-630.

Moniez, R. 1887. Sur les parasites nouveaux des Daphnies. C. R. Acad. Sci. Paris 104:183-185.

Mullins, J. T. 1961. The life cycle and development of *Dictyomorpha* gen. nov. (formerly *Pringsheimiella*), a genus of aquatic fungi. Amer. J. Bot. 48:377-387, 37 figs.

———, and A. W. Barksdale. 1965. Parasitism of the chytrid *Dictyomorpha dioica*. Mycologia 57:352-359.

Nagler, K. 1911. Studien über protozoen aus einem almtumpel. II. Parasitische Chytridiaceen in *Euglena sanquinea*. Arch. f. Protistenk. 23:262-268.

Nöller, W. 1920. Kleine Beobachtungen an parasitischen protozoen. Arch. f. Protistenk. 41:169-188, pls. 4-6.

Pascher, A. 1917. *Asterocystis* de Wildemann and *Asterocystis* Gobi. Beih. Bot. Centralb. 35: 578-579.

Penard, I. E. 1912. Nouvelles recherches sur les Amibes du groupe terricola. Arch. f. Protistenk. 28:74-140, 59 figs.

Perez, C. 1903. Sur un organisme nouveaus, *Blastulidium paedophthorum*, parasites des embryons daphnier. C. R. Soc. Biol. 55:715-716, figs. A-E.

———. 1905. Nouvelles observations sur le *Blastulidium paedophthorum*. C. R. Soc. Biol. 58: 1027-1029, 2 figs.

———. 1907. *Dermocystis pusula*, organisme nouveau parasite de la peau des Tritons Comp. Rendu Soc. Biol. 63:445-446.

———. 1908. Rectification de nomenclature a propos de *Dermocystis pusula*. Ibid. 64:738.

Perez Réyes, F. 1963. Algunas consideraciones sobre los parasitos de protozoarios incluidos en el genero *Sphaerita*. Rev. Soc. Mex. Hist. Nat. 24:1-6.

———, and E. Salaz Gomez. 1958. Euglenae de valle de Mexico. I. Algunas especies encontradas en el estangque de Capultepec. Rev. Latinoamer. Microbiol. 1:303-325.

———, ———. 1961. Euglenae de valle de Mexico IV. Descripcion de algunos endoparasitos. Ibid. 4:53-73.

Petersen, H. E. 1910. An account of Danish freshwater Phycomycetes, with biological and systematical remarks. Ann. Mycologici 8: 494-560, 27 figs.

Pumaly, A., De. 1927. Sur le *Sphaerita endogena* Dangeard. Chytridiacee Parasite des Euglenes.

Bull. Soc. Bot. France 74:472-476.

Rabenhorst, L. 1868. Flora Europaea algarum. Vol. 3, xx-561 pp. Leipzig.

Rieth, A. 1962. Beitrag zur Kenntnis der Phycomyceten. IV. *Pleotrachelus wildemanni* Petersen neue für Deutschland. Die Kultrupfl. 10:93-105, Figs. 1-5, pls. 1, 2.

Robertson, M. 1905. *Pseudospora volvocis* Cienkowski, Quart. J. Micro Sci. 49:213-230, pl. 12.

Rubcov, I. A. 1969. Rapports Réciproques entré l'hôte et le parasite. Response des Simulidae aux Microsporidies. Zh. Obshch. Biol. SSSR 27: 647-661.

Sahtiyance, S. 1962. Studien über einige wurzelparasitare Olpidiaceen. Arch. f. Mikrobiol. 41:187-228, 73 figs.

Sampson, K. 1932. Observations on a new species of *Olpidium* occurring in root hairs of *Agrostis*. Trans. Brit. Mycol. Soc. 17:182-194, 5 figs., Pls. 10-12.

Sawada, K. 1922. Descriptive catalogue of the Formosan fungi; Pt II, 172 pp.; pls. 1-7.

Scherffel, A. 1926a. Einiges über neue oder ungenügen bekannte Chytridineen. (Der "Beitrage zur Kenntnis der Chytridineen." Teil II.). Arch. f. Protistenk. 54:167-260, pls. 9-11.

________. 1926b. Beitrage zur Kenntnis der Chytridineen. Teil III. Ibid. 54:510-528, pl. 28.

Schneider, A. 1884. Development du *Stylophinchus longicollis*. Arch. Zool. Exp. Gen. 2 ser. 2:1-36.

Schroeter, J. 1885. Die Pilze Schlesiens. *In* Cohn, Kryptogamen-Fl. Schlesiens 3:180.

________. 1893-1897. Phycomycetes. *In* Engler und Prantl, Naturl. Pfanzenf. I(1):63-141.

Schwartz, E. J. and W. R. I. Cook. 1928. The life history and cytology of a new species of *Olpidium radicale* sp. nov. Trans. Brit. Mycol. Soc. 13:205-221, pls. 13-15.

Serbinow, J. L. 1907. Beitrage zur Kenntnis der Phycomyceten. Organisation und entwickelungsgeschichte einiger Chytridineen Pilze. Scripta Bot. Hortus Univ. Petropol. Fasc. 24:149-173, pls. 1-6.

Shcherban, Z. P. and A. M. Gol'berg. 1971. The pathogenic fungi *Coelomycidium* (Phycomycetes, Chytridiales) and *Coelomomyces* (Phycomycetes, Blastocladiales, in mosquitos *Culex* and *Aedes* (Family Culicidae, Diptera) from Uzbekistan Med. Parasit. i. parasit. bolezni 1:110-111, 2 figs. (Russ. Trans.)

Skvortzow, B. W. 1927. Über einige Phycomyceten aus China. Arch. f. Protistenk. 57:204-206, 10 figs.

Sörgel, G. 1952. Dauerorganbildung bei *Rozella allomycis* Foust, ein Beitrag zur Kenntnis der Sexualität der neideren Phycomyceten. Arch. f. Microbiol. 17:247, 5 figs.

Sparrow, F. K., Jr. 1933. Inoperculate chytridiaceous organisms collected in the vicinity of Ithaca, N. Y.; with notes on other fungi. Mycologia 25:513-535, 1 fig., pl. 49.

________. 1936. A contribution to our knowledge of the aquatic Phycomycetes of Great Britain. J. Linn. Soc. London (Bot.) 50:418-478, 7 figs., pls. 14-20.

________. 1948. Soil Phycomycetes from Bikini, Eniwetok, Rongerik and Rongelap atolls. Mycologia 40:445-453, 18 figs.

Stempell, W. 1903. Beitrage zur Kenntniss der Gattung *Polycaryum*. Arch. f. Protistenk. 2:343-363, pl. 9.

Strickland, E. H. 1913. Further observations on the parasites of *Simulium* larvae. J. Morph. 24:43-105, pls. 1-6.

Tomaschek, A. 1879. Uber Binnenzellen in der Grossen Zelle (Antheridiumzelle) des Pollens einiger Coniferen. Sitzungsber. K. Akad. Wiss. Math.-Naturw Klasse 78:197-212, 17 figs.

Tregouboff, G. 1913. Sur un Chytridiopside nouveau, *Chytridioides schizophyllii* n. g. n. sp., parasite de l'intestine de *Schizophyllum mediterraneum*. Arch. Zool. Exp. Gen. 52:25-31, 2 figs.

Uebelmesser, E. R. 1956. Uber einige neue Chytridineen aus Erdboden. *Olpidium, Rhizophydium, Phlyctochytrium* und *Rhizophlyctis*. Arch. f. Mikrobiol. 25:307-324, 7 figs.

Wagner, D. T. 1969. A monocentric holocarpic fungus in *Lemna minor* L. Nova Hedwigia 18:203-208, pls. 1, 2.

Weiser, J. 1947. Tri novi cezopasneci larevjepic. Vestnik Csl. Zool. Spolecnosti 11:297-303, 3 figs.

________. 1966. Nemoci hmyzu (Diseases of Insects). 554 pp. Academia, Prague.

________, and Zizka. 1973. The ultrastructure of the chytrid *Coelomycidium simulii* Deb. I. The vegetive thallus. Ceská Mycol. 28:159-162.

________ and ________. 1974. The ultrastructure of the chytrid *Coelomycidium simulii* Deb. II. Division of the thallus and structures of the zoospore. Ibid. 28:227-232, 7 figs.

Wildemann, E. de. 1893. Notes mycologiques I. Ann. Soc. Belge Micro. (Mem.) 17:5-30, pls. 1-3.

________. 1894. Notes mycologiques III. Ann. Soc. Belge Micro (Mem.) 18:135-161, pls. 4-6.

________. 1895. Notes mycologiques VI. Ann. Soc. Belge Micro (Mem.) 19:191-232, pls. 6-9.

Zopf, W. 1884. Zur Kenntniss der Phycomyceten. I. Zur Morphologie und Biologie der Ancylisteen und Chytridiaceen. Nova Acta Acad. Leop. - Carol 41:143-236, pls. 12-21.

________. 1892. Zur Kenntniss der Färbugsursachen niederer Organismem. Phys. Morph. Nied. Organismem 2:3-35, pls. 1, 2.

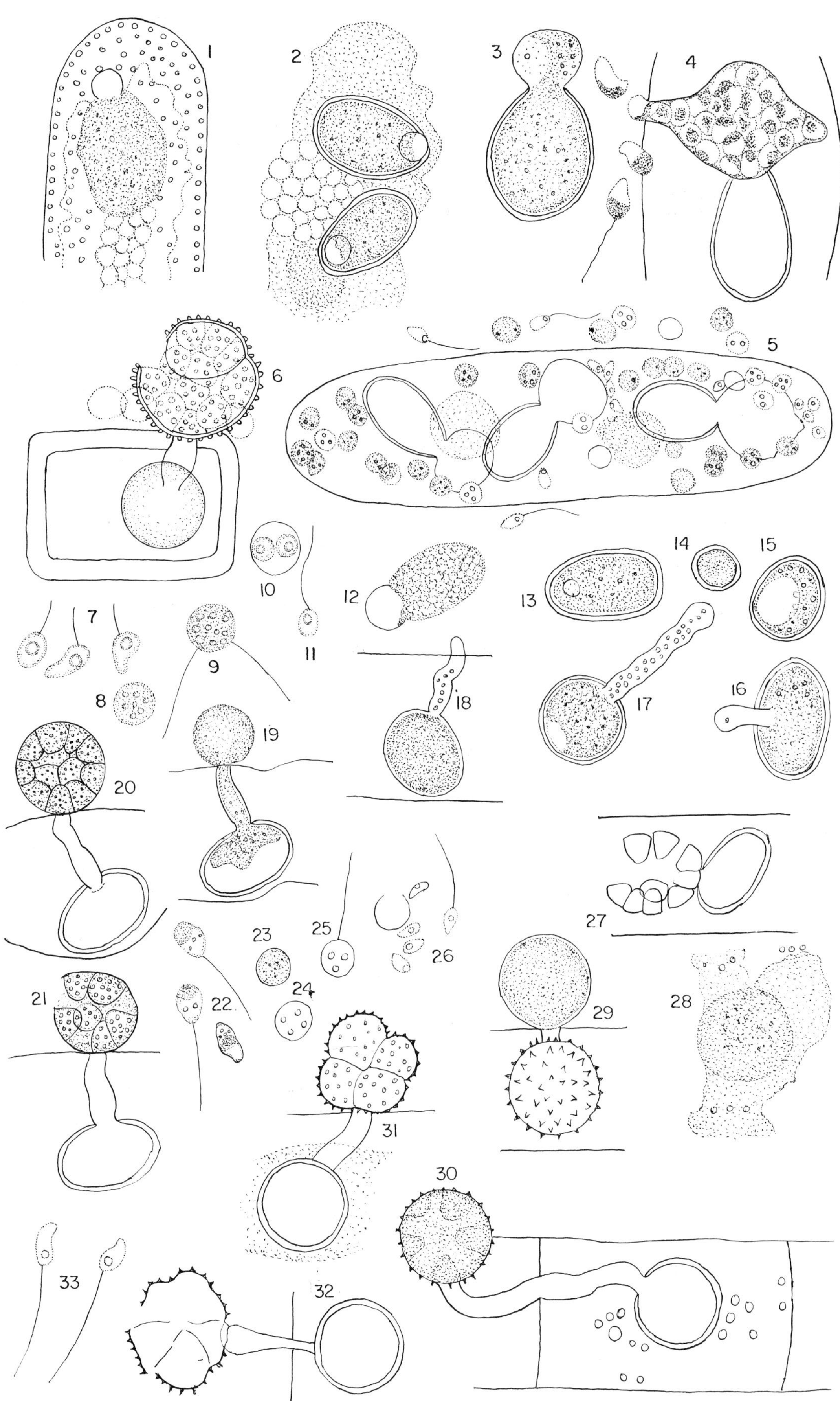

Endodesmidium, Micromycopsis

Chapter II

SYNCHYTRIACEAE

This is one of the large families of the Chytridiales and includes three or possibly four genera and more than two hundred species which are intracellular parasites of algae, mosses, ferns and flowering plants. Unlike in species of the previous family, the mature globular thallus is converted holocarpically into either a sorus, or an evanescent prosorus, or a resting prosorus, or a resting sporangium. Thus, the sequence and number of developmental stages vary in different species, and on the basis of such differences Karling (1964), following the procedures of previous workers, divided the genus into several subgenera. This classification is obviously tentative and will probably have to be modified as the developmental cycles of more species become known. Nevertheless, the present subgenera are illustrated separately to demonstrate and emphasize the known variations in types of development. The subgenus *Microsynchytrium* was established (Karling, 1953) to embrace the small aquatic species which were formerly placed in *Micromyces*, and these species were thus transferred to the genus *Synchytrium*. Later, it was discovered that many of the terrestrial species had the same type of development, and *Microsynchytrium* was, accordingly, extended to include both terrestrial and aquatic species. For purposes of comparison the life cycles and development of the aquatic and terrestrial species of this subgenus are illustrated separately below.

As reported in some species of the Olpidiaceae the motile cells are facultative and may function either as planospores or gametes. Thus, sexual reproduction has been reported to occur in several species of this family, particularly in the genus *Synchytrium*, and it occurs by the fusion of isomorphic gametes from the same or different zoosporangia (gametangia). In *S. endobioticum* and some other species both gametes are motile at the time of fusion; in *S. fulgens* one is non-flagellate and quiescent, and in *S. macrosporum* both gametes are quiescent and non-flagellate when they fuse. The zygote comes to rest on the host cell, infects it and develops into a resting prosorus. According to Kusano (1928, 1930) and Köhler (1930) sexuality in *Synchytrium* is expressed relative to the age of the gametes. When first discharged the gametes are neutral or indifferent to copulation, but after a short motile period the male potential develops. Such male gametes seek out and fuse with quiescent female gametes, but if fusion does not occur the male char-

PLATE 15

Figs. 1-5. *Endodesmidium formosum* Canter. (Canter, 1949.)

Fig. 1. Naked thallus or incipient prosorus with large anterior oil globule in *Netrium oblongum*.

Fig. 2. Mature prosori.

Fig. 3. Prosorus germinating to form a sorus.

Fig. 4. Sorus with exit papilla and emerging sluggish primary planospores (1 μ diam.); one with a flagellum.

Fig. 5. Three germinated prosori; primary planospores within and outside of host have encysted to become zoosporangia; three minute secondary planospores on the outside.

Figs. 6-11. *Micromycopsis cristata* var. *cristata* (Scherffel) Sparrow. (Scherffel, 1926.)

Fig. 6. Germinated prosorus with an exit canal in *Hyalotheca*; sorus of sporangia on the outside of the host.

Fig. 7. Amoeboid and sluggish primary planospores.

Figs. 8, 9. Encysted primary planospores or secondary zoosporangia; the one on the right has two flagella-like filaments.

Fig. 10. Secondary sporangium with two secondary planospores.

Fig. 11. Secondary planospore.

Figs. 12-27. *Micromycopsis fischeri* Scherffel. (Canter, 1949.)

Fig. 12. Naked thallus or prosorus with a large anterior oil globule.

Figs. 13-15. Variations in sizes and shapes of mature prosori.

Figs. 16-18. Germination stages of prosori in *Tetmemorus*.

Fig. 19. Migration of protoplasm through exit canal to form a sorus on the outside of the host.

Fig. 20. Cleavage in the sorus to form primary zoosporangia.

Fig. 21. Primary zoosporangia in ruptured sorus.

Fig. 22. Sluggish primary planospores, 2.4 × 4.3 μ diam.

Figs. 23-25. Encysted primary planospores developing into secondary zoosporangia; the one in fig. 25 has a flagellum-like filament.

Fig. 26. Secondary planospores, 2 μ diam., from a secondary zoosporangium.

Fig. 27. Germinated prosorus and empty sorus within the host cell; no exit canal has developed.

Figs. 28-33. *Micromycopsis intermedia* Canter. (Canter, 1949.)

Fig. 28. Naked thallus or prosorus in *Zygnema*.

Fig. 29. Germinated spiny prosorus.

Fig. 30. Germinated smooth-walled prosorus with an unusually long exit canal; sorus with echinulate wall.

Fig. 31. Division of sorus into four zoosporangia.

Fig. 32. Empty sorus.

Fig. 33. Primary planospores, 1.5 × 3.5 μ diam.; secondary planospores lacking or unknown.

acter or potential gradually decreases and vanishes. From this point on the female potential of the formerly male gamete begins to emerge and eventually reaches a climax at which time it may fuse with an actively motile male. If fusion does not occur the female characteristic decreases and gradually vanishes to the extent that the gamete finally becomes neutral again. Thus, as a gamete ages it may range in potential from neutral to male, to female, and neutral again, and during this period exhibits relative sexuality, according to Kusano.

This interpretation of sexuality has not been supported by subsequent investigations. Heim (1956) reported that no gametic fusions occur in *S. endobioticum*. Instead, the diplophase is established by fusion of small nuclei in the germinated resting prosorus (sporangium). The diploid nuclei divide after a pause, and these last two divisions prior to cleavage are meiotic, according to her account. These observations refute those of Kusano (1930) which indicated that meiosis in *S. fulgens* occurs during the first two divisions of the primary nucleus of the resting prosorus. B. T. Lingappa (1958) likewise did not confirm Kusano's and Köhler's views on relative sexuality. In *S. fulgens* he found that male and female gametes are differentiated shortly after discharge and fuse without undergoing changes from male to female.

Although it is not at all conclusive, there is some evidence which suggests an evolutionary series within the Synchytriaceae. In postulating relationships and phylogeny in this family on the basis of complexity of life cycle, the presence or absence of prosori or prosporangia, secondary sporangia, and secondary planospores, Karling (1954) suggested that *Micromycopsis cristata* and *M. fischeri* might be starting points in this evolutionary series. From such species suppression of secondary sporangia and zoospores may have occurred along several lines: (1) through *M. intermedia* and *M. mirabilis* to the subgenus *Microsynchytrium* by way of *Synchytrium (Micromyces) ovalis* which occasionally forms exit canals from the prosorus; (2) through *Endodesmidium* to *Microsynchytrium*; or (3) through *Micromycopsis*, in which only primary planospores are known, to *Micro-*

PLATE 16

Figs. 1, 2, 4, 5. Planospores of *Synchytrium (Micromyces) zygogonii, S. (Micromyces) grandis, S. (Micromyces) petersenii* (Scherff.) Karling and *S. (Micromyces) laevis*, (Canter) Karling, respectively. After Canter, 1949; Miller, 1955; Rieth, 1956.

Fig. 3. Paired planospores of *S. grandis*. (Miller, 1949.)

Fig. 6. Comma-like planospore cysts of *S. laevis* on surface of *Mougeotia* filament. (Rieth, 1956.)

Figs. 7-9. Infection by planospores of *S. zygonii* Dang., *S. longispinosus* Couch, and *S. laevis* Canter, respectively. (Canter, 1949; Couch, 1931; Rieth, 1956.)

Fig. 10. Minute parasite of *S. longispinosus* lying next to host nucleus. (Couch, 1931.)

Fig. 11. Amoeboid thallus of *S. zygogonii* lying next to host nucleus. (Canter, 1949.)

Fig. 12. Development of spines on the evanescent prosorus of *S. zygogonii*. (Canter, 1949.)

Fig. 13. Uninucleate spiny evanescent prosorus of *S. zygogonii*. (Dangeard, 1889.)

Fig. 14. Smooth evanescent prosorus of *S. zygogonii*. (Canter, 1949.)

Fig. 15. Early stage of germination of the evanescent prosorus of *S. zygogonii*; incipient sorus binucleate. (Dangeard, 1890-91.)

Fig. 16. Later stage, sorus multinucleate. (Dangeard, 1890-1891.)

Fig. 17. Uninucleate incipient zoosporangia of *S. zygogonii* from germinated evanescent prosorus. (Couch, 1937.)

Fig. 18. Cleavage in the sorus of *S. zygogonii*. (Couch, 1937.)

Fig. 19. Minutely spiny, germinated evanescent prosorus of *S. zygogonii*; and a sorus of five incipient zoosporangia. (Canter, 1949.)

Fig. 20. Hypertrophied and burst cell of *Zygogonium* containing germinated evanescent prosorus, zoosporangia, and planospores of *S. zygogonii*. (Dangeard, 1889.)

Fig. 21. Germinated smooth-walled evanescent prosorus of *S. petersenii* and a sorus of three incipient zoosporangia. (Canter, 1949.)

Fig. 22. Uninucleate evanescent prosorus of *S. longispinosus*. (Couch, 1937.)

Fig. 23 Zoosporangium of *S. longispinosus*. (Couch, 1937.)

Fig. 24. Planospores of *S. longispinosus*, 1 μ diam. (Couch, 1937.)

Figs. 25, 26. Paired planospores (planogametes) of *S. longispinosus*. (Couch, 1937.)

Fig. 27. Biflagellate zygote (?) of *S. longispinosus*. (Couch, 1937.)

Fig. 28. Rare germinated evanescent prosorus of *S. ovalis* Rieth with an exit canal and a sorus of four empty zoosporangia. (Rieth, 1950.)

Fig. 29. Resting prosori of *S. petersenii*. (Canter, 1949.)

Fig. 30. Uninucleate resting prosorus of *S. longispinosus*. (Couch, 1931.)

Fig. 31. Resting prosorus of *S. furcata* (Rieth) comb. nov. with reticulate outer wall bearing dichotomously branched spines. (Rieth, 1962.)

Fig. 32. Resting prosorus of *S. grandis*. (Miller, 1955.)

Fig. 33. Germination of resting prosorus of *S. longispinosus*; nucleus passing out into incipient sorus. (Couch, 1937.)

Figs. 34, 35. Stages in germination of resting prosori of *S. zygogonii* to form a single zoosporangium. (Rieth, 1956.)

Fig. 36. Empty prosorus and zoosporangium of *S. zygogonii*. (Rieth, 1956.)

Fig. 37. Germinated resting prosorus of *S. grandis* with sporangia. (Miller, 1955.)

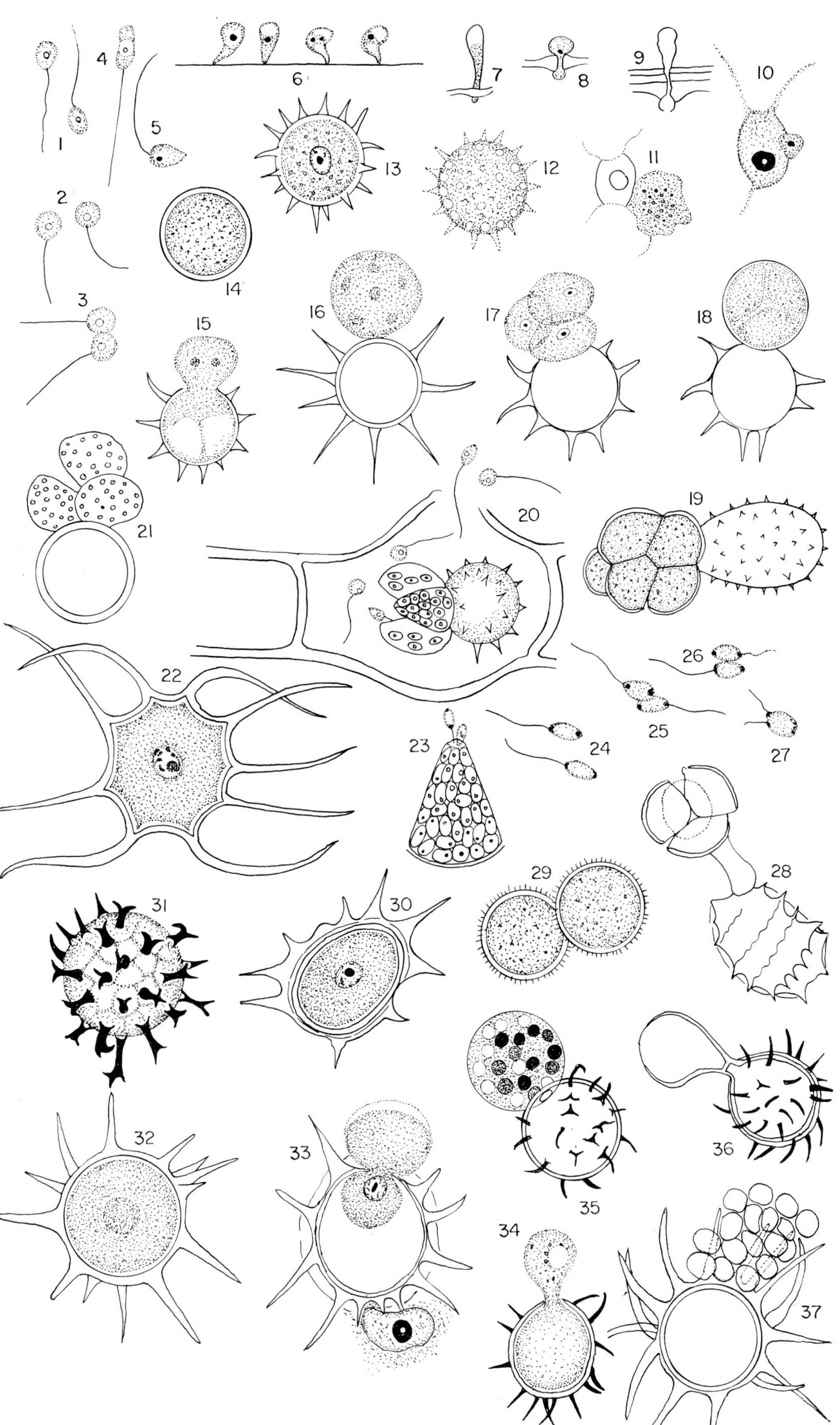
1
2
3
4
5
6
7
8
9
10
11
12
13
14
15
16
17
18
19
20
21
22
23
24
25
26
27
28
29
30
31
32
33
34
35
36
37

Micromyces

synchytrium. From the aquatic species (*Micromyces*) of this subgenus the complexity of life cycle may have retrogressed through the subgenera *Mesochytrium*, *Exosynchytrium*, *Synchytrium* (*Eusynchytrium*) to *Pycnochytrium* and *Woroninella* as the species left their watery habitat and began to parasitize terrestrial plants. Thus, according to this concept there has been a retrogression from a complex life cycle to one in which only resting prosori (*Pycnochytrium*) or sori (*Woroninella*) occur.

ENDODESMIDIUM Canter

Trans. Brit. Mycol. Soc. 22:72, 1949.

Plate 15, Figs. 1-5

This monotypic genus and its type species, *E. formosum* Canter, parasitizes *Netrium* and *Cylindrocystis* as a naked thallus (fig. 1) which develops into a fairly thick-walled smooth evanescent prosorus (fig. 2). So far, infection stages have not been described. The prosorus germinates within the host cell (figs. 3, 4) and forms an ovoid sorus with an exit papilla or a short canal (fig. 4). Its content divides into a large number of ovoid and subspherical cells which are mostly liberated within the host cell (fig. 5). Occasionally, they emerge to the outside through the exit canal (fig. 4) and become amoeboid or develop a flagellum. These may be regarded as primary planospores. They round up and encyst within a short time to become zoosporangia (fig. 5) and produce from two to five minute secondary planospores. Resting prosori or spores have not been observed in *Endodesmidium*.

MICROMYCOPSIS Scherffel

Arch. f. Protistenk. 54:202, 1926.

Plate 15, Figs. 6-33

At present *Micromycopsis* includes 7 fairly well-known and several imperfectly known parasites of

PLATE 17

Figs. 1-10, 12, 21-26, 29-40. *Synchytrium fulgens* Schroeter. Figs. 11-13, 17-20, 27, 28, 41 *S. endobioticum* Percival; figs. 14-16 *S. australe* Spegazzini. Figs. 7-10, 12, 34, 35 drawn from fixed section and stained preparations by Karling.

Fig. 1. Ovoid, 2.5-3.5 × 5-7 μ, planospores with an orange-yellow refractive globule. (Lingappa, 1958a.)

Fig. 2. Planospore flattened on the host cell. (Kusano, 1930a.)

Fig. 3. Young parasite within the host cell; refractive globules on the host cell wall. (Kusano, 1930b.)

Figs. 4, 5. Young thalli 1 and 2 days old, resp., lying next to the host nucleus. (Kusano, 1930a.)

Fig. 6. Three day old incipient prosorus; host nucleus compressed. (Kusano, 1930a.)

Fig. 7. Mature evanescent prosorus with a large primary nucleus; host nucleus below.

Fig. 8. Initial stage of "germination" of evanescent prosorus; papilla forming at the apex.

Fig. 9. Later stage; protoplasm flowing upward; primary nucleus pyriform.

Fig. 10. Primary nucleus entering incipient sorus.

Fig. 11. Similar stage in *S. endobioticum*. (Curtis, 1921.)

Fig. 12. Primary nucleus in prophase of mitosis in almost fully formed sorus.

Fig. 13. Division of primary nucleus in the sorus; division spindle intranuclear with 5 chromosomes. (Curtis, 1921.)

Fig. 14. Multinucleate sorus of *S. australe*; empty prosorus in the base of the host cell; dense plug between the sorus and empty prosorus. (Karling, 1955.)

Fig. 15. Cleavage of protoplasm into multinucleate zoosporangia. (Karling, 1955.)

Fig. 16. Zoosporangium discharging planospores. (Karling, 1955.)

Figs. 17-20. Fusion of isoplanogametes, biflagellate zygote, karyogamy, and diploid encysted zygote, resp. (Curtis, 1921.)

Figs. 21, 22. Pairing and fusion of gametes; female gamete sedentary and nonflagellate. (Kusano, 1930b.)

Figs. 23, 24. Uniflagellate, and quiescent zygotes, resp. (Kusano, 1930b.)

Figs. 25, 26. Fusion of gametes, anterior fusion ends amoeboid, and karyogamy, resp. (Lingappa, 1958a.)

Figs. 27, 28. Pairing of nuclei in a germinating resting sporangium (?) and a diploid nucleus, resp., in *S. endobioticum*. (Heim, 1956.)

Fig. 29. Infection by a zygote; refractive globule and shriveled zygote membrane on the outside of the host cell at the right. (Lingappa, 1958b.)

Fig. 30. Young zygote lying under the host nucleus. (Lingappa, 1958b.)

Fig. 31. Later developmental stage of the zygote or dormant prosorus. (Lingappa, 1958a.)

Fig. 32. Mature resting prosorus with a thick wall, large primary nucleus, and dense bodies in the cytoplasm. (Lingappa, 1958b.)

Fig. 33. Early stage in the germination of the resting prosorus; primary nucleus pyriform. (Lingappa, 1958b.)

Fig. 34. Primary nucleus entering incipient sorus. (Lingappa, 1958b.)

Fig. 35. Fully formed sorus with a dense plug between it and the empty prosorus; primary nucleus reduced in size and elongate. (Lingappa, 1958.)

Fig. 36. First meiotic division of the primary nucleus with 5 pairs of chromosomes. (Kusano, 1930a.)

Fig. 37. A secondary mitotic division. (Kusano, 1930a.)

Fig. 38. Multinucleate sorus; cleavage beginning at the periphery. (Lingappa, 1958b.)

Fig. 39. Sorus of zoosporangia. (Lingappa, 1958b.)

Fig. 40. Zoosporangium from a germinated resting prosorus discharging zoospores. Drawn from living material by Karling.

Fig. 41. Germinated resting prosorus of *S. endobioticum*; sorus containing a single incipient zoosporangium. Sketched from a photograph by Kole, 1965.

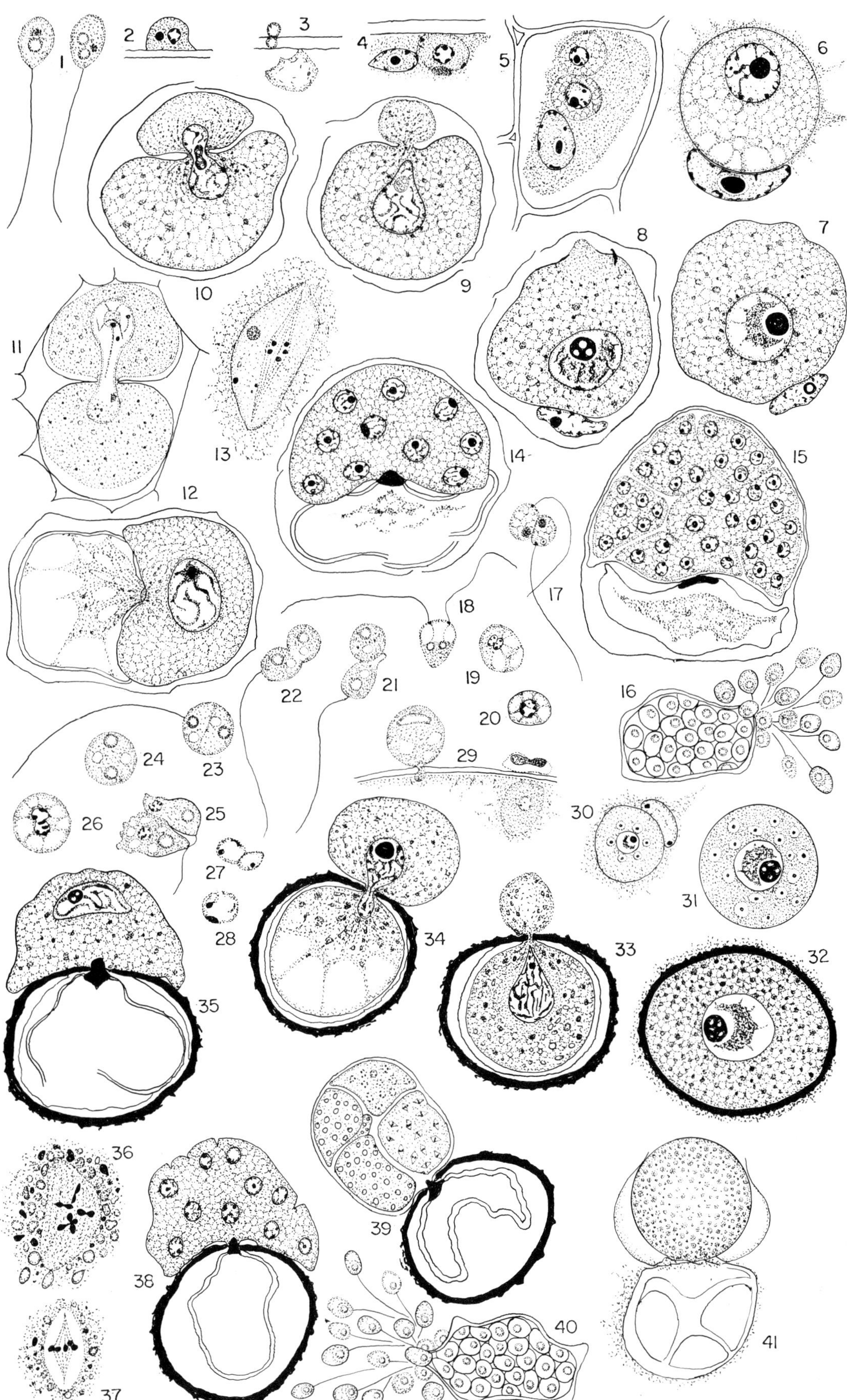

Subgenus *Microsynchytrium*

the Conjugales and *Oedogonium* in which the germinating evanescent prosorus usually develops an exit canal (figs. 16–19) through which its content passes to the outside of the host cell and develops into a sorus (figs. 6, 20, 21, 29–32). Rarely, no exit canal is formed, and the prosorus germinates and produces a sorus within the host cell (fig. 27) as in *Endodesmidium*. In *M. cristata* Scherffel and *M. fischerii* Scherffel sluggish primary planospores (figs. 7, 22) are produced by the primary zoosporangia, which soon encyst (figs. 8, 23) and in turn become secondary zoosporangia. Occasionally, these may bear flagellalike filaments (figs. 9, 25). These zoosporangia produce a small number of actively motile secondary planospores (figs. 11, 26) which emerge through a pore in the sporangial wall. In *M. intermedia* Canter and *M. oedogonii* Roberts, on the other hand, the secondary planospore stage apparently is suppressed, and the primary planospores (fig. 33) become actively motile. No resting prosori or spores have been observed in *Micromycopsis*.

The presence of sluggish primary planospores in this genus as well as in *Endodesmidium* which soon encyst, develop into zoosporangia, and produce a few minute actively motile secondary planospores is somewhat similar to developmental stages in *Allomyces monilioformis* and *Blastocladiella cystogena*. In the cystogenous groups of these two genera the germinating resting sporangium produces sluggish biflagellate planospores which soon encyst and later form four small posteriorly uniflagellate gametes which fuse in pairs. On the grounds of this similarity it has been postulated that the secondary planospores in *Endodesmidium* and *Micromycopsis* might be isogametes and that sexuality occurs in these genera. However, no fusions have been observed.

The distinctions between *Micromycopsis* and the aquatic species (*Micromyces*) of *Synchytrium* are not always sharply defined as is shown by the rare lack of an exit canal in *Micromycopsis fischeri* and the rare development of one in *S. ovalis* (Rieth) Karling. Obviously, *Micromycopsis* is closely related to the aquatic (*Micromyces*) species, and Canter (1949) and Rieth (1950, 1956) expressed the view that the genus *Micromycopsis* may prove to be superfluous. Sparrow (1960) concurred with this viewpoint and reduced *Micromycopsis* to a synonym of *Micromyces*.

MICROMYCES Dangeard

Le Botan. 1:55, 1889.

Plate 16

As noted earlier the species of *Micromyces* were transferred to the subgenus *Microsynchytrium* of *Synchytrium* by Karling (1953, 1954) on the grounds that their developmental cycle is identical with that of some species of *Synchytrium*. He emphasized that differences in habitat, hosts, size, and spinyness of prosori and sori do not merit generic distinction and that these characteristics are outweighed by the basic de-

Plate 18

Figs. 1-10. *Synchytrium desmodiae* Munasinghe. (Munasinghe, 1955.)

Fig. 1. Spherical, 1.4-6 μ diam. (?), planospore with a 2-6 μ long (?) flagellum.

Fig. 2. Paired encysted planospores.

Fig. 3. Biflagellate planospore.

Figs. 4, 5. Young and mature evanscent prosori.

Fig. 6. Formation of a sorus beneath the prosorus.

Fig. 7. Sorus with sporangial initials beneath the empty prosorus.

Fig. 8. Open sorus with sporangia.

Fig. 9. Thick-walled resting sporangium with planospore initials (?).

Fig. 10. Planospores (?) in a resting sporangium.

Figs. 11-40. *Synchytrium taraxaci* DeBary and Woronin.

Figs. 11, 12. Uninucleate and paired planospores. Drawn from living material.

Figs. 13-15. Biflagellate and amoeboid planospores. (DeBary and Woronin, 1863.)

Fig. 16. Binucleate planospore (plasmogamy ?). (Lowenthal, 1905.)

Fig. 17. Infection of epidermal cell (DeBary and Woronin, 1863.)

Fig. 18. Young parasite lying under the host nucleus. (Rytz, 1917.)

Fig. 19. Incipient sorus enveloped by host cytoplasm. (DeBary and Woronin, 1863.)

Figs. 20, 21. Uninucleate and multinucleate incipient sorus next to host nucleus. (Rytz, 1917.)

Fig. 22. Mature multinucleate sorus. (Drawn by Karling.)

Fig. 23. Cleavage. (Drawn by Karling.)

Figs. 24-28. Variations in sizes and shapes of zoosporangia. (Drawn by Karling.)

Figs. 29, 30. Papillate and dehiscing zoosporangia. (DeBary and Woronin, 1863.)

Figs. 31, 32. Discharge of planospores from small and large zoosporangia. (Drawn by Karling.)

Fig. 33. Young uninucleate incipient resting sporangium. (Lowenthal, 1905.)

Fig. 34. Similar stage with large primary nucleus. (Drawn by Karling.)

Fig. 35. Enlarged host cell with a mature, thick-walled resting sporangium lying beneath hypertrophied host nucleus. (Drawn by Karling.)

Fig. 36. Multinucleate resting sporangium. (Lowenthal, 1905.)

Fig. 37. Mature, living resting sporangium filled with golden-red granules. (DeBary and Woronin, 1863.)

Fig. 38. Resting sporangium prior to formation of planospore initials, filled with globules. (DeBary and Woronin, 1863.)

Figs. 39, 40. Planospores from germinated resting sporangium. (DeBary and Woronin, 1863.)

Figs. 41, 42. Amoeboid and giant planospores, resp. (DeBary and Woronin, 1863.)

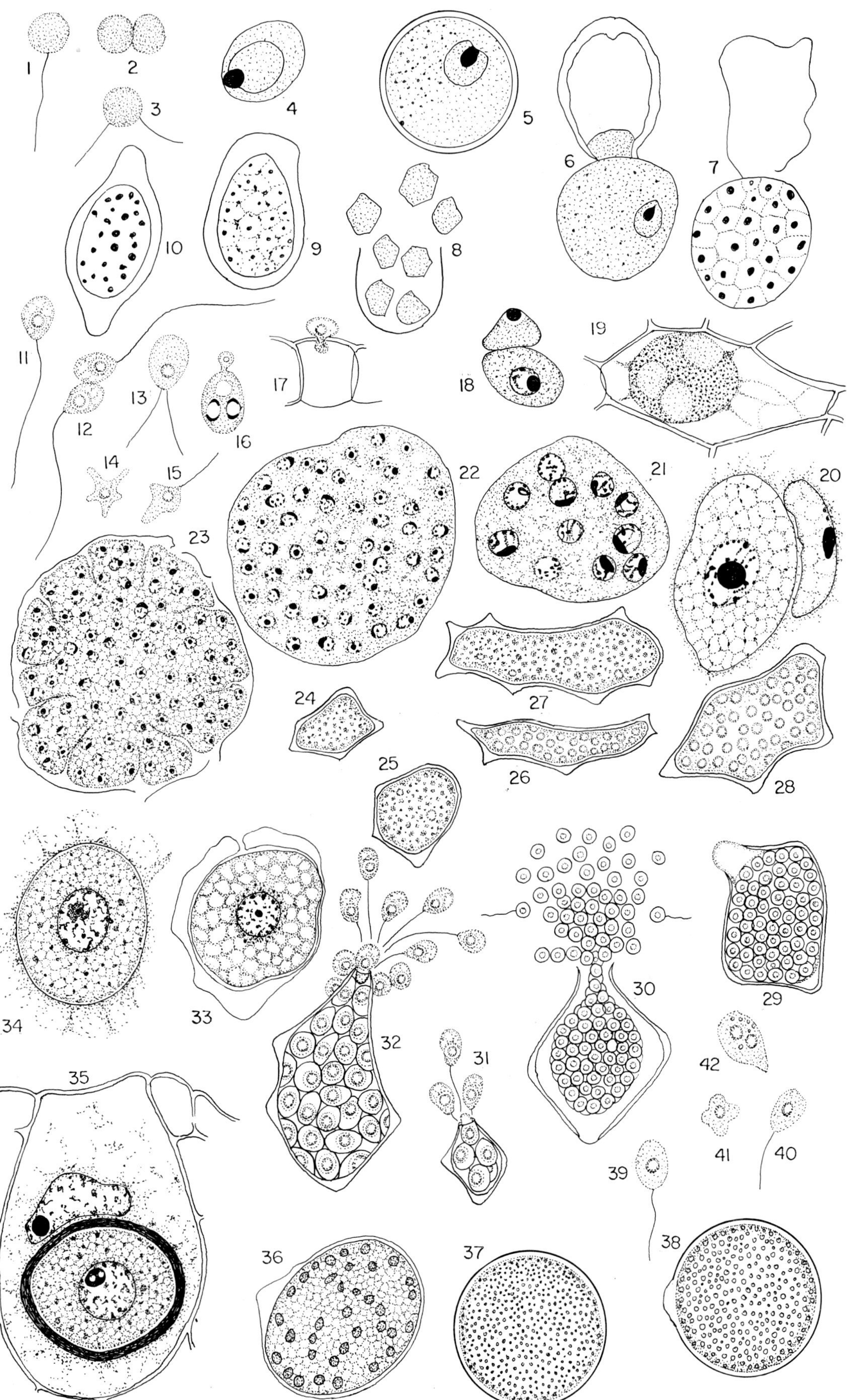

Subgenera *Mesochytrium* and *Synchytrium*

velopmental and structural similarities of the two genera. However, the genus and some of its species are illustrated here apart from *Micromycopsis* and the terrestrial species of *Synchytrium* to avoid personal interpretation as much as is possible and provide the reader with an opportunity for individual judgment. Nevertheless, the writer still considers the so-called *Micromyces* species to be members of *Synchytrium* (subgenus *Microsynchytrium*) and refers to them as such in the descriptions of the figures in Plate 6.

This group includes eight well-known and several other imperfectly known species, all of which parasitizes members of the Conjugatae and may cause marked local hypertrophy, bending, and elongation of the host cells. These parasites had been observed many years before Dangeard's time by algologists who described them as "asteridia" or "astrospheres."

The planospores (figs. 1–5) encyst on the host cell wall (fig. 6) and form infection tubes (figs. 7–9) through which their content passes into the host. The young parasite becomes amoeboid (fig. 11) and moves to the host nucleus (fig. 10). With further development it becomes globular and develops spines of varying lengths on its surface (figs. 12, 13, 22) or remains smooth-walled (figs. 14, 21). At this stage it is uninucleate (figs. 13, 22), and as in *Synchytrium* it is to be regarded as an evanescent prosorus. As it germinates the content emerges through a broad or small pore in the prosorus wall (fig. 15). Its nucleus apparently divides after emerging into the incipient sorus. The emerged protoplasm then develops a thin wall (fig. 16) and functions as a sorus as its content cleaves (fig. 18) into segments which become incipient zoosporangia (figs. 19, 21). These may vary in number from 4 to 54 per sorus in different species. They separate at maturity and produce planospores (figs. 20, 23) which escape through a single apical pore. In a single instance the evanescent prosorus of *S. ovalis* (Rieth) Karling was found (Rieth, 1950) to develop an exit canal (fig. 28) for the discharge of its content into a sorus as in *Micromycopsis*, and this species might possibly be regarded as a transition form between the latter genus and *Synchytrium*.

As the cultures of the parasites age thicker-walled, denser, resting prosori (figs. 29–32) develop. Apparently, these are uninucleate (fig. 30) at this stage like the evanescent prosori. At germination a pore is formed in the prosoral wall through which the cytoplasm and nucleus (fig. 33) pass to the outside and form a sorus of zoosporangia. The latter separate (fig. 37) and soon produce planospores.

The developmental cycle of these fully known aquatic species is identical with that of the terrestrial species of the subgenus *Microsynchytrium*. Occasionally, the germinating resting prosorus of *S. zygogonii* (Dang.) Karling develops only one zoosporangium (figs. 34–36) instead of a sorus, a type of development which has been shown (Kole, 1965) to occur in *S. endobioticum*.

As to the presence of sexuality in this group of aquatic species, Couch (1931, 1937) reported the pairing of planospores or isoplanogametes (figs. 25, 26) in *S. longispinosus* (Couch) Karling and "an apparent fusion" to form a biflagellate zygote (fig. 27) as in *S. endobioticum*. This has not been confirmed by subsequent workers, although Miller (1955) showed paired planospores (fig. 3) in *S. grandis*. (Miller) Karling.

SYNCHYTRIUM De Bary and Woronin

Ber. Verhandl. Naturf. Gesell. Freiburg 3:46, 1863.

Micromyces, Dangeard, Le Botan. 1:55, 1889.
Chrysophlyctis, Schiberszky, Ber. deut. Bot. Gesell. 14:36, 1896.
Pycnochytrium (De Bary) Schroeter, Engler and Prantl, Natürl. Pflanzenf. 1:73, 1893.
Woroninella Raciborski, Z. Pflanzenkr. 8:195, 1898.
Miyabella Ito and Homma, Bot. Mag. Tokyo 40: 110, 1926.

Plate 19

Figs. 1-26. *Synchytrium callirrhoe* Karling in *Callirrhoe involucrata*. (Karling, 1958.)

Figs. 1, 2. Uniflagellate, 3.8-4 × 4.6-5 μ, and biflagellate planospores, respectively.

Fig. 3. Pairing of isoplanogametes (?).

Fig. 4. Zygote (?).

Fig. 5. Infection of the epidermal cell.

Fig. 6. Young parasite lying above the host nucleus.

Fig. 7. Larger parasite below the host nucleus.

Fig. 8. Occasional division of host cell nucleus above the parasite.

Figs. 9, 10. Uni- and multinucleate incipient sori, respectively.

Fig. 11. Cleavage of a sorus.

Fig. 12. Sister incipient zoosporangia, nuclei dividing in one of them.

Fig. 13. Intranuclear division.

Figs. 14, 15. Variations in sizes of zoosporangia.

Fig. 16. Small sorus with four zoosporangia.

Figs. 17, 18. Monosporangiate sori in fixed and stained, and in living material, respectively.

Fig. 19. Discharge of planospores from a large zoosporangium.

Figs. 20, 21. Young and older incipient resting prosori, respectively.

Fig. 22. Mature resting prosorus surrounded by host residue in a simple unicellular gall.

Fig. 23. Early germination stage of resting prosorus in living material.

Fig. 24. Later stage, primary nucleus entering incipient sorus.

Fig. 25. Sorus of sporangia from a germinated resting prosorus.

Fig. 26. Discharge of planospores in a zoosporangium from a germinated resting prosorus.

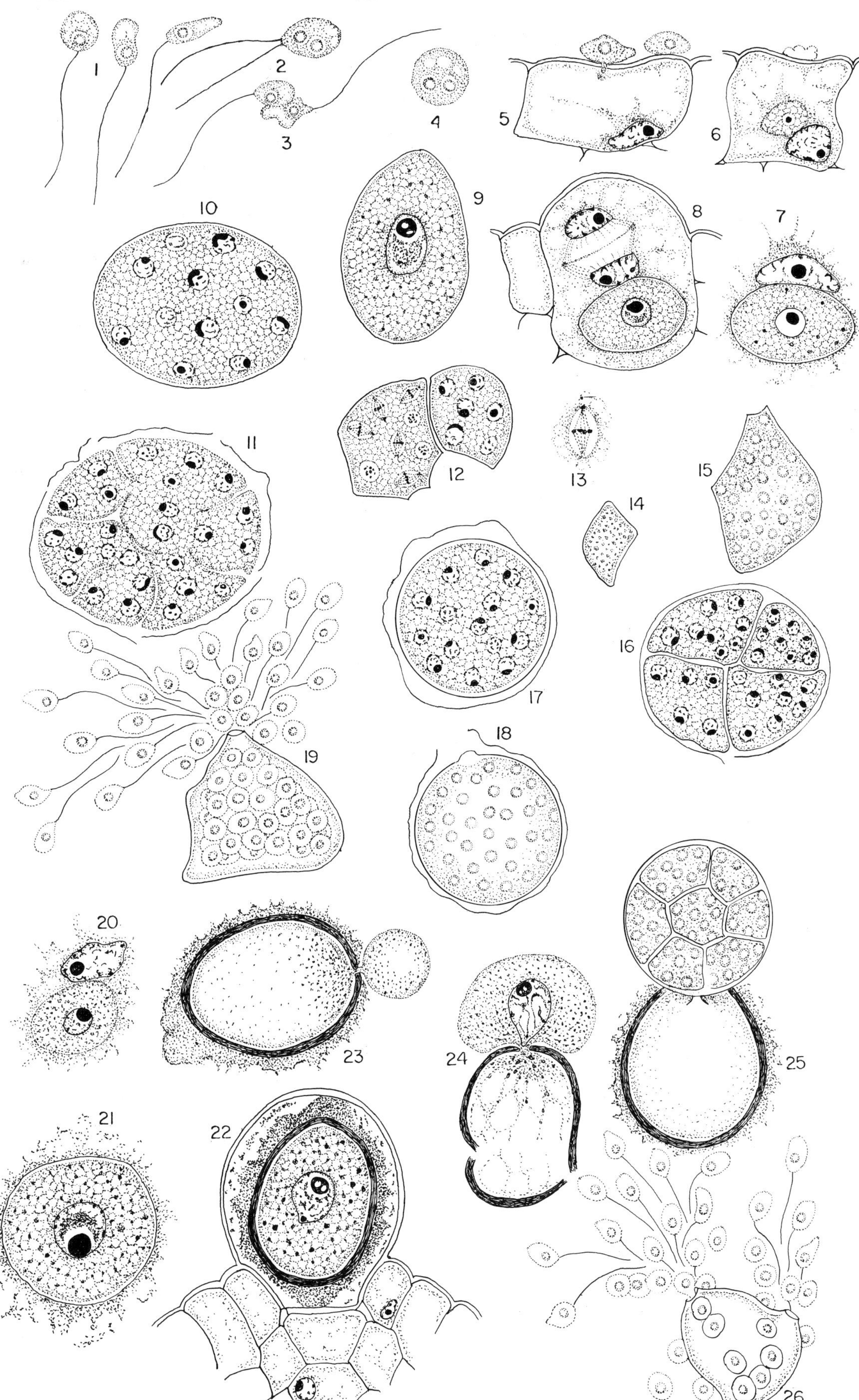

Subgenus *Exosynchytrium*

SUBGENUS MICROSYNCHYTRIUM Karling

Mycologia 45:279, 1953.

Terrestrial Species

Plate 17

The life cycles of these species are identical with those of the fully known aquatic (*Micromyces*) species, but their evanescent and resting prosori as well as sori and zoosporangia are markedly larger. Whereas in *S. grandis*, the largest of the aquatic species, the evanescent and resting prosori are only 40 to 48 μ and 16 to 32 μ in diameter, respectively, those of the terrestrial species may be up to 200 μ or more in diameter. Also, several of the latter species have been studied intensively from fixed, sectioned and stained material, and their cytology is better known than in the aquatic species. Like all terrestrial species of *Synchytrium* they induce the development of galls of various shapes, sizes and complexity on their hosts which are the result of cell enlargement or cell multiplication, or a combination of both. At present this subgenus includes thirty-three identified and seven unidentified species.

The planospores (fig. 1) with a whip-lash flagellum encyst on the host cell (fig. 2) and their contents enter it through a fine germ tube, leaving the refractive globules (fig. 3) and sometimes a shriveled membrane (fig. 3) on the outside. Within the host cell the young parasite migrates toward the host nucleus and eventually lies next to, above, or below it (figs. 4–8). In a few days its thallus increases markedly in size and is recognizable as an incipient evanescent prosorus (fig. 6). It attains maturity in 2 to 3 weeks, fills the host cell partly or completely, and possesses a remarkably large primary nucleus (fig. 7) which may be up to 26 μ in diameter. In contrast to the resting stage this thallus is an evanescent prosorus, and within a few weeks it gives rise to an incipient sorus on its surface (figs. 8, 9) into which the primary nucleus slowly flows (figs. 10, 11, 12). There, the nucleus divides (fig. 13), and with subsequent divisions of its daughter nuclei the sorus becomes multinucleate (fig. 14). This is followed by cleavage of the protoplasm into multinucleate segments (fig. 15) which mature into zoosporangia, are released by the breakdown of the soral membrane, and eventually produce planospores (fig. 16). These may function as planospores and form additional evanescent prosori, or fuse as gametes.

Plasmogamy is followed shortly by karyogamy (fig. 26). The quiescent zygote infects the host (fig. 29) and develops (figs. 30, 31) into a thick-walled uninucleate resting or dormant prosorus (fig. 32) which may germinate within a few weeks or remain dormant for many years. It germinates (figs. 33–35) in the same manner as the evanescent prosorus and forms an incipient sorus on its surface into which the primary nucleus moves and later divides. This division is meiotic, according to Kusano (1930a), with five pairs of chromosomes (fig. 36) in *S. fulgens* Schroeter. Subsequent mitoses (fig. 37) result in a multinucleate sorus (fig. 38) which cleaves into several segments or incipient zoosporangia (fig. 39). These eventually separate and produce planospores (fig. 40) which may function as zoospores or gametes. In *S. endobioticum* (Schilberszky) Perc. the sorus does not undergo cleavage, and its content is transformed into a single zoosporangium, according to Kole (1965).

Plate 20

Figs. 1-39. *Synchytrium macrosporum* Karling. (Karling, 1964.)

Figs. 1-5, 7. Variations in shapes and sizes of resting prosori; numerous bodies or granules in the cytoplasm; enveloping residue omitted.

Fig. 6. Highly magnified portion of the cytoplasm showing the structure of the bodies or granules.

Fig. 8. Composite, multicellular gall on *Ambrosia aptera* containing a mature resting prosorus surrounded by host cell residue.

Figs. 9, 10. Germination stages of a resting prosorus; expanded granules emerging into a relatively clear area in the incipient sorus.

Fig. 11. Division of the primary nucleus in the incipient sorus.

Figs. 12, 13. Bi- and multinucleate incipient prosori.

Fig. 14. Cleavage stage in the formation of uninucleate segments ("protospores"?).

Fig. 15. Sorus with expanded uninucleate segments ("protospores"?).

Fig. 16. Intranuclear mitosis in a uninucleate segment.

Fig. 17. Sorus of multinucleate incipient zoosporangia.

Figs. 18-22. Variations in sizes and shapes of zoosporangia.

Figs. 23-25. Zoosporangia just prior to dehiscence; outer wall has ruptured revealing a thin inner membrane.

Fig. 26. Discharge of planospores.

Figs. 27-29. Ovoid and elongate, 3-3.8 × 4-4.5 μ, planospores with a yellowish-orange refractive globule.

Fig. 30. Large abnormal biflagellate planospore.

Fig. 31. Aggregate of quiescent non-flagellate planospores and a few biguttulate zygotes (?); motile flagellate planospores darting through the aggregate.

Figs. 32, 33. Fusions.

Fig. 34. Amoeboid zygotes (?).

Figs. 35, 36. Fusion of refractive globules.

Fig. 37. Binucleate zygote (?).

Fig. 38. Completion of karyogamy.

Fig. 39. Young parasite lying in base of a host cell and next to the host nucleus.

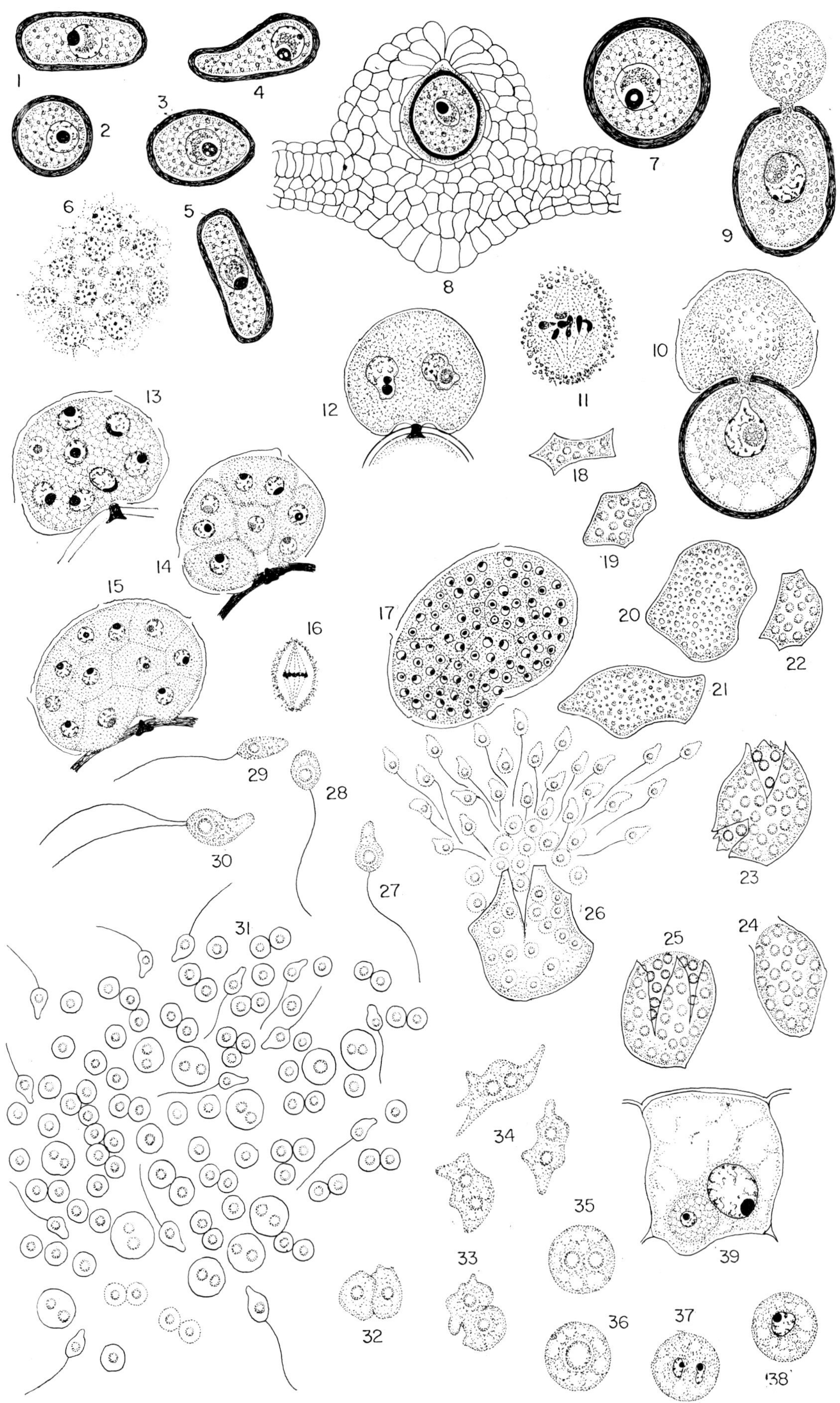

Subgenus *Pycnochytrium*

SUBGENUS MESOCHYTRIUM Schroeter

In Engler u. Prantl, Die Natürl. Pflanzenf. 1:73, 1892-1893 (sensu recent, Fitzpatrick, The lower fungi — Phycomycetes, pp. 80, 82, 1930)

Plate 18, Figs. 1-10

This subgenus, interpreted in the sense of Fitzpatrick (1930) and Karling (1964), became limited to only one species, *S. desmodiae* Munasinge, after Kole (1965) transferred *S. endobioticum* to *Microsynchytrium*. Its developmental cycle (figs. 1–8) is the same as that of *Microsynchytrium* up to the germination of the thick-walled resting stage or spore. Instead of functioning as a prosorus the content of the latter divides directly into planospores (figs. 9, 10) and it becomes thus a zoosporangium. Munasinghe's (1955) photographs and drawings of zoospore development in the resting sporangia are not clear, and it is not improbable that the spores may be found to function as prosori in germination. In that event this subgenus may be eliminated unless other species with this type of life cycle are discovered.

SUBGENUS SYNCHYTRIUM Karling

In *Synchytrium*, p. 193, 1964. (*Eusynchytrium* Schroeter, In Cohns Beitr. Biol. pflanz. 1:39, 1870)

Plate 18, Figs. 11-42

Synchytrium taraxaci De Bary and Woronin, the type species of the genus, is the only fully known (?) species which can be included in this subgenus at present, but further studies may show that additional species have the same developmental cycle. Its early developmental stages (figs. 12–20) are similar to those of the previously described subgenera, but the uninucleate initial cell of the thallus does not function as a prosorus. Instead of migrating out, the primary nucleus divides within the initial cell with the result that a multinucleate incipient sorus (figs. 21, 22) is formed directly. Eventually, its protoplasm cleaves (fig. 23) into segments which become the incipient zoosporangia. These vary markedly in shape and size (figs. 24–28), and the larger polyhedral ones may attain 112 μ in greatest diameter. Their hyaline walls are conspicuously thickened at the angles. One or more inconspicuous exit papillae are formed (fig. 29), and when these deliquesce the planospores swarm out in large masses (figs. 30–32). Large bi- and multiflagellate planospores may occur, and the uniflagellate ones may pair occasionally (fig. 12). However, no conclusive evidence of fusion of isogametes has been reported in *S. taraxaci*.

The so-called resting spores develop in the same manner (figs. 33–35) as the incipient sori and go into the dormant phase as thick-walled uninucleate bodies (fig. 35). In *S. taraxaci* these spores are usually much smaller than the incipient sori and do not induce as large galls as the former on the host. In germinating the primary nucleus and its derivatives divide whereby the spore becomes multinucleate (fig. 36) and is transformed directly into a zoosporangium (figs. 37, 38). So far only De Bary and Woronin have observed germination of such resting zoosporangia. The writer has attempted to induce germination of freshly collected material by treatment in various ways over a period of several years without success. In the event the resting spores are found to function as prosori this subgenus will probably be abandoned.

PLATE 21

Figs. 1-11, 16-25. *Synchytrium decipiens* Farlow. Figs. 1-11, 16-18, 25 after Karling, 1964; figs. 19-24 after Harper, 1899.

Fig. 1. Part of ruptured gall and sorus with extruding powdery zoosporangia. (Karling, 1964.)

Figs. 2-5. Variations in shapes and sizes of zoosporangia and successive stages in sporogenesis. Drawn from living material.

Fig. 6. Discharge of planospores. (Karling, 1964.)

Fig. 7. Fixed and stained planospore with a whiplash flagellum. (Karling, 1964.)

Fig. 8. Similar planospore, 3 × 4.5 μ, drawn from living material.

Fig. 9. Planospores on an epidermal cell of *Amphicarpaea bracteata*. (Karling, 1964.)

Fig. 10. Infection. (Karling, 1964.)

Fig. 11. Young parasite in the base of the host cell next to the host nucleus. (Karling, 1964.)

Figs. 12-15. *Synchytrium minutum* (Pat.) Gäumann. (Kusano, 1907.)

Figs. 12-14. Stages in the formation of a lysigenous cavity and a multinucleate symplast around the sorus by lysis of the host cell walls.

Fig. 15. Division of the primary nucleus of the sorus.

Fig. 16. Binucleate incipient sorus. (Karling, 1964.)

Fig. 17. Multinucleate incipient sorus drawn from a fixed and stained section.

Fig. 18. Cleavage into primary segments, drawn from a fixed and stained section.

Fig. 19. Primary multinucleate segment. (Harper, 1899.)

Fig. 20. Cleavage of primary segment, drawn from a fixed and stained section.

Figs. 21-23. Ultimate, secondary segments or protospores. (Harper, 1899.)

Fig. 24. Intranuclear mitosis in a protospore or incipient zoosporangium. (Harper, 1899.)

Fig. 25. Section through a group of multinucleate incipient zoosporangia. (Karling, 1964.)

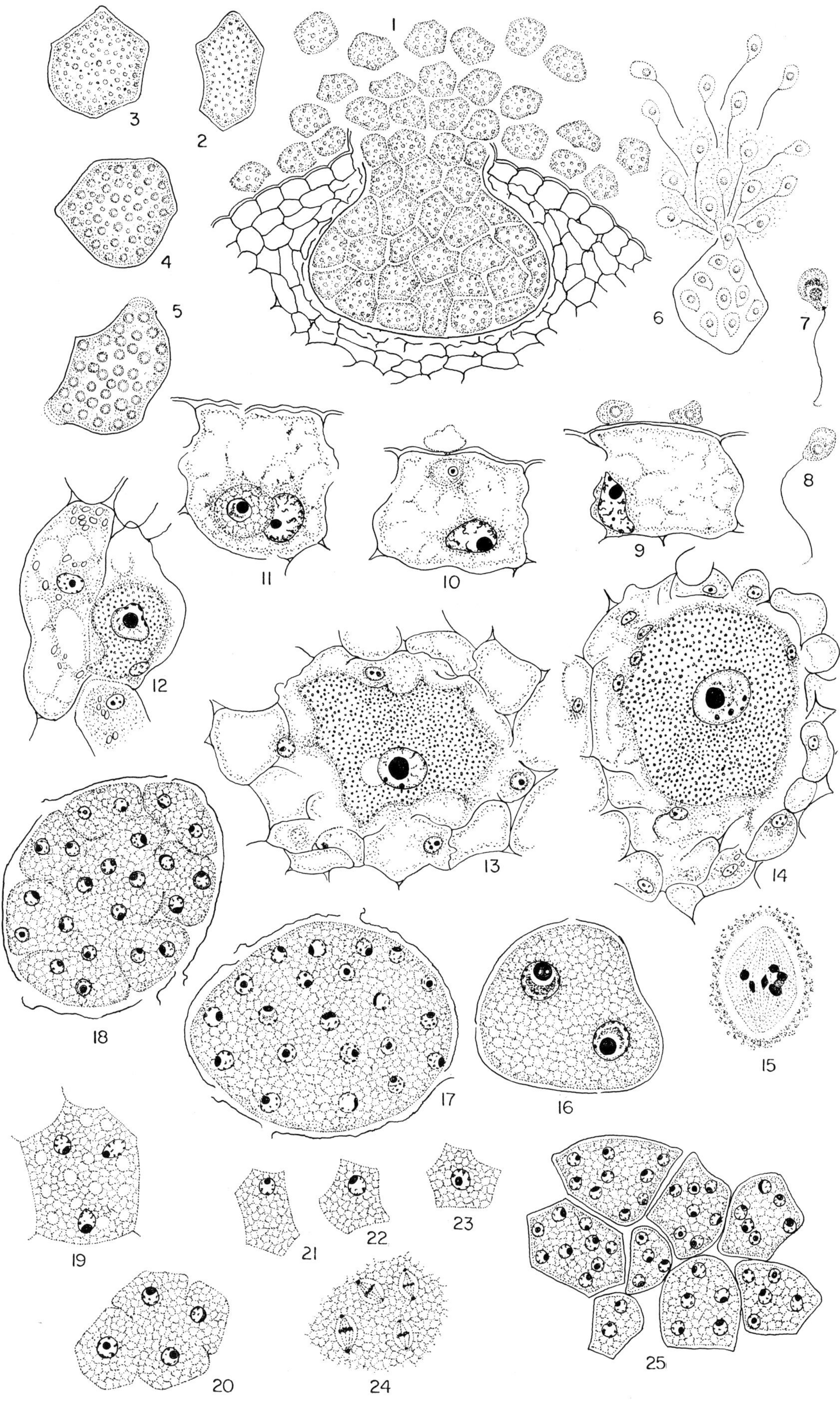

Subgenus *Woroninella*

SUBGENUS EXOSYNCHYTRIUM Karling

Mycologia 45:279, 1953

Plate 19

This subgenus was established to contrast it with *Endochytrium* Du Plessis (1933) and to include *Synchytrium fulgens* whose life cycle did not warrant inclusion in any of the other subgenera. Later it was found (Karling, 1956, 1958b) that *S. fulgens* has a *Microsynchytrium* type of development, and *Exosynchytrium* was temporarily abandoned. Subsequently, however, *S. callirrhoe* Karling (1958a) was found to have the type of development proposed originally for *Exosynchytrium*, and this subgenus was reestablished. *Synchytrium callirrhoe* is the only fully known species of this subgenus, but two other incompletely known ones are included tentatively in it.

Exosynchytrium has the same type of development as the subgenus *Synchytrium* except that the so-called resting spore functions as a prosorus in germination. The initial cell (figs. 6–11) derived from infecting planospores is transformed directly into a sorus at maturity and usually cleaves into a number of zoosporangia (figs. 11–16), but rarely it forms a single zoosporangium (figs. 17, 18). The planospores are facultative in *S. callirrhoe* and may infect the host as planospores (figs. 5, 6), or occasionally pair and fuse (figs. 3, 4). However, it has not been shown that the resting prosori (figs. 21–25) arise from such fusions and zygotes. In germination a pore is formed in the wall of the resting prosorus, and its content emerges to the outside (figs. 23, 24) to form a superficial sorus of zoosporangia (fig. 25) as in *Microsynchytrium*.

SUBGENUS PYCNOCHYTRIUM De Bary

Vergleich. Morp. u. Biol. d. Pilze, p. 180, 1884.

Plate 20

This is the largest of the subgenera with nearly sixty species, but of these only sixteen are fully known and can be definitely assigned to this group. As interpreted here *Pycnochytrium* is limited to species whose initial cell develops into a resting prosorus (figs. 1–7) which produces a superficial sorus of zoosporangia when it germinates (figs. 9–17). So far no so-called summer or evanescent prosori and sori have been reported in this subgenus. Two species, *S. aurem* Schroet. and *S. macrosporum* Karling, have a wide host range, and the latter species has been transferred experimentally by planospores to 1,483 species in 933 genera of 185 plant families which range from the liverworts to the Compositae (Karling, 1960, 1962, 1972, 1974).

In mounts of one or several dehiscing zoosporangia of *S. macrosporum* the planospores usually clump or aggregate in large dense masses (fig. 31), come to rest, and lose their flagella. Among such masses may be found many pairs as well as a few biguttulate zygotes. Occasional fusions (figs. 32, 33) occur after which the zygotes become amoeboid (fig. 34) for a while and then round up (fig. 35). Fusion of the refractive globules (fig. 36) may or may not take place, and at about the same stage karyogamy occurs (figs. 37, 38). Infection by zygotes and their development within the host cell into resting prosori have not been observed. If the zygotes develop into resting prosori meiosis, presumably, occurs during division of the primary nucleus (fig. 11) in the incipient sorus. Inasmuch as all developmental stages of *S. macrosporum* have been obtained from a monozoospore stock or inoculum this species, at least, is homothallic, but this has not been demonstrated in other species of the subgenus. The resting prosori of *S. macrosporum* have remained viable for 20 years under ordinary herbarium conditions at 24°C temperatures.

SUBGENUS WORONINELLA Gäumann

Ann. Mycol. 25:169, 1927

Plate 21

In this subgenus only sori, zoosporangia and planospores are developed, and so far no resting prosori or resting sporangia have been observed. The initial cell or thallus (figs. 11-17) is transformed directly into a sorus at maturity (fig. 18). It may be quite large, up to 1250 μ in greatest diameter in some species, and include as many as 2000 small zoosporangia. The galls and soral membrane burst at maturity (fig. 1) and expose powdery masses of brightly-colored zoosporangia which were often mistaken by the early mycologists for the aecial stage of rusts. At present *Woroninella* includes thirteen identified and approximately eighteen unnamed species, all of which parasitize only species of the Leguminosae. Although no thick-walled resting stages occur in this subgenus some of its species may persist over the winter months in temperature regions. According to Kusano (1932) and Karling (1936) the sori may remain viable during such periods on the underground portions and fruits of the host.

Occasional fusions of planospores have been reported to occur in *Woroninella* (Gaumann, 1927), but they do not appear to be significant sexually. Kusano (1930) regarded such fusions as an indication that the sex character had. not completely disappeared from the motile planospores or parthenospores.

REFERENCES TO SYNCHYTRIACEAE

Bary, A., de. 1884. Vergleichende Morphologie und Biologie der Pilze, Mycetozoa und Bakterien. xvi + 558 pp., 198 figs. Leipzig.

_______ and M. S. Woronin. 1863. Beitrag zur Kenntniss der Chytridieen. Ber. Verhandl. Naturf. Gesell. Freiburg. 3:22-61, pls. 1, 2.

Canter, H. M. 1949. Studies on British chytrids. VI. Aquatic Synchytriaceae. Trans. Brit. Mycol. Soc. 32:69-94, 13 text figs., pls. 7-11.

Couch, J. N. 1931. *Micromyces zygogonii* Dang., parasitic on *Spirogyra*. J. Elisha Mitchell Sci. Soc. 46:231-239, pls. 16-18.

_______. 1937. Notes on the genus *Micromyces*. Mycologia 29:592-596, 14 figs.

Curtis, K. M. 1921. The life history and cytology of *Synchytrium endobioticum* (Schilb.), Perc., the cause of wart disease in potato. Philos. Trans. Roy. Soc. London, Ser. B, 20:409-478, pls. 12-16.

Dangeard, P. A. 1889. Mémoirs sur les Chytridinées. Le Botan. 1:39-74, pls. 2, 3.

_______. 1890-1891. Mémoire sur quelques maladies des Alques et des Animaux. Le Botan. 2:231-268, pls. 16-19.

Du Plessis, S. J. 1933. Beskryrving van *Synchytrium cotulae* nov. spec op *Cotula coronopifolia* Linn. in Suid-Afrika. Ann. Univ. Stellenbosch 11 (reeks A, afl. 5):1-11, 8 figs.

Fitzpatrick, H. M. 1930. The lower fungi — Phycomycetes. vii + 331 pp. McGraw-Hill, New York.

Gäumann, E. 1927. Mykologische Mitteilungen. III. Ann. Mycol. 25:165-177.

Harper, R. A. 1899. Cell division in sporangia and asci. Ann. Bot. 13:467-525, pls. 24-26.

Heim, P. 1956. Remarques sur le développment les divisions nucléares et le cycle évolutiff de *Synchytrium endobioticum* (Schibb.) Perc. Rev. Mycol. 21:93, pls. 2, 3.

Ito, S. and Y. Homma. 1926. *Miyabella*, a new genus of the Synchytriaceae. Bot. Mag. Tokyo 40:110-113.

Karling, J. S. 1936. Overwintering of *Synchytrium decipiens* in New York. Bull. Torrey Bot. Club 63:37-40.

_______. 1953. *Micromyces* and *Synchytrium*. Mycologia 45:276-287.

_______. 1954. Possible relationships and phylogeny of *Synchytrium*. Bull. Torrey Bot. Club 81:353-362, 1 fig.

_______. 1955. The cytology of prosoral, soral and sporangial development in *Synchytrium australe*. Amer. J. Bot. 42:37-41, 31 figs.

_______. 1956. *Synchytrium fulgens* in relation to other species on onagraceous hosts. Ibid. 43:61-69, 11 figs.

_______. 1958a. *Synchytrium callirrhoe* sp. nov. Ibid. 45:327-330, 26 figs.

_______. 1958b. *Synchytrium fulgens*. Mycologia 50: 373-375.

_______. 1960. Inoculation experiments with *Synchytrium macrosporum*. Sydowia 14:138-169, 19 figs.

_______. 1962. Additional plants susceptible to *Synchytrium macrosporum*. Adv. Frontiers Pl. Sci. 1:55-71.

_______. 1964. *Synchytrium*. xviii + 470 pp., 13 pls. Academic Press, New York.

_______. 1972. Additional plants susceptible to *Synchytrium macrosporum*. II. Adv. Frontiers Pl. Sci. 1-29.

_______. 1974. Further induced infectivity by *Synchytrium macrosporum*. Bull. Torrey Bot. Club 101:311-316, 20 figs.

Köhler, E. 1930. Beobachtungen an zoosporen Aufschwemmungen von *Synchytrium endobioticum* (Schilb.) Perc. Zentralbl. Bakt. Parasitenk. Infekt abt. II, 82:1-10, 3 figs.

Kole, A. P. 1965. Resting-spore germination in *Synchytrium endobioticum*. Netherl. J. Pl. Path. 71:72-78, 7 figs.

Kusano, S. 1907. On the cytology of *Synchytrium*. Centralbl. Bakt. Parasit. Infekt. 19:538-542, 8 figs.

_______. 1909. A contribution to the cytology of *Synchytrium* and its hosts. Bull. Coll. Agric Imp. Univ. Tokyo 8:79-147, pls. 9-11.

_______. 1928. The relative sexuality in *Synchytrium*. Proc. Imp. Acad. (Tokyo) 4:497-499.

_______. 1930a. Cytology of *Synchytrium fulgens*. J. Coll. Agric. Imp. Univ. Tokyo 10:347-388, pls. 28, 29.

_______. 1930b. The life history and physiology of *Synchytrium fulgens* Schroet., with special reference to its sexuality. Jap. J. Bot. 5:35-132, 19 figs.

_______. 1932. Dormancy in the summer sorus of *Synchytrium*. J. Coll. Agric. Imp. Univ. Tokyo 11:427-439, 2 figs.

Lingappa, B. T. 1958a. Sexuality in *Synchytrium brownii* Karling. Mycologia 50:524-537, 78 figs.

_______. 1958b. The cytology of development and germination of resting spores in *Synchytrium brownii*. Amer. J. Bot. 45:613-620, 56 figs.

Lowenthal, W. 1905. Weitere Untersuchungen an Chytridiaceen. Arch. f. Protistenk. 5:221-239, pls. 7, 8.

Miller, C. E. 1955. *Micromyces grandis*, a new member of the aquatic Synchytriaceae. J. Elisha Mitchell Sci. Soc. 71:247-255, 29 figs.

Munasinghe, H. L. 1955. A wart disease of *Desmodium ovalifolium* caused by a species of *Synchytrium*. Quart. Circ. Ceylon Rubber Res. Inst. 31:22-28, 11 figs.

Raciborski, M. 1898. Pflanzenpathologische aus Java. Zeitschr. Pflanzenkr. 8:195-200.

Rieth, A. 1950. Beitrag zur Kenntnis der Gattung *Micromyces* Dangeard. I. *Micromyces ovalis* nov. spec. Oesterr. Bot. Zeitschr. 97:510-516, 11 figs.

_______. 1956. *Micromyces laevis* Canter in Deutschland nebst einigen Bemerkungen über die Wasserlebende algen-parasitare Gruppe der Synchytriaceae. Die Kulturpflanze 4:27-45, 25 figs., pl. 1.

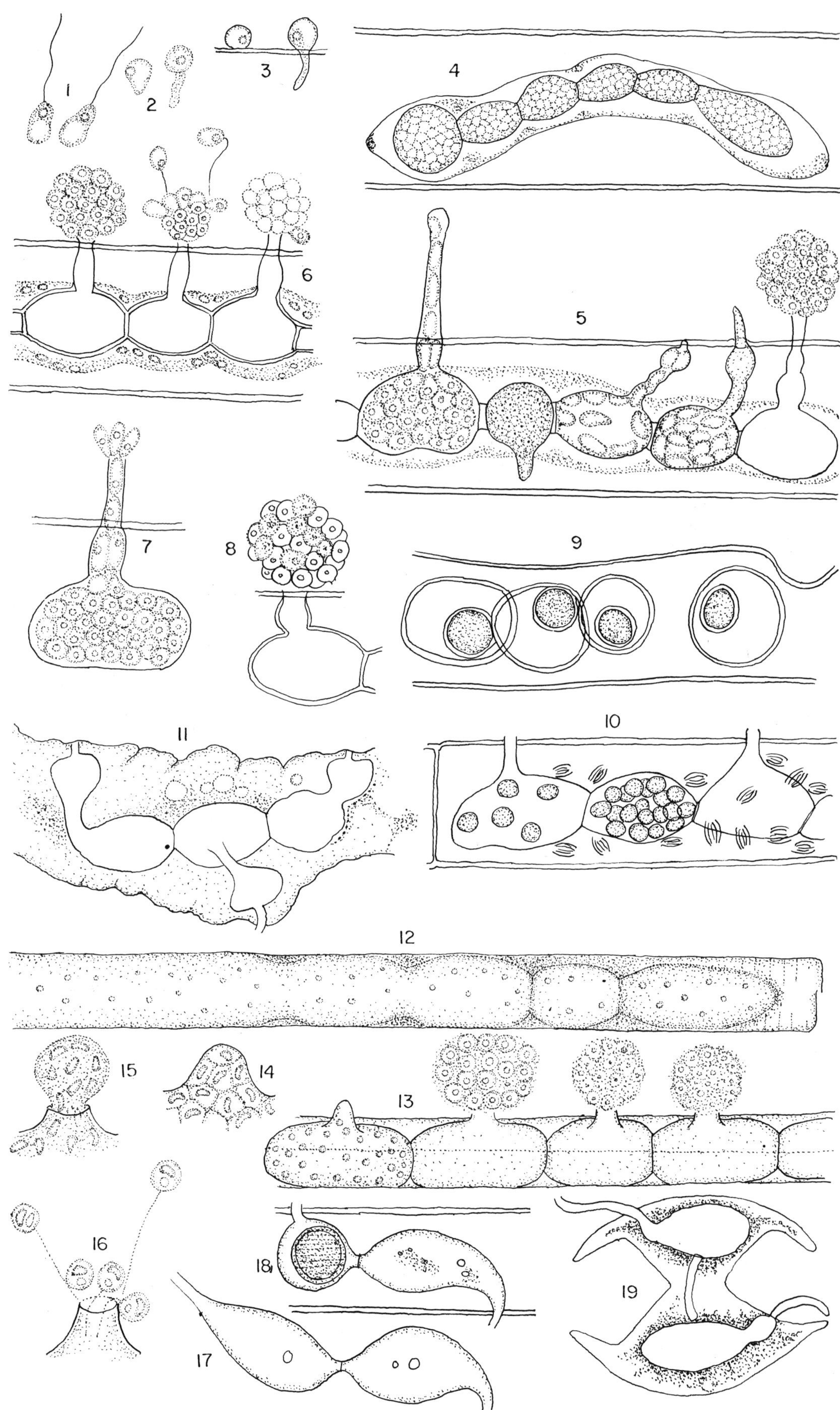

Achlyogeton, Septolpidium, Bicricium

________. 1962. *Micromyces furcata* n. sp., nebst einigen Bemerkungen über seinen Wirt. Die Kulturpflanze, Beihefte 3:286-295, 4 figs., pls. 1, 2.

Rytz, W. 1917. Beitrage zur Kenntniss der Gattung *Synchytrium*. Die cytologischen Verhaltnisse bei *Synchytrium taraxaci* de By. et Wor. Beiheft. Bot. Centralbl. 34:343-372, pls. 2-4.

Scherffel, A. 1926, Einiges über neue oder üngenugend bekannte Chytridineen (Der "Beitrage zur Kenntnis der Chytridineen," Teil II.) Arch. f. Protistenk 54:167-260, pls. 9-11.

Schilberszky, K. 1896. Ein neuer Schorfparasit der Kartoffelknollen. Ber deut. Bot. Gesell. 14:36-37.

Schroeter, J. 1870. Die Pilzparasiten aus der Gattung *Synchytrium*. Cohn's Beitr. Biol. Pflanz. 1:1-50, pls. 1-3.

________. 1892-1893. Phycomycetes. *In* Engler and Prantl's Naturl. Pflanzenf. 1, abt. 1:63-71.

Sparrow, R. K., Jr. 1960. Aquatic Phycomycetes. Univ. Mich. Sci. Ser. 15:xiii-1187, figs. 1-90.

Chapter III

ACHLYOGETONACEAE

This family was established by Sparrow (1942) to include three genera, *Achlyogeton*, *Septolpidium* and *Bicricium*, of imperfectly known species which are characterized by two- or many- celled, holocarpic, elongate, septate thalli whose segments become zoosporangia and produce posteriorly uniflagellate planospores. A fourth genus, *Myiophagus*, with more extensive thalli was tentatively assigned to this family by Karling in 1948. Thus, at present the family seems to be hardly more than a convenient grouping of species which may turn out to be unrelated. Resting sporangia occur in *Myiophagus*, but they are unknown in *Septolpidium* and have not been demonstrated with certainty in *Achlyogeton* and *Bicricium*.

The establishment of this family name from the genus *Achlyogeton* has been questioned by Karling (1942, 1948) on the grounds that this genus may not prove to be a chytrid at all when it is fully known. It is the only known chytrid genus, with the possible exception *Achlyella*, in which the zoospores encyst in clusters of cystospores at the exit orifices and later leave an empty vesicle behind as they germinate. Such behavior is characteristic of *Achlya* and *Aphanomyces* species in which the planospores are biflagellate. Schenk's (1859) observations of uniflagellate planospores in 1859a were made at a time when less importance was given to the number, position and relative lengths of the flagella as indications of identity and relationships, and it is possible that he may have overlooked a second flagellum in *Achlyogeton*. This is clearly suggested by the fact that he (1859b) illustrated the planospores of *Pythium* species as being uniflagellate, also. Although a few subsequent workers have reported clusters of cytospores at the exit orifices in *Achlyogeton*, none of them have shown actively motile uniflagellate planospores. It is thus possible that *Achlyogeton* may prove to be a genus with biflagellate planospores. Inasmuch as the

PLATE 22

Figs. 1-9. *Achlyogeton entophytum* Schenk. Figs. 1-7 after Schenk, 1859; figs. 8, 9 after Martin, 1927.

Fig. 1. Planospores with a minute refractive globule.

Fig. 2. Germination of planospores in water.

Fig. 3. Infection of the host.

Fig. 4. Multicellular thallus lying in the collapsed protoplasmic residue of *Cladophora*.

Fig. 5. Later stage, linear cells transformed into zoosporangia and discharging planospores.

Fig. 6. Discharged planospores in clusters; all planospores encysted as cystospores, 4 μ diam., in left-hand cluster; in the central one a planospore is emerging from the cysts, and in the right hand cluster all but one have emerged from the cysts.

Fig. 7. Reduced thallus consisting of one zoosporangium.

Fig. 8. Cluster of cystospores and empty cysts.

Fig. 9. Resting spores (?).

Fig. 10. *Achlyogeton salinum* Dang. Portion of a thallus in *Cladophora*, possibly a species of *Myzocytium* or *Sirolpidium*. (Dangeard, 1932.)

Fig. 11. *Achlyogeton rostratum* Sorokin. Reduced empty thallus in the body of *Anguillula*; exit tubes are greatly expanded before passing through body wall; possibly a species of *Myzocytium*. (Sorokin, 1876.)

Figs. 12-16. *Septolpidium lineare* Sparrow. (Sparrow, 1936.)

Fig. 12. Thallus in *Synedra* sp., right-hand portion segmenting into zoosporangia.

Fig. 13. Separate zoosporangia discharging planospores.

Figs. 14-16. Stages in deliquescence of exit papilla and discharge of planospores.

Fig. 17. *Bicricium lethale* Sorokin. Empty two-celled thallus in body of an eelworm. (Sorokin, 1883.)

Fig. 18. *Bicricium transversum* Sorokin. Two-celled thallus in *Cladophora*; resting .spore in one segment. (Sorokin, 1883.)

Fig. 19. *Bicricium naso* Sorokin. Two zoosporangia in *Arthrodesmus*. (Sorokin, 1883.)

name *Achlyogetonaceae* would then become invalid, Karling (1948) suggested a substitute name, *Septolpidiaceae*, for the family in that event.

ACHLYOGETON Schenk

Bot. Zeit. 17:398, 1859

Plate 22, Figs. 1-11

The thallus of this genus is strikingly similar to that of *Myzocytium*, and unless the number and position of the flagella on the planospores and resting spore development are observed it is difficult to differentiate between the two genera. *Achlyogeton* has often been placed in the Lagenidiales, but Sparrow (1942, 1960) regarded it as "unquestionably a member of the Chytridiales". Four species have been included in this genus, but three of these, *A. rostratum* Sorokin, *A. solatium* Sorokin, and *A. salinum* Dangeard, are so imperfectly known that it is impossible to classify them here as valid species. The type species, *A. entophytum* Schenk, also, is not fully known, and the resting spores described by Martin (1927) may not relate to this species at all. As described by Schenk the posteriorly uniflagellate planospores with a refractive globule (fig. 1) infect *Cladophora* (fig. 3), and the germ tube gives rise to an elongate, constricted, septate thallus (fig. 4). Occasionally, the thallus may be reduced and consists of a single cell or zoosporangium (fig. 7). In the elongate thalli the swellings between the septa and short isthmuses are transformed into zoosporangia with long exit canals (fig. 5) and through these the planospores emerge to the outside of the host. However, they do not swim directly away but encyst as clusters of cystospores (figs. 5, 6, 8) at the tips of the exit tubes as in *Achlya* and *Aphanomyces*. After a period of rest the cystospores develop a minute papilla and pore, and through these the planospores emerge and swim away, leaving an empty vesicle behind. Martin illustrated resting spores in large thick-walled vesicles in different *Cladophora* filaments from those which contained zoosporangia, but he was not certain that they belong to *A. entophytum*.

Two other named species, *A. rostratum* (fig. 11) and *A. salinum*, (fig. 10) are so poorly known that little can be said about their validity. Possibly, Sorokin's and Dangeard's (1932) drawings relate respectively to *Myzocytium* and *Sirolpidium*. In *A. solatium* Cornu (1870) reported that the branched filamentous thallus grows through several *Oedogonium* cells and forms irregularly-sized zoosporangia which produce 3 to 12 planospores of unknown size. These encyst in clusters at the exit orifices as in species of *Achlya*, *Aphanomyces*, and *Aphanomycopsis* species. Nothing additional is known about this species, and its inclusion in *Achlyogeton* is questionable.

SEPTOLPIDIUM Sparrow

Trans. Brit. Mycol. Soc. 18:215, 1933;
J. Linn. Soc. London (Bot.) 50:428, 1936

Plate 22, Figs. 12-16

This monotypic genus resembles *Achlyogeton* somewhat by its elongate cylindrical thallus which segments into a linear series of separate zoosporangia at maturity (figs. 12, 13), but differs from it by the lack of an encysted stage of the planospores at the exit orifice. Upon emerging the plano-

PLATE 23

Figs. 1-28. *Myiophagus ucrainica* (Wize) Sparrow. Figs. 1-18 after Karling, 1948; figs. 19, 22 after Wize, 1904; figs. 20, 21 after Sparrow, 1939; figs. 23-28 sketches by Thaxter drawn by Sparrow, 1939.

Fig. 1. Living, 3 × 8 μ, planospore with granules at anterior end.

Fig. 2. Fixed and stained spherical, 6 μ, diam., planospore with nuclear cap, nucleus, blepharoplast, rhizoplast and flagellum.

Figs. 3, 4. Germination of planospores in water.

Fig. 5. Early stage of infection of the body of *Lepidosaphes beckii.*

Figs. 6, 7. Bi- and 8-nucleate stages of young thalli; planospore cyst degenerating.

Fig. 8. Later stage of development, thallus constricting and branching.

Fig. 9. Later stage, thallus septate; isthmuses connecting incipient zoosporangia which contain numerous refractive globules.

Fig. 10. Five fully-grown zoosporangia shortly before separating from a portion of an extensive thallus connected by isthmuses; large refractive globules in centers.

Fig. 11. Separate sporangium with refractive globules becoming smaller and progressively dispersed.

Fig. 12. Synchronous intranuclear mitoses in incipient zoosporangium.

Figs. 13, 14. Enlarged views of prophase and metaphase, respectively.

Fig. 15. Uninucleate cleavage segments or planospore initials.

Fig. 16. Zoosporangium with exit papillae; granules aggregated in localized area.

Fig. 17. Early stage of dehiscence.

Fig. 18. Discharge of planospores.

Fig. 19. Incipient resting sporangium.

Fig. 20. Surface view of mature resting sporangium with a reticulate wall, lying in an envelope or sac.

Fig. 21. Optical section of the same.

Fig. 22. Resting sporangium lying in an envelope.

Fig. 23. Germination of resting sporangium; reticulate wall has burst and the zoosporangium has emerged.

Fig. 24. Emerged zoosporangium with 5 exit papillae.

Fig. 25. Almost empty zoosporangium with 4 exit papillae.

Fig. 26. Planospore with numerous granules.

Fig. 27. Paired planospores.

Fig. 28. Biflagellate planospore (zygote?.)

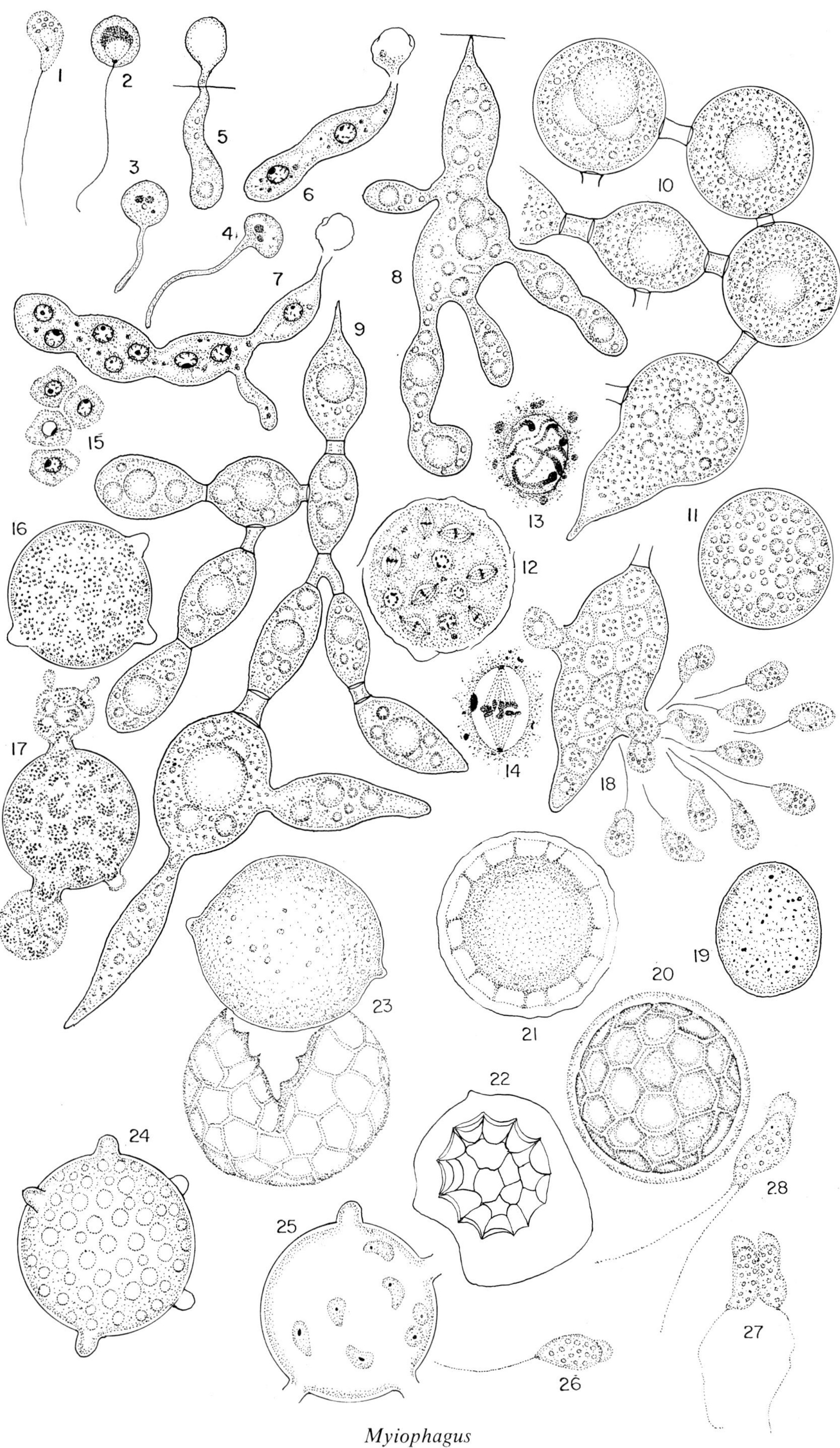

Myiophagus

spores remain quiescent for a while in a spherical motionless temporary cluster at the mouth of the discharge tube (figs. 13, 15), but they ultimately swim away without encysting. Sometimes they become motile directly after emergence (fig. 16). Resting spores have not been observed in this genus.

BICRICIUM Sorokin

Arch. Bot. Nord France 2:37, 1883

Plate 22, Figs. 17-19

This is a questionable genus of 4 imperfectly known species which parasitize eelworms and freshwater algae and which are characterized by a holocarpic, constricted, two-celled thallus. The cells are separated by a narrow septate isthmus, bear a single discharge tube, and function as zoosporangia, or one of them may contain a resting spore (figs. 17–19). Fischer (1892), Minden (1915) and Karling (1942) regarded these species as two-celled forms of *Myzocytium*, although Sorokin had described the planospores of *B. lethale* as being uniflagellate with an acuminate apex and a small basal globule. Planospores are unknown in the other species. Members of this genus have not been observed since Sorokin's report of them, and in the present state of knowledge *Bicricium* is a doubtful member of the Achlyogetonaceae.

MYIOPHAGUS* Thaxter

In Sparrow, Mycologia 31:443, 1939

Plate 23

This monotypic genus was classified tentatively by Karling (1948) in the family Achlyogetonaceae because of its septate holocarpic thallus whose zoosporangia separate at maturity and produce posteriorly uniflagellate planospores. He pointed out, however, that the structure of the planospores and the appearance of the content of the zoosporangia during sporogenesis are fairly similar to those of some members of the Blastocladiales. Possibly it may turn out to be a genus of this order when it becomes fully known, but it is retained provisionally in the Achlyogetonaceae for the time being. So far only one species, *M. ucrainicus* (Wize) Sparrow, has been described, and it is reported to parasitize pupa, larvae and bodies of dipterous and scale insects in the Ukraine U.S.S.R. (Wize, 1904), Maine (Thaxter, 1902) and Florida, U.S.A. (Fisher, 1950), Bermuda (Waterston, 1946; Karling, 1948), and Canada (Waterston, 1946). As a result of dense infection and the separation of the zoosporangia at maturity the interior of the body of the insect is reduced to an orange or reddish colored powdery mass of zoosporangia and resting sporangia.

The planospores come to rest on the host and develop a germ tube which develops into an elongate (figs. 6, 7), frequently constricted and branched thallus (figs. 8, 9). Subsequently it becomes septate at irregular intervals, and the swellings or enlargements between the septa develop into zoosporangia. These are usually connected by short isthmuses (figs. 9, 10) which disintegrate at maturity so that the zoosporangia separate and eventually lie free in the body of the insect. The zoosporangia develop 1 to several exit papillae (figs. 16–18), and at this stage the granular material gradually aggregates in localized areas and thus indicates the sites of the planospore initials. As the papillae tips deliquesce the planospores emerge and disperse (figs. 17, 18).

Resting sporangia were not observed in the Bermuda specimens and very little is known about their development. However, the brief accounts of Wize and Sparrow suggest that the content of globular enlargements contracts and develops an outer reticulated and an inner smooth wall (figs. 20, 21). Thus, the double-walled resting sporangium lies in a thin vesicle or envelope (figs. 21, 22). In germination the reticulated outer wall ruptures (fig. 23) and emits a thin-walled zoosporangium (endosporangium) that develops up to 5 exit papillae (fig. 24) and produces posteriorly uniflagellate planospores (fig. 26). Thaxter observed that these may pair occasionally (fig. 27), and his illustration, also, of large biflagellate planospores suggests that fusions might have occurred. However, this has not been proven and nothing conclusive is known about sexuality in *Myiophagus*.

*G. S. Torrey (Mycologia 37:161, 1945) noted that Sparrow's spelling of *Myrophagus* should be *Myiophagus*, which means "devourer of flies."

REFERENCES TO THE ACHLYOGETONACEAE

Cornu, M. 1870. Note sur une Saprolegniees, parasite d'une nouvelle espece *Oedogonium*. Bull. Soc. Bot. France 17:297-299.

Dangeard, P. A. 1932. Observations sur la famille des Labyrinthulees et sur quelques autres parasites des *Cladophora*. Le Botan. 24:217-258, pl. 24.

Fischer, A. 1892. Phycomycetes. Die Pilze Deutschlands, Oesterreichs and der Schweiz. Rabenh. Kryptogamen – Fl. 1(4):1-490. Leipzig.

Fisher, F. E. 1950. Entomogenous fungi attaching scale insects and rust mites on citrus fruits in Florida. J. Econ. Entomol. 43:305-309.

Karling, J. S. 1942. The simple holocarpic biflagellate Phycomycetes. x + 123 pp., 25 pls. New York.

________ . 1948. Chytridiosis of scale insects. Amer. J.

Bot. 35:246-254, 49 figs.
Martin, G. S. 1927. Two unusual water molds belonging to the family Lagenidiaceae. Mycologia 19:188-194, 4 figs.
Minden, M. von. 1911-1915. Chytridineae, Ancylistineae, Monoblepharidineae, Saprolegniineae. Kryptogamenfl. Mark Brandenburg 5:193-352, 353-496, 1911; 497-608, 1912; 609-630, 1915.
Petch, T. 1940. *Myrophagus ucrainicus* (Wize) Sparrow. A fungus new to Britain. The Naturalist 1940:68.
Schenk, A. 1859. *Achlyogeton*, eine neue Gattung der Mycophyceae. Bot. Zeit. 17:398-400, figs. A, 1-8, pl. 13.
Sorokin, N. W. 1876. Les Végétaux parasites des Anguillulae. Ann. Sci. Nat. Bot. VI, 4:62-71, pl. 3, figs. 1-45.
———. 1883. Aperçu systématique des Chytridiacées recoltées en Russie et dans l'Asie Centrale. Arch. Bot. Nord France 2:1-42, 54.
Sparrow, F. K., Jr. 1936. A contribution to our knowledge of the aquatic Phycomycetes of Great Britain. J. Linn. Soc. London (Bot.) 50:417-478, 7 figs., pls. 14-20.
———. 1939. The entomogenous chytrid *Myrophagus* Thaxter. Mycologia 31:439-444, 8 figs.
———. 1942. A classification of the aquatic Phycomycetes. Mycologia 304: 113-116.
Wize, M. C. 1904. Choroby komosnika buracznego (*Cleonus punctiventris*) powodowane przez grzby owodobojcze, ze szczlgolnem urvzglednieniem gattunkow nowych. Bull. Acad. Umiejetnosci Krakow 10:713-727, pl. 15, 11 figs.

Chapter IV

RHIZIDIACEAE

This is the largest family of the Chytridiales, and as interpreted here it includes approximately 45 genera and nearly 450 identified species. The families Phlyctidiaceae and Chytridiaceae are merged with it, and no familial distinctions are made on the presence or absence of an operculum on the zoosporangium for the reasons stated in the Introduction.

The thallus of species in this family is epi- endobiotic or extra- intramatrical, or largely extramatrical except for the tips of its rhizoids, eucarpic, and monocentric, although in some species it rarely becomes polycentric. The encysted planospore develops wholly or in part into a zoosporangium or a prosporangium. In exceptional instances, however, the zoosporangium of some species may develop from a swelling in the germ tube instead of directly from the encysted planospore. The germ tube usually develops into the absorbing vegetative system which varies from a fine, unbranched filament, peg, tapering, discoid or lobate haustorium, to a richly branched extensive rhizoidal system. It may arise from the base of the zoosporangium or at several points on the periphery of the latter, or the rhizoids may be centered on an apophysis which in some species functions as a prosporangium. In some species an apophysis without rhizoids constitutes the absorbing system, and in some genera several apophyses may develop in a catenulate series. In the polyphagus species the rhizoids are epibiotic or extramatrical except for their tips, and in non-polyphagus saprophytic species, also, part of the rhizoidal system may be extramatrical. The resting spores may be formed asexually or sexually, occur epibiotically or extramatrically as the zoosporangia, or develop endobiotically.

Sexual reproduction has been reported in a large number of genera and species, but the successive stage of sexual fusions are known in only a few species. In many species only the presence of a small empty cyst attached to a resting spore has been interpreted to be a "male" cell whose content has fused with that of another "female" cell. As reported, sexual reproduction varies considerably in this family and occurs by fusion of isoplanogametes, aplanogametes, fusion of the contents of gametangia through anastomosed rhizoids, or through long or short conjugation tubes, or through a pore. These variations will be described in detail for each genus in which they occur and need not be reported further at this point.

Primarily on the basis of the type of development exhibited by the genera and species this family is divided into 3 subfamilies, Rhizidioideae, Chytridioideae and Polyphagoideae, but the assignment of genera to any particular subfamily is arbitrary and provisional in many cases because their type of development is not fully known.

SUBFAMILY RHIZIDIOIDEAE

As interpreted here this subfamily includes most of the eucarpic, monocentric chytrids, and the thallus of its members is epi- endobiotic or extra- intramatrical. The encysted planospore is functional and develops partly or wholly into a zoosporangium so far as the developmental stages are known. In some species enlargement of the planospore into a zoosporangium is delayed until the absorbing system is well established, and an endo-exogenous type of development occurs. The asexually formed resting spores where known develop from the planospore and are borne like the zoosporangia. Likewise, the sexually developed resting spores are borne epibiotically or extramatrically although they may develop from fusion of aplanogametes or the contents of gametangia.

RHIZOPHYDIUM Schenk

Verhandl. Phys. Med. Gesell. Würzburg 8:245, 1858

Chytridium subgen. *Phlyctidium* Braun, 1856. Abhandl. Berlin. Akad. 1855:74.
Chytridium subgen. *Sphaerostildium* Braun, 1856. Ibid. 1855:75.
Phlyctidium Rabenhorst, 1868, Flora Europeae Algarum 3:280.
Rhizophyton Zopf, 1888. Nova Acta Acad. Leop.-Carol. 52:343.
Tylochytrium Karling, 1939. Mycologia 31:287.
Hapalopera Fott, 1942. Studia Bot. Cechica 5:170.

Plates 24-29

Rhizophydium is the second largest genus of the the inoperculate chytrids and together with the species formerly placed in *Phlyctidium* it includes at present 166 or more species, many of which are incompletely known. Also, several of them are doubtful, while others may prove to be identical. In addition a score or more unnamed specimens have been listed in this genus by numerous investigators who were unable to identify them accurately. Thirty-seven species or more have been added to the genus in the last 2 decades (see Ulken, 1972; Roane, 1973; Sparrow, 1973, 1974; Knox and Paterson, 1973; Johnson and Miller, 1974, and others). *Rhizophydium* species are worldwide in distribution and occur as parasites of freshwater and marine algae, various fungi, spores of ferns, roots of higher plants, pollen grains, microscopic animals, and liver fluke eggs, or as saprophytes in soil and water from which they may be isolated on various substrata. Many of the latter species have been grown in axenic cultures.

Phlyctidium is merged with *Rhizophydium* on the grounds that the differences between the endobiotic or intramatrical vegetative or absorbing system in some species of both genera are not sufficiently great to merit separate generic distinction. *Phlyctidium* has been generally regarded as having an endobiotic, unbranched rhizoid or a tubular, peg-like, clavate, knob-like, or discoid haustorium, while *Rhizophydium* is interpreted as a genus of species with a branched, sparse or abundant, and extensive endobiotic rhizoidal system—distinctions which may be highly variable and largely artificial. Otherwise, so far as they are known, the genera are identical in development and in the position of the zoosporangia and resting spores relative to the host or substratum. The artificiality of the distinctions is shown by the fact that in some species of *Rhizophydium*, i.e. *R. sphaerocarpum* (Zopf) Fischer, *R. haynaldii* (Schaarschmidt) Fischer, *R. michococci* Scherffel, *R. contractophilum* Canter, *R. anomalum* Canter, *R. tetragenum* Pongratz, etc., the rhizoid may be unbranched or only meagerly so as in some of the so-called *Phlyctidium* species. Also, some thalli of the same species may have an unbranched rhizoid while in other thalli it may be branched (*R. anomalum*). Furthermore, the distinctions are impractical in identifications because in many cases, particularly in parasitic species, the degenerating host protoplasm surrounds the rhizoid or absorbing system so closely that it is difficult or impossible to determine whether the rhizoid is branched or unbranched.

Another reason for rejecting *Phlyctidium* is that it is invalid. As Karling (1939) pointed out, this name was first used by Wallroth (1833) for a genus of Ascomycetes. Braun (1855) subsequently used it for a subgenus of *Chytridium*, and in 1868 Rabenhorst raised it to generic rank. Since that time several monographers, including Schroeter (1885), Serbinow (1907), Minden (1911) and Sparrow (1943, 1960), as well as research students, Karling included, have continued to use the name, although it was rejected by Fischer (1892), Schroeter (1893), and Scherffel (1926). Because this name is not valid Karling proposed a substitute one, *Tylochytrium*, as being more descriptive and appropriate, but it is rejected here as superfluous. Wallroth's genus was validly published, and according to the International Code, Edinb. ed. Art. 5, note 2, the subgenus *Phlyctidium* Braun and *Phlyctidium* Rabenhorst must be rejected unless specially conserved. Sparrow (1960) suggested that it should be regarded as a *nomen conservandum*, but conservation requires action of a special committee for fungi and of the Section on Nomenclature of an international congress. Such action has not been taken for *Phlyctidium* (Edinb. Code, pp. 240–244). Accordingly, it as well as the subgenera *Sphaerostilidium*, *Rhizophyton*, and the genera *Tylochytrium* and *Hapalopera* are regarded here as synonyms of *Rhizophydium*.

In merging *Phlyctidium* and other genera with this genus it is necessary to rename some species because their names are preempted by those of *Rhizophydium* species. Also, new combinations are established for several of the remaining species.

Accordingly, *Phlyctidium mycetophagum* Karling is renamed *Rhizophydium obpyriformis* (Karling) nom. nov., comb. nov.

Phlyctidium keratinophilum Ookubo and Kobayasi is renamed *R. ellipsoidium* (Ookubo and Kobayasi) nom. nov., comb. nov.

Phlyctidium globosum Skuja is renamed *R. skujai* (Skuja) nom. nov., comb. nov.

Phlyctidium marinum Karling is renamed *R. novae-zeylandiensis* (Karling) nom. nov., comb. nov.

Phlyctidium anatropum Braun becomes *R. anatropum* (Braun) comb. nov.

Phlyctidium tenue Sparrow becomes *R. tenue* (Sparrow) comb. nov.

Phlyctidium bumilleriae Couch becomes *R. bumilleriae* (Couch) comb. nov.

Phlyctidium brevipes var. *marinum* Kobayasi and Ookubo becomes *R. brevipes* var. *marinum* (Kobayasi and Ookubo) comb. nov.

Phlyctidium spinulosum Sparrow becomes *R. spinulosum* (Sparrow) comb. nov.

Phlyctidium olla Sparrow becomes *R. olla* (Sparrow) comb. nov.

Phlyctidium scenedesmi Fott becomes *R. scenedesmi* (Fott) comb. nov.

Hapalopera piriformis Fott becomes *R. piriformis* (Fott) comb. nov.

The remaining species placed by Sparrow (1943, 1960) and others in *Phlyctidium* have been classified as *Rhizophydium* species by other workers at one time or another and these names are retained as such.

As interpreted here *Rhizophydium* is characterized by an epibiotic or extramatrical zoosporangium and resting spore, and an endobiotic or intramatrical absorbing system which may consist of a tubular, peg-like, knob-like, discoid haustorium, or a single unbranched rhizoid or filament, or a sparse to richly branched and extensive rhizoidal system. The resting spores may be asexually or sexually formed. Because of the limitations of space only a few representatives of this genus are illustrated here to emphasize the variations in structure, development, and reproduction, and Plate 24 is devoted to several of the species which were formerly classified in *Phlyctidium*.

The type of development of *Rhizophydium* is relatively simple. The encysted planospore germinates on the host or substratum (figs. 6, 16, 38, 50, 54) and forms an endobiotic or intramatrical absorbing system of varied complexity, and after this has been established the planospore body enlarges to become a zoosporangium (figs. 6-9, 153-155, 162) or resting spore. In most species the latter is formed asexually, but in approximately 21 species at present it is reported to be formed sexually as will be described later.

Rhizophydium tetragenum Pongratz is a unique species in that as the encapsulated incipient cocoon-like zoosporangium develops from the planospore it segments into 2, 4 or more chambers (figs. 212b, c, 213) until it has the appearance of a *Sarcina* at maturity (figs. 214, 215). No exit papillae are developed, and the walls of the segments deliquesce completely (fig. 216), releasing the minute, approx 1 μ diam., planospores. The resting spores (fig. 212d), on the other hand, appear to develop like those of other *Rhizophydium* species. This is the only known chytrid in which this type of sporangial development has been observed, and it will likely prove to be representative of a new genus, in the event it is a chytrid. Pongratz did not report or figure the flagellum on the planospore nor its point of insertion. His study was made on fixed and stained material, and he did not follow the successive developmental stages of a single sporangial thallus.

The mature zoosporangium varies markedly from spherical (figs. 2, 46, 56, 60, 142, 167, 171, 174), ovoid (figs. 18, 26, 204), pyriform (figs. 160, 162, 165), fusiform, oblong (figs. 129, 133), irregular or gibbose (fig. 58), nodular (figs. 176, 189), angular (figs. 200, 202), and columnar to conical (fig. 165) in shape with a thin or relatively thick hyaline wall which may be smooth, echinate, or bear spines (fig. 52), hairs (fig. 84), or teeth (figs. 23, 25). Also, it varies from 3.3–6.7 μ in diameter in *R. achnanthis* Friedmann to 20–110 μ in *R. macrosporum* Karling and 30–140 μ in *R. macroporosum* Karling. The absorbing vegetative system varies from an unbranched peg-like (figs. 2, 17, 18, 23–25, 26), tubular (figs. 27, 28), clavate and discoid haustorium to a single unbranched filament or rhizoid (figs. 40, 98, 99, 120, 148), or a richly branched rhizoidal system (figs. 51, 167, 171, 202) which in *R. coronum* Hanson may extend for distances up to 500 μ. In *R. macrosporum* and *R. polystomum* Karling the main rhizoidal axes may be coarse, up to 9–10 μ in diameter at the base. In some thalli of these species the basal portion of the axis may be somewhat inflated or apophysis-like, or it may be inflated and constricted at irregular intervals (fig. 167). In species which parasitize planktonic organisms the absorbing system may consist of a single branched or unbranched filament (figs. 97, 98) or a few bushy rhizoids (figs. 142, 144), while in the saprophytic species from soil and water like *R. coronum*, *R. macroporosum*, *R. polystonum* and others the rhizoidal system is usually coarse extensive and richly branched.

In preparation for dehiscence the zoosporangia of most species form a broad, conspicuous, apical (figs. 24, 63) or lateral (fig. 1) exit papilla, or up to 28 to 30 low inconspicuous papillae (figs. 167, 193). In *R. macroporosum* and *R. novae-zealandiensis* the exit papilla may sometimes be up to 25 to 20 μ in diameter respectively, at its base and almost of the same diameter as that of the zoosporangium. In some species like *R. apiculatum* Karling (figs. 160, 165), *R. anomalum* Canter (fig. 123) and *R. condylosum* Karling (fig. 192) the papillae may be extended into short necks.

However, in 10 species no exit papillae or necks are developed for the discharge of planospores. In *R. contractophilum* Canter (figs. 101, 102), *R. collapsum* Karling, (fig. 113), *R. patellarium* Scholz (fig. 91), and *R. racemosum* Gaertner (figs. 204–206) part or most of the sporangial wall disintegrates or deliquesces to free the planospores. In *R. melosirae* Friedmann (figs. 89, 90), *R. achnanthis* (figs. 87, 88), *R. oblongum* Canter, *R. difficile* Canter, *R. sphaerocystidis* Canter and *R. tetragenum* Pongratz the wall deliquesces entirely leaving only the rhizoids intact. In *R. sphaerocystidis* the multinucleate sporeplasm is freed by the deliquescence of the wall somewhat as in *Nowakowskia* and later undergoes cleavage into planospores (figs. 148, 149). In *R. racemosum* Gaertner the mature zoosporangium (fig. 204) is packed so full of highly refractive bodies that the thin hyaline sporangial wall is scarcely perceptible, and when the apex is ruptured these bodies, surrounded by a thin layer of cytoplasm float out and away (fig. 205).

As the tip of the papilla or neck in the papillate species deliquesces the first planospores emerge in a

Plate 24

Figs. 1-14. *Rhizophydium laterale* (Braun) Rabenhorst. Figs. 1-3 after Braun, 1856; figs. 4-14 after Karling, 1938.

Fig. 1. Mature zoosporangium on *Ulothrix zonata* with a lateral exit papilla.

Fig. 2. Discharge of planospores through a subapical papilla; haustorium an unbranched, curved peg.

Fig. 3. Spherical, 2-3 μ diam., planospores.

Fig. 4. Spherical, 2.5-3.2 μ diam., planospore.

Fig. 5. Germination of a planospore in water.

Fig. 6. Infection of *Ulothrix zonata*.

Fig. 7. Young thallus; germ tube has developed into a haustorium.

Fig. 8. Later stage, delicate rhizoids have developed at the tipe of the haustorium within the host cell.

Fig. 9. Still later stage; delicate rhizoids more extensive.

Fig. 10. Haustorium with epibiotic and endobiotic swellings somewhat similar to those of *Physorhizophydium pachydermum* Scherffel.

Fig. 11. Mature zoosporangium with a lateral, conspicuous exit papilla; haustorium with a single short filament at its base.

Fig. 12. Discharge of planospores; haustorium is a long tapering, unbranched peg.

Fig. 13. Spherical, yellowish to amber, smooth-walled resting spore with a large central refractive globule; haustorium extending into host' cell and bearing delicate rhizoids at its tip.

Fig. 14. Resting spore lying free in a hyaline vesicle; formed possibly by the contraction of the content and its investment by a wall.

Figs. 15-21. *Rhizophydium anatropum* (Braun) Karling. (Couch, 1935.)

Fig. 15. Elongate, 2 × 5 μ, and amoeboid planospores.

Fig. 16. Infection of *Stigeoclonium* sp.

Fig. 17. Mature anatropus zoosporangium with a short basal haustorium.

Fig. 18. Erect, ovoid zoosporangium shortly before dehiscence with an apical exit papilla and a short peg-like haustorium.

Figs. 19, 20. Pairing of planospores (gametes.)

Fig. 21. Fusion?

Fig. 22. Spherical resting spore with a smooth, thick, faintly brownish wall and a slightly inflated basal haustorium.

Figs. 23-25. *Rhizophydium brebissonii* (Dang.) Fischer. (Dangeard, 1889.)

Fig. 23. Mature zoosporangium with an apical collarette of teeth and a curved isodiametric haustorium on *Coleochaete scutata*.

Fig. 24. Discharge of planospores.

Fig. 25. Empty zoosporangium with 7 teeth surrounding the exit orifice.

Fig. 26. *Rhizophydium brevipes* Atkinson; zoosporangium on *Spirogyra varians* discharging planospores. (Atkinson, 1909.)

Fig. 27. Unidentified species on *Cladophora* with a long tapering haustorium and a zoosporangium with radiating hairs. (Sparrow, 1933.)

Fig. 28. Spiny zoosporangium of *Rhizophydium spinulosum* (Sparrow) Karling on *Cladophora* sp. discharging planospores from subapical pore; haustorium unbranched and slightly inflated. (Sparrow, 1933.)

Figs. 29-34. *Rhizophydium piriformis* (Fott) Karling. (Fott, 1942.)

Fig. 29. Spherical, 2-3 μ diam., planospore with a small hyaline globule.

Fig. 30. Young thallus on *Characium ancora*.

Fig. 31. Stalked, pyriform zoosporangium.

Fig. 32. Sessile ovoid zoosporangium shortly before dissolution of the wall and release of the planospores.

Fig. 33. Sporangial stalk remaining after zoosporangia wall has dissolved.

Fig. 34. Stalked resting-spore thallus. (Drawing submitted by Fott to Canter, 1950.)

Figs. 35, 36. *Rhizophydium tenue* (Sparrow) Karling. (Sparrow, 1952.)

Fig. 35. Zoosporangium on *Zygnema* sp. discharging planospores; basal wall of zoosporangium thickened; haustorium isodiometric and unbranched.

Fig. 36. Resting spore with empty attached adnate male (?) cell.

Figs. 37-45. *Rhizophydium obpyriformis* Karling on the mycelium of a fungus imperfectus and *Rhizophydium ellipsoidium* (Ookubo and Kobayosi) Karling. (Karling, 1946.)

Fig. 37. Zoosporangium discharging planospores; absorbing system a short peg.

Fig. 38. An encysted planospore, a young thallus and a full grown zoosporangium with a basal peg.

Figs. 39-42. Variations in the structure of the absorbing system.

Fig. 43. Thick-walled resting body or spore (?) which fills the sporangium-like structure.

Fig. 44. Migration of protoplasm into the apex of the sporangium-like structure.

Fig. 45. Thick-walled resting body or spore in the apex.

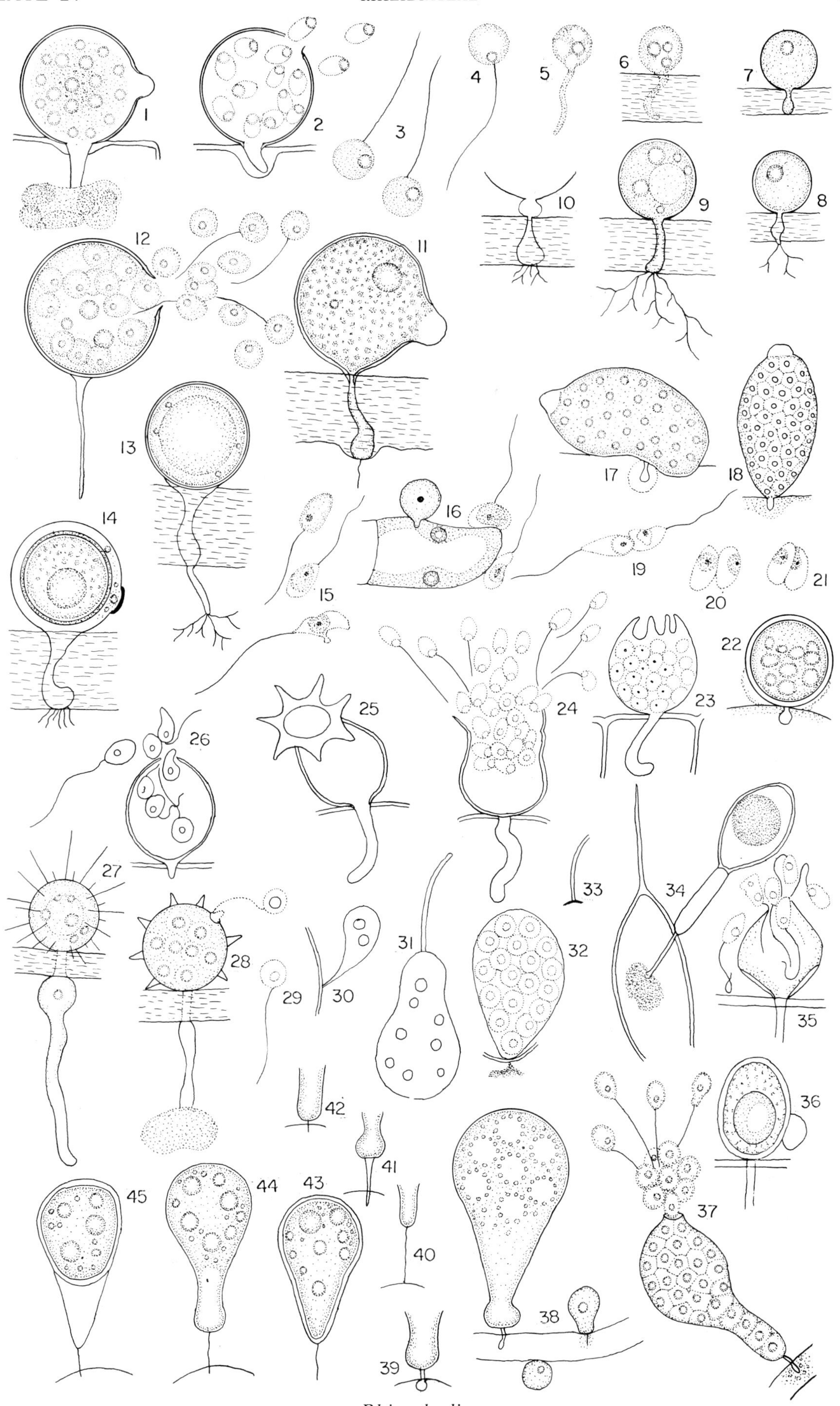

Rhizophydium

quiescent globular mass but soon disperse and become motile with a typical darting movement. In single-pored epibiotic species of this genus as well as in a few other similar genera Barr (1975) recently reported that they could be classified according to four well-defined methods of zoospore discharge. In *R. nobile* Canter, the globular mass is enveloped by a vesicular membrane (endosporangium?) which subsequently ruptures. The remnants of it remain attached to the edges of the exit orifice (fig. 147), and in *R. vaucheriae* de Wildemann (1931) the apex of the papilla rarely persists and is thrown back as a sort of an operculum by the emerging planospores. The planospores in most species are predominantly spherical to ovoid, vary from 1.5 μ to 7 μ in diameter with a posteriorly attached flagellum of varied lengths

PLATE 25

Figs. 46-52. *Rhizophydium globosum* (Braun) Rabenhorst. Fig. 46 after Cohn, 1853; fig. 47 after Braun, 1856; figs. 48, 49, 52 after Serbinov, 1907; figs. 50, 51, drawn from New Zealand specimens.

Fig. 46. Spherical zoosporangium with rhizoids on *Closterium*.

Fig. 47. Zoosporangium with a basal peg on *Oedogonium fonticola*.

Fig. 48. Spherical, 3 μ diam., planospore with a hyaline refractive globule.

Fig. 49. Germinated planospore with branched rhizoid infecting *Pleurotaenium trabecula*.

Fig. 50. Planospore infecting pollen grain of *pinus sylvestris*.

Fig. 51. Mature spherical zoosporangium on pollen grain with 3 low exit papillae and branched rhizoids.

Fig. 52. Spiny resting spore.

Figs. 53-57. *Rhizophydium pollinis-pini* (Braun) Zopf on pine pollen. (Zopf, 1887.)

Fig. 53. Spherical 2-4 μ diam., planospore with a hyaline refractive globule.

Fig. 54. Infection of pollen grain; drawn from N. Z. specimen.

Fig. 55. Young vacuolate zoosporangium with a branched rhizoidal axis.

Fig. 56. Mature zoosporangia on pine pollen with two fairly prominent exit papillae.

Fig. 57. Spherical, smooth-walled resting spore.

Fig. 58. *Rhizophydium gibbosum* (Zopf) Fischer; zoosporangium on rotifer egg with 3 swellings in rhizoidal axis; somewhat similar to *Physorhizophidium pachydermum*. (Zopf, 1888.)

Figs. 59, 60. *Rhizophydium sphaerotheca* Zopf. (Zopf, 1887.)

Fig. 59. Planospores, 2.5-3 μ diam., with a hyaline refractive globule.

Fig. 60. Spherical zoosporangium with 2 low exit papillae on a microspore of *Isoetes lacustris*.

Figs. 61-64. *Rhizophydium sphaerocarpum* (Zopf) Fischer. (Zopf, 1884.)

Fig. 61. Ovoid and spherical, 1.5-2 μ diam., planospores with a hyaline refractive globule.

Fig. 62. Infection of an algal (*Mougeotia*) cell.

Fig. 63. Dehiscing zoosporangium with a broad apical exit papilla.

Fig. 64. Resting spore with a smooth hyaline wall; rhizoid a long tapering unbranched peg.

Figs. 65-72. *Rhizophydium ovatum* Couch. (Couch, 1935.)

Fig. 65. Ovoid zoosporangium on *Stigeoclonium* discharging planospores.

Fig. 66. Male gamete at rest on the host cell.

Fig. 67. Germinated male gamete with an attached flagellate female gamete.

Figs. 68, 69. Stages in the development of a zygote at the apex of the male gamete by fusion with and enlargement of the female gamete.

Fig. 70. Incipient zygote with paired gametic nuclei.

Fig. 71. Later developmental stage of zygote with a large fusion nucleus.

Fig. 72. Zygote or resting spore atop an empty male gamete with a second female gamete attached at the side.

Figs. 73-79. *Rhizophydium granulosporum* Scherffel. Figs. 73, 75, 77, 78 after Scherffel, 1925; figs. 74, 76, 79 after Sparrow, 1939.

Fig. 73. Zoosporangium on *Tribonema bombycinum*.

Fig. 74. Empty zoosporangium on the same host.

Fig. 75. Ovoid planospores, 2-3 μ, with a hyaline, slightly eccentric refractive globule.

Fig. 76. Two paired thalli prior to conjugation, one of which has developed a fine rhizoid in the transverse wall of the host.

Fig. 77. Developing zygote in the female thallus on top of the empty germinated male thallus.

Fig. 78. Mature spiny zygote or resting spore at apex of a male thallus.

Fig. 79. Immature resting spore with two male thalli.

Figs. 80-82. *Rhizophydium goniosporum* Scherffel. (Scherffel, 1925.)

Fig. 80. Ovoid, 2-3 × 3-6 μ, planospores with an eccentric hyaline refractive globule.

Fig. 81. Ovoid zoosporangium on *Tribonema bombycinum* with 2 lateral exit papillae.

Fig. 82. Mature zygote or resting spore with an empty adnate male (?) cell.

Fig. 83. *Rhizophydium mischococci* Scherffel; zoosporangium on *Mischococcus confervicola* with an unbranched filamentous rhizoid. (Scherffel, 1926.)

Figs. 84, 85. *Rhizophydium chaetiferum* Sparrow; hairy zoosporangium and resting spore, resp., on *Cladophora* sp. (Sparrow, 1939.)

Figs. 86-88. *Rhizophydium achnanthis* Friedmann on *Achnanthes offinis*. (Friedmann, 1952.)

Fig. 86. Spherical zoosporangium.

Figs. 87, 88. Stages in the deliquescence or dissolution of the sporangium wall and freeing of the planospores.

Figs. 89, 90. *Rhizophydium melosirae* Friedmann on *Melosira varians*. (Friedmann, 1952.)

Fig. 89. Mature zoosporangium with planospores.

Fig. 90. Mass of planospores after the dissolution of the sporangium wall.

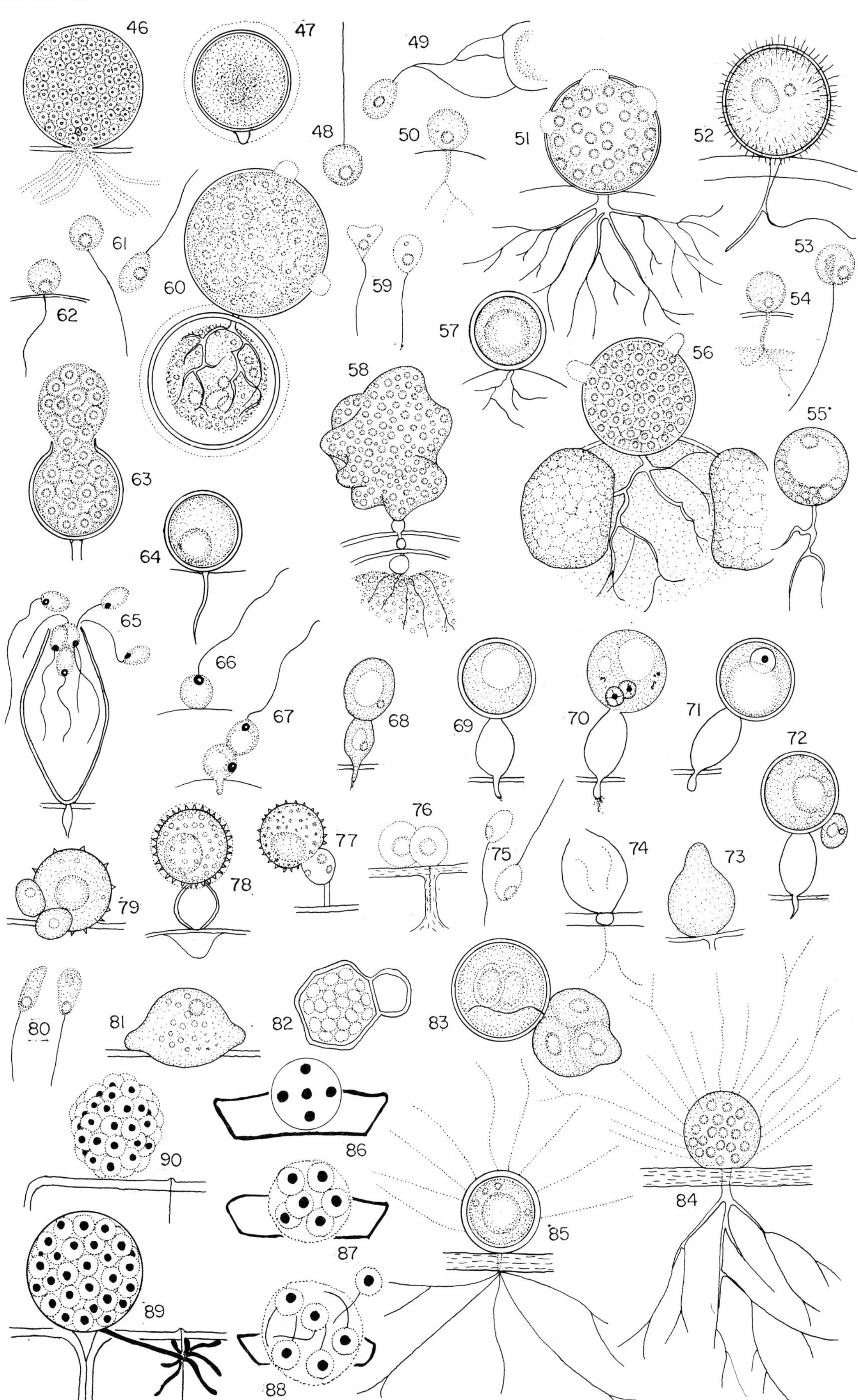

Rhizophydium

PLATE 26

Figs. 91-95. *Rhizophydium patellarium* Scholz. (Scholz, 1958.)

Fig. 91. Zoosporangium with ⅔s of the upper portion of its wall dissolved, exposing the planospores.

Fig. 92. Ovoid, 2.5-3 μ diam., planospore with an anterior hyaline refractive globule.

Fig. 93. Saucer-shaped remnant of basal portion of the zoosporangium after dispersal of the planospores.

Fig. 94. Spherical, smooth-walled resting spore.

Fig. 95. Germination of resting spore.

Figs. 96-111. *Rhizophydium contractophilum* Canter parastic on *Eudorina elegans* and *Eudorina* sp. (Canter, 1959.)

Fig. 96. Ovoid, 2 μ diam., planospores with a minute hyaline refractive globule.

Fig. 97. Germination of planospore and infection of the membrane of the algal colony by a fine unbranched rhizoid.

Fig. 98. Later stage; rhizoid has reached algal cell; planospore body enlarging to become the sporangial rudiment.

Fig. 99. Mature zoosporangium.

Fig. 100. Branching at end of rhizoidal filament.

Figs. 101, 102. Remnants of sporangial wall after irregular deliquescence.

Figs. 103-107. Stages in conjugation between a small male and a large female thallus (gametangia) and the development of the zygote or resting spore.

Fig. 108. Mature resting spore or zygote.

Fig. 109. Germination of zygote.

Figs. 110, 111. Male gametangia with short conjugation tubes which will probably develop into zoosporangia.

Figs. 112-115. *Rhizophydium collapsum* Karling on dead pollen of *Pinus sylvestris*. (Karling, 1964.)

Fig. 112. Spherical, 2.5-3 μ diam., planospore with a hyaline refractive globule.

Fig. 113. Collapse and breakdown of sporangium wall.

Fig. 114. Ovoid resting spore with a brown, verrucose wall.

Fig. 115. Germination of a spherical smooth-walled resting spore.

Fig. 116. *Rhizophydium bullatum* Sparrow; resting spore from pine pollen. (Sparrow, 1952.)

Figs. 117, 118. *Rhizophydium venustum* Canter parasitic on *Uroglena americana*. (Canter, 1963.) (This species has recently been found to be a species of *Zygorhizidium* by Canter, 1970.)

Fig. 117. Mature zoosporangium.

Fig. 118. Mature resting spore with empty attached male cell.

Figs. 119-127. *Rhizophydium anomalum* Canter parasitic on *Apiocystis brauniana*. (Canter, 1950.)

Fig. 119. Encysted planospore with a filamentous germ tube.

Fig. 120. Slightly elongated planospore with a long germ tube which has contacted the host cell.

Fig. 121. Elongate sporangial rudiment with a broadened germ tube.

Fig. 122. Young thallus; rhizoid has branched within the muscilage sheath of the host.

Fig. 123. Mature zoosporangium with a broad exit canal.

Fig. 124. Large male thallus (gametangium) with a small adherent female thallus (gametangium.)

Fig. 125. Fusion of minute female and large male thallus.

Fig. 126. Incipient zygote or resting spore near apex of elongate curved male thallus.

Fig. 127. Mature zygote or resting spore near the apex of large male thallus.

Fig. 128. Mature zygote or resting spore on top of male thallus.

Figs. 129-131. *Rhizophydium ephippium* Canter parasitic on *Stylosphaeridium stipitatum*. (Canter, 1950.)

Fig. 129. Dehiscing zoosporangium.

Fig. 130. Large female cell with a small attached male cell.

Fig. 131. Mature zygote with attached empty male cell.

Fig. 132-136. *Rhizophydium oblongum* Canter parasitizing *Dinobryon* spp. (Canter, 1954.)

Fig. 132. Planospores, 2.5 μ diam., with anteriorly attached and posteriorly directed flagellum.

Fig. 133. Oblong mature zoosporangium with 2 stubby rod-like rhizoids in body of host.

Fig. 134. Large female cell with a small male cell attached by a fine tube; rhizoid on the former cell omitted.

Fig. 135. Similar stage with a broader conjugation tube between gametangia.

Fig. 136. Mature resting spore or zygote with an adnate male cell or gametangium; rhizoid filamentous, short, and branched.

Figs. 137-140. *Rhizophydium uniguttulum* Canter parasitic on *Gmellicystis neglecta*. (Canter, 1954.)

Fig. 137. Zoosporangium discharging planospores.

Fig. 138. Male cell attached by a fine thread to a larger female cell.

Fig. 139. Male cell adnate to the female cell.

Fig. 140. Mature zygote or resting spore with a halo of mucilage and an attached empty male cell.

Figs. 141-144. *Rhizophydium difficile* Canter parasitic on *Staurastrum jaculiferum*. (Canter, 1954.)

Figs. 141, 142. Young and immature zoosporangia resp. with tufts of rhizoids.

Fig. 143. Small male cell attached directly to a larger female cell.

Fig. 144. Mature resting spore or zygote with a thick, smooth, brownish wall, and an adnate empty male cell; surrounding by a colorless muscilaginous halo with strands.

Figs. 145-147. *Rhizophydium nobile* Canter parasitic on resting spores of *Ceratium hirundinella*. (Canter, 1968.)

Fig. 145. Spherical, 4-5 μ diam., swimming planospore with a bent head and containing a lateral hyaline refractive globule and a thin nuclear cap.

Fig. 146. Mature zoosporangium with branched rhizoids.

Fig. 147. Empty zoosporangium with remnant of vesicular membrane attached to orifice.

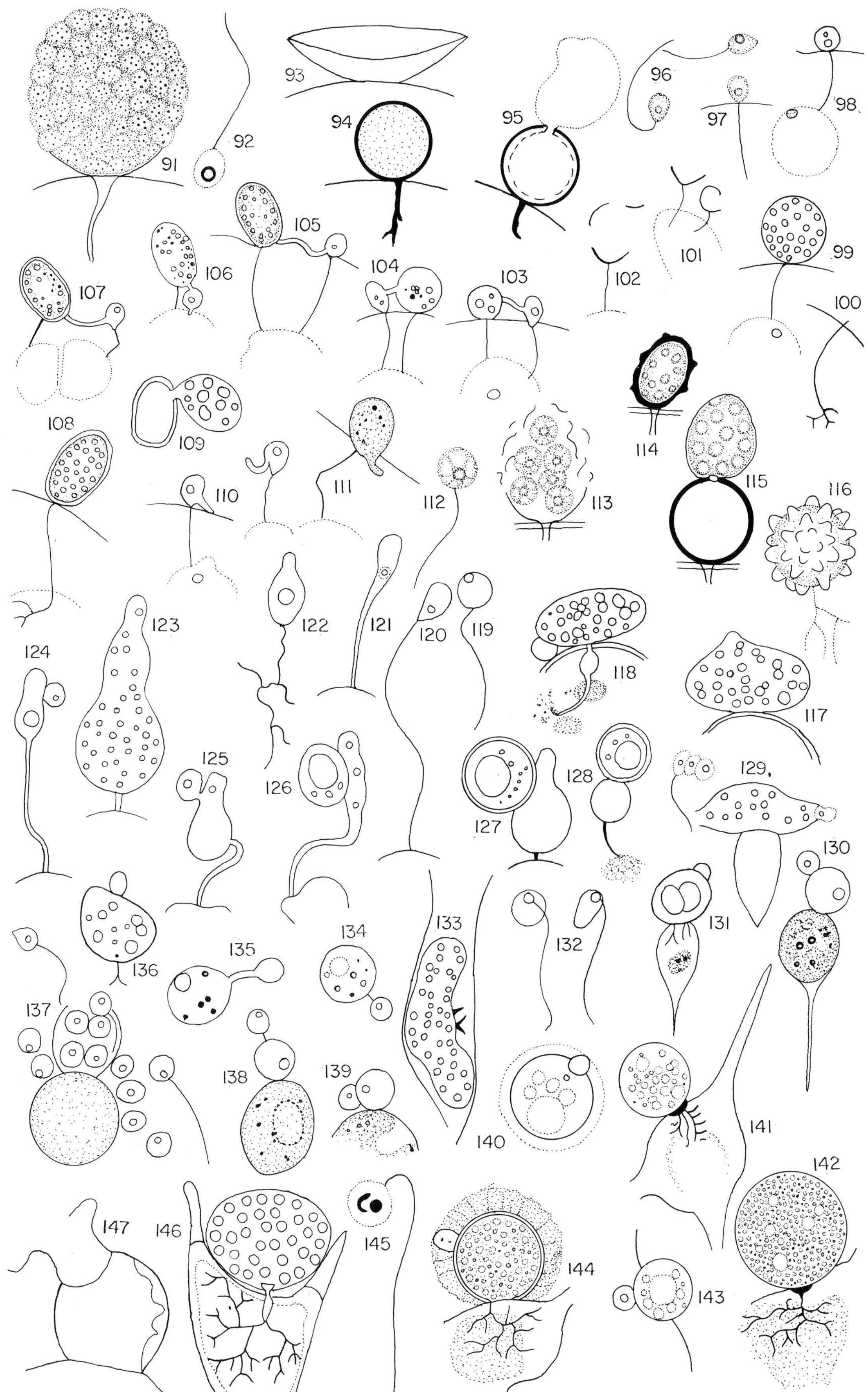

Rhizophydium

and contain a minute or large hyaline refractive globule. In *R. condylosum* and *R. rarotonganensis* Karling the single refractive globule is lacking and replaced by numerous granules (figs. 194, 199). In *R. oblongum* the globule lies at the anterior end, and the flagellum is anteriorly attached but posteriorly directed (fig. 132). In *R. racemosum*, according to Persiel (1963), two types of zoosporangia are developed. One is large and filled with huge refractive globules (fig. 207) and produces a few, 5–8 μ diam., planospores with an unusually large lipoidal globule (figs. 208, 209). These may be amoeboid and nonflagellate or flagellate. They develop into minute pygmaceous zoosporangia (fig. 210) which produce small, 3–4 μ diam., planospores. The large planospores may remain in the larger zoosporangium and become transformed directly into thick-walled small resting spores (fig. 211) which function as prosporangia when they germinate. Accordingly, the large planospores have 2 pathways of development. Persiel regarded these developments as analogous to the types exhibited by some of the aquatic Synchytriaceae.

In most species in which they have been observed the resting spores are formed asexually and borne like the zoosporangia (figs. 13, 57, 85, 114, 158, 172, 178, 186, 195). However, as noted earlier, they have been reported to be formed sexually in 22 species, and the process of sexual reproduction varies somewhat in these species. The motile cells or swarmers which function directly or indirectly in this process appear to be facultative. They may give rise to spo-

Plate 27

Figs. 148-151. *Rhizophydium sphaerocystidis* Canter. (Canter, 1950.)

Fig. 148. Two zoosporangia each with a filamentous rhizoid parasitizing *Sphaerocystis schroeteri*; wall of zoosporangium on the right has dissolved and released a multinucleate protoplasmic mass.

Fig. 149. Cleavage of protoplasmic mass into planospore initials.

Fig. 150. Spherical, 3μ diam., planospore with a small hyaline refractive globule.

Fig. 151. Resting spore with an empty adherent male (?) cell.

Figs. 152-159. *Rhizophydium coronum* Hanson. (Hanson, 1945.)

Fig. 152. Spherical, 3.7-4.5 μ diam., planospore with a large hyaline refractive globule.

Fig. 153. Germination of planospore.

Fig. 154. Surface view of sporangial rudiment on cellophane showing narrow enveloping coronum.

Fig. 155. Fixed and stained incipient uninucleate zoosporangium; coronum omitted.

Fig. 156. Mitosis with an intranuclear division spindle from a zoosporangium.

Fig. 157. Mature zoosporangium with a subapical exit papilla, surrounded by the coronum.

Fig. 158. Stained uninucleate resting spore with a golden, multilayered wall.

Fig. 159. Germination of resting spore; coronum omitted.

Figs. 160, 161. *Rhizophydium apiculatum* Karling. (Karling, 1946.)

Fig. 160. Protozoan infected by 1 mature and 3 incipient zoosporangia.

Fig. 161. Early germination stage of a hyaline, smooth-walled resting spore which was formed in an incipient zoosporangium.

Figs. 162, 163. *Rhizophydium mycetophagum* Karling. (Karling, 1946.)

Fig. 162. Zoosporangia and resting spores parasitizing *Choanephora* sp.

Fig. 163. Spherical, 3.5-5 μ diam., planospore with a hyaline refractive globule.

Fig. 164-166. *Rhizophydium conicum* Karling parasitizing *Netrium* sp. (Karling, 1946.)

Fig. 164. Subspherical, 2-2.5 μ diam., planospores with a hyaline refractive globule.

Fig. 165. Conical zoosporangium with 6 exit papillae.

Fig. 166. Conical, hyaline, smooth-walled resting spore.

Figs. 167, 168. *Rhizophydium polystomum* Karling. (Karling, 1967.)

Fig. 167. Empty zoosporangium on boiled corn leaf with 19 exit orifices surrounded by wrinkles; rhizoidal axis coarse, 10 μ diam., inflated and constricted at intervals.

Fig. 168. Spherical, 2.4-2.8 μ diam., planospore with a minute hyaline refractive globule.

Fig. 169-172. *Rhizophydium macrosporum* Karling. (Karling, 1938.)

Fig. 169. Spherical, 4.5-6 μ diam., planospore with a 3-4 μ diam., hyaline refractive globule.

Fig. 170. Germination of planospore in water.

Fig. 171. Mature zoosporangium growing on cooked beef with 3 exit papillae and filled with large hyaline refractive globules; rhizoidal axes coarse, 6 μ diam., extensive and branched.

Fig. 172. Spherical, large, 15-30 μ, hyaline, smooth-walled resting spore with a large central globule.

Figs. 173-175. *Rhizophydium macroporosum* Karling from boiled grass leaf. (Karling, 1967.)

Fig. 173. Broad exit papilla with a translucent conical plug of matrix protruding out of the orifice.

Fig. 174. Planospores emerging simultaneously from 3 orifices.

Fig. 175. Spherical, 3-4.2 μ diam., planospore with a minute hyaline refractive globule.

Figs. 176-178. *Rhizophydium elyensis* Sparrow on snake skin. (Karling, 1967.)

Fig. 176. Mature zoosporangium with 9 exit papillae surmounted with hyaline refractive plugs of matrix.

Fig. 177. Spherical, 2.6-3.5 μ diam., planospore with a hyaline refractive globule.

Fig. 178. Spherical and oblong, resp., hyaline and smooth-walled resting spores filled with angular refractive bodies.

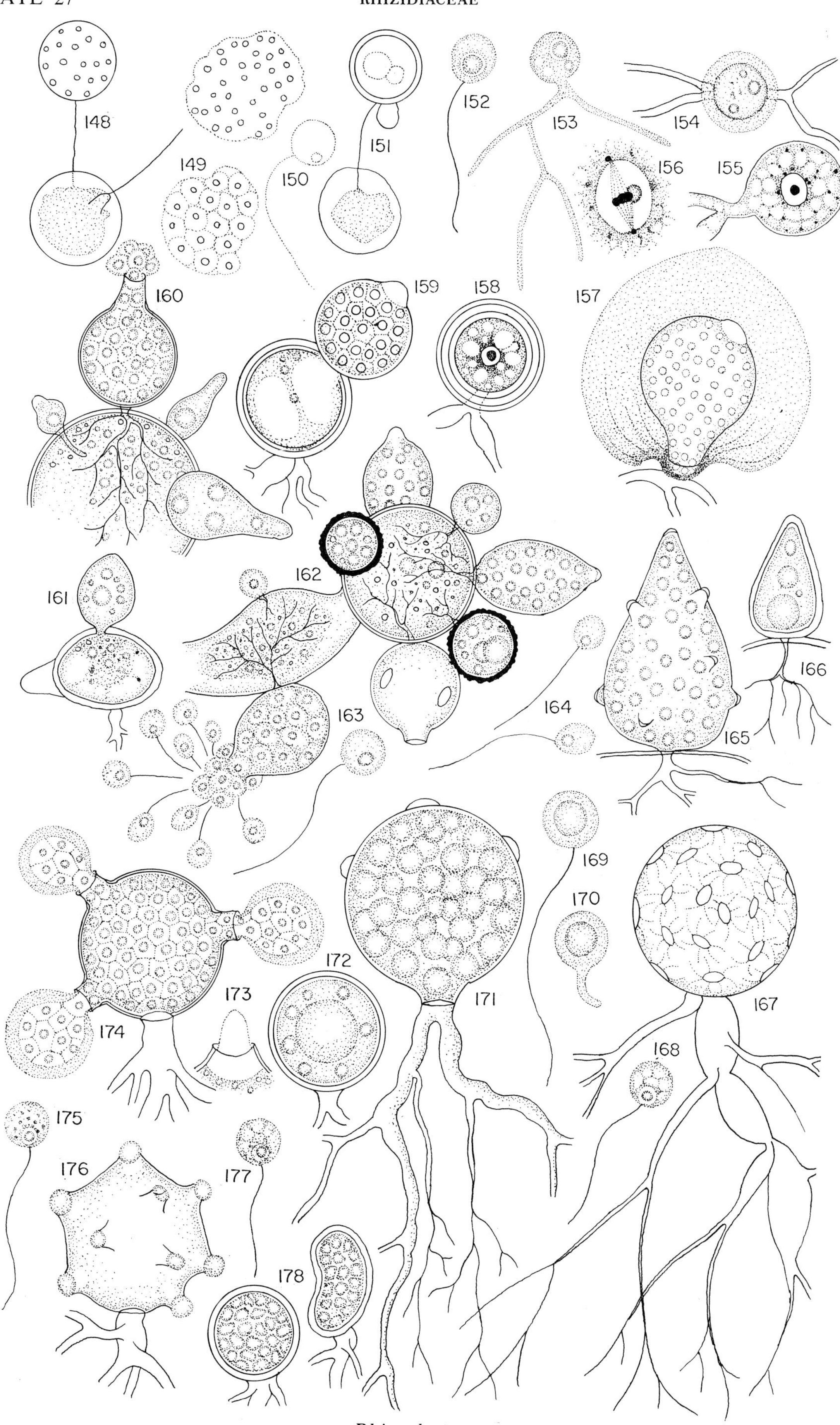

Rhizophydium

rangial thalli or fuse directly, or develop into male and female thalli (gametangia). This is borne out by the development into sporangial thalli by the male gametangia in *R. columnaris* and *R. contractophilum* even after the conjugation tube has developed (figs. 110, 111). So far no conclusive evidence of homo- or heterothallism has been reported in this genus.

In *R. beauchampii* Hovasse (1936) (*Phlyctidium eudorinae* Gimesi) Gimesi (1924) reported that isogametes are formed in zoosporangia as well as planospores. One, a female, comes to rest on the host and germinates after which a male gamete becomes attached at the apex or side of the female. The two fuse completely and develop into the zygote or resting spore. Particularly noteworthy is that the two gametes lose their identity in the fusion product as in some species of the Olpidiaceae and Synchytriaceae. Couch (1935) and Ledingham (1936) reported fusion of motile isogametes in *R. anatropum* and *R. graminis* Ledingham, but they did not observe the development of resting spores from such fusions. Instead Ledingham believed that fusions occurred through the rhizoids of a minute male thallus with those of a larger female thallus as in some species of *Siphonaria* and *Rhizoclosmatium*, but Barr (1973) was unable to confirm this. In *R. ovatum* Couch (figs. 67-72), *R. granulosporum* Scherffel (figs. 77, 78) and *R. anomalum* Canter (figs. 124-128) and *R. androdioctes* Canter, Couch, Scherffel, and Sparrow reported that a "male" planospore comes to rest on the host cell and germinates to form a small male thallus or gametangium after which a "female" planospore becomes attached to its apex or side. The two planospores are isogamous but they are designated as male and female on the basis of the subsequent development of the zygote. In *R. ovatum* the male gametangium receives nourishment from the host cell and may reach considerable size (fig. 70) while the female apparently is nourished by the male and has no contact with the host. A similar relationship occurs occasionally in *R. granulosporum* and in *R. androdioctes*. Eventually, the protoplasm of the male gametangium moves into the female which is subsequently transformed into the resting spore or zygote. According to Couch, both gametangia are uninucleate in *R. ovatum*, and karyogamy occurs in the young zygote (fig. 71). Occasionally, 2 female gametangia may be associated with one male (fig. 79) or 2 males with 1 zygote (fig. 72), suggesting polyandry.

PLATE 28

Figs. 179-188. *Rhizophydium keratinophilum* Karling. (Karling, 1946.)

Fig. 179. Portion of dead human hair infected by 13 zoosporangia and 2 resting spores.

Fig. 180. Vacuolate fully-grown zoosporangium with a conspicuous rhizoidal axis and one exit papilla.

Fig. 181. Uninucleate incipient zoosporangium.

Fig. 182. Simultaneous nuclear division in a fully-grown zoosporangium.

Fig. 183. Variations in single, bifurcated, and dichotomously branched spines, and hairs on the zoosporangia.

Fig. 184. Discharge of planospores from 2 exit papillae.

Fig. 185. Spherical, 2.5-3 μ diam., and amoeboid planospores with a minute hyaline refractive globule.

Fig. 186. Enlarged view of mature resting spore with a brown, prominently warted wall.

Figs. 187, 188. Germination of resting spores.

Fig. 189. *Rhizophydium nodulosum* Karling zoosporangia from dead human hair. (Karling, 1948.)

Figs. 190-196. *Rhizophydium condylosum* Karling from dead human hair. (Karling, 1968.)

Fig. 190. Mature zoosporangium with 2 basal rhizoidal axes and 15 visible exit papillae which give it a knobby or condylose appearance.

Fig. 191. Minute zoosporangium with 1 apical papilla.

Fig. 192. Zoosporangium on which the papillae have become extended into short necks.

Fig. 193. Simultaneous discharge of vermiform planospores from 5 exit papillae.

Fig. 194. Spherical, 2-2.8 μ diam., and vermiform planospores with evenly granular content.

Fig. 195. Mature, hyaline, smooth-walled resting spores with coarsely granular content.

Fig. 196. Resting spore with an aborted exit canal, apparently formed by the encystment of a zoosporangium.

Figs. 197-199. *Rhizophydium rarotonganensis* Karling. (Karling, 1968.)

Fig. 197. Mature zoosporangium with 3 exit papilla, causing marked local hypertrophy of the rhizomycelium of *Nowakowskiella profusa*; rhizoids reduced and sparse.

Fig. 198. Discharge of planospores.

Fig. 199. Spherical, 2.8-3.3 μ diam., planospore with finely and evenly granular content.

Figs. 200-203. *Rhizophydium angulosum* Karling. (Karling, 1968.)

Fig. 200. Cluster of 10 angular zoosporangia on a boiled corn leaf, formed as a result of the encystment and germination of planospores near the exit papilla of a mother zoosporangium.

Fig. 201. Spherical, 2-2.8 μ diam., and amoeboid planospores with a minute hyaline refractive globule.

Fig. 202. Angular zoosporangium on nutrient agar media showing extent and branching of rhizoids.

Fig. 203. Discharge of planospores and their feeble motility and encystment around the exit orifice.

Figs. 204-206. *Rhizophydium racemosum* Gaertner, on a pollen grain of *Pinus sylvestris*. (Karling, 1968.)

Fig. 204. Zoosporangium packed full of highly refractive angular and subspherical bodies.

Fig. 205. Rupture of sporangial wall releasing spherical bodies with an unusually large refractive body (planospores?.)

Fig. 206. Racemose cluster of resting spores released by the rupture of the sporangial wall.

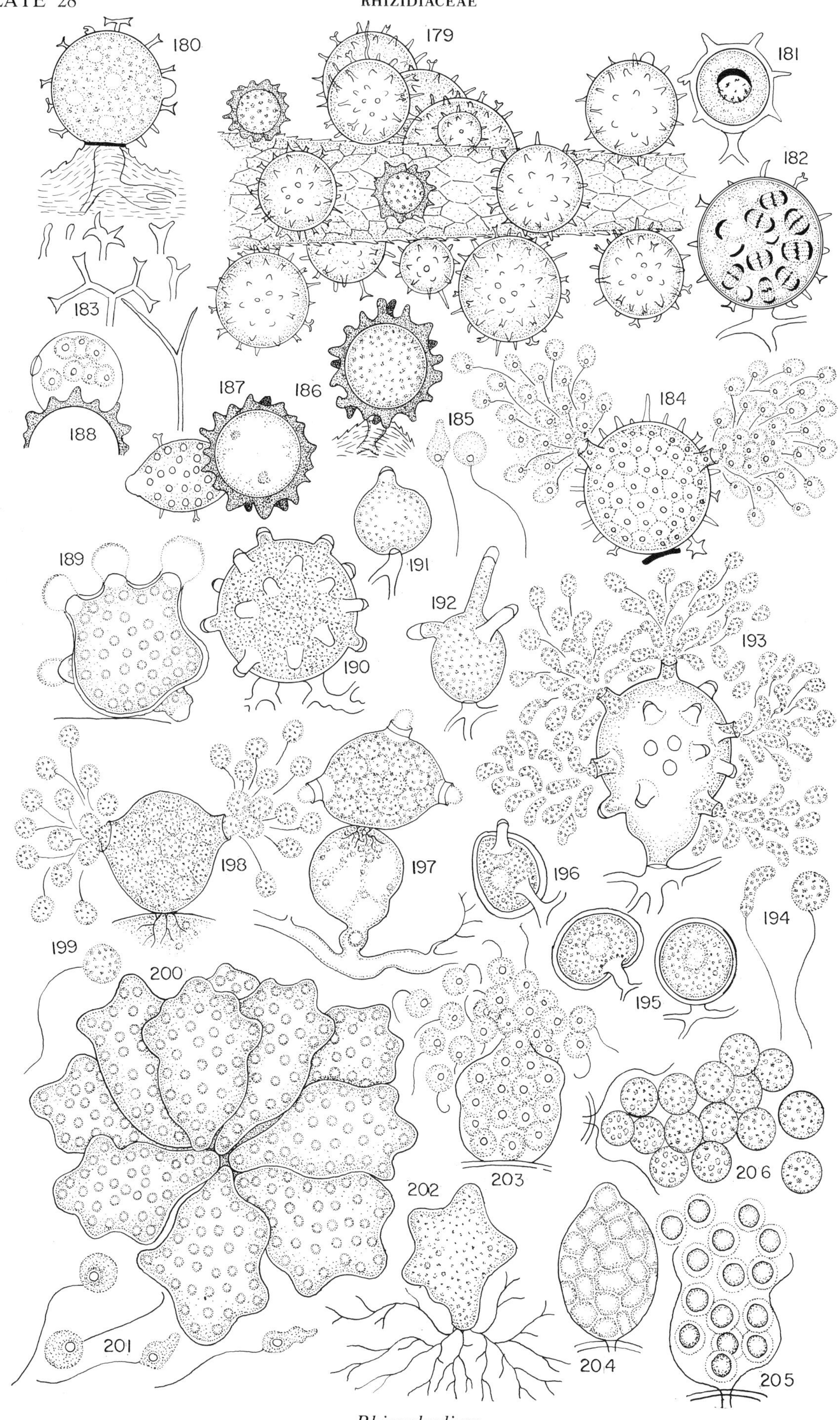

Rhizophydium

In *R. columnaris* Canter and *R. contractophilum* Canter the motile cells germinate on the host and develop into gametangia. One of the pair apparently grows faster and becomes larger, while the other one remains small and forms a conjugation tube which grows toward and fuses at its tip with the former gametangium (figs. 103–107). The length of the tube varies, and in *R. columnaris* it may attain a length up to 28 μ and be 2 μ in diameter. Occasionally, it is quite short, and the male gametangium is almost adnate to the female (fig. 106). After plasmogamy the zygote develops a thick wall (figs. 107, 108) and becomes dormant. The male gametangia with long or short conjugation tubes (fig. 110) and apparently female gametangia also, are capable of developing into zoosporangia if fusion does not occur. Sexual reproduction in these two species is, accordingly, strikingly similar to that of *Zygorhizidium*.

In *R. tenue* (fig. 36), *R. ephippium* Canter (figs. 130, 131), *R. uniguttulatum* Canter (figs. 138, 139), and *R. difficile* Canter (figs. 143, 144) the small male gametangium may be attached at the top or at the side of the larger female or mature resting spore. Quite probably the male gametangia in these species do not come into contact with the host, according to Sparrow's and Canter's drawings. In *R. couchii* Sparrow the motile cells which subsequently develop into gametangia are identical in size and shape, and after a period of swarming come to rest on the host in clumps of 4 to 10 or more. Each of an adjacent pair of encysted planospores develops an endobiotic rhizoidal system. The epibiotic body of one of them enlarges as its rhizoids increase in extent, while the other one (male gametangium) apparently ceases to grow. However, it develops a small tube which projects into the female gametangium. Its protoplasm flows through this tube into the female which develops into a zygote or resting spore while the male remains attached as an empty vesicle. A somewhat similar type of sexual reproduction was found by Valkanov (1964) in *R. simplex* Dangeard. The "male" and "female" cells are of equal size with the male resting on top of the female, and after fusion the female enlarges and develops into the resting spore while the empty male remains attached as an appendage. In *R. racemosum* clusters of resting spores are formed within or around zoosporangia with disintegrating walls, according to Gaertner (1954). These are formed from such bodies as are shown in figure 206 by their enlargement and the breaking up of the large refractive globules into small granules. Similar bodies were found by Karling (1968), but he was not certain that they were resting spores. In this regard he had overlooked a previous paper by Persiel (1963) who showed that these bodies are true resting spores which germinate as prosporangia in the zoosporangium.

The mature resting spores or zygotes vary in shape and size, and may be small or up to 36 μ in diameter as in *R. macrosporum* Karling. In most species the wall is hyaline and smooth, but in some they are bullate (fig. 116) or bear warts (fig. 114), spines (fig. 52) or hairs (figs. 84, 85). In other species the wall is light to dark brown and smooth (figs. 94, 95, 115). Germination of the spores has been observed in several species (figs. 95, 115, 161, 187, 188), and in these they function as prosporangia. In *R. ovatum*, however, the zygote or spore forms planospores directly, according to Couch, but in *R. contractophilum* Canter illustrated a zygote (fig. 109) which apparently functioned as a prosporangium during germination.

Plate 29

Figs. 207-211. *Rhizophydium racemosum* on pine pollen. (Persiel, 1963.)

Fig. 207. Large zoosporangium filled with huge lipoidal globules.

Fig. 208. Amoeboid, large, 5–8 μ diam., planospores; huge lipoidal globule with a thin enveloping layer of cytoplasm.

Fig. 209. Flagellate large planospore.

Fig. 210. Pygmy zoosporangia developed from the large planospores.

Fig. 211. Zoosporangium filled with small thick-walled resting spores which have developed from the large planospores; 2 have germinated as prosporangia.

Figs. 212-216. *Rhizophydium tetragenum* on *Astrionella formosa*. (Pongratz, 1966.)

Fig. 212. Five parasites on the host: a = infecting planospore; b = incipient zoosporangia which have segmented into 2 cells; c = tetroid zoosporangium; d = resting spore.

Fig. 213. Zoosporangium of 4 segments.

Figs. 214, 215. *Sarcina*-like stages of zoosporangia.

Fig. 216. Complete deliquescence of wall and release of planospores.

Figs. 217-223. *Rhizophydium fulgens* Canter on *Gemellicystis neglecta*. (Canter, 1951.)

Fig. 217. Spherical, 2.5 μ diam., planospore with an anterior globule.

Fig. 218. Encysted planospore.

Fig. 219. Germinated planospore with a rhizoidal filament infecting the alga.

Fig. 220. Oblong mature zoosporangium with a long rhizoidal filament which has grown through the muscilage surrounding the host cells.

Fig. 221. Empty zoosporangium with 2 subapical exit orifices.

Figs. 222, 223. Resting spore and germination of the same, resp.

Figs. 224-227. *Rhizophydium mougeotiae* Pongratz on *Mougeotia gracillima*. (Pongratz, 1966.)

Fig. 224. Spherical, 1.5-2 μ diam., planospore.

Fig. 225. Infected *Mougeotia* filament with a young, a mature and an empty zoosporangium.

Fig. 226. Mature zoosporangium with an apical exit papilla, and an empty zoosporangium.

Fig. 227. Resting spore.

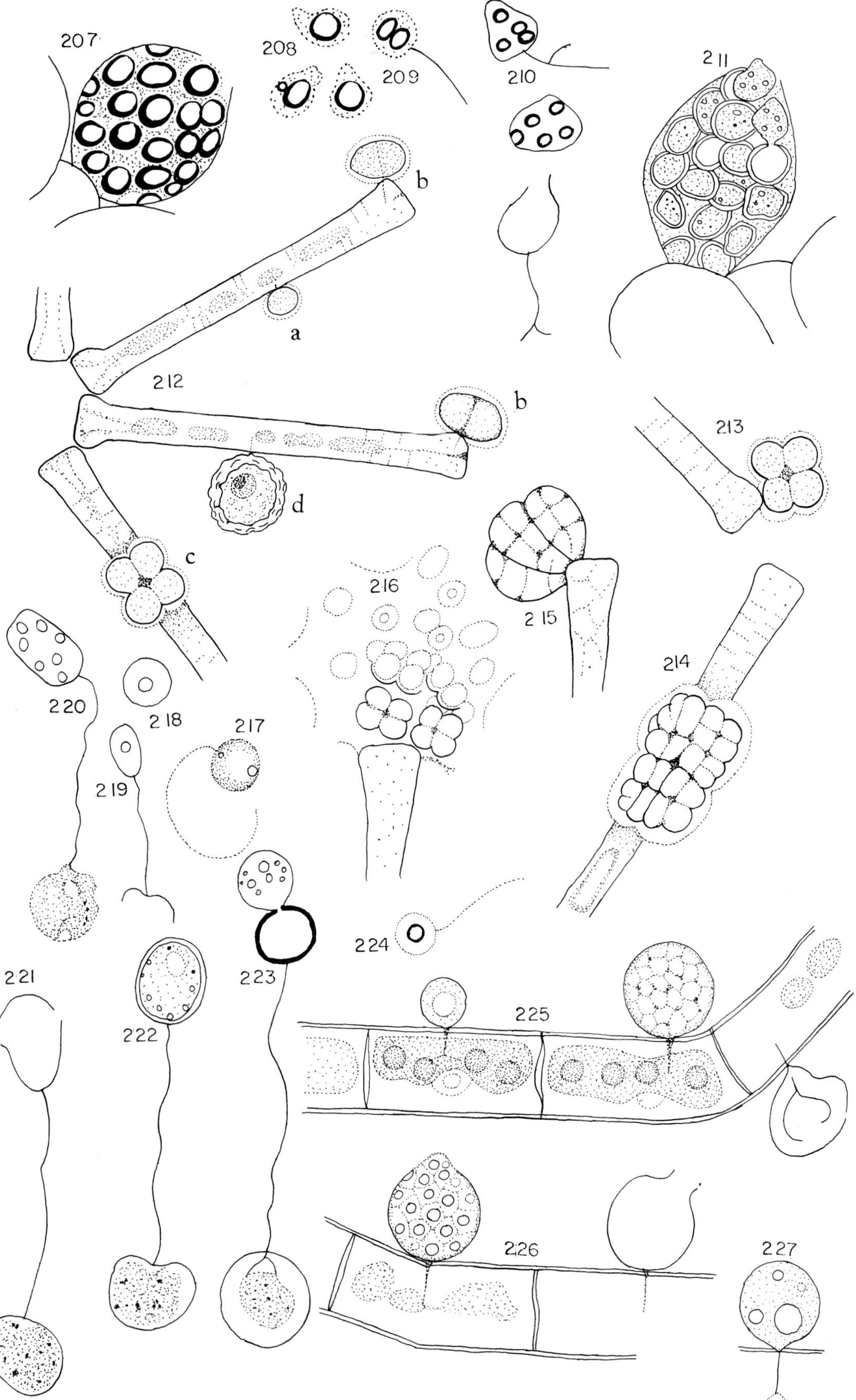

Rhizophydium

SEPTOSPERMA Whiffen ex Seymour

Mycologia 63:90, 1971
Septosperma Whiffen. 1942. Mycologia 34:352, *nomen nudum.*

Plate 30

Five parasites are included in this inoperculate genus at present, 3 of which occur on numerous chytrids and a 4th one on chrysophycean algae. The parasites on chytrids, *S. anomala* (Couch) Whiffen ex Seymour, *S. rhizophydii* Whiffen ex Seymour, *S. spinosa* Willoughby, and *S. irregularis* Dogma differ from species which were formerly included in *Phlyctidium* only by their septate resting spore thalli (figs. 2, 8–12, 32), but *S. multiforme* Canter on algae develops fairly extensive branched rhizoids (figs. 19, 20) like some species of *Rhizophydium.* Also, the flagellum on its planospores is not truly basal but sublateral in attachment (fig. 14), and the planospores contain 2 or 3 pulsating vacuoles. However, its resting spore thalli are septate (figs. 26–31) as in the other species. Such septate spores are formed in the same manner as those of *Sparrowia, Loborhiza, Nowakowskiella hemisphaerospora, Nephrochytrium stellatum, Podochytrium cornutum* and frequently in *Karlingia rosea* (Karling, 1968), and it is doubtful that this characteristic is generically distinctive. Probably, *Septosperma* will be merged with *Rhizophydium* in the future.

Parasitism of one chytrid upon another is not uncommon (Karling, 1942, 1960), and occasionally a species may parasitize itself (fig. 5). Accordingly, the occurrence of *S. anomala, S. rhizophydii* and *S. spinosa* on other chytrids is not exceptional, and a host zoosporangium may bear a large number of parasites (fig. 6).

During germination the planospore develops a fine germ tube which in *S. anomala, S. rhizophydii* and *S. spinosa* forms an unbranched or sparingly branched rhizoid, peg, or filament (figs. 1, 7, 13). In *S. multiforme* it branches fairly extensively with blunt ends (figs. 19, 20). Meanwhile, the planospore body enlarges to become the zoosporangium or resting spore. In *S. multiforme* the zoosporangia vary markedly in size and shape with the exit papillae occurring in various positions (figs. 17, 18, 21-25). The incipient resting spores in *S. anomala* and *S. rhizophydii* are usually more elongate (figs. 2, 8–13) than the young zoosporangia with denser, more coarsely granular content (fig. 8), and during maturation the content gradually aggregates in the upper portion (figs. 9–11). A cross wall then develops separating the empty basal part from the upper cell or spore (fig. 12). During germination the spore functions as a prosporangium (fig. 13).

LOBORHIZA Hanson

Amer. J. Bot. 31:169, 1944

Plate 31

This monotypic inoperculate genus includes a parasite, *L. metzneri*, of the reproductive cells of *Volvox carteri* and is characterized by an extracellular or epibiotic proliferating zoosporangium whose apex projects out of the gelatinous sheath of the coenobium, a polydigitate or polylobate endobiotic haustorium, and resting spores which are formed as a result of lateral fusion of isomorphic gametes. It differs primarily from *Rhizophydium* by its polylobate haustorium and true proliferation of the zoosporangium. Also, the loss of morphological identity and complete absorption of the gametes in the mature resting spore distinguishes it from nearly all members of the Rhizidiaceae except a few species of *Rhizophlyctis* and *Karlingia.*

Plate 30

Figs. 1, 2. *Septosperma anomala.* (Couch, 1932.)

Figs. 1, 2. Zoosporangium and resting spore, respectively, on *Rhizophydium bummilleriae.*

Figs. 3-13. *Septosperma rhizophydii.* Figs. 3-5, Whiffen, 1942; figs. 6-13 drawn from New Zealand material.

Fig. 3. Spherical, 1.6-2 μ diam., planospores with a hyaline refractive globule.

Fig. 4. Multiple infection of zoosporangium of *Rhizophydium macrosporum.*

Fig. 5. Self parasitism by *R. macrosporum.*

Fig. 6. Zoosporangium of *Karlingia rosea* parasitized by 10 zoosporangia and 6 resting spores.

Fig. 7. Discharge of planospores.

Figs. 8-12. Stages of resting spore development.

Fig. 13. Germination of resting spore.

Figs. 14-31. *Septosperma multiforme.* (Canter, 1963.)

Fig. 14. Planospores, 4 μ diam., with sublateral attachment of flagellum.

Fig. 15. Amoeboid planospores with 2 and 3 pulsating vacuoles.

Figs. 16, 17. Young thallus and mature zoosporangium on *Amphichrysis compressa.*

Fig. 18. Dehiscence of zoosporangium.

Fig. 19. Empty zoosporangium with branched, blunt-ended rhizoids.

Fig. 20. Similar rhizoids of a thallus in a cyst of *Mallomonas.*

Figs. 21-25. Variations in sizes and shapes of zoosporangia and positions of exit orifice.

Figs. 26-28. Development of resting spores on *Amphichrysis compressa.*

Figs. 29-31. Mature resting spores on *Mallomonas.*

Fig. 32. Spiny resting spore of *S. spinosa* on *Rhizophydium coronum.* (Willoughby, 1965.)

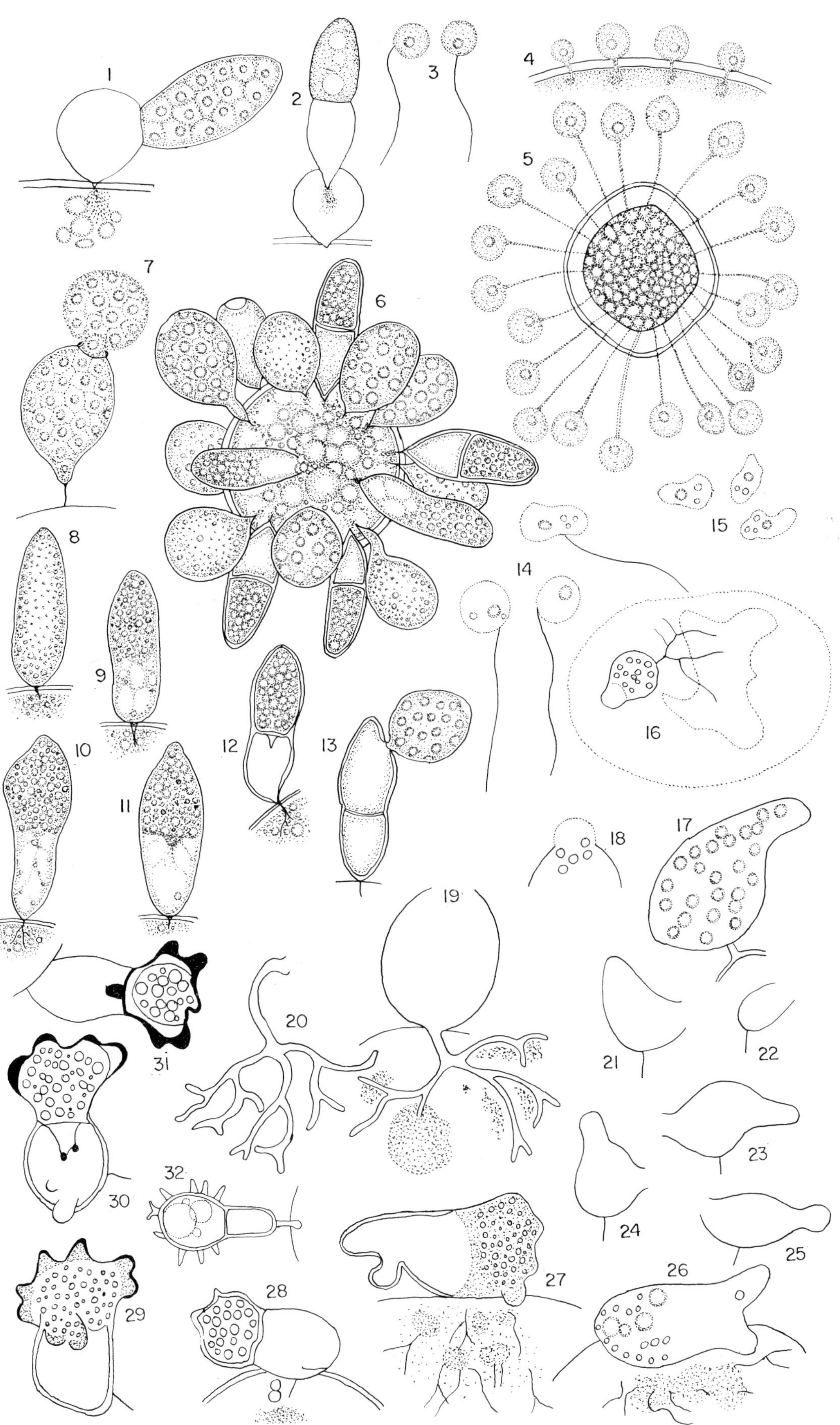

Septosperma

In the development of the sporangial thallus the planospore (fig. 1) develops a germ tube (fig. 3) which penetrates the gelatinous sheath of the coenobium and grows inward until it reaches a reproductive (vegetative or sexual) cell of the host (fig. 4). If a reproductive cell does not lie directly below the site of infection the tube will curve until it reaches its objective (figs. 4, 6). As it grows towards the host cell a small dilation develops at its tip which becomes the rudiment of the haustorium. After the dilated tip has penetrated the host cell digitations develop from the rudiment (fig. 8). The germ tube then increases almost uniformly in diameter (figs. 5, 6, 8), and the planospore cyst soon becomes indistinguishable from it. The cyst eventually becomes the apical portion of the zoosporangium which projects out of the gelatinous sheath. At first the incipient zoosporangium is almost cylindrical (figs. 7, 8), but later it enlarges from the tip downward (fig. 9) and becomes pyriform (figs. 10–12). After dehiscence, secondary and tertiary zoosporangia develop in the empty primary one (fig. 13).

In sexual reproduction isogametes come to rest in pairs on the host sheath, and one of them develops a tapering germ tube which anchors them to the sheath (figs. 14, 15). Possibly, as in *Rhizophlyctis* and *Karlingia* it is the planospore which came to rest first that develops the germ tube, but this has not been ascertained. As the germ tube of one gamete elongates, a conjugation papilla develops from the other one (fig. 16), and after contact between the gametes is achieved fusion through this papilla occurs (fig. 17). As the zygotic thallus develops and enlarges at its apex the respective gametes are gradually incorporated and lose their morphological identity. At first the apex of the thallus is lobed (figs. 18–20), indicating the partial outlines of the gametes, but this shape disappears in the mature resting-spore thalli. Karyogamy apparently is delayed so that the incipient resting spores remain binucleate for a time (figs. 18–20). Hanson did not observe karyogamy, but she found pairs of nuclei closely associated (fig. 20), and postulated that karyogamy might be delayed until shortly before germination. Also, she observed older spores with a single large nucleus (fig. 22) and believed that it might represent a fusion nucleus.

In the maturation of the predominantly clavate resting-spore thalli the protoplasm moves up into the apical portion (fig. 23) and becomes invested with a thick golden wall upon which blunt lobes develop (figs. 23–25). Thus, the resting-spore thalli become stalked and bear a lobate haustorium at the end of the stalk as in the sporangial thalli.

Occasionally, additional male thalli were found attached to incipient resting spores, but these appeared to be degenerating. Also, trinucleate incipient spores (fig. 21) were observed which suggests that more than one male gamete may have fused with the female. On the other hand, this may have been the result of division of one of the gametic nuclei. Germination of the resting spores has not been observed.

Sexual reproduction by fusion of isogametes in *Loborhiza* is similar to that of *Rhizophydium beauchampi* Hovasse, *Karlingia dubia* and species of *Rhizophlyctis*. Hanson did not note the origin of the gametes, whether they come from the same or different sporangia. If they came from the same zoosporangium the planospores are facultative. They may give rise to zoosporangia or fuse as gametes. Probably, *L. metzneri* will prove to be homothallic.

PHYSORHIZOPHIDIUM Scherffel

Arch. f. Protistenk, 54:181, 1926

Plate 32, Figs. 1-10

This inoperculate genus includes one species, *P. pachydermum* Scherffel, which parasitizes diatoms and is said to differ primarily from species of *Rhizophydium* by the development of an epibiotic secondary

PLATE 31

Figs. 1-25. *Loborhiza metzneri*. (Hanson, 1944.)

Figs. 1, 2. Motile, 3.7-3.9 μ diam., and amoeboid planospores with a large hyaline globule and several oscillating granules.

Figs. 3, 4. Germination of planospore on the gelatinous sheath of *Volvox carteri*; germ tube bending towards reproductive cells of host.

Fig. 5. Increase in diameter of germ tube and incorporation of the planospore cyst at the apex.

Fig. 6. Similar uninucleate stage from fixed and stained specimen.

Fig. 7. Multinucleate, fixed and stained incipient, elongate zoosporangium.

Fig. 8. Curved sporangial thallus with a basal digitate haustorium.

Figs. 9-11. Incipient and mature pyriform zoosporangia embedded in the gelatinous sheath except for their apices.

Fig. 12. Discharge of planospores.

Fig. 13. Proliferating sporangium.

Figs. 14-17. Stages in fusion of isogametes.

Fig. 18. Young binucleate zygotic thallus with lobed apex, indicating merging of gametes.

Fig. 19. Later stage, gametic nuclei separated by a large refractive globule.

Fig. 20. Gametic nuclei in close proximity.

Fig. 21. Trinucleate zygotic thallus.

Fig. 22. Fusion nucleus (?) in immature resting spore.

Fig. 23. Stalked resting-spore thallus with a digitate haustorium at the base and a resting spore in the apex; wall of spore with large pointed lobes which bear delicate threads.

Figs. 24, 25. Other mature resting-spore thalli with conspicuous refractive globules.

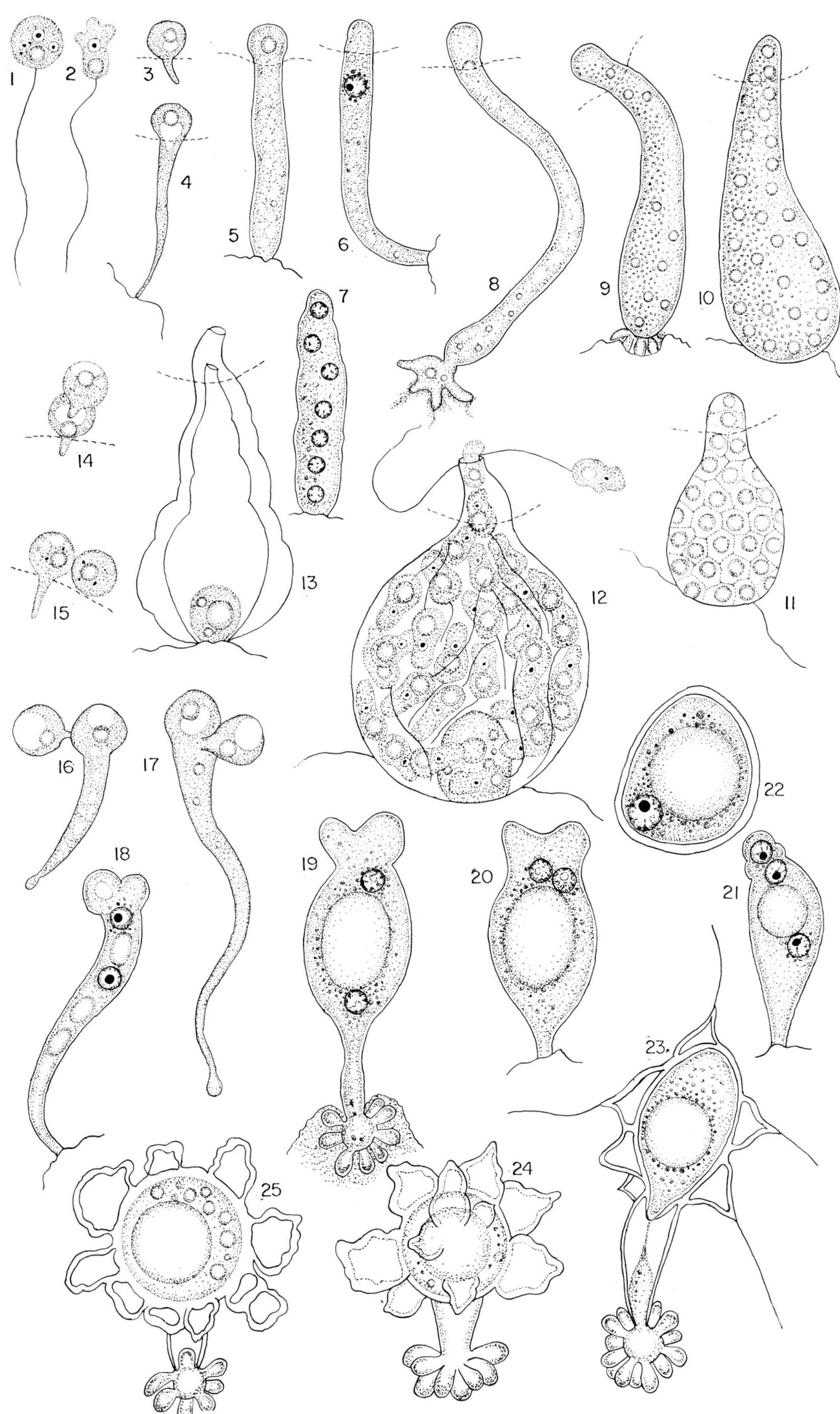

Loborhiza

knob-like swelling or appressorium on the host cell in addition to the endobiotic globular haustorium (fig. 7). Also, the zoosporangia are thick-walled, usually anatropus with an almost basal or lateral exit papilla (figs. 8, 9), and branched rhizoids are rarely developed (fig. 6). Whether or not these differences merit generic distinction is, obviously, open to serious question. Quite likely this genus will be merged with *Rhizophydium*.

In germination the encysted planospore (fig. 2) forms a germ tube which bores through the silicaceous host wall instead of entering through the girdle band, and its tip enlarges to become the endobiotic haustorium (fig. 3). However, in some thalli the haustorium and ectobiotic appressorium fail to develop, and in such instances an unbranched rhizoid (fig. 4) arises directly at the base of the incipient zoosporangium. In only one thallus (fig. 6) Scherffel figured a sparse, delicate, branched rhizoid arising from the base of the endobiotic haustorium, which suggests that the haustorium is usually the absorbing organ of this parasite. In the meantime the planospore body has enlarged usually laterally instead of equally in all directions so that the resulting elongate zoosporangium often lies parallel or almost so to the surface of the host (fig. 9). Sometimes, however, the encysted planospore enlarges almost equally in all directions to produce an erect pyriform zoosporangium (fig. 4). While this development has been going on a second swelling forms ectobiotically beneath the zoosporangium and becomes the appressorium.

The developing zoosporangium becomes thick-walled (figs. 5–9), and at maturity an exit papilla develops usually near the base of the zoosporangium (figs. 8, 10) or at the side (fig. 9). As a result the zoosporangia often have an anatropus shape. The planospores are liberated by the deliquescence of the exit papilla (fig. 9) and emerge singly in succession. Resting spores have not been observed in this genus.

PODOCHYTRIUM Pfitzer

Sitzungsber. Niederrhein. Gesell. Natur — und Heilkunde 1869:62, 1870

Septocarpus Zopf. 1888. Nova Acta Acad. Leop.-Carol. 52:348.
Rhizidiopsis Sparrow, 1933. Trans. Brit. Mycol. Soc. 18:216.

Plate 33

At present this inoperculate genus includes 4 parasites of diatoms and 2 others which occur as saprophytes on chitin and decaying leaf stalks. The genus is characterized by an epibiotic unexpanded portion of an encysted planospore (basal sterile cell) out of which the zoosporangium has developed apically, and an endobiotic rhizoidal system whose branches vary from fine and filamentous to coarse, broad, and strap-like.

In the development of the thallus the encysted planospore (fig. 1) of the parasitic species develops a fine infection tube which branches within the host to form the rhizoidal system (figs. 7, 14, 15, 16, 25, 35, 38). According to Friedmann (1952) the development in *P. emmanuelense* Sparrow and Paterson is endo-exogenous. The infection tube first develops an extensive irregular and lobed rhizoid (fig. 25), and after this organ has been established nourishment moves outward to expand the apical portion of the encysted planospore which becomes the rudiment of the zoosporangium (figs. 12, 13, 16). The lower portion of the encysted planospore remains unexpanded and eventually becomes thick-walled (figs. 3–6, 16–18, 45–47), or it may elongate slightly to become a short stalk (figs. 8, 9). Usually, it is delimited from the zoosporangium by a septum (figs. 8, 9, 12, 13, 16–18, 26). In *P. ellerbeckense* Willoughby (1963), a questionable species of *Podochytrium*, the thickened portion of the planospore cyst does not always remain as a sterile basal cell but is carried up by the developing zoosporangium to some distance from the rhizoid (fig. 38). The mature zoosporangium dehisces by the deliquescence of an inconspicuous, apical,

PLATE 32

Figs. 1-10. *Physorhizophydium pachydermum* Scherffel on *Amphora ovalis*. (Scherffel, 1926.)

Figs. 1, 2. Motile planospores, 2 μ diam., and infection of the host, resp.

Fig. 3. Young thallus with an endobiotic knob-like haustorium.

Fig. 4. Young and almost mature thallus with peg-like unbranched rhizoids; haustorium and epibiotic appressorium lacking.

Fig. 5. Irregular incipient zoosporangium with endobiotic haustorium.

Fig. 6. Incipient zoosporangium with an epibiotic appressorium and an endobiotic haustorium; branched rhizoid at base of haustorium.

Fig. 7. Similar zoosporangium without rhizoids.

Fig. 8. Mature zoosporangium with an almost basal exit papilla.

Fig. 9. Discharge of planospores from a lateral papilla.

Fig. 10. Empty zoosporangium showing thick, smooth persistent wall and basal exit pore.

Figs. 11-16. *Sporophlyctidium africanum* Sparrow on *Protoderma*. (Sparrow, 1938.)

Figs. 11, 12. Early stages in the development of the zoosporangia from the aplanospore body; germ tube forming the stalk.

Fig. 13. Empty zoosporangia.

Fig. 14. Aplanospores, 2 μ diam., at the exit orifice; also two empty zoosporangia.

Figs. 15, 16. Possible stalked resting spores with attached cysts, or companion, "male" cells.

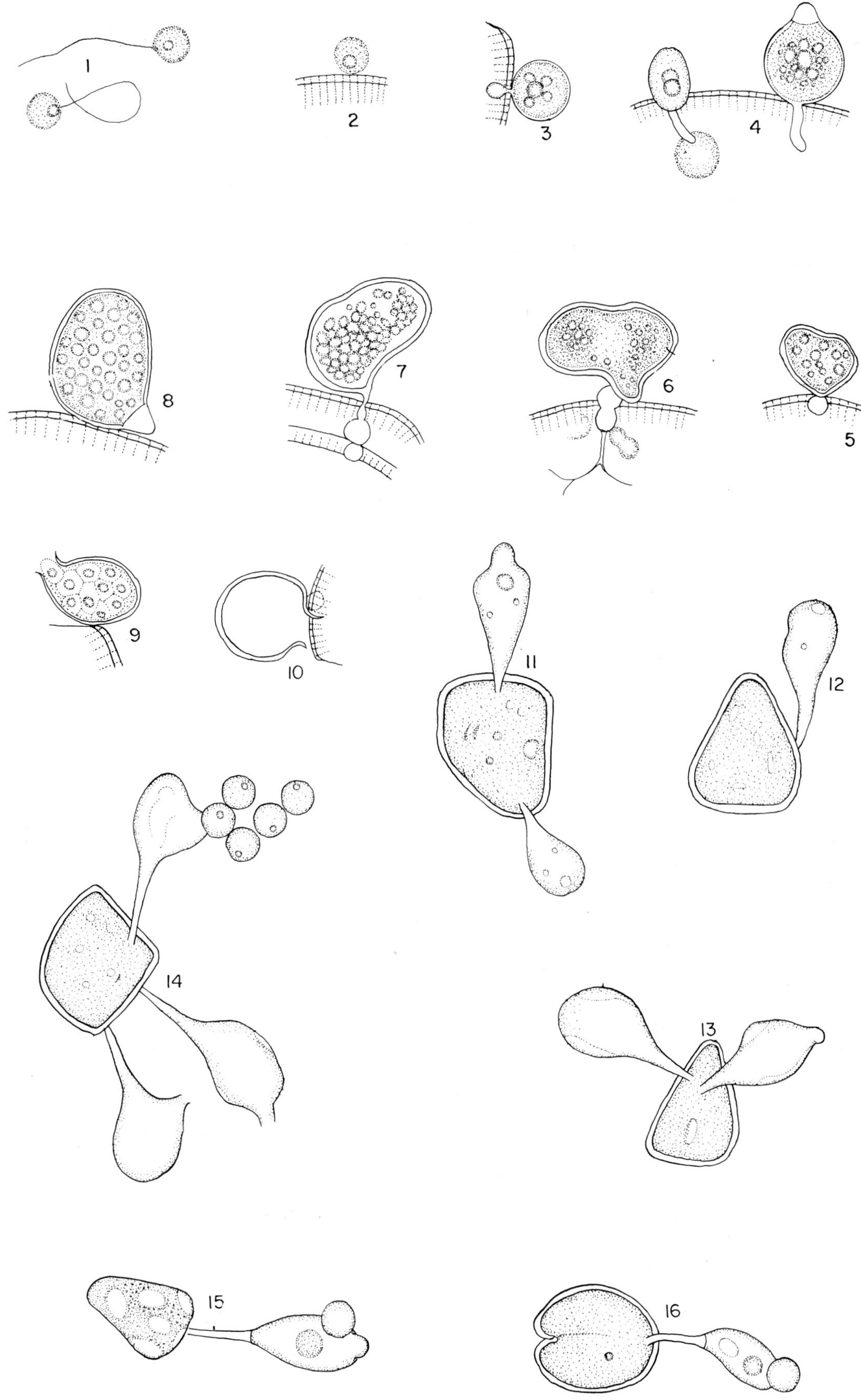

Physorhizophydium, Sporophlyctidium

broad or relatively narrow exit papilla. In *P. chitinophilum* Willoughby 2 subapical exit papillae are formed occasionally (fig. 43). In this species the discharged planospores swarm collectively (in a vesicle?) before dispersing, while in the other species they have been shown to disperse (figs. 9, 26) as they emerge from the zoosporangium.

Resting spores have been reported in *P. clavatum* Pfitzer, *P. emmanuelense* Sparrow, *P. cornutum* Sparrow and *P. ellerbeckense*. In the first named species and *P. cornutum*, according to Scherffel (1926) and Canter (1970), they develop in the upper portion of the thallus body (figs. 10, 19–21), while in *P. ellerbeckense* (fig. 48) they are borne like the zoosporangia. At maturity they have a thick, light brown, smooth wall with a basal or lateral cyst and contain numerous oil globules. In *P. emmanuelense* they are small, spherical to flattened, and surrounded by a dark-brown encrustation (figs. 27, 28). These apparently function as prosporangia in germination (figs. 29, 30).

PHLYCTOCHYTRIUM Schroeter

Engler u. Prantl, Naturl. Pflanzenf. 1(1):78, 1893.

Plates 34-37

Phlyctochytrium is reported to include 65 or more identified species and a large number of unidentified specimens at present, but the inclusion of some of these species here depends on one's interpretation of the genus. These species are worldwide in distribution (Ulken, 1972; Sparrow, 1973, 1974; Knox *et al*, 1973) and occur as parasites of freshwater and marine algae, fungi, eggs of microscopic animals, and

Plate 33

Figs. 1-10. *Podochytrium clavatum* Zopf. (Zopf, 1888.)

Fig. 1. Spherical, 3 μ diam., planospore with a hyaline refractive globule.

Fig. 2. Encysted planospore on the host wall.

Figs. 3-6. Successive developmental stages of the incipient zoosporangium out of the apex of the encysted planospore.

Fig. 7. Thallus with an epibiotic clavate zoosporangium and an endobiotic branched rhizoid; host cell omitted.

Fig. 8. Mature, stalked, septate zoosporangium.

Fig. 9. Discharge of planospores through a broad, apical exit orifice.

Fig. 10. Resting spores in upper part of the thallus.

Figs. 11-13. *Podochytrium lanceolatum* Sparrow on *Melosira varians*. (Sparrow, 1936.)

Fig. 11. Spherical, 3- 4 μ diam., planospore with a hyaline refractive globule.

Fig. 12. Fusiform zoosporangium with a bulbous base and septum; rhizoids sparse, delicate and branched.

Fig. 13. Discharge of planospores.

Figs. 14-21. *Podochytrium cornutum* Sparrow on *Stephanodiscus*. Figs. 16-18 after Sparrow, 1951; figs. 14, 15, 19-21 after Canter, 1970.

Figs. 14, 15. Germinated planospores on the host with coarse, strap-like branched rhizoids.

Fig. 16. Infection and developmental stages of thalli; incipient zoosporangia budding out of thick-walled planospore cysts; rhizoids coarse, branched and extensive.

Fig. 17. Elongate, clavate, septate zoosporangium with a bulbous base and planospores.

Fig. 18. Empty, curved zoosporangium.

Fig. 19. Early stage of resting spore development.

Fig. 20. Mature resting spore formed by the accumulation of the protoplasm in the apex of the thallus and its investment by a hyaline, thick wall.

Fig. 21. Stalked resting spore with a layer of external secretion.

Figs. 22-30. *Podochytrium emmanuelensis* Sparrow on *Melosira varians* and *Pinnularia viridis*. Figs. 22-24, 27-30 after Sparrow, 1936; figs. 25, 26 after Friedman, 1952.

Figs. 22-24. Encysted germinating planospore on *M. varians* and development of the incipient zoosporangium from the planospore cyst, resp.

Fig. 25. Almost mature zoosporangium on *Pinnularia viridis* connected by a fine infection tube to a coarse, irregularly branched rhizoid.

Fig. 26. Discharge of planospores from a broad apical orifice.

Figs. 27, 28. Surface and side views of resting spores on *Melosira varians*.

Figs. 29, 30. Germination of resting spores and an empty superficial zoosporangium, resp.

Figs. 31-35, 40-47. *Podochytrium chitinophilum* Willoughby on chitin. (Willoughby, 1961.)

Fig. 31. Ovoid, 4-5 × 3.5-4.5 μ, planospore with a hyaline refractive globule.

Fig. 32. Germination of encysted planospore.

Figs. 33, 34. Developmental stages of a sporangial rudiment out of apex of encysted planospore.

Fig. 35. Young sporangial thallus with branched rhizoids.

Fig. 40. Long unbranched rhizoid.

Fig. 41. Full grown zoosporangium with a sterile basal cell.

Fig. 42. Portion of mature zoosporangium with an apical exit papilla.

Figs. 43, 44. Empty zoosporangia with 2 and 1 exit orifices, resp.

Figs. 45-47. "Various types of thickenings of basal wall or sterile cell formation."

Figs. 36-39, 48. *Podochytrium ellerbeckense* on decaying leaf stalks. (Willoughby, 1963.)

Fig. 36. Large spherical quiescent planospore with a large yellow globule.

Figs. 37, 39. Germination stages.

Fig. 38. Young sporangial thallus; thickened portion of cyst wall lateral to sporangial rudiment.

Fig. 48. Resting spore with a basal cyst, filled with globules.

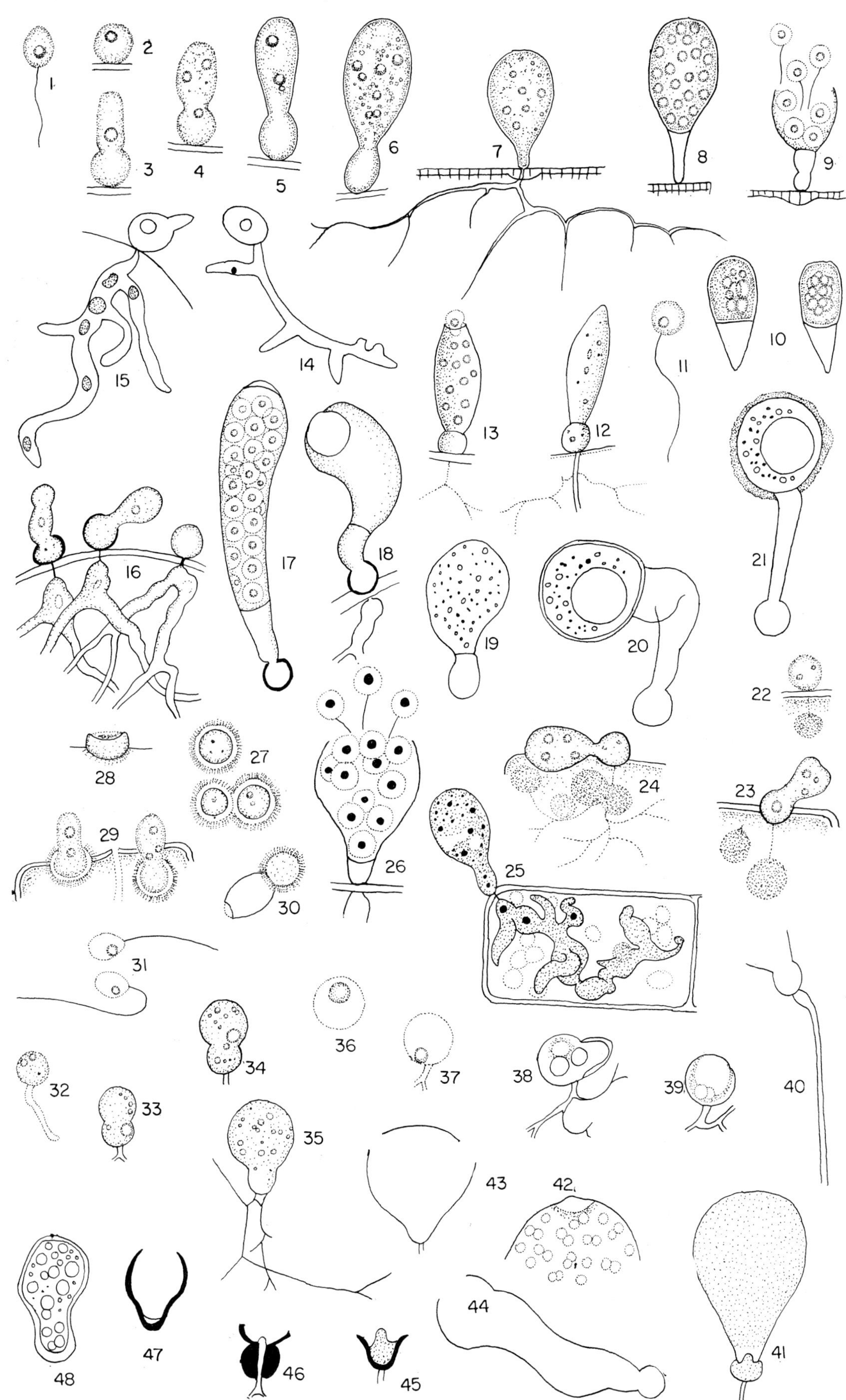

Podochytrium

pollen grains of higher plants, or as saprophytes in moribund algae, soil, and water from which they may be isolated on various types of substrata. Several of them have been cultivated successfully in axenic cultures.

The early-named species were classified in *Chytridium*, *Rhizidium* and *Rhizophydium*, but in 1893 Schroeter segregated the inoperculate apophysate species in a new genus, *Phlyctochytrium*, and recognized 2 subgenera: *Euphlyctochytrium* (*Nuda* E. Fischer) and *Dentigera* (Rosen). At present the genus is almost generally interpreted as including inoperculate species with an extramatrical, sessile or stalked zoosporangium or resting spore and an intramatrical or endobiotic apophysis and rhizoids. However, several of these characteristics may be so variable that it is difficult to classify some of the species accurately. The presence of an apophysis, for example, is not a constant and stable character, and in some thalli and species it may be lacking or consist only of a slight swelling in the rhizoidal axis, particularly when species are grown on agar media. Also, the apophysis (fig. 8) and rhizoids (figs. 59, 60) may sometimes be largely extramatrical, or rhizoids may be lacking (figs. 1, 40). Furthermore, in some zoosporangia dehiscence may appear to be operculate (fig. 72). It is to be noted in this connection that some thalli of *Rhizophydium*, *R. closterii* Karling, *R. macroporosum* Karling, and *R. polystomum* Karling, also, are reported to become apophysate occasionally (Karling, 1946, 1967) and that several species of *Rhizidium* develop a conspicuous intramatrical apophysis. Thus, some thalli of *Phlyctochytrium* may be similar to those of these genera, and the distinctions between them, particularly *Rhizophydium* and *Phlyctochytrium*, are often largely artificial as Koch (1957), Miller (1968), Barr (1969), and others have emphasized. More intensive studies on these species will doubtless support this view and might eventually lead to a merging of several genera. From the species which he studied Barr interpreted *Phlyctochytrium* as including the species which develop hollow, tubular rhizoids in pure culture, an interpretation accepted by Knox *et al* (1973), while those which develop thread-

Plate 34

Fig. 1. *Phlyctochytrium hydrodictyii* (Braun) Schroeter. Epibiotic zoosporangium on *Hydrodictyon utriculatum* and an intramatrical apophysis surrounded by a protective plug of host wall material; rhizoids lacking. (Braun, 1855.)

Figs. 2-9. *Phlyctochytrium zygnematis* Rosen. (Rosen, 1887.)

Fig. 2. Spherical, 3-4 μ diam., and ovoid planospores with a large hyaline refractive globule.

Fig. 3. Germination of planospore.

Fig. 4. Young thallus infecting *Zygnema*.

Fig. 5. Apex of fully-grown zoosporangium showing arrangement of solid bipartite deeply-incised teeth at the apex.

Fig. 6. Fully-grown zoosporangium with a large apophysis and a single rhizoidal axis penetrating adjacent cell wall.

Fig. 7. Mature biapophysate thallus with rhizoids arising from periphery of the intramatrical apophysis.

Fig. 8. Thallus with a large extramatrical apophysis and curved stalk.

Fig. 9. Discharge of planospores.

Figs. 10-12. *Phlyctochytrium quadricorne* (DeBary) Schroeter. Fig. 10 after Scherffel, 1926; figs. 11, 12 after Rosen, 1887.

Fig. 10. Enlarged view of planospore, 6 μ diam., with a hyaline refractive globule and granular content.

Fig. 11. Germination of planospore.

Fig. 12. Thallus on and in *Oedogonium rivulare*.

Figs. 13-15. *Phlyctochytrium planicorne* Atkinson. (Sparrow, 1938.)

Fig. 13. Early stages in thallus development on *Cladophora* sp.

Fig. 14. Mature thallus.

Fig. 15. Discharge of planospores.

Figs. 16, 17. *Phlyctochytrium bullatum* Sparrow. (Sparrow, 1938.)

Fig. 16. Mature thallus on *Oedogonium* sp. with 5 toothed bosses and an inner collarette of minute teeth on the zoosporangium.

Fig. 17. Top view of zoosporangium and 5 recently emerged planospores, 8 μ diam., with a 4-5 μ diam., hyaline refractive globule.

Figs. 18, 19. *Phlyctochytrium dentiferum* Sparrow. (Sparrow, 1938.)

Fig. 18. Mature thallus on *Oedogonium* sp.; zoosporangium with 2 apical concentric whorls of solid teeth.

Fig. 19. Discharge of planospores, 7 μ diam., with a hyaline refractive globule, 4 μ diam.

Fig. 20-22. *Phlyctochytrium urceolare* Sparrow. (Sparrow, 1938.)

Fig. 20. Young stage of stalked zoosporangium on *Cladophora* sp.

Fig. 21. Mature zoosporangium with an outer whorl of 6 and an inner whorl of 4 teeth.

Fig. 22. Discharge of planospores, 4 μ in diam., with a hyaline, 2 μ diam., refractive globule.

Figs. 23-25. *Phlyctochytrium aureliae* Ajello. (Ajello, 1945.)

Fig. 23. Spherical, 4-4.5 μ diam., planospore with a hyaline refractive globule.

Fig. 24. Young thallus with teeth developing at the apex of the incipient zoosporangium.

Fig. 25. Median optical view of zoosporangium discharging planospores.

Figs. 26, 27. *Phlyctochytrium mucronatum* Canter. (Canter, 1949.)

Fig. 26. Young thallus with a single apical spine on the zoosporangium.

Fig. 27. Top view of a zoosporangium with 4 whorls of spines.

Phlyctochytrium

like < 1.0 μ diam. rhizoids belong in *Rhizophydium*. However, these differences will probably prove to be as unstable as some of the other vegetative characters. In *R. macroporosum* and *R. polystomum*, for example, the main rhizoidal axis may be up to 9 μ in diameter and run out to fine points, while in *P. dichotomum* Umphlett and Olson they branch dichotomously with blunt ends (figs. 122, 123).

Nevertheless, for the time being and for the sake of convenience *Phlyctochytrium* is presented here in the generally interpreted but artificial sense as a genus of inoperculate species which usually develop an extramatrical or epibiotic zoosporangium or resting spore and an intramatrical apophysis and branched or unbranched rhizoids. Obviously, this interpretation might be applied equally well to the apophysate species of *Rhizidium* which do not develop free of the medium, substratum, or host.

On the basis of the presence or absence of ornamentation of the sporangial wall the known species may be placed in three groups: nude, dentigerous, and chaetiferous. But here again the ornamentations may vary markedly. In a dentigerous isolate which Miller grew on pollen, corn leaves, and purified chitin, the dentate character varied so greatly that he could not identify it with any of the other dentigerous species. Also, Ajello found that the ends of the teeth on *P. aureliae* Ajello sometimes grew out into long hairs. Koch likewise found that the zoosporangia of the nude *P. punctatum* Koch rarely formed coarse hairs or rhizoids from its periphery (fig. 69). However, in *P. furcatum* Sparrow (figs. 111–120), *P. multidentatum* Umphlett (fig. 124), and *P. circulidentatum* Koch (figs. 122–129) the distribution of the teeth and enations appear to be more specific and characteristic.

In the fairly general type of development the planospore germinates on or near the surface of the host or substratum and develops a germ tube which penetrates the surface and usually branches (figs. 3, 29, 34, 99). Subsequently a swelling develops in the germ tube above or at the juncture of the branches (figs.

Plate 35

Fig. 28. *Phlyctochytrium reinboldtae* Persiel. Mature thallus on pollen of *Pinus* sp.; zoosporangium with 6 short tapering necks. (Persiel, 1959.)

Figs. 29-32. *Phlyctochytrium chaetiferum* Karling. (Karling, 1937.)

Fig. 29. Germinated planospore with an apical hair on *Hydrodictyon reticulatum*.

Fig. 30. Mature hairy zoosporangium with a transverse intramatrical apophysis on *Oedogonium* sp.

Fig. 31. Spherical resting spore.

Fig. 32. Ovoid, stalked resting spore.

Figs. 33-36. *Phlyctochytrium hirsutum* Karling. (Karling, 1967.)

Fig. 33. Spherical, 3-4.5 μ diam., planospore with a nuclear cap.

Fig. 34. Germinated planospore on chitin with branched rhizoids.

Fig. 35. Young thallus with hairs beginning to develop on the periphery of the incipient zoosporangium; part of planospore wall unexpanded.

Fig. 36. Mature appendiculate zoosporangium with coarse hairs or rhizoids arising from its periphery; intramatrical apophysis transverse to zoosporangium.

Figs. 37, 38. *Phlyctochytrium hallii* Couch. (Couch, 1932.)

Fig. 37. Beginning of planospore discharge from a zoosporangium on *Spirogyra* sp.

Fig. 38. Bluntly-spiny resting spore.

Fig. 39. *Phlyctochytrium biporosum* Couch; zoosporangium with 2 exit orifices. (Couch, 1932.)

Figs. 40-43. *Phlyctochytrium synchytrii* Köhler. (Karling, 1960.)

Fig. 40. Stalked mature zoosporangium on the resting spore of *Synchytrium pilificum* with 13 exit papillae from which project curved and digitate masses of matrix; rhizoids lacking.

Fig. 41. Enlarged exit papilla with a curved projection of matrix.

Figs. 42, 43. Spherical, 3-3.4 μ diam., and tapering planospores with a relatively large, 2.3-2.6 μ diam., hyaline refractive globule.

Figs. 44-52. *Phlyctochytrium kniepii* Gaertner. (Gaertner, 1954.)

Fig. 44. Elongate planospore discharged on agar.

Fig. 45. Subspherical planospore, 3 × 4 μ, discharged in water with 2 hyaline refractive globules.

Fig. 46-49. Germination of planospores on agar and the development of thick, blunt-ending rhizoids.

Fig. 50. Young thallus growing on agar.

Fig. 51. Mature zoosporangium on agar discharging planospores.

Fig. 52. Resting spore on agar with 4 long projections from the exit papillae.

Figs. 52-57. *Phlyctochytrium africanum* Gaertner. (Gaertner, 1954.)

Fig. 53. Ovoid planospore with a hyaline refractive globule.

Fig. 54. Germination of planospore on agar.

Fig. 55. Young thallus on *Pinus* pollen.

Fig. 56. Incipient zoosporangium on *Pinus* pollen with a turnip-like apophysis.

Fig. 57. Discharge of planospores from a zoosporangium on agar.

Figs. 58-61. *Phlyctochytrium palustre* Gaertner. (Gaertner, 1954.)

Fig. 58. Germination of planospore on agar.

Fig. 59. Non-apophysate incipient zoosporangium with a single, slender, branched, curved, coiled and isodiametric, largely extramatrical rhizoidal axis attached to *Pinus* pollen.

Fig. 60. Largely extramatrical thallus on *Pinus* pollen.

Fig. 61. Resting spore.

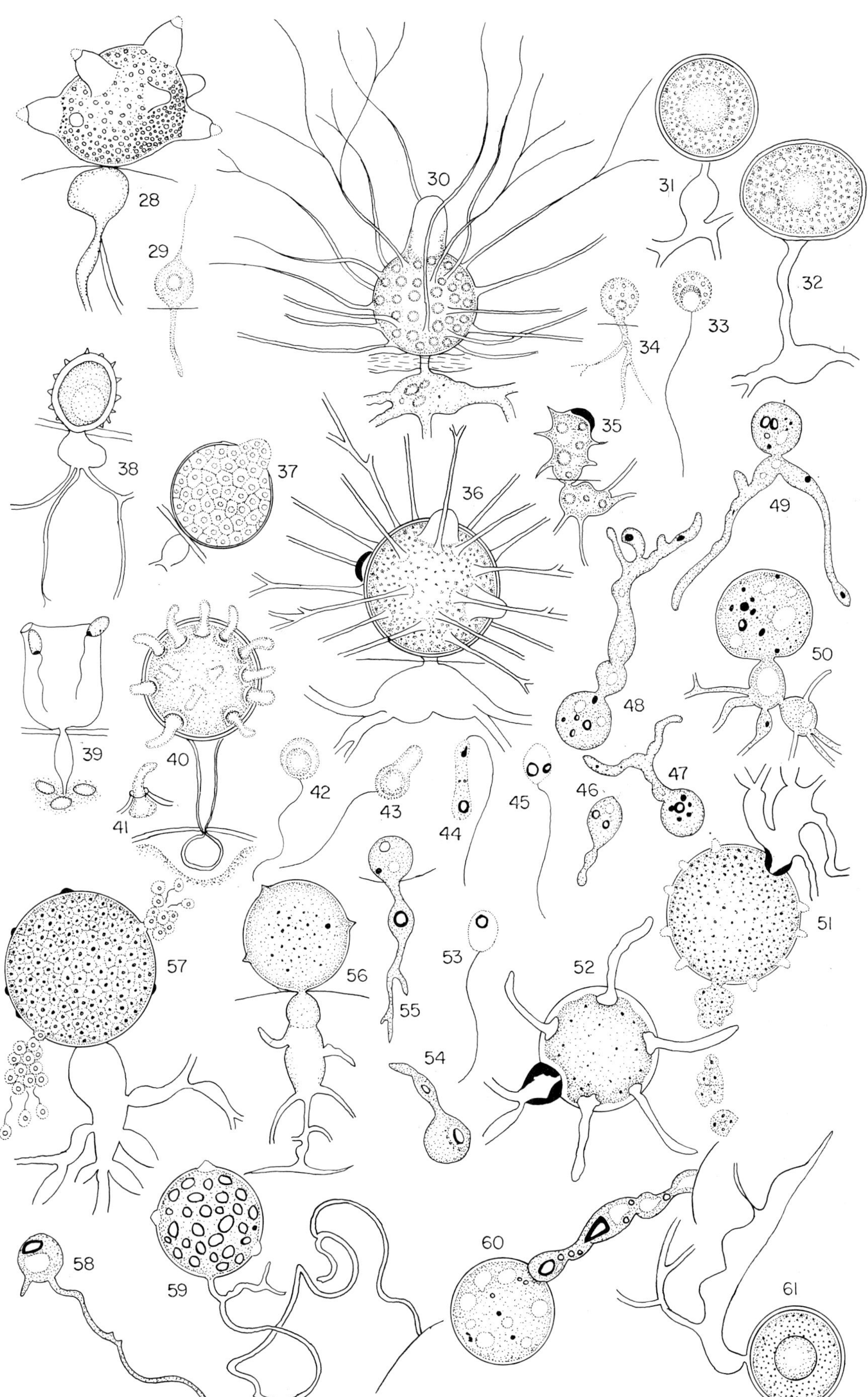

Phlyctochytrium

4, 13, 24, 35, 90, 100) which enlarges to become the apophysis. In some species in which rhizoids as such are lacking the apophysis apparently develops at the tip of an unbranched germ tube (figs. 1, 40). In instances when the zoospores germinate at some distance from the substratum or host the apophysis may develop extramatrically (fig. 8). Occasionally, more than one apophysis may form (figs. 7, 87, 115, 118). Also, some thalli on pollen may be nonapophysate (fig. 59), and when species are grown on nutrient agar the globular apophysis may be replaced by a broad tubular structure (figs. 60, 77).

After the intramatrical or endobiotic portion of the thallus is established the planospore cyst usually enlarges equally in all directions to become the zoosporangium, but in *P. hirsutum* Karling (fig. 35) a portion of it remains unexpanded and becomes thick-walled. Growth of the thallus is endo- exogenous, and in *P. desmidiacearum* Dangeard it is conspicuously so. According to Dangeard the planospore cyst does not begin to enlarge until the apophysis and rhizoids have reached nearly full size and complexity (fig. 107). As the incipient zoosporangium develops the rudiments of the teeth (figs. 13, 24, 113, 114) and setae (fig. 35) form at its apex or on the periphery.

Plate 36

Figs. 62-69. *Phlyctochytrium punctatum* Koch. (Koch, 1957.)

Fig. 62. Enlarged view of fixed and stained planospore showing whiplash flagellum, nucleus, extranuclear granules, and side body.

Figs. 63, 64. Elongate and ovoid planospores swimming on the surface of agar.

Figs. 65-67. Germination of planospore and development of thallus on YpSS agar.

Fig. 68. One of numerous exit papillae which give the zoosporangial wall a punctate appearance after dehiscence.

Fig. 69. Young zoosporangium on *Liquidambar* pollen developing broad hairs from its surface.

Figs. 70-72. *Phlyctochytrium irregulare* Koch. (Koch, 1957.)

Fig. 70. Enlarged view of spherical, 2.6-4.6 μ diam., planospore showing lipoid body and lobed nuclear cap.

Fig. 71. Thinning and deliquescence of apical papilla wall prior to planospore discharge.

Fig. 72. Papilla wall pushed up and back as a lid or operculum during planospore discharge.

Fig. 73. *Phlyctochytrium megastomum* Karling. Dehiscing zoosporangium on grass leaf; papilla wall torn, parts adhering to sporangial wall, and lying at the apex of extruding matrix. (Karling, 1968.)

Figs. 74-76. *Phlyctochytrium spectabile* Uebelmesser. (Uebelmesser, 1956.)

Fig. 74. Partly empty zoosporangium on *Pinus* pollen with 6 of the 15 to 17 exit papillae.

Fig. 75. Spherical, 6 μ diam., planospore with a hyaline refractive globule.

Fig. 76. Large spherical, 60-100 μ diam., resting spore with a large eccentric refractive globule.

Figs. 77-79. *Phlyctochytrium semiglobiferum* Uebelmesser. (Uebelmesser, 1956.)

Fig. 77. Developing thallus on agar; incipient zoosporangium with 5 large hemispherical exit papillae.

Fig. 78. Spherical, 6 μ diam., planospore with 2 refractive globules at the anterior end.

Fig. 79. Resting spore with 3 low papillae.

Figs. 80, 81. Zoosporangium and planospore, resp., of *Phlyctochytrium cladophorae* Kobayashi and Ookubo on *Cladophora* sp. (Kobayashi and Ookubo, 1954.)

Figs. 82, 83. Zoosporangium and resting spore, resp., of *Phlyctochytrium bryopsidis* Kobayashi and Ookubo on *Bryopsis* sp. (Kobayashi and Ookubo, 1954.)

Figs. 84-88. *Phlyctochytrium variabile* Rieth. (Rieth, 1954.)

Fig. 84. Young thallus on *Spirogyra* sp.

Fig. 85. Smooth zoosporangium.

Fig. 86. Zoosporangium with two undivided teeth.

Fig. 87. Empty zoosporangium with a collar of 6 divided teeth.

Fig. 88. Divided and undivided teeth from a zoosporangium.

Figs. 89-97. *Phlyctochytrium vaucheriae* Rieth on *Vaucheria*. (Rieth, 1956.) (Dogma has transferred this species to *Blyttiomyces.)*

Fig. 89. Spherical, 5 μ diam., planospore with a large hyaline refractive globule.

Fig. 90, 91. Young thalli on *Vaucheria*.

Figs. 92-97. Variations in the shapes of the zoosporangia.

Figs. 98-104. *Phlyctochytrium indicum* Karling on pollen of *Pinus*. (Karling, 1964.)

Fig. 98. Spherical, 2.5-3 μ diam., planospores with a hyaline refractive globule.

Figs. 99-101. Planospore germination and early developmental stages of thallus on dead pollen of *Pinus sylvestris*.

Fig. 102. Deliquescence of apical exit papilla and beginning of planospore discharge.

Fig. 103. Hyaline, smooth-walled resting spore with an irregular, large central refractive body surrounded by small globules.

Fig. 104. Early stage of resting spore germination.

Figs. 105-110. *Phlyctochytrium desmidiacearum* Dangeard. (Dangeard, 1937.)

Fig. 105. Spherical, 2-3 μ diam., planospore with a hyaline eccentric refractive globule.

Fig. 106. Germination of planospore on *Closterium ehrenbergii* with a protective host wall plug around the germ tube.

Fig. 107. Thallus with endobiotic portion fully formed before planospore begins to enlarge into a zoosporangium.

Fig. 108. Zoosporangium with 3 apical teeth.

Fig. 109. Discharge of planospores.

Fig. 110. Hyaline, smooth-walled resting spore with a large vacuole surrounded by small globules.

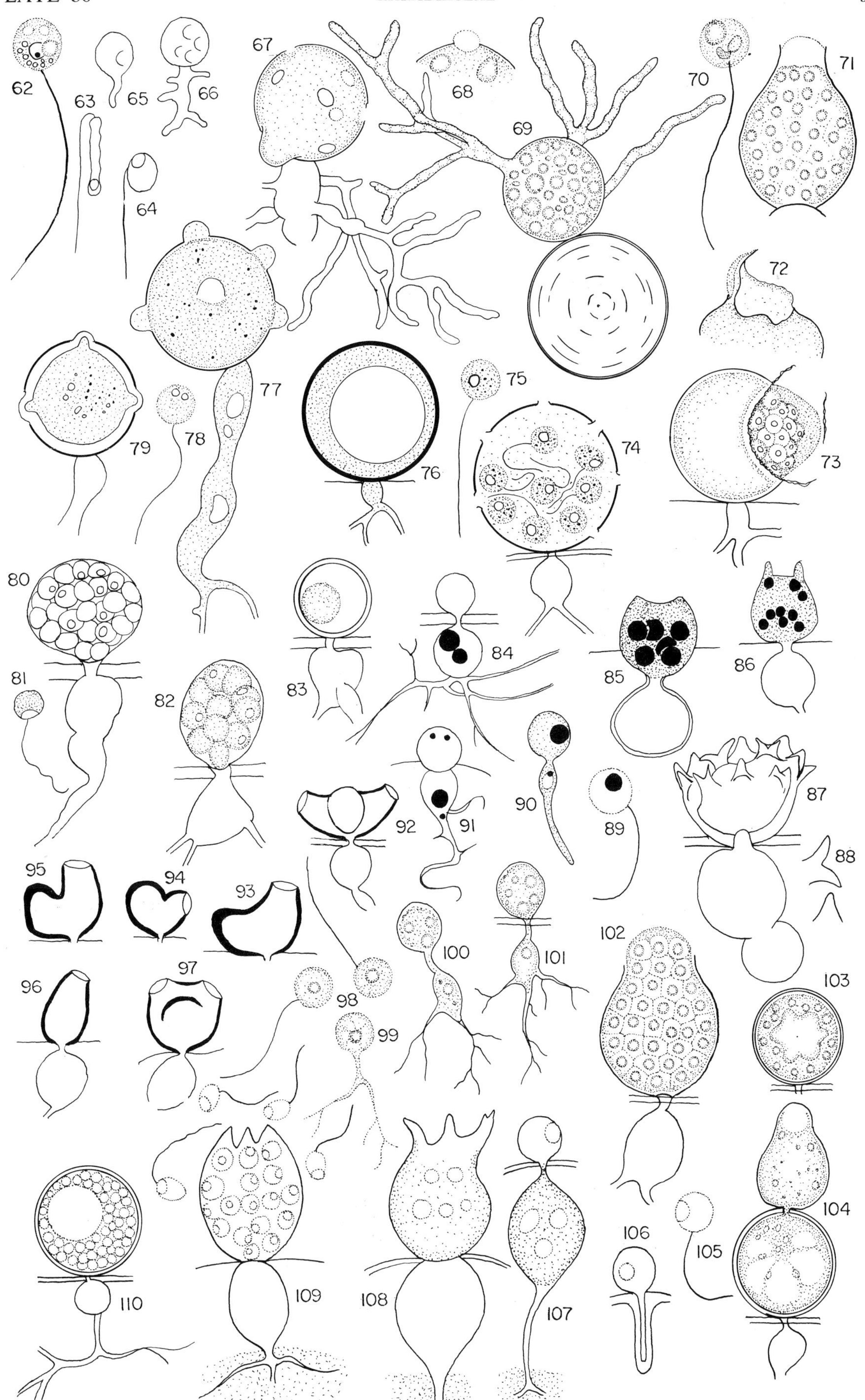

Phlyctochytrium

When the zoosporangia have attained full size they may vary markedly in shape and size in the different species. In most species they tend to be globular, ovoid, pyriform, obpyriform, urceolate or lobed, while in *P. vaucheriae* Rieth they are somewhat saccate (figs. 92–97) with 1 or 2 uptilted discharge papillae. When growing on a substratum or host the zoosporangia of most species rarely attain or exceed 50 μ in greatest diameter, but those of *P. semiglobiferum* Uebelmesser may attain diameters of 500 μ when grown on agar. In *P. palustre* Gaertner on pine pollen Persiel (1963) reported the development of two types of zoosporangia as in *Rhizophydium racemosum*. One type bears a few large, 5–8 μ diam., planospores whose refractive globule fills 2/3's of their volume. These germinate to produce the second type — minute zoosporangia in which develop a few small, 3–4 μ diam., planospores with a 1.5–1.7 μ diam. small refractive globule. At maturity the zoosporangia of other species develop 1 apical, subapical, lateral or up to 25 exit papillae in the various species. In *P. reinboldtae* Persiel (fig. 28) short tapering exit canals are formed, and in *P. synchytrii* Köhler curved and tapering horn-like and digitate masses of hyaline matrix project out of the papillae (figs. 40, 41).

The tips of the exit papillae or necks deliquesce prior to dehiscence in most species, but in *P. irregulare* Koch an operculum-like structure is occasionally lifted up (fig. 72). Also, in *P. megastomum* Karling and *P. circulidentatum* the torn wall of the papilla may adhere to the sides of the exit orifice or be carried up by the emerging zoospores (figs. 73, 126).

The planospores usually form a globular mass as they emerge and soon disperse (figs. 9, 57, 127). They vary from spherical, 2–7 μ diam., to ovoid and contain one minute or very large, or 2 hyaline refractive globules in the various species. Fixed and stained preparations of zoospores of *P. punctatum* (fig. 62) reveal a side body, nucleus, extranuclear granules, and a whip-lash flagellum. In *P. spectabile* the flagellum is reported to be 60 μ long. When discharged on agar the planospores of *P. kniepii* Gaertner and *P. punctatum* elongate markedly and creep around on the surface (figs. 44, 63) with the flagellum in a seemingly anterior placement.

Resting spores are known in only 18 species, and in 11 of these they are predominantly globular, subspherical or ovoid with a smooth outer wall (figs. 31, 32, 61, 76, 79, 83, 103, 110). In *P. spectabile* Uebelmesser the spores (fig. 76) are often as large as the zoosporangia and attain a diameter of 100 μ. In other species the outer wall may be bluntly spiny (fig. 38), papillate (fig. 79), or bear horn-like projections (fig. 52). So far as is known they are formed asexually and borne like the zoosporangia in most of these species. However, in *P. quadricorne* (DeBary) Schroeter and *P. planicorne* Atkinson, according to Sparrow (1933), Umphlett and Holland (1960) and Milanez (1967) the resting spore develops endogenously in a container cell or the epibiotic portion of a resting-spore thallus which has the same structure as a sporangial thallus. In *P. chaetiferum* Karling (fig. 104) and *P. synchytrii* Köhler they function as prosporangia in germination, but in *P. kniepii* Gaertner and *P. semiglobiferum* Uebelmesser Gaertner (1954) and Uebelmesser (1956) reported that the spores function directly as zoosporangia when they germinate. However, it is not improbable that what they described as germinating resting spores were zoosporangia which had become dormant.

POLYPHLYCTIS Karling

Sydowia 20:86, 1967.

Plate 38

This monotypic inoperculate genus was created for a chytrid which Paterson (1956) described as *Phlyc-*

PLATE 37

Figs. 111-120. *Phlyctochytrium furcatum* Sparrow on pine pollen. (Sparrow, 1966.)

Figs. 111-114. Stages in the growth and differentiation of the young zoosporangia.

Fig. 115. Nearly mature zoosporangium with furcate spines and external and internal apophyses connected by a stalk; rhizoids not shown.

Figs. 116, 117. Young zoosporangia with a single internal apophysis.

Fig. 118. Mature zoosporangium with an external apophysis connected with an internal one by an inflated stalk.

Fig. 119. A zoosporangium parasitized by 4 small zoosporangia of its own (cannibalism.)

Fig. 120. Discharge of spherical, 4 μ diam., planospores from 6 exit orifices.

Figs. 121-123. *Phlyctochytrium dichotomum* Umphlett and Olson. (Umphlett and Olson, 1967.)

Fig. 121. Planospore with a laterally attached flagellum.

Fig. 122. Germination of planospore with a broad germ tube.

Fig. 123. Incipient zoosporangium on agar with blunt incipient rhizoids.

Fig. 124. *Phlyctochytrium multidentatum* Umphlett and Olson thallus on *Liquidambar* pollen; rhizoids obscured by host, protoplasm. (Umphlett and Koch, 1969.)

Figs. 125-132. *Phlyctochytrium circulidentatum* Umphlett and Olson on *Closterium* and *Liquidambar* pollen. (Umphlett and Koch, 1969.)

Fig. 125. Mature thallus with a circular whorl of teeth above the equator of the zoosporangium; rhizoids obscured.

Fig. 126. Dehiscence of zoosporangium; apical wall largely deliquesced with a remnant at the side.

Fig. 127. Discharge of planospores.

Fig. 128. Three spherical, 3-4 μ diam., planospores with a hyaline refractive globule and nuclear cap.

Figs. 129-132. Various views of empty zoosporangia after discharging planospores.

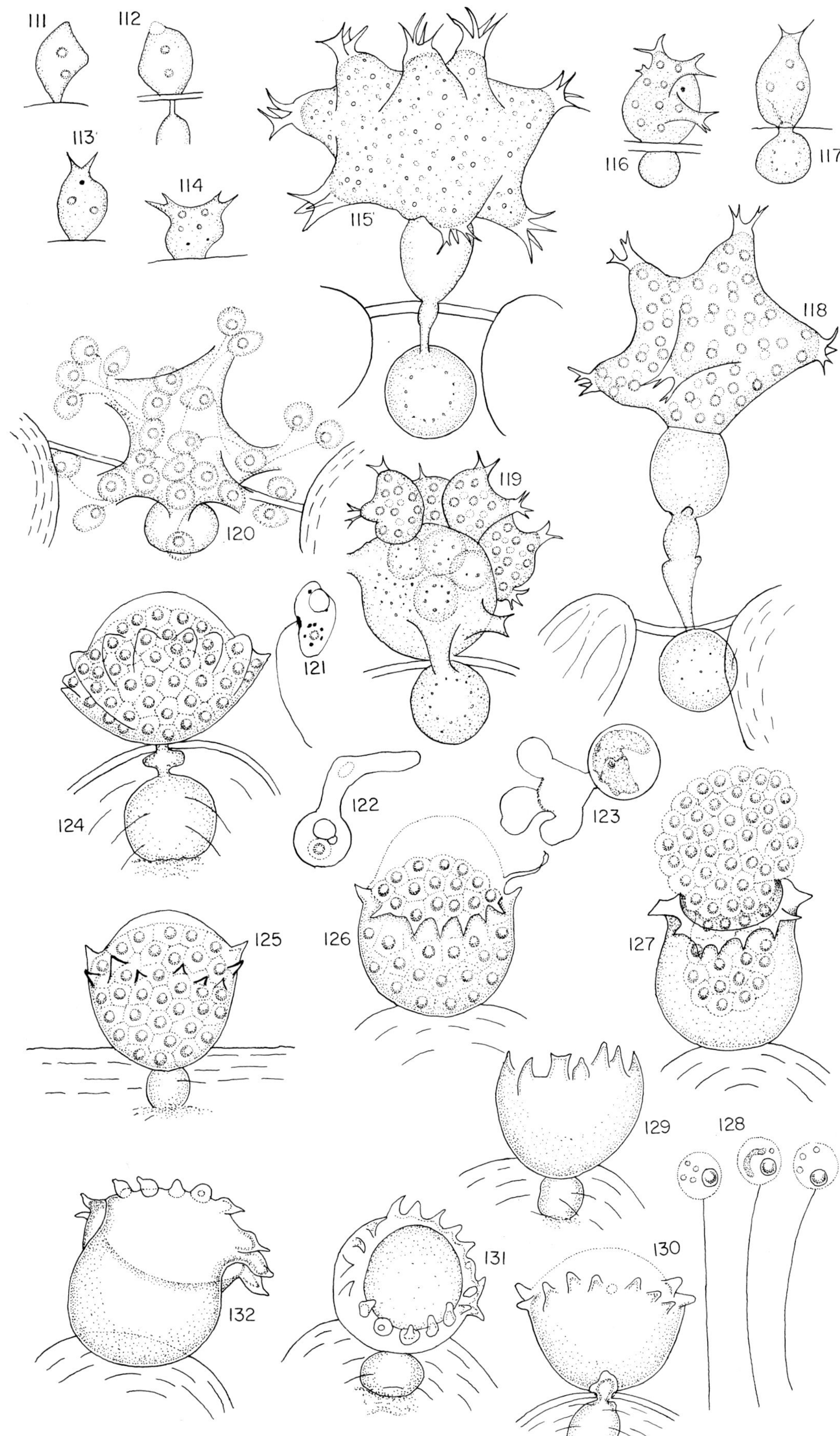

Phlyctochytrium

tochytrium unispinum. It is characterized by an extramatrical zoosporangium with 1–10 exit papillae which are surmounted by spines, knobs, pegs, or globules of gelatinous consistency, and usually bears the split halves of the planospore cysts as appendages, an intramatrical absorbing system consisting of 1–8 catenulate apophyses, and one unbranched rhizoid or an extensively branched rhizoidal system (fig. 1). It was reported first as a saprophyte in moribund cells of *Oedogonium* sp., *Zygnema* sp. and *Stigonema* sp., but later Willoughby and Townley (1961) and Karling (1967) isolated it on grass leaves from mud and soil samples. In algal cells Paterson described the intramatrical portion as consisting of a globular apophysis with an unbranched rhizoid (fig. 16), and the extramatrical zoosporangium as bearing a single apical spine (fig. 19). In boiled grass leaves the subsequent workers found that the zoosporangium may develop up to 10 exit papillae and gelatinous spines while the intramatrical portion may be more complex and extensive.

The development of *P. unispina* (Paterson) Karling is clearly endoexogenous but unusual with respect to the fate of the planospore cyst and sporangial dehiscence. The planospore encysts on the surface of the host cell or substratum, or at some distance from it, and develops a germ tube which penetrates the surface (fig. 5). In instances where the planospore germinates at some distance away from the host the germ tube may be fairly long and broad (fig. 7). The latter enlarges at its tip to form the primary apophysis (figs. 5–7, 9) from which subsequently a single rhizoid (fig. 16) or an extensively branched system develops (fig. 1). In the latter case a succession of swellings or apophyses (figs. 1, 10, 14) develop which bear rhizoids on their periphery or at their ends. The primary apophysis may become fairly thick-walled in old thalli. In exceptional thalli the rhizoids and apophyses may extend for distances up to 180 μ.

In the meantime the thick-walled planospore cyst has persisted, and after the endogenous development has been completed or nearly so, growth becomes exogenous. The protoplasm slowly flows outward into the planospore cyst which causes it to split vertically into halves (fig. 13). The sporangial rudiment is formed between the halves (figs. 14–17), and these are moved further apart as the rudiment enlarges and develops into a mature zoosporangium. Eventually they adhere as appendages at almost the opposite sides of the zoosporangium (figs. 1, 19–21). Occasionally, however, one or both may be shed and be close to the area where they were shed. Meanwhile, an apical (fig. 18) or up to 10 exit papillae develop on the zoosporangium and become surmounted by gelatinous, tapering spines, knobs, pegs, or globules (figs. 1, 30) which may vary considerably in size and shape (figs. 31–34).

In the process of dehiscence the tips of one (figs. 20–24), or more (fig. 35) exit papillae deliquesce, and the lower portion of the spine swells, or vesiculates (figs. 21–23). The planospores move up into this vesicle which soon bursts and disappears, freeing

PLATE 38

Figs. 1-36. *Polyphlyctic unispina* (Paterson) Karling. Figs. 1, 3, 5, 6, 10 drawn from New Zealand specimens; figs. 2, 15-18, 20, 21 after Paterson, 1956; figs. 4, 7-9, 11-14, 19, 22-36 after Willoughby and Townley, 1961.

Fig. 1. Portion of a sporangial thallus on boiled maize leaf with 8 intramatrical apophysal segments and branched rhizoids, and an extramatrical zoosporangium with 9 exit papillae capped by tapering gelatinous spines or globules; split halves of planospore cyst adhering as thickened appendages on the wall of the zoosporangium.

Fig. 2. Large spherical, 6-7.5 μ diam., planospore with a hyaline eccentric globule and several granules.

Figs. 3, 4. Smaller spherical, 3.5-4.5 μ diam., planospores from New Zealand and English specimens, resp.

Figs. 5, 6. Early germination stage and development of primary apophysis and rhizoidal axes, resp., on corn leaves.

Fig. 7. Germination of a planospore and formation of a broad tube at some distance from an algal cell; primary apophysis within host.

Fig. 8. Formation of primary and secondary apophyses.

Fig. 9. Elongate transverse primary apophysis.

Fig. 10. Portion of a young thallus with 6 incipient apophysal segments filled with dense protoplasm and refractive globules.

Fig. 11. Globular primary apophysis.

Fig. 12. Very large elongate primary apophysis and a smaller secondary one.

Fig. 13. Beginning of exogenous development; planospore cyst has split vertically as the protoplasm moved outward from the apophysis to form the sporangial rudiment between the halves.

Fig. 14. Later stage; halves of cyst have moved apart as sporangial rudiment enlarges; 4 apophysal segments within host.

Figs. 15, 16. Further stages of development of the zoosporangium in *Zygnema* sp., with the persistent halves of the planospore cyst.

Fig. 17, 18. Development of apical papilla.

Fig. 19. Mature zoosporangium and thallus on *Stigonema* sp. with one apical spine.

Fig. 20, 21. Two stages of dehiscence; basal portion of spine becoming vesicular; apical portion detached.

Figs. 22-24. Similar stages in dehiscence in which basal portion of gelatinous peg vesiculates while the apical part is detached.

Figs. 25-29. Stages in dehiscence in which the apical peg gradually deliquesces.

Fig. 30. Large mature zoosporangium with 9 exit papillae surmounted by globules of gelatinous material.

Figs. 31-34. Variations in shapes of the surmounting gelatinous material.

Fig. 35. Empty zoosporangium with halves of planospore cyst and 2 exit orifices.

Fig. 36. Germinated resting spore (?).

Polyphlyctis

the mass of planospores. The upper portion of the spine remains intact (fig. 21) or appears to be displaced (fig. 23) as the planospores swarm briefly in a localized area before dispersing. Willoughby and Townley described a second type of dehiscence in which the gelatinous peg or spine expanded, became nearly spherical (figs. 25–28) and "appeared to become completely absorbed in the spore mass as the latter left the sporangium". It is not certain that the fungus described by Paterson is the same species reported by Willoughby and Townley and Karling. Paterson reported the planospores to be 6–7.5 μ diameter (fig. 2) but in the English and New Zealand specimens they were only 3.5–4.5 μ in diameter (figs. 3, 4) a difference which seems to be significant at the present time.

Only one resting spore has been observed so far. It had a thick, smooth hyaline wall and appeared to be functioning as a prosporangium during germination (fig. 36).

Insofar as our knowledge of *P. unispina* and the chytrids in general goes *Polyphlyctis* appears to be a transitional genus which bears the same relations to *Phlyctochytrium* as *Catenochytridium* does to *Chytridium*.

DANGEARDIA Schröder

Ber. Deut. Bot. Gesell. 16:321, 1898.

Plate 39

This small inoperculate genus was reported to include 4 species and 3 unidentified specimens which parasitize colonial freshwater algae, but recently Batko, (1970) added another species, *D. echinulata* Batko, and transferred *D. sporapiculata* Geitler to *Dangeardiana*. However, the distinctions between these species are not very sharp, and are based primarily on the shape of the zoosporangia, the structure of the resting-spore wall, and the presence or absence of sexuality. As in species of *Rhizophydium* the thallus consists of epibiotic or extracellular zoosporangia and resting spores and sparse endobiotic, branched or unbranched rhizoids, or a haustorium. The planospores are spherical to ovoid with a hyaline refractive globule (figs. 1, 2, 19, 20, 32, 37), but in *D. mammillata* (Schröder) Canter (1946) reported two types of planospores, one ovoid with an oscillating granule (fig. 3) and another without the granule (fig. 2) which are produced in separate zoosporangia.

Plate 39

Figs. 1-18. *Dangeardia mammillata* Schröder. Figs. 1, 4, 5, 8, 11, 12, 15 after Schröder, 1898; figs. 2, 3, 6, 7, 9, 10, 13, 14, 16 after Canter, 1946.

Fig. 1. Ovoid planospore with a hyaline globule.

Fig. 2. Spherical, 2.5 μ diam., planospore with hyaline globule lying in a clear area.

Fig. 3. Second type of planospore, 1.8 × 3.5 μ, with an apical oscillating granule.

Fig. 4. Encysted, germinating planospore on the muscilagnous sheath of the coenobium of *Pandorina morum* (*Eudorina elegans*?) with a fine infecting filament or germ tube in contact with algae cell.

Fig. 5, 6. Elongation of planospore body and broadening of germ tube.

Fig. 7. Later stage, base of germ tube bearing two rhizoids and enlarging to become the lower portion of the incipient zoosporangium.

Fig. 8. Portion of a coenobium with algal cell and 4 parasites; almost mature zoosporangium embedded in the muscilaginous sheath except for its neck.

Figs. 9, 10. Mature zoosporangia with hyaline globules.

Fig. 11. Host cell and sporangium treated with concentrated sulphuric acid to show basal rhizoids.

Fig. 12. Discharge of planospores.

Fig. 13. Two young epibiotic resting spores with associated male (?) thalli.

Fig. 14. Smooth-walled resting spore with 2 male thalli (?), one of which still retains some of its content.

Fig. 15. Spiny resting spore.

Fig. 16. Spiny resting spore with broad, thick-walled planospore cyst and germ tube above, and an empty male thallus at right.

Fig. 17. Germination of smooth-walled resting spore; functioning as a prosporangium.

Fig. 18. Germinated smooth-walled resting spore with empty zoosporangium at left; broad, thick-walled planospore cyst and germ tube above, and empty male thallus at left.

Figs. 19-31. *Dangeardia laevis* Sparrow and Barr. (Sparrow and Barr, 1955.)

Figs. 19, 20. Spherical and ovoid planospores with a hyaline globule.

Fig. 21. Infection of (?) *Glenodinium* sp. by a long fine germ tube.

Figs. 22-25. Stages in the development of the zoosporangium; upper portion of planospore cyst becoming fairly thick-walled at the apex of exit canal.

Fig. 26. Stalked zoosporangium.

Figs. 27, 28. Dehiscence of zoosporangium.

Fig. 29. Almost empty zoosporangium with remnants of planospore cyst at apex of exit canal.

Figs. 30, 31. Thick-walled, smooth, hyaline resting spores with planospore cyst and broad germ tube persistent.

Figs. 32-38. *Dangeardia ovata* Paterson. (Paterson, 1958.)

Fig. 32. Spherical planospore, 3-3.5 μ diam., with a hyaline refractive globule.

Fig. 33. Infection of *Sphaerocystis schroeteri*.

Figs. 34-36. Stages in development of an ovate zoosporangium.

Figs. 37. Discharge of planospores.

Fig. 38. Resting spore with a thick, smooth wall and long persistent germ tube.

Figs. 39, 40. *Dangeardia* sp. on a colony of *Eudorina elegans*. (Paterson, 1958.)

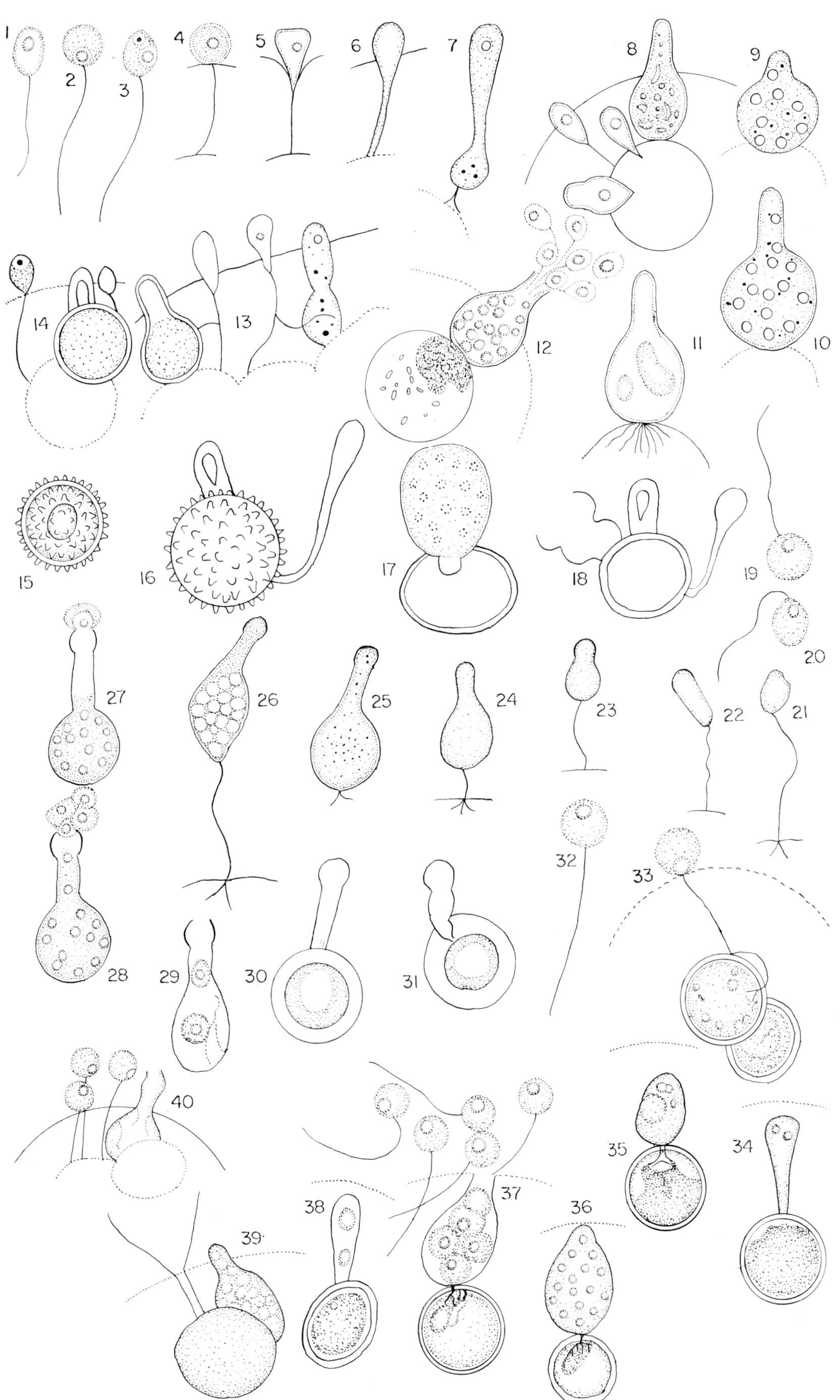
1
2
3
4
5
6
7
8
9
10
11
12
13
14
15
16
17
18
19
20
21
22
23
24
25
26
27
28
29
30
31
32
33
34
35
36
37
38
39
40

Dangeardia

The planospore comes to rest on the mucilaginous sheath of the host and produces a fine germ tube which grows down to the algal cell (figs. 4, 5, 21) within which it may branch or remain unbranched. The tube soon broadens (fig. 6) until the young thallus becomes almost cylindrical or spindle-shaped. The base in contact with the algal cell enlarges and eventually becomes the basal portion of the flask-shaped zoosporangium while the upper portion and planospore cyst become the neck, which usually protrudes slightly from the mucilaginous sheath at maturity (figs. 8–12, 37, 40). In *D. sporapiculata* (*Dangeardiana sporapiculata* Batko) Geitler (1963) reported that the zoosporangia develop in the gelatinous sheath and do not protrude beyond its surface. Accordingly, he did not regard them as being epibiotic but "extracellular" and release the planospores within the gelatinous sheath. The planospore cyst and upper portion of the germ tube, thus, function as an exit canal for the discharge of planospores in some species, but in *D. sporapiculata* a broad exit papilla develops adjacent to the thickened portion of the planospore cyst. In *D. laevis* Sparrow and Barr the wall of the upper portion of the planospore cyst at the apex of the neck becomes slightly thickened (figs. 25–29) and is clearly visible in dehiscing and empty zoosporangia. In *D. mammillata*, and *D. laevis* the rhizoids are unbranched and bushy (figs. 11, 24) while in *D. ovata* Paterson and *D. sporapiculata* they are branched and arise from a basal axis or stalk (fig. 26).

The resting spores are borne like the zoosporangia and at maturity bear the inflated germ tube and planospore cyst as appendages (figs. 16, 18, 30, 31, 38). Schröder illustrated them in *D. mammillata* as being endobiotic, but Canter described them as epibiotic, and her report has been confirmed by subsequent observations on *D. laevis* and *D. ovata*. In the latter two species the spores are smooth and thick-walled (figs. 30, 31, 38) while in *D. mammillata* Schrôder figured them as being spiny (fig. 15). However, Canter found that this species produces both spiny (fig. 16) and smooth (figs. 17, 18) resting spores. She, also, reported that they are formed sexually in this species. "In resting-spore formation, union occurs between a relatively large flask-shaped thallus (presumably the female), almost identical with a zoosporangium, and a relatively small one (presumably the male), resembling an early stage in the development of an asexual sporangium" (fig. 13). She believed that the male thalli might be derived from planospores with the oscillating granule, but she found no direct evidence to support this belief. "The germ tube of the male thallus makes direct contact with the swollen base of the female, and the empty male normally connected with each mature resting spore indicates that the contents of the male pass into the female" (figs. 16, 18). This type of sexuality most nearly resembles that of *Zygorhizidium willei* Lowenthal, in her opinion. No evidence of sexuality has been found in the development of the resting spores of *D. laevis* and *D. ovata*, and resting spores are unknown in *D. sporapiculata*.

In germinating the resting spores of *D. mammillata* function as prosporangia (fig. 17), and Canter believed that the zoosporangia from such spores produce planospores all of one type, "either with a single oil globule or with an additional oscillating granule."

PLATE 40

Figs. 1-7. *Dangeardia sporapiculata* Geitler. (Geitler, 1963.)

Fig. 1. Two encysted planospores, 2.8-3 μ diam., and a germinated one on the gelatinous sheath of the palmella stage of *Chlamydomonas* sp.

Fig. 2. Germinated planospore with the long germ tube attached to an algal cell.

Fig. 3. Later developmental stage; incipient zoosporangium forming as local enlargement of the germ tube.

Figs. 4, 5. Mature zoosporangia forming a broad exit papilla adjacent to the unexpanded, thick-walled portion of the planospore cyst or "sporapiculum."

Fig. 6. Empty collapsed zoosporangium with the sporangial stalk from whose base arise intramatrical rhizoids.

Fig. 7. Empty collapsed zoosporangium showing persistence of a portion of the planospore cyst.

Figs. 8-23. *Dangeardiana eudorinae* Valkanov. (Valkanov, 1964.)

Fig. 8. Thick-walled germinated planospores on the gelatinous sheath of a *Eudorina* colony; the filamentous germ tube *a*, has reached the oospore of the host; in *b* and *c* the germ tube is expanding to form the rudiment of the spindle-shaped zoosporangium in the gelatinous sheath.

Fig. 9. Later developmental stage; the planospore cyst and part of the germ tube are persistent, and the exit canal is forming at its right.

Fig. 10. Mature zoosporangium with a long exit canal projecting out of the gelatinous sheath, a tapering haustorium, and persistent planospore cyst and germ tube.

Fig. 11. Mature zoosporangium with planospores. Sketched from a photograph.

Fig. 12. Spherical, 2 μ diam., planospore with a hyaline refractive globule. Sketched from a photograph.

Figs. 13-16. Variations in the shapes of zoosporangia and lengths of the exit canal.

Fig. 17. Germinated female gametangium with an attached male cell.

Fig. 18. Later stage; germ tube of female gametangium enlarging to become the rudiment of the resting spore or zygote.

Fig. 19. Male cell has moved away from the planospore cyst of the female gametangium and formed a conjugation canal between them.

Figs. 20, 21. Later stages in the development of the zygote after plasmogamy.

Fig. 22. Mature zygote with a large central refractive globule and some smaller ones lying at the base of the gametangium; planospore cyst, male cell, and female gametangium and conjugation canal persistent. Sketched from a photograph.

Fig. 23. Similar zygote or resting spore formed by the fusion of the contents of 2 male cells with that of the female gametangium. Sketched from a photograph.

Dangeardia, Dangeardiana

DANGEARDIANA Valkanov

Arch. f. Microbiol. 48:245, 1964

Plate 40, Figs. 8-23

This inoperculate genus differs from *Dangeardia* primarily by the development of a long exit tube on the zoosporangia, a difference that does not merit generic distinction in this writer's opinion. *Dangeardiana eudorinae* Valkanov, which parasitizes the oospores of *Eudorina elegans*, will doubtless prove to be a species of *Dangeardia*, but it is illustrated and described separately so that readers may judge independently its validity as a representative of a new genus. Batko (1970) transferred *Scherffeliomyces appendiculatus* (Zopf) Sparrow, *S. leptorhizus* Johns, *Entophlyctis apiculata* (Braun) Fischer, and *Dangeardia sporapiculata* Geitler to this genus and interpreted it to be closely related to *Coralliochytrium*, *Scherffeliomyces*, and *Dangeardia*.

In the development of the sporangial thallus the planospore (fig. 12) encysts on the gelatinous sheath of the host, develops a thick wall, and forms a fine germ tube which grows inward until it reaches the algal oospore (fig. 8a). The germ tube enlarges locally, becomes narrowly spindle-shaped and develops into the rudiments of the zoosporangium with a tapering base or haustorium. It is not known whether or not rhizoids develop from the tip of the latter and within the host oospore. Meanwhile, the thick-walled planospore cyst and upper part of the germ tube persist (fig. 9) as an appendage. The young zoosporangium soon develops an exit canal at the side of the cyst and germ tube (fig. 9). This canal elongates further to the outside of the gelatinous sheath as a straight or curved tapering tube and may attain a length of 20 to 135 μ (figs. 10, 11, 13–16). The development of the zoosporangium proper is thus extracellular, and the exit canal is extramatrical or epibiotic. As the tip of the exit canal deliquesces the planospores emerge to the outside of the sheath. It appears that they are facultative. They may give rise to additional thalli or function as gametes (gametangia).

The resting spores are formed sexually by the fusion of the contents of a male cell and a female gametangium. A planospore encysts on the gelatinous sheath, develops a germ tube which grows inward to the host oospore, and develops as if it were to become a sporangial thallus. However, if a second planospore comes

PLATE 41

Figs. 1-11. *Rhizidium nowakowskii* Karling. Figs. 1, 2, 8-11 after Nowakowski, 1876; figs. 3-7 after Karling, 1944.

Fig. 1. Sporangial thallus with a pyriform zoosporangium and a sparsely branched rhizoidal axis, growing in the slime of *Chaetophora elegans*.

Figs. 2, 3. Planospores, with a large hyaline refractive globule.

Figs. 4, 5. Germination stages and development of central rhizoidal axis.

Fig. 6. Young pyriform zoosporangium.

Fig. 7. Mature ovoid zoosporangium with a lateral exit papilla.

Fig. 8. Discharge of planospores and their swarming in a vesicle which has floated away from exit orifice.

Fig. 9. Germination of a smooth-walled resting spore.

Fig. 10. Germination of a hairy resting spore; planospores swarming in a vesicle at the apex of the empty zoosporangium.

Fig. 11. Planospore from a germinated resting spore.

Figs. 12-15. *Rhizidium chitinophilum* Sparrow. Figs. 11-14 after Sparrow, 1937; fig. 15 after Karling, 1967.

Fig. 12. Young thallus with a central rhizoidal axis, in midges and mayflies.

Fig. 13. Apophysate incipient zoosporangium.

Fig. 14. Discharge of planospores which will swarm in a vesicle.

Fig. 15. Resting spore of *Rhizidium chitinophilum* (?) on chitin, from New Zealand.

Figs. 16-19. *Rhizidium ramosum* Sparrow. (Sparrow, 1937.)

Fig. 16. Ellipsoidal, 4 × 6 μ, planospore with a lateral hyaline refractive globule.

Fig. 17. Mature zoosporangium with an apical exit papilla; central rhizoidal axis lacking.

Fig. 18. Discharged planospores.

Fig. 19. Zoosporangium with a basal exit pore and two refractive nodules.

Figs. 20-33. *Rhizidium braziliensis* Karling. (Karling, 1945.)

Figs. 20, 21. Planospores, ovoid to elongate, 3-3.5 × 5-5 μ, with one and several refractive globules, resp.

Fig. 22. Early germination stage of planospore.

Fig. 23. Germinated planospore with 2 rhizoidal axes.

Fig. 24. Young thallus with apophysis forming in the germ tube beneath rudiment of zoosporangium.

Fig. 25. Later stage of the same thallus, zoosporangium and apophysis extramatrical.

Fig. 26. Young zoosporangium with 1 apical and 2 lateral exit papillae, apophysis intramatrical.

Fig. 27. Zoosporangium with 2 exit canals, apophysis and part of rhizoidal axis extramatrical.

Fig. 28. Discharged planospores swarming in a vesicle which has separated from the zoosporangium.

Fig. 29. Young resting spore with dense globular content.

Figs. 30-32. Variations in shape, wall structure, and content of resting spores.

Fig. 33. Germination of resting spore.

Figs. 34-36. *Rhizidium laevis* Karling. (Karling, 1945.)

Fig. 34. Smooth apophysate resting spore.

Fig. 35. Mature apophysate zoosporangium with 2 exit papillae, central rhizoidal axis lacking.

Fig. 36. Ovoid, 3 × 4 μ, planospore with a hyaline refractive globule.

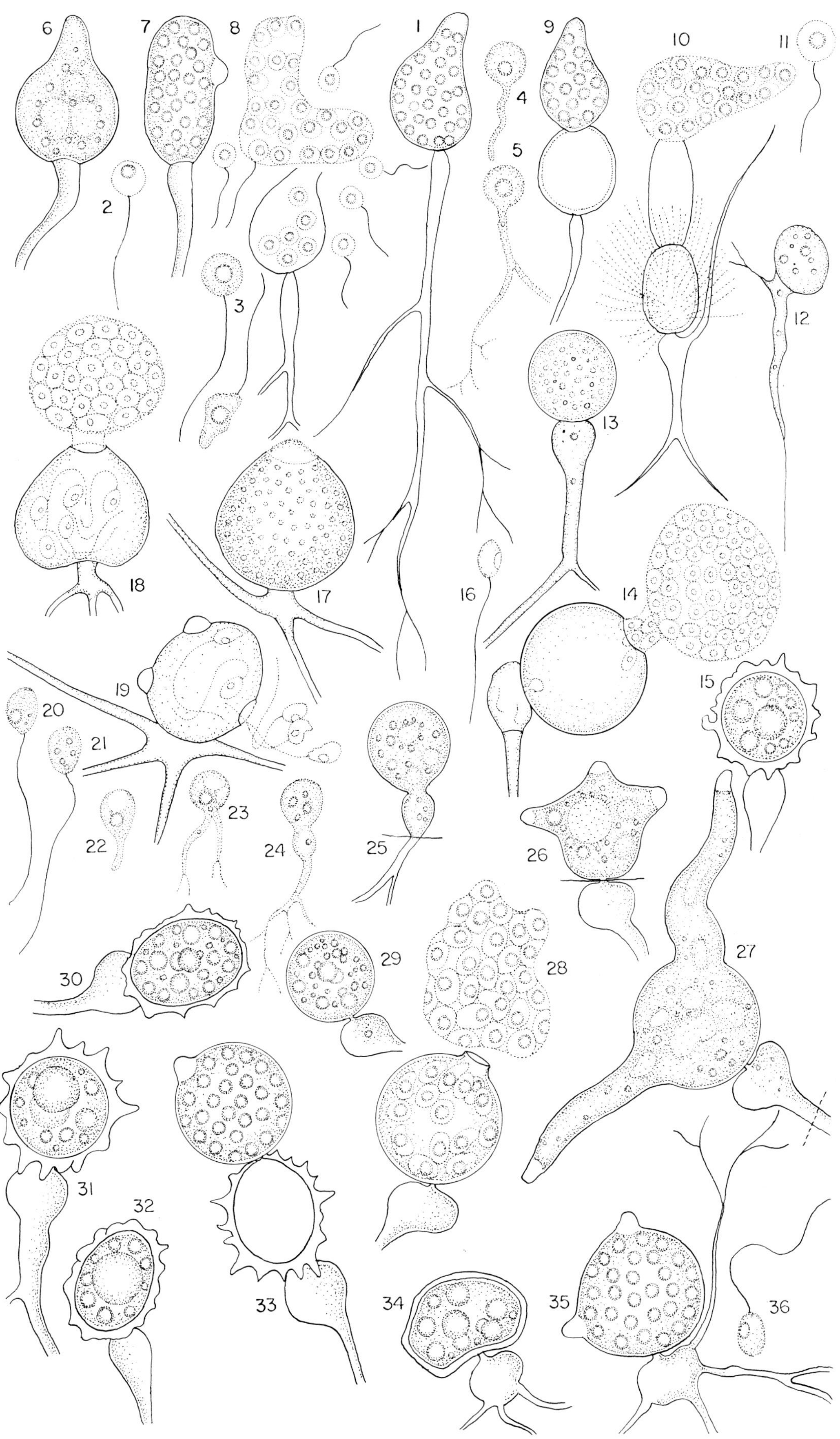
1
2
3
4
5
6
7
8
9
10
11
12
13
14
15
16
17
18
19
20
21
22
23
24
25
26
27
28
29
30
31
32
33
34
35
36

Rhizidium

to rest on it (fig. 17) it functions as a female gametangium and the second planospore functions as a male cell or gametangium. The two remain attached as the germ tube enlarges to form the rudiment of the resting spore (fig. 18). Shortly thereafter, the male cell moves a short distance away and develops a short conjugation tube which fuses with the planospore cyst of the female gametangium (fig. 19). Plasmogamy follows, and the resting-spore rudiment enlarges locally (figs. 20, 21) and becomes the zygote or resting spore (figs. 22, 23). The upper part of the gametangium and its cyst as well as the conjugation canal and male cell become thick-walled and persist. Occasionally, 2 male cells may fuse with the female gametangium (fig. 23). Germination of the zygote has not been observed.

Dangeardiana eudorinae is unusual in that it may complete its vegetative cycle and asexual reproduction within a single day.

Plate 42

Figs. 37-40. *Rhizidium braunii* Zopf. (Zopf, 1888.)

Fig. 37. Part of a young polyphagus thallus with tips of rhizoids parasitizing numerous thalli of *Pinnularia*.

Fig. 38. Young thallus with central rhizoidal axis.

Fig. 39. Mature zoosporangium.

Fig. 40. Part of a resting-spore thallus with a central rhizoidal axis and terminal resting spore.

Figs. 41-50. *Rhizidium verrucosum* Karling. (Karling, 1944.)

Figs. 41, 42. Ovoid, 3-3.5 × 5-5.5 μ, and spherical planospores with a reddish-brown refractive globule.

Fig. 43. Absorption of flagellum.

Figs. 44-46. Germination stages of planospores.

Fig. 47. Mature zoosporangium with dark outer verrucose wall.

Fig. 48. Similar but smaller zoosporangium with a long central rhizoidal axis.

Fig. 49. Discharge of planospores.

Fig. 50. Planospores swarming in a vesicle which has floated away from exit orifice.

Figs. 51-60. *Rhizidium elongatum* Karling. (Karling, 1949.)

Fig. 51. Ovoid, 4.8-5.2 × 5-6.2 μ, planospore with a hyaline refractive globule.

Fig. 52. Absorption of flagellum.

Fig. 53. Germination of a planospore and the rudiment of a central rhizoidal axis.

Fig. 54. Germination of a planospore and the rudiment of a branched rhizoidal system.

Figs. 55-58. Variations in the shapes of zoosporangia and the character of the rhizoidal system.

Fig. 59. Deliquescence of exit papilla and emergence of globule of matrix.

Fig. 60. Discharged planospores swarming in a vesicle which has floated away from the exit orifice.

Figs. 61-67. *Rhizidium windermerense* Canter. (Canter, 1950.)

Fig. 61. Encysted planospore with germ tube attached to cells of *Gemellicystis neglecta*, a colonial green alga.

Fig. 62. Branching of rhizoid and development of planospore into an incipient zoosporangium.

Fig. 63. Mature zoosporangium with tips of rhizoids attached to several algal cells.

Fig. 64. Discharge of planospores.

Fig. 65. Early stage of zygote (Z) developing at tips of fused rhizoids or from two gametangia (G) which are partly empty of content.

Fig. 66. Later stage of zygote development; gametangia empty.

Fig. 67. Mature zygote with thickened conjugation tubes.

Figs. 68-71. *Rhizidium variabile* Canter. (Canter, 1947.)

Fig. 68. Young thallus with central rhizoidal axis on a dead *Spirogyra* sp. cell.

Fig. 69. Discharge of planospores.

Figs. 70. Ovoid, 4-4.5 μ diam., planospore with a hyaline refractive globule.

Fig. 71. Early infection stage.

Figs. 72-77. *Rhizidium richmondense* Willoughby. Figs. 72-75 after Willoughby, 1956; figs. 76, 77 after Karling, 1967.

Fig. 72. Ovoid planospore, 3-3.3 × 2.5-3 μ, with a hyaline refractive globule.

Fig. 73. Very young thallus on epidermis of *Allium cepa*.

Fig. 74. Later stage of thallus development.

Fig. 75. Mature apiculate zoosporangium, apiculus developed from unexpanded portion of planospore.

Fig. 76. Discharge of planospores; apiculus and upper portion of zoosporangium wall lifted up like an operculum.

Fig. 77. Empty zoosporangium; apiculus and upper part of zoosporangium wall have flapped back into zoosporangium.

Figs. 78-81. *Rhizidium endosporangiatum* Karling. (Karling, 1967.)

Fig. 78. Spherical, 2.8-3 μ diam., planospore with large hyaline refractive globule.

Fig. 79. Germination of the planospore to form the rudiment of the central rhizoidal axis.

Fig. 80. Mature zoosporangium with 6 arms of the endosporangium extending out of deliquesced exit papillae.

Fig. 81. Swarming of planospores in the expanded arm of the endosporangium.

Figs. 82-85. *Rhizidium reniformis* Karling. (Karling, 1970.)

Fig. 82. Mature zoosporangium with a broad apical exit papilla and central rhizoidal axis on chitin.

Fig. 83. Ovoid zoosporangium with planospores swarming in a vesicle.

Fig. 84. Ovoid, 3-3.8 × 4-4.3 μ, planospore with a large refractive globule.

Fig. 85. Germination of the planospore to form the rudiment of rhizoidal axis.

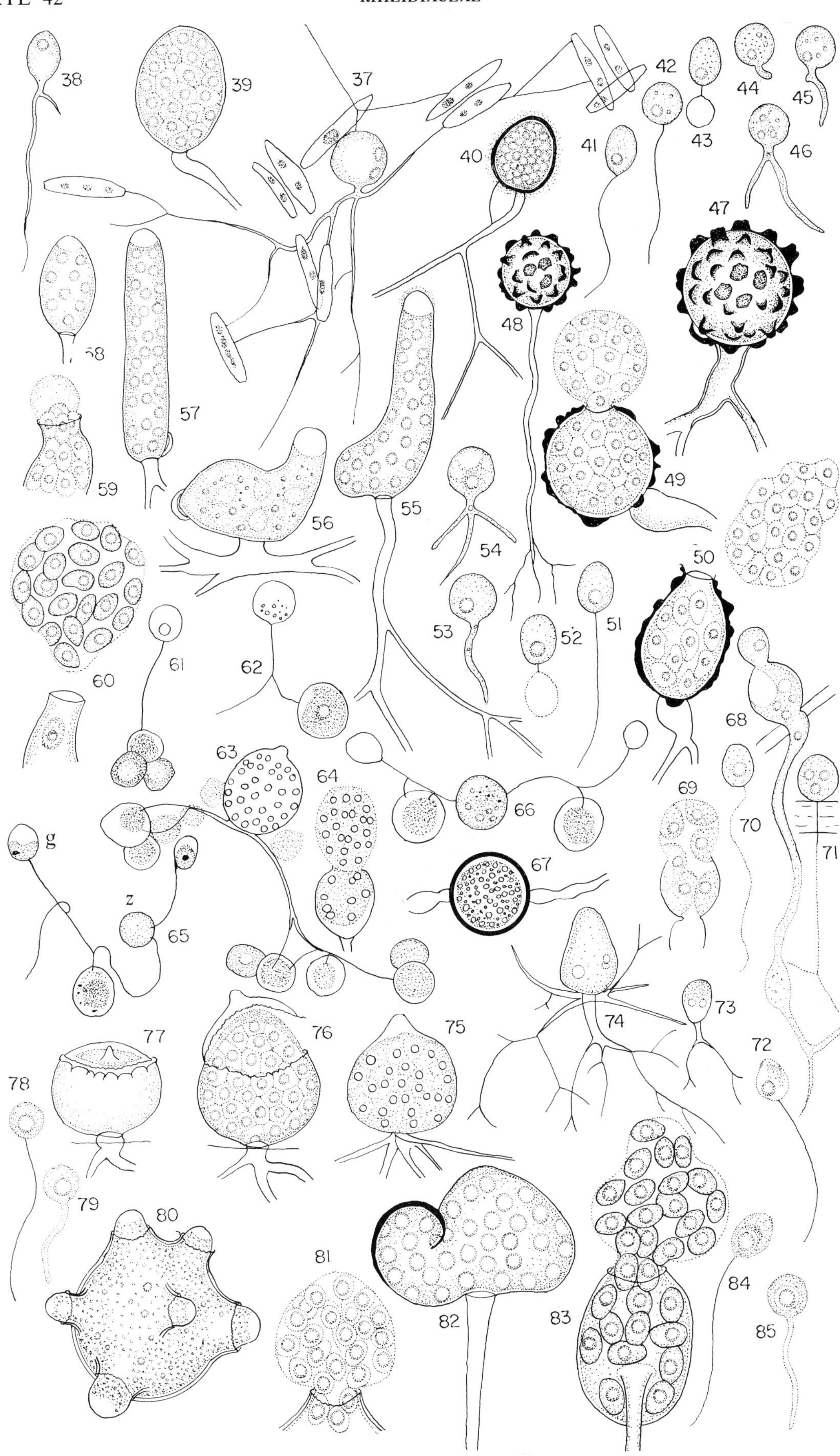
38
39
37
40
41
42
43
44
45
46
47
48
58
57
56
55
54
49
59
60
61
62
53
52
51
50
68
63
64
66
g
65
z
67
69
70
71
77
76
75
74
73
72
78
79
80
81
82
83
84
85

Rhizidium

SPOROPHLYCTIDIUM Sparrow

Trans. Brit. Mycol. Soc. 18:217, 1938

Plate 32, Figs. 11-16

This incompletely-known monotypic, inoperculate genus is characterized by an epibiotic, obpyriform to clavate, stalked zoosporangium with a small lateral exit papilla or pore, aplanospores, and possibly resting spores which are accompanied near the apex by an empty spherical cyst or "male" cell. *Sporophlyctidium africanum* Sparrow parasitizes *Protoderma* sp., and in its development the aplanospore produces a single unbranched, slightly inflated germ tube whose tip penetrates the algal cell but does not develop any rhizoids. Subsequently, the aplanospore body develops into a small zoosporangium (figs. 11–13) whose content eventually cleaves into 4 to 6 spores. These are liberated through a small lateral pore (fig. 14), and after remaining grouped for a while they separate and float away without becoming flagellate and mobile.

Several weeks after the occurrence of sporangial thalli Sparrow observed what he believed might be resting spores. These were similar in shape to the zoosporangia, but had a thicker wall and contained a large oil globule (figs. 15, 16). They were accompanied near the apex by an empty, spherical cyst "which undoubtedly functioned as a male or companion cell." Germination of such spores was not observed.

Sparrow emphasized the similarity of this genus to *Sporophlyctis rostrata* Sorokin by the presence of aplanospores, shape of the zoosporangium and its relationship to the host, and the position of the exit pore. Nevertheless, he (1943, 1960) classified it as the first genus in the Rhizidiaceae, as he interpreted this family, ahead of *Rhizidium*.

RHIZIDIUM Braun

Monatsber. Berlin Acad. 1856:591; Flora 14:599, 1856.

Plates 41-42

This inoperculate genus is reported to include 22 or more species at present, but 4 of these, *R. algaecolum* Zopf (1884), *R. equitans* Zopf (1885, 1890), *R. leptorhizum* Zopf (1884), and *R. spirogyrae* Fisch (1884), are merely mentioned by name or so inadequately described that it is impossible to establish their identity. *Rhizidium lignicola* Lindau (1899) appears to be an endooperculate chytrid, and *R. varians* Karling (1949) may prove to be a species of *Rhizophlyctis* because it may exhibit both the *Rhizidium* and *Entophlyctis* type of development. It has been transferred to *Rhizophlyctis*. Furthermore, the zoo-

PLATE 43

Figs. 1-15, 18-20. *Obelidium mucronatum* Nowakowski. Fig. 1 after Nowakowski, 1876; figs. 2, 3, 6, 8, 10-15, 18-20 after Karling, 1967; figs. 4, 5, 7, 9, after Sparrow, 1938.

Fig. 1. Thallus with apophysate stalked dehisced zoosporangium, planospores and branched rhizoids.

Figs. 2, 3. Spherical planospores, 2.5-3.5 μ diam., with a nuclear cap, and absorption of the flagellum, resp.

Figs. 4, 5. Germination of a planospore and development of the young thallus, resp.

Fig. 6. Cluster of young thalli on chitin agar.

Fig. 7. Bi-mucronate young thallus.

Fig. 8. Stalked apophysate, young zoosporangium with a large primary nucleus; upper part of apophysis wall thickened.

Fig. 9. Stalked thallus with 2 knob-like thick-walled regions at the base of the zoosporangium fundament.

Fig. 10. Full-grown zoosporangium with a large primary nucleus.

Fig. 11. Fixed and stained section of a primary nucleus.

Fig. 12. Division of the primary nucleus with an intranuclear spindle.

Fig. 13. Sessile zoosporangium with two lateral rhizoids near the thick-walled base; exit papilla subapical.

Fig. 14. Discharge of planospores from a free floating zoosporangium.

Fig. 15. Swarming of planospores in a vesicle which is continuous with the interior of the zoosporangium.

Figs. 16, 17. *Obelidium hamatum* Sparrow. (Sparrow, 1937.)

Fig. 16. Empty zoosporangium with intramatrical barbs on a larval case of midges.

Fig. 17. Planospore discharge from a basal papilla; planospores creeping out in amoeboid manner.

Figs. 18, 20. *Obelidium hamatum*-, and *Obelidium megarhizum*-like thalli of *O. mucronatum* on chitin agar with spines or barbs and extramatrical rhizoids on a stout rhizoidal axis.

Fig. 19. Rhizoidal branches of *O. mucronatum* thick-walled like the base of the zoosporangium.

Figs. 21-31. *Obelidium megarhizum* Willoughby. (Willoughby, 1961.)

Figs. 21, 22. Motile and encysted planospores, 3.5-4.5 μ diam., resp., with a nuclear cap.

Figs. 23, 24. Germination of a planospore and a young thallus, resp.

Fig. 25. Incipient zoosporangium with a large vacuole, (quite likely this is the primary nucleus instead of a vacuole.)

Fig. 26. Young stalked thallus with extramatrical rhizoids.

Fig. 27. Young zoosporangium with 2 apical spines, apophysis, and coiled rhizoidal axis.

Fig. 28. Swarming of planospores in what appears to be a vesicle.

Fig. 29. Empty zoosporangium with a sub-apical exit orifice.

Figs. 30, 31. Rhizoidal systems of mature thalli.

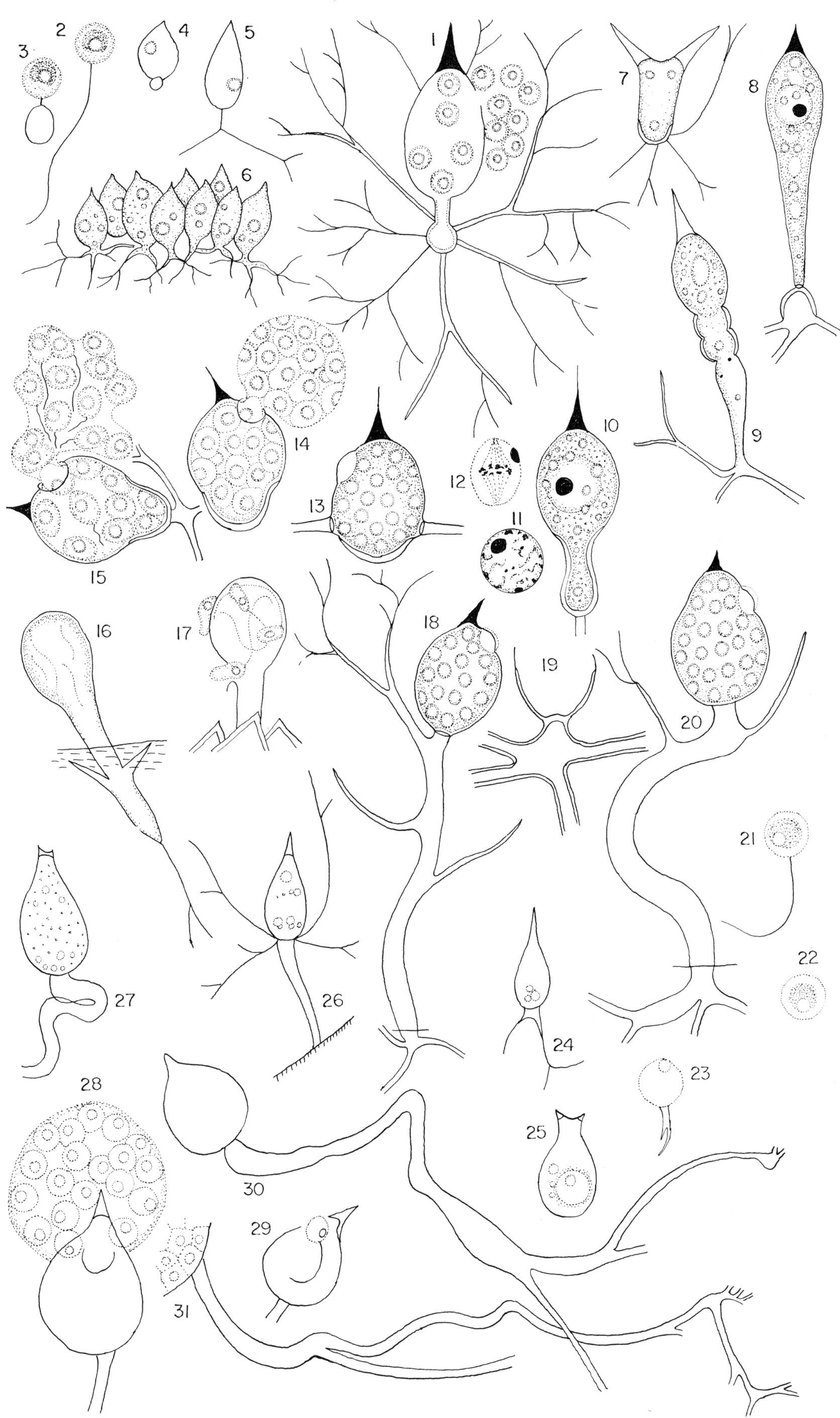
1
2
3
4
5
6
7
8
9
10
11
12
13
14
15
16
17
18
19
20
21
22
23
24
25
26
27
28
29
30
31

Obelidium

sporangium may bear several rhizoidal axes on its periphery, and approximately 4% of the thalli were observed to be polycentric. Also, *R. richmondense* Willoughby will probably prove to be a species of *Chytriomyces*. Two of the so-called valid species, *R. mycophilum* Braun and *R. nowakowskii* Karling, occur in the slime surrounding *Chaetophora elegans*; 2 species are polyphagus, *R. braunii* Zopf and *R. windermerense* Canter, and parasitize diatoms and green colonial algae, respectively, and the remaining ones occur on insect exuviae and purified chitin, and cellulosic substrata.

The concepts and limitations of *Rhizidium* are not sharply defined at present and have undergone several changes since the genus was established by Braun, probably because he failed to illustrate the type species, *R. mycophilum*, and was vague in his description of its characteristics. His description of "two-celledness" of the thallus apparently relates to the zoosporangium or resting spore as one cell and rhizoidal system as the second cell. On the other hand, it might relate to the zoosporangium and an occasional apophysis beneath it. The latter structure may sometimes be delimited from the stout tapering rhizoidal axis as in *R. chitinophilum* Sparrow, giving the thallus a two-celled appearance (fig. 14). Braun's concept of the genus was supported by Nowakowski (1876), but Fischer (1892) included in it the apophysate *Rhizophydium*-like species which are now placed in *Phlyctochytrium*. Dangeard (1889b) accepted this concept of the genus but included in it operculate and inoperculate species. Zopf (1884) apparently regarded it as including all monocentric inoperculate chytrids with a central tapering rhizoidal axis. Schroeter (1885, 1893) interpreted it more or less in the original sense of Braun, as did Sparrow (1943, 1960), who described the genus as including "all monocentric chytrids that develop free of the medium, form their sporangium from the enlarged body of the encysted zoospore, and have a definite taproot-like main rhizoidal axis of variable length from which arise most of the rhizoids." No distinction is made in this definition between operculate and inoperculate species, apparently an oversight. Antikajian (1949) proposed that *Rhizidium* should be enlarged to include *Phlyctochytrium* and *Asterophlyctis*. Although Sparrow's interpretation may serve as a convenient and perhaps workable concept at present, it does not encompass all species fully. So far only 3 known valid species (figs. 1, 37, 63) develop "free in the medium" (host, substratum?), and in others only the zoosporangium (figs. 26, 76, 77) is extramatrical while the rhizoidal axis, branches, and apophysis (figs. 26, 75–77) are intramatrical. Also, in several species the taproot-like rhizoidal axis may be lacking in many thalli, and in such cases the rhizoids may branch several times immediately beneath the zoosporangium (figs. 17–19, 56, 74–77) or arise from several points on the periphery of an apophysis (figs. 34, 35). Accordingly, it is often difficult to identify a chytrid as a species of *Rhizidium*.

In development the germinating encysted planospore produces a germ tube (figs. 4, 22, 38, 43–46, 53) which eventually elongates into a central axis with branches, or branches underneath the sporangial rudiment (figs. 52–54, 73) to become a branched rhizoidal system. Meanwhile, the encysted planospore body enlarges to become the incipient zoosporangium (figs. 12, 13, 23, 24, 62). At maturity the zoosporangium dehisces by the deliquescence of one apical (figs. 6, 17, 49, 55–59), or lateral (fig. 7), or basal (fig. 19), or 2 to 3 exit papillae. In *R. braziliensis* Karling (fig. 27) occasional long exit canals are formed, and in *R. richmondense* Willoughby (fig. 76) the apiculus and upper part of the sporangium wall may be lifted up as an operculum-like structure. In *R. megastomum* Sparrow (1965) the exit pore may be up to 15 μ in diameter, while in *R. endosporangiatum* Karling 1 to several exit papillae are formed from the periphery of the zoosporangium, which deliquesce as arms of an endosporangium protrude through them (fig. 80).

Characteristically, the emerging mass of planospores is usually enveloped by a clear layer of matrix which expands, thins out, and eventually becomes a vesicle in which the planospores swarm collectively

PLATE 44

Figs. 1-15. *Solutoparies pythii* Whiffen. (Whiffen, 1942.)

Fig. 1. Spherical, 4.5-5.6 μ diam., planospores with hyaline refractive globules.

Fig. 2. Germination of a planospore on agar.

Fig. 3. Germling on agar with branched central rhizoidal axis.

Figs. 4, 5. Infection of *Pythium* sp. filaments by tips of parasite's rhizoids; content of host filaments omitted; filament locally hypertrophied in fig. 4.

Fig. 6. Surface view of a zoosporangium.

Figs. 7, 8. Maturation stages of zoosporangium.

Fig. 9. Zoosporangium delimited by a wall from inflated rhizoidal axis; refractive material aggregating.

Fig. 10. Mature zoosporangium with planospore initials.

Figs. 11, 12. Stages in the dissolution of the zoosporangium wall and pegs.

Figs. 13, 14. Dispersal stages of planospores after dissolution of the wall.

Fig. 15. Remains of basal portion of zoosporangium wall.

Fig. 16-22. *Solutoparies* sp. (Dogma, 1969.)

Figs. 16, 17. Young zoosporangia.

Fig. 18. Mature zoosporangium with spine-free basal portion.

Figs. 19, 20. Remains of zoosporangium and spines after planospore discharge.

Figs. 21, 22. Surface view and optical section, resp., of mature resting spores.

Solutoparies

(figs. 8, 10, 28, 50, 60, 83). This vesicle with its swarming planospores is usually continuous with the interior of the zoosporangium (fig. 83), but it may often break loose and float free of the latter.

Resting spores have been reported in 6 species, and in 5 of these the spores are borne like the zoosporangia and develop by the enlargement of the planospore body and its investment by a thick wall. In germination they function as prosporangia (figs. 9, 10, 33). In *R. windermerense* Canter (1950) two spherical bodies or isogamous "sexual thalli" (gametangia) each develop a fine thread or conjugation tube which fuses or anastomoses at their tips. The contents of the gametangia pass through the tubes and form a globular swelling at their juncture (fig. 65) which is the incipient zygote. At the same time the conjugation tubes may form short accessory filaments or rhizoids which infect host cells (figs. 65, 66). Eventually the zygote becomes invested by a thick wall (fig. 67) while the conjugation tubes themselves remain as empty appendages.

The polyphagus nature of *R. windermerense* is similar to that of the genus *Polyphagus*, but the development of the zoosporangium differs markedly. Its type of sexual reproduction as reported by Canter, also, resembles somewhat that of *Siphonaria variabilis* and some species of *Polyphagus* by the subterminal development of the zygote, but in the *Polyphagus* species the conjugating thalli attain considerable size before fusion begins. Nothing is known about the origin of the 2 "sexual thalli" in *R. wildermerense*. Possibly, they are only facultative planospores which may give rise to sporangial thalli and also function as gametangia. Germination of the zygote has not been observed.

OBELIDIUM Nowakowski

Cohn. Beitr. Biol. Pflanzen 2:86, 1876.

Plate 43

Obelidium is an inoperculate genus which is reported to include three species, *O. mucronatum* Nowakowski, *O. hamatum* Sparrow, and *O. megarhizum* Willoughby. These are chitinophilic and occur on insect exuviae in nature, but *O. mucronatum* has been cultured on chitin agar. It is not certain whether *O. hamatum* and *O. megarhizum* are distinct species, or variants of, or identical with *O. mucronatum*. As shown by Sparrow (1938) and Karling (1967) *O. mucronatum* varies markedly in nature, and in cultures on chitin agar Karling observed thalli with spines or barbs and thick rhizoidal axes (figs. 18, 20) which resembled those of *O. hamatum* and *O. megarhizum*. However, in *O. hamatum* the zoospores are reported to be ellipsoidal and creep out as amoebae (fig. 17), and the zoosporangia lack an apical spine.

This genus is characterized by a subspherical, broadly or narrowly pyriform or clavate extramatrical zoosporangium which may be stalked, apophysate, or sessile, and bears usually one and sometimes two apical thick-walled spines in *O. mucronatum* and *O. megarhizum* (figs. 1, 7, 10, 13–15, 26–29). It usually rests in a thick-walled cup-like (figs. 13–15) or stalked base (figs. 1, 10), or is subtended by an apophysis (figs. 1, 8, 27) or stalk from which the intramatrical extensively branched main axes of the rhizoidal system radiate. However, some of the rhizoids may be extramatrical (figs. 18, 26), and in older thalli the axes and branches may become as thick-walled as the base of the zoosporangium (fig. 19). Also, sometimes the thickening of the stalk wall may be local (fig. 9). In development the planospore body enlarges to become the zoosporangial rudiment (figs. 4, 5, 23–25) while the fine germ tube gives rise to the stalk, or apophysis, and rhizoids.

So far resting spores have not been observed in *Obelidium*, although *O. mucronatum* has been maintained on sterilized shrimp chitin and chitin agar for many months. Possibly, this species is heterothallic, and compatible isolations must be brought together before resting spores will develop.

SOLUTOPARIES Whiffen

Mycologia 34:543, 1942

Plate 44

At the present time this inoperculate genus includes the type species, *S. pythii* Whiffen which parasitizes *Pythium* sp. and possibly a second species parasitic on *Karlingia rosea*. It is characterized primarily by an epibiotic thallus and the dissolution of all but the basal portion of the sporangial wall at maturity. The spherical planospore (fig. 1) of *S. pythii*

PLATE 45

Figs. 1-20. *Nowakowskia hormothecae* Borzi. (Borzi, 1885.)

Fig. 1. Several thalli with radial rhizoids attached to germinating planospores of *Hormotheca sicula*.

Figs. 2, 3. Stages in maturation of the zoosporangia.

Figs. 4, 5. Mature zoosporangium with planospores; wall deliquescing.

Figs. 6-8. Swarming colonies of planospores.

Figs. 9-11. Smaller swarming colonies of planospores derived as parts of larger colonies.

Fig. 12. Final separation of planospores.

Fig. 13. Isolated, elongate, 1 μ diam., darting planospores.

Fig. 14. Encysted planospores.

Figs. 15, 16. Germination of isolated planospores.

Fig. 17. Colony of encysted planospores enlarging.

Figs. 18, 19. Germination of colonies.

Fig. 20. Thalli in various stages of development.

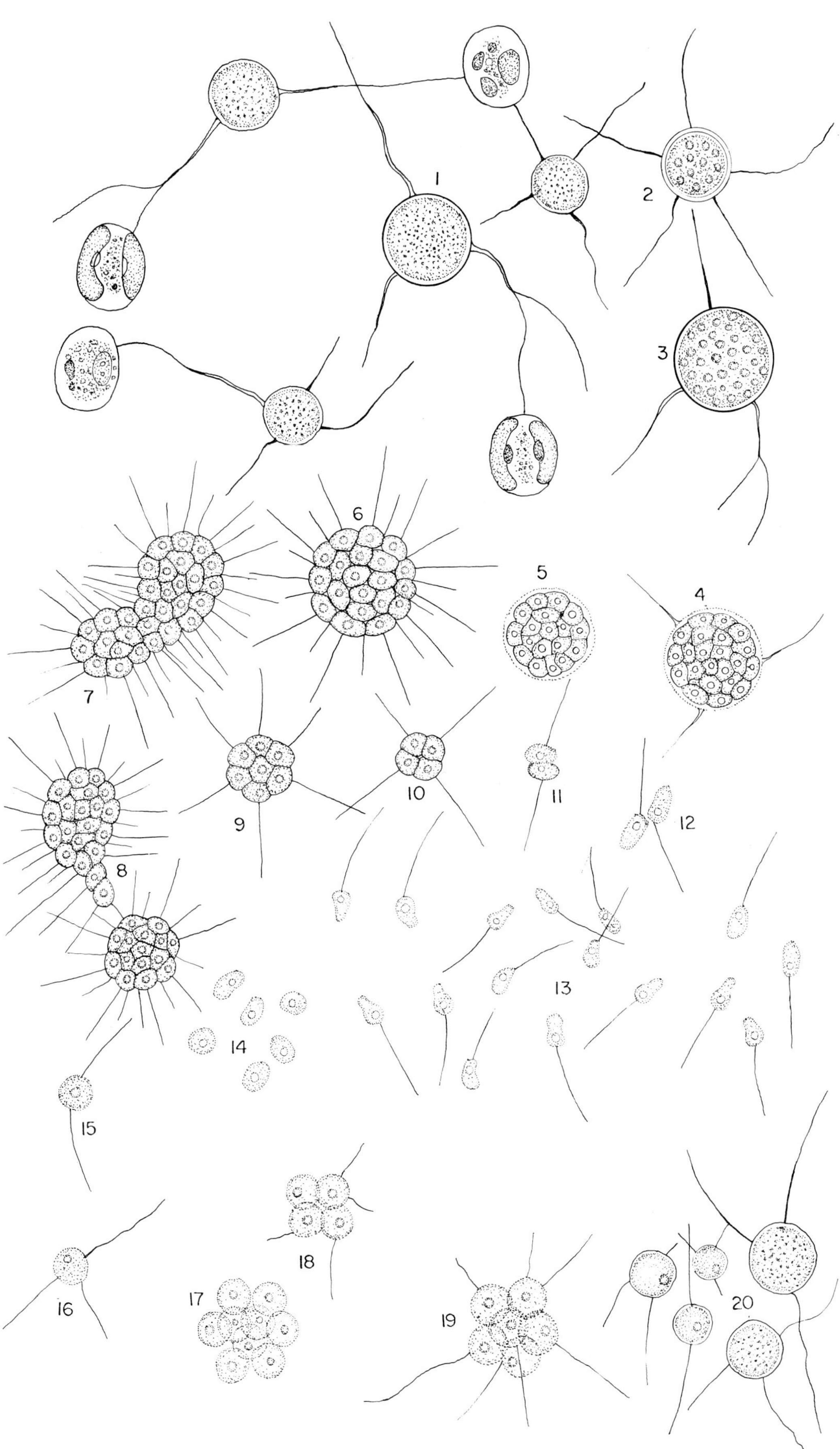

Nowakowskia

will germinate on agar and develop into a germling (figs. 2, 3) with a branched rhizoidal axis, but beyond this stage no further growth occurs unless *Pythium* hyphae are present. The parasite causes abnormal branching (fig. 5), snarling, and some local enlargement (fig. 4) of the hyphae. The rhizoidal tips do not enter the host filaments but merely become attached to or coiled around them so far as is known.

In further development of the sporangial thallus, the planospore body enlarges to become a spiny zoosporangium, and the single, basal rhizoidal axis becomes inflated and apophysis-like underneath the zoosporangium (fig. 6). After the planospore initials have been delimited (fig. 10) all but the basal portion of the sporangial wall thickens, swells, and gradually dissolves (figs. 11, 12), with the spines being the last to undergo dissolution. The gelatinous matrix around the initials then dissolves, and the planospores swim away one by one (figs. 13, 14), finally leaving behind the empty, irregular, bowl-like basal portion of the wall and attached apophysis (fig. 15) as in several species of *Rhizophydium*. Resting spores are unknown in this species.

Solutoparies sp. Dogma is reported to parasitize *Karlingia rosea*, but the relation of parasite to host is not known. It is similar to *S. pythii* in most respects but only rarely does a slight swelling of the rhizoidal axis underneath the zoosporangium occur. Its resting spores are spiny (figs. 21, 22) like the zoosporangia, but their method of germination is unknown.

Solutoparies is similar in some respects to *Nowakowskia hormothecae* Borzi (1885) whose rhizoidal tips penetrate several germinating zoospores of the alga *Hormotheca sicula* and in which dissolution of the sporangial wall occurs as in *Solutoparies*.

NOWAKOWSKIA Borzi

Bot. Centrabl. 22:23, 1885.

Plate 45

This monotypic genus parasitizes germinating planospores of *Hormotheca sicula* and causes an epidemic in cultures of this alga. It is characterized by an epibiotic globular zoosporangium from whose surface radiate several fine, branched or unbranched rhizoids (fig. 1), and in the latter respect it resembles somewhat some species of *Rhizophlyctis* and *Karlingia*. In germination the encysted planospore (fig. 14) enlarges and develops two to several fine rhizoids on its surface (figs. 15, 16). The tips of the rhizoids penetrate the algal cells (fig. 1) and as they derive nourishment from the host the planospore body enlarges to become an incipient zoosporangium (fig. 2) whose wall gives a *positive cellulose* reaction when treated with chloroiodide of zinc. After the planospores are delimited, they contract slightly *en masse* (figs. 4, 5), and within a short while thereafter the sporangial wall deliquesces or dissolves, freeing the globular mass of planospores. These develop flagella, and the whole mass swims away (fig. 6) somewhat like *Volvox*. It may undergo changes in shape, and after a while smaller groups of planospores break away from the larger masses (fig. 8) and swim away. By this process the masses become smaller and smaller (figs. 9–12) until individual minute planospores are isolated (fig. 13) which dart about with great rapidity. Occasionally, colonies of planospores remain

PLATE 46

Figs. 1-12. *Rhizoclosmatium globosum* Petersen. Figs. 1, 2 after Petersen, 1910; figs. 3-12 after Sparrow, 1937.

Fig. 1. Portion of a sporangial thallus with an apophysate zoosporangium.

Fig. 2. Planospore with a lateral attachment of the flagellum.

Fig. 3. Planospore with posterior attachment of flagellum.

Figs. 4-8. Developmental stages of thallus and incipient zoosporangium; apophysis fusiform and clavate.

Fig. 9. Discharge of planospores through a basal pore.

Fig. 10. Discharge of planospores through an apical pore.

Fig. 11. Early stage of conjugation of two thalli by rhizoidal anastomosis.

Fig. 12. Later stage; receptive thallus has been transformed into a resting spore.

Figs. 13-15. *Rhizoclosmatium aurantiacum* Sparrow. (Sparrow, 1937.)

Fig. 13. Ellipsoidal, 2 × 2.5 μ, planospore with a rusty-orange globule.

Fig. 14. Mature zoosporangium with a basal discharge papilla.

Fig. 15. Empty zoosporangium with a basal pore.

Figs. 16-18. *Rhizoclosmatium marinum* Kobayasi and Ookubo. (Kobayasi and Ookubo, 1954.)

Figs. 16, 17. Ellipsoidal, 5 × 7 μ, planospores.

Fig. 18. Mature sporangial thallus.

Figs. 19-31. *Rhizoclosmatium hyalinum* Karling. (Karling, 1967.)

Fig. 19. Spherical, 4.6-5 μ diam., planospore with a large hyaline globule.

Fig. 20. Young sporangial thallus.

Fig. 21. Incipient zoosporangium with the primary nucleus; apophysis elongate.

Fig. 22. Fully-grown zoosporangium with the large primary nucleus.

Fig. 23. Irregular zoosporangium with a basal exit papilla; apophysis and main rhizoidal axes thick-walled.

Fig. 24. Ovoid zoosporangium with a lateral papilla.

Fig. 25. Portion of a zoosporangium with a short neck or canal.

Fig. 26. Discharge of planospores which swarm in a vesicle.

Fig. 27. Young resting-spore thallus; content becoming coarsely granular in appearance.

Figs. 28-31. Variations in shapes of hyaline thick-walled resting spores; apophysis and main rhizoidal axes thick-walled.

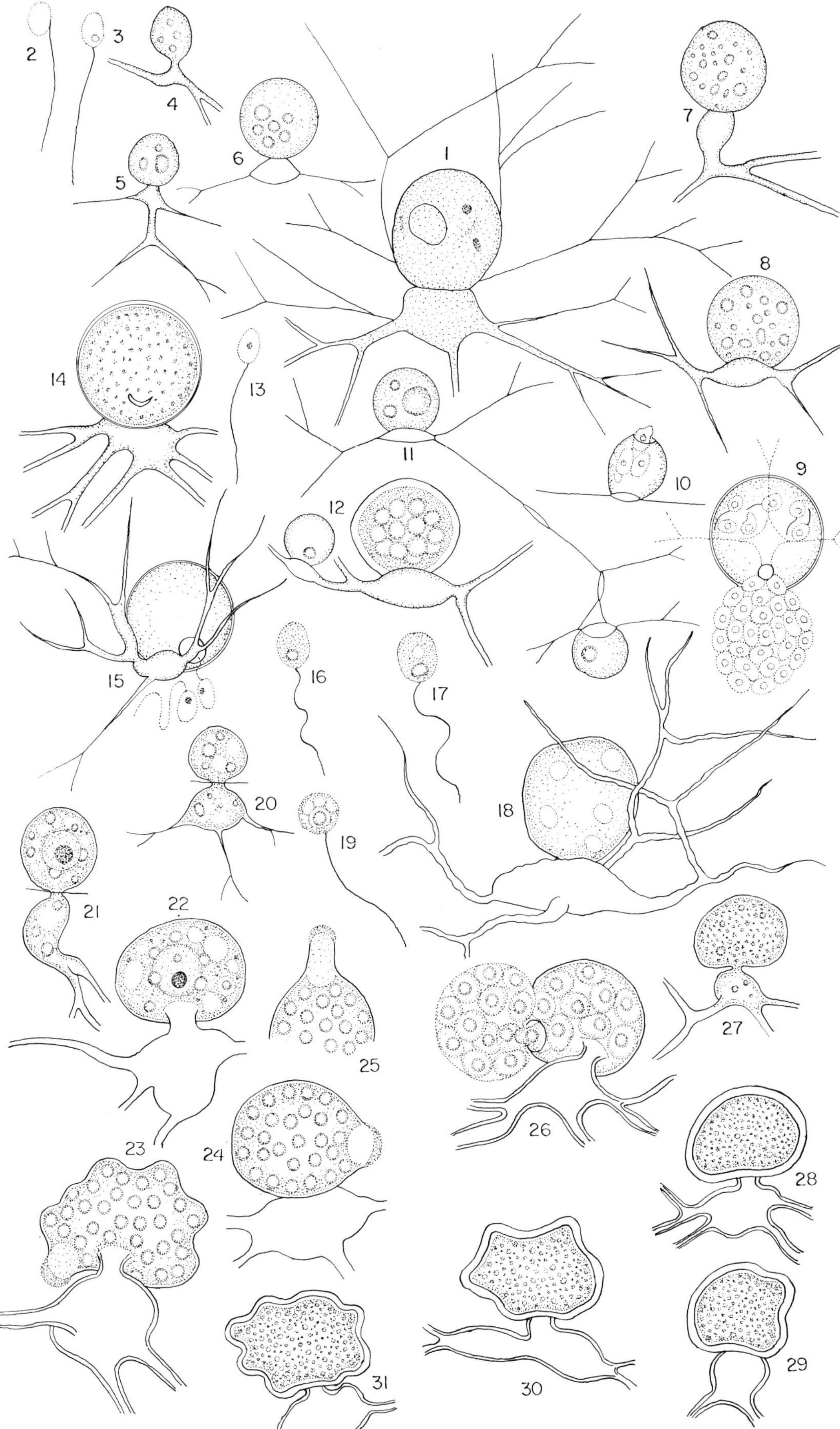

Rhizoclosmatium

together, encyst as colonies (fig. 17) and germinate as such (figs. 18, 19). Resting spores have not been observed.

Although its thallus resembles somewhat those of some species of *Rhizophlyctis* and *Karlingia*, dehiscence of the zoosporangium and the behavior of the globular mass of released planospores of *Nowakowskia* are quite unlike those of other genera. Fischer (1892) regarded this genus as doubtful and the behavior of the planospores as abnormal, but inasmuch as Borzi had an abundance of material at hand it is not likely that he described only the abnormal behavior of the planospores. Classification of this genus is uncertain because its life cycle is not fully known. Schroeter (1893) and Fitzpatrick (1930) placed it next to *Rhizophlyctis*, and this classification was followed by Sparrow (1943, 1960).

RHISOCLOSMATIUM Petersen

Journ. de Botan. 17:216, 1903.

Plate 46

At present this inoperculate genus is reported to include 4 species, *R. globosum* Petersen, *R. aurantiacum* Sparrow, *R. marinum* Kobayasi and Ookubo, and *R. hyalinum* Karling, but it has been suggested that *R. marinum* may be a species of *Diplophlyctis* because it is a parasite of the alga *Codium fragile*. The other 3 species are saprophytic, chitinophilic, and occur in insect exuviae in nature. The *Rhizoclosmatium* thallus consists of an extramatrical or epibiotic zoosporangium or resting spore and an intramatrical variously-shaped apophysis from whose surface arise delicate or coarse, branched rhizoids (fig. 1, 9, 10, 14, 15, 18, 22–24, 6–31).

In the development of the sporangial thallus the planospore body enlarges into a zoosporangium (figs. 4-7, 20, 21) while the germ tube gives rise to the apophysis and rhizoids. The tube branches shortly after germination, and soon thereafter an enlargement occurs in it immediately below the sporangial rudiment. This swelling enlarges to become the apophysis, and as a result the rhizoidal branches or axes are moved apart and become oriented on its periphery (figs. 1, 8, 14, 15, 18, 20- 24, 26). The apophysis is frequently fusiform and transversely elongate (figs. 6, 8, 12, 15, 18), but it may be elongately clavate (figs. 7, 21), or irregular in shape (figs. 14, 22, 26). The rhizoids are relatively delicate in *R. globosum* and *R. aurantiacum* but in *R. marinum* and *R. hyalinum* they are coarse. In *R. hyalinum* they are stiff-looking, sparsely branched, up to 8 μ in diameter, and may extend for distances up to 250 μ. In this species the primary nucleus is conspicuous (fig. 21) and does not divide until the zoosporangium is fully grown (fig. 22). At maturity the zoosporangium develops a basal (figs. 9, 14, 15, 23), or lateral (fig. 24), or apical (fig. 10) exit papilla. In *R. hyalinum* the papilla may be extended to form a short neck (fig. 25). The planospores are discharged through such papillae or necks and soon swarm in a vesicle outside of the zoosporangium (figs. 9, 26).

PLATE 47

Figs. 1-8. *Siphonaria variabilis* Petersen. Figs. 1, 6 after Petersen, 1910; figs. 2, 4, 5, 7, 8 after Sparrow, 1937; fig. 3 drawn from living material.

Fig. 1. Sporangial thallus with tubular thick-walled rhizoids.

Fig. 2. Pyriform, 2.5 $\times$ 5 μ, planospores with a lateral hyaline globule and a rust-colored body.

Fig. 3. Germination of planospore on shrimp chitin.

Fig. 4. Young thallus showing attachment of the rhizoidal axis.

Fig. 5. Discharge of planospores through a basal pore.

Fig. 6. Mature resting spore with empty male thallus or gametangium at right.

Fig. 7. Conjugation of gametangia through a rhizoid; content of small male gametangium passing into the larger female.

Fig. 8. Mature zygote or resting spore with two empty male gametangia.

Figs. 9-18. *Siphonaria petersenii* Karling. (Karling, 1945.)

Fig. 9. Spherical, 3-3.5 μ diam., planospores with a small golden-red globule.

Figs. 10-12. Germination of a planospore and early developmental stages of the sporangial thallus.

Fig. 13. Beginning of spine formation on the surface of a young zoosporangium.

Fig. 14. Mature sporangial thallus; spiny zoosporangium discharging planospores through a subapical pore; planospores later swarming in a vesicle.

Fig. 15. Conjugation of male and female gametangia through a rhizoid.

Fig. 16. Similar stage, gametangia close together.

Fig. 17. Mature zygote or resting spore; male gametangium attached to thick-walled rhizoidal axis.

Fig. 18. Mature zygote with an adnate male gametangium at the apex.

Figs. 19-28. *Siphonaria sparrowii* Karling. (Karling, 1945.)

Figs. 19, 20. Spherical, 5.5-6 μ diam., pyriform planospores with a large, 3-3.5 μ, hyaline refractive globule.

Figs. 21, 22. Encysted planospore and early germination stage, resp.

Fig. 23. Mature zoosporangium with an apical exit papilla; rhizoidal axis thick-walled.

Fig. 24. Empty male gametangium attached directly to young zygote by a conjugation papilla.

Fig. 25. Later stage; empty male gametangium attached to the rhizoid of the female gametangium; zygote enlarging.

Fig. 26. Portion of zygotic or resting spore thallus; mature zygote with a thick dark-brown wall; two empty male gametangia attached to the rhizoidal axis of zygote.

Fig. 27. Five mature zygotes enveloped by a common dark-brown wall; empty male gametangia adnate and partly enveloped by the common wall, or connected with the rhizoidal axis.

Fig. 28. Germination of a zygote.

Siphonaria

Resting spores are known only in *R. globosum* and *R. hyalinum*, and these are borne in the same manner as the zoosporangia (figs. 27-31). In *R. globosum* Sparrow (1937) reported that they are formed sexually by anastomosis of a rhizoid of a larger thallus with one from a smaller thallus (fig. 11). The content of the latter flows into the larger receptive thallus which subsequently develops into a resting spore (fig. 12), while the smaller one remains attached as an empty vesicle. No evidence of sexuality was observed in *R. hyalinum*, and its resting spores appear to develop asexually. Germination of the resting spores has not been observed in *Rhizoclosmatium*.

Petersen created the subfamily *Rhizoclosmatieae* for this genus and *Asterophlyctis*, and Sparrow (1943) placed *Rhizoclosmatium* in his subfamily *Obelidioideae* which he abandoned later. Whiffen (1944) placed it in the subfamily Rhizidioideae, and Sparrow (1960) followed this classification.

SIPHONARIA Petersen

Jour. de Bot. 17:220, 1903.

Plate 47

This inoperculate genus includes 3 species, *S. variabilis* Petersen, *S. petersenii* Karling and *S. sparrowii* Karling, which are chitinophilic in nature and occur in the exuviae of insects. So far only *S. petersenii* and *S. variabilis* have been cultured on chitin and nutrient agar, but the other species have been maintained on purified and sterilized shrimp chitin. *Siphonaria* is characterized by an extramatrical, apophysate or non-apophysate zoosporangium, an intramatrical coarse or fine, branched rhizoidal system, and resting spores or zygotes which are formed by the fusion of the contents of male and female thalli or gametangia. In *S. petersenii* the rhizoids are richly branched and the branches run out to fine filaments (fig. 14), while in the other two species they are less richly branched, coarser, more abruptly tapering and tube-like and may often form branches at right or obtuse angles (figs. 6, 25, 26). In all species, however, the main axes and primary branches usually become fairly thick-walled (figs. 1, 17, 23).

The developmental stages of the sporangial thalli are basically the same in all species. The planospore body enlarges to become the incipient extramatrical zoosporangium (figs. 4, 11-13) while the germ tube branches to form the intramatrical rhizoidal system. The apophysis when present is formed as a local swelling of the germ tube. The mature zoosporangium dehisces by the deliquescence of an exit papilla, and in *S. variabilis* the papilla is basal (fig. 5). In *S. petersenii* (fig. 14) and *S. sparrowii* (fig. 23) it is subapical, respectively. In all species, the present writer has observed that the emerged planospores swarm in a vesicle before dispersing.

PLATE 48

Figs. 1-30. *Asterophlyctis sarcoptoides* Petersen. Figs. 2-8, 13-15, 19-22, 24-27 after Antikajian, 1949; fig. 23 after Sparrow, 1937; figs. 16, 30-36 after Karling, 1967; figs. 1, 9-11, 18, 29 drawn from living and fixed and stained material.

Fig. 1. Portion of a thallus drawn from living material with branched, extensive rhizoids arising from an endobiotic apophysis, and an epibiotic zoosporangium with an exit canal.

Fig. 2. Ovoid, 2-3 $\times$ 5 μ, planospore with hyaline refractive globule.

Fig. 3. Germinating planospore.

Fig. 4. Later stage; planospore enlarging to become incipient zoosporangium; germ tube branching.

Fig. 5. Fully-grown zoosporangium in living material with large primary nucleus.

Fig. 6. Fixed and stained primary nucleus.

Fig. 7. Metaphase of primary nuclear division with an intranuclear spindle.

Fig. 8. Intranuclear spindle of a secondary mitosis.

Fig. 9. Section through a multinucleate zoosporangium.

Fig. 10. Mature zoosporangium with a basal exit papilla.

Fig. 11. Discharged planospores swarming in a vesicle.

Figs. 12-15. Stages in the development of incipient zoosporangia from an enlargement of the germ tube; nucleus of planospore has migrated into enlargement.

Fig. 16. Small zoosporangium developed from an enlargement of the germ tube; planospore cyst and germ tube persistent.

Fig. 17. Section through a fully-grown resting spore with large primary nucleus.

Fig. 18. Mature resting spore developed directly from encysted planospore.

Figs. 19-22. Successive stages of the development of a resting spore from a local enlargement of the germ tube.

Fig. 23. Sexually formed (?) resting spore.

Fig. 24. Early stage of resting spore germination; primary nucleus in prophase stage and migrating into the incipient zoosporangium.

Figs. 25, 26. Bi- and multinucleate incipient zoosporangia from germinated resting spore.

Figs. 27, 28. Profile and polar views of division of secondary nuclei.

Fig. 29. Discharge of planospores and swarming of them in a vesicle.

Figs. 30-38. *Asterophlyctis irregularis* Karling. (Karling, 1967.)

Fig. 30. Mature thallus.

Figs. 31-33. Variations of the thick-walled simple and bifurcate pegs on the zoosporangium.

Fig. 34. Discharge of planospores through a basal papilla and swarming of them in a vesicle.

Fig. 35. Spherical, 4.8 μ diam., planospore with a hyaline refractive globule.

Fig. 36. Germination of planospore.

Fig. 37. Young thallus, planospore body enlarging to become a zoosporangium.

Fig. 38. Incipient zoosporangium with primary nucleus developing from an enlargement of the germ tube.

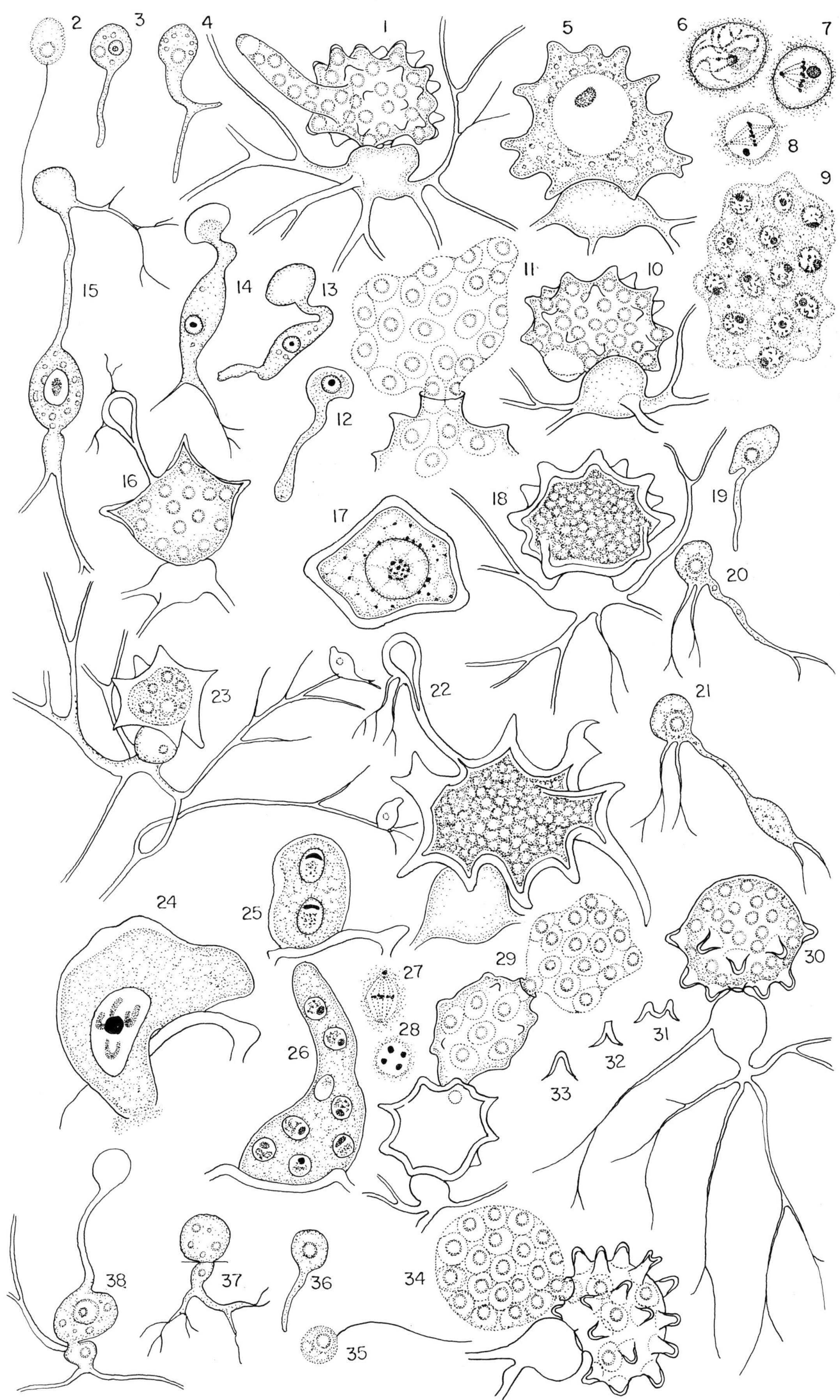

Asterophlyctis

In sexual reproduction the male and female thalli or gametangia develop as small isomorphic thalli, but in *S. petersenii* the male gametangium usually bears a small apical projection (figs. 15, 16). Contact between the two occurs by fusion of a rhizoid of each thallus (fig. 15), but frequently a fusion tube or rhizoid connects directly with the body of the female thallus. If the two gametangia are close together the conjugation tube may be very short (fig. 16). Also, the body of the male may connect directly with that of the female by a short tube or fine papilla (fig. 24). In such cases the empty male thallus becomes adnate (fig. 18) to the zygote as the latter enlarges and matures. The content of the male flows into the female (fig. 7) which sub-

PLATE 49

Figs. 1-10. *Zygorhizidium willei* Lowenthal. After Lowenthal, 1905; fig. 11 after Scherffel, 1926.

Figs. 1, 2. Side and top views of planospores, respectively, with an anterior globule.

Fig. 3. Infection of *Cylindrocystis brebissonii* by a planospore with an intact flagellum.

Figs. 4, 5. Uni- and multinucleate apophysate zoosporangia, respectively.

Fig. 6. Apophysate zoosporangium with refractive globules.

Fig. 7. Empty zoosporangium with the operculum close by.

Fig. 8. Male gametangium with a long conjugation tube.

Fig. 9. Conjugation.

Fig. 10. Binucleate zygote.

Fig. 11. Mature resting spore on *Mougeotia* sp. with attached conjugation tube and male gametangium.

Figs. 12-19. *Zygorhizidium verrucosum* Geitler. (Geitler, 1943.)

Fig. 12. Dorsiventrally flattened planospores, 1.5-2.2 × 3-3.5 μ, with the flagellum inserted *subapically* and projecting backward.

Figs. 13, 14. Germination of planospores on *Mesotaenium caldariorum*.

Figs. 15, 16. Young and mature zoosporangia.

Fig. 17. Conjugation between a smaller male and a larger female gametangium.

Fig. 18. Young resting spore with endobiotic basal haustorium and attached male gametangium and conjugation tube.

Fig. 19. Mature verrucose resting spore with the attached male gametangium which bears two non-functional conjugation tubes.

Figs. 20-28. *Zygorhizidium melosirae* Canter. Figs. 20-25, 28 after Canter, 1967; figs. 26, 27 after Paterson, 1958.

Fig. 20. Spherical, 3-3.5 μ diam., planospore.

Figs. 21, 22. Germinated planospores on *Melosira italica* with a short and long germ tube, resp., with inflated tips.

Fig. 23. Mature zoosporangium with an unbranched endobiotic rhizoid and a basal adnate swelling from the tip of the germ tube.

Fig. 24. Empty zoosporangium with a branched rhizoid.

Fig. 25. Endobiotic branched rhizoid.

Fig. 26. Discharged planospores from a zoosporangium on *Synedra* sp.

Fig. 27. Mature resting spore on *Asterionella* sp. with attached male gametangium and conjugation tube.

Fig. 28. Similar resting spore on *Melosira italica*.

Figs. 29-34. *Zygorhizidium planktonicum* Canter. Figs. 29-31, 33, 34 after Canter, 1967; fig. 32 after Paterson, 1956.

Fig. 29. Spherical, 2.5-3 μ diam., planospores.

Figs. 30, 31. Young and mature zoosporangium with branched rhizoids on *Synedra acus*.

Fig. 32. Mature zoosporangium on *Synedra* sp. with coarser rhizoids.

Fig. 33. Male gametangium which has developed into a zoosporangium; conjugation canal present.

Fig. 34. Mature resting spore with attached male gametangium and conjugation canal.

Figs. 35-38. *Zygorhizidium parvum* Canter. (Canter, 1950.)

Fig. 35. Spherical, 2 μ diam., planospore.

Fig. 36. Mature zoosporangium on *Sphaerocystis schroeteri*.

Fig. 37. Conjugation between germ tubes of male and female gametangia.

Fig. 38. Mature resting spore developed directly in the female gametangium with the male gametangium attached to the germ tube of the female.

Figs. 39-42. *Zygorhizidium parallelosede* Canter. (Canter, 1954.)

Fig. 39. Spherical, 2 μ diam., planospore.

Fig. 40. Immature zoosporangium with an endobiotic haustorial tube, lying parallel to the long axis of *Ankistrodesmus* sp.

Fig. 41. Conjugation between almost equal-sized male and female gametangia.

Fig. 42. Mature, detached, smooth, thick-walled resting spores with a large globule.

Figs. 43, 44. *Zygorhizidium cystogenum* Canter. (Canter, 1963b.)

Fig. 43. Immature zoosporangium on the envelope of *Dinobryon divergens* with a long unbranched thread or rhizoid penetrating the host cyst through the pore.

Fig. 44. Mature resting spore enveloped by a striated external secretion; conjugation tube from male gametangium has penetrated pore.

Figs. 45-48. *Zygorhizidium chlorophycidis* Canter. (Canter, 1963a.)

Fig. 45. Mature, almost hemispherical zoosporangium on *Gloeocystis planctonica* with a sparsely branched rhizoid.

Fig. 46. Spherical, 2.5 μ diam., planospore.

Fig. 47. Conjugation between male and female gametangia.

Fig. 48. Thick-walled resting spore with attached male gametangium and conjugation tube.

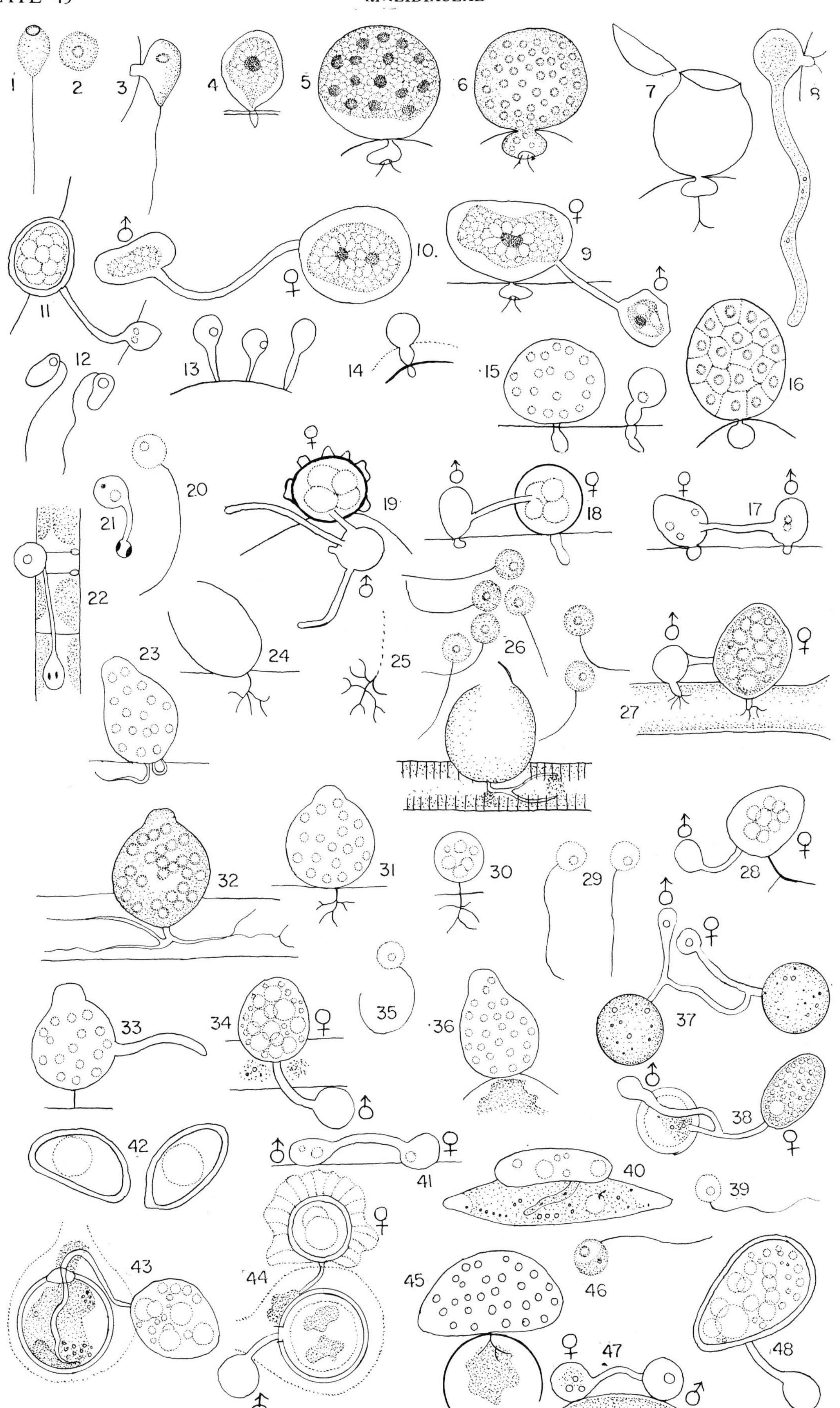

Zygorhizidium

sequently enlarges and becomes enveloped by a thick wall. Two empty male thalli may often be attached to the rhizoids of the female (figs. 8, 26), and in *S. sparrowii* as many as 6 empty male thalli have been found adnate to a mature zygote. In this species several adjacent zygotes may be enveloped by a common thick wall (fig. 27), and in germination they function as prosporangia (fig. 28). Dogma (1974d) emended Petersen's *S. variabilis* on the basis of his study of its sexual reproduction. He found fusions between gametangial rhizoids, but neither of the gametangia developed into the zygote. Instead, an outgrowth occurred at the point of fusion which subsequently developed into the zygospore or resting spore, and he compared this type of sexual reproduction to the scalariform-type found in the Zygomycetes and Conjugales. Essentially, this is the same type of reproduction reported to occur in *Canteria apophysata* (Canter) Karling (pl. 100), *Chytriomyces hyalinus* Karling (plate 54, fig. 112A), *Amphicypellus elegans* Ingold (pl. 56), and *Rhizidium windermerense* Canter (pl. 42).*

Karyogamy and meiosis have not been observed, and it is not known where sex segregation occurs. Karling postulated two types of life cycles for *Siphonaria*, one that genotypic segregation might occur at meiosis in the germinating zygote, and another that sex is phenotypically segregated at the close of the sporangial and planosporic period.

Petersen (1910) placed *Siphonaria* in his subfamily Siphonariae of the Rhizidiaceae together with *Obelidium*, and Sparrow (1943) included it in his subfamily Obelidioideae of the Rhizidiaceae. Later, he (1960) abandoned this subfamily, followed Whiffen's (1944) classification, and placed *Siphonaria* next to *Rhizidium* in the subfamily Rhizidioideae.

ASTEROPHLYCTIS Petersen

Jour. de Botanique 17:218, 1903.

Plate 48

This inoperculate genus is reported to include 2 species, *A. sarcoptoides* Petersen and *A. irregularis* Karling, which occur as saprophytes in insect exuviae in nature but may be cultured readily on chitin and synthetic media. However, Dogma (1974e) transferred the former species to *Diplophlyctis* as *D. sarcoptoides* and *D. irregularis* to *Septosperma* as *S. irregularis*. In the event his interpretations are confirmed *Asterophlyctis* becomes a synonym of *Diplophlyctis*. It is characterized by almost stellate and polyhedral epibiotic zoosporangia and resting spores with protuberances, conical pegs, warts, or blunt spines on the outer wall, and an endobiotic apophysis from which arise one to several, usually 2, branched rhizoidal axes (figs. 1, 18). Usually, the planospore body enlarges to become the incipient zoosporangium (figs. 4, 37) after germinating, while the germ tube usually branches dichotomously to form the rhizoidal axes and an apophysis at the juncture of rhizoids. However, in a careful and painstaking developmental and cytological study of monozoospore isolates of *A. sarcoptoides* from Brazil, Liberia, and Connecticut, U.S.A., Antikajian (1949) confirmed Karling's observations that the zoosporangia and resting spores may, also, develop as local enlargements of the germ tube. In a study of 1000 thalli in chitin and agar cultures of the Connecticut isolation she found that 30% developed from the germ tube, while in the Brazilian and Liberian isolations only 9 and 11%, respectively, develop in the same manner. Karling (1967) found that a few zoosporangia develop from the germ tube in *A. irregularis* (fig. 38), also. Thus, while *Asterophlyctis* develops predominantly as do other members of the Rhizidiaceae it also exhibits some of the developmental characteristics of the Entophlyctaceae. Dogma (1974e) found in his pure culture isolates of *A. sarcoptoides* on agar media that the zoosporangia develop from a swelling in the germ tube as in *Entophlyctis*, and on these grounds as well as on the type of sexual

*See Dogma (1976), Philipp. I. Biol. 5:121–142.

PLATE 50

Figs. 49-63. *Zygorhizidium venustum* (*Rhizophydium venustum*) Canter parasitic on the cysts of *Uroglena americana*. (Canter, 1963, 1970.)

Fig. 49. Encysted planospore on the host.

Figs. 50, 51. Mature zoosporangia; rhizoid has entered through the pore of the cyst.

Figs. 52-55. Empty zoosporangia.

Fig. 56. Pair of encysted gametes on the cyst.

Fig. 57. Fusion; content of 1 gamete has flowed into the other.

Fig. 58. Immature resting spore (zygote) with an attached male (?) cell.

Figs. 59-62. Mature resting spores (zygotes) with attached male (?) cells.

Fig. 63. Mature resting spores (zygotes) with 2 attached male (?) cells.

Figs. 64-79. *Zygorhizidium affluens* Canter parasitic on *Astrionella formosa*. (Canter, 1969.)

Fig. 64. Spherical, 3.5 μ diam., planospores with a globule and nuclear cap.

Figs. 65, 66. Encysted planospores on the host, and infection, resp.

Fig. 67. Young sporangial thallus; rhizoid largely epibiotic.

Fig. 68. Immature zoosporangium with a branched rhizoid.

Figs, 69, 70. Mature zoosporangia.

Figs. 71-74. Empty zoosporangia with opercula.

Fig. 75. Minute operculate zoosporangium.

Fig. 76. Young resting spore (zygote?) with an empty attached male (?) cell.

Fig. 77. Later developmental stage of resting spore (zygote?.)

Figs. 78, 79. Mature resting spores (zygotes?) with 2 and 6 attached male (?) cells, resp.

Zygorhizidium

reproduction exhibited he transferred Petersen's species to *Diplophlyctis*. The types of development are associated with the distribution of the nucleus of the zoospore. If it remains in the zoospore body the latter enlarges to become a zoosporangium, but if it migrates into the germ tube a local enlargement forms at the site of the nucleus and becomes a zoosporangium or resting spore (figs. 12-15).

As in species of *Obelidium* and *Chytriomyces* the primary nucleus of the fully-grown zoosporangium or resting spore (figs. 5, 17) is quite large and readily visible in living specimens. It remains undivided until shortly before sporogenesis, and in preparation for this process it and its derivatives divide (figs. 6–8) with an intranuclear spindle. In germinating resting spores it moves out during the prophases of mitosis (fig. 24) into the incipient zoosporangium being formed on the surface of the spore.

The resting spores are reported by some investigators to be borne in the same manner as the zoosporangia, but Sparrow (1937, 1943, 1960, Dogma, 1974e) reported that they are formed sexually as the result of fusion of male and female thalli through a rhizoid as in *Siphonaria* and *Rhizoclosmatium*. However, in a critical study of this developmental phase Antikajian found planospore cysts and germ tubes attached to the resting spores which developed as enlargements of the germ tube (figs. 19–22), but in no instances were resting spores observed with cysts attached to the rhizoids as figured by Sparrow. Also, in a cytological study of developing resting spores she never found any but uninucleate ones, and from this direct developmental and indirect cytological evidence she concluded that in her material, at least, the resting spores develop asexually. The tendency to interpret all resting spores with attached cysts and tubes as being sexually developed is not warranted unless the developmental and fusion stages are actually observed, and in many species the presence of such appendages on the resting spores apparently has no sexual significance. In dense infections of chitin by *Asterophlyctis* it is not uncommon to find many zoospores and small thalli lying over the rhizoids of developing resting spores, and these may give the appearance of "male" thalli which are contributing to the development of the resting spore without doing so. On the other hand, Dogma (1974e) found in his pure isolates that the resting spores are sexually formed as a result of rhizoidal fusions in the same manner as he described for *Siphonaria variabilis* H. E. Petersen.

Petersen (1910) classified *Asterophlyctis* in the subfamily Rhizoclosmatieae of the Rhizidiaceae, and Sparrow (1943) first placed it in the subfamily Obelidioideae. Later (1960) he abandoned this subfamily and agreed with Whiffen's (1944) classification of *Asterophlyctis* in the subfamily Rhizidioideae. Previously, Karling (1945) had suggested that if the observations on sexuality in *Rhizoclosmatium* and *Asterophlyctis* are confirmed these two genera should be merged with *Siphonaria* on the basis of similarity of types of sexual reproduction. Antikajian concurred with Whiffen's classification, but she did not believe that *Asterophlyctis* can be retained as a valid genus and should be merged with *Phlyctochytrium* and *Rhizidium* in one genus, *Rhizidium*.

Plate 51

Figs. 80-97. *Zygorhizidium* (?) *vaucheriae* Rieth on *Vaucheria*. (Rieth, 1967.)

Fig. 80. Spherical, 4.5-5.0 μ diam., planospores with a colorless refractive globule and an *anteriorly* attached and directed, 10-13 μ long, flagellum.

Fig. 81. Germinating planospore.

Fig. 82. Infection of *Vaucheria*; host filament beginning to branch at the left because of the infection.

Fig. 83. "Witches-broom" branching of host filament; primary zoosporangium with 3 exit papillae.

Fig. 84. Secondary sporangial thalli with endobiotic apophyses which do not cause "witches broom" effects in the host.

Fig. 85. Young male gametangium with a long conjugation tube growing towards young female gametangium; no branching of the host filament.

Fig. 86. Male and female gametangia; no branching of the host.

Fig. 87. Stage in fusion of the contents of the gametangia; protoplasm accumulating in one portion of the female gametangium.

Fig. 88. Later developmental stage; zygote or resting spore delimited in upper part of female gametangium.

Fig. 89. Still later stage; brown zygote developing sharp pegs on its upper surface.

Fig. 90. Mature brown zygote in upper part of female gametangium.

Fig. 91. Zygote as seen from above.

Fig. 92. Zygote formed by direct fusion of gametangia without the development of a conjugation tube; male gametangium at the right did not develop its own apophysis or rhizoid.

Fig. 93. Similar type of fusion; both gametangia developed an endobiotic apophysis.

Fig. 94. Zygotic thallus of a form on *Vaucheria woroniniana* with a long rhizoid.

Fig. 95. Zygote of a form on *V. woroniniana* f. *quadripora*.

Fig. 96. Empty zoosporangium with 3 exit orifices showing the persistence of the wall.

Fig. 97. Diagram illustrating the life cycle of *Z. vaucheriae* with some hypothetical suggestions:

A = Large primary zoosporangium, causing "witches broom" effect on the host, whose planospores may develop into other large primary zoosporangia (I), or smaller secondary zoosporangia (III), or into gametangia (II).

B = Gametangia.

C = Small secondary zoosporangia whose planospores theoretically develop into gametangia.

D = Zygotic thallus; germination of zygote unkown.

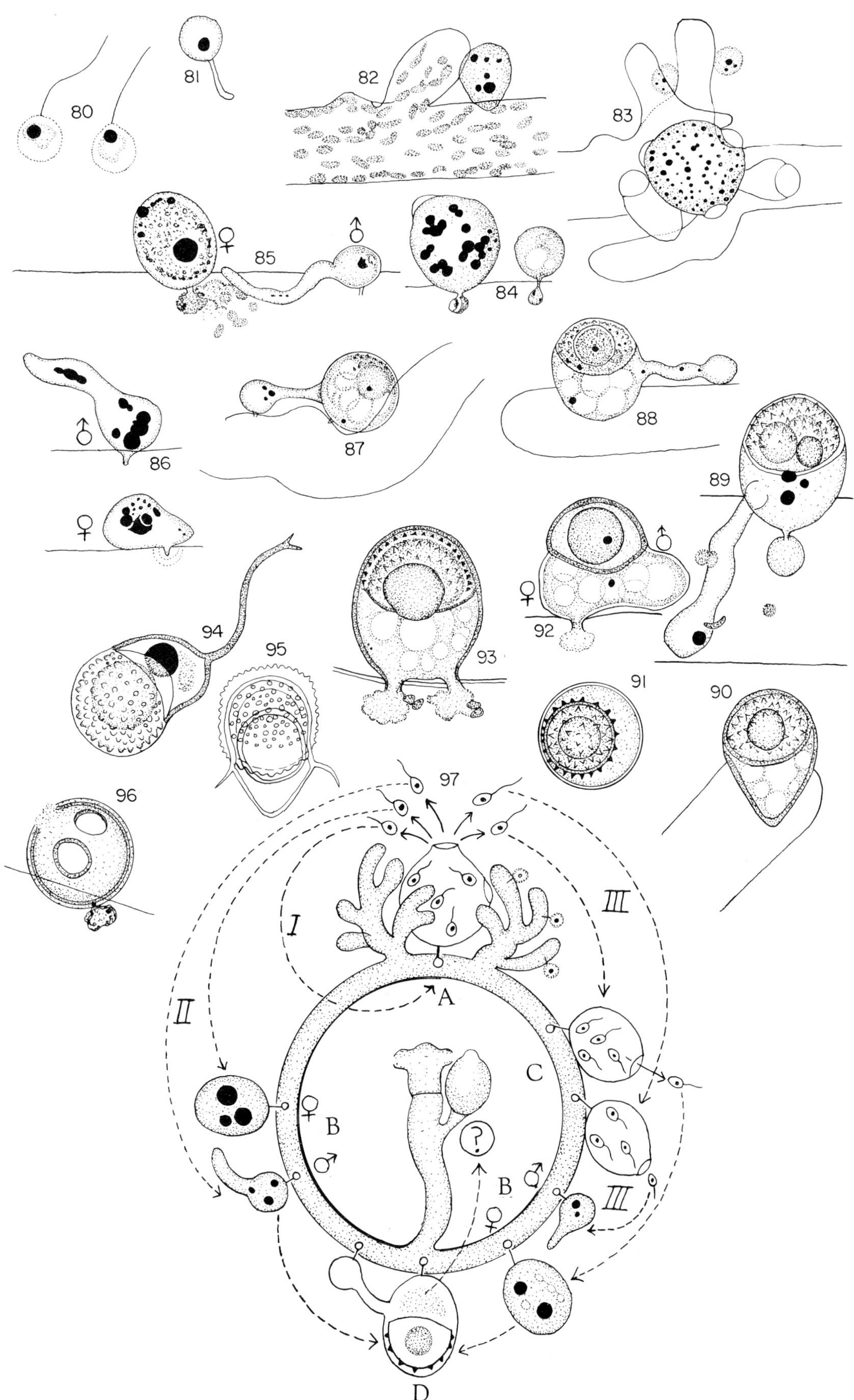

Zygorhizidium

ZYGORHIZIDIUM Lowenthal

Arch. f. Protistenk. 5:228, 1905.

Ectochytridium Scherffel, 1925. Arch. f. Protistenk. 53:7.

Plates 49-51

This operculate genus includes 11 species at present which parasitize green and chrysophycean algae and diatoms. Another species, *Z. vaucheriae* Rieth (1967), which has the same type of sexual reproduction as other species of *Zygorhizidium*, produces *anteriorly, uniflagellate* planospores and has been included in this genus, but its identity as a member of *Zygorhizidium* is questionable. The other species are characterized by epibiotic, subspherical, pyriform, elongate, subhemispherical operculate, apophysate or non-apophysate zoosporangia, small planospores, an endobiotic sparse, branched or unbranched rhizoid or haustorium, and epibiotic resting spores or zygotes which are formed by fusion of the contents of male and female gametangia through a conjugation tube. The planospores are small, asymmetrically ovoid (fig. 1), spherical (figs. 20, 26, 29, 35, 39, 46, 64) with a posteriorly attached flagellum, or reniform with the flagellum attached subapically but projecting backward (fig. 12). In germination they form a knob-like structure (fig. 3), or a short (figs. 13, 21) or a long germ tube (fig. 22) which in *Z. melosirae* Canter may be inflated at its tip (figs. 21, 22). From these structures develop an endobiotic peg or knob (figs. 5, 6), or a haustorium (figs. 15, 18), or a branched (figs. 24, 25, 27, 30–32, 60, 69, 70) or unbranched rhizoid (figs. 23, 43, 67, 71). In the meantime, the encysted planospore, and sometimes part of the germ tube, enlarges to become the zoosporangium (figs. 4–6, 15, 16, 31, 67–70) which usually develops 1 apical or 2 lateral operculate papillae for discharge of the planospores.

Sexual reproduction has been reported in all species of this genus, and although it may vary somewhat in some species it is basically similar. Usually a smaller gametangium develops a long (fig. 8) or short (figs. 19,

PLATE 52

Figs. 1-10. *Chytriomyces aureus* Karling. (Karling, 1945.)

Fig. 1. Portion of a thallus with an apophysate operculate zoosporangium, planospores swarming in a vesicle; rhizoids frequently branched and extensive.

Figs. 2, 3. Ovoid, 3-3.5 × 5 μ, planospore with a golden-red globule, and early germination stage on chitin, resp.

Fig. 4. Young apophysate thallus.

Fig. 5. Living, full-grown non-apophysate zoosporangium with the large primary nucleus.

Fig. 6. Division of the primary nucleus with an intranuclear spindle.

Figs. 7, 8. Progressive stages in the development of resting spores.

Fig. 9. Mature resting spore with dark thick wall, coarsely globular content, and a hyaline area in the center.

Fig. 10. Germination of resting spore.

Figs. 11-15. *Chytriomyces hyalinus* Karling. (Karling, 1945.)

Figs. 11, 12. Ovoid, 3-3.5 × 5-5.5 μ, planospore with a hyaline refractive globule, and early germination stage on chitin, resp.

Fig. 13. Young thallus.

Fig. 14. Mature operculate, non-apophysate zoosporangium with a coarse thick-walled rhizoidal axis.

Figs. 15, 16. Mature apophysate resting spore, and germination of the same.

Figs. 17, 18. *Chytriomyces hyalinus* var. *granulatus* Karling. (Karling, 1967.)

Fig. 17. Spherical, 5-6 μ diam., planospore with numerous granules.

Fig. 18. Mature resting spore.

Figs. 19-24. *Chytriomyces parasiticus* Karling. (Karling, 1947.)

Fig. 19. Minute, ovoid, 2.5–3 μ diam., planospore with minute hyaline refractive globule.

Fig. 20. Infection of the mycelium of *Aphanomyces laevis*.

Fig. 21. Later stage, mycelium of host enlarged locally.

Fig. 22. Local enlargement and branching of the host mycelium; zoosporangium partly enveloped.

Fig. 23. Living, full grown zoosporangium with large primary nucleus.

Fig. 24. Planospores swarming in a vesicle outside of zoosporangium.

Figs. 25-38. *Chytriomyces appendiculatus* Karling. (Karling, 1947.)

Fig. 25. Ovoid, 4-5 × 6-6.5 μ, planospore with a large hyaline refractive globule.

Figs. 26, 27. Early and late germination stages of planospores, resp.

Fig. 28. Partial thickening of incipient sporangium wall.

Fig. 29. Young zoosporangium with wall thickened at one end, and primary nucleus within.

Fig. 30. Older zoosporangium with uniformly thickened wall, and primary nucleus within.

Fig. 31. Mature thick-wall zoosporangium with coarse, thick-walled rhizoidal axis; sac-shaped area of hyaline matrix underneath the operculum.

Fig. 32. Emergence of globule of hyaline matrix.

Fig. 33. Emerging planospores enveloped by a layer of the hyaline matrix.

Fig. 34. Planospores bursting out of the matrix and dispersing.

Figs. 35-37. Some variations in the shapes of mature resting spores.

Fig. 38. Germination of resting spore, and planospores bursting out of hyaline matrix.

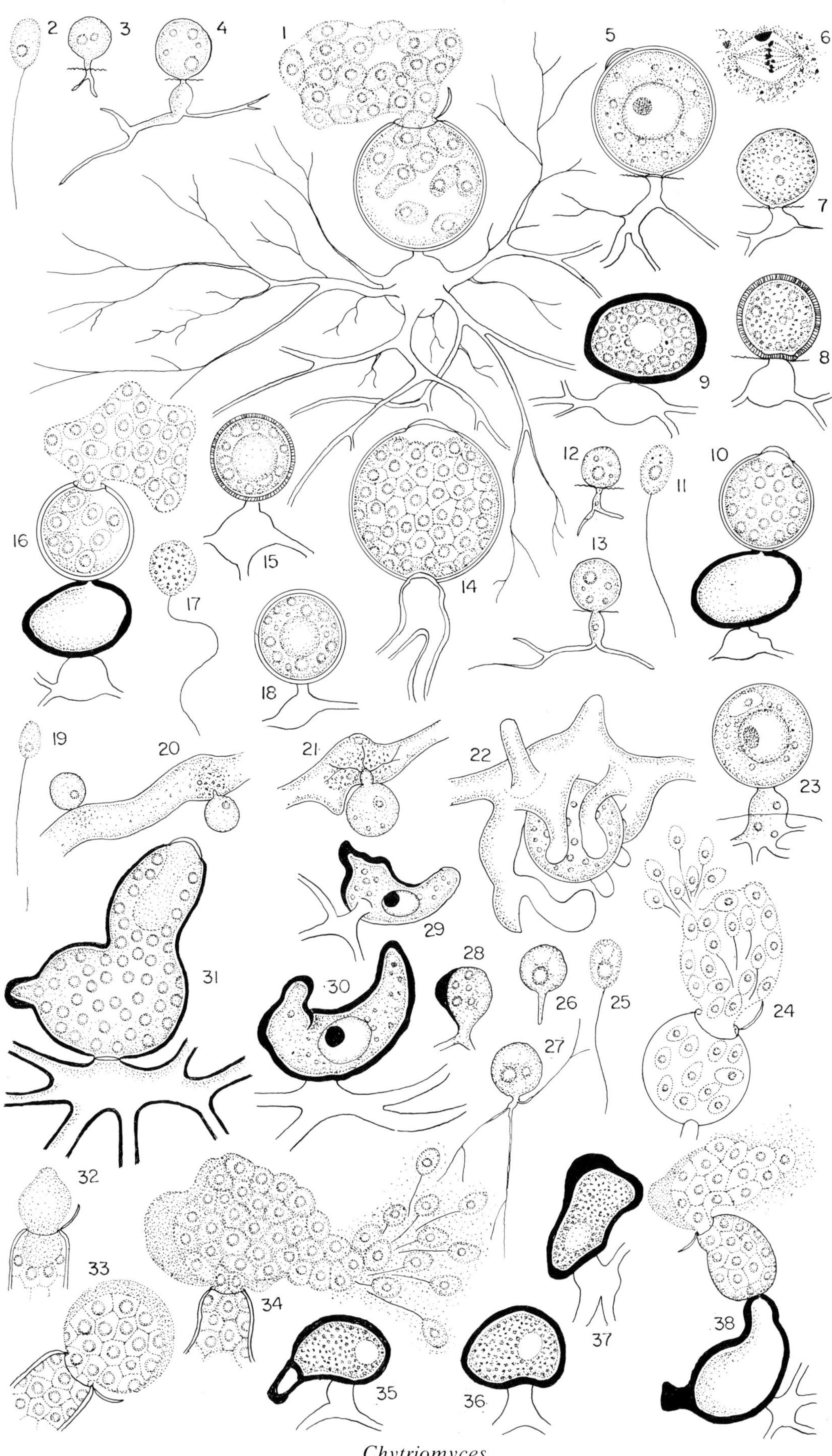

Chytriomyces

27) conjugation tube which grows towards and fuses at its tip with the surface of a larger gametangium (figs. 9, 17). The content of the former flows into the latter, and plasmogamy occurs. In *Z. willei* Lowenthal the gametangia are known to be uninucleate (fig. 9) at the time of plasmogamy, but karyogamy is delayed ap-

PLATE 53

Figs. 39-43. *Chytriomyces stellatus* Karling. Drawn from living material by Karling.

Fig. 39. Sparsely spiney, apophysate zoosporangium with two exit canals which are filled with hyaline matrix.

Fig. 40. Zoosporangium with two exit papillae.

Fig. 41. Ovoid, 3.5-4 × 5-5 μ, planospore with a hyaline refractive globule.

Figs. 42, 43. Mature, hyaline somewhat stellate resting spores.

Figs. 44-48. *Chytriomyces spinosus* Fay. (Fay, 1947.)

Figs. 44, 46. Ovoid and spherical, 3-4.5 μ diam., planospores with a hyaline refractive globule.

Fig. 45. Spiny zoosporangium; hyaline matrix fills upper ⅓ of zoosporangium.

Fig. 47. Mature hyaline resting spore with spines on the upper surface.

Fig. 48. Germination of a resting spore with the planospores beginning to swarm in a vesicle.

Figs. 49-51. *Chytriomyces closterii* Karling. (Karling, 1949.)

Fig. 49. Spherical, 2-2.5 μ diam., and ovoid planospores with hyaline refractive globules.

Fig. 50. A portion of parasitized *Closterium rostratum* with three young thalli and one mature dehiscing zoosporangium.

Fig. 51. Mature hyaline resting spore.

Figs. 52-54. *Chytriomyces lucidus* Karling. (Karling, 1949.)

Fig. 52. Mature zoosporangium filled with large brilliantly refractive globules; rhizoidal axis thick-walled.

Fig. 53. Ovoid, 5.8-6.2 μ, planospores with a large refractive globule.

Fig. 54. Mature hyaline resting spore filled with polyhedral refractive bodies.

Figs. 55-66. *Chytriomyces fructicosus* Karling. (Karling, 1949.)

Figs. 55, 56. Ovoid, 3.8 × 4.2 μ, planospore with a hyaline refractive globule, and early germination stage, resp.

Figs. 57-59. Early developmental stages of mono-, bi- and tri-apophysate thalli, resp.

Fig. 60. Incipient zoosporangium with the primary nucleus developing in a local enlargement of the long germ tube.

Fig. 61. Similar, later stage of a zoosporangium developing from a shorter germ tube.

Fig. 62. Mature papillate zoosporangium.

Fig. 63. Mature zoosporangium with an exit canal on which is an exit papilla.

Fig. 64. Mature zoosporangium with the planospore cyst and germ tube persistent as appendages; rhizoids have developed on the planospore cyst.

Fig. 65. Spiny hyaline resting spore.

Fig. 66. Warty resting spore with planospore cyst and germ tube persistent as an appendage.

Figs. 67, 68. *Chytriomyces poculatus* Willoughby and Townley. (Willoughby and Townley, 1961.

Fig. 67. Zoosporangium with crenulated wall and vacuolate cytoplasm from keratin substratum.

Fig. 68. Encysted planospores.

Fig. 69. *Chytriomyces mortierellae* Persiel. (Persiel, 1960); multioperculate non-apophysate zoosporangium on *Mortierella hygrophila.*

Fig. 70-73. *Chytriomyces reticulatus* Persiel. (Persiel, 1960.)

Fig. 70. Empty cup-shaped zoosporangium with collapsed operculum on *Pythium proliferum.*

Fig. 71. Spherical, 2.5-3.5 μ diam., planospore with a hyaline refractive globule.

Fig. 72. Resting spore with thick, reticulated, ridgėd wall.

Fig. 73. Germinated resting spore.

Fig. 74-77. *Chytriomyces mammilifer* Persiel. (Persiel, 1960.)

Fig. 74. Mature, warty resting spore on pine pollen.

Fig. 75. Partly dehisced zoosporangium with a large operculum.

Fig. 76. Spherical, 3.5-4.5 μ diam., planospore with a hyaline refractive globule.

Fig. 77. Mature zoosporangium before dehiscence.

Figs. 78, 79. *Chytriomyces willoughbyi* Karling. (Karling, 1968); mature zoosporangium and resting spore, resp., on *Phytophthora* sp. with intramatrical peg.

Figs. 80-83. *Chytriomyces cosmaridis* Karling. (Karling, 1967.)

Fig. 80. Dehiscing zoosporangium enveloped by a halo on *Cosmarium* sp; endobiotic apophysis continuous with a peg which projects up into zoosporangium; rhizoids lacking.

Fig. 81. Spherical, 2-2.6 μ diam., planospore with golden-orange refractive globule.

Fig. 82. Germinating planospore, forming rudiment of the endobiotic apophysis.

Fig. 83. Mature, apophysate resting spore surrounded by a halo.

Figs. 84-86. *Chytriomyces vallesiacus* Persiel. (Persiel, 1960.)

Fig. 84. Mature resting spore on pine pollen.

Fig. 85. Subspherical, 3-3.5 μ diam., planospore with a hyaline refractive globule.

Fig. 86. Empty zoosporangium with large collapsed operculum.

Figs. 87-91. *Chytriomyces rotoruaensis* Karling. (Karling, 1970.)

Fig. 87. Mature smooth hyaline resting spore on a chitinic substratum.

Fig. 88. Ovoid, 2.8-3.2 × 3-3.6 μ, planospore with a hyaline refractive globule.

Figs. 89, 90. Minute elongate and irregular zoosporangia.

Fig. 91. Dehiscing zoosporangium with a thin, barely perceptible operculum.

Chytriomyces

parently for some time. At least, Lowenthal noted that the zygote remained binucleate up until the time it developed a thick wall. Germination of the zygote has not been observed with certainty, although Lowenthal illustrated a zygote with small pointed zoospores within. Some variations of the above account have been reported by Canter. In *Z. parallelosede* Canter the gametangia may be almost equal in size at the time of conjugation (fig. 41). In *Z. parvum* Canter the infection tube of the male gametangium may branch and produce a conjugation tube which fuses with the infection tube of the female gametangium (fig. 37), or its infection tube may fuse directly with the female gametangium (fig. 38). In either event the female gametangium becomes a thick-walled zygote or resting spore, presumably after the contents of the 2 gametangia have fused. Canter's drawing (1963b, fig. 10) suggests that conjugation in *Z. cystogenum* may occasionally occur through the infection tube. In *Z. affluens* Canter 1 to many "male" cells may be adnate to the mature resting spores (figs. 78, 79).

Male gametangia with conjugation tubes may develop into zoosporangia in the event the tubes do not reach a female gametangium (figs. 8, 33). Presumably, this occurs with the latter, also, if contact with a male gametangium is not effected. This suggests strongly that the planospores of *Zygorhizidium* are facultative and may develop into either zoosporangial thalli or gametangia.

Zygorhizidium vaucheriae differs in several respects from the other species of this genus, and it may turn out to be representative of a new genus. Its planospores (fig. 80) have an *anteriorly inserted* and directed flagellum; the zoosporangia dehisce by the swelling of 1 to 5 exit papilla, and the resting spores or zygotes develop in the upper part of the female gametangium (figs. 88–94). Rieth did not observe whether the zoosporangia are operculate or inoperculate, but he described the papillae as swelling just before planospore discharge. The initially infecting planospores form a short germ tube (fig. 81) which penetrates the host wall and develops into a peg or knob-like haustorium while the planospore body develops into the so-called primary zoosporangium. This infection causes marked branching of the host filaments at the point of infection (figs. 82, 83) with the result that a "witches broom"-like effect is produced, similar to that produced by *Chytriomyces parasiticus* on *Aphanomyces* (pl. 52, fig. 22). Planospores from the primary zoosporangia may repeat this phase (fig. 97A, I) or give rise to smaller (fig. 84) secondary zoosporangia (fig. 97, III), which do not cause malformation of the host, or give rise to gametangia (fig. 97B, II).

Sexual reproduction in this species is basically similar to that of the other species of *Zygorhizidium*. The gametangia, like the secondary zoosporangia, do not cause malformation of the host. The smaller male gametangium develops a conjugation tube which elongates towards the female gametangium (fig. 85), and its tip fuses with the body of the latter. As the content of the former moves into the female the fused protoplasm aggregates or accumulates in the upper portion of the female gametangium (fig. 87). There it eventually becomes invested with a brown wall (fig. 88) whose upper or outer surface develops sharp pegs or short spines (figs. 89–91). The zygote, thus, occupies only a portion of the female gametangium. Occasionally, adjacent gametangia may fuse directly (figs.

PLATE 54

Figs. 1-12. *Chytriomyces annulatus* Dogma. (Dogma, 1969.)

Fig. 1. Spherical 4.7-5.4 μ diam., planospores with a hyaline refractive globule; from zoosporangia on pure pollen.

Fig. 2. Spherical, 6.0-6.5 μ diam., planospore with a large hyaline refractive globule from zoosporangia on *Liquidambar* pollen.

Fig. 3. Young stalked apophysate thallus.

Fig. 4. Young non-apophysate thallus with interbiotic branched rhizoids.

Fig. 5. Immature stalked apophysate thallus; collar-like undulations beginning to appear on the lower portion of the zoosporangium.

Fig. 6. Later developmental stage; zoosporangium filled with globules; annular undulations conspicuous on lower half of the zoosporangium.

Fig. 7. Zoosporangium with planospores.

Fig. 8. Discharge of planospores.

Fig. 9. Empty zoosporangium with a torn flap-like operculum.

Figs. 10-12. Variations in the rhizoid axes.

Fig. 12 A. Fusion through the rhizoids of the contents of 2 gametangia to form a central incipient zygote with a diploid fusion nucleus in *C. hyalinus*. (Moore and Miller, 1973.) Sketched from a photograph.

Figs. 13-23. *Pseudopileum unum* Canter on *Mallomonas*. (Canter, 1963.)

Fig. 13. Mature zoosporangium with an apical endo-operculum (?) and a basal rhizoidal stalk which has entered the pore of the host cyst.

Fig. 14. Young zoosporangium with an apical refractive cap.

Fig. 15. Later stage; refractive cap displaced; endo-operculum (?) lying underneath a muscilaginous papilla or plug.

Fig. 16. Later stage just before the disappearance of the muscilaginous papilla.

Fig. 17. Extent of the rhizoids in the host cyst.

Fig. 18. Empty operculate zoosporangia.

Fig. 19. Immature resting spore.

Figs. 20, 21. Mature resting spores with male (?) thalli adjacent.

Fig. 22. Resting spore with an enveloping mucilaginous secretion and possibly a male thallus; germ tubes of both thalli converging on the pore of the host cyst.

Fig. 23. Resting spore with a refractive cap and surrounded by a mucilaginous secretion; germ tube of male (?) thallus in contact with the resting spore.

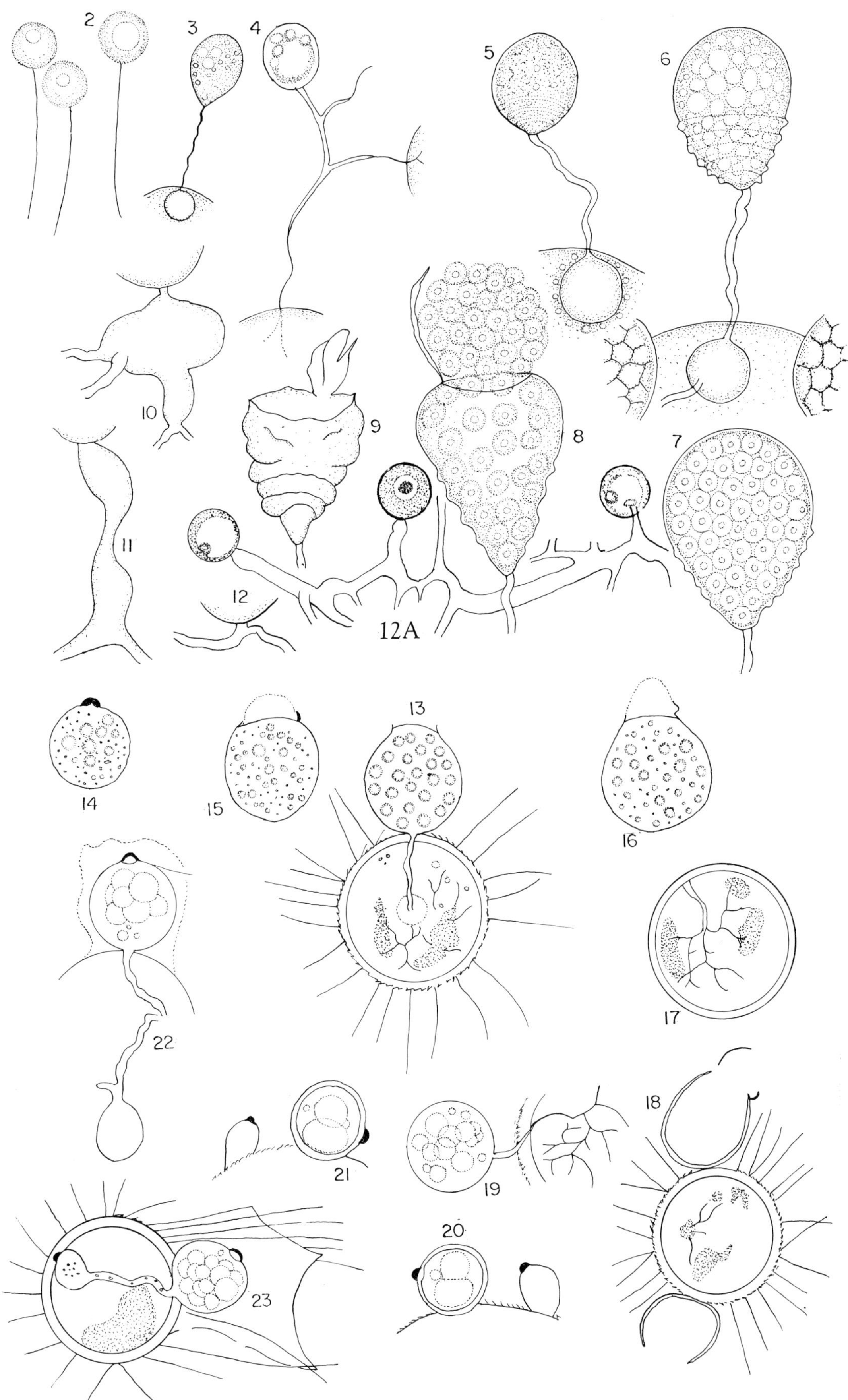

Chytriomyces, Pseudopileum

92, 93) without the development of a conjugation canal by the male. Germination of the zygotes has not been observed.

The origin of the planospores (andro- and gynoplanospore) which give rise to the gametangia is unknown, but Rieth postulated that they may originate from the smaller secondary zoosporangia (fig. 97C, III) or the larger primary zoosporangia (fig. 97A, II). Accordingly, the planospores of the latter appear to be facultative. Pertinent here is the observation that the secondary zoosporangia and gametangia do not cause "witches broom" symptoms in the host.

In the event the zoosporangia of this species are inoperculate and the anterior flagellum on the planospores proves to be tinsillate, *Z. vaucheriae* will be classified as a species of the Hyphochytriales. Minden (1911) classified the other species of *Zygorhizidium* in the subfamily Obelideae, but Sparrow (1943) created a separate subfamily, Zygochytrioideae, in his Operculatae for *Zygorhizidium*. Later (1960) he merged this subfamily with the Chytridioideae and placed this genus next to *Amphicypellus*. Whiffen (1944) classified it in the Rhizidioideae, and it is retained here tentatively in this subfamily of the Rhizidiaceae.

CHYTRIOMYCES Karling

Amer. J. Bot. 32:363, 1945.

Plates 52-54, Figs. 1-12

Twenty-three or more species, one variety, and several unidentified specimens are included in this operculate genus at present. Its distinction as a genus separate from *Chytridium* has been questioned on the grounds that occasional epibiotic resting spores may occur in the latter genus, also. Some of its species occur in the soil as saprophytes and can be trapped readily on chitinic, keratinic, and cellulosic subtrata and pollen grains and grown in axenic culture, while others parasitize diatoms, green algae, fungi, and Heliozoa. This genus is reported to be characterized by an apibiotic, smooth or spiny, apophysate or nonapophysate zoosporangium or resting spore, an endobiotic apophysis, if present, and an absorbing vegetative system which varies from fine, sparsely branched, or coarse, richly branched and extensive rhizoids to reduced pegs, digitations, or a single filament. In *C. appendiculatus* Karling and *C. lucidus* Karling the main rhizoidal axes may be up to 18 μ in diameter, and thick-walled, and their branches often extend for distances up to 400 μ. In *C. tabellariae* Canter (1949, 1951) the rhizoids may consist of a single branched or unbranched filament, and in *C. heliozoicola* Canter (1966) the infecting filament is very fine, but branches and becomes blunt and digitate within the host. In *C. willoughbyi* Karling the rhizoids may be reduced to a few sparsely branched filaments or a peg (figs. 78, 79), while in *C. cosmaridis* Karling they appear to be lacking (figs. 80, 83). In this species absorption of nutriments apparently occurs through an apophysis. Also, in *C. verrucosus* Karling (1960) and *C. suburceolatus* Willoughby (1964) the rhizoidal system is sparse. In *C. gilgaiensis* Willoughby (1965) on *Nowakowskiella crassa* it is replaced by a small, endobiotic, lobed haustorium.

The operculum on the mature zoosporangium, also, may vary. In *C. hyalinus* Karling on chitin it is usually sharply defined, saucer-shaped (fig. 14) and up to 15 μ in diameter, but when this species is grown in a nutrient solution the operculum may be incompletely delimited and flap-like, according to Bostlick (1968). Miller (1968), also, reported that some strains of this species are operculate and others inoperculate. In *C. verrucosus* the operculum may be up to 20 μ in diameter, but it is quite thin and flap-like, and becomes wrinkled.

The encysted germinated planospore develops into an epibiotic zoosporangium or resting spore, and the germ tube gives rise to an endobiotic apophysis and absorbing vegetative system (fig. 3, 4, 12, 26, 27). In *C. fructicosus* Karling, however, the zoosporangium occasionally develops as a local enlargement of the germ tube (figs. 60, 61), and in such cases the planospore cyst and germ tube may persist as appendages (fig. 64). Also, rare bi- and tri-apophysate zoo-

PLATE 55

Figs. 1-23. *Sparrowia parasitica* Willoughby. After Karling, 1970.

Fig. 1. Spherical, 2.5 -3 μ diam., and oblong planospores with a hyaline refractive globule.

Fig. 2. Amoeboid planospores.

Fig. 3. Infection of oogonium of *Pythium debaryanum*.

Fig. 4, 5. Sporangial rudiments budding out of the side of thick-walled planospore cyst; rhizoid filamentous and unbranched.

Figs. 6-8. Stages in the development of zoosporangia.

Figs. 9-15. Variations in sizes and shapes of mature zoosporangia and their position in relation to the oogonial wall and oospore of the host; rhizoid extending down into oospore.

Fig. 16. Discharge of planospore; operculum bowl-shaped.

Fig. 17. Dispersal of planospores from a zoosporangium with an exit canal.

Figs. 18-20. Developmental stages of young clavate incipient resting-spore thalli at the side of thick-walled planospore cysts.

Fig. 21. Migration of the protoplasm into the apex of the resting-spore thallus.

Fig. 22. Investment of the protoplasm in the apex by a hyaline wall.

Fig. 23. Variations in sizes and shapes of resting spores.

Figs. 24, 25. Planospore, resting spore, and a zoosporangium of *Rhizophydium pythii* de Wildemann which was frequently present with *Sparrowia parasitica*.

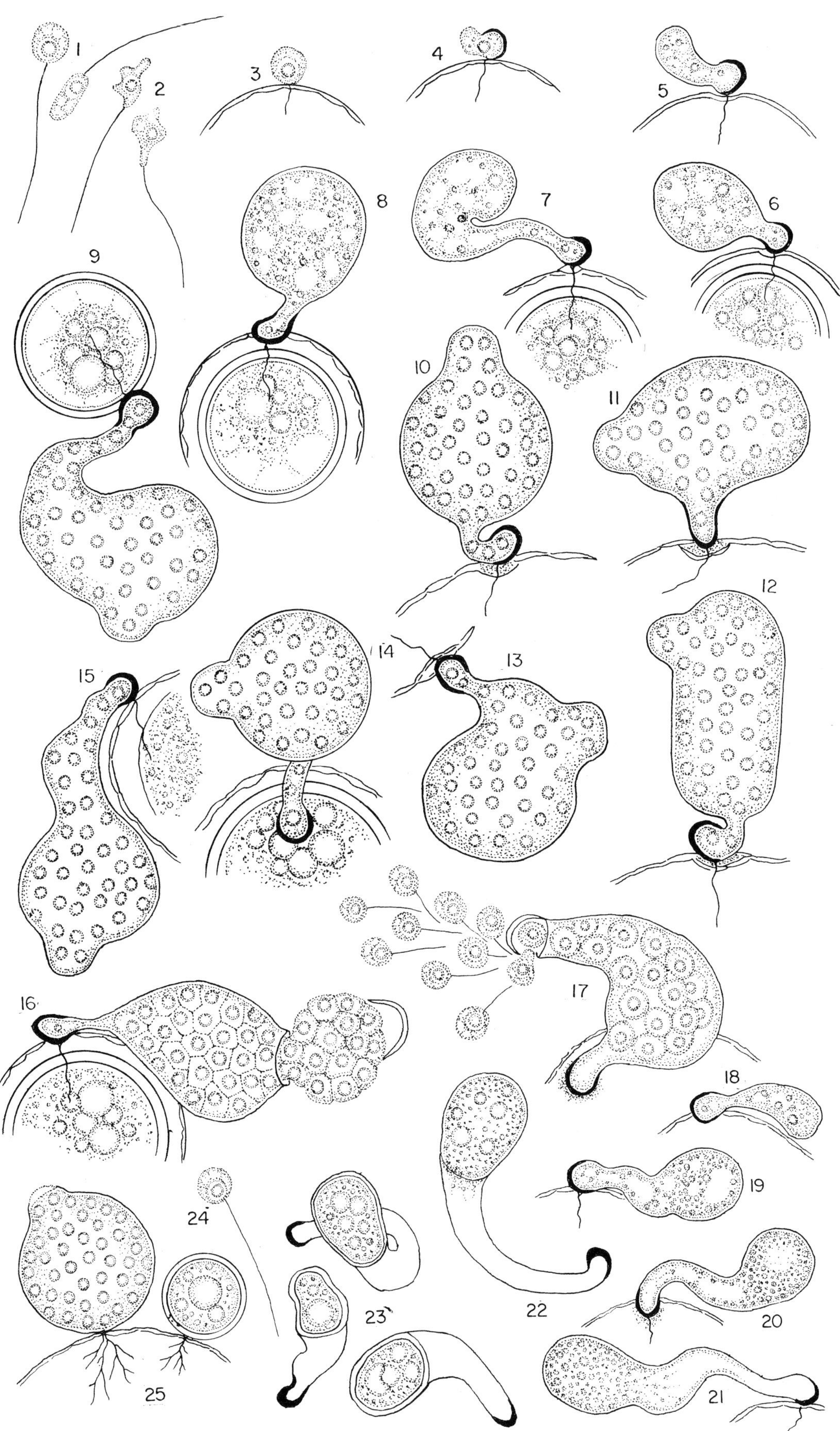

Sparrowia

sporangia may develop in this species (figs. 58, 59). A striking feature of the young (figs. 29, 30, 60, 61) and fully grown zoosporangia (figs. 5, 23), as in *Obelidium* and *Asterophlyctis*, is the presence of a large primary nucleus which is visible in living material as a subhyaline globule with a darker nucleolus. The zoosporangia remain uninucleate up to the time of zoosporogenesis, and in preparation for this process the nucleus divides with an intranuclear spindle (fig. 6). At maturity the zoosporangia dehisce through an apical or subapical operculate exit papilla (figs. 1, 14, 24, 80, 91), but in *C. stellatus* Karling, *C. fructicosus* and *C. appendiculatus* exit canals as well as papillae may be formed (figs. 32, 33, 39, 63). Occasionally, an exit papilla may form on an exit canal (fig. 63), and in *C. mortierellae* Persiel the sporangia are multioperculate (fig. 69). Underneath the operculum is a small biconvex (figs. 14, 40, 62, 64) copious (figs. 31, 45) mass of hyaline matrix which exudes as a globule (figs. 32, 33) as the operculum is pushed off. The planospores emerge into (fig. 33) and expand it to a thin layer. Apparently, the periphery becomes a membrane so that the planospores eventually swarm in a pliable vesicle which is continuous with the zoosporangium (figs. 1, 16, 24) before dispersing. In some species swarming in a vesicle does not occur, and the planospores disperse as they burst out of the enveloping matrix (figs. 34, 38).

The resting spores are borne in the same manner as the zoosporangia and appear to be formed asexually in most species. However, if *Chytridium confervae* (Wille) Minden, *C. chaetiferum*, and *C. cornutum* Braun turn out to be species of this genus, as will be pointed out under *Chytridium*, sexually formed resting spores apparently occur in *Chytriomyces*. In these 3 species the resting spores are accompanied by small empty cells which Scherffel and Canter interpret to be male cells which have fused with a female. In agar cultures of *Chytriomyces hyalinus* Karling, Koch (1959) reported "that the resting spore thalli form sexually at the tips of the rhizoids of contributing thalli." His observations have been confirmed by Moore and Miller (1971, 1973) in hanging drop cultures. "Practically the entire protoplast of each contributing sporangium flows through the common rhizoidal system into the incipient resting spore which enlarges rapidly at this time. These resting spores then become enveloped by a thickened and pigmented wall" (pl. 54, fig. 12A). The above reports indicate that sexual reproduction in this species is basically similar to that of *Rhizoclasmatium globosum, Sirhonaria petersenii* and *S. sparrowii*, all inoperculate species.

In *C. fructicosus*, however, resting spores occasionally bear a thick-walled tube and planospore cysts as appendages (fig. 66) which might suggest that the spores developed by the fusion of the contents of two gametangia or contributing thalli. On the other hand, such spores probably develop as an enlargement in the germ tube after which the planospore cyst and germ tube persist. The fact that similar appendages may occur on the zoosporangia (fig. 64), also, supports the latter view. However formed, the resting spore functions as a prosporangium during germination and forms a zoosporangium on its surface (figs. 10, 16, 38, 48). Such germination has been reported in 13 of the known species.

SPARROWIA Willoughby

Nova Hedwigia 5:336, 1963.

Plate 55

This operculate genus of 2 species, *S. parasitica* Willoughby and *S. subcrusiformis* Dogma, parasitizes the oogonia and oospores of *Pythium* sp., *P. debaryanum* and *Aphanomyces laevis* and the zoosporangia of *Rhizophydium coronum, Rhizophydium* sp., *Chytriomyces poculatus, C. annulatus* and *Karlingia rosea*. It is characterized by epibiotic, sessile or stalked, clavate, pyriform to irregular zoosporangia, filamentous or blunt-ending, branched or unbranched endobiotic rhizoids, and epibiotic resting spores borne in the apex or center of the thallus.

In *S. parasitica* the planospores encyst on the oogonial wall and infect the oospore by a fine germ tube (fig. 3) or rhizoid, and as the latter absorbs nourishment the sporangial rudiment buds out at right angles from the side of the thick-walled cyst (figs. 4, 5) as in

PLATE 56

Figs. 1-18. *Amphicypellus elegans* Ingold. Figs. 1, 5, 8 after Ingold, 1944; figs. 2, 9-11 after Paterson, 1958; figs. 3, 6, 7, 12-18 after Canter, 1961. Hosts not included.

Fig. 1. Unusually large thallus with a globular zoosporangium, apophysis, and extensive, branched rhizoids.

Figs. 2, 3. Spherical, 3.5-4.5 μ diam., planospores with a hyaline refractive globule.

Fig. 4. First stage in germination; sketch based on Paterson's description.

Fig. 5. Young thallus.

Figs. 6, 7. Variations in sizes of mature zoosporangia.

Fig. 8. Operculate zoosporangium with delimited planospores.

Fig. 9. Protoplasm with refractive globules emerging into a vesicle outside of the operculate zoosporangium.

Fig. 10. Planospores after being delimited simultaneously in the vesicle.

Fig. 11. Empty operculate zoosporangium.

Fig. 12. Early stage in resting spore formation with one small associated body or thallus at the right.

Figs. 13-16. Later stages in resting spore development, and mature spores; two small apophysate thalli associated with the resting spores.

Fig. 17. Resting spore mounted in India ink showing surrounding mucilaginous secretion.

Fig. 18. Smooth-walled resting spores.

Amphicypellus

some species of *Chytridium*. Accordingly, the sporangial rudiment often lies almost parallel to the surface of the oogonium. The rudiment usually elongates so that the mature sporangia often become stalked with the remaining portion of the planospore cyst at the base (figs. 6, 15) to which the rhizoid is attached. The zoosporangia develop a broad apical (fig. 10) or lateral exit papilla, and in some cases it may become extended into a short neck (fig. 17). At dehiscence a broad bowl- or shallow cup-shaped operculum is pushed off (fig. 16) or remains attached at the side of the orifice (fig. 17). The planospores emerge in a globular mass (fig. 16) but soon separate and disperse. Those remaining in the zoosporangium emerge one by one (fig. 17). In *S. subcrusiformis* Dogma early developmental stages begin as in the former species, but the developing zoosporangia are usually erect and develop 1 or 2 operculate exit papillae. The short rhizoidal branches are usually blunt-ended.

The resting-spore thalli apparently develop in the same manner as the sporangial thalli in *S. parasitica*, but may be recognized in the early stages by their smaller size, elongate shape, and more coarsely granular content (figs. 18, 19). As they elongate and enlarge the protoplasm moves up in the inflated apical portion (figs. 20, 21) and becomes invested by a hyaline wall (fig. 22) which thickens as the spore matures (fig. 23). Accordingly, the resting spore is formed in the apex of the thallus as in species of *Septosperma*. In *S. subcruciformis* the resting-spore thallus becomes somewhat cruciform in shape, and as its content contracts the resting spores form in the center. Germination of the spores has not been observed.

Willoughby regarded *Sparrowia* as the operculate counterpart of *Septosperma*, but as Karling (1970) pointed out it will probably prove to be a species of *Chytriomyces*.

PSEUDOPILEUM Canter

Trans. Brit. Mycol. Soc. 46:309, 1963.

Plate 54, Figs. 13-23

This incompletely known monotypic operculate genus parasitizes the cysts of *Mallomonas* sp. and is characterized by endooperculate zoosporangia, minute planospores, a fairly long rhizoidal axis or germ tube which branches after it has entered the host, and epibiotic resting spores with a refractive cap. The young zoosporangia, also, possess a refractive cap (fig. 14), but it is displaced by the development of a mucilaginous papilla or plug as the zoosporangia mature (figs. 15, 16).

The planospore encysts on the envelope of the host, and in germinating it forms a long germ tube which grows towards and enters the pore (fig. 13) of the cyst and branches within (fig. 17). In the meantime, the planospore body enlarges directly and eventually becomes the zoosporangium. As it matures it develops a mucilaginous apical papilla, and an endooperculum forms underneath (figs. 15, 16) which is pushed out (fig. 18) as the planospore mass emerges.

The resting spores are borne in the same manner as the zoosporangia and Canter believed that they are formed by the fusion of the contents of a small male and a larger female gametangium through a conjugation tube (fig. 22). She did not observe the actual contact of the so-called gametangia nor fusion, but association of mature resting spores with small empty thalli (figs. 20, 21) on the same cysts led her to the belief that a sexual process is involved in resting spore development. Germination of the resting spores was not observed.

Whether or not this chytrid represents a new genus is open to question. So far as it is known it is basically similar to *Chytriomyces*, and it may prove to be an endooperculate species of this genus. Probably, the refractive cap on the young zoosporangia and resting spores is a small unexpanded portion of the planospore cyst which became thick-walled and refractive. Endooperculation is fairly common among chytrids, and sometimes a few endoopercula may be found in species which are predominantly exooperculate, or vice versa, i.e. *Nowakowskiella macrospora*, *N. sculptura*, *N. multispora*, *Chytridium oedogonii*, etc. (Karling, 1945, 1961, 1964, 1966, resp.). The interpretation of

PLATE 57

Figs. 1-23. *Rhopalophlyctis sarcoptoides* Karling. (Karling, 1945.)

Fig. 1. Mature sporangial thallus on mayfly exuviae with an extramatrical, stalked, operculate zoosporangium, and intramatrical rhizoids.

Fig. 2. Ovoid, 5 × 6.5-7.5 μ, planospore with a large hyaline refractive globule.

Fig. 3. Absorption of flagellum.

Figs. 4, 5. Germination stages of planospore on mayfly exuviae.

Figs. 6, 7. Later stages of development of the incipient zoosporangium from the planospore body, with a basal peg and rhizoids entering the lumen of the insect case.

Fig. 8. Small zoosporangium which has broken loose, leaving a small hole in the integument.

Figs. 9, 10. Later stages of zoosporangial development.

Fig. 11. Clavate zoosporangium shortly before cleavage into zoospores.

Figs. 12, 13. Minute zoosporangia.

Figs. 14, 15. Stages in the discharge of planospores.

Fig. 16. Planospores swarming in a vesicle.

Figs. 17, 18. Continuous and septate empty zoosporangia.

Figs. 19, 20, 22. Uni-, tetra- multinucleate incipient zoosporangia.

Fig. 21. Simultaneous intranuclear mitoses.

Fig. 23. Fixed and stained planospore with nucleus, nuclear cap, and a basal granule.

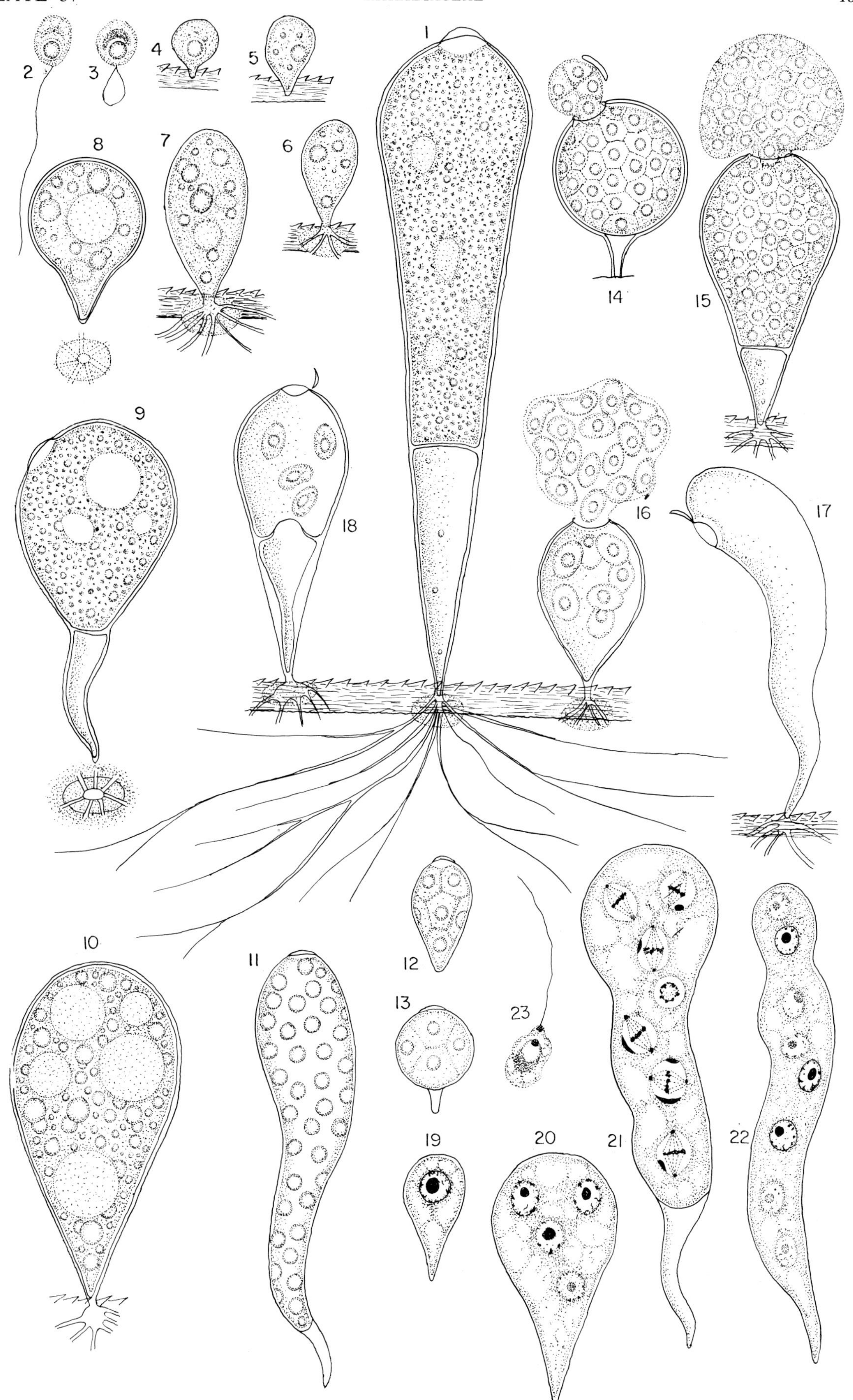

Rhopalophlyctis

endoopercula as pseudoopercula is not warranted by the data in the literature, and in light of this *P. unum* Canter should be regarded as an operculate species, in this writer's opinion. Canter believed that further studies may indicate that Persiel's (1960) endooperculate *Chytridium* sp. no 2307 on pollen grains should be transferred to *Pseudopileum*.

AMPHICYPELLUS Ingold

Trans. Brit. Mycol. Soc. 27:96, 1944.

Plate 56

This operculate monotypic and apophysate genus includes a single species, *A. elegans* Ingold, which has been found growing on the theca of two dinoflagellates, *Ceratium hirundinella* and *Peridinium* sp. in Scotland, Ireland, Denmark, Sweden, England, and Italy, and on *C. hirundinella* in Michigan, U.S.A. With the exception of the tips of its rhizoids this species is epibiotic, and Ingold believed that it might be saprophytic.

The planospores (figs. 2, 3) first produce a basal bud (fig. 4) in germination, according to Paterson (1958), which eventually becomes an apophysis, and from it develops lateral rhizoids (fig. 5). These branch and may become fairly extensive (fig. 1), frequently trailing out into the water and forming a fringe around the host. In the meantime, the planospore body enlarges to become the zoosporangium (figs. 1, 6, 8) which may be quite small (fig. 6) and produce as few as 2 planospores or become fairly large (figs. 7, 8). According to Canter (1961), it as well as the resting spore is surrounded by a mucilaginous secretion (fig. 17). From preserved material Ingold illustrated the planospores as being delimited within the operculate zoosporangium (fig. 8), but in living material Paterson reported that the protoplasm escapes in a vesicle to the outside of the zoosporangium (fig. 9) and in which the planospores are delimited simultaneously (fig. 10).

The resting spore thalli develop in a manner similar to that of the sporangial thalli, but in the formation of the resting spores Canter found 2 (figs. 13–16) and rarely 1 (fig. 12) bodies attached at some point to the main rhizoidal axis. Sometimes such bodies were minutely apophysate and bore a rhizoid. Since no resting spores were found without such companion cells or bodies she concluded that they might be contributing thalli such as those which are associated with resting spore formation in *Siphonaria* and *Rhizoclosmatium*. If this is confirmed, the resting spores of *A. elegans* are formed as a result of the conjugation of the contents of thalli (gametangia). The mature resting spores are usually spiny (figs. 14–17), but exceptional smooth ones (fig. 18) may occur, also. Germina-

PLATE 58

Figs. 1-4. *Rhizophlyctis mastigotrichis* (Nowak.) Fischer. (Nowakowski, 1876.)

Fig. 1. Zoosporangium with 4 radiating rhizoidal axes, one of which is infecting a filament of *Mastigothrix* (*Calothrix*) *aeruginea*.

Figs. 2, 3. Ovoid, 5 × 8 μ, and amoeboid zoospores with a large ovoid, hyaline refractive globule in the base.

Fig. 4. Monorhizoidal zoosporangium discharging planospores from a broad apical exit orifice.

Figs. 5, 6. *Rhizophlyctis tolypothricis* Zukal. (Zukal, 1893.)

Fig. 5. Zoosporangium with 3 rhizoids infecting filaments of *Tolypothrix lanata*.

Fig. 6. Oblong, 3-3.5 × 5-6 μ, planospores.

Figs. 7, 8. *Rhizophlyctis borneensis* Sparrow; zoosporangia with branched rhizoids infecting diatoms. (Sparrow, 1938.)

Figs. 9-14. *Rhizophlyctis petersenii* Sparrow. (Sparrow, 1937.)

Fig. 9. Amoeboid, and spherical, 5.2 μ diam., planospores with a minute orange-brown globule and several refractive granules.

Fig. 10. Young zoosporangium with several stout rhizoidal axes in empty larval cases of aquatic insects.

Fig. 11. Mature zoosporangium with an exit canal.

Fig. 12. Discharged planospores in a large globular mass at the exit orifice.

Figs. 13, 14. Ovoid and irregular resting spores with orange-brown content.

Figs. 15-20. *Rhizophlyctis petersenii* var. *appendiculata* Karling on bleached corn leaves and shrimp chitin. (Karling, 1967.)

Fig. 15. Spherical, 4.8-5.2 μ diam., planospore with an orange-brown globule and several refractive granules.

Figs. 16, 17. Germination of planospore and young thallus, resp., part of planospore case unexpanded and thickened.

Fig. 18. Young appendiculate zoosporangium with numerous large faintly yellowish to orange refractive globules; rhizoidal axis arising only from the base of the zoosporangium.

Fig. 19. Mature appendiculate zoosporangium with a long curved exit canal.

Fig. 20. Discharge of planospores.

Figs. 21-27. *Rhizophlyctis harderi* Uebelmesser on pine pollen grains. (Uebelmesser, 1956.)

Fig. 21. Spherical, 4 μ diam., planospore.

Fig. 22. Young zoosporangium on a pollen grain with 5 rhizoidal axes.

Fig. 23. Dehisced urn-shaped zoosporangium with planospores; apical exit orifice up to 40 μ diam.

Fig. 24. Germinated planospore.

Fig. 25. Mature smooth and thick-walled resting spore with a large refractive globule; bearing numerous long apical hairs.

Fig. 26. Germination of a resting spore (?).

Fig. 27. Spherical, 5.5 μ diam., larger planospore from a germinated resting spore.

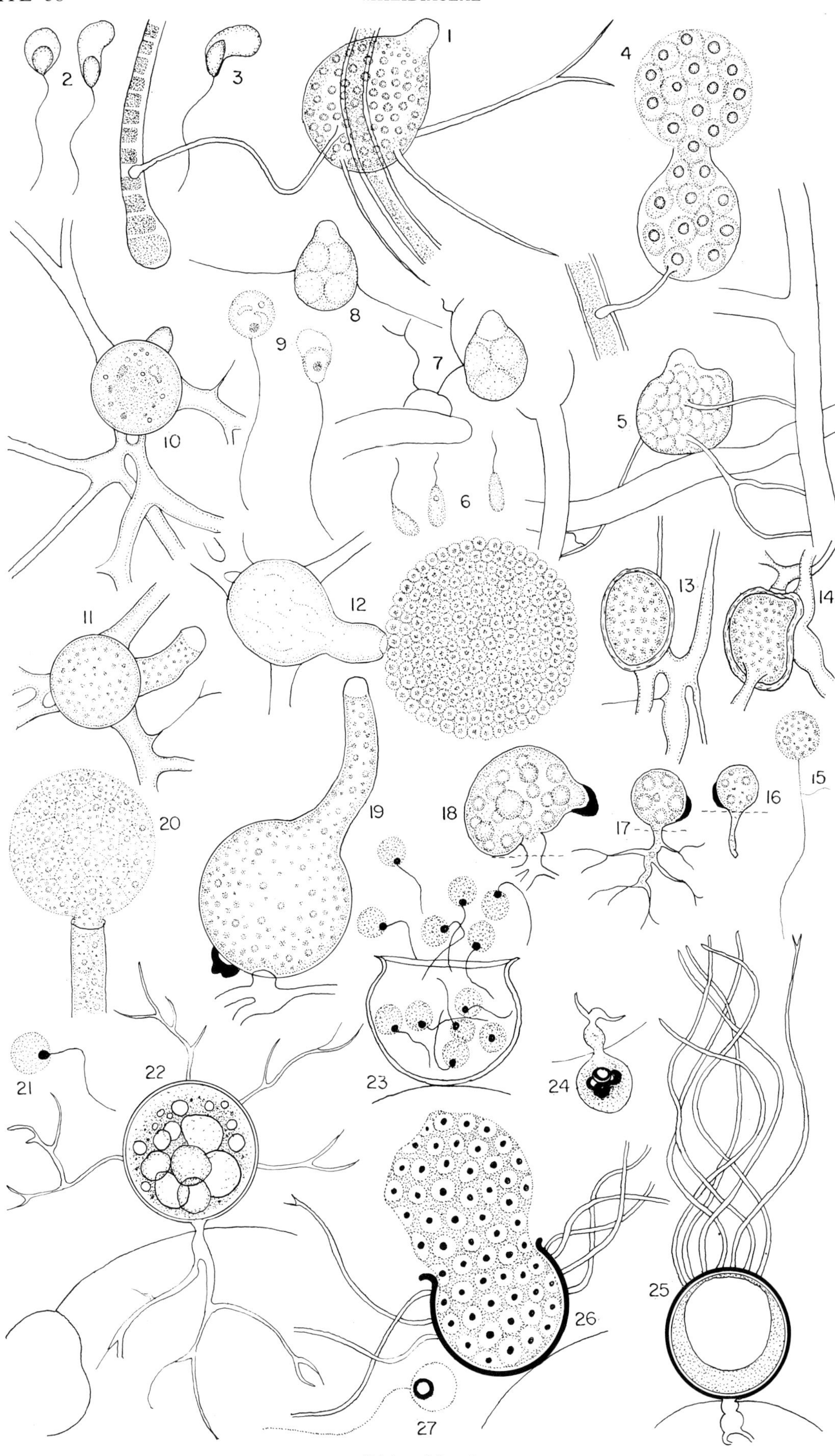

Rhizophlyctis

tion of the spores has not been observed.

Because of its epibiotic, operculate and apophysate zoosporangia Paterson and Sparrow (1960) compared *Amphicypellus* with *Chytriomyces*, but Canter believed that it resembles more closely *Siphonaria, Asterophlyctis* and *Rhizoclosmatium* by its manner of resting spore development. Pertinent to this point as noted before is Koch's (1959) report, confirmed by Moore and Miller (1971, 1973), that in agar cultures of *Chytriomyces hyalinus* Karling "the resting spore thalli form sexually at the tips of the rhizoids from two contributing thalli."

RHOPALOPHLYCTIS Karling

Amer. J. Bot. 32:363, 1945.

Plate 57

This monotypic operculate genus occurs as a saprophyte on the exuviae of mayflies in nature and it has been cultured sparingly on purified shrimp chitin and chitin agar. So far as the type species, *R. sarcoptoides* Karling, is known, its thallus consists of a predominantly clavate extramatrical, stalked or sessile

PLATE 59

Figs. 28-37. *Rhizophlyctis boneseyi* Sparrow on chitin. Figs. 28-36 after Sparrow, 1965; fig. 37 drawn from New Zealand material.

Fig. 28. Spherical, 7 μ diam., and ovoid 5 $\times$ 7 μ, planospores with a basal pale lemon-yellow globule and a hyaline nuclear cap.

Figs. 29, 30. Germination of planospore and young thallus, resp.

Fig. 31. Young elongate thallus with a thick-walled hemispherical cyst at the apex and rhizoidal axes at its ends.

Fig. 31 A. Fully-grown zoosporangium with the hemispherical cyst and 3 rhizoidal axes.

Figs. 32, 33. Initial stages of planospore discharge.

Figs. 34, 35. Two paired planospores (isogametes?) and their subsequent fusion.

Fig. 36. Resting spore with a thick sheath or wall and a possible attached male thallus.

Fig. 37. Germination of resting spore.

Figs. 38-59. *Rhizophlyctis oceanis* Karling on chitin. (Karling, 1969.)

Fig. 38. Spherical, 3-5 μ diam., and elongate planospores with a conspicuous hyaline refractive globule.

Figs. 39, 40. Germination of a planospore and the formation of a single basal rhizoidal axis, resp.

Fig. 41. Germination of a planospore in water and the formation of 3 peripheral rhizoids.

Fig. 42. Appearance of exit papilla and mass of matrix beneath it shortly before dehiscence.

Fig. 43. Initial stage of dehiscence.

Fig. 44. Later stage; remnants of papilla wall attached to edges of the orifice.

Fig. 45. Still later stage of dehiscence; mass of planospores enveloped by a layer of matrix.

Fig. 46. A flagellate motile cell approaching an encysted non-flagellate one.

Figs. 47-49. Successive stages in the movement of the flagellate cell around the encysted one; flagellum extended and beating.

Fig. 50. Same cells after flagellum has been lost.

Fig. 51. Beginning of fusion.

Fig. 52. Fusing cells with a rhizoid developing on the cell which came to rest first and encysted.

Fig. 53. Rhizoid developing from isthmus between two fusing cells.

Fig. 54. Spherical fusion product or zygote 26 hours old; fusing cells have lost their identity.

Fig. 55. Fusion of 3 cells.

Fig. 56. Large triflagellate cell partly enveloping a smaller encysted one.

Figs. 57, 58. Spherical and ovoid brown resting spores with spiney and verrucose outer walls, resp.

Fig. 59. Germination of resting spore.

Figs. 60-69. *Rhizophlyctis varians* (Karling) comb. nov. on cellulosic substrata. (Karling, 1949.)

Fig. 60. Spherical, 3.5-4 μ diam., planospore with a hyaline refractive globule.

Fig. 61. Germination of a planospore to form a single basal rhizoidal axis.

Fig. 62. Young broadly fusiform thallus with rhizoids at its ends.

Fig. 63. Young stage in the development of a polycentric thallus.

Fig. 64. Mature zoosporangium with 3 exit papillae and 3 basally-inflated rhizoidal axes.

Fig. 65. Early stage of planospore discharge from a monorhizoidal zoosporangium; planospores enveloped by a layer of matrix.

Fig. 66. Planospores swarming in a vesicle.

Fig. 67. Resting spore with a smooth light-brown wall and a single central *Rhizidium*-like rhizoidal axis.

Fig. 68. Three resting spores in a polycentric thallus.

Fig. 69. Germination of a resting spore.

Fig. 70. *Rhizophlyctis* sp. Sparrow; lobed zoosporangium with 6 exit papillae and several rhizoidal axes. (Sparrow, 1953.)

Fig. 71. *Rhizophlyctis* sp. immature thallus on snake skin. (Sparrow, 1957.)

Figs. 72, 72 A. *Rhizophlyctis* sp. on keratinic substratum. (Persiel, 1960.)

Fig. 72. Zoosporangium with 6 exit papillae which may become elongated into 5-15 μ long necks.

Fig. 72 A. Spherical resting spore with a brown, smooth, thick wall and 4 radiating rhizoidal axes.

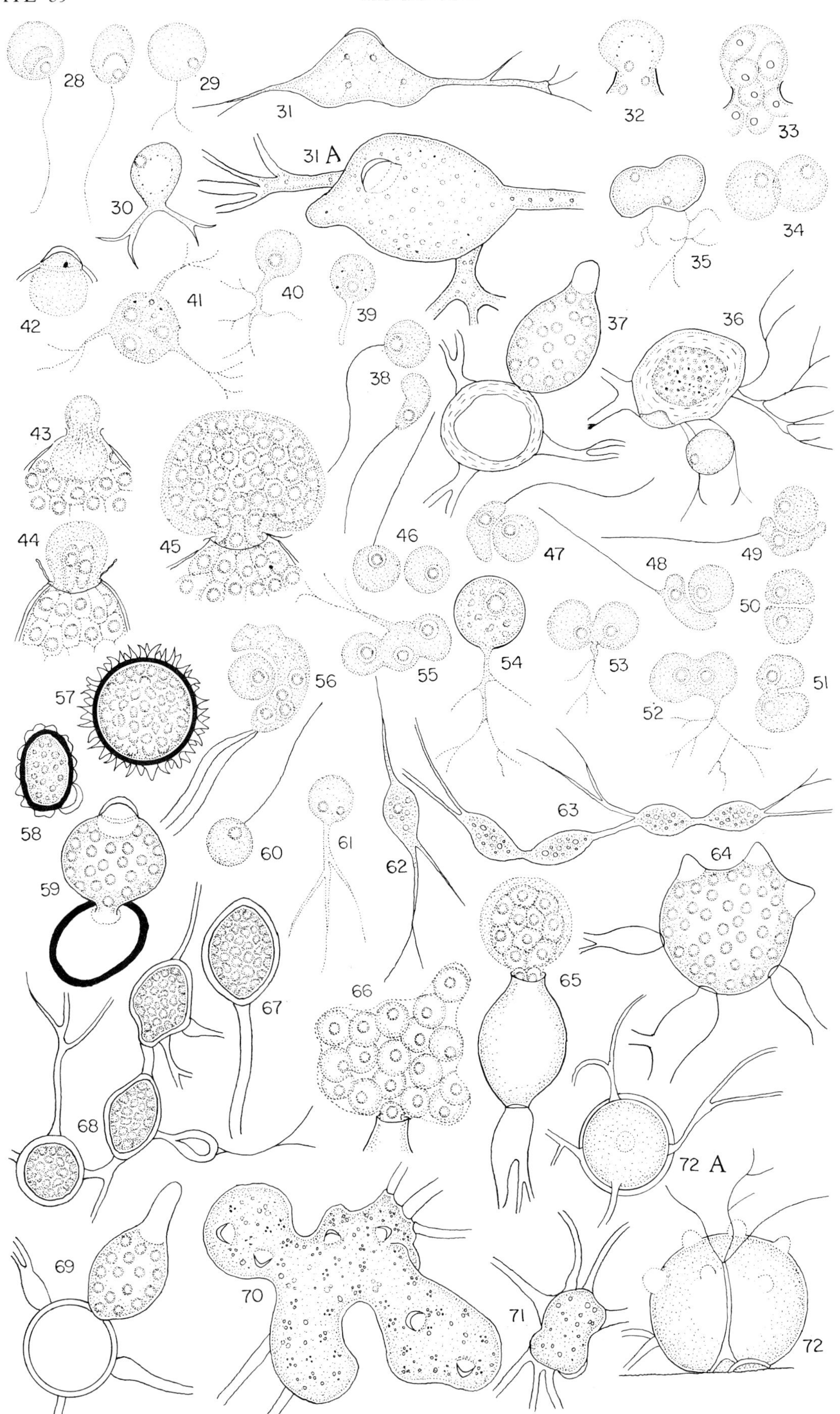

Rhizophlyctis

zoosporangium and an intramatrical peg or holdfast from which arise the branched and stiff-looking rhizoids (fig. 1). Resting-spore thalli and resting spores are unknown.

The ovoid planospores (fig. 2) absorb their flagellum (fig. 3), and after encysting produce a short germ tube which penetrates the insect integument (figs. 4, 5) and becomes the peg or holdfast from which the rhizoids arise. As they spread they usually separate the layers of the integument in a local circular area (figs. 8, 9). Meanwhile, the planospore body enlarges, usually becomes obpyriform or ovoid, and develops into a zoosporangium. Although the latter becomes predominantly clavate and stalked at maturity (figs. 1, 9, 14, 15, 18), it may, also, be sessile and continuous, obpyriform, ovoid, and subspherical (figs. 8, 10, 13, 22). Some zoosporangia may be quite small (figs. 12, 13) and produce as few as 8 planospores. In the maturation of stalked zoosporangia the protoplasm in the basal portion becomes vacuolate as it moves upward, and eventually the base becomes empty. A cross septum is then formed which separates the upper fertile portion from the base. Quite often mature zoosporangia are broken off from the rhizoids and substratum in mounting them for study and float free in the surrounding water (figs. 8, 9, 11). The operculum is usually apical, but sometimes it is subapical (figs. 9, 14, 17) or almost lateral. At dehiscence it is pushed off or aside as the planospores emerge in a globular mass (figs. 14, 15), which is enveloped by a thin layer of translucent matrix. This layer expands as more planospores emerge, and eventually its periphery becomes a vesicle in which the planospores swarm collectively (fig. 16) before dispersing.

The zoosporangia are uninucleate in the early developmental stages (fig. 19) but become multinucleate as they increase in size (figs. 20, 22). The nuclei divide simultaneously with an intranuclear spindle and conspicuous nucleolus (fig. 21). Fixed and stained planospores reveal a conspicuous nucleus and nucleole surmounted by a nuclear cap, and a basal granule on which the flagellum appears to be oriented.

Plate 60

Figs. 73-88. ***Rhizophlyctis*** *ingoldii* Sparrow on chitin. Figs. 73, 74 after Sparrow, 1957; figs. 75, 77, 78, 86, 87 after Willoughby, 1961; figs. 76, 79-85, 88 after Karling, 1970.

Fig. 73. Developing zoosporangium with a broad apical papilla and crusty wall.

Fig. 74. Early stage of planospore discharge.

Fig. 75. Mature intercalary zoosporangium with a subapical papilla under which is a U-shaped area of translucent matrix; wall with branched hairs.

Fig. 76. Stalked, hairy zoosporangium with 2 prominent exit papillae filled with matrix; wall of stalk and rhizoids greatly thickened.

Fig. 77. Discharge of planospores enveloped by matrix.

Fig. 78. Spherical, 4.5 μ diam., planospore with a minute hyaline refractive globule.

Fig. 79. Pairing of a flagellate and an encysted nonflagellate cell.

Fig. 80. Flagellate cell moving over surface of encysted cell.

Fig. 81. Pairing of cells after the loss of the flagellum.

Fig. 82. Fusion; rhizoid developing from the cell which came to rest first.

Figs. 83, 84. Later stages; globules have fused and the rhizoid has elongated and branched.

Fig. 85. Fusion completed before the development of a rhizoid.

Figs. 86, 87. Mature hairy resting spores with a yellow wall; thickened rhizoidal wall in fig. 86.

Fig. 88. Germination of resting spore.

Figs. 89-94. *Rhizophlyctis lovetti* Karling on fibrin film. (Karling, 1964.)

Fig. 89. Ovoid, 2.5-2.8 × 3.5-4 μ, and oblong planospores with a conspicuous nuclear cap and no refractive globule.

Fig. 90. Germinating planospore.

Fig. 91. Young multirhizoidal thallus.

Fig. 92. Free-lying zoosporangium near the edge of fibrin film with 8 coarse, curved and coiled rhizoidal axes.

Fig. 93. Elongate zoosporangium with a long neck.

Fig. 94. Discharge of planospores.

Figs. 95-101. *Rhizophlyctis fusca* Karling on fibrin film. (Karling, 1964.)

Fig. 95. Ovoid, 2-2.5 × 3.8-4.5 μ, and spherical planospores with a conspicuous nuclear cap and no refractive globule.

Fig. 96. Germination of planospore.

Fig. 97. Sporangial rudiment developing within the substratum at the end of the germ tube.

Fig. 98. Young extramatrical thallus with 4 rhizoids.

Fig. 99. Hyaline thick-walled zoosporangium with an appendage and 3 coarse rhizoidal axes.

Fig. 100. Discharge of planospores; exit canal has broken out of the thick wall.

Fig. 101. Dehiscence of a thick-walled reddish-brown dormant zoosporangium.

Figs. 102-105. *Rhizophlyctis hirsutus* Karling on bits of boiled hemp seed. (Karling, 1964.)

Fig. 102. Ovoid, 3.8-4.2 × 6-8 μ, or a slightly spindle-shaped planospore almost filled by a refractive body.

Fig. 103. Young thallus of a group, filled with large refractive bodies.

Fig. 104. Hirsute zoosporangium with a long exit canal.

Fig. 105. Discharge of planospores.

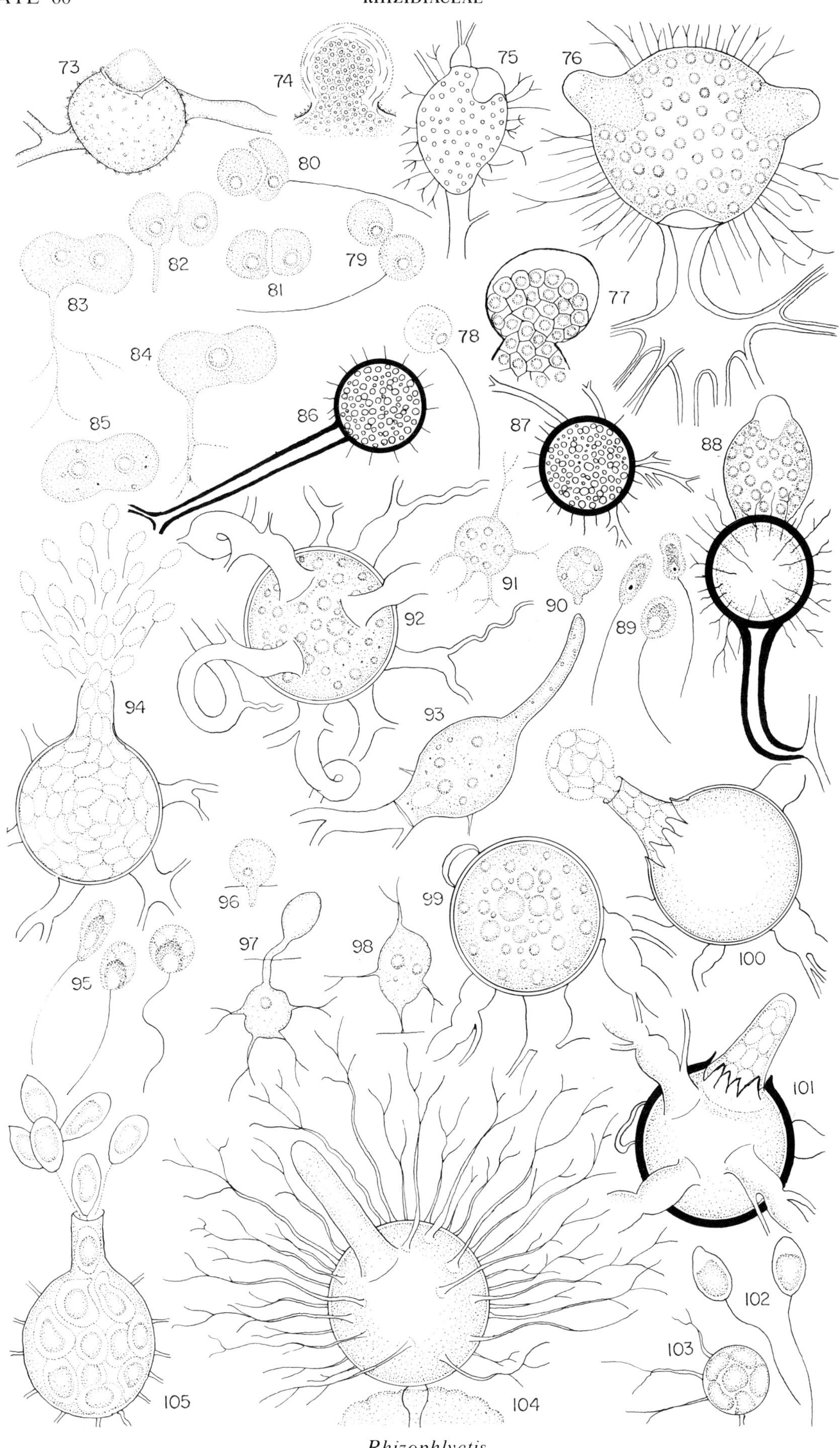

Rhizophlyctis

RHIZOPHLYCTIS Fischer

Rabenhorst Kryptogamen-Fl. 1:119, 1892.
(sensu recent Minden, Kryptogamenfl. Mark Brandenburg 5:374, 1911)

Plates 58-61

This genus was established by Fischer to include all *Rhizidium* species whose development is epibiotic except for the tips of the rhizoids and which form several rhizoidal axes that radiate from the periphery of the zoosporangia and resting spores. However, only 3 species (figs. 1–8) are known which fit exactly into this interpretation of the genus. Schroeter (1893) retained the genus in this sense but did not distinguish it clearly from *Rhizidium*. Minden (1911) emphasized the presence of a single rhizoidal axis in *Rhizidium* and interpreted *Rhizophlyctis* to include species with several rhizoidal axes. This distinction has been accepted almost generally but it is not always clearly defined. Some thalli of *Rhizophlyctis* may develop only one central axis (figs. 65, 67, 114), while some thalli of *Rhizidium* may be largely extramatrical and develop several axes. Accordingly, it is not always easy or possible to classify some thalli of either genus, and Antikajian (1949) proposed that *Rhizophlyctis* as well as other fairly similar genera be merged with *Rhizidium*. Usually the thalli are sessile on the host or substratum with some or many of the rhizoids embedded therein and do not develop free of the host. Also, it may be noted here that some zoosporangia of *Rhizophlyctis* may be partly intramatrical, or develop from a swelling in or at the tip of the germ tube (fig. 97) as in *Entophlyctis* and *Endochytrium*. Furthermore, the thalli may sometimes become polycentric as in *Rhizophlyctis varians* (Karling) comb. nov. (figs. 63, 68).

Rhizophlyctis is retained here provisionally in the sense of Minden and limited to 16 presently known inoperculate species, 2 varieties and several unidentified specimens. The morphologically similar endo-operculate species which Sparrow (1960) included in the genus has been replaced in *Karlingia*. Most of the known species are saprophytic and have been isolated from soil, water and debris on cellulosic, chitinic, keratinic substrata and fibrin film. However, *Rhizophlyctis mastigothricis* (Nowa.) Fischer (figs. 1–4), *R. polythricis* Zukal (fig. 5), *R. palmellacearum* B. Schroder, and *R. borneensis* Sparrow (figs. 7, 8) are reported to be parasitic on *Mastigothrix*, *Tolypothrix*, a palmellaceous green alga, and diatoms, respectively. *Rhizophlyctis harderi* Uebelmesser (figs. 21–27) was isolated on pine pollen in sea and freshwater deposits, *R. hirsuta* Karling (figs. 102- 105) on bits of boiled hemp seed, and *R. lovetti* Karling (figs. 89–94), *R. fusca* Karling (figs. 95–101) on fibrin film, and *R. boninensis*, Kobayasi and Konno (1969) from soil. Recently, Dogma (1974b) isolated and described several new species on chitinic substrata.

In the development of the sporangial thallus the encysted planospore develops 1 (figs. 16, 39, 61) or several (figs. 31A, 41) germ tubes which eventually become the rhizoidal axes. In some mature thalli these may number from 1 to 12, attain a diameter up to 30 μ at their base, branch profusely, and extend for distances of 500 μ or more. The extramatrical peripheral rhizoids in the water surrounding a zoosporangium may become curved, coiled and contorted (fig. 92). In New Zealand specimens of *R. ingoldi* Sparrow the main axis was usually thick-walled (fig. 76) and developed tyloses and trabeculae (Karling, 1970). In these specimens the rhizoidal system was so abun-

PLATE 61

Figs. 106-140. *Rhizophlyctis* sp. on chitin. After Karling, 1970.

Fig. 106. Spherical, 3.6-4.2 μ diam., planospores with a hyaline refractive globule and granular content.

Fig. 107. Large triflagellate planospore.

Figs. 108-112. Variations in young thalli with one or more rhizoids; part of planospore wall thickened and persistent as an appendage.

Fig. 113. Elongate and constricted thallus with 3 rhizoidal axes, 2 of which have expanded and given rise to several radiating branches.

Fig. 114. Oblong zoosporangium in the "granular" stage with a single, *Rhizidium*-like, central, intermittently inflated, thick-walled rhizoidal axis.

Fig. 115. Irregular, lobed appendiculate zoosporangium with the granular protoplasm in the "balled up" stage.

Fig. 116. Stalked, spherical, appendiculate zoosporangium with 3 prominent papillae which are clear of granular protoplasm.

Fig. 117. Papilla extended into a neck.

Fig. 118. Part of a zoosporangium and exit papilla with hyaline refractive globules, shortly before dehiscence.

Fig. 119. Simultaneous discharge of planospores from two short necks.

Fig. 120. An encysted non-flagellate cell approached by a flagellate one.

Fig. 121. Flagellate cell creeping over the encysted one.

Fig. 122. Same cells paired and at rest after loss of the flagellum.

Figs. 123-125. Stages in fusion before the development of a rhizoid.

Fig. 126. Spherical, zygote-like thallus resulting from fusion of two cells.

Figs. 127-129. Stages in fusion during the development of a rhizoid by the cell which encysted first.

Fig. 130. Fusion of 3 cells.

Figs. 131-135. Successive stages in the formation of a clump of cells.

Figs. 136-139. Variations in sizes and shapes of smooth-walled, light-brown resting spores with coarsely granular content and a clear area in the center.

Fig. 140. Germination of a resting spore.

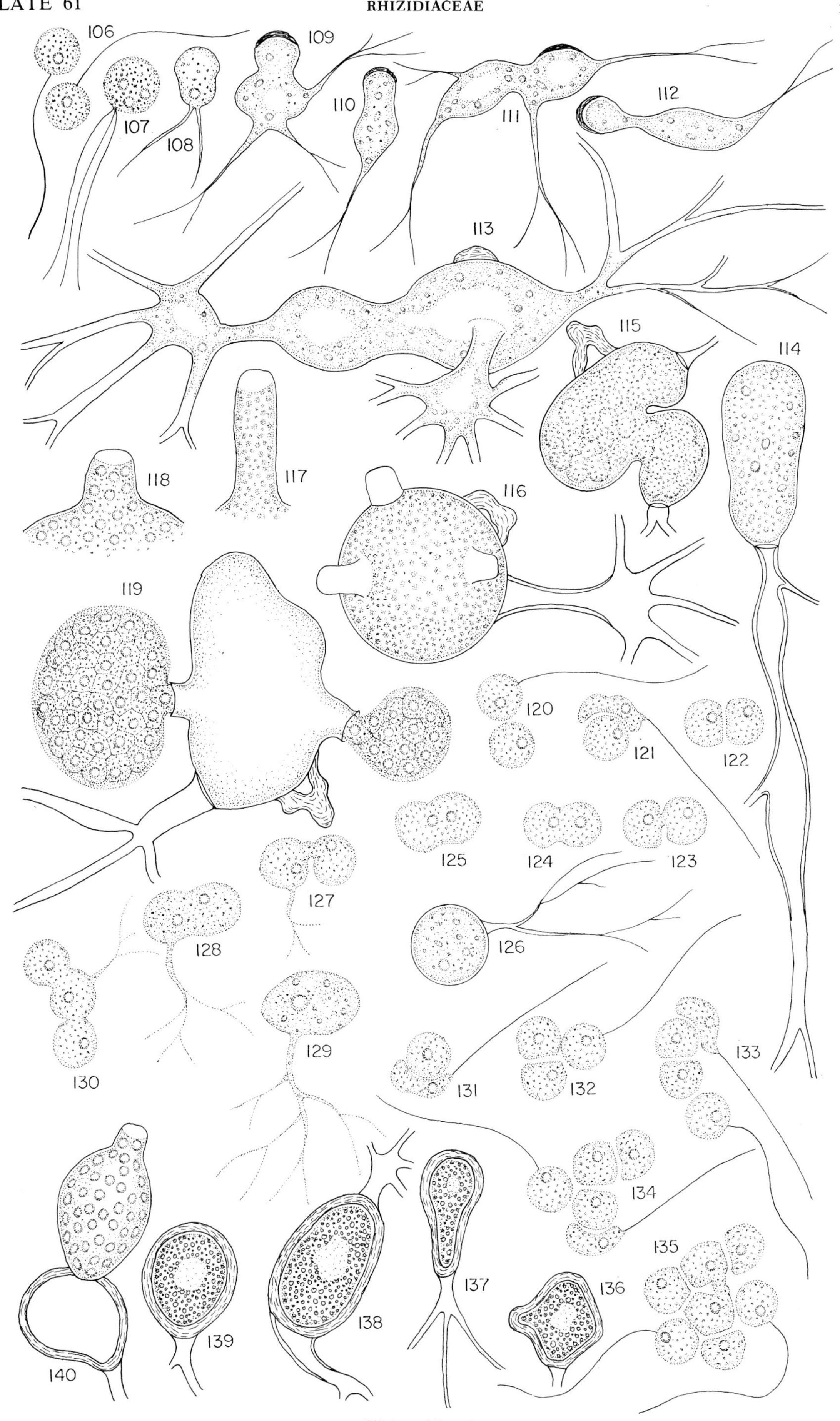

Rhizophlyctis

dant, dense, extensive, brown and richly-branched at the edges of the chitin that the zoosporangia were largely obscured.

The planospore body usually enlarges directly into a zoosporangium (figs. 10, 16–19), but as noted before the latter may develop in rare cases from an enlargement of the germ tube (fig. 97) in the substratum. The mature zoosporangia vary markedly in size, 3 to 290 μ diam., and shape from spherical (figs. 10, 19, 94, 100, 104, 116), subspherical (figs. 72, 73), ovoid (figs. 18, 31A, 65), oblong (fig. 114), irregular and lobed (figs. 70, 115, 119) to pyriform (figs. 1, 4, 7) and somewhat urn-like (fig. 23). In *R. ingoldii* Sparrow and *R. hirsuta* Karling they bear numerous hairs on their peripheries (figs. 75, 76, 104) in addition to the rhizoids. Usually 1 to a few exit papillae are formed before dehiscence, but in *R. oceanis* Karling up to 20 may develop in rare cases. These may be low and inconspicuous (figs. 42, 70) or prominent and protruding (figs. 64, 76) with a bowl- or U-shaped area of translucent matrix beneath (figs. 42, 64, 76). In *R. harderi* Uebelmesser the diameter of the exit orifice may be up to 40 μ and almost as great as that of the zoosporangium (fig. 23). In some cases the papillae may become extended into fairly long, up to 50 μ or more in length, necks (figs. 11, 12, 19, 104, 117).

As the tip of the papillae or necks deliquesce the hyaline matrix underneath oozes out (figs. 43–45, 77) and envelopes the initial mass of emerging planospores. Occasionally, in old zoosporangia the upper part of the papillae wall is pushed off like an operculum (Karling, 1969) or is split so that parts of it adhere to the edge of the exit orifice (fig. 44). As the mass of planospores expands (figs. 45, 119) the enveloping layer of matrix thins out and the planospores begin to disperse. In *R. varians* Karling, however, they swarm in a vesicle (fig. 66) before dispersing, and in *R. harderi* Uebelmesser reported that they are enveloped for a time by the gelatinized upper part of the zoosporangium wall.

The planospores vary considerably in size in dif-

Plate 62

Figs. 1-28. *Karlingia rosea* (DeBary and Woronin) Johanson. Figs. 1, 2, 5, 8 after Johanson, 1944; figs. 3, 7 after Ward, 1939; figs. 4, 9 after DeBary and Woronin, 1863; figs. 10, 16, 18 drawn from New Zealand specimens; figs. 6, 12-15, 22-24, 26, 28 after Karling, 1947; fig. 27 after Karling, 1968; fig. 25 after Willoughby, 1958; figs. 24A, 24B after Dogma, 1974a.

Figs. 1, 2. Ovoid, 4-5.3 μ, and amoeboid planospores, resp., with several salmon-pink granules.

Fig. 3. Elongate planospore with a large hyaline refractive globule.

Figs. 4, 5. Germination of planospores.

Fig. 6. Young intramatrical thallus developed from an enlargement of the germ tube in boiled grass leaf.

Figs. 7, 8. Young thalli with 4 and 6 rhizoidal axes, resp.

Fig. 9. Full-grown endooperculate zoosporangium with a plug of hyaline material in the short neck.

Fig. 10. Small zoosporangium with an elongate endooperculate neck.

Fig. 11. Surface of a zoosporangium with thickened remnants of planospore cyst and a plug of hyaline material above a low exit papilla.

Figs. 12-15. Stages in the deliquescence of the wall at the tip of the exit papilla, and the formation of a plug of hyaline material and an endooperculum beneath.

Fig. 16. Portion of a sporangial thallus with a subspherical zoosporangium bearing 10 exit papillae and 7 rhizoidal axes.

Fig. 17. Zoosporangium wall with short setae; possibly a variety of *K. rosea.*

Fig. 18. Zoosporangium at eroded edge of cellophane discharging planospores.

Figs. 19-21. Detached endoopercula, some drawn in optical section as seen in emptying sporangia.

Fig. 22. Spherical resting spore with 3 rhizoidal axes and a brown wall.

Fig. 23. Elongate resting spore in eroded cellophane developed intramatrically from a swelling in the germ tube.

Fig. 24. Young resting spore partially intramatrical.

Figs. 24A, 24B. Stages in the fusion of encysted gametes (gametangia?) in the formation of resting spores.

Fig. 25. Irregular resting spore with several large globules.

Fig. 26. Two small resting spores developed in a resting-spore thallus.

Fig. 27. Resting spore developed by contraction of the content and its investment by a wall in the apex of a resting-spore thallus.

Fig. 28. Germination of a similar resting spore.

Figs. 29-43. *Karlingia granulata.* (Karling, 1947.)

Fig. 29. Spherical planospores, 5.5-6.5 μ, with numerous refractive granules.

Fig. 30. Germination of a planospore.

Fig. 31. Young thallus with 2 rhizoidal axes and the remnant of a zoospore cyst at apex.

Fig. 32. Portion of a zoosporangium with thickened portion of a planospore cyst.

Fig. 33. Fully-grown zoosporangium with 2 exooperculate exit papillae, granular content, and 3 rhizoidal axes.

Fig. 34. Minute endooperculate zoosporangium with 4 planospore initials.

Fig. 35. Deliquescence of the wall at the tip of exit papilla and formation of an endooperculum.

Fig. 36. Large zoosporangium with 2 exo- and 1 endooperculate exit papillae, rhizoidal axes irregularly constricted with trabeculae.

Fig. 37. Shapes of opercula.

Fig. 38. Discharge of planospores.

Figs. 39, 40. Young resting spores.

Figs. 41, 42. Mature resting spores filled with numerous globules; outer wall brown and crusty.

Fig. 43. Germination of resting spore.

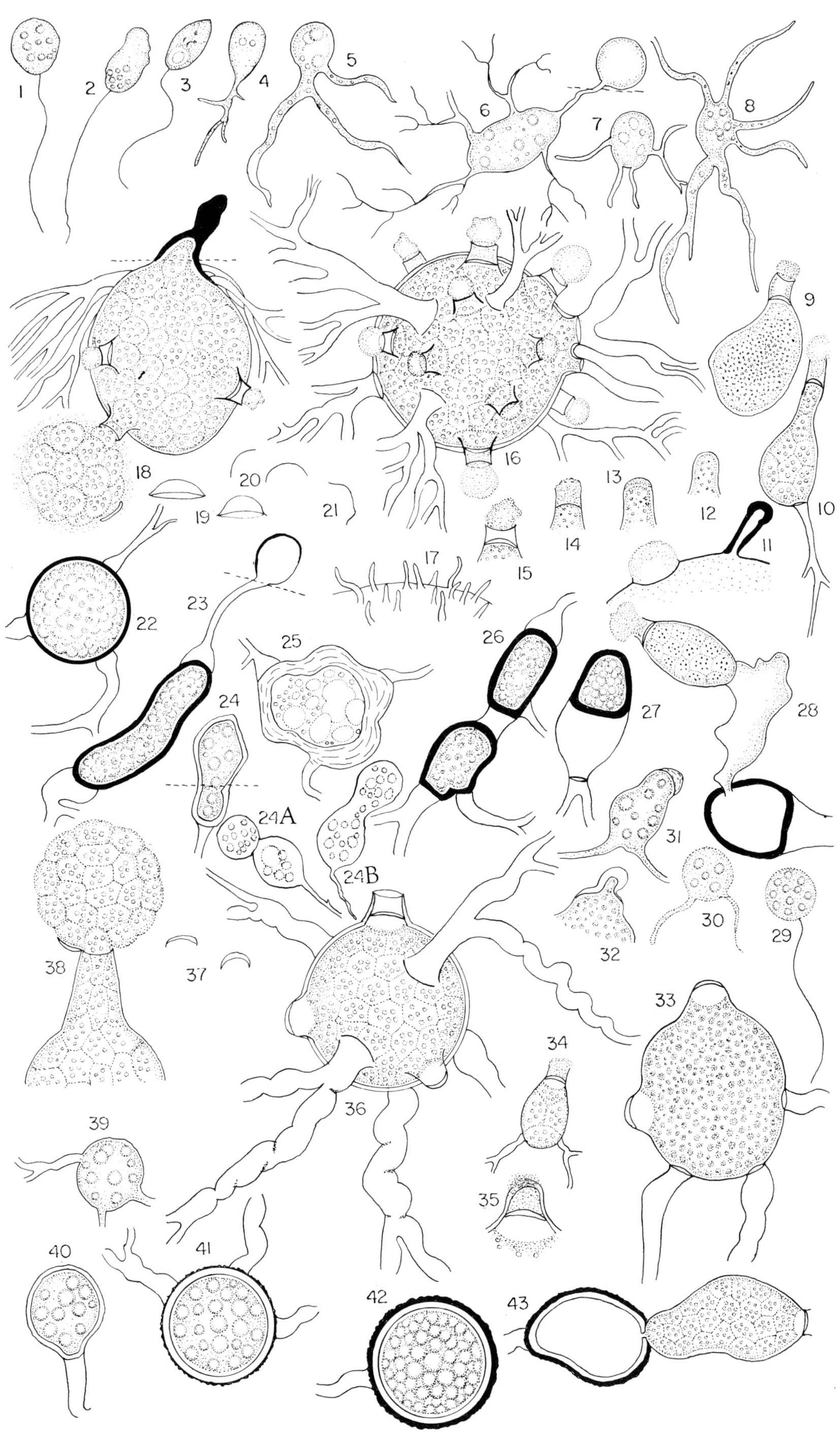
1
2
3
4
5
6
7
8
9
10
11
12
13
14
15
16
17
18
19
20
21
22
23
24
24A
24B
25
26
27
28
29
30
31
32
33
34
35
36
37
38
39
40
41
42
43

Karlingia

ferent species and also in the same species (*R. oceanis* Karling, 1969) and in most members contain a hyaline refractive globule and nuclear cap. The planospores of *R. petersenii* Sparrow include a minute orange-brown and several small refractive granules (fig. 9), and those of *R. petersenii* var. *appendiculata* Karling have a greyish granular content (fig. 15). In *R. hirsuta* planospores the refractive globule is so large (figs. 102, 105) that it almost fills the planospore body, and in *R. bonseyi* Sparrow the globule is lemon-yellow (figs. 28, 29).

Resting spores are known in all but 4 species, and in most of these they are reported to be borne like the zoosporangia and develop asexually. They vary considerably in size and shape with smooth, hyaline (fig. 72A) or light-brown to brown (figs. 13, 14, 25, 67, 68), spiny or verrucose (figs. 57, 58) outer walls. In *R. ingoldi* they may bear fine hairs (figs. 86–88), and in *R. harderi* tufts of long rhizoids occur at the upper surface (fig. 25). During germination they function as prosporangia (figs. 37, 59, 69, 88, 140) in most species where the process has been observed, but in *R. harderi* (fig. 26) Uebelmesser reported that they form planospores endogenously. Probably, figure 26 relates to a thick-walled dormant zoosporangium which formed planospores in the same manner as a thin-wall zoosporangium. Such zoosporangia and their germination (fig. 101) occur fairly often in *R. fusca* Karling.

In *R. boneysi* (figs. 34, 35), *R. oceanis, R. ingoldi* and *Rhizophlyctis* sp. the resting spores appear to be formed by the fusion of isomorphic gametes, although it has not been proven conclusively that the fused cells develop into mature spores. The 3 last named species are homothallic, and their motile cells or planospores are reported facultative. They may function as planospores and give rise to sporangial thalli, or fuse in pairs or more to form zygote-like thalli. In cases of fusions of pairs, one cell loses its flagellum and encysts and is approached by a flagellate one (figs. 46, 79, 120). The latter elongates and creeps over the encysted one (figs. 47–49, 80, 121) with its flagellum beating back and forth. Eventually, it loses its flagellum and comes to rest adjacent to the previously encysted cell. At this stage the two cells look like adjacent hemispheres (figs. 50, 81, 122). After a while an isthmus or bridge develops between them as fusion begins (figs. 51, 82, 123), and at the same time the cell which encysted first develops a rhizoid (figs. 52, 83, 127). However, the latter may not develop until fusion has been completed (figs. 85, 124, 125) or it may develop on the isthmus between the fusing cells (fig. 53). The fusion cell is oblong at first (figs. 84, 85, 128) but becomes spherical or subspherical (figs. 54, 126). Following fusion the refractive globules may fuse (figs.

PLATE 63

Figs. 44-51. *Karlingia spinosa* Karling. (Karling, 1947.)

Figs. 44, 45. Spherical planospores, 3.3-4.4 μ diam., with several golden-brown refractive granules or bodies.

Fig. 46. Germinated planospore on decaying vegetable debris with 3 rhizoidal axes.

Fig. 47. Partially intramatrical thallus in vegetable debris.

Fig. 48. Young intramatrical thallus in decaying vegetable debris; planospore cyst persistent.

Fig. 49. Young intramatrical irregular spiny resting spore in decaying vegetable debris with persistent planospore cyst and germ tube.

Fig. 50. Mature extramatrical resting spore with numerous golden-brown globules.

Fig. 51. Germination of spiny resting spore.

Figs. 52-56. *Karlingia hyalina* Karling. (Karling, 1947.)

Fig. 52. Central spherical zoosporangium of a polycentric thallus with 5 endooperculate exit papillae and 4 rhizoidal axes; content with large hyaline refractive globules.

Fig. 53. Shapes of endoopercula.

Fig. 54. Spherical, 4-5.5 μ diam., and pyriform planospores with a large hyaline refractive globule.

Fig. 55. Rare intramatrical young thallus in decaying vegetable debris.

Fig. 56. Usual type of planospore germination on surface of decaying vegetable debris.

Figs. 57-70. *Karlingia chitinophila* Karling. Figs. 57-66, 68-70 after Karling, 1967; fig. 67 after Karling, 1949.

Fig. 57. Ovoid and spherical planospores, 3.1-3.7 μ, with greyish granular content.

Figs. 58, 59. Germination of planospores at edge of chitin.

Fig. 60. Germination of planospores with 4 rhizoidal rudiments at the edge of the chitin.

Figs. 61, 62. Rare intramatrical development of thallus within the eroded chitin.

Fig. 63. Irregular zoosporangium with 2 endooperculate exit papillae and an appendage.

Fig. 64. Endooperculate exit papilla shortly before dehiscence.

Fig. 65. Discharge of planospores.

Fig. 66. Young resting spore with appendage.

Figs. 67-69. Smooth mature greenish-brown resting spores; appendage in fig. 69 bears rhizoids.

Fig. 70. Germination of resting spore.

Figs. 71-80. *Karlingia asterocystia* Karling. (Karling, 1949; figs. 79, 80 drawn from New Zealand specimens.)

Fig. 71. Spherical planospore, 4.2-4.6 μ diam., with a hyaline refractive globule and nuclear cap.

Fig. 72. Irregular zoosporangium with 3 low exooperculate exit papillae; large hemispherical zone of hyaline material under operculae.

Fig. 73. Initial stage of dehiscence.

Fig. 74. Discharge of planospores in a vesicle in which they will swarm collectively.

Fig. 75. Escape of planospores from vesicle.

Figs. 76-78. Variations in mature, greenish-brown resting spores.

Figs. 79, 80. Germination of resting spores.

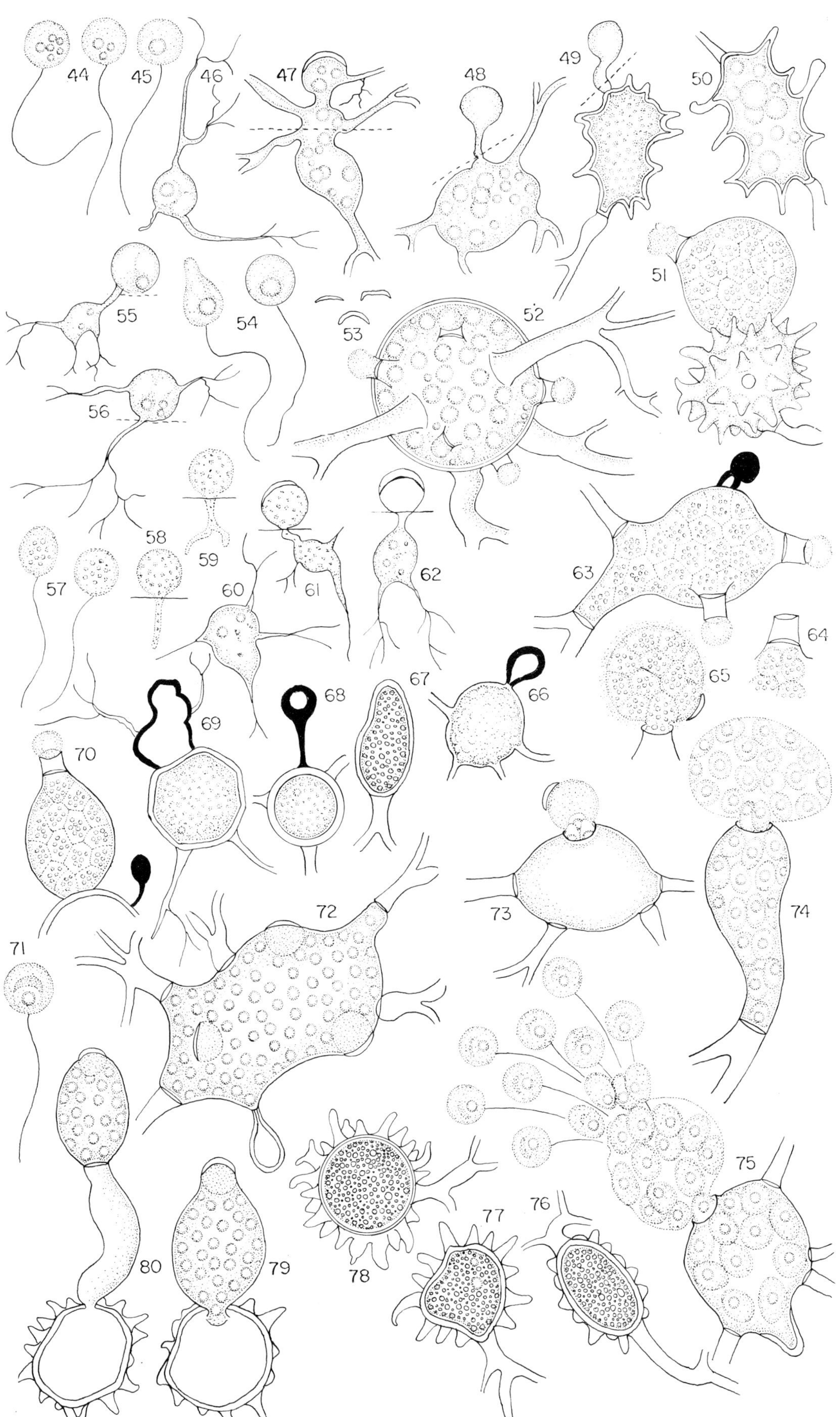
44
45
46
47
48
49
50
51
52
53
54
55
56
57
58
59
60
61
62
63
64
65
66
67
68
69
70
71
72
73
74
75
76
77
78
79
80

Karlingia

84, 129) or remain separate. In *R. oceanis*, and *Rhizophlyctis* sp. fusions may occur between 3 to 5 cells (figs. 55, 130), and a small cell may fuse with a large multiflagellate one (fig. 56) or vice versa. Occasionally 2 giant multiflagellate cells may fuse and form a monstrous zygote. Possibly, some of the very large resting spores in *R. oceanis* may be the result of such or multiple fusions, but this has not been demonstrated. The type of fusions described above appear to be common in *Rhizophlyctis* and *Karlingia*. Dogma (1974b) found it to occur in *R. aurantiaca* Dogma, *R. serpentinus* Dogma, *R. reynoldsii* Dogma, and *Rhizophlyctis* sp. D1.

As noted earlier, the fusion product has not been brought to maturity so far in *R. oceanis*, and it is not definitely known that the mature resting spores develop from it as in *Karlingia dubia*. Nevertheless, the fusions in these species has been interpreted to be gametic, but they might be regarded as gametangial in cases where the initially encysted cell develops a rhizoid before or during fusion. Particularly significant is the isomorphy of the fusing cells, their loss of morphological identity after fusion, and their complete merging in the fusion product. In these respects the fusions are similar to those of some species of the *Olpidiaceae* and *Synchytriaceae*, particularly *Synchytrium fulgens*, and unlike those of nearly all species of the Rhizidiaceae where one of the empty gametes or gametangia remain attached to the zygote as an appendage.

The motile cells of these *Rhizophlyctis* species may, also, clump together in small to large clusters (figs. 131–135) as in *Synchytrium endobioticum* and *S. fulgens*, and in these clumps fusions may occur (fig. 135), or the cells develop into sporangial thalli or degenerate.

Plate 64

Figs. 81-85. *Karlingia curvispinosa* Karling. (Karling, 1949.)

Fig. 81. Spherical planospore, 3.8-4.2 μ diam., with a small hyaline refractive globule and short flagellum.

Fig. 82. Small broadly pyriform mature zoosporangium with an exooperculate exit papilla.

Fig. 83. Discharge of planospores from 2 exit papillae.

Fig. 84. Dark-amber to brown resting spore with predominantly curved spines.

Fig. 85. Row of spines showing their variations in size and shapes.

Figs. 86-109. *Karlingia dubia* Karling. Figs. 86, 91-94, 96, 97, 99, 100 after Karling, 1949; figs. 87, 90, 95, 98, 101-109 after Willoughby, 1957.

Fig. 86. Spherical, 6-6.5 μ diam., and ovoid planospores with a hyaline refractive globule and numerous minute granules.

Fig. 87. Spherical and ovoid planospores with a hyaline refractive globule and vacuole.

Figs. 88-90. Germination of planospore and development of thallus, resp., on chitin; wall of incipient zoosporangium locally thickened.

Fig. 91. Portion of a thick-walled and irregularly constricted rhizoidal axis.

Figs. 92, 93, 95. Exit papillae showing relation of exooperculum to remainder of zoosporangium.

Fig. 94. Mature zoosporangium with exooperculate exit papilla and constricted rhizoidal axes.

Figs. 96, 97. Initial and advanced stages, resp., of planospore discharge.

Fig. 98. Empty operculate zoosporangium.

Figs. 99, 100. Mature dark-brown, encrusted and slightly verrucose resting spores with coarsely granular content and a central vacuole.

Figs. 101-104. Stages in the fusion of isomorphic facultative gametes; a rhizoid is developed by one gamete.

Fig. 105. Occasional fusion of 3 gametes.

Fig. 106. Young zygote.

Figs. 107-109. Mature zygotes or resting spores.

Figs. 110-118. *Karlingia lobata* Karling. (Karling, 1949.)

Fig. 110. Spherical planospore, 5.2-6.1 μ diam., with numerous hyaline refractive granules.

Fig. 111. Mature zoosporangium on cellulosic substrata with endo- and exooperculate exit papilla, thick, wrinkled wall, and constricted rhizoidal axes.

Fig. 112. Portion of a thick-walled constricted rhizoidal axis.

Fig. 113. Deliquescing tips of exit papilla with endooperculum beneath.

Fig. 114. Stalked endooperculate zoosporangium.

Fig. 115. Streams of planospores escaping from a globular mass above exit orifice.

Figs. 116-118. Variations in sizes and shapes of lobed hyaline resting spores.

Figs. 119-125. *Karlingia marylandia* Karling. (Karling, 1949.)

Fig. 119. Mature endooperculate zoosporangium on cellulosic substrata with a long contorted exit canal.

Fig. 120. Zoosporangium which formed 2 opercula.

Fig. 121. Appendiculate, endooperculate zoosporangium with constricted rhizoidal axes.

Figs. 122, 123. Initial and advanced stages, resp., of planospore discharge.

Figs. 124, 125. Abnormally small and normal-sized, 5.5-6 μ diam., planospores, resp., with a hyaline refractive globule.

Fig. 126. *Karlingia* sp. (*Rhizophlyctis* sp.) salmon pink zoosporangium with 10 exit papillae and 8 detached endoopercula being swirled around within the sporangium by the swarming zoospores. (Sparrow, 1953.)

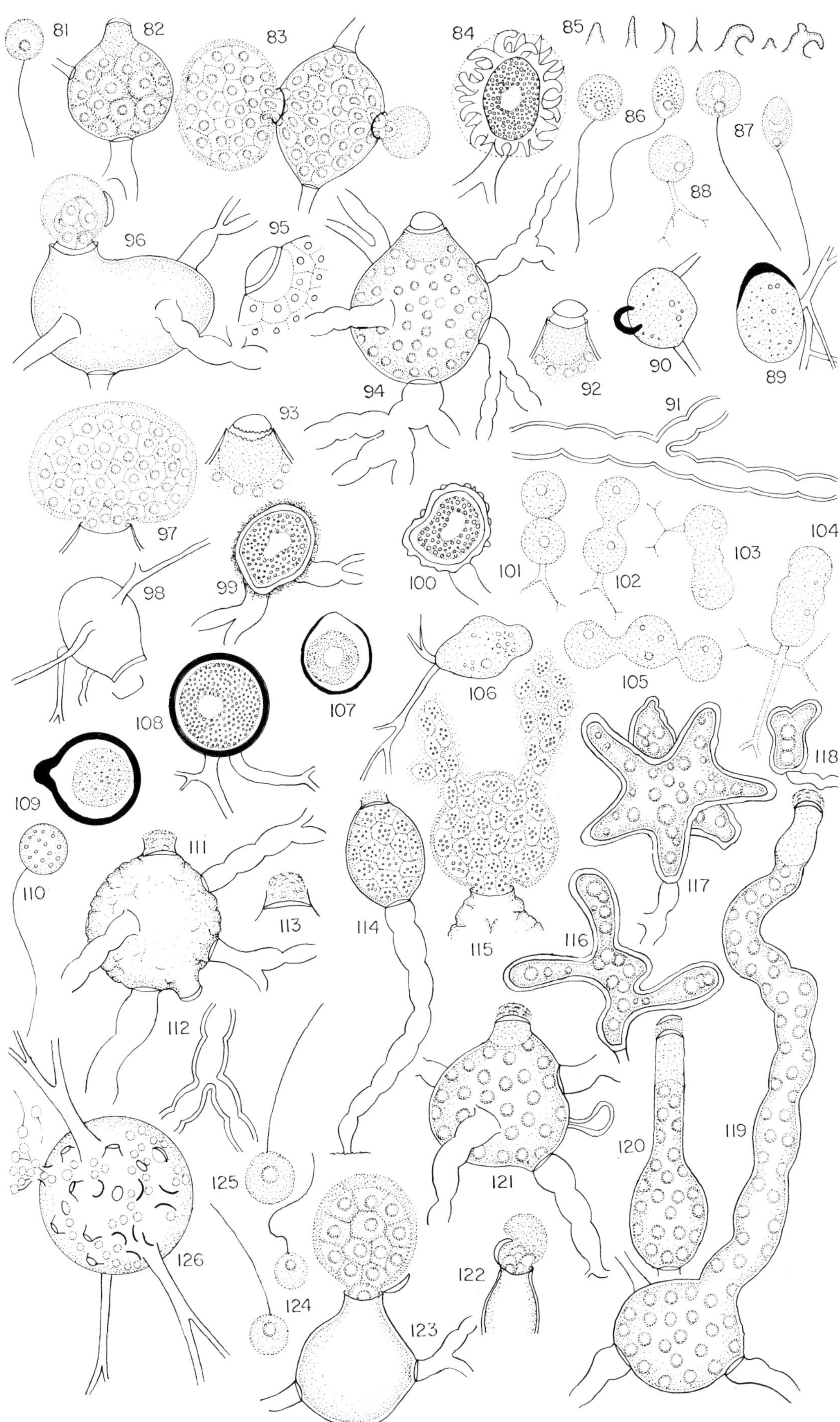
81
82
83
84
85
86
87
88
89
90
91
92
93
94
95
96
97
98
99
100
101
102
103
104
105
106
107
108
109
110
111
112
113
114
115
116
117
118
119
120
121
122
123
124
125
126

Karlingia

KARLINGIA Johanson

Amer. J. Bot. 31:397, 1944.

Karlingiomyces Sparrow, Aquatic Phycomycetes. 2nd. ed., p. 559, 1960.

Plates 62-64

Karlingia is the operculate counterpart of *Rhizophlyctis* and includes 11 or possibly 12 species and several unidentified specimens which occur as saprophytes in soil and can be readily trapped or isolated on various substrata. The salmon-pink or rosy type species, *K. rosea* (De Bary and Woronin) Johanson appears to be the most common of all chytrids and has been isolated from soil from most parts of the world. It is generally regarded as the universally-occurring chytrid. It has been grown extensively on known synthetic media by Stanier (1942), Quantz (1943), Haskins and Weston (1950) and others, and its environmental and nutritional requirements are well-known. It, as well as K. *granulata* Karling, K. *spinosa* Karling, K. *hyalina* Karling, *K. lobata* Karling and *K. marylandia* Karling occur in nature on cellulosic substrata, while K. *chitinophila* Karling, K. *asterocysta* Karling, K. *curvispinosa* Karling and K. *dubia* Karling are chitinophilic and have been isolated only on chitnic substrata. *Karlingia asterocysta* has been found to be an obligate chitinophile by Murray and Lovett (1966), which has an absolute requirement for chitin or preformed N-acetyl-D-glucosamine. Dogma (1974a and c) has made an extensive study of 93 isolates of the *Karlingia rosea*-complex as well as other species of this genus and grown them in axenic cultures. Only 30 of the 93 isolates of *K. rosea* developed resting spores, and in isolates HK_2, HK_4, and $N1Cr_4$ they were formed sexually as in *K. dubia* and *Rhizophlyctis oceanis* Karling. Probably, similar fusion will be found to occur in *Allochytridium expandens* Salkin, also.

As interpreted here the genus includes endo- and exooperculate species as well as those which may be both endo- and exooperculate. Sparrow (1960) assigned the endooperculate species to *Rhizophlyctis* because he believed that endoopercula are abnormal and develop only in old zoosporangia where dehiscence is delayed. A new genus, *Karlingiomyces,* was created for the exooperculate species. Sparrow's view of endooperculation is based primarily on the inadequate observations by Haskins (1948, 1950) on *K. rosea* and *Diplophlyctis sexualis*, but other investigators of *K. rosea*, Karling (1947, 1967); Willoughby, (1958); Chambers and Willoughby (1964), and Dogma (1974a) have shown that opercula are always formed in the exit papillae, regardless of age. Sparrow (1952), also, described and illustrated (fig. 126) endooperculate dehiscence in a salmon-pink species in Cuba, which apparently is *K. rosea*, and showed that several endoopercula are retracted into the zoosporangium and swirled about by the zoospores in the manner described by Johanson and Willoughby.

As in species of *Rhizophlyctis* the eucarpic thallus of *Karlingia* consists usually of an extramatrical highly variable zoosporangium or resting spore from whose peripheries or bases arise several (figs. 16, 36, 72, 94, 121) or one (figs. 10, 74, 114, 120) rhizoidal axes which spread and branch richly in the substratum and also may extend extramatrically for great distances in the surrounding water. Such axes may be almost straight, curved, coiled or contorted, thick-walled and irregularly constricted (figs. 36, 91, 111, 112, 114), and attain a diameter up to 38 μ with the branches extending for distances up to 1500 μ. Occasionally, the thallus may become polycentric and bear more than one zoosporangium. The zoosporangia may vary greatly in size and shape, but they are predominantly subspherical, ovoid, pyriform or oblong, and may be minute (figs. 10, 34), 6.3 μ, or up to 400 μ in diameter. In *K. rosea* (De Bary and Woronin) Johanson they may bear one (figs. 9, 10), or up to 24 exit papillae, 2–12 μ high by 5–18 μ diam. at the base, whose apices are filled by a gelatinous plug of hyaline material (figs. 9, 16, 18).

Plate 65

Figs. 1-22. *Allochytridium expandens* on agar and onion skin. Sketched from photographs by Salkin, 1970. Figs. 4-17 illustrate the primary pathway of development of the zoosporangium.

Fig. 1. Spherical, 6-8 μ diam., planospores with a large, 2.4-3 μ diam., hyaline refractive globule and a 35-40 μ long flagellum.

Fig. 2. Encysted planospore, colorless at first but later becoming orange.

Fig. 3. Germination of the planospore.

Figs. 4, 5. Later stages, germ tube enlarging locally to form the rudiment of the zoosporangium, and branching.

Fig. 6. Content of cyst has passed into the expanding spherical sporangial rudiment; cyst adnate to the rudiment.

Figs. 7, 8. Further expansion of the sporangial rudiment which gradually expands and fills the portion of the germ tube between it and the cyst until the latter become adnate.

Fig. 9. A later stage of confluence of the cyst and the incipient zoosporangium.

Figs. 10, 11. Still later stages; cyst incorporated in the zoosporangium; rhizoids omitted from sketches.

Fig. 12. Ampulliform zoosporangium developing an exit tube.

Figs. 13, 14. Mature zoosporangia on agar with several constricted segments of catenulate, rhizoidal axes arising from their periphery; adnate cyst in fig. 14 with a rhizoid.

Fig. 15. Discharge of planospores which are enveloped by a vesicular membrane; operculum at the right.

Figs. 16, 17. Irregular, elongate zoosporangia on agar and onion skin, resp.

Figs. 18-22. Stages in the secondary pathway of development of the zoosporangium directly from the enlargement of the planospore cyst, according to Salkin.

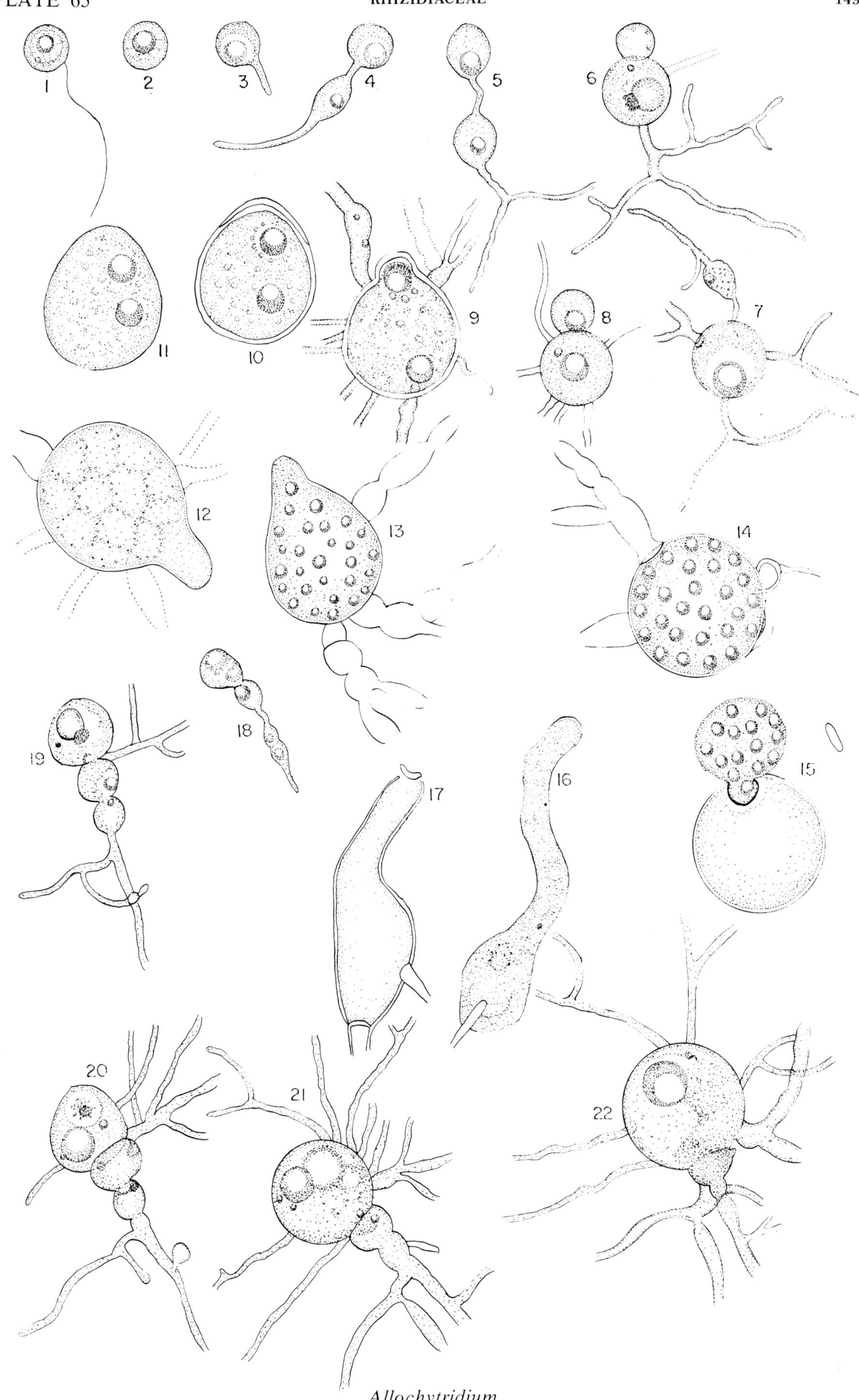

Allochytridium

Similar plugs occur in *K. hyalina* Karling (fig. 52), and *K. chitinophila* Karling (figs. 63, 70). In small zoosporangia of *K. rosea* the papillae may become fairly long necks (fig. 10), and in *K. marylandia* Karling long and contorted necks (fig. 119) may be formed. In the exooperculate species the papillae may be fairly prominent (figs. 82, 94), or low and inconspicuous (fig. 72) and usually have a conspicuous hyaline matrix under the operculum.

In the development of the thallus the planospore develops one or more germ tubes (figs. 4, 5, 7, 8, 58, 60), which ultimately become the rhizoidal axes while the planospore body usually enlarges to become the incipient zoosporangium. However, in rare cases it has been reported that the sporangial and resting spore rudiments develop as an enlargement in the germ tube (figs. 6, 23, 48, 49, 55, 61, 62) whereby the zoosporangium and resting spore become partly or fully intramatrical. However, should *Entophlyctis aurea* Haskins (1946) prove to be a species of *Karlingia*, possibly *K. rosea*, this type of development may be quite common. The zoosporangia and resting spores may frequently be partly intramatrical, particularly when infection occurs at the eroded, degenerating edges of the substratum. Fairly often the whole or part of the planospore cyst remains unexpanded (figs. 11, 18, 63, 66, 68–70, 89, 90) and persists as a thick-walled appendage.

In the maturation of exit papillae which become endooperculate the apical wall of the papilla gradually deliquesces (figs. 13, 14, 35, 113), and underneath it a mass of hyaline gelatinous material expands and extrudes out of the orifice as plugs, and these may be quite conspicuous in *K. rosea, K. hyalina*, and *K. chitinophila* (figs. 11, 16, 18, 28, 52, 63, 70) and usually disappear before dehiscence (fig. 64). In dehiscence in *K. rosea* the endooperculum and a globular mass of planospores are quickly extruded (fig. 18), and this

PLATE 66

Figs. 1-11. *Chytridium olla*. Figs. 1, 2, 4, 5 after Braun, 1856; figs. 6, 7 after Sorokin, 1872; figs. 3, 8, 9 after Sparrow, 1936; figs. 9, 10, 11 after DeBary, 1884.

Fig. 1. Two ovoid zoosporangia on an oogonium of *Oedogonium rivulare* with elongate haustoria attached to the oospore of the host.

Fig. 2. Spherical, 3-3.5 μ diam., planospore with a hyaline refractive globule.

Fig. 3. Amoeboid planospore.

Fig. 4. Detached zoosporangium and haustorium.

Fig. 5. Umbonate operculum.

Figs. 6, 7. Early infection stages.

Fig. 8. One of a group of empty zoosporangia on an oogonium of *Nitella tenuissima* with branched rhizoids arising from the tip of the haustorium or stalk.

Fig. 9. Resting spore in the same host with a tubular attachment running through the host wall; possibly the germ tube.

Fig. 10. Enlarged view of germination of a resting spore with a long germination canal.

Fig. 11. Dispersal of planospores; germination canal elongated to form an epibiotic zoosporangium on the surface of the oogonium of the host.

Figs. 12, 13. *Chytridium sphaerocarpum* Dangeard. (Sparrow, 1933.)

Fig. 12. Narrowly pyriform zoosporangium on *Spirogyra* with a filamentous branched rhizoid.

Fig. 13. Discharge of planospores from a broader zoosporangium.

Figs. 14-17. *Chytridium versatile* Scherffel. Fig. 14 after Scherffel, 1926; figs. 15-17 after Sparrow, 1933.

Fig. 14. Mature zoosporangium with planospore initials and a short stalk or rhizoid on *Cymatopleura solea*.

Fig. 15. Initial stage of planospore discharge on *Navicula* sp.

Fig. 16. Spherical, 3-5 μ diam., planospore with a hyaline refractive globule.

Fig. 17. Large zoosporangium with a short stalk and branched rhizoids in *Navicula*.

Fig. 18. *Chytridium versatile* var *podochytrioides* Friedmann; zoosporangium on *Navicula oblonga* with a knob-like sterile base and branched rhizoids. (Friedmann, 1952.)

Fig. 19. *Chytridium epithemiae* Nowakowski; zoosporangium with a prominent apiculus on *Epithemia zebra*. (Nowakowski, 1876.)

Fig. 20. *Chytridium nodulosum* Sparrow. Discharged planospores with coiled flagella. (Sparrow, 1933.)

Figs. 21-28. *Chytridium coleochaetes* Nowakowski. Figs. 21, 22 after Nowakowski, 1876; figs. 23-28 after Canter, 1960.

Fig. 21. Narrowly fusiform zoosporangium with a curved foot and planospores in the oogonia of *Coleochaete pulvinata*.

Fig. 22. Planospores.

Fig. 23. Elongate thallus with a markedly curved foot.

Fig. 24. Spherical, 2.5 μ diam., planospores with a hyaline refractive globule.

Figs. 25, 26. Variations in shapes of the rhizoidal sac.

Fig. 27. Apex of a zoosporangium with an operculum.

Fig. 28. Almost empty zoosporangium with an operculum far above the exit orifice.

Fig. 29. *Chytridium curvatum* Sparrow; zoosporangium epiphytic on *Stigeoclonium* (?) with a cup-like base and a short stalk. (Sparrow, 1933.)

Fig. 30. *Chytridium appressum* Sparrow; zoosporangium on *Melosira* with a cylindrical tube which has penetrated the host wall. (Sparrow, 1953.)

Figs. 31-34. *Chytridium surirellae* Friedmann. (Friedmann, 1953.)

Fig. 31. Encysted planospore on *Surirella ovata*.

Fig. 32. Young thallus with a branched rhizoid.

Fig. 33. Mature zoosporangium with refractive oil globules and branched rhizoids.

Fig. 34. Empty zoosporangium with a thin, wrinkled flap-like operculum.

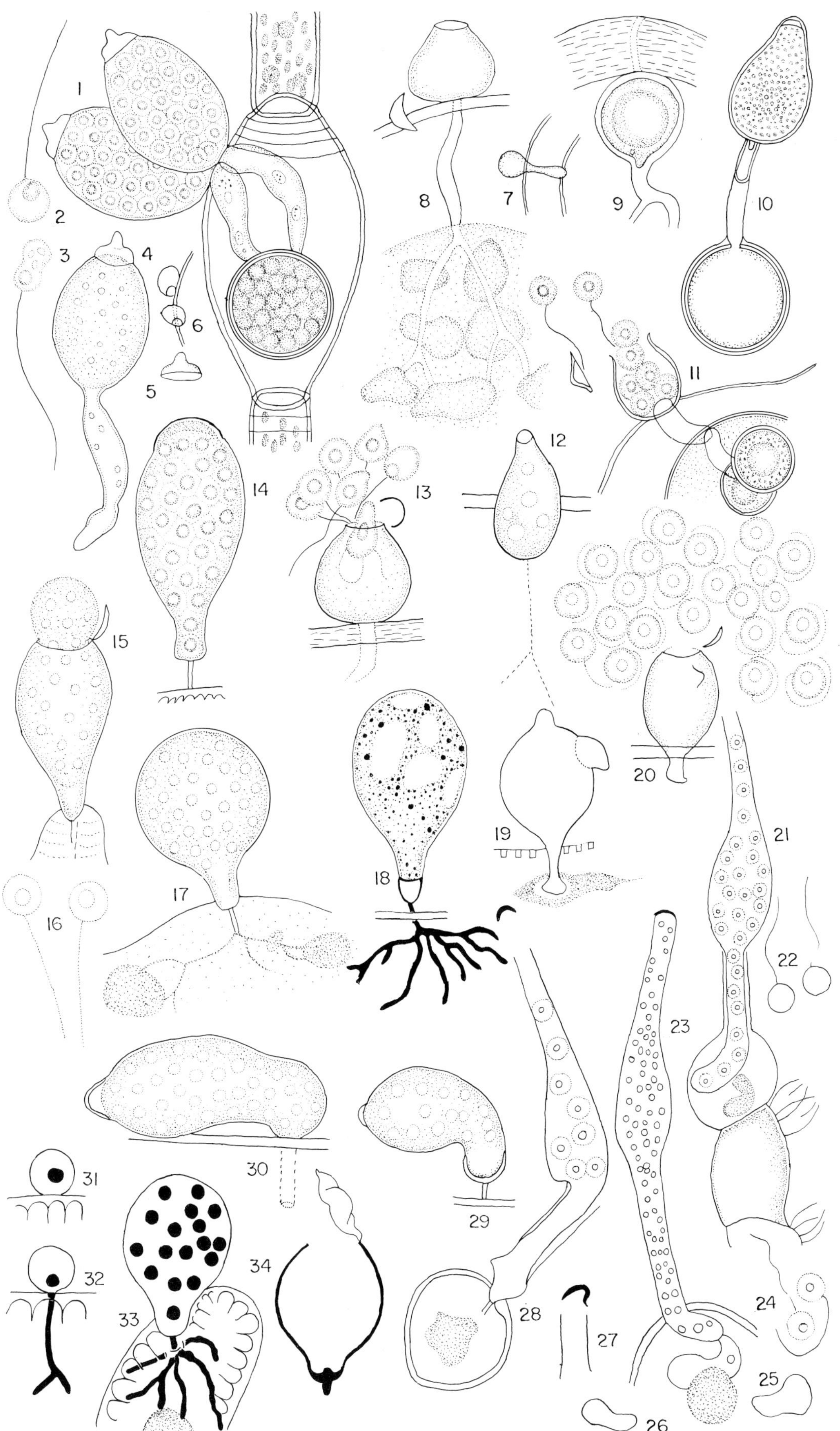
1
2
3
4
5
6
7
8
9
10
11
12
13
14
15
16
17
18
19
20
21
22
23
24
25
26
27
28
29
30
31
32
33
34

Chytridium

process occurs so rapidly that unless one observes the instant of dehiscence the extrusion of the endooperculum may be overlooked. Discharge of planospores may occur from one or more papillae, and presumably as the pressure within the zoosporangium is suddenly reduced several endoopercula are retracted or displaced inward and swirled around within the zoosporangium by the swarming planospores (fig. 126). In the exooperculate species the operculum is pushed up or aside by the emerging mass of planospores (figs. 38, 65, 73, 96, 123), and in *K. asterocysta* Karling the planospores swarm in a vesicle before dispersing.

Resting spores are known in all species except *K. hyalina* and *K. marylandia*, and these are borne in the same manner as the zoosporangia in most of the species. They, also, vary markedly in size, shape and ornamentation of the outer wall (figs. 23, 27, 41, 42, 49, 50, 76–78, 84, 99, 100, 116–118). In rare instances 2 spores may develop in a thallus (fig. 26), or they may be formed in the upper part of the thallus by contraction of the content and its investment by a thick wall (fig. 27). In germinating they function as prosporangia

PLATE 67

Figs. 35, 36. *Chytridium* (?) *chaetophilum* Scherffel. (Scherffel, 1925.)

Fig. 35. Hairy zoosporangium on the setae of *Bulbochaete* sp. with an endobiotic peg.

Fig. 36. Questionable spiny, epibiotic resting spore of *C. chaetophilum* with an attached male (?) cell.

Figs. 37-39. *Chytridium perniciosum* Sparrow on *Navicula*. (Sparrow, 1933.)

Fig. 37. Two empty zoosporangia with branched rhizoids and a solid operculum.

Fig. 38. Abnormal zoosporangium with a horn-like operculum.

Fig. 39. Endobiotic spherical, hyaline, smooth-walled resting spore.

Figs. 40, 41. *Chytridium megastomun* Sparrow on *Striaria attenuata*. (Sparrow, 1934.)

Fig. 40. Zoosporangium with planospores and a branched rhizoid.

Fig. 41. Empty zoosporangium with a large subapical orifice.

Figs. 42, 43. *Chytridium polysiphoniae* Sparrow on *Ceramium fructiculosum*. (Sparrow, 1934.)

Fig. 42. Mature zoosporangium with a peg-like rhizoid and an apical operculum.

Fig. 43. Discharge of planospores.

Fig. 44. *Chytridium papillatum* Sparrow; zoosporangium on *Stigeoclonium* with a branched rhizoid. (Sparrow, 1933.)

Figs. 45-48. *Chytridium rhizophydii* Karling on *Rhizophydium nodulosum*. (Karling, 1948.)

Fig. 45. Spherical, 2-2.8 μ diam., planospores with a minute hyaline refractive globule.

Fig. 46. Young stalked thallus.

Fig. 47. Mature zoosporangium with fine, branched rhizoids.

Fig. 48. Discharge of planospores; rhizoid needle-like and unbranched.

Figs. 49-52. *Chytridium microcystidis* Rohde and Skuja on *Microcystis* spp. (Skuja, 1948.)

Fig. 49. Young thallus with some of the rhizoids epibiotic.

Fig. 50. Mature zoosporangium with rhizoids arising from several points on its basal periphery.

Fig. 51. Zoosporangium with planospores and a strongly convex operculum.

Fig. 52. Spherical, 3 μ diam., planospore with a hyaline refractive globule.

Figs. 53-56. *Chytridium lagenula* Braun on *Tribonema bombycina*. Figs. 53, 54 after Braun, 1855; figs. 55, 56 after Scherffel, 1926.

Fig. 53. Young zoosporangium.

Fig. 54. Zoosporangium with planospores.

Fig. 55. Zoosporangium with a needle-like basal filament; host has formed a protective plug around the base of the filament.

Fig. 56. Nearly spherical endobiotic hyaline, smooth-walled resting spore with a large oil globule.

Figs. 57-61. *Chytridium lecythii* Goldie-Smith on the rhizopod, *Lecythium hyalinum*. Fig. 57 after Ingold, 1941; figs. 58-61 after Goldie-Smith, 1946.

Fig. 57. Two young thalli with branched rhizoids.

Fig. 58. Mature zoosporangium.

Fig. 59. Spherical planospore with an eccentric refractive globule.

Fig. 60. Empty zoosporangium with an attached wrinkled, flap-like operculum.

Fig. 61. Detached operculum.

Figs. 62-65. *Chytridium cocconeidis* Canter on *Cocconeis pediculus*. (Canter, 1947.)

Fig. 62. Zoosporangium with an unbranched epibiotic rhizoid.

Fig. 63. Almost mature zoosporangium with a branched epibiotic rhizoid.

Fig. 64. Zoosporangium with planospores.

Fig. 65. Dehisced zoosporangium and spherical, 2-3 μ diam., planospores with a hyaline refractive globule; operculum in surface view at and above the exit orifice.

Figs. 66-72. *Chytridium proliferum* Karling; saprophytic on various substrata. (Karling, 1967.)

Fig. 66. Spherical, 2-3 μ diam., and ovoid planospores with a hyaline refractive globule.

Fig. 67. Germination of planospore.

Fig. 68. Young thallus with a branched rhizoid.

Fig. 69. Sessile zoosporangium with a thickened basal portion and branched rhizoids.

Fig. 70. Mature stalked zoosporangium.

Fig. 71. Upper portion of a zoosporangium showing initial stage of planospore discharge.

Fig. 72. Proliferation of zoosporangia.

Fig. 73. *Chytridium lateoperculatum* Scherffel; urn-shaped zoosporangium on the zoocyst of *Vampyrella pendula*. (Scherffel, 1926.)

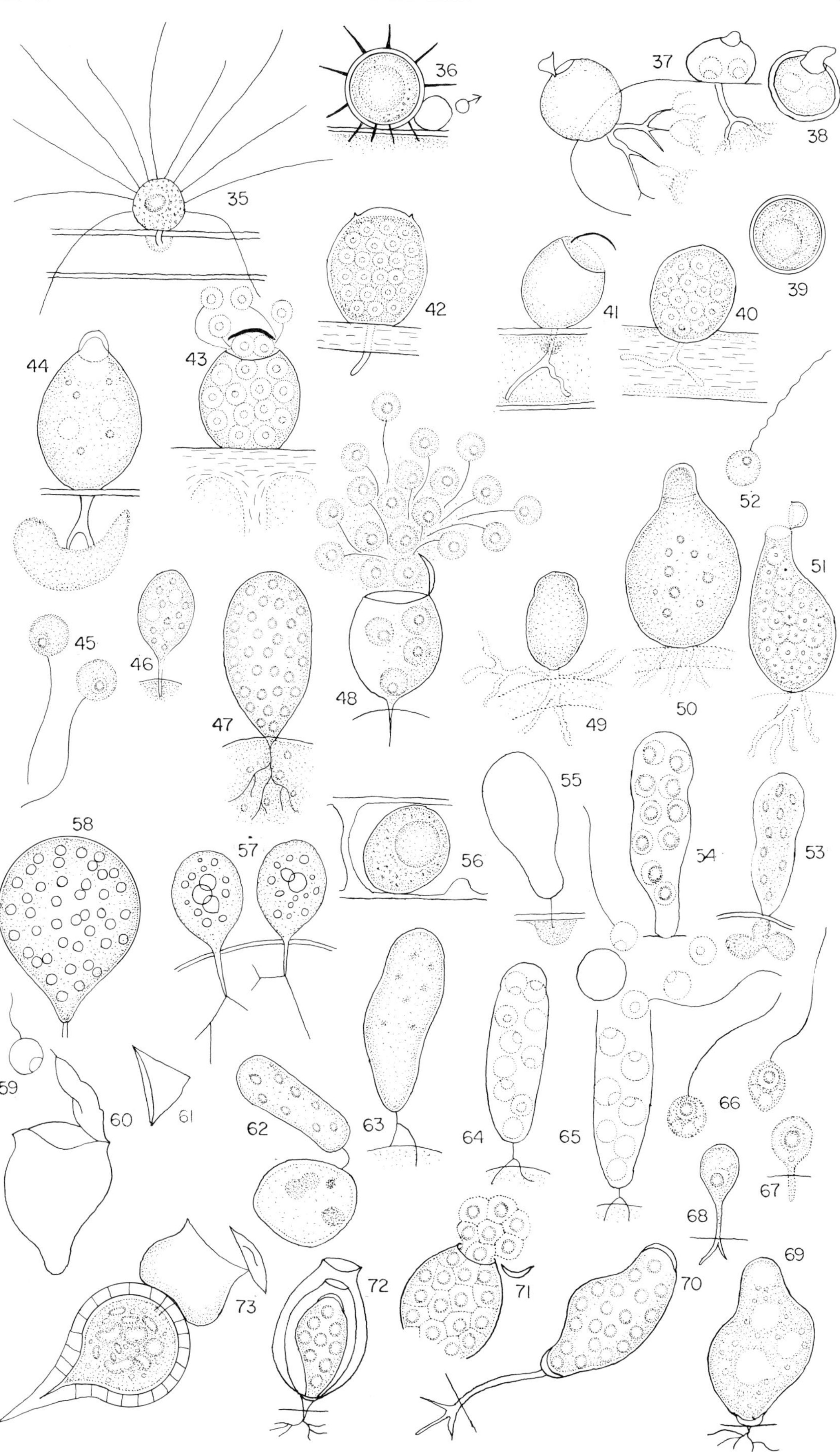
35
36
♂
37
38
39
40
41
42
43
44
45
46
47
48
49
50
51
52
53
54
55
56
57
58
59
60
61
62
63
64
65
66
67
68
69
70
71
72
73

Chytridium

(figs. 28, 43, 51, 70).

As to the occurrence of sexual reproduction in *Karlingia*, Couch (1939) reported that *K. rosea* is heterothallic and that sporangial strains of the opposite sex must be brought together before resting spores will develop. He did not observe how or if fusions occur. In *K. dubia* Karling as well as in other species, Karling (1937, 1939) reported that the resting spores develop asexually, but Willoughby (1957) found that they are formed by fusion of gametes. According to him, the planospores appear to be facultative and may give rise to vegetative thalli or fuse as gametes. He followed the successive fusion stages in hanging drops and cultures and showed that the zygotes develop into resting spores (figs. 107–109). After discharge the gametes become distinguishable by their greatly reduced motility and often occur close together in groups. Pairs of gametes, apparently with entangled flagella, come to rest on the substratum in contact with each other and encyst. A rhizoid develops from one of them (fig. 101) without an increase in diameter of both gametes, and after a while an isthmus develops between them (fig. 102). This bridge widens (fig. 103) until an ovoid zygote (fig. 104) is formed which subsequently increases in size in all directions (fig. 106) and eventually develops into a mature resting spore. Occasionally, triple

PLATE 68

Figs. 1-16. *Diplochytridium lagenaria* Karling. Figs. 1-12 after Karling, 1936; figs. 13-16 after Sparrow, 1936.

Fig. 1. Slightly ovoid, 3-5.5 μ diam., planospore with a hyaline refractive globule.

Fig. 2. Germinated planospore on *Oedogonium* sp.

Fig. 3. Later stage, apophysis or prosporangium developing in the germ tube above the branches.

Fig. 4. Fully-formed prosporangium or apophysis; planospore cyst persistent.

Fig. 5. Enlargement of planospore cyst as exogenous growth begins.

Fig. 6. Incipient zoosporangium developing from planospore cyst as protoplasm moves outward.

Fig. 7. Ovoid zoosporangium shortly before cleavage.

Fig. 8. Globular mass of planospores above exit orifice.

Fig. 9. Young resting spore formed from apophysis or prosporangium; planospore cyst persistent.

Fig. 10. Mature, smooth-walled, hyaline resting spore with a large central refractive globule.

Fig. 11. Early stage in resting spore germination; protoplasm moving outward from prosporangium or apophysis and expanding the planospore cyst.

Fig. 12. Mature epibiotic zoosporangium formed from a large prosporangium; rhizoids omitted.

Fig. 13. Flagellate planospore at rest atop a germinated one in a hanging drop culture.

Figs. 14, 15. Fusions at the point of contact leading to a marked rejuvinescence of the thallus from the germinated planospore.

Fig. 16. Empty zoospore cyst atop a rejuvinated thallus after fusion.

Figs. 17-28. *Diplochytridium sexuale* Karling on *Vaucheria germinata*. (Koch, 1951.)

Fig. 17. Spherical, 3.2-4.5 μ diam., planospore with a hyaline refractive globule.

Fig. 18. Enlarged view of a fixed and stained planospore showing whiplash flagellum, nuclear cap, colorless nucleus, and lipoidal globule.

Fig. 19. Incipient clavate zoosporangium which has budded out from the apex of the persistent planospore cyst; with numerous lipoidal globules; rhizoids lacking on the globular apophysis or prosporangium.

Fig. 20. Discharge of planospores from a somewhat curved clavate zoosporangium.

Fig. 21. Enlarged view of exit orifice and operculum.

Fig. 22. Germinated female gametangium on the host connected with a male by a short conjugation tube.

Fig. 23. Three males on a female gametangium.

Fig. 24. Apophysate female gametangium with a fusing male and a supernumerary flagellate male (?) at the apex.

Fig. 25. Completion of fusion, protoplasm moving down (?) into the apophysis or prosporangium.

Fig. 26. Later stage of resting spore or zygote development.

Fig. 27. Immature zygote formed in enlarged apophysis (?), supernumerary male cell attached to empty female gametangium.

Fig. 28. Mature, hyaline zygote with a large central refractive globule surrounded by smaller ones, and a slightly warted outer wall.

Figs. 29-40. *Diplochytridium stellatum* Karling on *Zygnema* sp. (Koch, 1951.)

Fig. 29. Spherical, 4.2-5.5 μ diam., planospores with a hyaline refractive globule.

Fig. 30. Germination of planospore.

Figs. 31, 32. Stages showing the development of the apophysis or prosporangium as an enlargement in the germ tube.

Fig. 33. Young thallus with rhizoids attached at the base of the prosporangium.

Fig. 34. Sporangial rudiment budding out of the side of the thick-walled zoospore cyst; thallus stalked and largely epibiotic.

Fig. 35. Small, mature zoosporangium with an empty prosporangium.

Fig. 36. Discharge of planospores from the same zoosporangium 4 hrs. later.

Fig. 37. Early stage of resting-spore thallus (?); upper part of planospore cyst collapsed.

Fig. 38. Later stage in resting spore development from the apophysis or prosporangium.

Fig. 39. Still later stage, spines forming on the outer wall.

Fig. 40. Mature resting spore with a large central globule surrounded by smaller ones; wall 2-layered, glistening outer layer bearing spines, lobes, and excrescences.

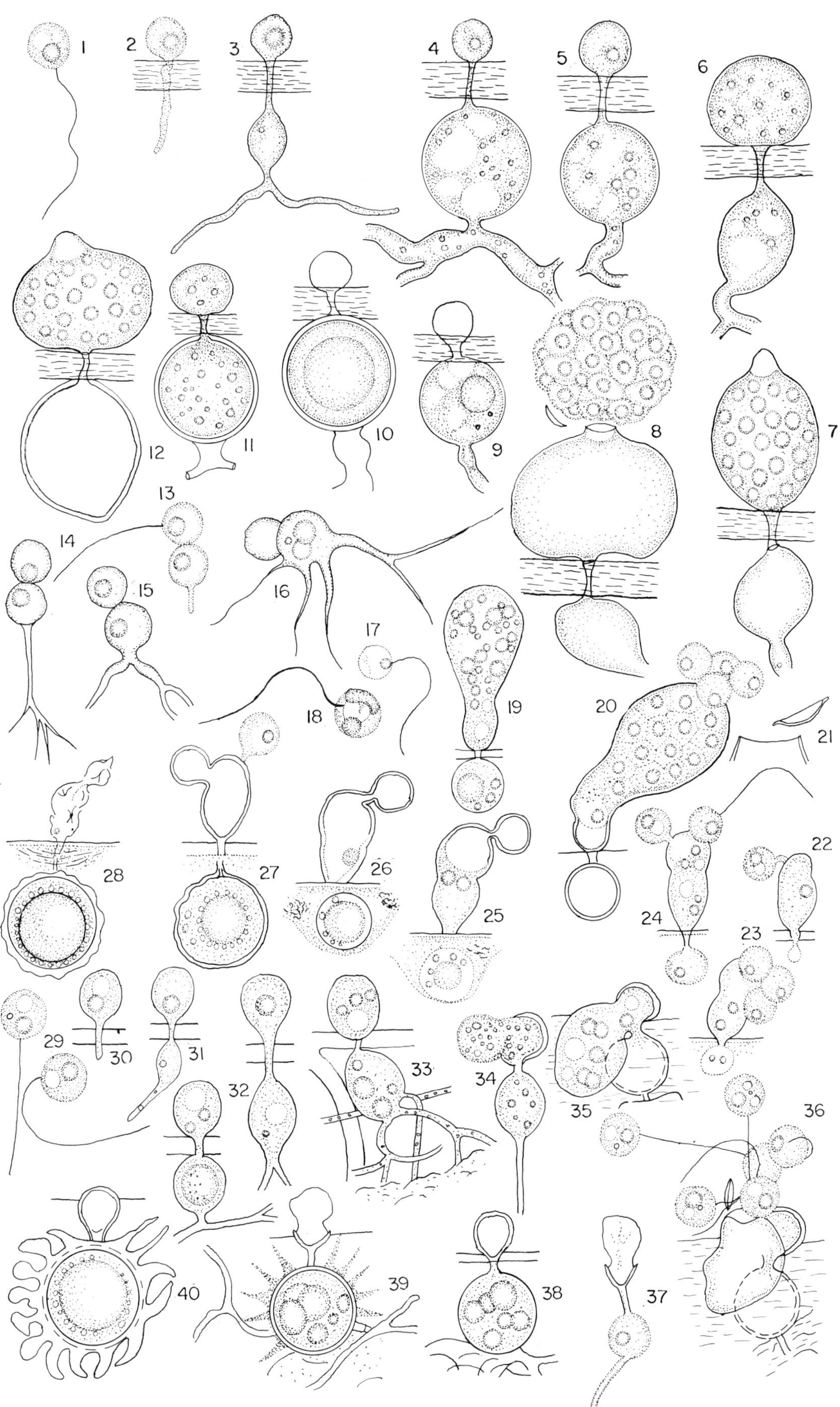

Diplochytridium

fusions occur (fig. 105). Dogma (1974c) confirmed Willoughby's observations in most details and noted the development of mature resting spores and their germination from such fusions. He (1974a) also found that essentially the same type of fusions occurred in some isolates of the *Karlingia rosea*-complex. Particularly noteworthy, as stressed by Willoughby, is the loss of morphological identity of the gametes during fusion and their incorporation in the resting spore. Only in a single instance did Willoughby find a spore with an empty thick-wall cyst or appendage. This type of sexual reproduction is unlike that of nearly all members of the Rhizidiaceae and more like that of species of the Olpidiaceae and Synchytriaceae. Also, its fundamental similarity to the fusions reported in *Rhizophlyctis oceanis, R. ingoldii* and *Rhizophlyctis* sp. (Karling, 1970) shows that *Karlingia* and *Rhizophlyctis* are closely related genera in this respect.

ALLOCHYTRIDIUM Salkin

Amer. J. Bot. 57:655, 1970.

Plate 65

This operculate genus of one species, *A. expandens* Salkin, was isolated from a roadside puddle and grown axenically on a defined agar medium and onion skin, and in defined nutrient media in a perfusion chamber. Under these conditions it exhibited a primary and

PLATE 69

Figs. 41-49. *Diplochytridium aggregatum* Karling on *Oedogonium* sp. (Karling, 1938.)

Fig. 41. Localized aggregation of encysted planospores on the surface of an algal cell.

Fig. 42. Germinated planospore with branched rhizoid; incipient prosporangium or apophysis forming in the germ tube.

Fig. 43. Incipient zoosporangium budding out of the brown thick-walled planospore cyst as exogenous growth or movement of the protoplasm out of the prosporangium begins.

Fig. 44. Later stage of sporangial development; half of planospore cyst persistent; prosporangium elongate and constricted.

Fig. 45. Mature zoosporangium with the planospore cyst near the base.

Fig. 46. Globular mass of discharged planospores above the exit orifice.

Fig. 47. Spherical, 3-4.5 μ diam., planospore with a hyaline refractive globule.

Figs. 48, 49. Spherical and oblong hyaline resting spores with several large refractive globules; planospore cyst and germ tube persistent.

Fig. 50. *Diplochytridium gibbosum* Karling; empty apophysate zoosporangium on *Cladophora fracta*. (Scherffel, 1926.)

Figs. 51-56. *Diplochytridium schenkii* Karling. Figs. 51-54 after Scherffel, 1926; figs. 55, 56 after Sparrow, 1932.

Fig. 51. Germinated planospore on *Oedogonium* sp. with an incipient apophysis or prosporangium and a branched rhizoidal axis.

Fig. 52. Later stage of thallus development; prosporangium large; planospore cyst persistent.

Fig. 53. Zoosporangium budded out of planospore cyst, exogenous growth completed.

Fig. 54. Discharged globular mass of planospores above exit orifice.

Fig. 55. Spherical, hyaline, smooth-walled resting spore.

Fig. 56. Germination of resting spore.

Figs. 57-65. *Diplochytridium oedogonii* Karling on *Oedogonium* sp. Figs. 57, 62 after Couch, 1938; figs. 58-61, 63-65 after Canter, 1950.

Fig. 57. Recently discharged spherical, 4-5.5 μ diam., planospore with a hyaline refractive globule.

Fig. 58. Germinated planospore, apophysis forming in the germ tube.

Fig. 59. Young thallus with a large prosporangium or apophysis bearing 2 blunt-ended rhizoids which have penetrated 3 cell walls; planospore cyst persistent.

Fig. 60. Sporangial rudiment budding out of planospore cyst as exogenous growth begins.

Fig. 61. Later stage or sporangial development.

Fig. 62. Zoosporangium discharging planospores.

Fig. 63. Young resting spore developing from the prosporangium or apophysis; planospore cyst persistent.

Fig. 64. Mature resting spore with a large and numerous small refractive globules and a smooth hyaline to yellowish-brown wall.

Fig. 65. Oblong resting spore with refractive globules.

Figs. 66-68. *Diplochytridium cejpii* Karling on *Vaucheria sessilis*. (Fott, 1950.)

Fig. 66. Subspherical 3-4 μ diam., planospore with a hyaline refractive globule.

Fig. 67. Dehisced zoosporangium with planospores which has budded out of apex of planospore cyst; rhizoids lacking (?) on prosporangium.

Fig. 68. Incipient zoosporangium budding out of apex of planospore cyst.

Fig. 69. *Diplochytridium inflatum* Karling on *Cladophora*; zoosporangium discharging planospores; rhizoids lacking on prosporangium or apophysis. (Sparrow, 1933.)

Figs. 70-73. *Diplochytridium kolianum* Karling on *Spirogyra*. (Domjan, 1936.)

Fig. 70. Young thallus with a large prosporangium and branched rhizoids; planospore cyst persistent.

Fig. 71. Incipient zoosporangium which budded out of planospore cyst as exogenous growth occurred; planospore cyst moved upward by expanding zoosporangium.

Fig. 72. Fully-grown zoosporangium with a sublateral planospore cyst.

Fig. 73. *Diplochytridium mucronatum* Karling on *Oedogonium* sp.; mucronate zoosporangium discharging planospores. (Sparrow and Barr, 1955.)

Fig. 74. *Diplochytridium citriforme* Karling; mature zoosporangium on pollen of *Pinus caribaea* with empty prosporangium and branched rhizoids. (Sparrow, 1952.)

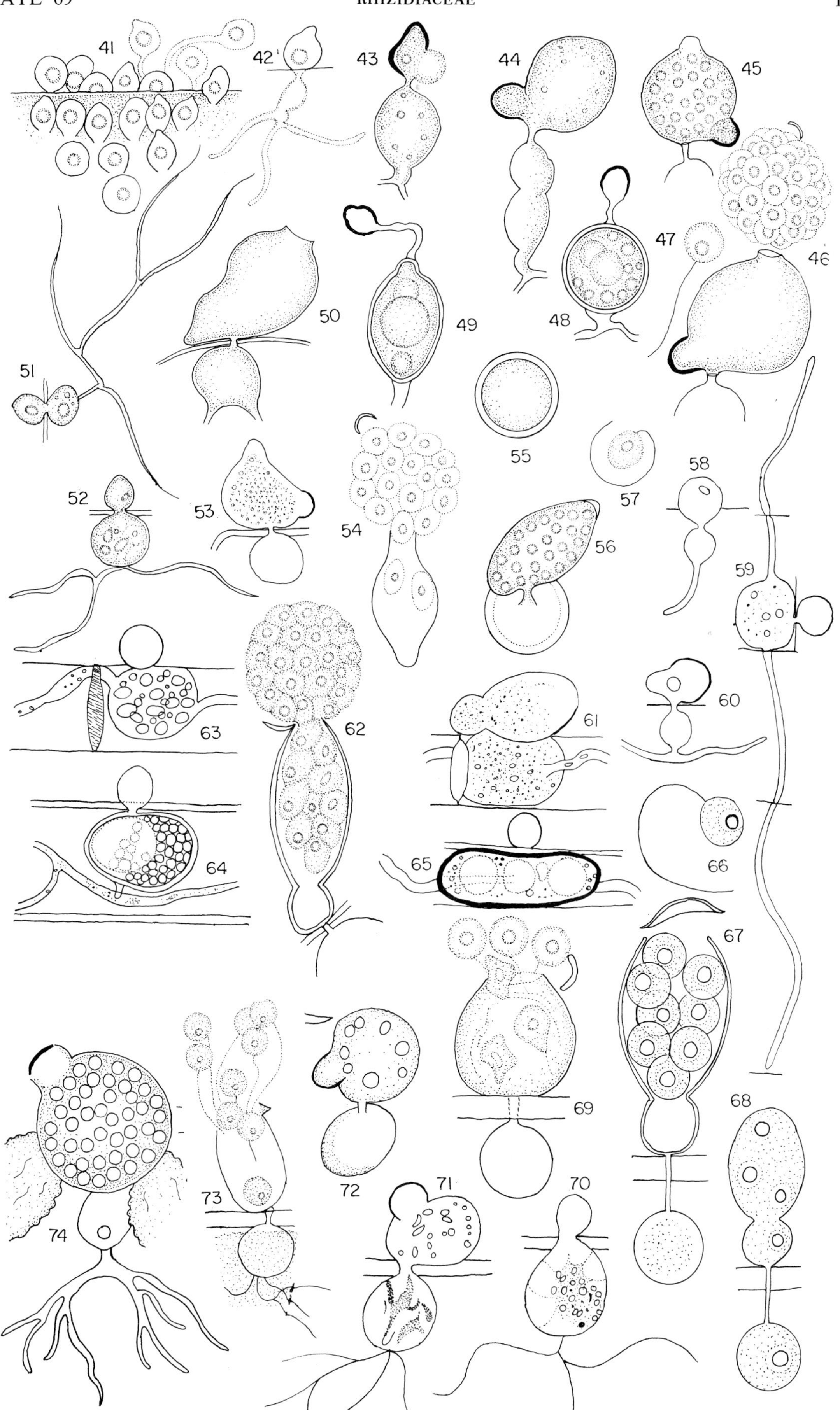

Diplochytriaium

secondary pathway of development of the zoosporangium. In the former, the zoosporangium developed from an enlargement of the germ tube (figs. 4–17), and this primary type was characteristic for 70–75% of the 500 thalli tabulated. The remainder of the thalli followed the secondary pathway of development in which the zoosporangium was formed directly by the enlargement of the planospore cyst (figs. 18–22).

In the primary type Salkin described the planospore cyst as fusing with the zoosporangium (figs. 7–10), but use of the word fusion here is misleading. Apparently, what occurs is the gradual expansion of the sporangial rudiment whereby the portion of the germ tube between the former and the cyst gradually becomes incorporated in the incipient zoosporangium. As enlargement of the sporangial rudiment continues the planospore cyst itself may become adnate to (figs. 8, 14), or finally confluent with the zoosporangium.

In either type of development the resulting mature thallus consists of an ampulliform, globose, or subglobose, hyaline, smooth, operculate zoosporangium with several rhizoidal axes arising from its periphery (figs. 9, 12–14, 20–22) as in *Karlingia*, or occasionally from the base (fig. 16). In onion skin and agar plate cultures the rhizoidal axes are more often inflated and constricted at irregular intervals (figs. 13, 14) as in *Cylindrochytridium, Truitella,* and *Karlingia*, for example, and have a catenulate appearance. In such cultures the zoosporangia may be irregular in shape with multiple, or elongate exit tubes (figs. 16, 17). At dehiscence the operculum is lifted off by the emerging planospores (fig. 15) which escape into a vesicle and remain motionless for 1–2 minutes before becoming motile and gradually dispersing. However, they apparently do not swarm in a vesicle, according to Salkin's account, which suggests that the so-called vesicular membrane may be only a surrounding layer of viscid material.

It is worth noting here that the 2 types of development described above occurred on agar and onion skin cultures and in a perfusion chamber with a defined nutrient medium free of bacteria and other contaminants. This may be a very unnatural environment for *A. expandens* compared with the roadside puddle water in which it was initially growing. Unfortunately, no tabulations of the dominant pathway of development were made from thalli growing in the natural non-enriched environment of ditch water containing bacteria and other organisms.

Until this is done and resting spores have been found and germinated the validity and relationships of *Allochytridium* will remain uncertain. By its monocentric eucarpic thalli and operculate zoosporangia with predominantly radiating rhizoidal axes it is strikingly similar to *Karlingia*. In the latter genus the zoosporangia may develop sometimes from an enlargement in the germ tube and the radiating rhizoids may be distinctly constricted at irregular intervals (pl. 64, figs. 91, 94, 111, 114, 121) in several species. Development of the zoosporangium from the germ tube occurs in fairly high percentages in other chytrids, also, i.e., *Asterophlyctis sarcoptoides* and *Chytriomyces fructicosus*, for example, according to Antikajian (1949), Karling (1949), Dogma (1974).

Thus, it remains to be seen how significant taxonomically type of development is, although it is presently used as a convenience to distinguish families and subfamilies. *Allochytridium* is herewith placed next to *Karlingia* with which it will probably prove to be identical.

Plate 70

Figs. 1-6, 8-11, 18-24, 27. *Blyttiomyces spinulosus* Bartsch. Figs. 1, 5, 8, 27, 18-24 after Bartsch, 1939; fig. 6 after Rieth, 1956; figs. 10, 11 after Scherffel, 1926.

Fig. 1. Zygote of *Spirogyra weberi* infected by 3 young thalli, a resting spore, and an empty spiny zoosporangium.

Fig. 2. Spherical, 4.2 μ, planospores with a large clear refractive globule seen in 2 views.

Fig. 3. Germination of planospore.

Figs. 4, 5. Young thalli with apophyses and rhizoids developing within the host zygospore.

Fig. 6. Multi-apophysate young thallus.

Fig. 8. Mature apiculate zoosporangium with 2 endobiotic apophyses.

Fig. 9. Spiny apiculate zoosporangium with planospores.

Fig. 10. Empty zoosporangium showing irregular subapical orifice.

Fig. 11. Planospore.

Figs. 18-24, 26. Stages in the development of the endobiotic resting spores; planospore cyst and germ tube persistent.

Fig. 27. Early stages of resting spore germination.

Figs. 7, 17, 25. *Blyttiomyces* sp. Persiel on pollen grains. (Persiel, 1960.)

Fig. 7. Planospore body expanded equally in all directions to form an incipient zoosporangium; apiculus lacking.

Fig. 17. Early stage of resting spore development as an enlargement in the germ tube.

Fig. 25. Mature resting spore with planospore cyst as an appendage.

Figs. 12-14. *Blyttiomyces helicus* Sparrow and Barr on pine pollen. Figs. 12, 13 after Sparrow and Barr, 1955; fig. 14 after Rieth, 1956.

Fig. 12. Zoosporangium with helical bands.

Fig. 13. Spherical, 4-4.8 μ diam., planospore with a large hyaline refractive globule.

Fig. 14. Resting spore.

Figs. 15, 16. *Blyttiomyces laevis* Sparrow on *Zygnema*. (Sparrow, 1952.)

Fig. 15. Discharge of planospores from 2 subapical orifices.

Fig. 16. Empty apiculate zoosporangium.

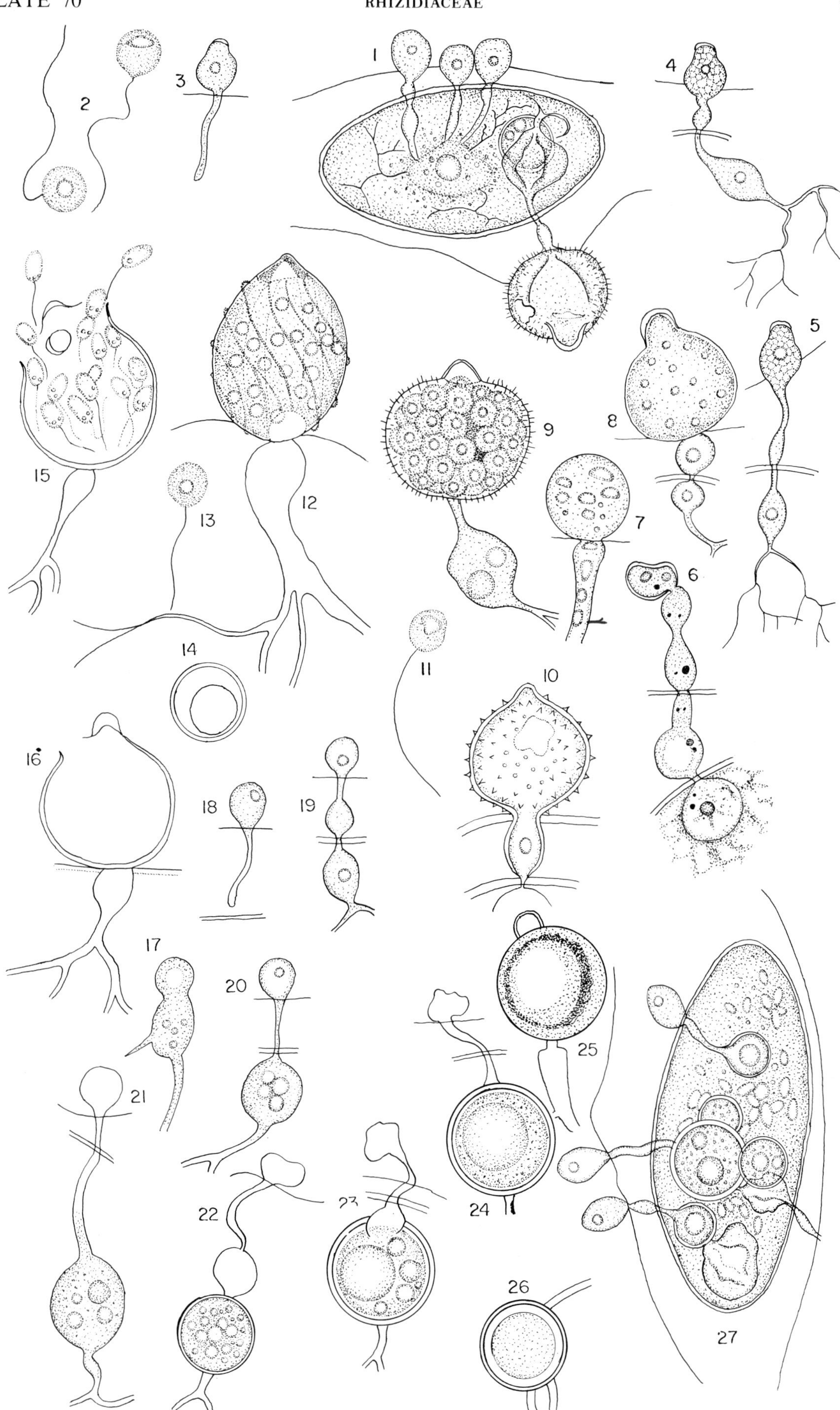

Blyttiomyces

SUBFAMILY CHYTRIDIOIDEAE

As interpreted here this subfamily includes *Chytridium, Diplochytridium, Catenochytridium,* and *Blyttiomyces*. The thallus of these genera is epi-endobiotic, and in some species the encysted planospore develops directly in part or wholly into a zoosporangium as in *Rhizophydium*, while other species have an endo- exogenous type of development. In the latter, development of the persistent planospore cyst is delayed until the endobiotic absorbing system is fully developed. The apophysis functions as a prosporangium in the sense that its accumulated protoplasm flows outward slowly and expands the planospore cyst wholly or partly into a zoosporangium. The resting spores are endobiotic and are formed asexually or sexually, usually in the enlarged apophysis or haustorium. Where known these function as prosporangia in germination. Possibly, *Canteria* might be included in this subfamily because it appears to develop endo-exogenously and produces endobiotic, sexually formed resting spores, but it is listed among the monocentric genera of doubtful affinity.

CHYTRIDIUM Braun

Betrachtungen über die Erscheinung der Verjüngung in der Natur. —, p. 198.
Leipzig, 1851; Monatsber. Berlin Akad. 1855:378.
(Sensu recent Karling, Arch. Mikrobiol. 76:127, 1971)

Plates 66, 67

As presently defined by nearly all chytridologists this operculate genus includes approximately 57 species, few of which are fully known and several are doubtful members. In addition several unidentified specimens in which the presence of an operculum is unknown have been assigned to this genus. However, this interpretation of *Chytridium* is inconsistent in that it includes some apophysate species with an endo-exogenous type of growth and development and other non-apophysate members which develop like *Rhizophydium* species insofar as their zoosporangia and absorbing or rhizoidal system are concerned. Several workers including Whiffen (1944) and Karling (1948) have noted this inconsistency and suggested that these operculate chytrids might be best accommodated in separate genera. This had not been done, and this writer (1971) proposed the establishment of a new genus, *Diplochytridium*, for the species with an endobiotic or intramatrical prosporangium or apophysis, of which some have the endo- exogenous type of development.

Chytridium is, accordingly, restricted to approximately 25 known species and several incompletely known members which lack an apophysis and develop like species of *Rhizophydium*, insofar as the zoosporangium and absorbing system or rhizoids are concerned, and form predominantly endobiotic or intramatrical resting spores. Knox (1970) and Roane (1973) however, reported occasional epibiotic resting spores in some species of this genus. Nonetheless, the above interpretation corresponds fairly closely to Braun's diagnosis of the type species. Even so, the distinctions between so-called apophysate and non-apophysate species are not always sharply defined, and it is difficult to determine whether a fairly thick, tubular or sac-like endobiotic haustorium is to be regarded as an apophysis or not. Furthermore, the successive developmental stages in most of such species are not

PLATE 71

Figs. 28-34. *Blyttiomyces* sp. Willoughby on *Rhizophydium coronum*. (Willoughby, 1965.)

Fig. 28. Apiculate zoosporangium, young thallus, and encysted planospore lying within the coronum and parasitizing the host zoosporangium.

Fig. 29. Two infecting planospores and a young zoosporangium.

Fig. 30. Maturing zoosporangium with apiculus and germ tube.

Fig. 31. Two young zoosporangia with an elongate distal end and apiculus.

Fig. 32. Mature zoosporangium with apiculus and 3 exit papillae or pores.

Figs. 33, 34. Empty apiculate zoosporangia.

Figs. 35-51. *Blyttiomyces rhizophlyctidis* Dogma and Sparrow on *Karlingia* (*Rhizophlyctis*) *rosea*. (Dogma and Sparrow, 1969.)

Fig. 35. "Resting spore and sporangial thalli in various stages of development on the host sporangium."

Fig. 36. Amoeboid and spherical, 2.5 μ diam., planospores with pale-yellowish, clustered, orange refractive globules.

Figs. 37-40. Stages in the development of sporangial thalli.

Fig. 41. Apiculate zoosporangium and thallus developing in the host exit plug and sporeplasm.

Fig. 42. Immature zoosporangium with its protoplasm at the "accumulation stage" of development.

Fig. 43. Mature zoosporangium; apiculus capping a discharge tube.

Fig. 44. Discharge of planospores from 3 orifices.

Figs. 45, 46. Empty zoosporangia.

Figs. 47-49. Young stages of resting spore development.

Figs. 50, 51. Mature hyaline resting spores formed by the encystment of the apophysis; planospore cyst and germ tube persistent; rhizoids omitted; shrunken fat globule pale yellowish-orange.

Figs. 52-54. *Blyttiomyces aureus* Booth on pollen grains. (Booth, 1969.)

Fig. 52. Germinating planospore cyst with an apiculus on the end of the germ tube.

Fig. 53. Young vacuolate zoosporangium with the apiculus on the sporangial wall.

Fig. 54. Irregularly-shaped planospores with granular protoplasm and one posterior minute refractive globule.

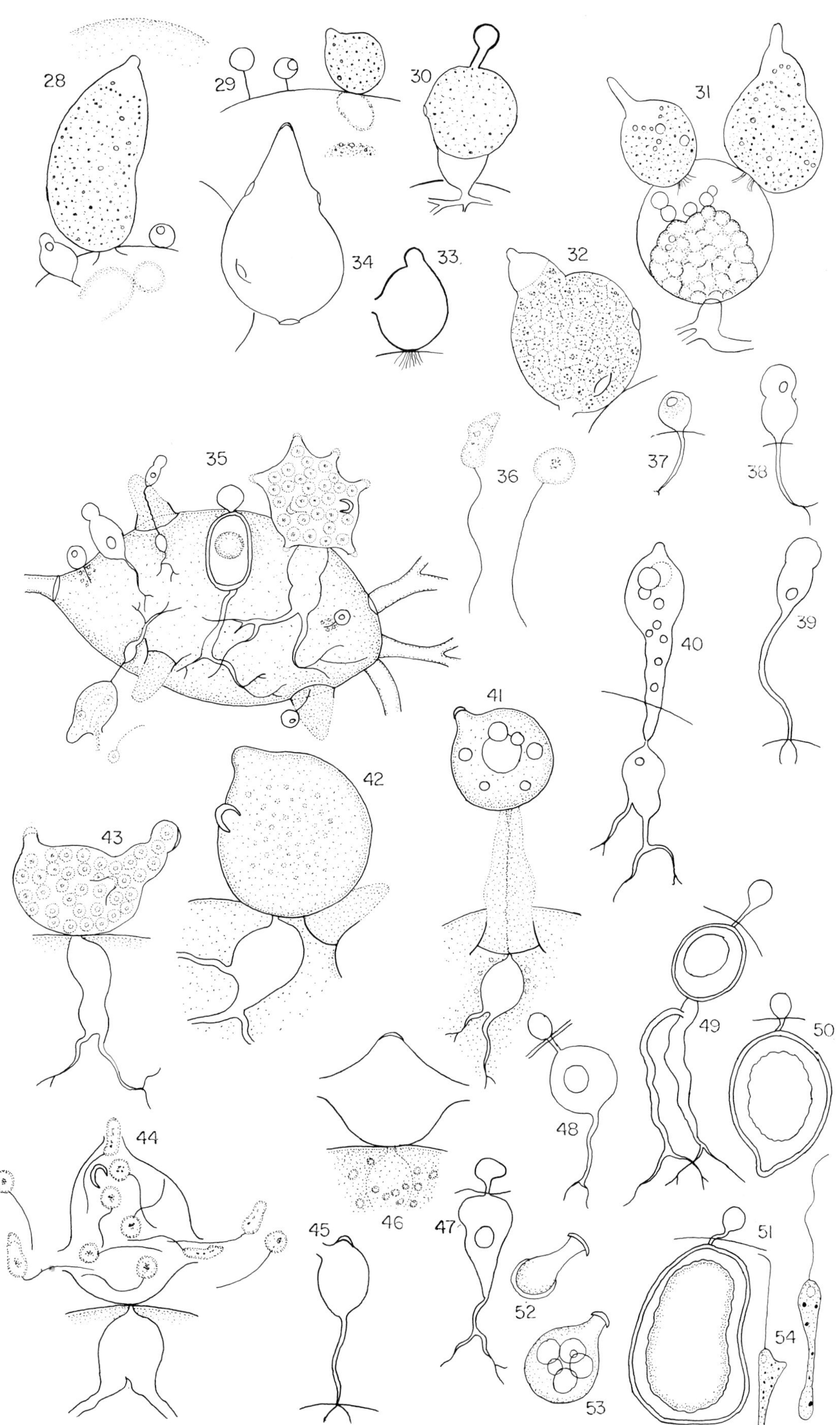

Blyttiomyces

fully known, stages which would show whether or not the haustorium functions as a prosporangium. Therefore, it remains to be seen from intensive studies on many incompletely-known species if this segregation into 2 genera will be justified. Nevertheless, it eliminates some inconsistencies and appears to be a practical solution insofar as our knowledge of these chytrids goes at present. Also, it may be argued that if the presence or absence of an apophysis is acceptable as a generic distinction between *Phlyctochytrium* and *Rhizophydium*, it should be recognized as such for the various species generally included in *Chytridium*.

In the restricted sense noted above, species of *Chytridium* are characterized by epibiotic or extramatrical operculate sessile or stalked zoosporangia, endobiotic haustoria or branched or unbranched rhizoids, and predominantly endobiotic resting spores. Such species are almost worldwide in distribution and parasitize freshwater and marine algae, fungi, rhizopods and protozoa. *Chytridium proliferum* Karling apparently occurs as a saprophyte in acid, pH 4.7, soil and has been isolated on dead pollen grains, bleached corn leaves, snake skin and fibrin film, and it may occur as a weak parasite of moribund zoosporangia of *Chytriomyces* and the aeciospores of a rust.

The zoosporangia of these species vary markedly in shape (figs. 1, 12, 15, 17–21, 29, 20, 51, 64, 70) and size and are usually sessile on the host although stalked ones may occur (figs. 12, 29, 70). In *C. pilosum* Kobayasi and Konno (1970) they are densely spinose and frequently deeply lobed. Likewise, the endobiotic absorbing system varies markedly from a tubular (figs. 1, 4, 11, 28), sac-like (figs. 21–23), or lobed haustorium which may bear rhizoids at its tip (fig. 8) to a single unbranched (figs. 42, 48, 55) or branched (figs. 8, 17, 33, 47, 44, 57, 69, 72) rhizoidal axis. In some thalli and species the unbranched rhizoid may be needle-like (figs. 46, 48) or filamentous (fig. 62). The rhizoidal axis usually arises at one point at the base of the zoosporangium, but in *C. microcystidis* Skuja they may arise at several points on the basal periphery (fig. 50). An exception to the endobiotic position of the rhizoid occurs in *C. curvatum* Sparrow where the tip of the subsporangial stalk does not penetrate beyond the gelatinous sheath around the *Stigeoclonium* filament (fig. 29). Also, in *C. cocconeidis* Canter (figs. 62–65), *C. microcystidis* (fig. 49) and *C. schenkii* (Schenk) Scherffel the rhizoidal branches may be partly epibiotic.

In development, the encysted planospore develops a germ tube which usually penetrates the host wall or substratum and branches (figs. 32, 67, 68) or remains unbranched (figs. 46, 58), and eventually becomes the absorbing system. Meanwhile, the planospore body enlarges to become the epibiotic zoosporangium which bears an apical or subapical operculum. In *Chytridium* sp. Schulz (1923) two convex opercula occur at the apex of the zoosporangium. The opercula in various species may be umbonate (figs. 1, 4, 5) shallow saucer-shaped (figs. 15, 20, 41, 43), disc-like, convex (figs. 11, 51), or horn-like and solid (figs. 37, 38). In *C. lecythii* (Ingold) Goldie-Smith (fig. 60) and *C. surirellae* Friedmann (fig. 34) the operculum may be quite thin and become wrinkled and flap-like after dehiscence.

During dehiscence the operculum is pushed off (figs. 11, 13, 20, 43) or remains attached at the edge of the orifice (figs. 15, 51, 48, 71) and the planospores usually emerge slowly to form a motionless globular mass which loosens up as the planospores become active and disperse (figs. 13, 20). However, in *C. rhizophydii* Karling (fig. 48) dehiscence occurs with almost explosive force, and as a mass of planospores is shot out the operculum is pushed up or back with such speed that its movement is usually invisible. The recoil of the zoosporangium may be so great that it is torn loose from the rhizoidal axis and host.

Resting spores have been reported in only 3 of the distinctly operculate species assigned here to *Chytridium*. These develop endobiotically, but their successive developmental stages are not known. In *C. olla* Braun they appear to be formed in an expanded haustorium. In *C. chaetophilum* Scherffel the questionable spiny resting spore (fig. 36) is reported to be formed sexually and epibiotically, but if this spore belongs to the species with the hairy operculate (?) zoosporangium shown in fig. 35 *C. chaetophilum* should be excluded from *Chytridium* and transferred to *Chytriomyces*. Germination of the resting spores has been observed only in *C. olla* (figs. 10, 11) and in this process a canal is formed for the emergence of the protoplasm and the formation of a zoosporangium on the surface of the host.

Probably, some of these incompletely-known *Chytridium* species will be found to develop epibiotic resting spores (*C. neopapillatum* Kobayasi, *et al*, 1971) for instance, and in that event they should be transferred to *Chytriomyces*, like *C. suburceolatus* Willoughby (1964) and *C. parasiticum* Willoughby (see Karling, 1968). *Chytridium confervae* (Wille) Minden, also, develops epibiotic resting spores, according to Scherffel (1925) and Rieth (1951), and if their reports are confirmed this species should be excluded from *Chytridium* and assigned to *Chytriomyces*. According to Canter (1963) a species which she identified as *Chytridium cornutum* Braun forms resting spores sexually and epibiotically in sporangia-like container cells, and in the event her identification and account of development are correct this species, also, should be placed in *Chytriomyces*. *Chytridium* (?) *spirotaeniae* Scherffel with its epibiotic sexually formed resting spores may be a species of *Chytriomyces* if its zoosporangia are found to be operculate.

In relation to *Chytridium* it may be noted that Koch (1957) described a species on pollen grains and grass leaves in which plasmogamy occurs at the tip of rhizoids of 2 contributing thalli. "A short tube grows out spirally from the point of fusion, and the tip of this tube swells or buds out. The nuclei of the two contributing thalli move out of the old spore cysts, and karyogamy occurs in the swelling. This initiates the formation of a new rhizoidal system. The swelling continues to enlarge and develops into a typical chytrid resting spore with an orange wall." Koch did not identify this species, but his description of the type of sexuality suggests that it may be a species of *Chytriomyces*.

DIPLOCHYTRIDIUM Karling

Arch. f. Microbiol. 76:128, 1971.

Non *Diplochytrium* Tomaschek, Sitzungsber. Acad. Wiss. Wien (Math. Nat. Cl.) 78:198, 1879.

Plates 68, 69

As defined here *Diplochytridium* includes approximately 21 species which develop epibiotic or extramatrical zoosporangia, endobiotic prosporangia or apophyses with or without rhizoids, and endobiotic resting spores. Some of the species have been shown to grow and develop endo- exogenously, but in others the developmental stages are too incompletely known to determine their type of development.

In addition to the species illustrated in plates *68* and *69*, *D. langenaria* var. *japonense* (Kobayashi and Ookubo, 1953) Karling, *D. chlorobotytris* (Fott, 1952) Karling, *D. scherffelii* (Sparrow) Karling, *D. codicola* (Zeller) Karling, *D. isthmiophilum* (Canter, 1960) Karling, *D. mallamonadis* (Fott) Karling and *D. deltanum* (Masters) comb. nov. are included in this genus. *Chytridium neochlamydococci* Kobayashi and Ookubo (1954) is a doubtful *Diplochytridium* species because its so-called apophysis is so minute and obscure that it may be interpreted as a rhizoidal axis bearing a few filaments at its base.

Species of *Diplochytrium* are primarily parasites of algae, but *D. citriforme* (Sparrow) Karling (fig. 74) is saprophytic on pine pollen grains. *Diplochytrium lagenaria* (Schenk) Karling apparently is a weak parasite of moribund algal cells, and it has been grown as a saprophyte on dead algae. As in *Chytridium* the epibiotic zoosporangia vary considerably in shape from subspherical to clavate, pyriform and citriform and bear an apical or subapical operculum of various sizes and shapes. Several species are reported to lack rhizoids on the endobiotic apophysis (figs. 19, 20, 67–69), and this organ apparently absorbs nutriments from the hosts. In *D. oedogonii* (Couch) Karling the rhizoids become very extensive and blunt-ended (fig. 59), and may pass through the walls of several host cells or even to the outside of the host.

The weakly parasitic or saprophytic *D. lagenaria* may be taken as an illustration of the endo- exogenous type of growth and development. After encysting on the moribund or dead algal cell the planospores form a germ tube, which penetrates the host wall (fig. 2) and branches. Shortly thereafter a swelling develops in the germ tube (fig. 3) and becomes the incipient prosporangium or apophysis. The rhizoids continue to branch and extend in the host and may reach a diameter up to 9 μ. At the same time the prosporangium continues to enlarge as protoplasm accumulates in it until it reaches the sizes shown in figs. 4–6, and 43. Meanwhile the planospore cyst persists unchanged and is connected to the prosporangium or apophysis by the germ tube. Accordingly, growth up to this stage has been inward or endogenous. Then, after the endobiotic system has become well established, growth and development is outward or exogenous. Protoplasm of the endobiotic system slowly moves out of the prosporangium and expands the zoospore cyst (figs. 5, 6) until it becomes the epibiotic zoosporangium (figs. 7, 8). In some species (figs. 19, 34, 35, 43, 44, 53, 60, 61, 68, 71, 72) part of the zoospore cyst fails to expand, and the sporangial rudiment buds out of its side or apex. The unexpanded portion of the cyst persists on the zoosporangium as a thick-walled appendage (figs. 35, 45, 46, 71, 72).

The same type of growth occurs in the development of the resting-spore thalli. After the rhizoidal system and apophysis have become well-established by endogenous growth, protoplasm accumulates in the latter (fig. 9) which eventually becomes a thick-walled resting spore (figs. 9, 10, 48, 49, 63–65) and may remain dormant for a long time. Meanwhile, the planospore cyst and germ tube persist, and when germination begins the protoplasm moves outward and expands the cyst (fig. 11) into a zoosporangium (fig. 12).

The endo- exogenous type of development has been observed in *D. stellatum* (Koch) Karling (figs. 30–40) and *D. aggregatum* Karling (figs. 41–49), and possibly it occurs also in *D. schenkii* (Scherffel) Karling (figs. 51–56), *D. oedogonii* (Couch) Karling (figs. 58–65), *D. kolianum* (Domjan) Karling (figs. 70–72), *D. gibbosum* (Zopf) Karling (fig. 50), *D. citriforme* (Sparrow) Karling, and *D. mucronatum* (Sparrow and Barr) Karling (fig. 73). Whether or not it occurs in the apophysate species without rhizoids such as *D. sexuale* (Koch) Karling (figs. 19–28), *D. cejpii* (Fott) Karling (figs. 67, 68), *D. inflatum* (Sparrow) Karling (fig. 69), *D. mallomanadis* (Fott, 1957) Karling and *D. deltanum* (Masters) Karling remains to be seen from more intensive studies of the successive developmental stages. In *D. isthmiophilum*, Canter (1960) reported that the zoosporangium develops at the same time as the apophysis, so that endo- exogenous growth is not evident.

In *D. lagenaria, D. schenkii, D. oedogonii, D. aggregatum, D. stellatum* and *D. chlorobotrytis* the endobiotic resting spores apparently are formed asexually. However, in *D. lagenaria*, Sparrow (1936) reported that in pure water mounts a mobile planospore may fuse with another one which had germinated (figs. 13–15), but such fusions did not lead to the development of a zygote or resting spore. Instead it resulted in a remarkable rejuvenescence of vegetative development of the germinated planospore or thallus (figs. 15, 16). In *D. sexuale* on the other hand, the resting spores develop as the result of the fusion of the contents of a "male" planospore with a small "female" thallus or gametangium. A female planospore germinates on the host after which a male planospore comes to rest on it and forms a short conjugation tube (fig. 22) through which its protoplasm flows into the female (fig. 24). At the same time one or more (figs. 23, 24) male planospores may come into contact with a female gametangium, but it is not known if multiple fusions occur. The protoplasts of the "male" and "female" fuse (figs. 25, 26), and the empty male cyst persists on the female gametangium. Presumably, the fused protoplasts move down into the apophysis, but this has not been demonstrated conclusively. Koch (1951) found that the mature resting spore (figs. 27, 28) is markedly larger than the apophyses shown in figs. 25 and 26, and he questioned that this increase in size of the resting spore is due to the physical addition of the small male protoplast. Possibly, the apophysis with the fused protoplast continues to absorb nutriment from the host and thus increases in size. In *D. isthmiophilum* Canter reported that the endobiotic resting spores are formed sexually "by terminal fusion of two isogametes 3 μ diam. (?), one of which had previously come to rest and germinated." After fusion presumably, the resting spore is formed by an enlargement of the germ tube of the planospore which had germinated, although Canter did not observe germination of one of the planospores. Much of the same type of sexual reproduction has been found by Masters (1971) in *D. deltanum*. A male gametangium (gamete?, 1.5–2.5 μ diam.) encysts near a germinated female gametangium (gamete?, 2.5–3 μ diam.) and develops a short conjugation tube through which its protoplasm passes into the female. The fused protoplasts then pass through the infection tube of the female and presumably its tip within the host enlarges to become a resting spore. As in *D. lagenaria* several, 2–10, male gametangia may cluster around a female gametangium.

Germination of the sexually formed resting spores has not been observed, but in those developed asexually in *D. lagenaria* and *D. schenkii* they function as prosporangia when they germinate (figs. 12, 56).

BLYTTIOMYCES Bartsch

Mycologia 31:559, 1939.

Plates 70, 71

Although some previous workers had noted that the type species, *B. spinulosus* (Blytt) Bartsch, of this

PLATE 72

Figs. 1-12, 22. *Catenochytridium carolineanum* Berdan. Fig. 1 after Berdan, 1939; figs. 2-12 after Berdan, 1941; fig. 22 after Hanson, 1946.

Fig. 1. Sporangial thallus with intramatrical compound catenulate apophysis and rhizoids, an extramatrical operculate zoosporangium, and a globular mass of discharged zoospores.

Fig. 2. Spherical, 5-6 μ diam., planospores with hyaline refractive globules.

Fig. 3. Early germination stage.

Fig. 4. Later stage with the incipient primary segment of the apophysis formed at juncture of germ tube branches.

Fig. 5. Still later stage with the primary segment of the apophysis developing above the branches.

Figs. 6, 7. Stages in the development of the compound apophysis indicating endogenous growth of the intramatrical portion of thallus; constricted segments filled with refractive globules; planospore cyst persistent.

Fig. 8. Sporangial rudiment budding out at side of planospore cyst as exogenous growth begins; protoplasm moving outward.

Fig. 9. Similar stage; planospore cyst at apex of sporangial rudiment.

Fig. 10. Mature zoosporangium with broad exit papilla and unexpanded portion of planospore cyst.

Fig. 11. Mature intramatrical resting spore with a pale- to dark-brown smooth wall and filled with refractive globules; formed from one segment of compound apophysis.

Fig. 12. Germinated resting spore with layered wall.

Figs. 13-28. *Catenochytridium laterale* Hanson. (Hanson, 1946.)

Fig. 13. Spherical planospore, 2.9-4.5 μ diam., from living material, with a large and 2 smaller hyaline refractive globules.

Fig. 14. Fixed and stained planospore with a large nuclear cap enveloping the clear nucleus.

Fig. 15. Early germination stage.

Fig. 16. Later stage, primary segment of the apophysis developing above the rhizoidal branches.

Fig. 17. Fixed and stained extramatrical planospore cyst showing nucleus in the cyst.

Figs. 18, 19. Sporangial rudiment budding out of thick-walled planospore cyst.

Fig. 20. Later stage, fixed and stained, of sporangial rudiment with primary nucleus.

Fig. 21. Prophase nucleus in a binucleate sporangial rudiment.

Fig. 22. Nuclear division in sporangial rudiment.

Fig. 23. Anaphase in an incipient zoosporangium.

Fig. 24. Fully grown extramatrical zoosporangium with a conspicuous exit papilla, a thickened portion of the planospore cyst at the base, and primary and secondary segments of the apophysis.

Fig. 25. Initial stage of zoospore discharge.

Figs. 26-28. Mature intramatrical spherical and irregular resting spores with smooth golden-colored walls and filled with refractive globules; developed asexually by encystment of the primary segment of apophysis, thick-walled portion of planospore persistent as an appendage.

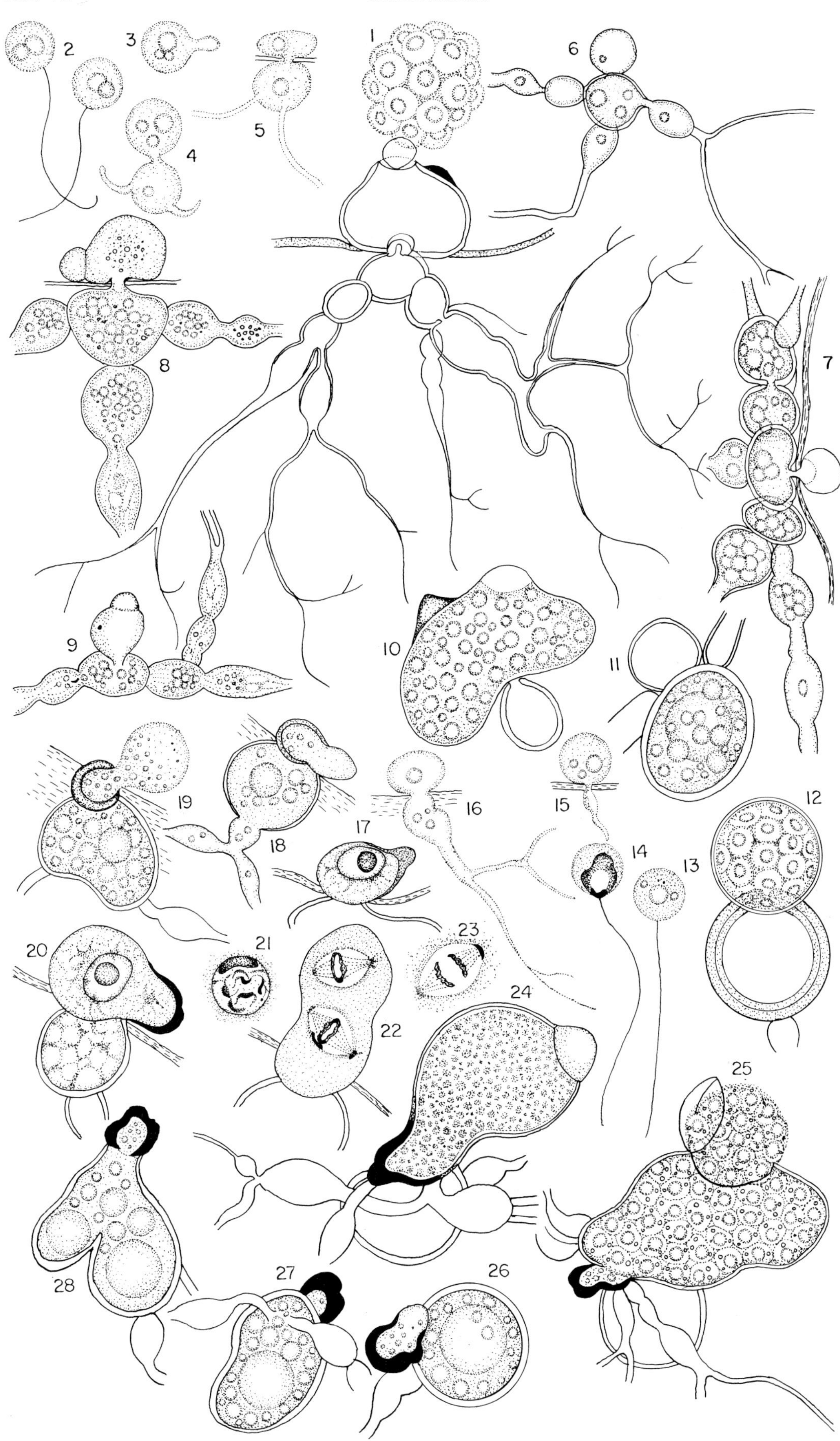

Catenochytridium

genus is inoperculate it was still classified as a species of *Chytridium* until Bartsch showed conclusively that dehiscence of the zoosporangium occurs by the deliquescence of a subapical papilla. At the present time *Blyttiomyces* includes 2 and possibly 3 parasites of conjugate algae, possibly 3 saprophytes of pollen grains, and 1 or 2 more hyperparasites of chytrids. Another species has been added (Dogma and Sparrow, 1969) by transferring *Phlyctochytrium vaucheriae* Rieth to this genus. *Blyttiomyces* sp. Persiel (1960) on pollen grains has been isolated and grown on synthetic media, and *B. harderi* Sparrow and Dogma (1973) parasitizes *Karlingia rosea.*

Five of the species are characterized primarily by apiculate epibiotic zoosporangia, and endobiotic apophyses, rhizoids and resting spores, and their type of growth and development is strikingly similar to that of some species of *Diplochytridium*. The zoosporangia of *Blyttiomyces* sp. Persiel lack the apiculus, and its resting spores may develop both epi- and endo-biotically. Also, its zoosporangia on pollen grain cultures may be inter-biotic occasionally with rhizoids arising at several points on the periphery of the zoosporangium as in *Rhizophlyctis*.

In *B. spinulosus* (fig. 3) and *B. rhizophlyctidis* Dogma and Sparrow (figs. 37–40) the planospores produce a long germ tube (fig. 39) which penetrates the host and develops swellings. These later become one or more apophyses in tandem (figs. 4–6). The germ tube tip branches to form rhizoids which may be relatively reduced in *B. spinulosus* but more coarse in *B. helicus* Sparrow and Barr, *B. laevis* Sparrow, and *Blyttiomyces* sp. Persiel (figs. 7, 12, 15), and bushy in *Blyttiomyces* sp. Willoughby (figs. 31, 33). In the meantime, the planospore body begins to enlarge except at the apex where the wall becomes markedly thickened (figs. 3–5). This portion of the wall is the primordium of the future cucullate apiculus which persists at the apex of the zoosporangia (figs. 8–10, 12, 15, 16, 28–34). In *Blyttiomyces* sp. Persiel, as noted before, no apiculus is present (fig. 7). The remaining portion of the planospore body enlarges to become the zoosporangium whose outer wall is spiny in *B. spinulosum* (figs. 9, 10), smooth in *B. laevis* (figs. 15, 16), *Blyttiomyces* sp. Persiel (fig. 25) and *Blyttiomyces* sp. Willoughby (figs. 31, 32), and traversed by spiral thickenings in *B. helicus* Sparrow (fig. 12). In *B. aureus* Booth the wall is smooth as in *B. laevis* but faintly brown, and the apiculus is hemispherical and golden brown. At maturity the zoosporangia discharge planospores through one (fig. 10) or more subapical, apical, or lateral orifices (fig. 15, 32, 35, 44).

In contrast to the zoosporangia, the resting spores of *B. spinulosus* and *B. rhizophlyctidis* develop endobiotically by the enlargement of an apophysis or swelling, an accumulation of refractive globules within it (figs. 18–22), and a thickening of the wall. The empty planospore cyst and germ tube persist as appendages to the smooth-walled resting spores (figs. 23, 24, 47–51). During germination the spore develops a tube, or possibly utilizes the empty germ tube, which expands at its tip outside of the host and develops into a zoosporangium (fig. 27).

Accordingly, the type of growth and development is endo- exogenous as in *Diplochytridium lagenaria* (Karling, 1936). Bartsch regarded *Blyttiomyces* as closely related to *Chytridium*, but other workers have placed it next to *Phlyctochytrium*, presumably because of its apophysate thalli and inoperculate dehiscence. Disregarding its type of dehiscence it is obviously close to *Diplochytridium*.

CATENOCHYTRIDIUM Berdan

Amer. J. Bot. 26:460, 1939.

Plates 72, 73

This operculate genus includes 4 saprophytic species, and one variety, *C. carolineanum* var. *marinum* Kobayashi and Ookubo, which parasitizes *Cladophora japonica*. This variety appears to be a distinct species because of its marine habitat and parasitic nature, and it is renamed here as *C. marinum* (Kobayashi and Ookubo) nom. nov. *Catenochytridium oahuensis* Sparrow resembles *Septochytrium* and *Truitella* in several respects and will probably prove to belong to another genus. The sporangial thallus of this genus consists of an extramatrical operculate zoosporangium and an intramatrical compound apophysis which consists of a linear series of constricted, catenulate segments subtended by extensive richly branched rhizoids (figs. 1, 29, 30, 32). Occasionally, the thallus may become polycentric (fig. 33). The rhizoids may extend for distances of 55 to 800 μ and are richly branched at the extremities. The resting-

PLATE 73

Fig. 29. Thallus of *Catenochytridium marinum* on *Cladophora japonica*. (Kobayashi and Ookubo, 1953.)

Fig. 30. Thallus of *Catenochytridium kevorkiana* Sparrow with a citriform operculate zoosporangium, thin-walled catenulate segments, and basal planospore cyst. (Sparrow, 1952.)

Fig. 31. Discharge of planospores. *C. kevorkiana*. (Sparrow, 1952.)

Figs. 32-36. *Catenochytridium oahuensis* Sparrow. (Sparrow, 1965.)

Fig. 32. Usual monocentric thallus with a globular zoosporangium and cylindrical constricted rhizoidal segments.

Fig. 33. Unusual polycentric *Septochytrium*-like thallus with 3 zoosporangia and occasionally septate, cylindrical, constricted rhizoidal segments.

Fig. 34. Planospore, 4-5 μ diam., with a minute hyaline refractive globule.

Figs. 35, 36. Discharge of planospores.

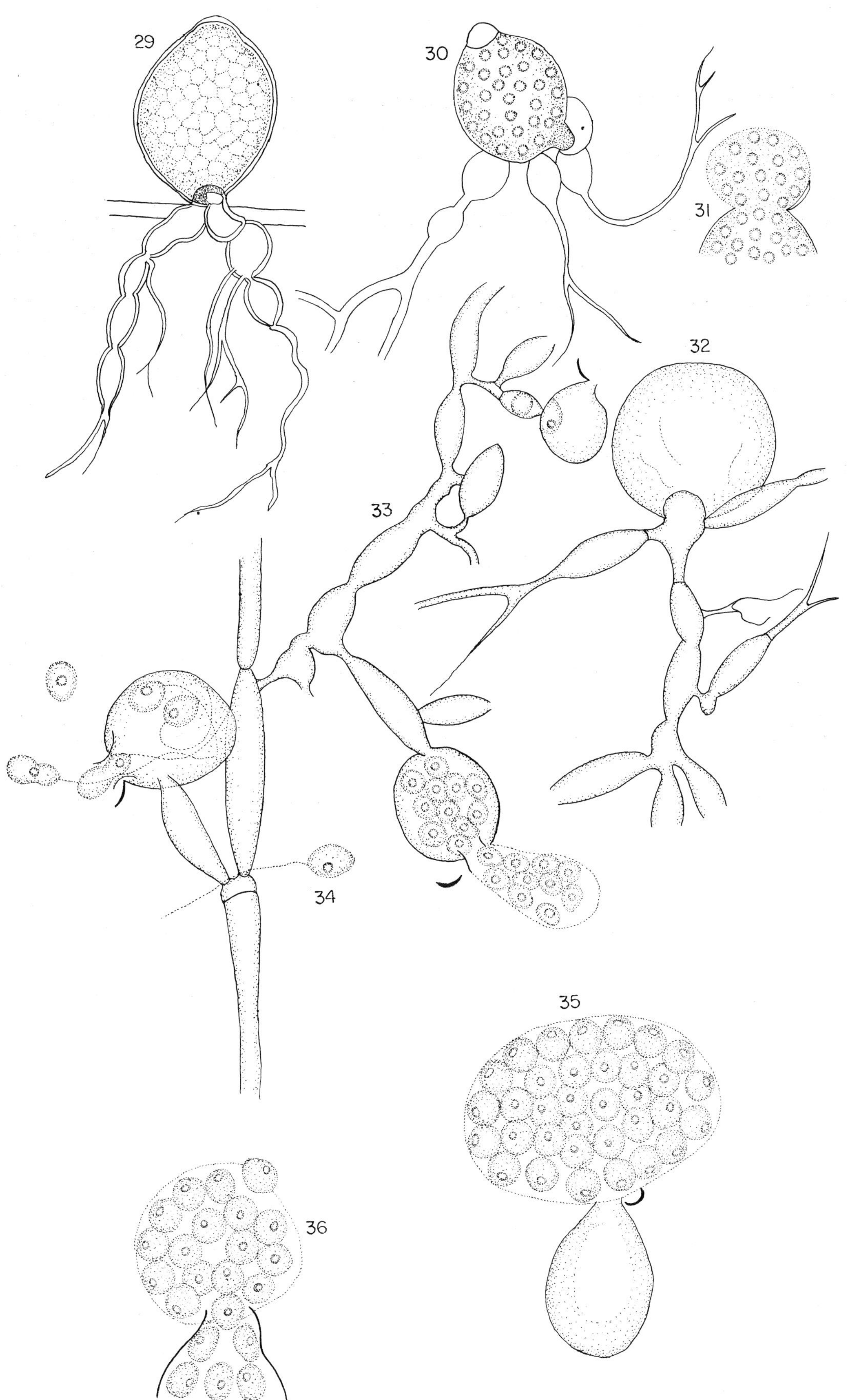

Catenochytridium

spore thalli are similar to the sporangial thalli but usually smaller, and in *C. carolineanum* Berdan the resting spore may be occasionally borne like the zoosporangia and be extramatrical.

The development of the sporangial thallus is typically endo- exogenous as in *Diplochytridium lagenaria.* The encysted zoospore develops a germ tube (figs. 3, 5), which branches (figs. 4, 5, 16) in the substratum, and shortly thereafter the primary segment of the apophysis appears as a swelling above, below, or at the juncture of the branches (figs. 4, 5, 16). The planospore cyst persists on the surface of the substratum in continuity with the primary segment, and its nucleus remains in the persistent planospore case (fig. 17). As intramatrical growth continues additional inflated segments usually develop in the branches of the germ tube (figs. 6, 7) whereby series of catenulate constricted segments are formed. These are arranged in 1 to 4 linear series, number from 1 to 30, and become filled with refractive globules (figs. 7, 8). Occasionally, only one apophysal segment is formed in *C. laterale* Hanson (fig. 18) so that the thallus is monoapophysate like species of *Diplochytridium.*

After completion of the intramatrical development growth becomes exogenous. The protoplasm flows outward into the thick-walled zoospore cyst, and as a result the rudiment of the zoosporangium buds out as a blister or vesicle at one side (figs. 8, 18–20), above (fig. 30), or on the lower portion of (fig. 9) the cyst. The remainder of the cyst persists as a thick-walled appendage on the developing zoosporangium (figs. 1, 10, 18–20, 24, 30). As the latter enlarges the primary nucleus (fig. 21) and its daughter nuclei (fig. 22) divide by intranuclear division spindles (fig. 23) and the sporangium becomes multinucleate and eventually filled with hyaline refractive globules (fig. 10). At dehiscence the apical (figs. 1, 30), or subapical (fig. 10) or lateral (figs. 24, 25) operculum is lifted up, and the zoospores ooze out in a globular mass (figs. 1, 25) and soon disperse.

Resting spores are known only in *C. carolineanum* and *C. laterale*, and as noted before they may occasionally be borne like the zoosporangia in the former species. Generally, they are intramatrical and develop asexually in and from the primary segment of the apophysis (figs. 11, 26–28). Rarely, in *C. carolineanum* both a resting spore and a zoosporangium may be borne on the same thallus. The spores of *C. carolineanum* function as prosporangia (fig. 12) when they germinate.

Berdan (1941) regarded *Catenochytridium* as being closely related to *Chytridium* by its method of development and growth and thought that it showed an evolutionary tendency towards *Catenaria*, then regarded as a chytrid. Hanson (1946), on the other hand, believed that there is but little reason for separating *Catenochytridium* and *Chytridium*. Sparrow (1960) placed it in the subfamily *Chytridiodeae* next to *Karlingomyces* (*Karlingia*).

SUBFAMILY POLYPHAGOIDEAE

This subfamily includes the genera *Polyphagus, Sporophlyctis, Endocoenobium* and *Arnaudovia* and approximately 15 polyphagus species which parasitize freshwater algae and microscopic aquatic animals. In these species the encysted planospore develops into a prosporangium as the radially oriented rhizoids develop and derive nourishment from the host. At maturity the zoosporangium buds out and develops from the prosporangium. The resting spores or zygospores are formed by the fusion of the contents of male and female thalli through a long or short conjugation tube.

So far as it is known *Arnaudovia* is basically similar in development and reproduction, asexually and sexually, to *Polyphagus* and should be merged with it. *Saccomyces* is included tentatively in this sub-

Plate 74

Figs. 1, 2, 4, 6, 14, 16-19. *Polyphagus euglenae* Nowakowski. Figs. 1, 2, 4, 6, 15, 16, 18, 19 after Nowakowski, 1876; figs. 3, 5, 14 after Wager, 1913; fig. 17 after Dangeard, 1900.

Fig. 1. Large thallus with rhizoids attached to 10 encysted *Euglena* cells.

Figs. 2, 3. Living 3-5 × 6-13 μ, and fixed and stained planospores, resp.

Fig. 4. Germinated planospore with 5 radially oriented rhizoids, one of which is attached to a *Euglena* cell.

Fig. 5. Same stage, fixed and stained.

Fig. 6. Young thallus, planospore cyst has enlarged to become a prosporangium.

Figs. 7-10. *Polyphagus serpentinus* Canter. (Canter, 1963.)

Fig. 7. Elongate thallus parasitizing *Spondylosum planum*.

Fig. 8. Dwarf thalli; zoosporangia forming on calvate prosporangia.

Fig. 9. Monoaxial germination of planospore and infection of *S. planum*.

Fig. 10. Young thallus on *S. planum*.

Figs. 11, 12. *Polyphagus elegans* Canter. (Canter, 1963.)

Fig. 11. Portion of thallus parasitizing *Ulothrix mucosum*; sporangial rudiment beginning as a bud from the prosporangium.

Fig. 12. Later stage of zoosporangium development.

Figs. 13, 15. *Polyphagus laevis* Bartsch. Early and late stages of zoosporangium development from prosporangia, resp. (Nowakowski, 1876.)

Fig. 14. Nucleus of prosporangium passing into sporangial rudiment.

Fig. 16. Curved elongate zoosporangium.

Fig. 17. Simultaneous intranuclear mitoses in the incipient zoosporangium. Possibly, this figure relates to *P. laevis*.

Figs. 18, 19. Discharge of planospores, and planospore enlarged, resp.

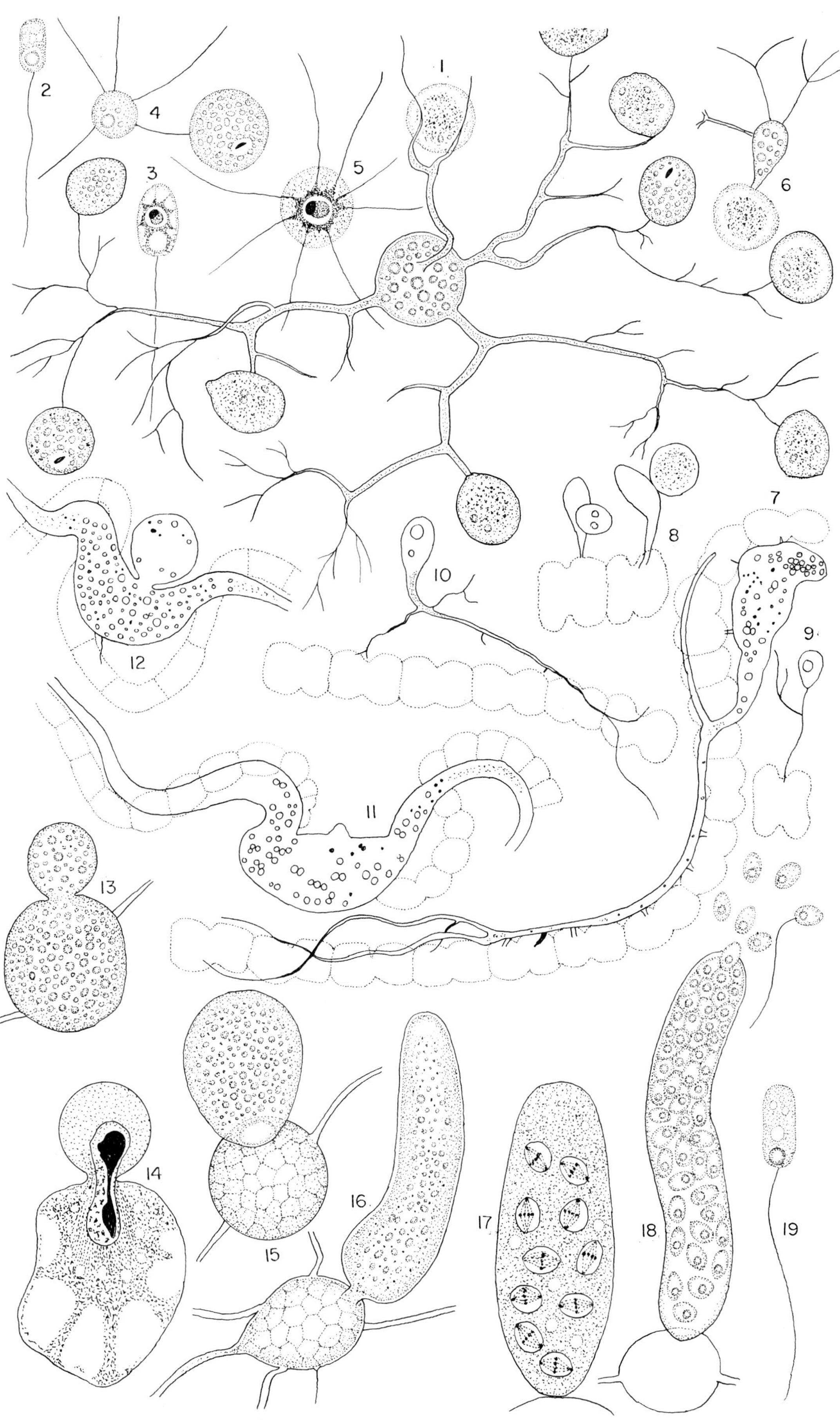

Polyphagus

family, but this classification is highly questionable and will likely prove to be incorrect.

POLYPHAGUS Nowakowski

Cohn, Beitr. Biol. Pflanzen 2:203, 1876.

Plates 74, 75

At present this genus includes 11 or possibly 12 inoperculate species which parasitize a large number of algae. So far as it is known *Polyphagus* is almost worldwide in distribution and usually develops abundantly when susceptible hosts are present. *Polyphagus parasiticus* Nowakowski and *P. ramosus* Jaag and Nipkov (1951) may prove to be host specific, and *P. starrii* Johns infects only species of the Volvocales. Also, studies by Johns (1964) have shown that *P. laevis* Bartsch is limited to *Euglena* hosts, while *P. euglenae* Nowakowski has a wide host range. *Polyphagus forminii* Milovtzova and *P. asymmetricus* Valkanov have been found only on *Botrydiopsis*. In addition to such differences in host range, the various species are differentiated primarily on the basis of size and shape of the planospores and zoosporangia, number and orientation of the predaceous rhizoids, size, shape and position of the zygospores relative to the receptive thallus and conjugation tube, and the ornamentation of the outer wall of the zygospore.

Valkanov (1963) in particular has stressed the symmetrical or asymmetrical arrangement of the primary rhizoids on the young prosporangium and the hyponeustic habit of the thallus as distinguishing characteristics. Unfortunately for present purposes several of the species are illustrated only by photographs, and it is accordingly difficult to picture them here. Nevertheless, sketches have been made from a few of these photographs to show the fundamental types of development and reproduction. *Polyphagus euglenae* Nowakowski is the best known and most common species which has been studied cytologically, and it will be used to a large extent to illustrate the genus.

Its oblong planospores (figs. 2, 3, 19) round up, encyst, and develop radially oriented rhizoids (figs. 4, 5) whose tips penetrate the host cells. In *P. starrii*, however, the rhizoids first make contact with and adhere to the flagella of the algal cells, after which infection apparently occurs. The rhizoids gradually absorb the content of the host cells, and as further growth occurs they usually branch and become more extensive. As a result large thalli (fig. 1) are formed which may capture and adhere to as many as 10 or more host cells. In the meantime the planospore body enlarges to become the prosporangium. In *P. serpentinus* Canter and *P. elegans* Canter germination of the planospore is monoaxial (fig. 9) which results eventually in a unilateral elongate prosporangium (figs. 7, 10–12). Dwarf thalli or prosporangia are formed occasionally (fig. 8), and these produce minute zoosporangia which may bear 1 to a few planospores.

Cytologically, the prosporangium remains uninucleate until maturity. Then, as the sporangial rudiment buds out from it (figs. 11–13) the nucleus enters the incipient zoosporangium (fig. 14). As the protoplasm flows out of the prosporangium the incipient zoosporangium enlarges in size and becomes ovoid (fig. 15), to pyriform, almost cylindrical and sometimes curved (figs. 16–18) or subspherical. During this growth the nuclei divide simultaneously with intranuclear division spindles (fig. 17) so that the zoosporangium becomes multinucleate and eventually produces planospores which emerge as the tip of the exit papilla deliquesces (fig. 18). Occasionally, the uninucleate prosporangium encysts and becomes thick-walled (figs. 20, 21), and when it germinates it develops a zoosporangium in the same manner as the evanescent prosporangium.

PLATE 75

Figs. 20-32. *Polyphagus euglenae* Nowakowski. Figs. 20, 21 after Dangeard, 1900; figs. 22-32 after Wager, 1913.

Figs. 20, 21. Encysted uninucleate thick-walled prosporangia.

Figs. 22, 23. Conjugation of a small thallus with a larger female thallus; incipient zygospores formed at the tip of the male filament or conjugation tube.

Fig. 24. Later stage; nuclei moving towards large incipient zygospore; part of male thallus shortened in drawing.

Fig. 25. Male nucleus in incipient zygospore.

Fig. 26. Entrance of female nucleus into incipient zygospore.

Fig. 27. Later stage of plasmogamy; gametic nuclei unequal in size.

Fig. 28. Mature binucleate zygospore.

Fig. 29. Germinating zygospore; equal sized nuclei have migrated into sporangial rudiment.

Fig. 30. Karyogamy in incipient zoosporangium.

Figs. 31, 32. Zoosporangium with planospores and their discharge, resp.

Fig. 33. *Polyphagus laevis* Bartsch. Small smooth zygospore adnate to female thallus. (Bartsch, 1945.)

Figs. 34, 35. *Polyphagus serpentinus* Canter zygospores with attached male and female thalli. (Canter, 1963.)

Figs. 36-38. *Polyphagus parasiticus* Scherffel. (Scherffel, 1925.)

Fig. 36. Zygospore with a large attached female thallus and part of the filamentous male thallus.

Fig. 37, 38. Germination of zygote, and planospores, resp.

Figs. 39-41. *Polyphagus elegans* Canter. (Canter, 1963.)

Fig. 39. Stage in the formation of the zygospore.

Fig. 40. Male and female thalli equal in size; zygospore between them.

Fig. 41. Male thallus minute.

Fig. 42. *Polyphagus* sp. Rieth zygospore on *Tolypothrix lanata* with a facet-structured outer wall. (Rieth, 1954.)

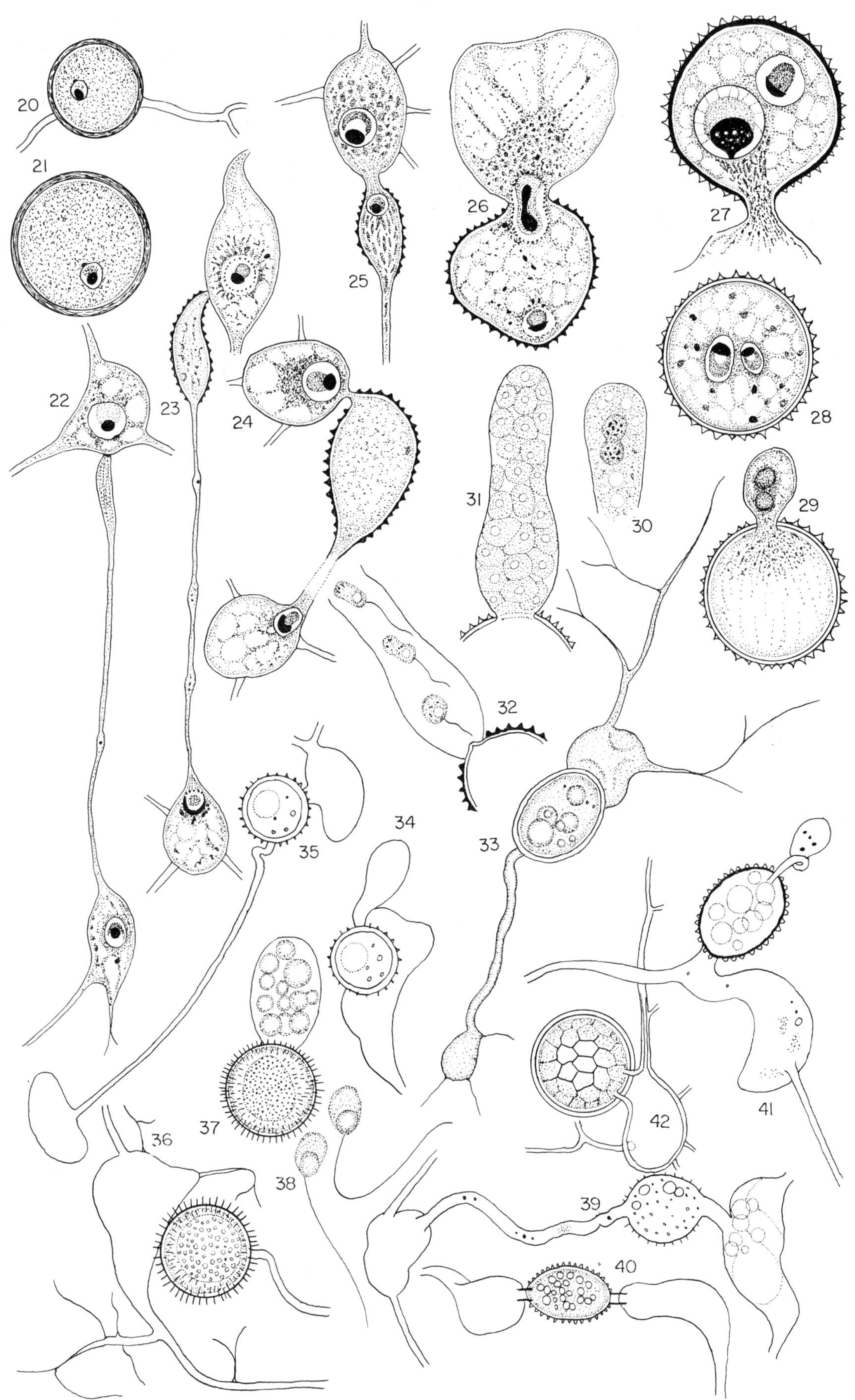

Polyphagus

Sexual reproduction occurs by the fusion of the contents of a male and female thallus through a conjugation tube and the formation of a zygote or so-called zygospore. The male thallus is usually smaller than the female thallus (figs. 22, 23, 34, 35, 41) but not always so (fig. 40) and in some instances it may be the larger. It develops a conjugation tube which comes into contact with the female thallus. Its distal part at the point of contact then begins to enlarge (figs. 22–24) and become the incipient zygote. The contents of the uninucleate thalli or gametangia and the two nuclei migrate into this swelling (figs. 25, 26) which becomes invested with a fairly thick wall. At first, the female nucleus is much larger than the male (figs. 27, 28), but the latter nucleus enlarges until it attains the size of the former. The two nuclei do not fuse but move apart in the mature zygospore and remain separate until the latter germinates. In germination the zygospore functions as a prosporangium and produces a zoosporangium into which the two nuclei migrate (fig. 29) and fuse (fig. 30). The diploid nucleus soon divides, and its derivatives multiply until the zoosporangium becomes multinucleate, after which its content cleaves into planospores (fig. 31). Occasionally, two normal zygospores are associated with a single receptive thallus in *P. ramosus* and *P. starrii*, and in *P. asymmetricus* Valkanov a zygospore may be formed by double fertilization, according to Valkanov (1963).

Although meiosis has not been observed it is presumed to occur during the first division of the fusion nucleus. On the basis of this presumption and the fact that Nowakowski found nearly equal numbers of male and female thalli in his cultures of *P. euglenae*, Kniep (1928) suggested that this species is dioecious and produces sexually differentiated planospores. However, in *P. starrii* Johns found that all of more than 60 clonal isolates were homothallic.

From studies on single isolated zygospores and their germination Bartsch (1945) showed that Nowakowski's studies on *P. euglenae* relate to 2 species, one with spiny-walled zygospores and another with smooth zygospores which he named *P. laevis* (fig. 33). In the latter species the zygospore is formed terminally in the conjugation tube and becomes adnate to the female thallus at maturity. *Polyphagus nowakowski* Raciborski (1900), also, develops smooth zygospores, and in *Polyphagus* sp, Rieth the outer wall has a facetted structure (fig. 42). In *P. starrii* the zygospores are spherical with elongate spines and develop terminally in the conjugation tube as in *P. laevis*. The zygospores of the other species are warty or bear relatively short delicate conical spines.

ARNAUDOVIA Valkanov

Arch. f. Protistenk. 106:562, 1963.

Plate 76, Figs. 50-57

This monotypic genus was created for a *Polyphagus*-like parasite of species of planktonic algae, *Trachelomonas, Phacotus* and *Strobomonas,* which develops a specialized and complex apparatus for the capture of its prey. The young hyponeustic thallus or broadly pyriform rudiment of the prosporangium is attached to the underside of a water film by 5 short anchoring rhizoids (fig. 50). Five long primary rhizoids arise from the periphery and curve down into the water like ribs of an open umbrella, and a 6th one develops from the terminal pole of the prosporangium and extends straight down like the handle of an umbrella. These 6 primary rhizoids are spirally twisted and bear numerous short, sharp setae which are the organelles of capture (fig. 50). Thus, the chances of capturing prey are increased many-fold, and in one case Valkanov noted the capture of 63 algae by one thallus.

Otherwise, the life cycle and development are similar to those of *Polyphagus*. After coming to rest and encysting on the underside of a mouldy water film the spherical, 7 μ diam., planospore enlarges and develops the specialized apparatus noted above. After capturing and absorbing the contents of the host

PLATE 76

(Figs. 43, 50-53 are diagrams; remaining figures are sketches from photographs.)

Fig. 43. Surface view of young, star-shaped, radially symmetrical thallus of *Polyphagus forminii* Valkanov with 5 primary rhizoids. (Valkanov, 1964.)

Figs. 44-47. *Polyphagus asymmetricus* Valkanov. (Valkanov, 1963.)

Fig. 44. Subspherical thick-walled zoosporangium and prosporangium.

Figs. 45, 46. Zygospores photographed at different focal levels.

Fig. 47. Zygospore formed as the result of double fertilization.

Figs. 48, 49. *Polyphagus starrii* Johns. (Johns, 1964.)

Fig. 48. Zoosporangium attached to empty prosporangium and showing refractive globules of planospores.

Fig. 49. Spiny zygospore with attached conjugation tube and male thallus.

Figs. 50-57. *Arnaudovia hyponeustonica* Valkanov. (Valkanov, 1963.)

Fig. 50. Diagram of a young thallus attached to the underside of a mouldy water film by 5 primary anchoring rhizoids. Five curved sublateral and one polar rhizoid extend down in the water and bear numerous short sharp predaceous setae; two host species, *Trachelomonas* and *Phacotus*, have been captured.

Figs. 51, 52. Asymmetrical and symmetrical arrangements of 5 anchoring rhizoids, resp.

Fig. 53. Free-floating young thallus which has been torn loose from the water film.

Fig. 54. Thallus with 2 captured *Trachelomonas volvocina* cells; sporangial rudiment forming.

Fig. 55. A fully-developed zoosporangium.

Fig. 56. Conjugation of male and female thalli and an early stage in zygospore formation.

Fig. 57. Mature zygospore.

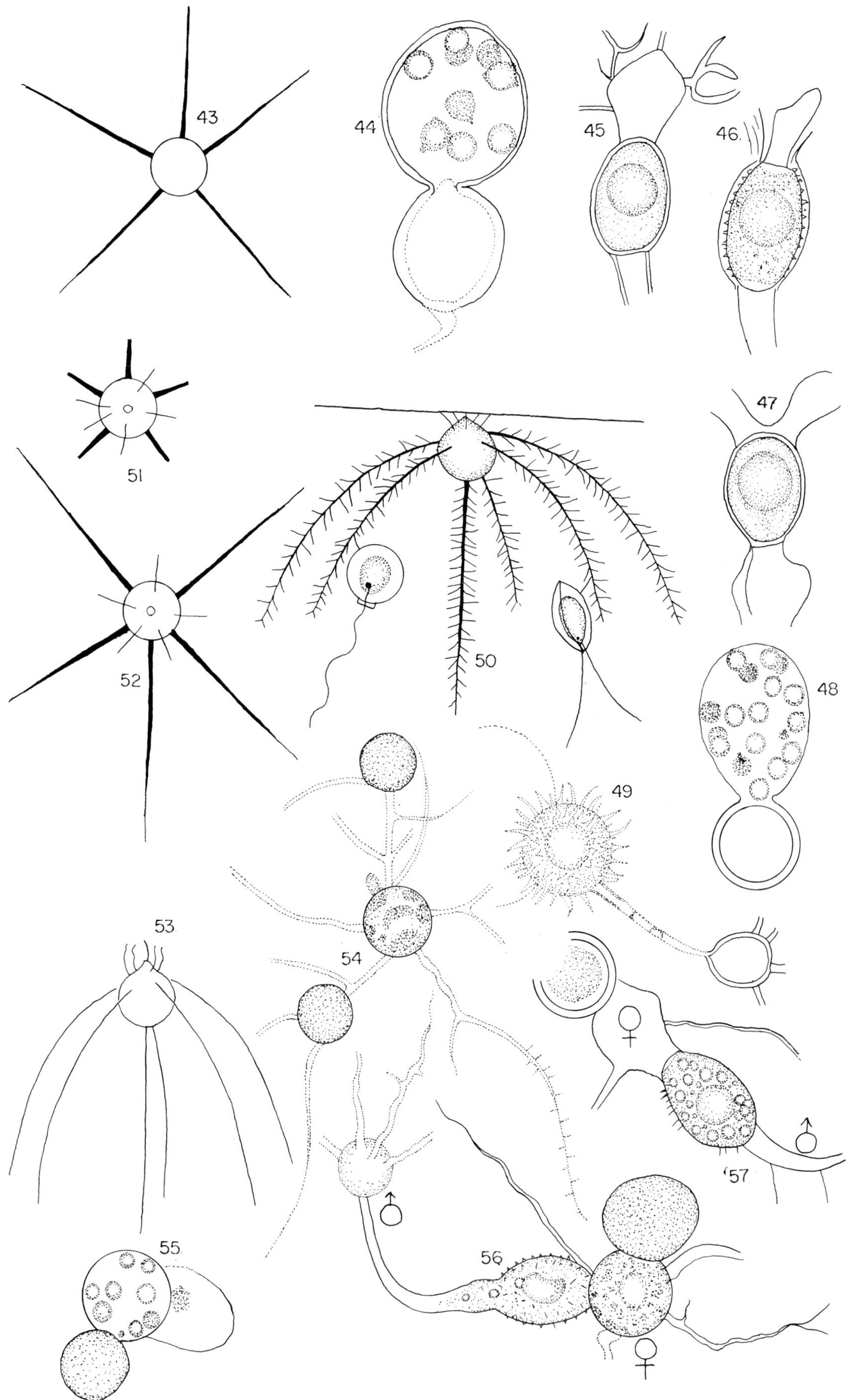

Polyphagus, Arnaudovia

cells the primary rhizoids branch so that the thallus becomes asymmetrical (fig. 54). At the same time the planospore body enlarges to become the prosporangium, 25–35 μ diam., from which the zoosporangium eventually grows out and produces usually up to 20 planospores (fig. 55).

Sexual reproduction, likewise, is identical with that of *Polyphagus*. A rhizoid or tube of a small male thallus grows toward a female thallus. When its tip reaches the latter it enlarges to become the rudiment of zygote (fig. 56). The contents of the two thalli or gametangia move into this rudiment which then enlarges and matures into a spiny zygospore (fig. 57).

On the basis of its close similarity in type of development, asexual and sexual reproduction *Arnaudovia* should be merged with *Polyphagus*, and the name *P. hypneustonica* (Valkanov) nov. nom. is proposed for Valkanov's species. Nevertheless, the genus is presented separately here so that readers may judge independently its validity as a separate generic taxon. However, its hyponeustic habit and the development of a specialized and more complex apparatus for the capture of its prey are specific characters, in this writer's opinion, and do not merit generic rank. Similar variations in the organs of capture occur in *Zoophagus* and other predacious fungi, and these are generally regarded as specific variations. In *Arnaudovia* such variations are outweighed taxonomically by the similarity of type of development, asexual, and sexual reproduction to those of *Polyphagus*.

SPOROPHLYCTIS Serbinov

Scripta Bot. Horti Univ. Imper. Petropl. 24:164, 1907.

Plate 77

This polyphagus inoperculate genus is reported to include two species which parasitize the alga *Draparnaldia* in Russia, Denmark (Petersen, 1909), the U.S.A. (Graff, 1928), and China (Shen, 1944). *Sporophlyctis rostrata* Serbinov has an ovoid or broadly fusiform apiculate prosporangium and produces aplanospores in a spherical thin-walled zoosporangium, while *S. chinensis* (Shen) Sparrow develops an elongate or clavate non-apiculate prosporangium and forms typical chytrid planospores in an ovoid zoosporangium. In both species the aplanospore or planospore (figs. 2, 11, 17, 18) enlarges to become a prosporangium as the tips of the rhizoids (figs. 1, 19) derive nourishment from the host. At maturity a sporangium or zoosporangium buds out from the prosporangium (figs. 5–8, 20, 21) and forms aplanospores (figs. 8, 9) or planospores (fig. 16).

Sexual reproduction is very similar to that of *Polyphagus* and occurs by conjugation at the apical portion of a small, spiny, slender, and a larger ovoid or clavate thallus (figs. 12, 22, 23). As the two thalli come into contact part of the wall between them deliquesces to form a narrow or broad pore (figs. 22, 23). At this stage the thalli in *S. rostrata* are uninucleate, and part or most of the protoplasm of the larger thallus moves into the small one which increases in size and spinyness (figs. 13, 23). Inasmuch as movement of the protoplasm is from the larger to the smaller thallus Serbinov regarded the former as female and the latter as male in *S. rostrata*. On this basis the same interpretation may be made for *S. chinensis*, although Shen did not designate the thalli specifically as male and female. Fusion of the protoplast results in the development of a spiny zygote (figs. 14, 15, 24, 25) whose germination has not been observed. Although meiosis has not been observed, it is assumed to occur during the 1st division of the zygote nucleus. Shen suggested that the planospores produced in *S. chinensis* might be of three kinds, plus, minus, and neutral, and that sexual differentiation is delayed in the neutral ones.

Shen believed that aplanospore development in *S. rostrata* might be related to certain environmental conditions, and that his species is identical with Serbinov's fungus. Since he did not regard aplanospore development as a dependable criterion for separating

PLATE 77

Figs. 1-15. *Sporophlyctis rostrata* Serbinov. (Serbinov, 1907.)

Fig. 1. Two young thalli parasitizing filaments of *Draparanaldia glomerata*.

Fig. 2. Germinated aplanospore.

Figs. 3, 4. Uni- and multinucleate prosporangia.

Fig. 5. Incipient apiculate sporangium budding out of multinucleate prosporangium.

Figs. 6, 7. Later stages in sporangium development.

Fig. 8. Sporangium filled with aplanospores.

Fig. 9. Aplanospores beginning to germinate within sporangium.

Fig. 10. Germinated aplanospores bursting out of sporangium.

Fig. 11. Colony of sporelings.

Fig. 12. Conjugation between a small spiny female and a larger smooth uninucleate thallus.

Fig. 13. Binucleate spiny zygote, fixed and stained.

Figs. 14, 15. Mature spiny zygotes with large refractive bodies, living material.

Figs. 16-25. *Sporophlyctis chinensis* Sparrow. (Shen, 1944.)

Fig. 16. Spherical planospores, 8-8.5 μ diam., with hyaline refractive bodies.

Figs. 17, 18. Germination of planospore and early developmental stage of the thallus.

Fig. 19. Mature thallus on a filament of *Draparnaldia* sp.

Fig. 20. Incipient zoosporangium budding out of apex of a nonapiculate prosporangium.

Fig. 21. Mature zoosporangium prior to cleavage.

Figs. 22, 23. Early and late stages of conjugation of two thalli.

Figs. 24, 25. Mature zygotes.

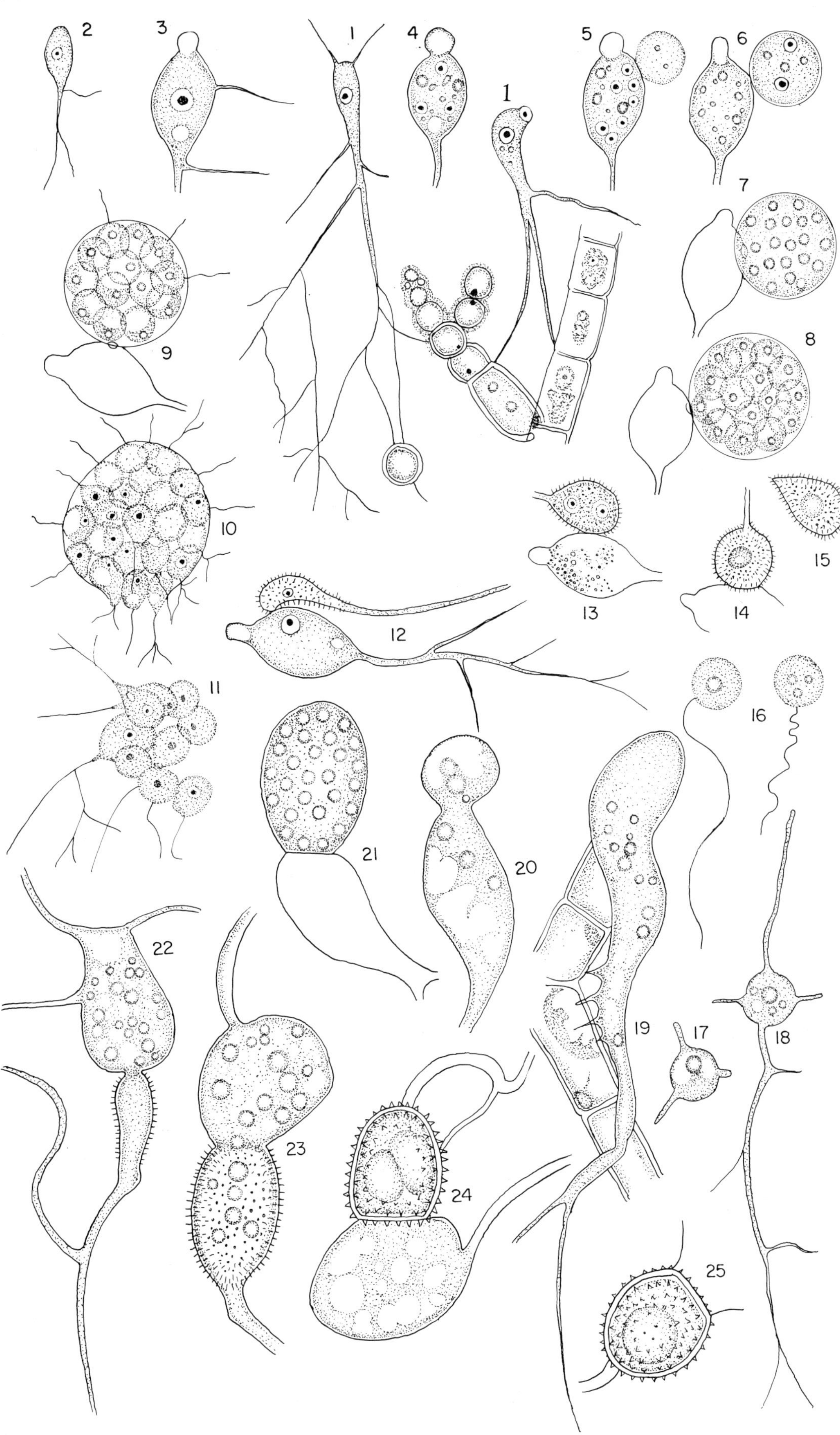

Sporophlyctis

genera he merged *Sporophlyctis* with *Polyphagus* and renamed both species *P. rostratus*. Sparrow (1960), however, maintained the two genera as separate taxa and renamed Shen's species *S. chinensis*. He believed that there is a fundamental difference in the types of sexual reproduction: "in *Polyphagus* the zygote is formed in or near the tip of a special conjugation tube, whereas in *Sporophlyctis* there is no such tube, and the receptive structure, while smaller than the contributing thallus, is the thallus itself." Such differences may not prove to be generically significant, and Shen's interpretation may prove to be the correct one.

ENDOCOENOBIUM Ingold

New Phytologist 39:97, 1940.

Plate 78

This genus includes one identified species, *E. eudorinae* Ingold, which parasitizes the motile coenobium of *Eudorina elegans* Ehrenb. in England, Sweden and Switzerland and possibly *Sphaerocystis schroeteri* Chod. in the U.S.A. Paterson (1958) found the latter host to be parasitized by a species whose young stages were similar to those described by Ingold, but it is too little known to identify it with *E. eudorinae*.

The planospores (fig. 8) come to rest on the coenobium, encyst, and form an appressorium which attaches it to the host (fig. 1). A tube develops from the appressorum which penetrates the coenobium, branches to form rhizoids, and gradually enlarges to become the incipient prosporangium (figs. 2-5). At the same time the planospore cyst becomes thick-walled and persists as an appendage on the surface of the coenobium. The tips of the rhizoids grow towards and penetrate the algal cells, causing them to change color and shrink (fig. 5). As the prosporangium matures a zoosporangium begins to bud out of it at the side of the planospore cyst (fig. 6), and as the protoplasm passes into it the incipient zoosporangium elongates to the outside of the coenobium and is delimited by a septum (fig. 7). It dehisces by an apical tear while the host is still motile.

Resting spores (zygospores) are reported to develop indirectly as the result of fusion of two thalli (fig. 9) within the coenobium, although the initial fusion stages have not been observed. Possibly, the planospores which give rise to these thalli are facultative, and the evidence at hand suggests that they may form prosporangial thalli, or other thalli which fuse. From the fusion thallus the incipient resting spore buds out (fig. 10) in the same manner as the zoosporangium, but this development occurs within the coenobium with the result that the spore is endobiotic. As the content of the fusion thallus moves into the bud, it enlarges, becomes delimited by a septum (fig. 11), and eventually develops a fairly thick wall covered with small spines or pegs (figs. 12, 13). Germination of these spores (zygospores) has not been observed.

The development of the so-called zygospores as an outgrowth of a fusion thallus is unique for most of the Chytridiales but similar to the formation of zygospores in other Phycomycetes such as *Syncephalis nodosa* and *Entomophthora americana*, as Ingold pointed out.

The proper classification of this genus is problematical and uncertain. Sparrow (1943, 1960) placed it in the subfamily Polyphagoideae, but Whiffen (1944) classified it in her subfamily Diplophlyctoideae on the grounds that the prosporangium is formed from the germ tube instead of directly as an enlargement of the planospore as in *Polyphagus*. This is a basic developmental difference but it is outweighed to a degree by a similarity in type of sexual reproduction. Accordingly, *Endocoenobium* is classified here only tentatively in the Polyphagoideae.

SACCOMYCES Serbinov

Scripta Bot. Horti Univ. Petropol. 24:162, 1907.

Plate 79

This inoperculate genus parasitizes cysts of *Euglena* and is characterized in the type species by an endobiotic deeply-lobed or digitate, sac-like haustorium which may or may not be centered on an endobiotic apophysis, and a relatively small epibiotic, pyri-

PLATE 78

Figs. 1-13. *Endocoenobium eudorinae* Ingold. Figs. 1-9, 12 after Ingold, 1940; figs. 10, 11, 13 after Canter, 1961. Parts of coenobium omitted to avoid obscuring the parasite.

Fig. 1. Encysted planospore forming an appressorium on the coenobium of *Eudorina elegans*.

Fig. 2. Young endobiotic thallus with 3 rhizoids attached to the host cells; planospore cyst external to coenobium.

Fig. 3. Later developmental stage of a thallus in a small coenobium.

Figs. 4, 5. Later stages, incipient prosporangium becoming almost globular.

Fig. 6. Incipient zoosporangium budding out of the prosporangium and becoming external to the coenobium.

Fig. 7. Mature zoosporangium delimited from the prosporangium by a septum and external to the coenobium.

Fig. 8. Spherical, 4-5 μ diam., planospores.

Fig. 9. Fusion thallus formed by fusion of two thalli with attached planospore cysts.

Fig. 10. Incipient resting spore or zygospore budding out of fusion thallus within the coenobium.

Fig. 11. Young resting spore (zygospore) delimited by a septum from the fusion thallus within the coenobium.

Figs. 12, 13. Mature resting spores (zygospores).

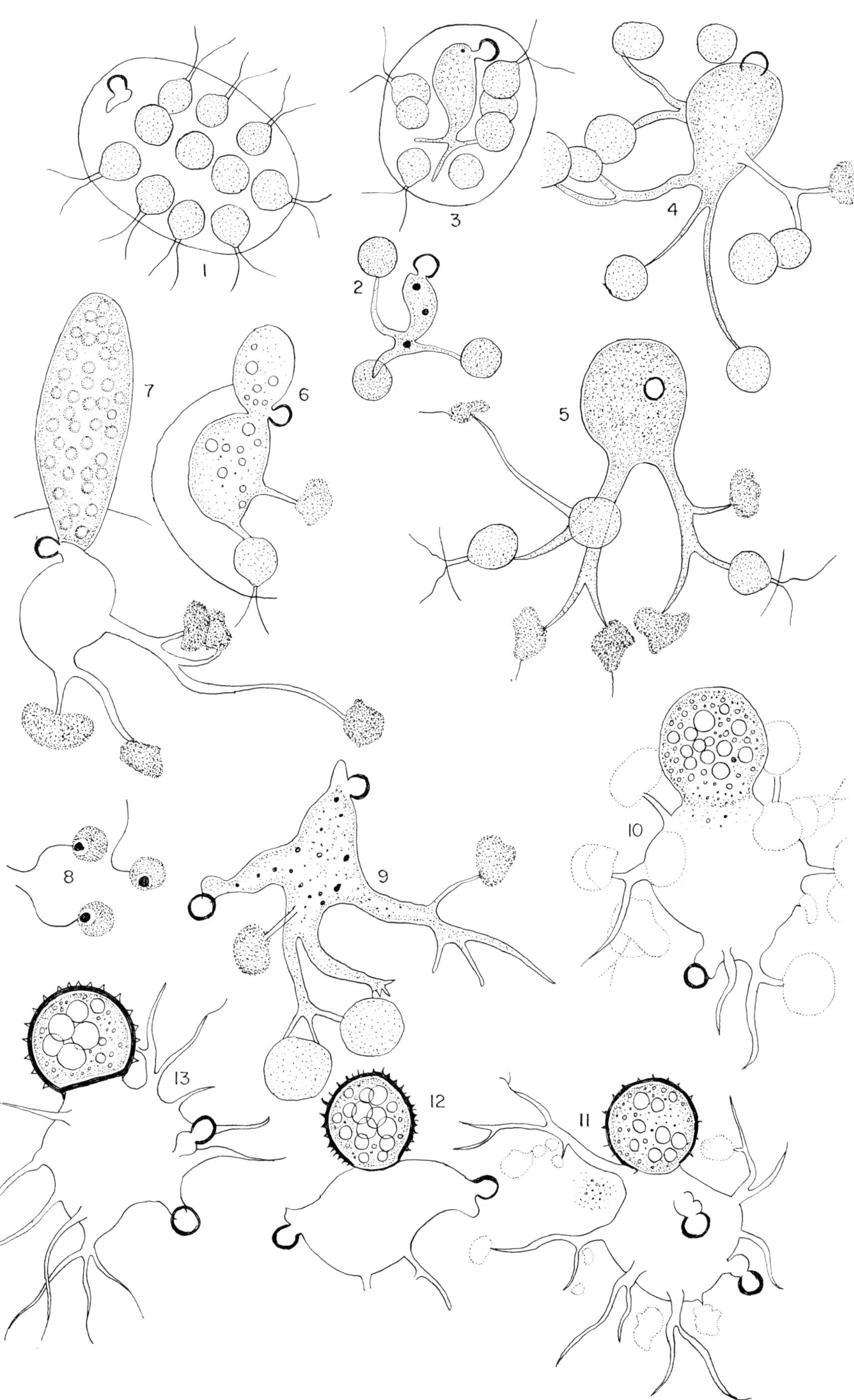

Endocoenobium

form, apiculate prosporangium out of which an almost cylindrical or ovoid zoosporangium develops (figs. 1, 7, 8). In another parasite (figs. 10–17) which Nowakowski (1878) described as *Polyphagus endogena* Nowakowski, the haustorium is oblong and slightly curved, and the epibiotic prosporangium is lacking, according to his figures 108 to 110, and 112. Sparrow (1943) regarded this parasite as identical with *Saccomyces dangeardii* Serbinov and renamed the latter *S. endogena* (Nowa.) Sparrow. Apparently, Nowakowski's species does not relate to *Polyphagus*, and it is referred to here as *Saccomyces* sp. That it is identical with Serbinov's species is open to question, and it is unlikely that a careful observer like Nowakowski would have failed to see a digitate haustorium and epibiotic prosporangium.

Although the early developmental stages are not known in *S. dangeardii* the digitate haustorium and apophysis apparently develop from the germ tube of the germinating planospore (fig. 6) while the epibiotic planospore body develops into the prosporangium. The latter remains relatively small (figs. 3–5), but the 2–4 digitations of the haustorium become relatively large and sac-like. The incipient zoosporangium buds out apically or laterally from the prosporangium as in species of *Polyphagus* and becomes elongate to almost cylindrical (figs. 1, 7), or ovoid in shape (fig. 8). After the planospores are mature the thin wall of the zoosporangium bursts, and the planospores are liberated (fig. 2).

The spiny resting spores of this species (fig. 9) are formed in the same manner as the zoosporangia from prosporangia and no evidence of sexuality has been observed. Germination of such spores has not been observed.

According to Nowakowski's figures, no prosporangium is formed in *Saccomyces* sp. unless the knob at the end of the short stalk in fig. 15 is interpreted as such. Possibly, the elongated and curved endobiotic haustorium serves as a prosporangium.

The method of sporangial development and planospore formation in *S. dangeardii* are similar to those of *Polyphagus*, but its resting spores are not formed sexually as in the latter genus so far as is known. Also, it differs by the non-polyphagus nature of its thallus and its confinement to a single host cell. For these reasons *Saccomyces* has been classified in several subfamilies and families. Minden (1911) placed it in the subfamily Rhizidieae next to *Rhizidium*, but Sparrow (1943, 1960) included it in the family Phlyctidiaceae next to *Podochytrium*. However, Whiffen believed that it belongs in the Polyphagoideae because its planospore becomes a prosporangium and gives rise to the zoosporangium as in other species of this subfamily. It is classified temporarily in this subfamily for want of a more apt group in which to place it.

REFERENCES TO THE RHIZIDIACEAE AND GENERA OF DOUBTFUL AFFINITY

Ajello, L. 1945. *Phlyctochytrium aureliae* parasitized by *Rhizophydium chytridiophagum*. Mycologia 37:109-119, 28 figs.

Antikajian, G. 1949. A developmental, morphological, and cytological study of *Asterophlyctis* with special reference to its sexuality, taxonomy, and relationships. Amer. J. Bot. 36:245-262, 78 figs.

Atkinson, G. F. 1909. Some fungus parasites of algae. Bot. Gaz. 48:321-337, 9 figs.

Bahnweg, G., and F. K. Sparrow, Jr. 1972. *Aplanochytrium kerguelensis* gen. nov., sp. nov., a new Phycomycete from subantartic marine waters. Arch. f. Mikrobiol. 81:45-49, 8 figs.

Barr, D. J. S. 1969. Studies on *Rhizophydium* and *Phlyctochytrium* (Chytridiales). I. Comparative Morphology. Canad. J. Bot. 47:991-997, fig. 1, pls. 1-8.

———. 1973. *Rhizophydium graminis* (Chytridiales): Morphology, host range, and temperature effect. Can. Pl. Dis. Survey 53:191-193, 3 figs.

———. 1975. Morphology and zoospore discharge in single-pored epibiotic Chytridiales. Canad. J. Bot. 53:164-178, 65 figs.

PLATE 79

Figs. 1-9. *Saccomyces dangeardii* Serbinov. (Serbinov, 1907.)

Fig. 1. Parasitized cyst of *Euglena viridis* with two empty prosporangia and 2 zoosporangia filled with planospores.

Fig. 2. Dispersal of planospores after bursting of sporangium wall.

Figs. 3-5. Variations in the digitate sac-like endobiotic haustoria; apophysis lacking in figs. 3, 5.

Fig. 6. Infection by an encysted planospore.

Fig. 7. Four prosporangia, and one from which an incipient zoosporangium has developed.

Fig. 8. Mature zoosporangium before cleavage into planospores.

Fig. 9. Spiny resting spore developed from a prosporangium.

Figs. 10-17. *Saccomyces* sp. [*S. endogena* (Nowa.) Sparrow, pro parte]. (Nowakowski, 1878.)

Figs. 10, 11. Spherical, 2 μ diam., planospores with a large hyaline refractive globule.

Fig. 12. Young epibiotic zoosporangium and a curved non-digitate endobiotic apophysis or prosporangium (?) on and in *Euglena*.

Figs. 13, 14. Clavate and lobed zoosporangia.

Fig. 15. Ovoid zoosporangium with a short epibiotic stalk.

Fig. 16. Branched zoosporangium.

Fig. 17. Resting spore.

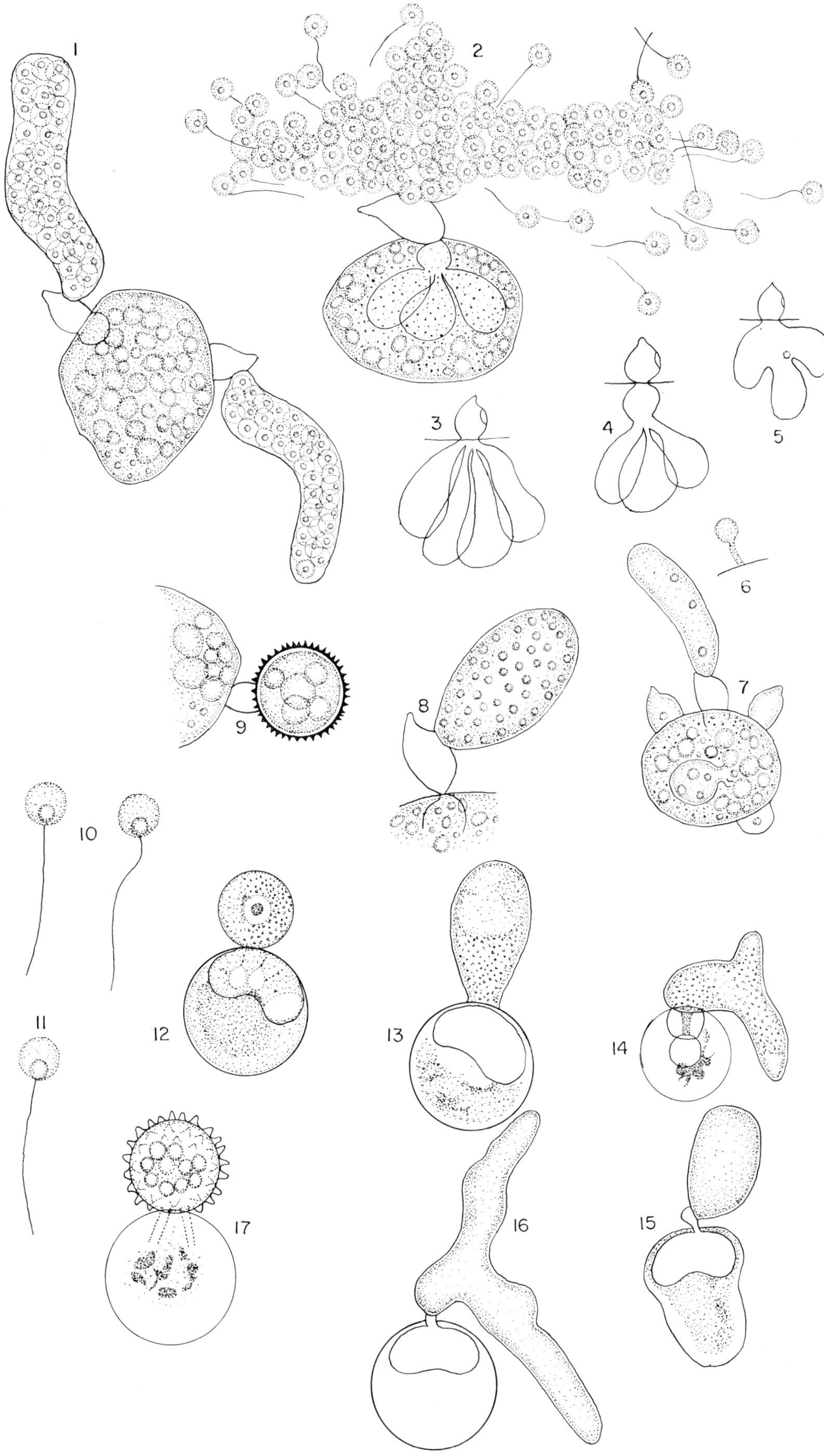

Saccomyces

Bartsch, A. F. 1939. Reclassification of *Chytridium spinulosm* with additional observations on its life history. Mycologia 31:558-571, 24 figs.

———. 1945. The significance of zygospore character in *Polyphagus euglenae*. Ibid. 37:553-570, 23 figs.

Batko, A. 1970. A new *Dangeardia* which invades motile chlamydomonadaceous monads. Acta Mycol. 12:407-436, 27 figs.

Berdan, H. 1939. Two new genera of operculate chytrids. Amer. J. Bot. 26:459-463, 2 figs.

———. 1941. A developmental study of three saprophytic chytrids. Ibid. 28:901-911, 72 figs.

Booth, T. 1969. Marine fungi from British Columbia. Monocentric chytrids and chytridiaceous species from coastal and halomorphic soils. Syesis 2: 141-161, 74 figs.

Borzi, A. 1885 *Nowakowskia*, eine neue Chytridiee. Bot. Centralb. 22:23-26, pl. 1, figs. 1-10.

Bostick, L. R. 1968. Studies on the morphology of *Chytriomyces hyalinus*. J. Elisha Mitchell Sci. Soc. 84:94-99, 7 figs.

Braun, A. 1855. Ueber *Chytridium*, eine Gattung einzelliger Schmarotzergewäsche auf Algen and Infusorien. Monatsber. Berlin Akad. 1855:378-384.

———. 1856. Ueber *Chytridium*, eine Gattung einzelliger Schmarotzergewäsche auf Algen and Infusorien. Abhandl. Berlin Akad. 1855:21-83, pls. 1-5.

———. 1865a. Ueber einige neue Arten der Gattung *Chytridium* und die damit verwandte Gattung *Rhizidium*. Monatsber. Berlin Akad. 1856: 587-592.

———. 1956b. Untersuchungen über einige mikroskopische Schmarotsgewäsche...Flora (n.s.) 14:599-600.

Butler, E. J. 1928. Morphology of the chytridiacean fungus, *Caternaria anguillulae*, in liver fluke eggs. Ann. Bot. 42:813-821, 19 Figs.

Canter, H. M. 1946. Studies on British chytrids. I. *Dangeardia mamillata* Schröder. Trans. Brit. Mycol. Soc. 29:128-134, 5 figs., pl. 7.

———. 1947a. Studies on British Chytrids II. Some new monocentric species. Ibid. 31:94-105, 8 figs., pls. 9, 10.

———. 1947b. Studies on British chytrids. III. *Zygorhizidium willei* Lowenthal and *Rhizophydium columnaris* n. sp. Ibid. 31:128-135, 4 figs., plate 11.

———. 1949. Studies on British chytrids IV. *Chytriomyces tabellariae* (Schroter) n. comb. parasitized by *Septosperma anomalum* (Couch) Whiffen. Ibid. 32:16-21, 3 Figs.

———. 1949. Studies on British chytrids. VII. On *Phlycotochytrium mucronatum* n. sp. Ibid. 32: 236-240, 2 figs.

———. 1950a. Studies on British chytrids. VIII. On *Rhizophydium anomalum* n. sp. New Phytol. 49:98-102, 2 figs.

———. 1950b. Fungal parasites of the Phytoplankton. I. (Studies on British chytrids, X.) Ann. Bot. n. s. 54:263-389, 16 figs., l. 8.

———. 1950c. Studies on British chytrids. XI. *Chytridium oedogonii* Couch. Trans. Brit. Mycol. Soc. 33:354-358, 3 figs., pls. 28, 29.

———. 1951. Fungal parasites of phytoplankton. II. (Studies on British chytrids. XII). Ann. Bot. n. s. 15:129. 14 figs., pls. 8-11.

———. 1954. Fungal parasites of the phytoplankton. III. Trans. Brit. Mycol. Soc. 37:111-133, 9 figs., pls. 3-5.

———. 1959. Fungal parasites of the phytoplankton. IV. *Rhizophydium contractophilum* sp. nov. Ibid. 42:185-192, 4 figs.

———. 1960. Fungal parasites of the phytoplankton. V. *Chytridium isthmiophilum* sp. nov. Ibid. 43: 660-664, 1 fig.

———. 1961a. Studies on British chytrids. XVII. Species occurring on planktonic desmids. Ibid. 44:163-176, 4 figs., pl. 12.

———. 1961b. Studies on British chytrids XVIII. (Further observations on species invading planktonic algae). Nova Hedwigia 3:73-78, pls. 38-41.

———. 1961c. Studies on British chytrids. XIX. On *Phlyctidium apophysatum* Canter emend. Trans. Brit. Mycol. Soc. 44:522-523, 3 figs.

———. 1963a. Fungal parasites of phytoplankton. VI. *Zygorhizidium chlorophycidis* sp. nov. Ibid. 5:1-6, pls. 1, 2.

———. 1963b. Fungal parasites of the phytoplankton VII. Ibid. 5:419-428, pls. 71-76.

———. 1963c. Concerning *Chytridium cornutum* Braun. Ibid. 46:208-212, 2 figs.

———. 1963b. Studies on British chytrids. XXIII. New species on chrysophycean algae. Ibid. 46: 305-320, 7 figs.

———. 1966. Studies on British chytrids XXV. *Chytriomyces heliozoicola* sp. nov., a parasite of *Heliozoa* in the plankton. Ibid. 49:633-638, 1 fig.

———. 1967. Studies on British chytrids. XVI. A critical examination of *Zygorhizidium melosirae* Canter and *Z. planktonicum* Canter. J. Linn. Soc. (Bot.) London 60. no. 38:85-97, 4 figs., pls. 1-4.

———. 1968. Studies on British Chytrids XXVIII. *Rhizophydium nobile* sp. nov. parasitic on the resting spore of *Ceratium hirundinella* O. F. Mull. from plankton. Proc. Linn. Soc. London 179: 197-201 pls. 1, 2, figs. 1.

———. 1969. Studies on British chytrids XXIX. A taxonomic revision of certain fungi on the diatom *Asterionella*. Ibid. 62:267-278, 3 figs., pls. 1-3.

———. 1970. Studies on British chytrids. XXX. On *Podochytrium cornutum* Sparrow Ibid. 63:47-52, 2 figs., pls. 1, 2.

———. 1971. Studies on British chytrids. XXXI. *Rhizophydium androdiocytes* sp. nov., parasitic on *Dictyosphaerium pulchellum* Wood from the plankton. Trans. Brit. Mycol. Soc. 56:114-120,

fig. 1, pl. 13.
Chambers, T. C. and L. G. Willoughby. 1964. The fine structure of *Rhizophlyctis rosea*, a soil Phycomycete. J. Roy. Micro. Soc. 83:355-364, pls. 152-158.
Cienkowski, L. 1857. *Rhizidium confervae-glomeratae*. Bot. Zeit. 15:233-237, pl. 5A, figs. 16.
Cohn, F. 1853. Untersuchungen über die Entwickelungsgeschichte der Mikroskopischen Algen and Pilze. Nova Acta Acad. Leop-Carol. 24:101-256, pls. 15-20.
Cook, P. W. 1963. Host range studies of certain Phycomycetes parasitic on desmids. Amer. J. Bot. 50: 580-588.
________. 1966. *Entophlyctis reticulospora* sp. nov., a parasite of *Closterium*. Trans. Brit. Mycol. Soc. 49:545-550, pls. 28, 29.
Couch, J. N. 1932. *Rhizophydium, Phlyctochytrium* and *Phlyctidium* in the United States. J. Elisha Mitchell Sci. Soc. 47:245-360, pls. 14-17.
________. 1935. An incompletely known chytrid: *Mitochytridium ramosum*. Ibid. 51:293-295, pl. 62.
________. 1935. New or little known Chytridiales, Mycologia 27:160-175, 21 figs.
________. 1938. A new species of *Chytridium* from Mountain Lake, Virginia. J. Elisha Mitchell Sci. Soc. 54:256-259, pl. 24.
________. 1938. A new chytrid on *Nitella*: *Nephrochytrium stellatum*. Amer. J. Bot. 25:507-511, 34 figs.
________. 1939a. Technic for collection, isolation, and culture of chytrids. J. Elisha Mitchell Sci. Soc. 55:208-214.
________. 1939b. Heterothalliam in the Chytridiales. Ibid. 55:409-414, pl. 49.
Crasemann, J. M. 1954. The nutrition of *Chytridium* and *Macrochytrium*. Amer. J. Bot. 41:302-310, 5 figs.
Dangeard, P. A. 1888. Les Péridiniens et leurs parasites. J. de Botan. 2:126-132, pl. 5.
________. 1889. Mémoire sur les Chytridinees. Le Botaniste 1:39-72, pls. 2, 3.
________. 1900. Recherches sur las structure *Polyphagus euglenae* Nowakowski et sa reproduction sexualle. Ibid. 7:213-257, 2 figs. 6, 7.
________. 1911. Un nouveau genre de Chytridiacées. Bull. Soc. Mycol. France. 27:200-203, 1 fig.
________. 1932. Observations sur la familee des Labyrinthulées et sur quelques autres parasites des *Cladophora*. Le Botaniste 24:217-258, pl. 24.
________. 1937. Sur un nouveau moyen de défense très curieux de certaines Desmidiées contra les parasites de la familee des Chytridiacées. Ibid. 28:187-200, pl. 19.
DeBary, A. 1884. Vergleichende Morphologie and Biologie der Pilze, Mycetozoen und Bacterien. XVI + 558 pp., 198 figs. Leipzig.
________, and M. Woronin. 1863. Beitrag zur Kenntniss der Chytridieen. Ber. Verh. Nat. Gesell. Freiberg, 3:22-61, pls, 1,2.
Dogma, I. J., Jr. 1969a. Observations on some cellulosic chytridiaceous fungi. Arch. f. Mikrobiol. 66:203-219, 70 figs.
________. 1969b. Additions to the Phycomycete flora of the Douglas Lake region. VIII. *Chytriomyces annulatus* sp. nov., and notes on other zoosporic fungi. Nova Hedwigia 18:349-365, pls. 1, 2.
________. 1970. Additions to the Phycomycete flora of the Douglas Lake region. IX. On the genus *Sparrowia* Willoughby, Chytridiales. Ibid. 19: 503-508, 16 figs.
________. 1974a. Developmental and taxonomic studies of rhizophlyctoid fungi, Chytridiales. II. The *Karlingia* (*Rhizophlyctis*) *rosea*-complex. Nova Hedwigia 25:1-49, 77 figs.
________. 1974b. Developmental and taxonomic studies on rhizophlyctoid fungi, Chytridiales. III. Chitinophilic *Rhizophlyctis* with resting spores of sexual origin. Ibid. 25:51-87, 9 pls.
________. 1974c. Developmental and taxonomic studies on rhizophlyctoid fungi, Chytridiales. IV. *Karlingia granulata, Karlingia spinosa*, and *Karlingiomyces dubius*. Ibid. 25:91-105, 3 pls.
________. 1974d. Studies on chitinophilic *Siphonaria, Diplophlyctis*, and *Rhizoclosmatrium*. I. *Siphonaria variabilis* H. E. Petersen emend. with emphasis on sexuality. Ibid. 25:107-110, 2 pls.
________. 1974e. Studies on chitinophilic *Siphonaria, Diplophlyctis*, and *Rhizoclosmatium*. II. *Asterophlyctis sarcroptoides* H. E. Petersen: a *Diplophlyctis* with a sexual phase. Ibid. 25:121-141, 4 pls.
Domjan, A. 1936. "Visigombas"—Adatok Szeged es Tihany Videkerol (Wasserpils"—daten aus der Umgebung von Szeged and Tihanyl). Folio crytogam. 2:8-50, pl. 1.
Fay, D. J. 1947. *Chytriomyces spinosus* nov. sp. Mycologia 39:152-157, 2 figs.
Fisch, C. 1884. Beitrage zur Kenntniss der Chytridiaceen. Sitzungsber. Phys. Med. Soc. Erlangen 16: 29-72, pl. 1, 39 figs.
Fischer, A. 1892. Phycomycetes, Die Pilze Deutschlands, Oesterreichs und der Schweiz. Rabenhorst Kryptog.—Fl. 1:1-490. Leipzig.
Fitzpatrick, H. M. 1930. The lower fungi-Phycomycetes. XI-331 pp. McGraw-Hill, New York.
Fott, B. 1942. Über eine auf den Protocaccalenzellen parasitierende Chytridiacee. Studia Bot. Cechica 5:167-170, 5 figs.
________. 1950. New chytrids parasitizing on algae. Z. vestnik Kral ces spol. Nauk, 1 Tr. Matem.—Privo. Roc. 4:1-10, 20 figs. 1 pl.
________. 1952. Midroflora oravskych raselin. Preslia 24:189-209, 4 figs.
________. 1957. Taxonomic drobnokledne flory masich vod. Ibid. 29:278-319, 11 figs., pl. 19.
________. 1967. *Phlyctidium scenedesmi* spec. nova, a new chytrid destroying mass cultures of algae. Zeitschr. Allgem. Mikrobiol. 7:97-102.
Friedmann, I. 1952. Über neue and wenig bekannte

auf Diatomeen parasitierende Phycomyceten. Oesterr. Bot Zeitschr. 99:1 8-217, 8 figs.

———. 1953. Eine neue Chytridiale, *Chytridium surirellae* n. sp. Ibid. 100:5–7, 1 fig.

Gaertner, A. 1954. Beschreibung drier neuer Phlyctochytrien und eines *Rhizophydium* (Chytridiales) aus Erdoden. Arch. f. Mikrobiol. 21:112-126, 7 figs.

Geitler, L. 1943. Eine neue Chytridiale *Zygorhizidium verrucosum* n. sp. und ihre Wirkung auf den Wirtzellen. Arch. f. Protistenk. 96:109-118, 2 figs.

——— 1962a. Entwickelungsgeschichte der Chytridiale *Entophlyctis apiculata* auf der Protococcale *Hypnomonas lobata*. Oesterr. Bot. Zeitschr. 109:138-149, 4 figs.

———. 1962b. Entwickelung und Beziehung zum Wirt der Chytridiale *Scherffeliomycopsis coleochaetis* n. spec. Ibid. 109:250-275, 8 figs.

———. 1963. *Dangeardia sporapiculata* n. sp., der Begriff "apikulus" und die Gattungs abgrenzung bei einigen Chytridialen. Sydowia 16:324-330, 2 figs.

Gimesi, N. 1924. Hydrobiologiai Tanulmanyok (Hydrobiologische Studen) II. *Phlyctidium eudorinae* Gim, n. sp. (Adatok A. Phycomycesek Ismeretehez). Novenytani Szakosztalyanak 124:1-5. Nemetul 6-8 (1 tabla, 1 rojz).

Goldie-Smith, E. K. 1946. *Chytridium lecythii* (Ingold) n. comb. Trans. Brit. Mycol. Soc. 29:68-69, 1 fig.

Graff, P. 1928. Contributions to our knowledge of western Montana fungi. II. Phycomycetes. Mycologia 20:158-179.

Hanson, A. M. 1944a. Three new saprophytic chytrids. Torreya 44:30-33.

———. 1944b. A new chytrid parasitizing *Volvox*: *Loborhiza metzneri* gen. nov., sp. nov. Amer. J. Bot. 31:166-171, 31 figs.

———. 1945. A morphological, developmental, and cytological study of four saprophytic chytrids. II. *Rhizophydium coronum* Hanson. Ibid. 32: 479-487, 61 figs.

———. 1946a. A morphological, developmental and cytological study of four saprophytic chytrids. III. *Catenochytridium laterale* Hanson. Ibid. 33:389-393, 31 figs.

———. 1946b. A morphological, developmental, and cytological study of four saprophytic chytrids. IV. *Phlyctorhiza endogena* gen. nov., sp. nov. Ibid. 33:732-739, 49 figs.

Harant, H. 1931. Les ascidies et leurs parasites. Ann. L'inst. Oceanographie 8:231-389, 61 figs.

Haskins, R. H. 1946. New chytridiaceous fungi from Cambridge. Trans. Brit. Mycol Soc. 29:135-140, 21 figs.

———. 1948. Studies in the lower Chytridiales. Thesis, Harvard University.

———. 1950. Studies in the lower Chytridiales II. Endo-operculation and sexuality in the genus *Diplophlyctis*. Mycologia 42:772-778, 10 figs.

———, and W. H. Weston, Jr., 1950. Studies in the lower Chytridiales I. Factors affecting pigmentation, growth, and metabolism of a strain of *Karlingia* (*Rhizophlyctis*) *rosea*. Amer. J. Bot. 37: 739-750, 11 figs.

Hillegas, A. B. 1938. Two new operculate chytrids. Mycologia 30:302-312, 37 figs.

———. 1940. The cytology of *Endochytrium operculatum* (de Wildeman) Karling in relation to its development and organization. Bull. Torrey Bot. Club. 67:1-29, pls. 1-7.

Höhnk, W. von. 1962. Über die Phycomyceten der Insel Madeira. Veroff. Inst. Meeresforsch in Bremerhaven 8:99-108.

Hovasse, R. 1936. *Rhizophydium beauchampi* sp. nov., Chytridinée parasite de la Volvocinée *Eudorina* (*Pleodorina*) *illinoisensis* (Kofoid). Ann. Protistol. 5:73-81, 4 figs.

Ingold, C. T. 1940. *Endocoenobium eudorinae* gen. et sp. nov., a chytridiaceous fungus parasitizing *Eudorina elegans* Ehrenb. New Phytologist 39: 97-103, 4 Figs. pl. 2.

———. 1941. Studies on British chytrids. I. *Phlyctochytrium proliferum* sp. nov. and *Rhizophydium lecythii* sp. nov. Ibid. 25:41-48, 3 figs. pl. 4.

———. 1944. Studies on British chytrids II. A new chytrid on *Ceratium* and *Peridinium*. Trans. Brit. Mycol. Soc. 27:93-96, 3 figs. pl. 9.

Jaag, O. and F. Nipkow. 1951. Neue and wenig bekannte parasitische Pilze auf planktonorganismen schweizerischer Gewasser, 1. Ber. Schweiz. Bot Gesell. 61:478-498, pls. 11-16.

Johanson, A. E. 1944. An endo-operculate chytridiaceous fungus: *Karlingia rosea* gen. nov: Amer. J. Bot. 31:397-404, 37 Figs.

Johns, R. M. 1956. Additions of the Phycomycete flora of the Douglas Lake Region. III. A new species of *Scherffeliomyces*. Mycologia 48:433-438, 12 figs.

———. 1964. A new *Polyphagus* in algal culture. Ibid. 56:441-451, 6 figs.

Johnson, T. W. 1968a. Aquatic fungi of Iceland: Introduction and preliminary account. J. Elisha Mitchell Sci. Soc. 84:179-183, 1 fig.

———. 1968b. A note on *Macrochytrium botrydiodies* Minden. Arch. f. Mikrobiol. 63:292-294, 13 figs.

———. 1969. Aquatic fungi of Iceland: *Phlyctochytrium* Schroeter. Ibid. 64:357-368, 25 figs.

———. 1971. Aquatic fungi of Iceland: *Chytriomyces* Karling emend Sparrow. J. Elisha Mitchell Sci. Soc. 87:200-205, 49 figs.

———, and C. E. Miller. 1974. Two unusual chytridiaceous fungi. Mycologia 66:859-867, 31 figs.

Karling, J. S. 1928a. Studies in the Chytridiales I. The life history and occurrence of *Entophlyctis helimorpha* (Dang.) Fischer. Amer. J. Bot 15:32-42, pl. 1.

———. 1928b. Studies in the Chytridiales. II. Contribution to the life history and occurrence of *Diplophlyctis intestina* (Schenk) Schroeter in cells of American Characeae. Ibid. 15:204-214, pl. 14.

———. 1930. Studies in the Chytridiales. IV. A further study of *Diplophlyctis intestina* (Schenk) Schroeter. Ibid. 17:770-778, 2 figs., pls. 46-49.

———. 1931. Studies in the Chytridiales. V. A further study of species of the genus *Entophlyctis*. Ibid. 18:443-464, 54 text-figs., pls. 35-38.

———. 1936. The endo-exogenous method of growth and development of *Chytridium lagenaria*. Ibid. 23:619-627, 2 figs.

———. 1936b. Germination of the resting spores of *Diplophlyctis intestina*. Bull. Torrey Bot. Club 63:467-471, 8 figs.

———. 1937a. A new species of *Phlyctochytrium* on *Hydrodictyon reticulatum*. Mycologia 29: 178-186, 3 figs.

———. 1937b. The structure, development, identity, and relationship of *Endochytrium*. Amer. J. Bot. 24:352-264, 53 figs.

———. 1938a. A new chytrid genus: *Nephrochytrium*. Ibid. 25:211-215, 2 figs.

———. 1938b. A large species of *Rhizophydium* from cooked beef. Bull. Torrey Bot. Club 64: 439-452, pls. 20, 21.

———. 1938c. Studies on *Rhizophydium*. II. *Rhizophydium laterale*. Ibid. 65:615-624, pl. 31.

———. 1938d. Two new operculate chytrids. Mycologia 30:302-312, 37 figs.

———. 1939a. Studies on *Rhizophydium*, III. Germination of the resting spores. Bull. Torrey Bot. Club 66:281-286, pl. 6.

———. 1939b. A note on *Phlyctidium*. Mycologia 31:286-288.

———. 1941a. *Cylindrochytridium johnstonii* gen nov. et sp. nov., and *Nowakowskiella profusum* sp. nov. Bull. Torrey Bot. Club 68:381-387, 16 figs.

———. 1941b. Texas chytrids. Torreya 41:105-108.

———. 1942. Parasitism among chytrids. Amer. J. Bot. 29:24-35, 47 figs.

———. 1944a. Brazilian chytrids. II. New species of *Rhizidium*. Ibid. 31:254-261, 72 figs.

———. 1944b. Brazilian chytrids. III. *Nephrochytrium amazonensis*. Mycologia 36:351-357, 28 figs.

———. 1945a. Brazilian chytrids. V. *Nowakowskiella macrospora* n. sp., and other polycentric species. Amer. J. Bot. 32:29-35, 51 figs.

———. 1945b. Brazilian chytrids. VI. *Rhopalophlyctis* and *Chytriomyces*., two new chitinophilic operculate genera. Ibid. 32:362-369, 61 figs.

———. 1945c. Brazilian chytrids. VII. Observations relative to sexuality in two new species of *Siphonaria*. Ibid. 32:580-587, 53 figs.

———. 1946a. Brazilian chytrids. IX. Species of *Rhizophydium*. Ibid. 33:328-344, 37 figs.

———. 1946b. Keratinophilic chytrids. I. *Rhizophydium keratinophilum* n. sp., a saprophyte isolated on human hair, and its parasite, *Phlyctidium mycetophagum* n. sp. Ibid. 33:751-757, 60 figs.

———. 1947a. Brazilian chytrids. X. New species with sunken opercula. Mycologia 39:56-70, 56 figs.

———. 1947b. Keratinophilic chytrids. II. *Phyctorhiza variabilis* n. sp. Amer. J. Bot. 34:37-32, 48 figs.

———. 1947c. New species of *Chytriomyces*. Bull. Torrey Bot. Club. 74:334-344, 48 figs.

———. 1948. Keratinophilic chytrids. II. *Rhizophydium nodulosum* sp. nov. Mycologia 40:328-335, 20 figs.

———. 1949a. Three new species of *Chytriomyces* from Maryland. Bull. Torrey Bot. Club. 76:352-362, 59 Figs.

———. 1949b. Two new eucarpic inoperculate chytrids from Maryland. Amer. J. Bot. 36:681-687, 48 Figs.

———. 1949c. *Truittella setifera* gen. nov., et sp. nov., a new chytrid from Maryland. Amer. J. Bot. 36:454-460, 44 figs.

———. 1949d. New monocentric eucarpic operculate chytrids from Maryland. Mycologia 41:505-522, 78 Figs.

———. 1960. Parasitism among chytrids. II. *Chytriomyces verrucosus* sp. nov. and *Phlyctochytrium synchytrii*. Bull. Torrey. Bot. Club. 87: 326-336, 29 Figs.

———. 1961. *Nowakowskiella sculptura* sp. nov. Trans. Brit. Mycol. Soc. 44:453-457, 24 figs.

———. 1964a. Indian chytrids. I. Eucarpic monocentric species. Sydowia 17:285-396, 31 figs.

———. 1964b. Indian chytrids. III. Species of *Rhizophlyctis* isolated on human fibrin film. Mycopath. et Mycol. Appl. 23:215-222, 26 figs.

———. 1964c. Indian chytrids. IV. *Nowakowskiella multispora* sp. nov. and other polycentric species. Sydowia 17:314-319, 8 figs.

———. 1965. *Catenophlyctis*, a new genus at the Catenaeriaceae. Amer. J. Bot. 52:133-138, 12 figs.

———. 1966. The chytrids of India with a supplement of other zoosporic fungi. Beih. z. Sydowia VI:1-125.

———. 1967a. Some zoosporic fungi of New Zealand. III. *Phlyctidium, Rhizophydium, Septosperma, and Podochytrium*. Sydowia 20:74-85, pls. 12-14.

———. 1967b. Some zoosporic fungi of New Zealand. IV. *Polyphlyctis* gen. nov., *Phlyctochytrium* and *Rhizidium*. Ibid. 20:86-95, pls. 15, 16.

———. 1967c. Some zoosporic fungi of New Zealand. V. *Asterophlyctis, Obelidium, Rhizoclosmatium, Siphonaria* and *Rhizoplyctis*. Ibid. 20: 96-107, pls. 17-19.

———. 1967d. Some zoosporic fungi of New Zealand. VI. *Entophlyctis, Diplophlyctis, Nephrochytrium* and *Endochytrium*. Ibid. 20:109-118, pls. 20-23.

———. 1967e. Some zoosporic fungi of New Zealand.

VII. Additional monocentric species. Ibid. 20: 119-128, pls. 23.

———. 1968a. Zoosporic fungi of Oceania. III. Monocentric chytrids. Arch. f. Mikrobiol. 61: 112-127, 3 figs.

———. 1968b. Zoosporic fungi of Oceania. IV. Additional monocentric species. Mycopath. et Mycol. Appl. 36:165-178; 61 figs.

———. 1969. Zoosporic fungi of Oceania. VII. Fusions in *Rhizophlyctis*. Amer. J. Bot. 56:211-221, 108 figs.

———. 1970. Some zoosporic fungi of New Zealand. XIV. Additional species. Arch. f. Mikrobiol. 70:266-287, 6 figs.

———. 1971. On *Chytridium* Braun, *Diplochytridium* N. G., and *Canteria* N. G. (Chytridiales). Ibid. 76:126-131.

Kniep, H. 1928. Die sexualität der niederen pflanzen. IV + 544 pp. Jena.

Knox, J. S. 1970. Biosystematic studies of aquatic Phycomycetes; Chytridiales and Blastocladiales. Ph.D. Thesis, Virginia Poly. Inst. and State Univ.

———, and R. A. Paterson. 1973. The occurrence and distribution of some aquatic phycomycetes on Ross Island and the dry valleys of Victoria Land, Antarctica. Mycologia 65:373-387, 25 figs.

Koch, W. J. 1951. Studies in the genus *Chytridium* with observations on a sexually producing species. J. Elisha Mitchell Sci. Soc. 67:267-278, 2 figs., pls. 19-21.

———. 1957a. Two new chytrids in pure culture, *Phlyctochytrium punctatum* and *Phlyctochytrium irregulare*. Ibid. 73:108-121, 24 figs.

———. 1957b. Further studies in the genus *Chytridium*. Ibid. 73:239-240.

———. 1959. The sexual stage of *Chytriomyces*. Ibid. 75:66.

Kobayashi, Y., and M. Ookubo. 1953. Studies on marine Phycomycetes. (1). Bull. Nat. Sci. Mus. (Tokyo) 33:53-65, 9 figs.

———. 1954a. Studies on marine Phycomycetes. II. Ibid. 2:63-71, 9 figs.

———. 1954b. Studies on the aquatic fungi of the Osegahara moor. Rept. Osegahara Gen. Sci. Survey Comm. 1954:561-575, 18 figs.

———, and K. Konno. 1969. Enumeration of the watermoulds found in Chickijima, Bonin Islands. Bull. Nat. Sci. Mus. Tokyo 12:725-733, 5 figs.

———, and K. Konno. 1970. Watermoulds isolated from soil in Tsusuima Island (1). Jap. J. Bot. 45:325-337, 2 figs.

———, N. Hiratsuka, Y. Otani, K. Tubaki, S. Ubagawa, J. Sugi-Yama, and K. Konno. 1971. Mycological Studies of the Angmagssalik Region of Greenland. Ibid. 14:1-96, 58 figs., pls. 1-6.

Kono, K. 1968. Studies on Japanese lower aquatic Phycomycetes. I. On *Blyttiomyces spinulosus* (Blytt) Bartsch. Trans. Mycol. Soc. Japan 8: 130-135, 2 figs.

———. 1969. Studies on Japanese lower aquatic Phycomycetes. III. *Rhizidium tomiyamanum*. J. Jap. Bot. 44:315-317, 8 figs.

———. Studies on Japanese lower aquatic Phycomycetes. Sci. Rept. Tokyo Kyoiku Daigaku, ser. B., 14:227-292, 13 pls.

Lagerheim, 1890. *Harpochytrium* and *Achlyella*, zwei neue Chytridiaceen Gattungen. Hedwigia 29: 142-145, pl. 2.

———. 1892. *Mastigochytrium* eine neue Gattung der Chytridiaceen. Ibid. 31:185-189. pl. 18.

Ledingham, G. A. 1936. *Rhizophydium graminis* n. sp., a parasite of wheat roots. Canad. J. Res. (c) 14:117-121, 15 figs.

Lindau, G. 1899. *Rhizidium lignicola* nov. spec., eine holzbewohnende Chytridiacee. Verh. Bot. Verein Brandenburg 41: XXVII-XXXIII, 12 figs.

Lowenthal, W. 1905. Weitere Untersuchungen an Chytridiaceen. I. *Synchytrium anemones* Woronin. II. *Olpidium Dicksonii* (Wright) Wille. Arch. f. Protistenk. 5:221-239, pls. 7, 8.

Masters, M. J. 1971. *Chytridium deltanum* n. sp. and other Phycomycetes on *Oocystis* spp. in the Delta Marsh, Manitoba. Canad. J. Bot. 49:471-481, 6 figs., pl. 1.

Milanez, A. I. 1967. Resting spores of *Phlyctochytrium planicorne* on Saprolegniaceae. Trans. Brit. Mycol. Soc. 50:679-681, 1 fig.

Miller C. E. 1968. Observations concerning taxonomic characteristics in chytridiaceous fungi. J. Elisha Mitchell Sci. Soc. 84:100-107, 44 figs.

Milovtzova, M. 1938. *Polyphagus fominii* sp. nov. Sbirnik praz frisvjaceni pamjeti akad. O. V. Fomina. Acad. Nauk. U. S. S. R. pp. 146-148.

Minden, M. von. 1902. Ueber Saprolegnüneen. Centralb. f. Bact. Parasitk. u. Infekt. Abt. 2, 8:805-810, 821-825.

———, 1911. Chytridineae, Ancylistineae, Monoblepharidienae Saprolegniieae Kryptogamenfl. Mark Brandenburg 5:193-352, pl. 2.

———. 1916. Beitrage zur Biologie und Systematik einheimischer submerser Phycomyceten. In Falck. Mykolog. Untersuch. Berichte 2:146-255, 24 figs., pls. 1-8.

Moore, E. D., and C. E. Miller, 1971. Observations on sexual fusions in *Chytriomyces hyalinus*. Amer. J. Bot. 58:474.

——— and ———. 1973. Resting body formation by rhizoidal fusion in *Chytriomyces hyalinus*. Mycologia 65:145-154, 14 figs.

Murray, C. L., and J. S. Lovett. 1966. Nutritional requirements of the chytrid *Karlingia asterocysta*, an obligate chitinophile. Amer. J. Bot. 53:469-476.

Nemec, B. 1912. Zur Kenntniss der niederen Pilze. IV. *Olpidium brassicae* Wor. und zwei *Entophlyctis*- Arten. Bull. Intern. Acad. Sci. Boheme 1912: 1-11, pls. 1, 2, 1 fig.

Nowakowski, L. 187a. Beitrag zur Kenntniss der Chytridiaceen. *In* Cohn, Beitr. Biol. Pflazen 2:73-100,

pls. 4-6.
———. 1876b. Beitrag zur Kenntniss der Chytridiaceen II. *Polyphagus euglenae*, eine Chytridiacee mit geschlechtlicher Fortpflanzung. Ibid. 2: 203-216, pls. 8, 9.
———. 1878. Przyczynek do morphologii i systematyki Skozckow (Chytridiaceae). Akad. Umiejnosci Krakowie. Wydziat mat. - przyod., Pamistnik 4:174-198, pls. 7-10.
Ookubo, M. 1954. Studies on the aquatic fungi in the moor and ponds of Hakkoda. Nagaoa 4:48-60, 47 figs.
——— and Y. Kobyasi, 1955. Studies on the water molds on kerantinized materials. Ibid. 5:1-10, 6 figs.
Paterson, R. A. 1956. Additions to the Phycomycete flora of the Douglas Lake region. II. New chytridiaceous fungi. Mycologia 48:270-277, 2 figs.
———. 1958a. Parasitic and saprophytic Phycomycetes which invade planktonic organisms. I. New taxa and records of chytridiaceous fungi. Ibid. 50: 85-96, 2 figs.
———. 1958b. Parasitic and saprophytic Phycomycetes which invade planktonic organisms. II. A new species of *Dangeardia* with notes on other lacustrine fungi. Ibid. 50:453-468, 27 figs.
———. 1958c. On the planktonic chytrids *Zygorphizidium melosirae* Canter and *Z. planktonicum* Canter. Trans. Brit. Mycol. Soc. 41:457-460, 1 fig.
Persiel, I. 1959. Über *Phlyctochytrium reinboldtae* n. sp. Arch. f. Mikrobiol. 32:411-415, 1 fig.
———. 1960. Beschreibung neuer Arten der Gattung *Chytriomyces* und einiger seltener niederer Phycomyceten. Ibid. 36:283-305, 14 figs.
———. 1963. Über klein- und grosporige sporangien bei *Phlyctochytrium palustre* Gaertner und *Rhizophydium racemosum* Gaertner. Ibid. 46:343-346, 1 fig.
Petersen, H. E. 1903. Note sur les Phycomycètes observés dans les téguments vides des nymphes de Phryganées avec description trois espéces nouvelles de Chytridinées. J. de Bot. 17:214-222, 17 figs.
———. 1909. Studier over Ferskvands-Phycomyceter. Bidrag til Kundskaben om de submerse Phykomyceter's Biologi og Systematik, samt om deres Udredelse i Danmark. Bot. Tidsskr. 29:345-440, 27 figs.
———. 1910. An account of Danish freshwater Phycomycetes, with biological systematical remarks. Ann. Mycol. 8:494-560, Figs. I-XXVII.
Pfitzer, E. 1870. Über weiterer Beobachtungen...auf Diatomaceen parasitischen Pilze aus der Familie Chytridieen. Sitzungsber. Niederrhein. Gesell. Natur- und Heilkunde 27:62.
Pongratz, E. 1966. De quelques champignons parasites d'organismes plantoniques de Leman Schweitz Zeitschr. f. Hydrologie 28:104-132, pls. 1-5, fig. 1.
Quantz, L. 1943. Untersuchungen über die Ernährungsphysiologee einiger niederer Phycomyceten (*Allomyces kniepii, Blastocladia variabilis* und *Rhizophlyctis rosea*). Jahrb. Wiss. Bot. 91:120-168.
Rabenhorst, L. 1868. Flora Europeae algarum—. Vol. 3, XX + 461 pp., Leipzig.
Raciborski, M. 1900. Parasitische Algen and Pilze Javas 1:1-39. Batavia, Java.
Richards, M. 1951. The life-history of *Diplophlyctis laevis*. Trans. Brit. Mycol. Soc. 34:483-488, 19 figs., pl. 24.
Rieth, A. 1951. Zur Phycomycetenflora Wurtenbergs. I. Teil. Natursch. in Wurtemberg-Hohenzollern, 1950, 259-271, 14 figs.
———. 1954. Ein weiterer Beitrag zur kenntnis algenparasitarer Phycomyceten. Die Kulturpflanze 2: 164-184, 9 figs.
———. 1956. Beitrag zur Kenntnis der Phycomyceten. III. Ibid. 4:181-186, pl. 6, 2 figs.
———. 1967. Beitrage zur Kenntnis algenparasitarer Phycomyceten. Biol. Zentralb. 86:435-448, 4 figs.
Roane, M. K. 1973. Two new chytrids from the Appalachian Highlands. Mycologia 65:531-538, 20 figs.
Rosen, F. 1887. Ein Beitrag zur Kenntnis der Chytridiaceen. *In* Cohn, Beitr. Biol. Pflanz. 4:253-266, pls. 13, 14.
Salkin, J. F. 1970. *Allochytridium expandens n.* gen. et sp. n. Growth and morphology in continuous culture. Amer. J. Bot. 57:649-658, 49 figs.
Schaarschmidt, J. 1883, *Phlyctidium haynaldii* n. sp. Magyar Novenytani Lapok. Kolozsvar 1883:58-63, pl. 2.
Schenk, A. 1858. Ueber das Vorkommen contractiler Zellen im Pflanzenreich. 20 pp., 15 figs.
———. 1858. Algologische Mittheilungen. Verhandl. Phys-Med. Gesell. Wurzburg, A. F. 8:235-259, pl. 5.
Scherffel, A. 1925. Zur Sexualität der Chytridineen (Der "Beitrage zur Kenntnis der Chytridineen" Teil I). Arch. f. Protistenk. 53:1-58, pls. 1, 2.
———. 1926. Einiges über neue oder ungenügend bekannte Chytridineen (Der "Beitrage zur Kenntnis der Chytridineen". Teil II). Arch. f. Protistenk. 54:167-260, pls. 9-11.
Scholz, E. 1958. Über Morphologische Modifikationen bei neiderer Erdphycomyceten und Beschreibung zweir neuer Arten von *Rhizophydium* and *Traustochytrium*. Arch. f. Mikrobiol. 29:354-362, 4 figs.
Schröder, B. 1898. *Dangeardia*, ein neues Chytridianeen Genus auf *Pandorina morum* Bory. Ber. deut. Bot. Gesell. 16:314-321, pl. XX.
Schroeter, J. 1885. Die Pilze Schlesiens. *In* Cohn, Kryptogamenfl. Schlesiens 3:1-814.
———. 1893. Phycomycetes. *In* Engler u. Prantl, Natürl. Pflanzenf. 1:63-141.
Schulz, P. 1923. Kurze Mitteilungen über Algenparasiten. Schrft. f. Süsswasser u. Meeresk. 2:178-181, 14 figs.
Serbinov, J. L. 1907. Kenntniss der Phycomyceten. Organisation u. Enwickelungsgeschichte einiger

Chytridineen Pilze (Chytridinae Schroeter). Scripta Bot. Horti Univ. Imper. Petropol. 25:1-173, pls. 1-6.

Seymour, R. L. 1971. Studies on mycoparasitic chytrids. I. The genus *Septosperma*. Mycologia 63: 83-93, 29 figs.

Shen, San-chun. 1944. A form of *Sporophlyctis rostrata* with ciliated spores. Amer. J. Bot. 31:229-233, 21 figs.

Skuja, H. 1948. Taxonomie des Phytoplanktons einiger seen in Uppland, Schweden. Symbol. Bot. Upsaliensis 9:1-399, pls. 1-39.

Sorokin, N. W. 1872. Aperçu systématique du groupe des Siphomycètes. Bull. Soc. Nat. Kazan 4:1-26, pls. 1, 2 (In Russian).

_______. 1883. Apercu systématique des Chytridiacées recoltées en Russie et dans l'Asie Centrale. Arch. Bot. Nord France 2:1-42, 54 figs.

Sparrow, F. K. Jr., 1932. Observations on the aquatic fungi of Cold Spring Harbor. Mycologia 24: 268-303, 4 figs., pls. 7, 8.

_______. 1933a. New chytridiaceous fungi. Trans. Brit. Mycol. Soc. 18:215-217.

_______. 1933b. Observations on operculate chytridiaceous fungi collected in the vicinity of Ithaca, New York. Amer. J. Bot. 20:63-77, 2 figs., pl. 20.

_______. 1934a. *Scherffeliomyces*. Mycologia 26:377.

_______. 1934b. Observations on marine fungi collected in Denmark. Dansk. Bot. Ark. 8:1-24, 4 pls.

_______. 1936a. Evidences for the possible occurrence of sexuality in *Diplophlyctis*. Mycologia 28: 321-323, 2 figs.

_______. 1936b. A contribution to our knowledge of the aquatic Phycomycetes of Great Britain. J. Linn. Soc. London (Bot.) 50:417-478, 7 figs., pls. 14-20.

_______. 1937. Some chytridiaceous inhabitants of submerged insect exuviae. Proc. Amer. Philos. Soc. 78:23-53, 2 figs., pls. 104.

_______. 1938a. Some chytridiaceous fungi from North Africa and Borneo. Trans. Brit. Mycol. Soc. 11:145-151, 2 figs.

_______. 1938b. The morphology and development of *Obelidium mucronatum*. Mycologia 30:1-14, 44 figs.

_______. 1938c. Chytridiaceous fungi with unusual sporangial ornamentation. Amer. J. Bot. 25: 485-493, 41 figs.

_______. 1939a. Unusual chytridiaceous fungi. Papers Mich. Acad. Sci., Arts and Letters 24:121-126, pls. 1, 2.

_______. 1943. The aquatic Phycomycetes...XIX × 785 pp., 634 figs. Univ. Mich. Press, Ann Arbor.

_______. 1950. Some Cuban Phycomycetes. J. Wash. Acad. Sci. 40:50-55, 30 figs.

_______. 1951. *Podochytrium cornutum* n. sp., the cause of an epidemic on the planktonic diatom *Stephanodiscus*. Trans. Brit. Mycol. Soc. 34:170-173, 1 fig.

_______. 1952a. A contribution to our knowledge of the Phycomycetes of Cuba. Rev. Soc. Cubana Bot. 9:34-40, 10 figs.

_______. 1952b. Phycomycetes from the Douglas Lake region in northern Michigan. Mycologia 44: 759-772, 1 fig.

_______. 1957. A further contribution to the Phycomycete flora of Great Britain. Trans. Brit. Mycol. Soc. 40:523-535, 2 figs.

_______. 1960. The aquatic Phycomycetes...2nd ed. VII - 1187 pp., 91 figs. Univ. Mich. Press, Ann Arbor.

_______. 1965. The occurrence of *Physoderma* in Hawaii, with notes on other Hawaiian Phycomycetes. Mycopath. et. Mycol. Appl. 25:119-143, pls. 1-7.

_______. 1966. A new bog chytrid. Arch. f. Mikrobiol. 3:178-180, 9 figs.

_______. 1973. Three monocentric chytrids. Mycologia 65:1331-1336, 20 figs.

_______. 1974a. Chytridiomycetes, Hyphochytridiomycetes: Chap. 6, Vol. IV B, in The Fungi. An advanced treatise Edited by G. C. Ainsworth, F. K. Sparrow, and A. S. Sussman. Academic Press. New York, N.Y.

_______. 1974b. A zoosporic tribute from the Delphi Sibl. Proc. Iowa Acad, Sci. 81:2-5, 14 figs.

_______, and M. E. Barr. 1955. Additions to the Phycomycete flora of the Douglas Lake region. I. New taxa and records. Mycologia 47:546-556, 27 figs.

_______, and L. J. Dogma. 1973. Zoosporic Phycomycetes from Hispaniola. Arch. f. Mikrobiol. 89: 177-204, 7 figs.

_______, and R. A. Paterson, 1955. A note concerning *Rhizidiopsis* and *Podochytrium*. Ibid. 47: 272-274.

Stanier, R. Y. 1942. The cultivation of and nutrient requirements of a chytridiaceous fungus, *Rhizophlyctis rosea*. J. Bact. 43:499-520.

Tomaschek, A. 1879. Über Binnenzellen in der grossen Zelle (Antheridiumzelle) des Pollens einiger Conifern. Sitzungsber. Acad. Wiss. Wien (Math-Nat. Cl.) 78:197-212, 17 figs.

Uebelmesser, E. R. 1956. Über einige neue Chytridineen aus Erdboden (*Olpidium*, *Rhizophydium*, *Phlyctochytrium* und *Rhizophlyctis*). Arch. f. Mikrobiol. 25:307-324, 7 figs.

Ulken, A. 1972. Physiological studies on a Phycomycete from a mangrove swamp at Cataneia, São Paulo, Brazil. Veroff Inst. Meeresforsch. Bremerhaven 13:217-250.

Umphlett, C. J. and M. M. Holland, 1960. Resting spores in *Phlyctochytrium planicorne*. Mycologia 52: 429-434, 15 figs.

_______, and W. J. Koch, 1969. Two new dentigerate species of *Phlyctochytrium* (Chytridiomycetes).

Ibid. 61:1021-1030, 17 figs.
———, and L. W. Olson. 1967. Cytological and morphological studies on a new species of *Phlyctochytrium*. Ibid. 59:1085-1096.
Valkanov, A. 1963. Über zwei hyponeustische *Polyphagus*-Arten. Arch. f. Protistenk. 106:565-568, 2 figs., pl. 14.
———. 1964. *Dangeardiana eudorinae* n. g. n. sp. ein neuer vertreter der Algenpilze. Arch. f. Mikrobiol. 48:239-246, 13 figs.
Voss, L. R. 1969. Morphology and life cycle of a new chytrid with aerial sporangia. Amer. J. Bot. 56: 898-909, 56 figs.
———, and L. S. Olive. 1968. A new chytrid with aerial sporangia. Mycologia 60:730-733, 5 figs.
Wager, H. 1913. The life history and cytology of *Polyphagus euglenae*. Ann. Bot. 27:173-202, pls. 16-19.
Wallroth, F. G. 1833. Flora Cryptog. Germ. 2nd sect. 4:416.
Ward, M. W. 1939. Observations on *Rhizophlyctis rosea*. J. Elisha Mitchell Sci. Soc. 55:353-360, pls. 32, 33.
Whiffen, A. J. 1941. A new species of *Nephrochytrium*: *Nephrochytrium aurantium*. Amer. J. Bot. 28: 41-44, 30 figs.
———. 1942. Two new chytrid genera. Mycologia 24:543-557, 52 figs.
———. 1944. A discussion of taxonomic criteria in the Chytridiales. Farlowia 1:583-589.
Wildemann, E. de. 1895. Notes mycologiques. V. Ann. Soc. Belge Micro. (Mem.) 19:85-117, pls. 3, 4.
———. 1896. Notes mycologiques. XXII. Champignons parasites des oogones des Characées. Ibid. 20:109-131, pls. 6-12.
———. 1931. Sur quelques Chytridinées parasites d'algues. Bull. Acad. Roy. Belg. (Sci.) V, 17: 281-298, 3 figs., 2 pls.
Willoughby, L. G. 1956. Studies on soil chytrids. I. *Rhizidium richmondense* sp. nov. and its parasites. Trans. Brit. Mycol. Soc. 39:125-141, 9 figs.
———, 1957. Studies on soil chytrids, II. On *Karlingia dubia* Karling. Ibid. 40:9-16, 5 figs.
———. 1958. Studies on soil Chytrids III. On *Karlingia rosea* Johanson and a multi-operculate chytrid on *Mucor*. Ibid. 41:309-319, pl. 17, 4 figs.
———. 1961a. Two new saprophytic chytrids from the Lake District. Ibid. 44:177-184, 3 figs., pls. 13, 14.
———. 1961b. The ecology of some lower fungi at Esthwaite Water. Ibid. 44:305-332, 17 figs., pls. 22, 23.
———. 1961c. Chitinophilic chytrids from lake muds. Ibid. 44:586-592, 2 figs., pl. 37.
———. 1961d. New species of *Nephrochytrium* from the English Lake District. Nova Hedwigia 3: 439-444, pls. 112-116.
———. 1962. The ecology of some lower fungi in the English Lake District. Trans. Brit. Mycol. Soc. 45:121-136, 5 figs.
———. 1963. A new genus of the Chytridiales from soil and a new species from freshwater. Nova Hedwigia 5:335-340. pls. 52, 53.
———. 1964. A study of the distribution of some lower fungi in soil. Ibid. 7:123-150, pls. 17-26.
———. 1965. A study of Chytridiales from Victorian and other Australian soils. Arch. Mikrobiol. 52: 101-131, 12 figs.
———, and P. J. Townley, 1961. A further contribution to our knowledge of *Phlyctochytrium unispinum* Paterson. J. Roy. Micro. Soc. 80:131-136, figs. 1-52, pl. 18.
Zopf, W. 1884. Zur Kenntniss der Phycomyceten. I. Zur Morphologie und Biologie der Ancylisteen und Chytridiaceen. Nova Acta Acad. Leop. - Carol. 47:143-236, pls. 12-21.
———. 1885. Die Pilzthiere oder Schleimpilze. Encyklop. d. Naturwiss. 174 pp., 51 figs. Breslau.
———. 1887. Ueber einige niedere Algenpilze (Phycomyceten) und eine neue Methode ihre Keime aus den wasser zu isolieren. Abhandl. Naturf. Gesell. Halle 17:77-107. 2 pls.
———. 1887. Zur Kenntnis der Infections-Krankheiten niederer Thiere und Pflanzen. Nova Acta Acad. Leop-Carol. 52:313-376, pls. 17-23.
———. 1890. *See* A. Schenk, Handbuch d. Bot. Encyklop. I. Naturwiss 4:271-781, Breslau.
Zukal. H. 1893. Mykologische Mittheilungen. Oestrr. Bot. Zeitschr. 43:310-314, pl. 11.

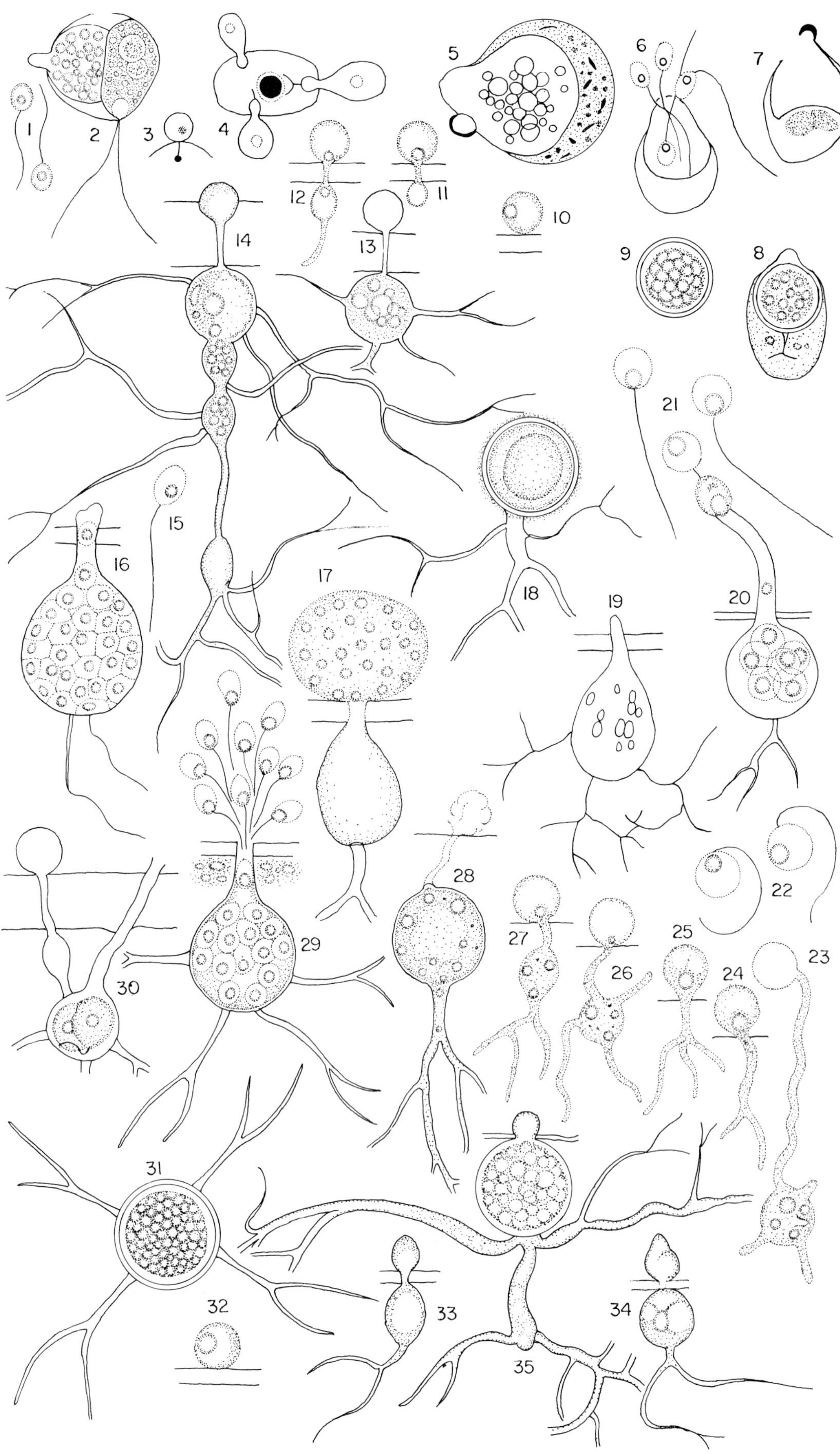

Entophlyctis

Chapter V

ENTOPHLYCTACEAE

This family includes the monocentric, eucarpic operculate and inoperculate chytrids in which the planospore cyst is usually non-functional but may persist with a portion of the germ tube as an appendage on the apophysis, the zoosporangium, or resting spore. The zoosporangium develops either directly from an enlargement of the germ tube or the enlargement becomes an apophysis or prosporangium out of which the zoosporangium develops later. However, as noted earlier, in exceptional thalli of *Asterophlyctis, Chytriomyces* and *Karlingia* of the family Rhizidiaceae the zoosporangia may develop from an enlargement of the germ tube. Thus, the limits of these two families are not always sharply-defined and it remains to be seen whether or not the family Entophlyctaceae will stand as a distinct taxon. Occasionally, the thalli of some species may become polycentric. On the basis of whether the zoosporangium and resting spore develop directly from the germ tube or later from an apophysis or prosporangium the family was divided into 2 subfamilies, Entophlyctoideae and Diplophlyctoideae, respectively, by Whiffen (1944), and this division is followed here. But it must be noted again that the types of development for each division are not always well-defined. It is not always certain that the apophysis functions as a prosporangium in *Diplophlyctis*, and in one species, *D. nephrochytrioides*, several types of development may occur. In other genera the apophysis clearly functions as a prosporangium.

The mature thallus of some genera is wholly endobiotic and consists of a zoosporangium or resting

PLATE 80

Figs. 1-9. *Entophlyctis apiculata* Fischer. (*Dangeardiana apiculata*. Batko, 1970) Figs. 1, 8, 9 after Zopf, 1884; fig. 2 after Braun, 1859; figs. 3-7 after Geitler, 1962.

Fig. 1. Planospores with a hyaline refractive globule.

Fig. 2. Motile cell of *Gleococcus* (*Chlamydomonas*) *pulviscus* parasitized by a mature zoosporangium.

Fig. 3. Infection of *Hypnomonas lobata*.

Fig. 4. Young thalli on *H. lobata*.

Fig. 5. Large zoosporangium with a broad exit papilla in *H. lobata*; planospore cyst persistent and thick-walled.

Fig. 6. Discharge of planospores.

Fig. 7. Empty zoosporangium with apical thick-walled planospore cyst at the side of the exit orifice.

Figs. 8, 9. Mature resting spores in and from *Gleococcus mucosus*, resp., wall smooth and hyaline, content globular.

Figs. 10-14, 17, 18. *Entophlyctis cienkowskiana* Fischer. Figs. 10-14, 18 after Zopf, 1884; fig. 17 after Sparrow, 1936.

Figs. 10-12. Encysted planospore on wall of *Cladophora* sp., germination of the planospore, and formation of sporangial rudiment at end of germ tube, resp.

Fig. 13. Incipient zoosporangium with rhizoids arising from its periphery.

Fig. 14. Sporangial thallus with 3 swelling in the main rhizoidal axis; zoospore cyst and germ tube persistent.

Fig. 17. Discharge of planospore mass.

Fig. 18. Mature resting spore with a golden-smooth wall and a large central globule.

Figs. 15, 16. *Entophlyctis confervae-glomeratae* Sparrow. (Cienkowski, 1857.)

Figs. 15, 16. Ovoid, 3 μ diam., planospore with a hyaline refractive globule and a zoosporangium with planospores in *Cladophora glomerata*, resp.

Figs. 19-21. *Entophlyctis rhizina* Minden. Fig. 19 after Domjan, 1936; figs. 20, 21 after Schenk, 1858.

Fig. 19. Zoosporangium in *Spirogyra* with 3 rhizoidal axes.

Fig. 20. Zoosporangium with a long neck discharging planospores.

Fig. 21. Spherical planospores with a large reddish-yellow globule.

Figs. 22-31. *Entophlyctis helioformis* Ramsbottum. Figs. 22, 29, 31 after Dangeard, 1888; figs. 23, 30 after Karling, 1928; figs. 24-28 drawn from New Zealand material.

Fig. 22. Spherical, 3-4 μ diam., planospores with a hyaline refractive globule.

Fig. 23. Sporangial and rhizoid rudiments forming at tip of a long germ tube in *Nitella*.

Fig. 24, 25. Germ tubes branching before the formation of the sporangial rudiment in *Nitella*.

Fig. 26. Sporangial rudiment developing at juncture of 3 rhizoidal branches.

Fig. 27, 28. Sporangial rudiment developing above rhizoid branches.

Fig. 29. Discharge of planospores from polyrhizoidal zoosporangium in *Nitella*.

Fig. 30. Small zoosporangium in *Nitella* with planospore cyst and germ tube persistent.

Fig. 31. Mature resting spore with a hyaline, smooth wall, globular content, and 4 radiating rhizoids.

Figs. 32-35. *Entophlyctis bulligera* Fischer. (Zopf, 1884.)

Fig. 32. Encysted planospore on wall of *Spirogyra crassa*.

Figs. 33, 34. Young thalli, planospore cysts persistent.

Fig. 35. Part of an almost mature sporangial thallus with a coarse rhizoidal axis whose ends bored through the cross wall of the host.

spore, a haustorium, bushy, reduced, or richly-branched, or an extensive rhizoidal system which is usually centered on the base of the zoosporangium or an apophysis. In some species, however, the rhizoids may arise from more than one point on the periphery of the zoosporangium or resting spore as in *Rhizophlyctis*. In other genera the thallus is interbiotic in the sense that the zoosporangium and resting spore are extracellular or epibiotic in the gelatinous sheath of the host, and the rhizoids or haustorium are endobiotic. In *Rhizosiphon* and *Mitochytridium*, which are included only provisionally in this family, the resting spores are endobiotic.

Sexual reproduction of the resting spores has been reported to occur in *Diplophlyctis* and *Rhizosiphon*. In the former genus fusion of the contents of male and female thalli is said to occur through anastomosed rhizoids. In *Rhizosiphon* pairs of planospore cysts attached to resting spores have been interpreted to be gametes whose contents have fused to produce resting spores at the end of a long or short tube. In other genera where such spores are known they are apparently formed asexually and borne like the zoosporangia. However, in *Mitochytridium* they appear to be borne on a polycentric thallus. Where known they function as prosporangia during germination.

SUBFAMILY ENTOPHLYCTOIDEAE

In this subfamily the zoosporangium and resting spore usually develop from an enlargement within or at the tip of a short or long germ tube, but in *Scherffeliomycopsis* the secondary zoosporangia borne within the primary one develop directly from the planospore body. Also, this genus develops aflagellate amoebospores instead of planospores as in the other genera. Its thallus, like that of *Scherffeliomyces* and *Coralliochytrium*, is interbiotic with an extracellular or epibiotic zoosporangium or resting spore in the gelatinous sheath of the host and an endobiotic

PLATE 81

Figs. 36, 40-43. *Entophlyctis vaucheriae* Fischer. Figs. 37-39 [*E. spirogyrae* (?)]. All after Fisch., 1884.

Fig. 36. Planospore with a hyaline refractive globule.

Figs. 37-39. Infection of *Spirogyra* cell and early stages of thallus development.

Fig. 40. Young vacuolate zoosporangium in *Vaucheria* with a stout rhizoidal axis.

Fig. 41. Mature zoosporangium.

Fig. 42. Mature resting spore with smooth brown outer wall and lustrous endospore.

Fig. 43. Germination of resting spore; outer wall cracked, and subspherical protruding endosporangium.

Fig. 44. *Entophlyctis characearum* de Wildemann resting spore from an oogonium of the Characeae. (de Wildemann, 1896.)

Fig. 45. *Entophlyctis tetraspora* de Wildemann zoosporangia in *Rhinconema* (*Spirogyra*) sp. (Sorokin, 1883.)

Fig. 46. *Entophlyctis maxima* Dangeard zoosporangium in *Cladophora glomerata*. (Dangeard, 1932.)

Figs. 47-56. *Entophlyctis texana* Karling in boiled maize leaves. (Karling, 1967.)

Fig. 47. Spherical, 4-4.5 μ diam., planospore with a deep-red brilliantly refractive globule.

Fig. 48. Germinating planospore.

Fig. 49. Sporangial rudiment developing above germ tube branches.

Fig. 50. Sporangial rudiment developing at juncture of branches of germ tube.

Fig. 51. Young polyrhizoidal zoosporangium.

Fig. 52, 53. Pyriform and irregularly lobed mature polyrhizoidal zoosporangia with long exit tubes.

Fig. 54. Discharge of planospores from a basally monorhizoidal zoosporangium.

Fig. 55. Mature spherical resting spore with a smooth rust colored wall and a large central golden-red or rust colored globule surrounded by smaller ones.

Fig. 56. Germination of an ovoid resting spore.

Fig. 57. *Entophlyctis* sp. Willoughby knobby zoosporangium. (Willoughby, 1961b.)

Fig. 58-67. *Entophlyctis lobata* Willoughby in chitin. (Willoughby, 1961a.)

Fig. 58. Ovoid, 3.5-4 × 2.5-3 μ, planospore with an orange colored refractive globule.

Figs. 59-61. Young thalli with persistent planospore cyst.

Fig. 62. Young flattened thallus with 7 rhizoids and persistent planospore cyst.

Fig. 63. Flattened and lobed thallus with evenly distributed orange globules and a central region of clear protoplasm; persistent planospore cyst shown as a black globule.

Fig. 64. Enlarged view of central clear area; "inner circle represents exit pore rim and the outer the position where the exit tube has broken through the substratum surface."

Fig. 65. Ball of discharged planospores shortly before collective swarming.

Figs. 66, 67. Irregular mature resting spores with yellow-layered walls and large globules.

Figs. 68-74. *Entophlyctis crenata* Karling in epidermal cells of *Vallisneria* sp. (Karling, 1967.)

Fig. 68. Slightly ovoid, 3.8-4.3 μ, planospore with a large hyaline refractive globule.

Fig. 69. Germination of planospore.

Fig. 70. Sporangial rudiment developing in one branch of germ tube.

Fig. 71. Fusiform sporangial rudiment developing at juncture of 2 rhizoidal axes.

Fig. 72. Later developmental stage of incipient zoosporangium.

Fig. 73. Intercalary cylindrical zoosporangium discharging planospores.

Fig. 74. Mature resting spore with a hyaline crenate wall and filled with refractive globules.

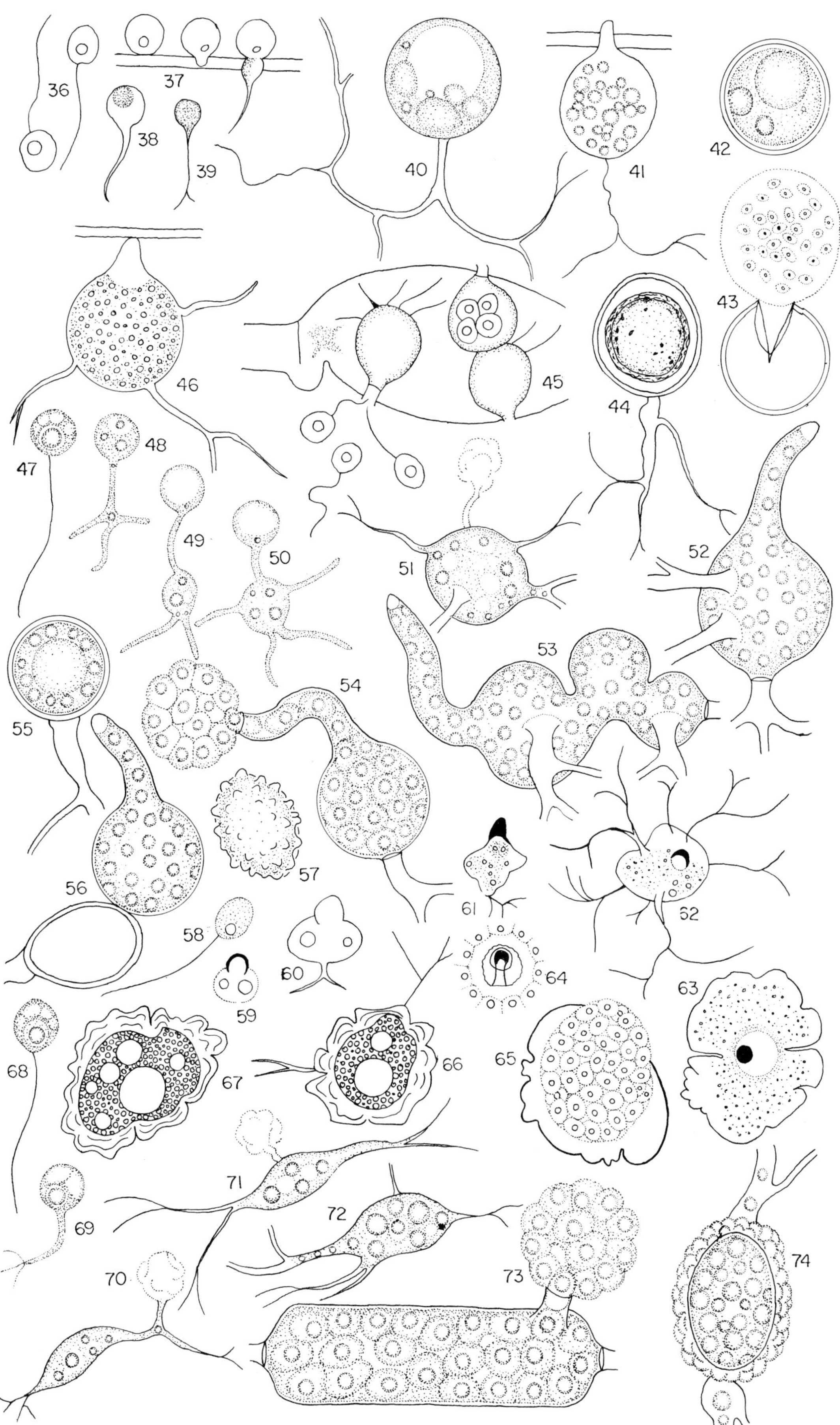

Entophlyctis

absorbing system. The thalli of the other genera are endobiotic, but in disintegrating substrata they may be partly intra- and extramatrical. The thallus of *Mitochytridium* is wholly endobiotic, elongate, tubular and branched and may occupy 2 host cells.

ENTOPHLYCTIS Fischer

Rabenhorst Kryptogamen-Fl. 1:114, 1892. (sensu recent Schroeter, 1893, Phycomycetes, in Engler u. Prantl, die Naturl. Pflanzenfam. 1:75)

Plates 80, 81

The earlier-named species of this genus were placed in *Chytridium* and *Rhizidium*, but Fischer (1892) established *Entophlyctis* for all so-called *Rhizidium* species whose thalli develop within the host cell. In 1893, however, Schroeter limited it to such species with non-apophysate zoosporangia. It is a relatively small inoperculate genus, as interpreted here, but to which 23 species, 1 variety and several unidentified specimens have been assigned from time to time. One species, *E. willoughbyi* Bradley, has been described as an Eocene fossil (Bradley, 1971). *Entophlyctis* is usually characterized by intramatrical or endobiotic sporangial and resting-spore thalli with one or more, reduced or coarse, extensive and richly branched rhizoidal axes arising from the base or periphery of the zoosporangia or resting spores. In *E. reticulospora* Cook (1966), however, rhizoidal axes are lacking, and the zoosporangium periphery bears numerous long hairs, which apparently function as the absorbing system. Many of the reported species are not sharply defined nor completely known and may prove to be identical or invalid. *Entophylctis aurea* Haskins, for example, appears to be a species of *Rhizophlyctis* or *Karlingia*; *E. cienkowskiana* (Zopf) Fischer and *E. vaucheriae* (Fisch) Fischer may be identical with *E. confervae-glomeratae* (Cienkowski) Sparrow; *E. confervae-glomeratae* (Cienkowski) Sparrow f. *marina* Kobayashi and Ookubo likely belongs in another genus, while *E. spirogyrae* Fisch, *E. characearum* de Wildemann, E. *tetraspora* Sorokin, *E. woronichinii* Jaczewski, and *E. maxima* Dangeard are only partially known, which makes their identification very difficult. *Entophlyctis brassicae* Nemec and *E. salicorniae* Nemec each appear to be a combination of more than one fungus. Accordingly, Barr (1971a) believed that the genus includes only 7 species.

Most species are weakly parasitic or saprophytic in freshwater green algae and members of the Characeae. *Entophlyctis apiculata* Braun *(Dangeardiana apiculata* Batko, 1970) parasitizes the thalli and motile cells of *Gleococcus, Chlamydomonas* and *Hypnomonas*, while *E. texana* Karling may occur in moribund cells of *Elodea, Eriocaulon* and *Vallisneria*, or as a saprophyte in the soil where it can be trapped in boiled grass leaves. *Entophlyctis lobata* Willoughby, also, occurs in soil as a chitinophile and can be isolated on chitin. *Entophlyctis brassicae* and *E. salicorniae* occur in the roots of *Brassica* and *Salicornia*, and *E. reticulospora* parasitizes species of *Closterium*.

The marked variations which species of this genus exhibit when grown on synthetic media is evident from Booth's (1971) study of 10 isolates. A comparison of the variations common to all isolates relate them to *E. apiculata, E. crenata,* and *E. reticulospora* as well as to different genera, subfamilies, families, and series. Such variations make current taxonomic dispositions and concepts of little value, according to

PLATE 82

Figs. 1-22. *Endochytrium operculatum* Karling. Fig. 1 after de Wildemann, 1895; figs. 2-4, 7-9, 20-22 after Hillegas, 1940; figs. 5, 6, 10-16, 18, 19 after Karling, 1937.

Fig. 1. Mature zoosporangium with rhizoids arising at 2 places on the periphery.

Fig. 2. Living planospore, 3-5 μ diam., greatly enlarged, with a large hyaline globule.

Fig. 3. Fixed and stained planospore with a conspicuous nuclear cap over the clear nucleus.

Fig. 4. Early germination stage of a fixed and stained planospore.

Fig. 5. Germination of the planospore on *Cladophora.*

Fig. 6. Later stage, rudiment of zoosporangium forming as a swelling in the germ tube.

Fig. 7. Nucleus of germinated planospore has moved down into sporangial rudiment.

Fig. 8. Young tetranucleate zoosporangium.

Fig. 9. Prophase nucleus with centrosomes and astral rays.

Fig. 10. Young zoosporangium dissected out of *Nitella* with planospore cyst attached, one rhizoidal axis at the base.

Fig. 11. Mature zoosporangium in *Cladophora* with 3 rhizoidal axis at the base.

Fig. 12. Discharged mass of planospores.

Fig. 13. Almost empty zoosporangium with rhizoidal axes arising from 4 places on its lower periphery.

Figs. 14-16. Some variations in the shapes of zoosporangia.

Fig. 17. Young uninucleate resting spore.

Figs. 18, 19. Mature warty and smooth-walled resting spores, resp., with a large refractive central globule surrounded by smaller ones.

Fig. 20. Binucleate vacuolate resting spore with an undulating outer wall.

Fig. 21. Early stage of resting spore germination, incipient zoosporangium binucleate.

Fig. 22. Later stage of resting spore germination.

Figs. 23-27. *Endochytrum ramosum* Sparrow. (Sparrow, 1933.)

Fig. 23. Portion of a *Cladophora* filament with 3 zoosporangia and coarse rhizoids.

Figs. 24-26. Stages in discharge of planospores.

Fig. 27. Zoospore dispersal; zoospores enlarged out of proportion to the size of the zoosporangium.

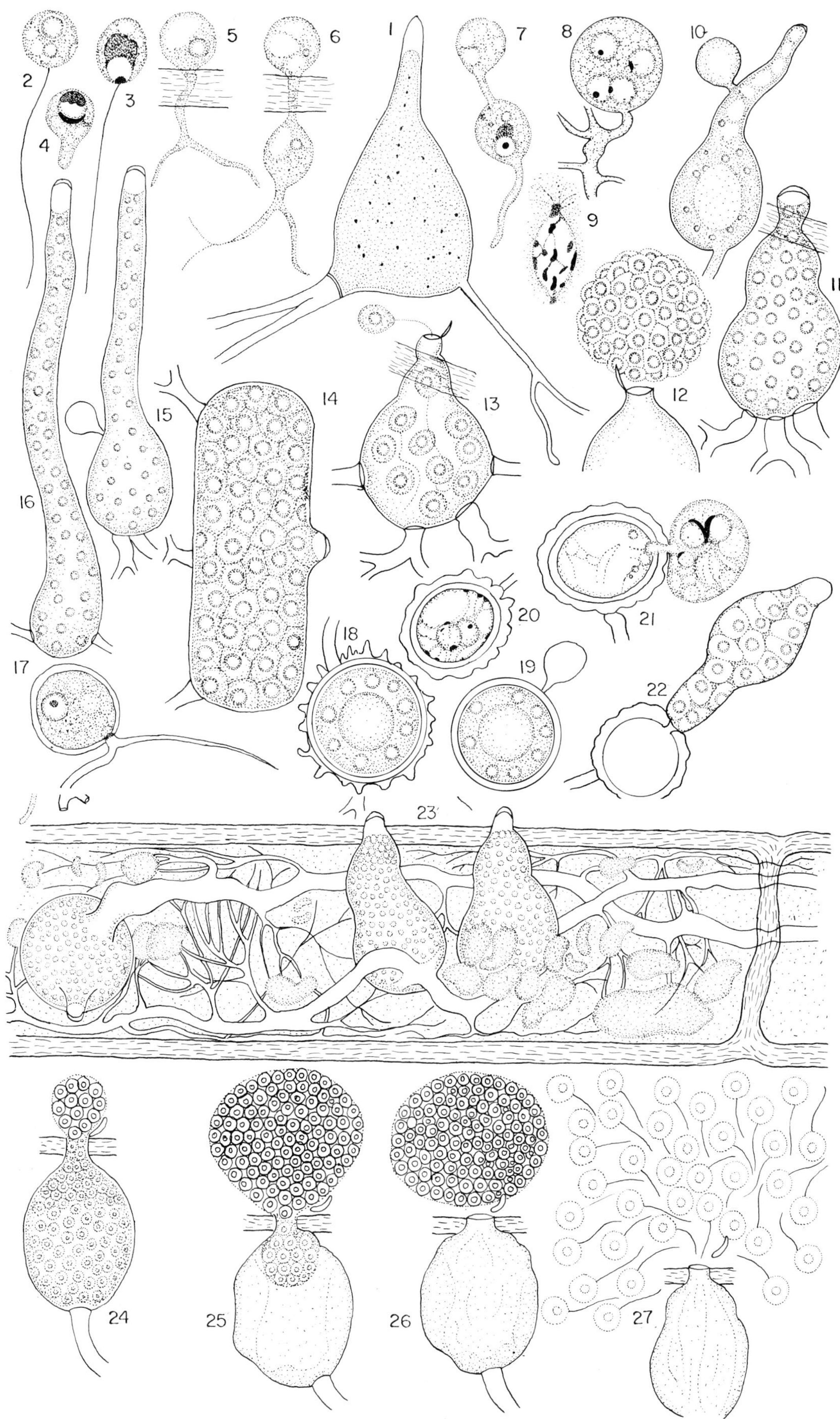

Endochytrium

Booth. Barr (1971b), also, found that a species, *E. confervae-glomeratae*, varied markedly in size and general morphology in different hosts, substrata, and in axenic culture.

The planospores vary from spherical to ovoid in shape and form 1.5 μ diam. in *E. pygameae* (Serbinow) Sparrow to 4.5 μ in *E. texana*, and up to 7 μ in *Entophlyctis* sp. Karling (1931). In *E. rhizina* (Schenk) Minden, *E. aurea*, and *E. lobata* the refractive globule in the planospore is orange or yellowish in color, while in *E. texana* it is brilliantly red. In the other species it is hyaline or colorless. The germ tube of the planospore penetrates the host cell wall (figs. 3, 12, 37) or surface of the substratum (figs. 24, 25, 37) and may become quite long (fig. 23) or remain barely perceptible (fig. 61). Its tip is reported to enlarge (figs. 12, 13, 23) and become the rudiment of the zoosporangium before branches are formed, but in more recent studies it has been shown that the germ tube usually branches (figs. 24, 25, 48, 69) before the sporangial rudiment develops. In any event, the rudiment develops as an enlargement in the germ tube. This enlargement may develop above (figs. 27, 49) or at the juncture of the branches (figs. 26, 50). In the former event the ensuing zoosporangium or resting spore usually has one or more basal rhizoidal axes (figs. 17, 18, 35, 40, 44, 54–56). In the latter the germ tube branches are moved apart as the sporangial rudiment enlarges with the result that the rhizoidal axes arise at several points on the periphery of the zoosporangium or resting spore (figs. 13, 14, 19, 29, 31, 46, 52, 53, 62, 72). In *E. apiculata* Fischer (figs. 2, 5, 7), *E. cienkowskiana* Fischer (figs. 13, 14), *E. bulligera* (Zopf) Fischer (figs. 33–35), *E. helioformis* Ramsbottum and *E. lobata* Willoughby the planospore cyst may be persistent as well as the germ tube. The latter may occasionally function as the exit canal, and in *E. apiculata* Braun and Zopf re-reported that the planospore cysts serve as the exit papilla. Geitler, however, found that the exit papilla develops close by but independent of the thick-walled cyst (figs. 5, 7). In *E. lobata*, also, the exit papilla develops close by the planospore cyst (figs. 63, 64).

The zoosporangia vary markedly in size and shape in the same as well as in different species with the result that it is often difficult to differentiate the species on these grounds. In *E. helioformis* the sporangia are reported to be up to 20 μ in diameter, but in the New Zealand specimens they were sometimes up to 60 μ. In *E. texana* they may be subspherical, pyriform to deeply-lobed (figs. 52–54), and occasionally bud out and occupy parts of 6 substratum cells. In *E. crenata* Karling (fig. 73) they usually fill the host cell and become cylindrical, while in *E. lobata* they may be flattened and deeply-lobed (figs. 62, 63, 65). As noted before they discharge their planospores through inconspicuous papillae (figs. 2, 7, 65), or 1 or more long necks (figs. 29, 52–54, 56). In *Entophlyctis* sp. Karling (1931) these may branch and extend for distances up to 160 μ. As the tip of the tube or papilla deliquesces the planospores emerge to form a globular mass (figs. 17, 54, 73) and usually disperse in a short while (fig. 29), but in *E. lobata* they swarm collectively (fig. 65), presumably in a vesicle, before dispersing.

The rhizoidal axes, also very markedly in number, degree of branching and extensiveness. They may vary from 1 basal axis to 7 around the periphery of the zoosporangium or resting spore. In *E. apiculata* the axis is quite small and sparingly branched (figs. 3, 4,

Plate 83

Figs. 28-34. *Endochytrium pseudodistomum* Karling. (Domjan, 1936.)

Fig. 28. Zoosporangium and a part of the rhizoids in a *Spirogyra* cell.

Fig. 29. Planospores, 5-7.5 μ diam., with a hyaline refractive globule.

Figs. 30, 31. Mature zoosporangia with long and short exit neck, resp.; planospore cyst persistent.

Fig. 32. Tip of exit canal with attached operculum.

Fig. 33. Young resting spore forming apparently by contraction of the contents of a sporangium-like vesicle; planospore cyst persistent.

Fig. 34. Mature resting spore formed endogenously in a sporangium-like vesicle; outer wall covered with thick reflexed scales.

Figs. 35-49. *Endochytrium digitatum* Karling. Figs. 35-48 after Karling, 1938; fig. 49 after Willoughby, 1961.

Fig. 35. Spherical, 4.4-5.5 μ diam., and amoeboid planospores with a hyaline refractive globule.

Fig. 36. Germinated planospore with a long, branched germ tube.

Fig. 37. Rudiment of zoosporangium forming in the germ rube above the branches.

Figs. 38-40. Young irregular and digitate zoosporangia with irregular apophysis-like swellings in figs. 38 and 40.

Fig. 41. Elongate irregular zoosporangium with a long irregular and coiled exit canal and an irregular apophysis; content evenly granular.

Fig. 42. Ovoid zoosporangium with a straight tapering exit canal.

Fig. 43. Deeply-lobed mature zoosporangium with a short neck.

Fig. 44. Irregular end of an exit canal.

Fig. 45. Discharge of planospores from an irregular zoosporangium with a long coiled exit canal; part of rhizoidal system shown.

Figs. 46-48. Young and mature resting spores with a smooth wall and a large refractive globule surrounded by a few smaller ones.

Fig. 49. Mature resting spore with a large hyaline globule.

Figs. 50, 51. *Endochytrium oophilum* Sparrow. (Sparrow, 1933.)

Fig. 50. Two holocarpic (?) operculate zoosporangia in a rotifer egg.

Fig. 51. Two empty operculate zoosporangia with 2 emerging planospores.

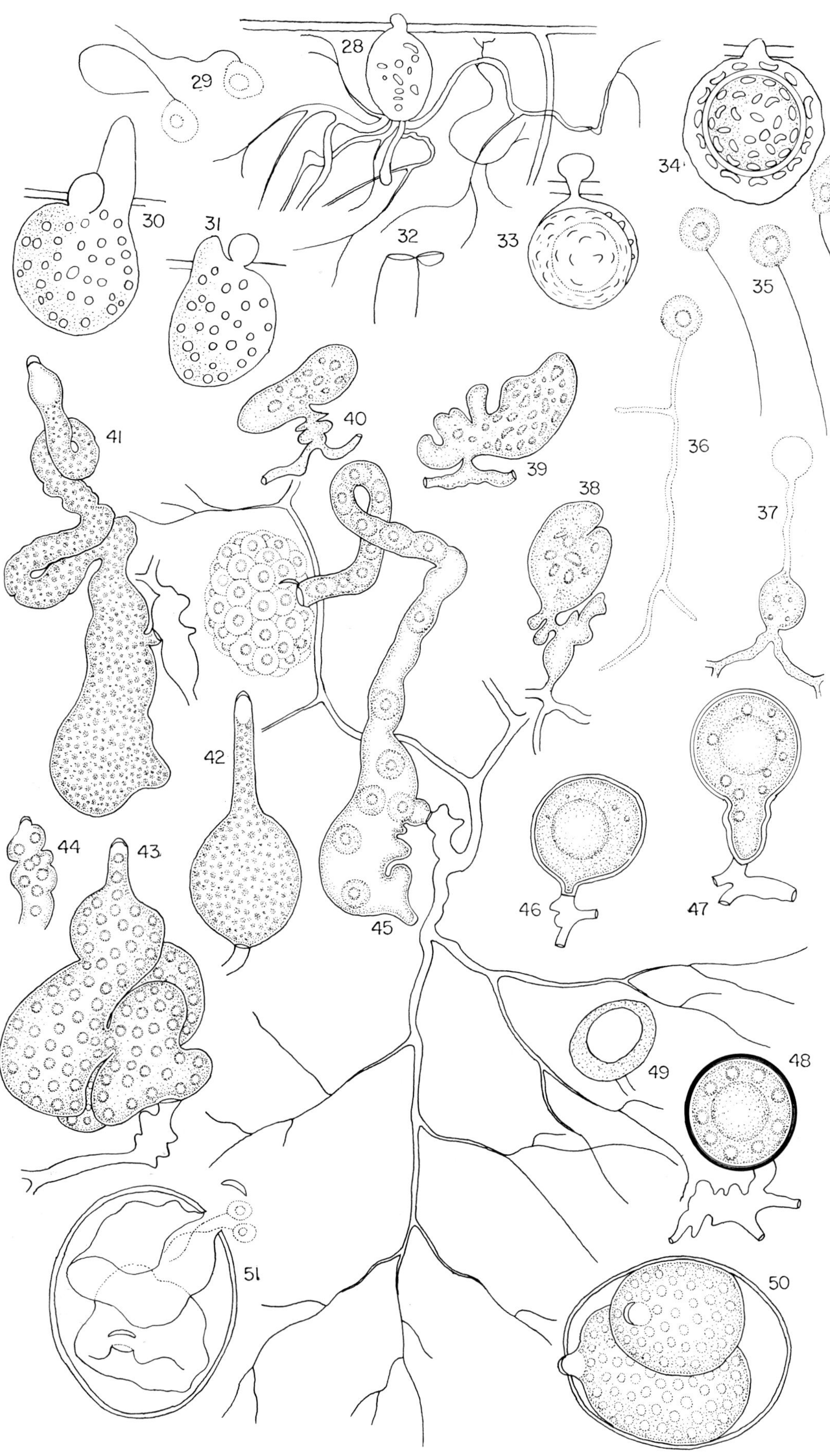

Endochytrium

8), while in *E. texana* they may be up to 7 μ in diameter, 600 μ in length, and richly branched at the extremities.

Resting spores are known in *E. apiculata* (figs. 8, 9), *E. cienkowskiana* (fig. 18), *E. heliformis* (fig. 31), *E. vaucheriae* (fig. 42), *E. characearum* (fig. 44), *E. pygmaea, E. texana* (figs. 55, 56), *E. lobata* (figs. 66, 67), *E. crenata* (fig. 74) and *E. confervae-glomeratae*, and these are borne in the same manner as the zoosporangia. No evidence of sexuality has been reported in relation to their development. In germination Fisch described the exospore of *E. vaucheriae* as cracking and protruding an endospore (fig. 43) in which the planospores developed. However, Karling (1941) and Haskins (1946) showed that the resting spores of *E. texana* (fig. 56) and *E. aurea*, respectively, function as prosporangia in germinating.

ENDOCHYTRIUM Sparrow

Amer. J. Bot. 20:71, 1933. (Sensu recent Karling, Amer. J. Bot. 24:353, 1937)

Plates 82-84

Endochytrium is the operculate counterpart of *Entophlyctis* and is reported to include 6, or possibly 7, species which are weakly parasitic in green algae or saprophytic in cellulosic substrata and eggs or cysts of microscopic animals. Possibly, other species earlier described as members of *Entophlyctis* may prove to belong in this genus. *Endochytrium ramosum* Sparrow will likely be shown to be identical with *E. operculatum* (de Wildemann) Karling, and *E. oophilum* Sparrow in which no rhizoids or vegetative system have been found may prove to be the operculate counterpart of *Olpidium* and a representative of a new genus. *Endochytrium multiguttulatum* Dogma resembles somewhat *Cylindrochytridium endobioticum* and species of *Nephrochytrium*. One species, *E. operculatum*, has been grown on synthetic media.

In creating *Endochytrium* for *E. ramosum* Sparrow seems to have been uncertain about the monocentricity of the genus and he avoided any mention of its polycentric or monocentric organization in his diagnosis. Also, his figure G, plate 2, shows 3 zoosporangia attached to 1 rhizoidal axis. The genus was emended by Karling (1937) from his intensive study of *E. operculatum*, and subsequent studies have confirmed his observations that the thallus is predominantly monocentric.

The intramatrical thallus is similar to that of *Entophlyctis* and consists usually of an intramatrical zoosporangium or resting spore from whose base or periphery arise 1 (figs. 8, 10, 20–22, 42, 45–47) to several (figs. 1, 11, 13, 14) course, up to 10 μ diam., rhizoidal axes which generally branch profusely and extend for long distances (fig. 23). Accordingly, some thalli may resemble those of *Rhizidium* with one basal central rhizoidal axis (fig. 10) or those of *Rhizophlyctis* and *Karlingia* with several axes arising from the periphery (figs. 1, 13, 14). In *E. multiguttulatum*, however, the rhizoids are largely extramatrical and inserted locally near the persistent portion of the planospore case (figs. 58–64), and in rare cases they may be lacking and are replaced by fine hairs attached to the planospore case (fig. 67). In *E. cystarum* Dogma the rhizoids are depauperate and consist of a few short, blunt-ending, simple or branched filaments (figs. 72–77).

The operculate zoosporangia may vary markedly in size and shape (figs. 1, 11, 14–16, 38–43, 45, 75, 76) and develop long straight (figs. 15, 16, 42, 63), coiled and contorted (figs. 41, 45), or short (figs. 11, 13, 31, 43) necks, or low papillae (figs. 14, 50, 75, 76) for discharge of the planospores. Occasionally, the necks may branch once to several times. In all but one species the planospores contain one hyaline refrac-

PLATE 84

Figs. 52-69. *Endochytrium multiguttulatum* Dogma. (Dogma, 1969.)

Fig. 52. Active spherical, 6-7 μ diam., planospores with *laterally* inserted but posteriorly directed flagellum, and multiguttulate content.

Fig. 53. Amoeboid planospore with anterior pseudopodia.

Fig. 54. Encysted planospore.

Fig. 55, 56. Early germination stages.

Fig. 57. Later stage with a broad germ tube; upper part of planospore cyst wall thickening.

Figs. 58-60. Later development of extramatrical rhizoids in relation to thickened portion of planospore cyst; germ tube continuous with incipient zoosporangia.

Figs. 61, 62. Zoosporangia with conspicuous, golden-brown, thick-walled persistent germ tubes.

Figs. 63, 64. Zoosporangia with a tapering neck and exit papilla, resp.; content greyish, coarsely granular.

Fig. 65. Discharge of planospores.

Fig. 66. Opercula.

Fig. 67. Empty, non-rhizoidal (?) sporangium with hairs arising from thickened portion of planospore cyst.

Figs. 68, 69. Surface view and optical section, resp., of resting spores.

Figs. 70-83. *Endochytrium cystarum* Dogma. (Dogma, 1969.)

Fig. 70. Spherical, 2.2-2.6 μ diam., planospores with a minute hyaline refractive globule.

Figs. 71-73. Developmental stages of zoosporangia by local enlargement of the germ tube; depauperate rhizoids simple; planospore cyst and germ tube persistent.

Figs. 74-76. Mature zoosporangia with 1 and 2 exit papillae, resp., and the persistent planospore cyst and germ tube, with depauperate simple or branched rhizoids.

Fig. 77. Empty zoosporangium with 2 exit canals and 1 papilla.

Fig. 78-80. Developmental stages of resting spores with persistent planospore case and germ tube.

Figs. 81-83. Mature resting spores; the one in fig. 82 has 2 empty planospore cysts, sexually formed (?).

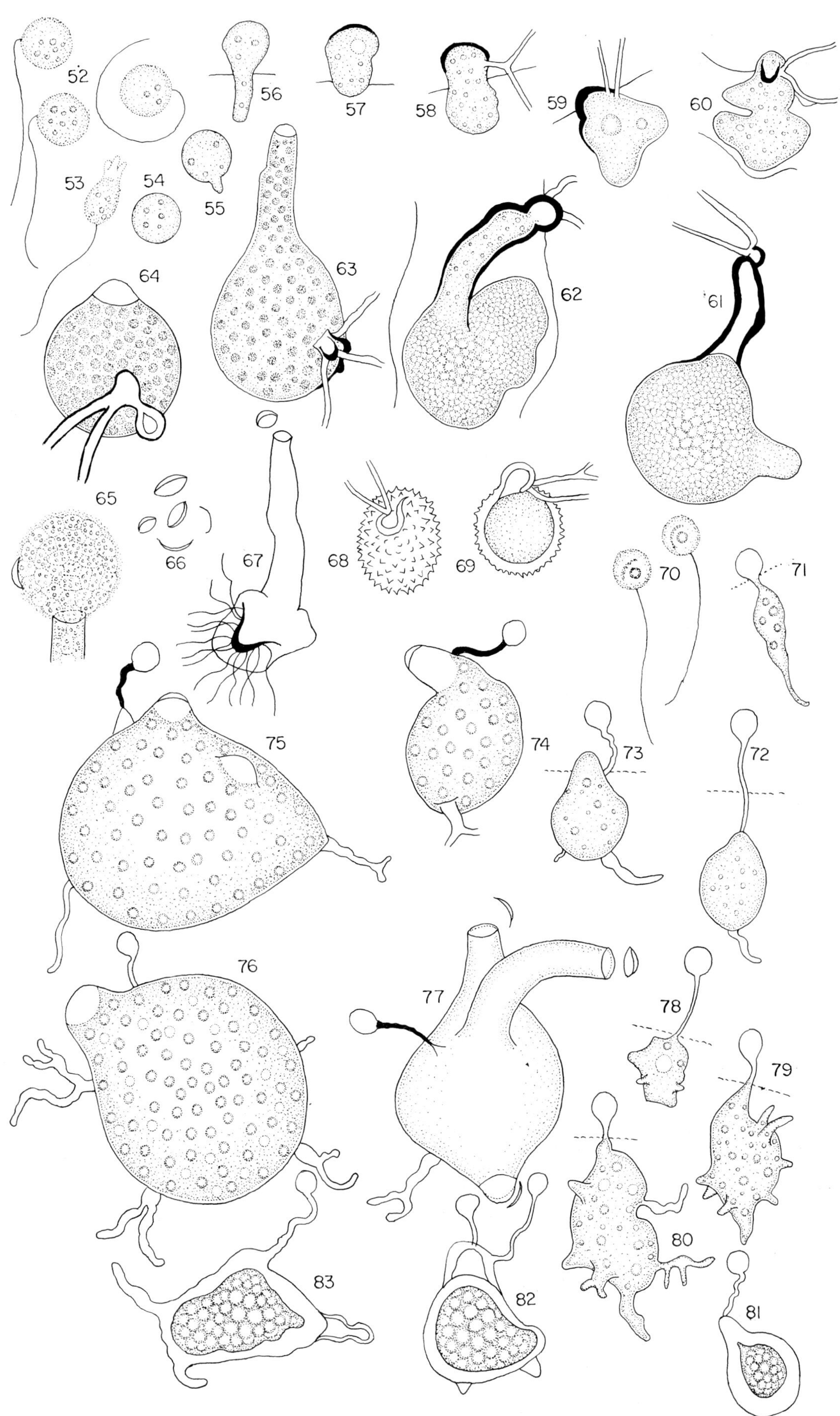

Endochytrium

tive globule, but those of *E. multiguttulatum* include 5 to 10 minute globules (fig. 52). Furthermore, the flagellum is inserted *laterally* instead of posteriorly, but it is directed backward during motility.

The development of the thallus is similar to that of *Entophlyctis*. The germinating planospore forms 1, sometimes 2, germ tubes which penetrate the host cell or substratum and branches (figs. 5, 6). In *E. multiguttulatum* the germ tube may be so broad (figs. 57–59) that it is almost indistinguishable from the sporangial rudiment. In other species a swelling develops in the germ tube into which the planospore nucleus migrates (fig. 7). This swelling is the rudiment of the zoosporangium, and as its nucleus divides (fig. 9) it becomes multinucleate and eventually develops into the mature zoosporangium. As in *Entophlyctis* the sporangial rudiment may occur considerably above the branches of the germ tube or at their juncture, resulting either in a mono- or polyrhizoidal zoosporangium. In *E. operculatum, E. pseudodistomum* (Scherffel) Karling, *E. multiguttulatum* and *E. cystarum* the planospore cyst and germ tube may persist as appendages on the zoosporangium (figs. 10, 15, 30, 31, 61–63, 67) or the resting spore (figs. 19, 33, 68, 69, 81–83).

In all species in which they have been found the resting spores are formed and borne like the zoosporangia, but in *E. multiguttulatum* Dogma found a spore with 2 attached cysts (fig. 82) which he suggested might have developed sexually. The outer wall of the spores varies from smooth (figs. 19, 46–49), warty (fig. 18), bullate (figs. 68, 69), spiny and tuberculate (figs. 82–83), or undulating (figs. 20, 21) and may be hyaline, golden-brown (figs. 68, 69) or light-brown (figs. 81–83). In *E. pseudodistomum* Karling, however, the spores develop in a sporangium-like structure, apparently by the contraction of the content and its investment by a thick scaly wall (figs. 33, 34). In *E. operculatum* the resting spore functions as a prosporangium during germination (figs. 21, 22), but in the other species germination has not been observed.

CYLINDOCHYTRIDIUM Karling

Bull. Torrey Bot. Club 68:382, 1941.

Siphonochytrium Couch, nom. nud. J. Elisha Mitcell Sci. Soc. 55:208, 1939.

Plates 85, 86

This operculate genus is reported to include two saprophytic species, *C. johnstonii* Karling and *C. endobioticum* Willoughby, which may be isolated on cellulosic substrata from soil and water. *Cylindrochytridium endobioticum* varies considerably from the type species and is a doubtful member of the genus. The thallus of *C. johnstonii* may be partly intra- and extramatrical, depending on the condition of the substratum, and consists of a predominantly cylindrical or clavate, stalked or continuous zoosporangium with one to several rhizoidal axes which are usually inflated at irregular intervals (figs. 1, 8–11, 13). When the grass substratum is soft and at the stage of disintegration the thallus may be wholly intramatrical. Sometimes the zoosporangium may vary considerably in size and shape from cylindrical (fig. 11) to clavate (figs. 1, 8), spherical (fig. 12) and irregular.

In the development of the thallus, the planospores germinate on or partly within the softened substratum with a germ tube which soon branches (figs. 4, 5). Later, an elongate swelling develops in the tube above the branches (fig. 6), and this enlargement is the rudiment of the sporangium. If the germ tube is short, the sporangial rudiment is continuous with the planospore cyst which, also, may elongate outward and become part of the zoosporangium (figs. 7, 8). As the latter matures its protoplasm slowly moves up into the apical portion (figs. 8, 9), with the result that the base becomes highly vacuolate and eventually empty. At the same time a septum develops and delimits the base from the upper fertile portion so that the zoosporangia usually become stalked (figs. 1, 11). Meanwhile, the intramatrical rhizoid branches further and becomes extensive in a radius of 100 to 1200 μ. Also, catenulate swelling develops in the branches at irregular intervals which may be up to 12×30 μ, and sometimes has an appearance similar to the compound apophysis of *Catenochytridium*. In cases where a single rhizoidal axis is formed the portion immediately beneath the zoosporangium may be inflated so that the zoosporangium appears to be uni-apophysate (fig. 10). The fertile portion of the zoosporangium usually projects partly out of the substratum, and as the apical

PLATE 85

Figs. 1-14. *Cylindrochytridium johnstonii* Karling. (Karling, 1941.)

Fig. 1. Mature thallus with dehiscing stalked septate, operculate sporangium and fusiform swellings in rhizoids.

Figs. 2. 3. Motile, spherical, 5.6-7 μ diam., and amoeboid planospores, respectively, with a large hyaline refractive globule.

Figs. 4, 5. Germination of planospores.

Fig. 6. Incipient zoosporangium developing as a swelling of the germ tube.

Fig. 7. Young thallus; incipient sporangium developing partly from the germ tube and elongating zoospore cyst.

Figs. 8, 9. Developmental stages of zoosporangia, basal portions becoming vacuolate.

Fig. 10. Later stage of development, protoplasm coarsely granular in appearance.

Fig. 11. Cylindrical septate zoosporangium shortly before cleavage.

Figs. 12-14. Some variations in shapes and sizes of empty zoosporangia.

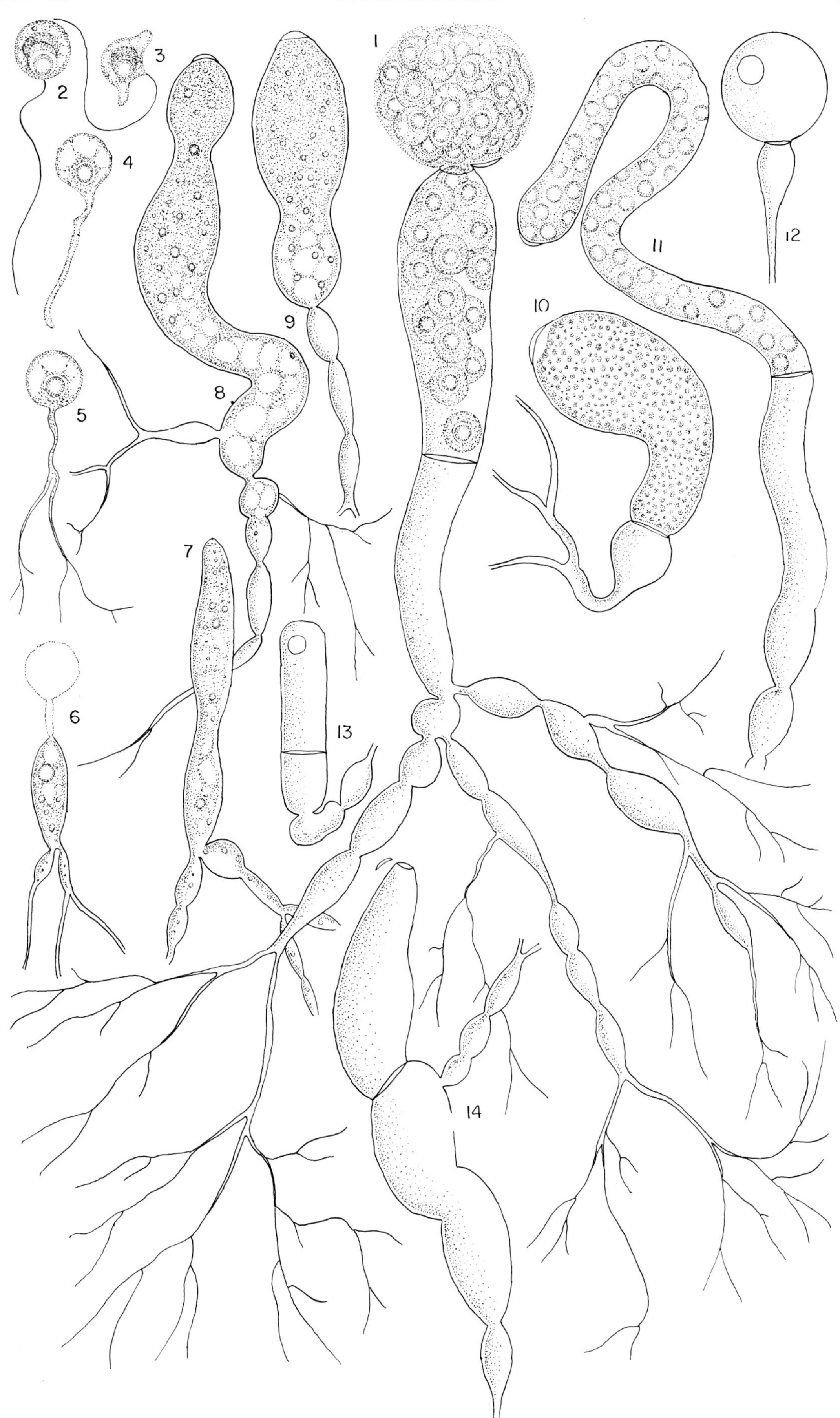

Cyclindrochytridium

or lateral operculum is pushed off or aside the planospores emerge slowly and form a globular mass at the exit orifice.

Resting spores have been observed only in *C. johnstonii*, and according to Shanor (1944), who did not illustrate them, they are nearly spherical, thick-walled and smooth with light-brown or amber-colored walls and a large yellow globule in the contents.

The thallus of *C. endobioticum* is partly intra- and extramatrical like *C. johnstonii*, but its thickened planospore cyst, part of the stalk, and most of the rhizoids are extramatrical while the zoosporangium proper is usually intramatrical. In this respect the relative positions of the thallus parts of the two species appear to be reversed, according to Willoughby's drawings. In the development of the thallus the planospore cyst elongates downward towards or into the substratum as the upper portion of its wall thickens (figs. 17, 18), and the rhizoids develop in close proximity to it (figs. 18, 20), usually on the surface of or partly within the substratum. The elongate portion of the cyst eventually develops into the stalk and zoosporangium proper (figs. 21–25). The partly intra- and extramatrical rhizoids branch further and become extensive and thick-walled. Also, the walls of the planospore cyst and the stalk become markedly thickened (figs. 22–29), and in mature thalli the zoosporangium becomes invested with the finer branches of the rhizoids. Discharge of planospores occurs through a broad, conspicuous exit papilla (fig. 28), and their release is dependent apparently on erosion of the substratum.

PHLYCTORHIZA Hanson

Amer. J. Bot. 33:732, 1946.

Plate 87

At the present time this inoperculate genus includes one species, *P. endogena* Hanson, which occurs as a saprophyte within the basement membrane of insect integuments. Formerly, two other species, *P. variabilis* Karling and *P. peltata* Sparrow, were included in this genus, but they have been transferred to *Catenophlyctis* (Karling, 1965) of the family Catenariaceae. The development of the thallus and zoosporangia of *P. endogena* is unique in several respects. In germination the planospore cyst or body is persistent, and the thick germ tube penetrates the integument and branches (fig. 2) within the basement membrane. As these branches extend and develop further their dorsal surface vesiculates (figs. 3, 4, 5) to form the rudiment of the zoosporangium, while the ventral portion of the rhizoids remains unexpanded. At this stage the incipient sporangium is flat and irregular or almost stellate in shape and has the appearance of a thin flattened extension between the main rhizoidal branches. Superficially, the thalli in this stage resemble somewhat a bat's wing or a duck's foot with the incipient zoosporangia corresponding to the web or membrance between the digits and the ventral portion of the rhizoids corresponding to the digits. As the zoosporangia mature they acquire an ovoid, oblong, subspherical or irregular shape (figs. 7–11), and the rhizoids frequently anastomose to form a network of filaments (fig. 6). A low broad exit papilla develops more or less centrally in the ventral wall as the zoosporangium matures, but the papilla may occasionally develop dorsally or laterally. It usually forms in the region of the main rhizoidal axis (fig. 11), and in some instances the axis may be interrupted in the region of the papilla (fig. 11). As the papilla deliquesces the planospores emerge in a globular mass (fig. 10), and after becoming motile they swarm actively within a thin hyaline enveloping vesicle for a few minutes before bursting out and dispersing.

In some instances the terminal branches of the rhizoids vesiculate (fig. 11), and if nuclei from the incipient zoosporangium migrate into such vesicles or swellings they may develop into secondary zoosporangia. Accordingly, polycentric thalli with numerous secondary zoosporangia (fig. 12) may be found rarely within the basement membrane. Resting spores are reported to develop asexually in the same manner as the zoosporangia and may have thick smooth, tuberculate (figs. 13, 14, 15) or undulate light-golden or

PLATE 86

Figs. 15-29. *Cylindrochytridium endobioticum* Willoughby. (Willoughby, 1964.)

Fig. 15. Flagellate and encysted planospores, 4.5-5.5 μ diam., resp., with numerous hyaline refractive globules.

Fig. 16. Thickening of upper portion of wall of planospore cyst.

Fig. 17. Young thallus with the thickened wall of the planospore cyst.

Figs. 18, 20. Young thalli showing the relation of the planospore cyst to the rhizoids.

Fig. 19. Young, partly intramatrical thallus eroding edge of the cellophane strip.

Figs. 21, 22. Later development stages of thalli relative to stomata of the grass substratum; planospore cyst thick-walled and extramatrical together with the rhizoids and upper portion of incipient zoosporangium.

Figs. 23, 24. "Young thalli with extensive thickening in the region of the sterile stalk."

Fig. 25. Almost mature vacuolate zoosporangium; planospore cyst, rhizoidal axes, and upper portion of sterile stalk extramatrical.

Fig. 26. Upper portion of a thallus showing relation of thickened planospore cyst to the rhizoidal system.

Fig. 27. Mature zoosporangium delimited from the stalk by a septum.

Fig. 28. Mature zoosporangium with a broad exit papilla.

Fig. 29. Empty, stalked, operculate zoosporangium.

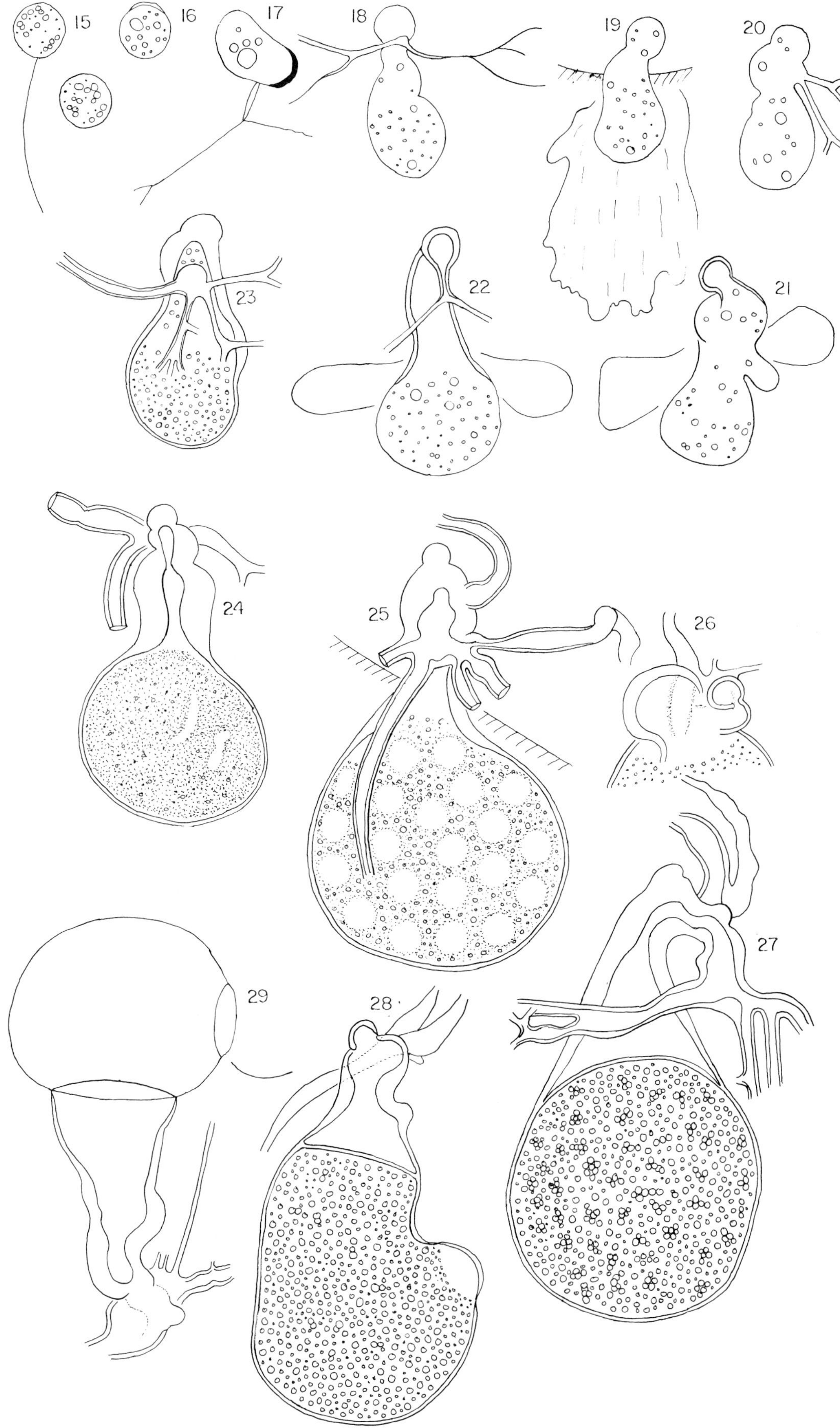

Cylindrochytridium

deep amber walls. They function as prosporangia during germination and form a thin-walled zoosporangium on their surface (fig. 15).

Superficially, *Phlyctorhiza* resembles somewhat *Rhizophlyctis*, but on the grounds that its zoosporangia and resting spores develop endogenously from the germ tube and its branches, it is classified tentatively in the family Entophlyctaceae.

TRUITELLA Karling

Amer. J. Bot. 36:454, 1949.

Plate 88

This monotypic operculate genus is characterized by an intramatrical, ovoid, spherical, oblong, elongate, lobed, irregular zoosporangium, locally constricted and inflated, dichotomously branching rhizoids which usually arise at several points on the periphery of the zoosporangium, a persistent planospore cyst which bears 1 to several simple or branched setae, and spherical, elongate, constricted and irregular resting spores which develop in a segment of the rhizoids (figs. 16-19). Occasionally, the thalli become polycentric and develop more than one zoosporangium (fig. 14) or resting spore.

Three variations of development may occur in the formation of the sporangial thalli of *T. setifera* Karling. In the first type the germinating planospore on onion skin or grass leaves (fig. 4) form a fairly long germ tube which branches once to several times (fig. 5) and forms the rudiments of the absorbing system. Shortly thereafter a swelling forms, usually at the juncture of the branches, in the germ tube, and as it enlarges the branches are moved apart (fig. 6). This enlargement becomes the rudiment of the zoosporangium. Meanwhile, the rhizoidal axes elongate, branch dichotomously, and enlarge in diameter. At the same time setae develop as small protuberances on the planospore cyst (figs. 6, 9) and eventually elongate into slender filaments (figs. 10, 11).

A more common variation of development is shown in figures 7 to 10. The germ tube (fig. 7) is quite broad, and after its tip has penetrated the substratum it enlarges and forms an irregular or lobed structure (figs. 8, 9) out of which the rhizoidal rudiments develop as thick branches or tubes (figs. 9, 10). These elongate, branch, and become constricted at irregular intervals, while the center of the lobed structure enlarges to become the incipient zoosporangium. Thirdly, the germ tube may be quite short, and in such cases the zoosporangium and rhizoids are formed close to the surface of the substratum and lie partly exposed, particularly if the substratum is eroded and disintegrating. Oftentimes, a rhizoidal segment adjacent to the zoosporangium becomes markedly inflated (fig. 13) and looks like an apophysis. At dehiscence the operculum is pushed off or aside, and the planospores emerge slowly to form a globular mass at the exit orifice (fig. 1).

The resting spores so far observed were formed asexually in segments of the rhizoids by the concentration of the coarsely granular protoplasm in such areas and its investment by a smooth, dark, reddish-brown wall, and at maturity the spores vary markedly in shape (figs. 16–19). So far germination has not been observed.

In its type of development *Truittella* is basically similar to that of species of the Entophlyctaceae, and it is classified provisionally in the subfamily Entophlyctoideae. By its constricted rhizoids and inflated apophysis-like segments as well as the formation of resting spores in the segments of the rhizoids it resembles somewhat *Catenochytridium*, but it differs by the origins of rhizoids from several points on the zoosporangium as in *Rhizophlyctis* and *Karlingia*, and the development of the zoosporangium from an enlargement of the germ tube instead of from the planospore body.

PLATE 87

Figs. 1-15. *Phlyctorhiza endogena* Hanson. (Hanson, 1946.)

Fig. 1. Ovoid, 2.2-2.9 × 2.9-3.7 μ, planospore.

Fig. 2. Germination of planospore with a branched germ tube.

Fig. 3. Ventral view of dorsally unilateral, vesiculating rhizoidal branches to form an incipient zoosporangium in the basement membrane of insect exuviae.

Fig. 4. Later stage of thallus development.

Fig. 5. *In toto*- stained uninucleate thallus with persistent planospore cyst, germ tube, and incipient zoosporangium expanding by continued dorsal vesiculation of rhizoidal branches.

Fig. 6. Anastomosis of rhizoids in a mature thallus.

Figs. 7, 8. Bi- and 8-nucleate thalli.

Fig. 9. Profile view of zoosporangium with basal exit papilla penetrating basement membrane; planospore cyst and rhizoids omitted.

Fig. 10. Discharge of planospores through a basal papilla.

Fig. 11. Mature primary zoosporangium with large exit papilla; terminal branches of rhizoids vesiculating dorsally to form incipient secondary zoosporangia.

Fig. 12. Portion of a sectioned and stained polycentric thallus with 7 secondary zoosporangia; primary zoosporangium empty.

Figs. 13, 14. Dorsal views of smooth-walled and tuberculate resting spores; rhizoids vesiculate.

Fig. 15. Germination of resting spore.

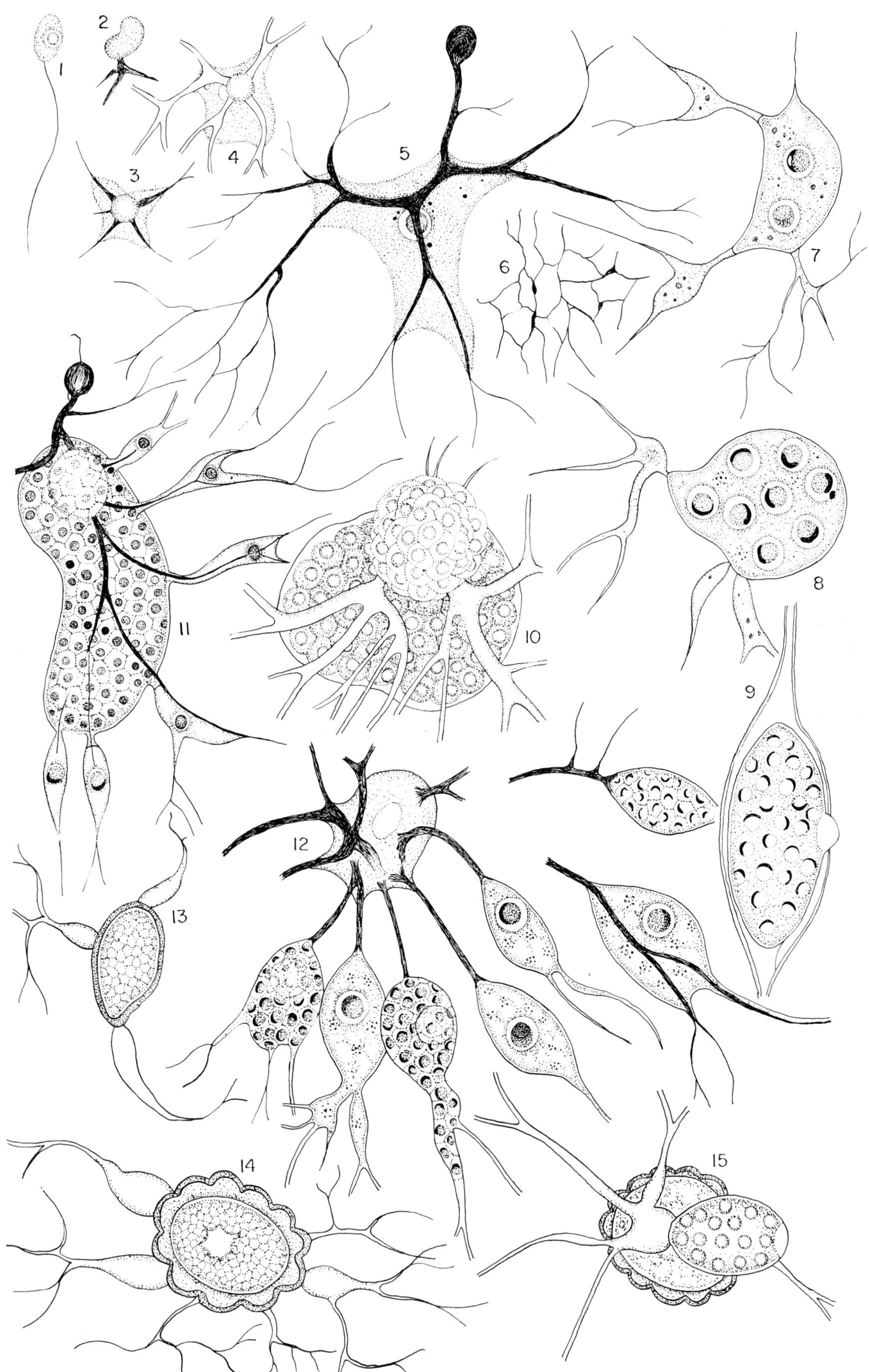

Phlyctorhiza

SCHERFFELIOMYCES Sparrow

Mycologia 36:377, 1934; J. Linn. Soc. London (Bot) 50:446, 1936.

Scherffelia Sparrow, 1933. Trans. Brit. Mycol. Soc. 18:216; non *Scherffelia* Pascher 1912, Hedwigia 52:281.

Plate 89, Figs. 1-19

This inoperculate genus includes 3 species which parasitize *Euglena, Chlamydomonas* and *Zygnema* in Hungary, Germany, Great Britain, and the U.S.A. It is characterized by partly or wholly epibiotic or extracellular zoosporangia, an endobiotic vegetative absorbing system which may be rhizoidal or haustorial, and epibiotic resting spores. In these respects it resembles some species of *Rhizophydium*, but its method of development is more like that of *Entophlyctis* in that the zoosporangium develops as an enlargement of the germ tube instead of the encysted planospore.

The planospore (figs. 1–3) encysts and germinates at a short or long distance from the host cell (figs. 4, 5, 11) or in the gelatinous sheath surrounding it (figs. 4–7, 10), producing a short or long germ tube. When it comes into contact with the host cell its tip enlarges (figs. 4, 5), or an enlargement develops locally within it (fig. 8), and this enlargement becomes the rudiment of the zoosporangium which develops epibiotically (figs. 11–13) or partly within the gelatinous sheath surrounding the host (figs. 9, 10, 16). At the same time digitate haustoria develop endobiotically at the base of the incipient zoosporangium (figs. 5, 11), or branched rhizoids form at the base of an endobiotic peg (figs. 7–9). During this development the empty planospore cyst remains attached to the zoosporangium as an appendage (figs. 14–16), or is connected with it by a short or long persistent germ tube (figs. 10–12). At maturity of the zoosporangium the apical or subapical papilla deliquesces, and the planospores are usually discharged in a globular mass (figs. 12, 13, 15).

Resting spores are formed asexually in the same manner as the zoosporangia, and in *S. parasitans* Sparrow and *S. leptorhizus* Johns (*Dangeardiana leptorhiza*, Batko, 1970) (figs. 18, 19) the whole swelling of the germ tube or sporangium-like structure is converted into a spore. In the latter species the spore has an exceptionally thick wall (fig. 19) and the shape of a sporangium. In *S. appendiculatus* Johns, however, the content of an incipient spore apparently contracts and becomes enveloped by a thick wall so that the spore lies in a sporangium-like vesicle (fig. 17). Germination of the resting spores has not been observed.

The classification and relationships of *Scherffeliomyces* are uncertain at present. As noted previously the thalli of some species resemble those of *Rhizophydium* in that the zoosporangia and resting spores are partly or fully epibiotic and the haustoria or rhizoids are endobiotic. But the method of development is different and more like that of *Entophlyctis*, and for this reason it is classified tentatively in the subfamily Entophlyctoideae. Perhaps this genus might be regarded as a transition from the largely epibiotic to the wholly endobiotic genera like *Entophlyctis*.

CORALLIOCHYTRIUM Domjan

Folio Cryptogam 2:22, 1936.

Plate 89, Figs. 20-38

This monotypic inoperculate genus which parasitizes *Zygnema* in Hungary has the same type of development as *Scherffeliomyces*, and it is not necessary to describe it in detail. It differs, however, by its coralloid absorbing system (fig. 27), apophysate zoosporangia (figs. 30, 33), and discharge of zoosporangia through 1 to 5 exit papillae (figs. 30–37). These differences appear to be specific instead of generic

PLATE 88

Figs. 1-19. *Truittella setifera* Karling. (Karling, 1949.)

Fig. 1. Portion of a sporangial thallus with a spherical operculate zoosporangium discharging planospores, a setiferous thick-walled planospore cyst, and dichotomously branched, constricted and blunt-ending rhizoids (fig. 1A).

Fig. 2. Enlarged drawing of a primary transversely-oriented rhizoidal segment.

Fig. 3. Spherical, 6 μ, and 4.2 μ diam., planospores with a large hyaline refractive globule.

Figs. 4, 5. Germination of planospores with a fairly long and branched germ tube, respectively.

Fig. 6. Young thallus, planospore case persistent with developing setae, and the sporangial rudiment developing in the germ tube at the juncture of the rhizoidal branches.

Fig. 7. Germination of a planospore with a short blunt germ tube.

Fig. 8. Later stage of the same planospore; germ tube enlarging and forming an irregular body.

Figs. 9-11. Later successive stages of the same thallus; planospore case persistent and developing setae; previously irregular body of thallus enlarging and giving rise to constricted segments of the rhizoids.

Fig. 12. Median view of a subtriangular-shaped zoosporangium.

Fig. 13. Ovoid zoosporangium with a somewhat bell-shaped basal apophysis-like swelling.

Fig. 14. Portion of a rare polycentric sporangial thallus.

Fig. 15. Irregular zoosporangium with branched setae arising from the planospore cyst.

Figs. 16-19. Variations in sizes and shapes of dark, reddish-brown resting spores formed in segments of the rhizoids.

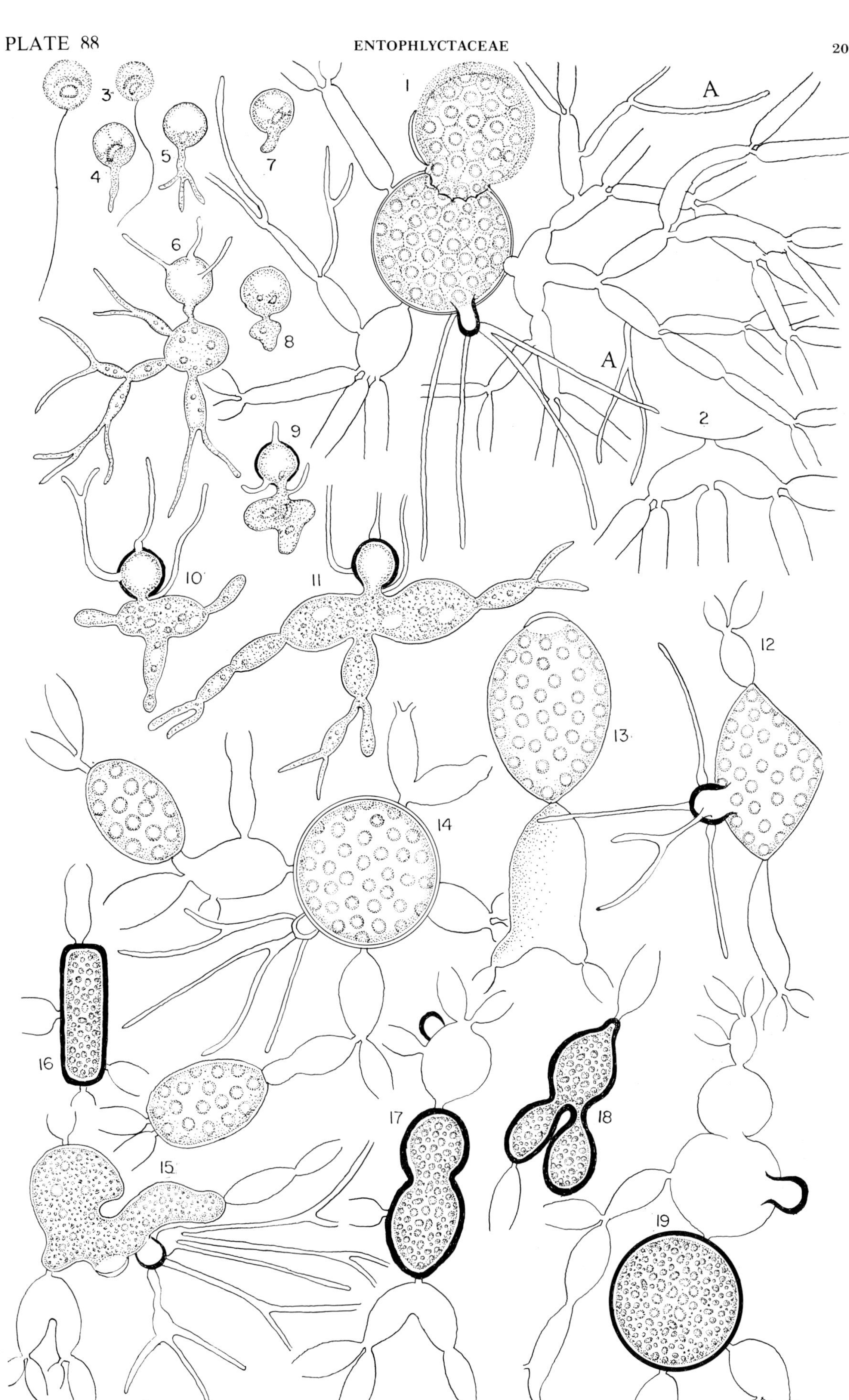

Truittella

inasmuch as the type of development is identical with that of *Scherffeliomyces*. Johns (1936) merged the two genera, and his viewpoint seems fully justified. Nevertheless, *Coralliochytrium* is illustrated separately (figs. 20–38) so that chytridologists may judge its identity without personal bias or interpretation.

SCHERFFELIOMYCOPSIS Geitler

Oesterr. Bot. Zeitschr. 109:272, 1962

Plate 90

This inoperculate genus includes one species, *S. coleochaetis* Geitler, which parasitizes the terminal vegetative cells of *Coleochaete soluta* and *C. irregularis* in southern Austria without causing much harm to its host. The morphology and development of the primary thalli and sporangia are almost identical with those of *Sherffeliomyces*, but its sporangia produces amoebospores instead of planospores. Also, the thallus reacts *positively* for cellulose in its walls when tested with choroiodide of zinc. Other differences involve the development of secondary and tertiary thalli differently from the primary ones, and the formation of the primary sporangia in the outer membrane of the host, according to Geitler.

The content of the mature sporangium emerges to the outside as a naked globular mass of protoplasm and soon divides into amoebospores. These possess a hyaline anterior end (figs. 1, 2) which develops into a lobopodium, and a posterior end with numerous globules. As they creep about 2 pulsating vacuoles become visible near the center (figs. 3, 4), and later pseudopods develop at the anterior end (figs. 5, 6) which retracts periodically. After a while the amoebospores round up, encyst, and later germinate in the gelatinous envelope of *C. soluta* (fig. 7) or directly on the wall of *C. irregularis*, which lacks the envelope.

Depending on whether primary thalli and sporangia or secondary and tertiary ones are formed this genus exhibits two types of development. In the former the encysted and germinating amoebospores develop a long germ tube which penetrates the host cell and develops a tuft of rhizoids or a digitate haustorium at the tips (fig. 8). The lower portion of the germ tube within the outer layer of the host wall enlarges to form the incipient sporangium which becomes multinucleate (figs. 15, 16) and exhibits wide variations in size and shape (figs. 15–18). The amoebospore cyst and germ tube persist as appendages to the sporangia (figs. 15–26).

Secondary and tertiary thalli develop from amoebospores which do not emerge from the primary sporangium (fig. 21). In their development no cyst and germ tube are formed, and the encysted amoebospore develops directly into a sporangium with haustoria at its base (figs. 22–26). Resting spores are unknown in this genus.

This genus is difficult to classify at the present time, although its morphology and development of the primary thalli and zoosporangia are almost identical with those of *Scherffeliomyces*, and on the basis of this similarity it is placed only provisionally in the subfamily Entophlyctoideae. However, its lack of flagellate planospores, positive reactions to tests for cellulose, and the formation of secondary zoosporangia directly by enlargement of the amoebospore body makes this classification very uncertain.

PLATE 89

Figs. 1, 10, 16, 17 after Zopf, 1884; figs. 4, 5, 11-13, 18 after Sparrow, 1936; figs. 6-9, 14, 15, 19 after Johns, 1956; figs. 20-38 after Domjan, 1936.

Fig. 1. Planospore. *Scherfelliomyces appendiculatus* Sparrow.

Fig. 2. Minute motile and amoeboid planospores. *S. parasitans* Sparrow.

Fig. 3. Large, 4-5 μ diam., planospore. *S. leptorrhizus* Johns.

Figs. 4, 5. Infection of *Euglena* cysts, *S. parasitans*.

Figs. 6-8. Germination of a planospore and development of the thallus in the gelatinous sheath and cell of *Zygnema*; branched rhizoids attached to basal peg, *S. leptorrhizus*.

Fig. 9. Incipient zoosporangium developing partly in gelatinous sheath, with a basal peg and rhizoids in host cell, *S. leptorrhizus*. (*Dangeardiana leptorhiza* Batko, 1970.)

Fig. 10. Incipient appendiculate zoosporangium in the gelatinous sheath of the palmella stage of *Chlamydomonas*, *S. appendiculatus*.

Fig. 11. Zoosporangium detached from the host cell with basal digitate haustoria, and the planospore cyst and long germ tube, *S. parasitans*.

Figs. 12, 13. Dehiscence and discharge of planospores, *S. parasitans*.

Figs. 14, 15. Mature zoosporangium and discharge of planospores, *S. leptorrhizus*.

Fig. 16. Discharge of planospores, *S. appendiculatus*.

Figs. 17-19. Resting spores of *S. appendiculatus*, *S. parasitans*, and *S. leptorrhizus*, respectively.

Figs. 20-38. *Coralliochytrium scherffellii* Domjan. (Domjan, 1936.)

Figs. 20-22. Motile, encysted, and germinating planospores, respectively, on *Zygnema*.

Figs. 23-26. Developmental stages of thalli and incipient epibiotic zoosporangia.

Fig. 27. Typical coralloid haustorium oriented on an apophysis within the host cell.

Figs. 28, 29. Variations in rhizoid or haustorium.

Figs. 30-35. Variations in sizes and shapes of zoosporangia, number of exit papillae or tubes, and the lengths of the attached germ tubes and planospore cysts.

Fig. 36. Discharge of planospores from two basal papillae.

Fig. 36a. Planospore.

Fig. 37. Median view of empty zoosporangium with 5 exit orifices.

Fig. 38. Resting spore in a sporangium-like vesicle.

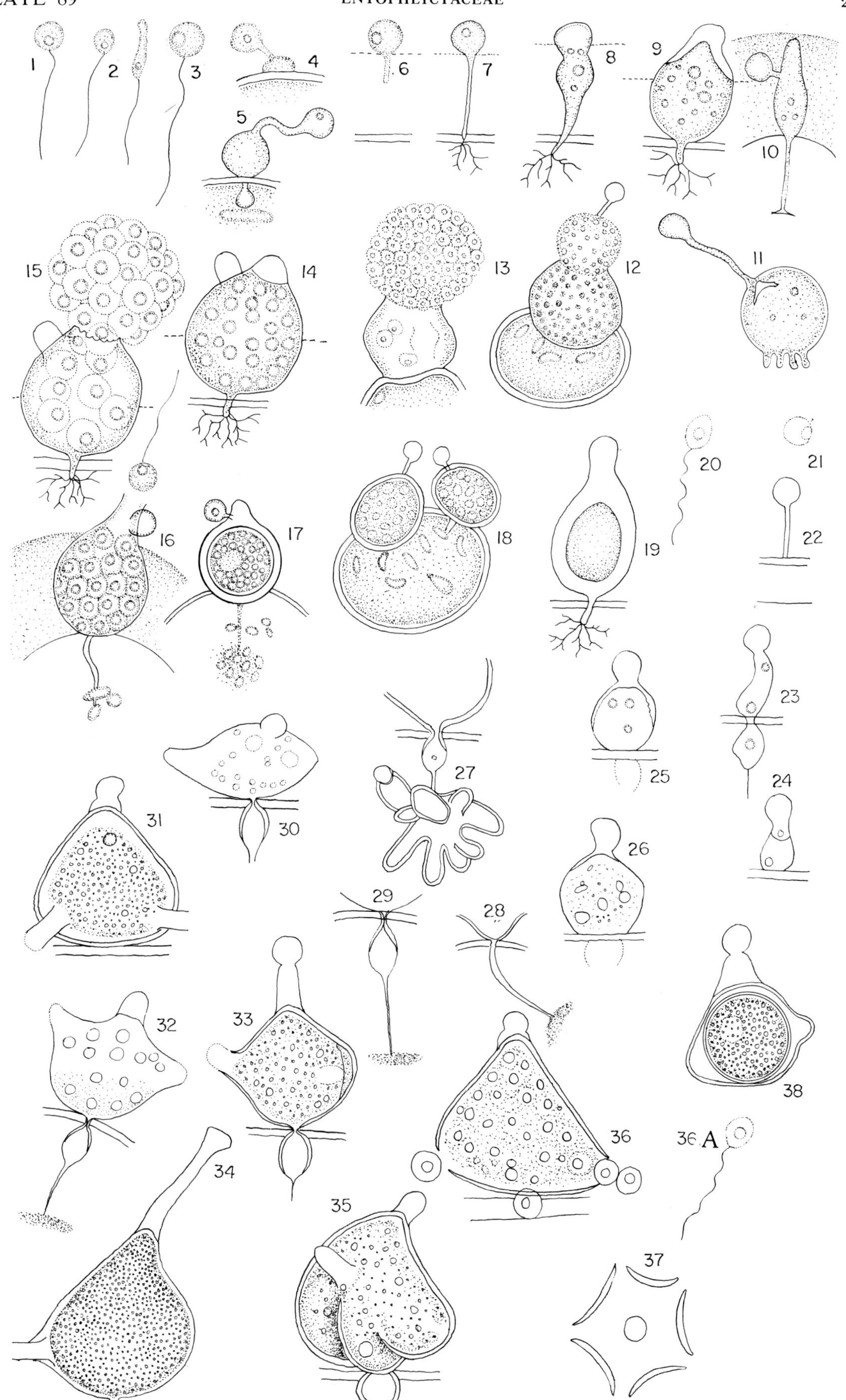

Scherffelomyces, Corallochytrium

MITOCHYTRIDIUM Dangeard

Bull. Soc. Mycol. France 27:202, 1911.

Plate 91

This is a monotypic inoperculate genus whose type species, *M. ramosum* Dangeard, parasitizes desmids and develops entirely within the host cell as an elongate, tubular, curved, lobed, and branched thallus with rhizoids at irregular intervals along its periphery (figs. 1, 8). So far it has been reported only three times, once in France and twice in the U.S.A. (Cook, 1963). The sporangial thallus is monocentric, and at maturity the unicellular tubular portion is transformed into a zoosporangium and discharges planospores through one or more papillae or short tubes. The planospores (fig. 2) infect the host, and the short (fig. 5) or long (fig. 6) penetration or germ tube may enlarge at its tip and elongate to the full length, sometimes 660 μ, of the host cell to form the zoosporangium. Usually, however, the thallus is shorter (fig. 6). In some instances the rudiment of the zoosporangium appears to be laid down (fig. 3) before the rhizoids appear, but in other cases (fig. 4) the rhizoids may be formed fairly early.

According to Dangeard and Couch (1935) the resting spore thallus appears to be polycentric in contrast to the monocentric sporangial thallus. At least, several resting bodies are reported to be formed endobiotically on a common filamentous and rhizoidal system (fig. 9). These bodies bear rhizoids which disappear as the spores mature, leaving a few spines and protuberances (figs. 10, 11). Germination of the thick-walled spores has not been observed.

The development of a monocentric sporangial thallus and a polycentric resting spore thallus is unique among the chytrids except for species of *Physoderma*, and *Mitochytridium* stands alone in this respect insofar as our knowledge of its developments goes. Several monocentric chytrids occasionally develop polycentric thalli, but such occurrences are exceptional.

The classification of *Mitochytridium* is uncertain at present because its life cycle is not fully known. Dangeard pointed out the resemblance of its thallus to those of some species of *Lagenidium* and *Myzochytium*, but because of the presence of rhizoids he regarded it as intermediate between the Chytridiaceae and what was then known as the Ancylistaceae. Butler (1928) believed that it should be placed in the Cladochytriaceae close to *Catenaria* which at that time was believed to be a chytrid. Couch, also, thought that it is closer to the Chytridiales than to the Lagenidiales (Ancylistales) and probably belongs in the Cladochytriaceae. However, his discovery that the thallus wall gives a *positive* cellulose reaction when tested with chloroiodide of zinc as in the Lagenidiaceae confuses the classification even more. Sparrow (1960) believed that *Mitochytridium* is more nearly related to *Entophlyctis* than to any of the Cladochytriaceae, and it is retained here in the subfamily Entophlyctoideae only tentatively.

SUBFAMILY DIPLOPHLYCTOIDEAE

This subfamily includes the monocentric eucarpic species in which the enlargement of the germ tube usually becomes an apophysis and which functions as a prosporangium in most species of *Nephrochytrium* and in *Rhizosiphon*. But in some species of *Diplophlyctis* it has not been clearly shown that this type of development occurs. The thallus in *Nephrochytrium* and *Diplophlyctis* is wholly endobiotic except for a persistent planospore cyst and part of the germ tube in some species, but in *Rhizosiphon* the zoosporangium is epibiotic. Further details on the structure and development of the thallus and reproduction will be presented below for each genus.

PLATE 90

Figs. 1-26. *Scherffeliomycopsis coleochaetis* Geitler. (Geitler, 1962.)

Figs. 1, 2. Newly emerged amoebospores with a hyaline end and a posterior end with globules.

Figs. 3, 4. Creeping amoebospores with an anterior lobopodium, two contracting vacuoles, and an attenuated posterior end with globules.

Figs. 5, 6. Amoebospores with pseudopodia at anterior end.

Fig. 7. Uninucleate encysted germinating amoebospores on the gelatinous envelope of *Coleochaete soluta*.

Fig. 8. Germinated amoebospore with a long germ tube whose tip has penetrated the host cell and developed a tuft of rhizoids.

Figs. 9-11. Stages in the development of incipient sporangia.

Figs. 12-14. Young sporangia.

Figs. 15, 16. Multinucleate sporangia.

Fig. 17. Mature sporangium with exit papillae shortly before dehiscence; amoebospore initials indicated by aggregated globules.

Fig. 18. Mature zoosporangium on *Coleochaete irregularis*.

Figs. 19, 20. Deliquescence of exit papilla or canal.

Fig. 21. Encysted amoebospores in primary sporangium.

Fig. 22. Germination of encysted amoebospores in a primary sporangium.

Figs. 23-26. Various stages in the development of secondary sporangia within a primary sporangium; encysted amoebospores developing directly into sporangia without a sporocyst and a long germ tube.

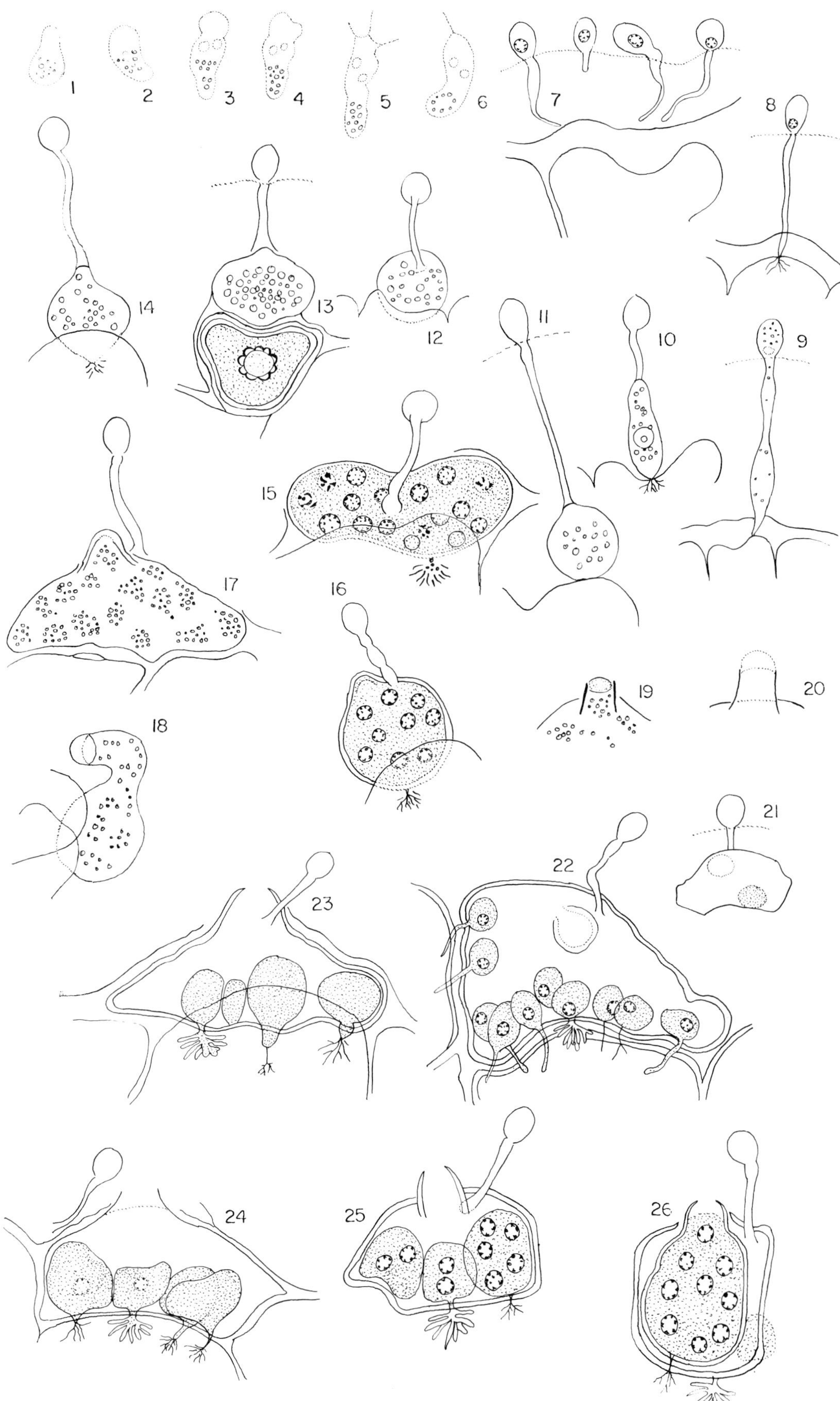

Scherffelomycopsis

DIPLOPHLYCTIS Schroeter

Engler u. Prantl, Naturl. Pflanzenf. 1:78, 1892

Plates 92, 93

Several species have been included in this intramatrical inoperculate genus up to the present time, but some of these are questionable members. *Diplophlyctis amazonensis* (Karling) Sparrow is obviously endooperculate and should be removed from the genus. *Diplophlyctis sexualis* Haskings, also, appears to be endooperculate (figs. 23, 24) and will probably prove to belong to the operculate genus *Nephrochytrium* with more intensive study. On the other hand, *Nephrochytrium complicatum* Willoughby is inoperculate, and for this reason, Dogma (1974b) transferred it to *Diplophlyctis* and renamed it *D. complicatus* (Willoughby) Dogma. These chytrids are saprophytic or weakly parasitic in moribund cells of the Characeae and Chlorophyceae and saprophytic in vegetable debris, cellophane and chitin.

The thallus of this genus is intramatrical except for the persistent planospore cyst which is present in some species, and it consists usually of an apophysate zoosporangium or resting spore, and richly branched extensive rhizoids which arise usually from a basal or lateral apophysis. It develops from the germ tube of the planospore, and in the type species, *D. intestina* (Schenk) Schroeter two types of development have been reported. Schenk (1858), Zopf (1884) and Karling (1928) reported that the rudiment of the zoosporangium is formed first at the tip of the germ tube. However, in a more intensive study Karling (1930) found that the branches of the rhizoid are formed first (fig. 4), after which the sporangial rudiment develops at one end of the branches (figs. 5, 6) while the apophysis forms later as an enlargement beneath the zoosporangium. Sparrow (1936) and Haskins (1950), also, reported that the zoosporangium and apophysis develop from a branch of the germ tube in *D. intestina* and *D. sexualis* Haskins. In *D. nephrochytrioides* Karling three types of development have been reported, as follows:

Firstly, the germ tube may branch almost at right angles (fig. 33), and these 2 branches enlarge to form two lateral bladder-like vesicles (fig. 34). At the same time a 3rd branch arises at the juncture of these vesicles and becomes the rudiment of the rhizoids. The 2 vesicles may enlarge equally for a while (fig. 35), but later one becomes larger as the content of the other moves into it (figs. 36, 37). Thus, it eventually becomes the zoosporangium, and the other empty vesicle remains as a lateral apophysis (fig. 38) which may be lobed (fig. 39), digitate, elongate (fig. 40), and sometimes almost as large as the zoosporangium.

Secondly, only one vesicle may be formed from the germ tube near or at the rhizoidal rudiment (fig. 41), and with further development it enlarges considerably (fig. 42). A bud is subsequently formed at its side or apex (fig. 43), which increases in size as the content of the initial vesicle or swelling moves into it, and eventually becomes a zoosporangium (fig. 44). Thus, as in *Nephrochytrium appendiculatum* Karling the zoosporangium develops out of the initial swelling or apophysis which functions as a prosporangium. In this type of development the apophysis lies at the base or side of the zoosporangium.

Thirdly, a swelling or vesicle may develop in the germ tube (fig. 45) as in *Entophlyctis*, usually above the rudiment of the rhizoid, and it gradually enlarges to become a non-apophysate zoosporangium (fig. 46). In all of the types of development noted above in *D. nephrochytrioides* the planospore cyst and germ tube are persistent and thick-walled, and are attached directly to the surface of the zoosporangium (figs. 38, 39, 40, 46) or resting spore (figs. 50–52), or the base of the apophysis (fig. 44), or the main rhizoidal axis (fig. 40). In *D. chitinophila* Willoughby the persistent planospore cyst and germ tube may be attached to the zoosporangium or resting spore (figs. 57–59, 62). In *D. complicatus* it may become thick-walled and almost solid (figs. 66–67), and persists at one end of the zoosporangium or is attached to the rhizoidal axis of the resting spores (fig. 70).

At maturity the zoosporangia of *D. intestina* may vary markedly in shape and size and usually bear one apical, subapical or lateral exit canal which may be relatively short or up to 7 times as long as the diameter of the zoosporangium (Karling, 1930). Sometimes as many as 6 canals may be formed, and in exceptional cases an exit canal may branch several times. Occasionally, the canals are reduced to papillae (figs. 40, 46). The planospores are released after the tip of the canal or papilla deliquesces (figs. 8, 47), but in *D. laevis* Sparrow the tip may deliquesce before the planospores are fully formed (fig. 17).

The resting spores are borne in the same manner as the zoosporangia and may be smooth (figs. 10, 19, 20, 51–53), spiny or warty (figs. 9, 11, 13, 70, 72), hairy or spiny (fig. 27) or verrucose with a hyaline, yellowish, or dark wall and coarsely to finely globular content.

PLATE 91

Figs. 1-11. *Mitochytridium ramosum* Dangeard. Fig. 1 after Dangeard, 1911; figs. 2-11 after Couch, 1935.

Fig. 1. Curved and branched zoosporangium with zoospores in *Docidium ehrenbergii*.

Fig. 2. Planospores.

Figs. 3-7. Early stages in development of the thallus and zoosporangia.

Fig. 8. Elongate curved, lobed and branched thalli in *Docidium* sp., planospores emerging from 3 zoosporangia.

Fig. 9. Incipient resting spores connected by filaments.

Figs. 10, 11. Mature and empty resting spores with spines and protuberances.

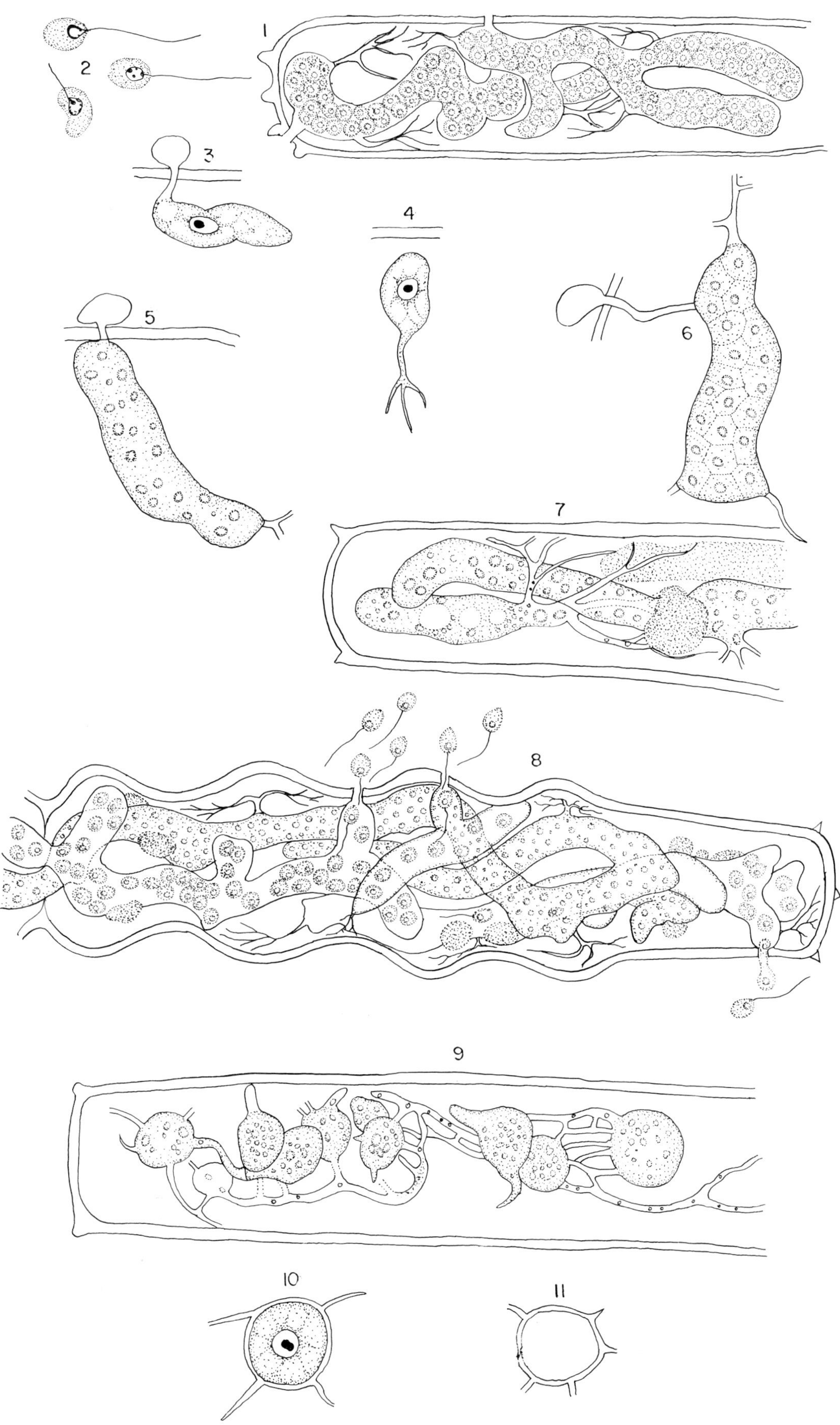

Mitochytridium

In *D. intestina* spiny, warty, verrucose, and occasionally smooth spores are formed which raises the question of whether *D. verrucosa* Ookuba (figs. 31), occurring in internodes of the Characeae, might be identical to the former species or a variant of it.

Sparrow (1936) and Haskins (1950) reported evidence that the resting spores are formed sexually by the fusion of the contents of two thalli through anastomosed rhizoids in *D. intestina* and *D. sexualis*. Sparrow showed anastomosis of lateral rhizoids of two small thalli (fig. 12) and the movement of the contents of a smaller thallus into a larger one which finally developed, presumably into a resting spore with the smaller one attached to the rhizoids as an empty vesicle (fig. 13). In *D. sexualis* Haskins described and illustrated small, thin-walled "male thalli" enclosing one to several cysts which were connected to outside resting-spore thalli by a fine tube or rhizoid (fig. 27), and he regarded this as evidence that the rhizoid of the cyst had anastomosed with the rhizoids of a receptive female (?) thallus to produce the resting spore. He, also, recorded asexually developed resting spores. It seems obvious to the present writer that what he illustrated as sexual reproduction may be planospores which have encysted and germinated within a zoosporangium, producing a long germ tube at whose end a resting-spore thallus developed. At least, the present writer has observed germination *in situ* of planospores in *D. intestina* and the formation of resting-spores as well as sporangial thalli outside in relation to the germ tube. Sparrow (1960) suggested that the type of sexuality reported by Haskins might be a process of "rejuvinescence." The recent observations on the occurrence of persistent planospore cysts and germ tubes attached to the zoosporangia, resting spores, apophyses, and rhizoidal axes in *D. chitinophila* and *D. nephrochytrioides* cast some doubt on the view that the presence of an empty vesicle on a resting-spore thallus always relates to sexual reproduction. Unless the successive development stages of resting spores are followed, the persistent planospore cysts and germ tubes on spores such as shown in figures 50–53, 62, and 70 might be interpreted as small male thalli and anastomosed rhizoids. However, in *D. complicatus* Dogma (1974) found clear evidence of sexual reproduction of the type described for *Siphonaria variabilis* where the zygote or zygospore forms as an outgrowth at the point of fusion of rhizoids.

Zopf reported that the resting spores of *D. intestina* form and discharge planospores in the same manner as the zoosporangia, but Karling (1936) found that they function as prosporangia in germination (fig. 11). This type of germination has been confirmed by Richards (1951) in *D. laevis* (fig. 21).

PLATE 92

Figs. 1-13. *Diplophlyctis intestina* Schroeter. Figs. 2-6, 8, 10 after Karling, 1928, 1930, 1936; figs. 7, 9 after Zopf, 1884; figs. 12, 13 after Sparrow, 1936.

Fig. 1. Portion of mature sporangial thallus in a *Nitella* cell showing a zoosporangium with an exit canal, apophysis, and rhizoids. Drawn from living material.

Fig. 2. Spherical and ovoid, 4-6 μ diam. planospores with a hyaline refractive globule.

Figs. 3-6. Successive stages of planospore germination and the formation of the sporangial rudiment, apophysis and rhizoidal branches.

Fig. 7. Sporangial rudiment with remains of the germ tube, rudiment of apophysis, and rhizoidal branches.

Fig. 8. Discharge of planospores through a long exit canal.

Figs. 9, 10. Spiny and smooth resting spores, resp.

Fig. 11. Germination of resting spore.

Fig. 12. Anastomosis of lateral rhizoids and fusion (?) of the contents of two small thalli.

Fig. 13. Mature resting spore connected by its rhizoid with that of a minute thallus.

Figs. 14-20. *Diplophlyctis laevis* Sparrow. (Sparrow, 1939; figs. 15, 21 after Richards, 1951.)

Fig. 14. Large spherical, 7 μ diam. planospore with a hyaline refractive globule.

Fig. 15. Germinating planospore.

Fig. 16. Young thallus in *Cladophora* sp., showing planospore cyst, germ tube, rhizoids, apophysis, and rudiment of the zoosporangium.

Fig. 17. Mature zoosporangium with lateral apophysis; tip of exit tube deliquesced.

Fig. 18. Discharge of planospores.

Figs. 19, 20. Smooth resting spores with lateral apophyses.

Fig. 21. Germination of resting spore.

Figs. 22-27. *Diplophlyctis sexualis* Haskins in decaying vegetable debris. (Haskins, 1950.)

Fig. 22. Germinated planospore on boiled maize leaf showing long germ tube, rhizoid branches, and rudiments of zoosporangium and apophysis beneath the branches of the rhizoids.

Fig. 23. Almost mature zoosporangium, apophysis, and part of rhizoids.

Fig. 24. Plug of gelatinous material in exit canal.

Fig. 25, 26. Endooperculate exit canals.

Fig. 27. Five spiny to hairy resting spores with apophyses connected by filaments to male gametes (?) in a male (?) thallus.

Fig. 27A. *Diplophlyctis complicatus* (Willoughby) Dogma. Incipient resting spore developing as an outgrowth at the point of fusion of the rhizoids of 2 gametangial thalli. (Dogma, 1974f.)

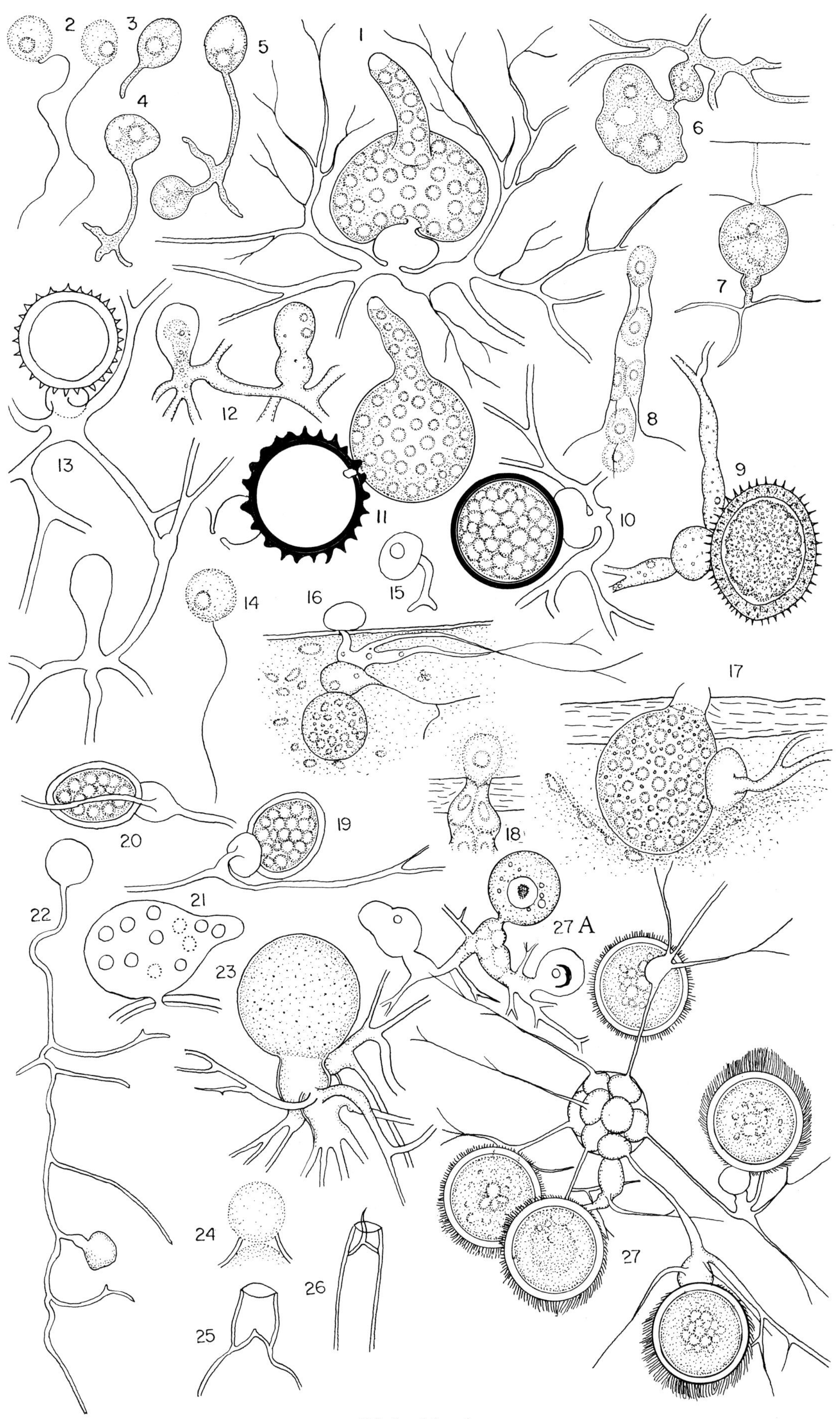

Diplophlyctis

NEPHROCHYTRIUM Karling

Amer. J. Bot. 25:211, 1938.

Plates 94, 95

Nephrochytrium is interpreted herewith to include 5 operculate intramatrical species which occur as saprophytes in soil and water and can be trapped on grass leaves, vegetable debris, cellophane and chitin. As noted earlier *Diplophlyctis sexualis* Haskins, also, may prove to be a species of this genus, and *Nephrochytrium complicatum* Willoughby was transferred to *Diplophlyctis* by Dogma. As diagnosed originally for *N. appendiculatum* Karling and confirmed by the reports on *N. stellatum* Couch and *N. aurantium* Whiffen, *Nephrochytrium* included operculate species whose zoosporangia bud out as discrete entities from an apophysis or prosporangium, but it is emended to include other operculate species such

PLATE 93

Figs. 28-31. *Diplophlyctis verrucosa* Kobayashi and Ookubo in *Chara*. (Ookubo, 1954.)

Fig. 28. Ovoid planospores with a hyaline refractive globule.

Figs. 29, 30. Zoosporangia in *Chara fragilis* with exit canals, basal apophyses, and part of the rhizoids.

Fig. 31. Mature verrucose resting spore with a large globule.

Figs. 32-53. *Diplophlyctis nephrochytrioides* Karling in *Nitella*. (Karling, 1967.)

Fig. 32. Spherical, 4-4.8 μ diam., planospore with a hyaline refractive globule.

Fig. 33. Early germination stage on a moribund cell of *Nitella* sp. with one central and two lateral branches.

Fig. 34. Later stage, lateral branches enlarged to form two bladder-like outgrowths; central branch has branched to form rhizoids.

Fig. 35. Still later stage, lateral bladders have enlarged; planospore cyst and germ tube persistent.

Fig. 36. Same thallus later, one bladder slightly larger than the other.

Fig. 37. Similar thallus in later stage of development; content of smaller bladder flowing into the larger; planospore cyst and germ tube wall becoming thick and dark.

Fig. 38. Mature zoosporangium and portion of a thallus formed in the manner described above; smaller bladder has become an empty lateral apophysis; planospore cyst and germ tube are attached to the zoosporangium.

Fig. 39. Mature zoosporangium with a lobed lateral apophysis and persistent thick-walled planospore cyst.

Fig. 40. Irregular zoosporangium with a basal exit papilla; planospore cyst, germ tube, and apophysis attached to the rhizoidal axis.

Fig. 41. Germinated planospore forming a single lateral bladder, or incipient apophysis, or prosporangium above rhizoidal branches.

Fig. 42. Later stage of development.

Fig. 43. Rudiment of zoosporangium budding out of prosporangium or apophysis.

Fig. 44. Portion of a mature thallus formed in the manner described in fig. 41.

Fig. 45. Germinated planospore with the rudiment of the zoosporangium forming in the germ tube above the rhizoidal branches; rudiment of apophysis lacking.

Fig. 46. Non-apophysate zoosporangium formed in the manner described in fig. 45.

Fig. 47. Discharge of planospores.

Figs. 48-50. Stages in the development of resting spores; similar to those in the formation of some zoosporangia.

Fig. 51. Mature resting spore with a thick, dark-brown wall and a large central globule surrounded by smaller ones; apophysis attached to the central rhizoidal axis and bearing rhizoidal filaments at one end; planospore cyst and germ tube attached to the base of the spore.

Fig. 52. Mature resting spore; apophysis has become thick-walled and a part of resting spore.

Fig. 53. Ovoid non-apophysate resting spore; formed in the same manner as the zoosporangium in fig. 46.

Figs. 54-63. *Diplophlyctis chitinophila* Willoughby on chitin. (Willoughby, 1962.)

Figs. 54, 55. Ovoid planospores, 5-5.2 × 3.5-3.7 μ, with a hyaline refractive globule and nuclear cap.

Fig. 56. Early germination stage on chitin.

Fig. 57, 58. Young thalli with planospore cyst and germ tube persistent on sporangial rudiments.

Fig. 59. Reduced planospore cyst and long germ tube attached to the sporangial rudiment.

Fig. 60. Mature apophysate zoosporangium with a short apical exit canal.

Fig. 61. Tip of long exit canal with clear area of gelatinous material above the planospore globules.

Fig. 62. Mature smooth hyaline resting spore with the attached reduced planospore cyst and germ tube.

Fig. 63. Irregular resting spore.

Figs. 64-73. *D. complicatus* Dogma on chitin. (Willoughby, 1962.)

Fig. 64. Encysted planospores.

Figs. 65, 66. Germinated planospores and minute thalli; planospore cysts and upper part of sporangial rudiment, resp., thickened and almost solid.

Fig. 67. Mature zoosporangium with a long exit canal; planospore cyst and germ tube persistent and thickened; part of sporangial wall thickened.

Fig. 68. Tip of curved exit canal with hyaline matrix above planospore globules.

Fig. 69. Small empty zoosporangium with an almost solid planospore cyst.

Fig. 70. Young resting spore; persistent planospore cyst and germ tube attached to the rhizoidal axis.

Fig. 71. Young resting spore before development of spines on the surface.

Fig. 72. Optical view of mature resting spore; planospore cyst and germ tube attached at base of apophysis.

Fig. 73. Surface view of mature resting spore with blunt spines.

Diplophlyctis

as *N. amazonense* Karling and *N. buttermerense* Willoughby in which formation of the zoosporangium out of an apophysis is not sharply defined. In some morphological and developmental aspects the thallus of the latter two species is similar to that of *Diplophlyctis* species, and they might be regarded as operculate counterparts of the latter genus. Sparrow (1960) and Dogma (1969), however, classified them as species of *Diplophlyctis*.

In the development of the thallus of most species of *Nephrochytrium* the germinating planospore (figs. 2–4, 41–43, 55–55, 69) forms a germ tube which branches in the substratum and develops an apophysis at (fig. 4) or above (fig. 43) the juncture of two primary branches. In *N. buttermerense* the germ tube may become greatly elongate and "transformed into a rhizoid" (figs. 69–71), and in *N. stellatum* Couch (1938) reported that the rudiment of the apophysis forms first at the tip of the germ tube before it branches (figs. 19–21). However, when the planospores germinate within a zoosporangium they form long and branched germ tubes (fig. 27). As the apophysis attains mature size it may be elongate and constricted (figs. 4, 5, 7), ovoid or slightly irregular (figs. 22, 23), or spherical (figs. 44, 45, 47). The branches of the germ tube develop into the rhizoidal system which usually becomes quite extensive in the substratum and is oriented on the apophysis (figs. 5, 28, 37, 48). In *N. appendiculatum, N. aurantium, N. amazonense* and *N. buttermerense* the terminal branches run out to fine filaments and points (figs. 6, 46, 55), but in *N. stellatum* the rhizoids are tubular and end bluntly (fig. 39). The planospore cyst and germ tube persist as appendages and may become thick-walled. In *N. buttermerense* and *N. amazonense* the cyst remains thin-walled (figs. 69–74, 76), and in the former species the germ tube may be unusually long and rhizoid-like.

In 3 species the rudiment of the zoosporangium buds out, usually at the juncture of the apophysis and germ tube, as a globular body (figs. 8, 23, 48) from the fully-grown apophysis, and in these species the latter structure functions as a prosporangium. In the other species (figs. 55, 56, 69) the sporangial rudiment is closely associated with an apophysis, but it has not been shown conclusively that it develops out of the apophysis. In *N. buttermerense* the wall of the apophysis may become greatly thickened (figs. 73–76) by the time the zoosporangium is mature. The bud from the apophysis in the 3 previously mentioned species enlarges to become the zoosporangium (figs. 9, 11, 24, 28, 49–51) which at maturity discharges planospores through a long or short exit canal (figs. 13, 14, 58–60), or a short cone-shaped papilla (fig. 51). Two species are endooperculate (figs. 58–60, 75, 76), and the remainder are exooperculate. The operculum is pushed off or out by the emerging planospore mass (figs. 13, 25, 51, 60), but in *N. buttermerense* remnants of it may persist in the exit canal (figs. 79, 81).

Resting spore are known in 3 species (figs. 16–18,

PLATE 94

Figs. 1-18. *Nephrochytrium appendiculatum* Karling in *Nitella*. After Karling, 1938; figs. 12, 13 drawn from New Zealand material.

Fig. 1. Spherical, 3.5-4.5 μ diam., planospore with a large hyaline refractive globule.

Figs. 2, 3. Germination of planospore; germ tube branching dichotomously.

Figs. 4, 5, 7. Formation of the constricted apophysis or prosporangium at the juncture of rhizoidal branches.

Fig. 6. Portion of a rhizoidal branch; terminal filaments fine and running out to fine points.

Fig. 8. Rudiment of zoosporangium budding out of apophysis or prosporangium.

Fig. 9. Almost fully grown zoosporangium developing out of the apophysis or prosporangium.

Fig. 10. Occasional fusiform swelling in rhizoids.

Fig. 11. Mature zoosporangium with 2 short exit tubes; planospore cyst thick-walled and dark.

Fig. 12. Tip of operculate exit canal.

Fig. 13. Initial discharge of planospores; operculum at side of emerged planospores.

Fig. 14. Later stage of planospore discharge; planospore mass expanding.

Fig. 15. Young resting spore budded out of an apophysis or prosporangium.

Figs. 16-18. Variations in shapes and sizes of mature, smooth, dark-walled resting spores.

Figs. 19-39. *Nephrochytrium stellatum* Couch in *Nitella*. (Couch, 1938.)

Figs. 19-21. Early stages in planospore germination; rudiment of apophysis formed first.

Fig. 22. Empty planospore cyst and young apophysis or prosporangium.

Fig. 23. Rudiment of zoosporangium budding out of apophysis or prosporangium near point of union of apophysis and rhizoid.

Fig. 24. Fully grown zoosporangium with an exit papilla in surface view near the planospore cyst; apophysis empty.

Fig. 25. Discharge of planospores from a small zoosporangium; operculum carried off by expanding mass of planospores.

Fig. 26. Planospores, 5 μ diam., with a hyaline refractive globule.

Fig. 27. Germination of planospore *in situ* with long branched tubes.

Fig. 28. Large mature zoosporangium with a slit-like basal papilla.

Figs. 29, 30. Appearances of the apex of exit papillae after discharge of operculum and planospores.

Fig. 31. Portion of resting-spore thallus with apophysis, rhizoid, and empty planospore cyst.

Fig. 32. Apophysis elongating; basal portion vacuolute.

Figs. 33, 34. Formation of a smooth-walled resting spore in the upper end of an elongated apophysis.

Figs. 35, 36. Similar stages in the development of the resting spore with an undulate outer wall.

Figs. 37, 38. Mature, almost stellate resting spores.

Fig. 39. Terminal portion of tubular rhizoids with blunt ends.

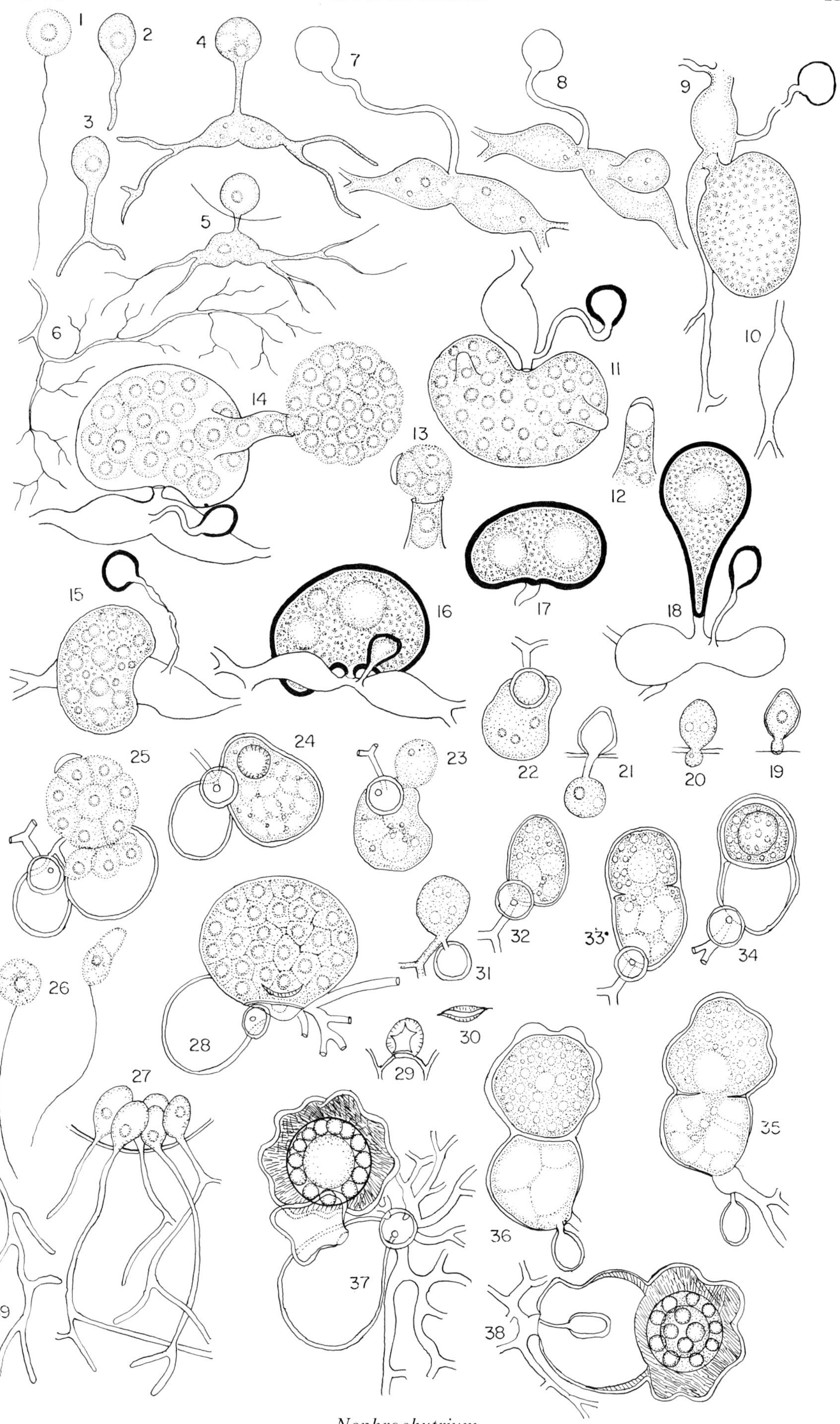

Nephrochytrium

33–38, 62–65) and in these they are borne in the same manner as the zoosporangia. No evidence of sexuality has been reported in relation to their development. In *N. appendiculatum* they are formed as an outgrowth from the apophysis, but in *N. stellatum* they develop in the upper part of the apophysis. In this process the apophysis elongates (fig. 32) as its protoplasm moves into it and aggregates in one end. Subsequently, a septum develops across the equator of the apophysis (figs. 33–36) delimiting a spore in one end and leaving the other end empty (figs. 37, 38). Couch emphasized the similarity of this type of resting spore development to azygospore formation in certain Mucorales. Germination of the spore has been observed only in *N. amazonense* where it functions as a prosporangium.

RHIZOSIPHON Scherffel

Arch. f. Protistenk. 54:189, 1926.

Plate 96

This inoperculate genus includes 2 well known species, *S. crassum* Scherffel and *S. anabaenae* (Rodhe and Skuja) Canter, and a third incompletely known one, *S. akinetum* Canter, which parasitize bluegreen algae in Hungary, Sweden, Czechoslovakia, and Great Britain. It is characterized by an endobiotic prosporangium from which may (figs. 7, 9, 13) or may not (fig. 20) arise a tubular vegetative system or haustorium that may penetrate several host cells, an epibiotic

PLATE 95

Figs. 40-51. *Nephrochytrium aurantium* Whiffen in grass leaves. (Whiffen, 1941.)

Fig. 40. Spherical, 4-4.8 μ diam. planospore with orange-colored refractive globule.

Fig. 41. Germinated planospore with a long, fairly thick germ tube.

Fig. 42. Dichotomous branching of germ tube; small rhizoid at apex.

Fig. 43. Formation of the apophysis or prosporangium.

Figs. 44, 45. Stages in the formation of a spherical apophysis or prosporangium with a dark-walled empty planospore cyst on its surface.

Fig. 46. Terminal portion of a rhizoid with the branches running out to fine filaments and points.

Fig. 47. Fully formed spherical apophysis or prosporangium filled with refractive globules.

Fig. 48. Same prosporangium 5½ hours later with the sporangial rudiment budding from the surface.

Fig. 49. Same vacuolate prosporangium 17 hours later; vacuolate zoosporangium fully formed with an exit papilla in surface view.

Fig. 50. Same empty prosporangium 19 hrs. later; zoosporangium fully formed and ready to discharge planospores, operculum in center of discharge papilla.

Fig. 51. Discharge of planospores from the conical exit papilla; operculum at the side of the expanding planospore mass.

Figs. 52-66. *Nephrochytrium amazonense* Karling in decaying vegetable debris. (Karling, 1944.)

Fig. 52. Spherical, 5-6.5 μ diam. and elongate planospores with a hyaline refractive globule.

Figs. 53, 54. Germination of planospores, and the formation of rhizoidal and apophysis rudiments.

Fig. 55. Later stage; sporangial rudiment forming above apophysis.

Fig. 56. Portion of a young sporangial thallus.

Fig. 57. Young zoosporangium; exit canal forming at apex.

Fig. 58. Tip of an exit canal with a plug of matrix and an endooperculum beneath it.

Fig. 59. Small zoosporangium shortly before dehiscing; plug of matrix has disappeared.

Fig. 60. Discharge of planospores; endooperculum on surface of expanding mass of planospores.

Fig. 61. Variations in shapes of discharged endoopercula.

Figs. 62-65. Variations in the shapes, sizes, and character of the outer wall of mature resting spores.

Fig. 66. Germination of the resting spore.

Figs. 67-81. *Nephrochytrium buttermerense* Willoughby on cellophane. (Willoughby, 1962.)

Figs. 67, 68. Spherical, 5-6 μ diam., planospores with a hyaline refractive globule and nuclear cap.

Fig. 69. Germination of a planospore with a long germ tube, rhizoidal branches, and the rudiments of the apophysis and zoosporangium.

Fig. 70. Young zoosporangium continuous with the broad apophysis; persistent planospore cyst bearing a rhizoid at one end and connected to the main rhizoidal axis.

Fig. 71. Later stage; apophysis wall beginning to thicken.

Fig. 72. Later developmental stage; incipient zoosporangium still continuous with the apophysis.

Fig. 73. Zoosporangium delimited from the thick-walled apophysis; planospore cyst connected to the rhizoidal axis and bearing a long, curved, branched rhizoid.

Fig. 74. Fully grown, small zoosporangium with a short exit tube before deliquescence of its tip.

Fig. 75. Two mature zoosporangia borne on a branched apophysis; endoopercula in exit canals.

Fig. 76. Mature endooperculate zoosporangium with a large thick-walled apophysis; planospore cyst attached to the rhizoidal axis by a long branched rhizoid.

Fig. 77. Small empty zoosporangium with an unusually long neck.

Fig. 78. Empty zoosporangium with 2 necks.

Fig. 79. Tip of exit canal with a beaked endooperculum.

Fig. 80. Empty tip of exit canal with reflexed wall material at its rim and portion of the endooperculum within.

Fig. 81. Similar tip of exit canal with outer rim of an endooperculum.

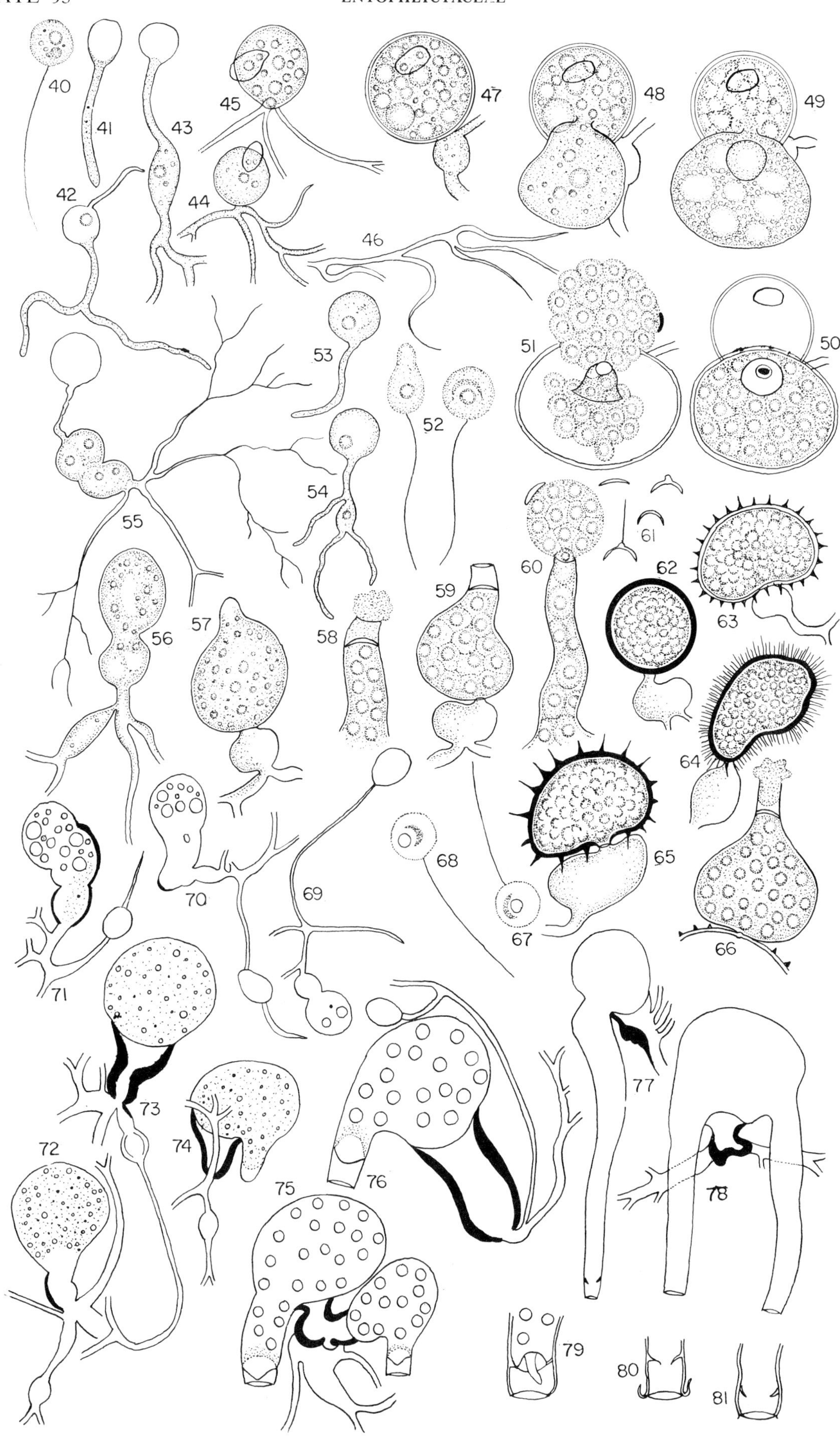

Nephrochytrium

zoosporangium which buds out of the prosporangium (figs. 1, 21, 32), planospores with several refractive granules (fig. 3), and an endobiotic resting spore (figs. 15, 30, 37, 38).

In *S. crassum* the planospore usually germinates at a long distance from the host cell and forms a long infection tube (figs. 4–6, 8, 9) which enlarges at its tip after entering the host. In *S. anabaenae* and *S. akinetum* the germ tube is usually shorter, and in the former of these species the planospores are attracted to and become attached to the heterocyst of the host. However, it does not infect this cell but forms a lateral germ tube which infects a neighboring cell (fig. 18). The enlarged tip of the germ tube becomes the incipient prosporangium which develops a tubular and irregular vegetative system (figs. 6, 7) from its periphery. *Rhizosiphon anabaenae* and *R. akinetum* lack a vegetative system or haustorium, and the prosporangium apparently functions directly as a nutriment absorbing system (figs. 18–20, 32–34). The zoosporangium buds out of the prosporangium (figs. 8, 9, 21, 11) and becomes epibiotic. In *R. crassum* and *B. anabaenae* short necks are formed for discharge of the planospores (figs. 10, 24), but in *S. akinetum* a refractive apical papilla develops (figs. 33, 34).

Endobiotic resting spores are formed by the transformation of the prosporangium into a thick-walled cell (figs. 13, 15, 28–30, 37, 38), and these apparently function as prosporangia in germination (fig. 16). Canter (1954) found an empty resting spore without an attached sporangium in *S. akinetum*, and she suggested that it might have functioned directly as a sporangium in germination. Whether or not sexuality is involved in the development of the resting spores has not been proven conclusively. Canter (1951, 1954) described and illustrated pairs of encysted planospores (gametes?) attached to the resting spores (figs. 13, 27–30, 36–38). She believed that such gametes make contact directly (fig. 27) or laterally by a short tube (fig. 29) in *R. anabaenae* after which infection

PLATE 96

Figs. 1-16. *Rhizosiphon crassum* Scherffel. Figs. 1, 2, 15, 16 after Scherffel, 1926; figs. 3-14 after Canter, 1951.

Fig. 1. Epibiotic vacuolate zoosporangium and an elongate endobiotic prosporangium in *Filarzkya* sp.

Figs. 2, 3. Spherical, 3 μ diam., planospores with one refractive globule and others with several granules.

Fig. 4. Infection of *Anabaena* by a long germ tube.

Fig. 5. Tip of germ tube enlarging in the host cell to form the rudiment of the haustorium and prosporangium.

Fig. 6. Later stage, incipient prosporangium forming a bud or tube into adjacent cell.

Fig. 7. Ovoid prosporangium with rhizoidal tubes extending into several cells; planospore cyst and germ tube persistent.

Figs. 8, 9. Incipient zoosporangium budding out of of the prosporangium near the tip of the persistent germ tube.

Fig. 10. Highly vacuolate zoosporangium with an empty prosporangium beneath.

Figs. 11, 12. Deliquescence of the exit papilla.

Fig. 13. Immature resting spore associated with the germ tubes of two empty planospores (gametes?).

Fig. 14. Mature resting spore

Fig. 15. Resting spore with coarsely granular content.

Fig. 16. Germination of a resting spore to form a vacuolate superficial zoosporangium.

Figs. 17-30. *Rhizosiphon anabaenae* Canter in *Anabaena*. Figs. 17-22, 25-30 after Canter, 1951; fig. 23 after Fott, 1951; fig. 24 after Skuja, 1948.

Fig. 17. Germinated planospore attached to a heterocyst of *Anabaena*.

Fig. 18. Production of a lateral germ tube and infection of a neighboring cell.

Fig. 19. Content of germinated planospores passing into the host cell through the germ tube and forming the rudiment of the prosporangium.

Fig. 20. Prosporangium enlarging in the hypertrophied host cell.

Fig. 21. Incipient zoosporangium budding out of the prosporangium.

Fig. 22. Mature zoosporangium with a mucilaginous apex at right angles to the infected cell and empty prosporangium.

Fig. 23. Almost erect zoosporangium.

Fig. 24. Similar zoosporangium.

Figs. 25, 26. Mature zoosporangium and planospores, respectively.

Fig. 27. Possible early fusion stage of gametes.

Figs. 28, 29. Young endobiotic resting spores in association by a broad tube with pairs of empty planospore (gametes?).

Fig. 30. Mature resting spore with an undulate wall connected by a tube to 2 empty planospores (gametes?).

Figs. 31-38. *Rhizosiphon akinetum* Canter in *Anabaena*. (Canter, 1954.)

Fig. 31. Empty encysted planospore on akinete of *Anabaena*.

Fig. 32. Empty endobiotic prosporangium and incipient epibiotic zoosporangium; thick walled planospore cyst adnate to zoosporangium.

Figs. 33, 34. Mature zoosporangia with empty prosporangia and thick walled planospore cysts attached by short and long filaments respectively; apical exit papillae.

Fig. 35. Dehisced zoosporangium with planospores; planospore cyst adnate; distal part of sporangial wall thicker and covered with short spines or bristles.

Fig. 36. Views of gametes (?) associated with resting spores.

Fig. 37. Young resting spore associated with two planospores (gametes?) at one end of the host akinete.

Fig. 38. Similar mature resting spore; highly refractive triangular area connecting gametes (?) with the spore.

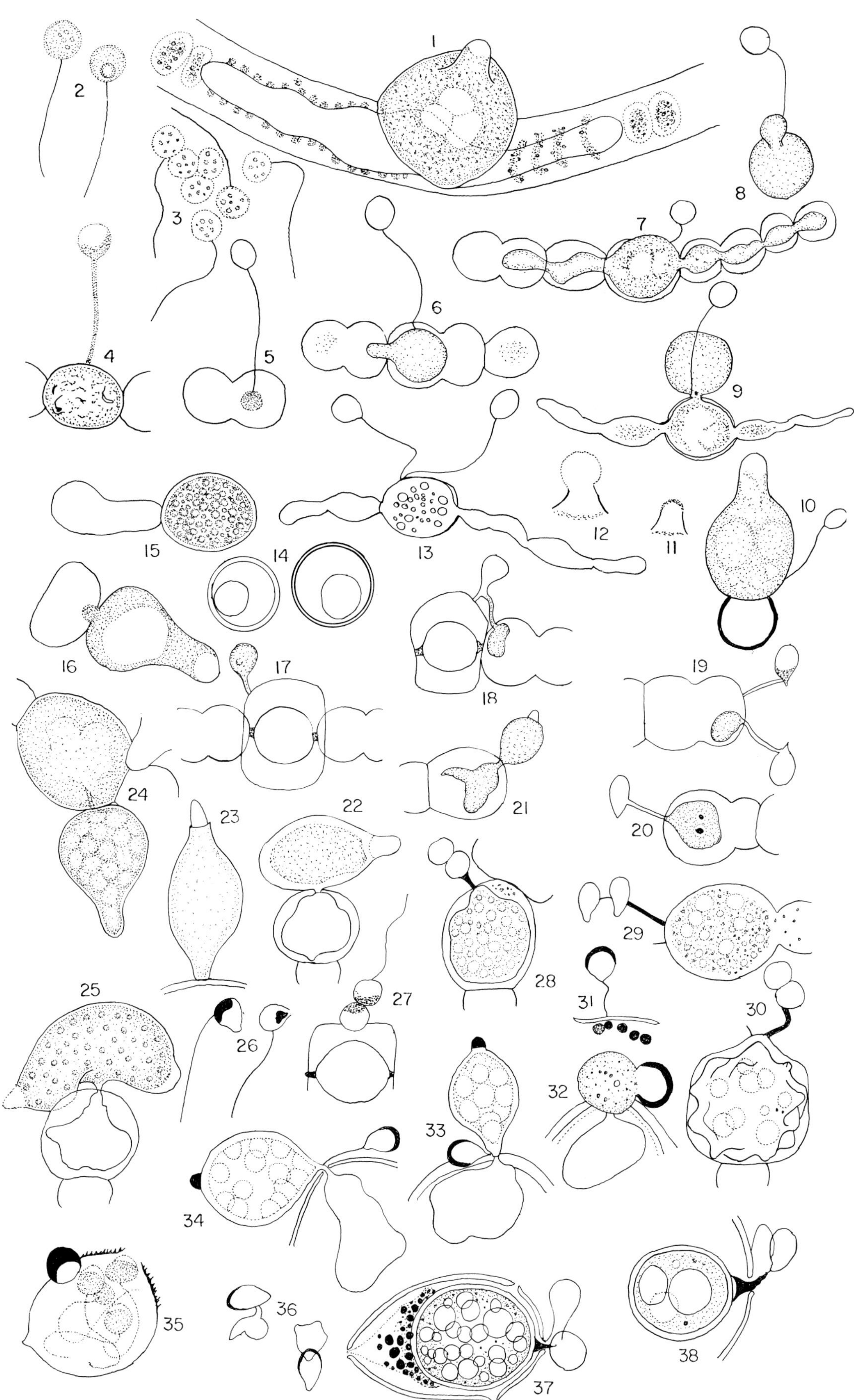

Rhizosiphon

of the host occurred by means of a fairly broad tube (figs. 28–30). Her studies of *R. anabaenae* and *R. akinetum* were made on preserved material contributed by other workers on these species, and her observations and interpretations of sexuality in *Rhizosiphon* need confirmation from study of living material.

Scherffel (1926) believed that this genus was allied to *Polyphagus* and *Saccomyces* because of the outgrowth of the zoosporangium from prosporangium, and for the same reason Whiffen (1944) classified it in the subfamily Polyphagoideae. Sparrow (1943, 1960) included it in the subfamily Entophlyctoideae, but Canter (1951) placed it in the subfamily Entophlyctoideae on the grounds that the prosporangium develops from the germ tube. Obviously, the relationships of this genus are obscure. It is included arbitrarily in this subfamily, but its exact taxonomic position is doubtful.

REFERENCES TO THE ENTOPHLYCTACEAE

Barr, D. J. S. 1971a. Morphology and taxonomy of *Entophlyctis confervae-glomeratae* (Chytridiales) and related species. Canad. J. Bot. 49:2215-2222, 19 figs.

———. 1971b. *Entophlyctis confervae-glomeratae* (Chytridiales): physiology. Ibid. 49:2223-2225.

Batko, A. 1970. A new *Dangeardia* which invades motile chlamydomonaceous monads. Acta Mycol. 6:407-435, 27 figs.

Bradley, W. H. 1967. Two aquatic fungi (Chytridiales) of Eocene age from the Green River formation of Wyoming. Amer. J. Bot. 54:577-582, 9 figs.

Booth, T. 1971. Problematical taxonomic criteria in the Chytridiales: comparative morphology of 10 *Entophlyctis* sp. isolates. Canad. J. Bot. 49:977-987, 12 figs., pls. 1-3.

Braun, A. 1856. Ueber *Chytridium*, eine Gattung einzelliger Schmarotzergewäsche auf Algen and Infusorien. Abhandl. Berlin Akad. 1855:28-83. pls. 1-5.

Butler, E. J. 1928. Morphology of the chytridiacean fungus *Catenaria anguilluae* in liver fluke eggs, Ann. Bot. 42:813-821, 19 figs.

Canter, H. M. 1951. Fungal parasites of Phytoplankton. II (Studies on British chytrids XII). Ann. Bot. n.s. 25:129-156, 13 figs. pls. 8-11.

———. 1954. Fungal parasites of the phytoplankton. III. Trans. Brit. Mycol. Soc. 37:111-133, 9 figs., 3 pls.

Cienkowski, L. 1857. *Rhizidium confervae-glomeratae.* Bot. Zeit. 15:233-237, pl. 5A, figs. 1-6.

Cook, P. W. 1966. *Entophlyctis reticulospora* sp. nov. A parasite of *Closterium*. Trans. Brit. Mycol. Soc. 49:545-550, *pls.* 28, 29.

Couch, J. N. 1935. An incompletely know chytrid: *Mitochytridium ramosum*. J. Elisha Mitchell Sci. Soc. 51:293-296, p. 62.

———. 1938. A new chytrid on *Nitella*: *Nephrochytrium stellatum*. Amer. J. Bot. 25:507-511, 34 figs.

Dangeard, P. A., 1888. Les Peridiniens, et leurs parasites. J. de Bot. 2:126-132, pl. 5.

———. 1911. Un nouveau genere de Chytridiacées. Bull. Soc. Mycol. France 27:200-203, fig. 1.

———. 1932. Observations sur la famille des Labyrinthulées et sur quelques autres parasites des *Cladophora*. Le Botan. 24:217-258, pls. 22-24.

Dogma, J. J., Jr. 1969. Observations on some cellulosic chytridiaceous fungi. Arch. f. Mikrobiol. 66:203-219, 70 figs.

———. 1974. Studies on chitinophilic *Siphonaria*, *Diplophlyctis*, and *Rhizoclosmatium*, Chytridiales. III. *Nephrochytrium complicatum* Willoughby, another *Diplophlyctis* with a sexual phase. Nova Hedwigia 25:143-159, 52 figs.

Domjan, A. 1936. "Visigombás"-Adtok Szeged es Tihany Vidékérol ("Wasserpilzdaten aus der Umgebung von Gzeged und Tihany). Folio cryptogam. 2:8-59, pl. 1.

Fisch, C. 1884. Beträge zur Kenntniss der Chytridiaceen. Sitzungsb. Phys. Med. Soc. Erlangen 16'. 29-72, pl. 1, 39 figs.

Fischer, A. 1892. Phycomycetes. Die Pilze Deutschlands, Oesterreichs und der Schweiz. Rabenh. Kryptog. -Fl. 1:1-490 Leipzig.

Fott, B. 1951. New chytrids parasiting on algae. Z. Vestnik Kral ces Spol. Nauk. Tr. Matem.-Privo Roc. 4:1-10, 20 figs., pl. 1.

Geitler, L. 1962a. Entwickelungsgeschichte der Chytridiale *Entophlyctis apiculata* auf der Protococcale *Hypnomonas lobata*. Oesterr. Bot. Zeitschr. 109:138-149.

———. 1926b. Entwickelung und Beziehung zum Wirt der Chytridiale *Scherffeliomycopsis*. Ibid. 109:250-275, 8 figs.

Hanson, A. M. 1946. A morphological, developmental and cytological study of four saprophytic chytrids. IV. *Phlyctorhiza endogena*. Amer. J. Bot. 33:732-739, 49 figs.

Haskins, R. H. 1946. New chytridiaceous fungi from Cambridge. Trans. Brit. Mycol. Soc. 29:135-140, pl. 8, 15 figs.

———. 1950. Studies in the lower Chytridiales. II. Endo-operculation and sexuality in the genus *Diplophlyctis*. Mycologia 42:772-778, 10 figs.

Hillegas, A. B. 1940. The cytology of *Endochytrium operculatum* (de Wildemann) Karling in relation to its development and organization. Bull. Torrey Bot. Club. 67:1-29, pls. 1-7.

Johns, R. M. 1956. Additions to the Phycomycete flora of the Douglas Lake Region. III. A new species of *Scherffeliomyces*. Mycologia 48:433-438, 12 figs.

Karling, J. S. 1928a. Studies in the Chytridiales. I. The life history and occurrence of *Entophlyctis heliomorpha* (Dang.) Fischer. Amer. J. Bot. 15:

32-42, pl. 1.

———. 1928b. Studies in the Chytridiales. II. Contribution to the life history and occurrence of *Diplophlyctis intestina* (Schenk) Schroeter in cells of American Characeae. Ibid. 15:204-214, pl. 14.

———. 1930. Studies in the Chytridiales. IV. A further study of *Diplophlyctis intestina* (Schenk) Schroeter. Ibid. 17:770-778, 2 figs., pls. 46-49.

———. 1931. Studies in the Chytridiales. V. A further study of species of the genus *Entophlyctis*. Ibid. 18:443-464, 54 figs. pls. 35-38.

———. 1936. Germination of the resting spores of *Diplophlyctis intestina*. Bull. Torrey Bot. Club, 63:467-471, 8 figs.

———. 1937. The structure, development, identity and relationship of *Endochytrium*. Amer. J. Bot. 24:352-364, 53 figs.

———. 1938a. A new chytrid genus: *Nephrochytrium*. Ibid. 25:211-215, 2 figs.

———. 1938b. Two new operculate chytrids. Mycologia 30:302-312, 37 figs.

———. 1941. *Cylindrochytridium johnstonii* gen. nov., and *Nowakowskiella profusum* sp. nov. Bull. Torrey Bot. Club. 68:381-387, 16 figs.

———. 1944. Brazilian chytrids. III. *Nephrochytrium amazonensis*. Mycologia 36:351-357, 28 figs.

———. 1947. Keratinophilic chytrids. II. *Phlyctorhiza variabilis* n. sp. Amer. J. Bot. 34:27-32, 48 figs.

———. 1949. *Truittella setifera* gen. nov., sp. nov., a new chytrid from Maryland. Ibid. 36:454-460, 44 figs.

———. 1965. *Catenophlyctis*, a new genus of the Catenariaceae. Ibid. 52:133-138, 12 figs.

———. 1967. Some zoosporic fungi of New Zealand. VI. *Entophlyctis, Diplophlyctis, Nephrochytrium* and *Endochytrium*. Sydowia 20:109-118, pls. 20-23.

Minden, M. von. 1916. Beitrage zur Biologie und Systematik einheimischer submerser Phycomyceten. In Falck, Mykolog. Untersuch. Berichte 2:146-255, 24 figs., pls. 1-8.

Ookubo, M. 1954. Studies on the aquatic fungi in the moors and ponds of Hakkoda. Nagaoa 4:48-60, 47 figs.

Richards, M. 1951. The life history of *Diplophlyctis laevis*. Trans. Brit. Mycol. Soc. 34:483-488, pl. 24.

Schenk, A. 1858. Algologische Mittheilungen. Phys.-Med. Gesell. Würzburg. A. F. 8:235-259, pl. 5.

Scherffel, A. 1926. Einiges über neue oder ungenügend bekannte Chytridineen (Der "Beitrage zur Kenntnis der Chytridineen" Teil II). Arch f. Protistenk. 54:167-260, pls. 9-11.

Schroeter, J. 1893. Phycomycetes. *In* Engler u. Prantl, Natürl. Pflanzenf. 1:63-141.

Skuja, H. 1948. Taxonomie des Phytoplanktons einiger Seen in Uppland, Schweden. Symbol. Bot. Upsaliensis 9:1-399, pls. 1-39.

Sorokin, N. W. 1883. Apercu systématique des Chytridiacées recoltées en Russie et dans l'Asie Centrale. Arch Bot. Nord France. 2:1-42, 54 figs.

Sparrow, F. K. Jr., 1933. Observations on operculate chytridiaceous fungi collected in the vicinity of Ithaca, New York. Amer. J. Bot. 20:63-77, 2 figs., pl. 20.

———. 1936. Evidence for the possible occurrence of sexuality in *Diplophlyctis*. Mycologia 28:321-323, 2 figs.

———. 1939. Unusual chytridiaceous fungi. Papers Mich. Acad. Sci. Arts, Letters 24:121-126, pls. 1, 2.

———. 1943. The aquatic Phycomycetes...XIX-785 pp., 634 figs. Univ. Mich. Press. Ann Arbor.

———. 1960. The aquatic Phycomycetes...2nd. ed. VII-1187 pp., 91 figs. Univ. Mich. Press. Ann Arbor.

Whiffen, A. J. 1941. A new species of *Nephrochytrium*: *Nephrochytrium aurantium*. Amer. J. Bot. 28:41-44, 30 figs.

———. 1944. A discussion of taxonomic criteria in the Chytridiales. Farlowia L:583-589, 2 figs.

Wildemann, E. de. 1895. Notes mycologiques. V. Ann. Soc. Belge Micro. (Mém.) 19:85-117, pls. 3, 4.

———. 1896. Notes mycologiques. XXII. Ibid. 20:109-131, pls. 6-12.

Willoughby, L. G. 1961a. Chitinophilic chytrids from lake muds. Trans. Brit. Mycol. Soc. 44:586-592, 2 figs., pl. 37.

———. 1961b. New species of *Nephrochytrium* from the English Lake District. Nova Hedwigia 3:439-444, pls. 112-116.

———. 1962. The ecology of some lower fungi in the English Lake district. Trans. Brit. Mycol. Soc. 45:121-136, 5 figs.

———. 1964. A study of the distribution of some lower fungi in soil. Ibid. 7:123-150, pls. 17-26.

Zopf, W. 1884. Zur Kenntniss der Phycomyceten. I. Zur Morphologie und Biologie der Ancylisteen und Chytridiaceen. Nova Acta Acad. Leop.-Carol. 47:143-236, pls. 12-21.

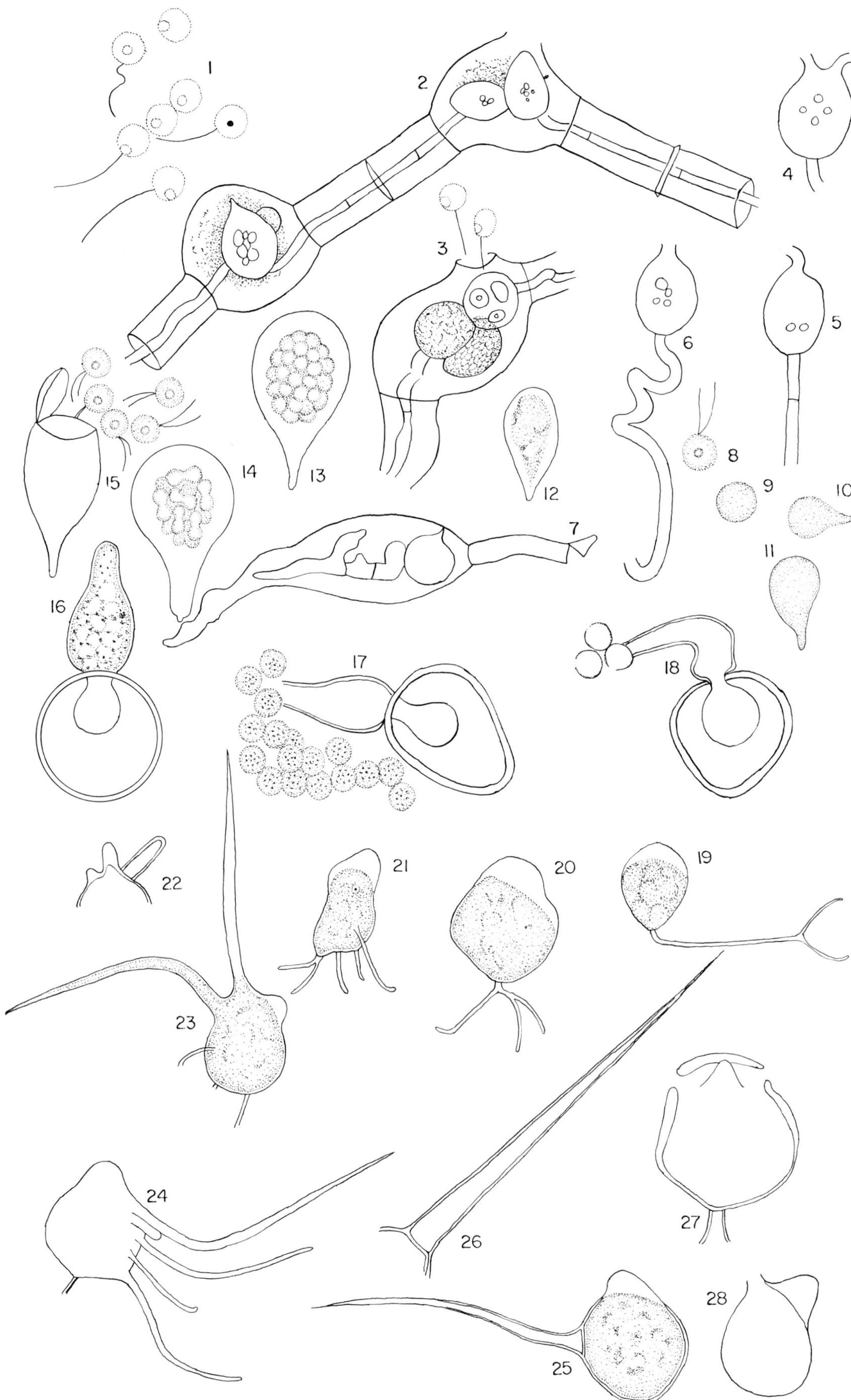

Aphanistis, Haplocystis, Achlyella, Mastigochytrium

Chapter VI

MONOCENTRIC, EUCARPIC GENERA OF DOUBTFUL AFFINITY*

APHANISTIS Sorokin

Arch. Bot. du Nord, France 2:35, 1883

Plate 97, Figs. 1-7

This monocentric eucarpic and inoperculate genus parasitizes *Oedogonium* and includes 1 or possibly 2 species which have not been seen since their discovery. *Aphanistis oedogoniarum* Sorokin develops pyriform to avoid zoosporangia in the oogonia of the host. These bear a tubular straight or curved, septate, isodiometric branched or unbranched hypha (figs. 2–6), which grows through the walls of several cells. One or two exit papilla or short necks develop on the zoosporangia (figs. 4–6), and the planospores are liberated through the fertilization pore of the host (fig. 3). The posteriorly uniflagellate planospores (fig. 1) are typically chytrid-like. Resting spores are not known. *Aphanistis pellucida* Sorokin is unknown except for one zoosporangium and an attached tubular and irregular hypha, (fig. 7) and is a questionable species. Sorokin was not certain of its identity and affinity with *A. oedogoniarum*.

The planospores of the latter species are chytrid-like in structure and appearance, and *A. oedogoniarum* is to be included among the chytrids. However, its taxonomic position and relationship are uncertain. Sparrow (1943, 1960) included it in the subfamily Entophlyctoideae next to *Rhizosiphon*, and this classification seems to be the best that can be attempted on the basis of present-day knowledge of *Aphanistis*.

HAPLOCYSTIS Sorokin

Bull. Soc. Nat. Kazan 4:11, 1872.

Plate 97, Figs. 8-15

This monotypic, monocentric, eucarpic, operculate genus occurs on submerged wood and is characterized by an obpyriform zoosporangium which is superficially attached to the substratum by a tapering base. No rhizoids nor resting spores have been observed. In development the content of the mature zoosporangium divides by successive cleavage into 32 portions (figs. 13) which rotate within the zoosporangium and later conjugate in pairs (fig. 14). The fusion cells develop flagella, and 16 are discharged from the operculate zoosporangium as biflagellate planozygotes (figs. 15). After coming to rest and encysting they germinate (figs. 9–11) and develop into incipient zoosporangia (fig. 12).

This remarkable genus has not been seen since its discovery, and Sorokin's account of its unusual life history needs confirmation.

*Literature references to the genera are given in the bibliography of the Rhizidiaceae.

PLATE 97

Figs. 1-6. *Aphanistis oedogoniarum* Sorokin in *Oedogonium*; fig. 7 *A. pellucida*. (Sorokin, 1883.)

Fig. 1. Spherical planospores with a hyaline refractive globule and a posterior flagellum.

Fig. 2. Parasitized *Oedogonium* filament with the sporangia in the oogonia; hyphae isodiamentric, septate and growing through several cross walls of the host.

Fig. 3. Planospores liberated through the fertilization pore of the host oogonium.

Figs. 4-6. Zoosporangia with 2 and 1 short exit canals or papillae, resp.

Fig. 7. Thallus of *A. pellucida* in a sporeling of *Oedogonium*.

Figs. 8-15. *Haplocystis mirabilis* Sorokin on submerged wood. (Sorokin, 1874.)

Fig. 8. Spherical, 11 μ diam., biflagellate planozygote.

Figs. 9-12. Encystment of zygote, germination, and stages of development into an incipient zoosporangium.

Fig. 13. Zoosporangium after progressive cleavage of content into 32 cells.

Fig. 14. Fusion of cells within the zoosporangium.

Fig. 15. Discharge of biflagellate planozygotes from the operculate zoosporangium.

Figs. 16-18. *Achlyella flahaulti* Lagerheim on pollen of *Typha*. (Lagerheim, 1890.)

Fig. 16. Mature epibiotic zoosporangium with an endobiotic apophysis; rhizoids lacking.

Fig. 17. Discharged mass of cystospores; contents finely granular.

Fig. 18. Three empty germinated cystospores at the exit orifice.

Figs. 19-28. *Mastigochytrium sacardiae* Lagerheim on the perithecia of *Saccardia durantae*. (Lagerheim, 1892.)

Fig. 19. Young epibiotic zoosporangium with a single branched rhizoidal filament or axis.

Fig. 20. Older zoosporangium with basal rhizoids.

Fig. 21. Zoosporangium with rhizoids arising from several points on its basal periphery.

Fig. 22. Beginning of hair or spine formation on the zoosporangium.

Figs. 23-25. Zoosporangia with 2, 4 and 1 spine, resp.

Fig. 26. Enlarged hair or spine showing its thick, double-contoured wall.

Figs. 27, 28. Large and small empty zoosporangia.

ACHLYELLA Lagerheim

Hedwigia 29:144, 1890.

Plate 97, Figs. 16-18

This monotype, eucarpic, inoperculate genus is a doubtful member of the Chytridiales at present because the position and number of flagella on the planospores are unknown. *Achlyella flahaultii* parasitizes the pollen grains of *Typha*, and at maturity its thallus consists of an epibiotic zoosporangium and an endobiotic apophysis or swelling which lacks rhizoids (fig. 16). Lagerheim presumed that the planospore enters the pollen grain and develops into a globular cell or thallus as in species of *Olpidium*. When it attains a certain size it breaks through the host wall and forms the flagon-shaped epibiotic zoosporangium which contains finely-granular protoplasm. As the tip of the exit canal deliquesces the protoplasm apparently undergoes cleavage, and the non-flagellate portions or cells emerge and remain quiescent near the exit orifice (fig. 17). These develop a thin wall and are transformed into cystospores as in *Achlya, Achlygeton* and occasionally in *Caulochytrium*. After a while the cystospores develop a minute papilla, germinate and presumably release a planospore, leaving an empty vesicle behind near the exit orifice (fig. 18). Lagerheim did not observe motile planospores, the presence, number and position of the flagella, and resting spores. For these reasons it is not certain that this fungus is a chytrid, and its taxonomic position is doubtful. Hohnk (1962) reported it from Maderia, but did not contribute any additional data on its development and morphology.

Lagerheim regarded it as showing relationship with the Chytridiales and Saprolegniales, and Fischer (1892) placed it in his family Sporochytriaceae (Rhizidiaceae). Schroeter (1893) placed it in the genus *Rhizidiomyces* which was believed to be a chytrid genus at that time. Minden (1911) included it in the subfamily Rhizophidiae of the Rhizidiaceae between *Asterophlyctis* and *Rhizidiomyces*, and Sparrow (1943, 1960) listed it as an imperfectly known genus of the Phlyctidiaceae. Hohnk placed it in this family, also.

Possibly, the planospores from the germinated cystospore may be uniflagellate as has been described for *Achlyogeton*, and in that event there may exist among the chytrids other monocentric, eucarpic, epi-endobiotic species that form masses of cystospores at the orifice of the zoosporangia, which later germinate as in *Achlya*. Accordingly, it may become necessary in the future to establish a separate family apart from the *Achlygetonaceae* for such eucarpic, epi-endobiotic, monocentric species.

MASTIGOCHYTRIUM Lagerheim

Hedwigia 3:188, 1892

Plate 97, Figs. 19-28

This *Rhizophydium*-like, monotypic, monocentric, eucarpic, inoperculate genus, which parasitizes the perithecia of an ascomycete, *Saccardia durantae*, is of doubtful affinity at present because its planospores and resting spores are unknown. Its type species, *M. saccardiae* Lagerheim, consists of an epibiotic zoosporangium which bears 1 to 7 tapering long, stiff hairs or spines on its periphery (figs. 23–25) and one to several branched or unbranched endobiotic rhizoids (figs. 19–21) from its base or side. The hairs or spines are tapered and solid at the apex, and have a double-contoured wall (fig. 26), and are delimited from the zoosporangium by a cellulose plug and later fall off. Lagerheim did not observe dehiscence and planospores, but on empty zoosporangia he found up to 4 exit orifices which were 4 μ in diameter.

He regarded this genus as being close to *Rhizophydium*, but Minden (1911) discussed it in relation to *Rhizophlyctis*, presumably on the grounds that several rhizoids may arise at more than one point on the periphery of the zoosporangium (fig. 21). Sparrow (1943, 1960) believed that it might be referable to *Rhizophydium*.

PLATE 98

Figs. 1-16. *Macrochytrium botrydioides* Minden from fruits in stagnant water. Figs. 1, 2, 4-6, 9, 10, 12, 13 after Minden, 1916; figs. 3, 7, 8, 11, 15, 16 after Johnson, 1968.

Fig. 1. Spherical planospore with a colorless refractive globule.

Fig. 2. Large biflagellate planospore.

Fig. 3. Spherical, 2.5-3.5 μ diam., planospores with a hyaline colorless globule.

Fig. 4. Amoeboid planospores.

Fig. 5. Young thallus; zoosporangium budding out from the side of the broad germ tube near the apex.

Fig. 6. Almost mature thallus with an ovoid zoosporangium, a central axis, and coarse branched rhizoids.

Figs. 7, 8. Young thalli.

Fig. 9. Possibly a resting spore or resistant sporangium.

Fig. 10. Mature ovoid, laterally attached zoosporangium.

Fig. 11. Young zoosporangium with circumferential flange.

Fig. 12. Beginning of planospore discharge.

Fig. 13. Ovoid zoosporangium with planospore initials.

Fig. 14. Discharge of planospores.

Figs. 15, 16. Thalli with empty zoosporangia.

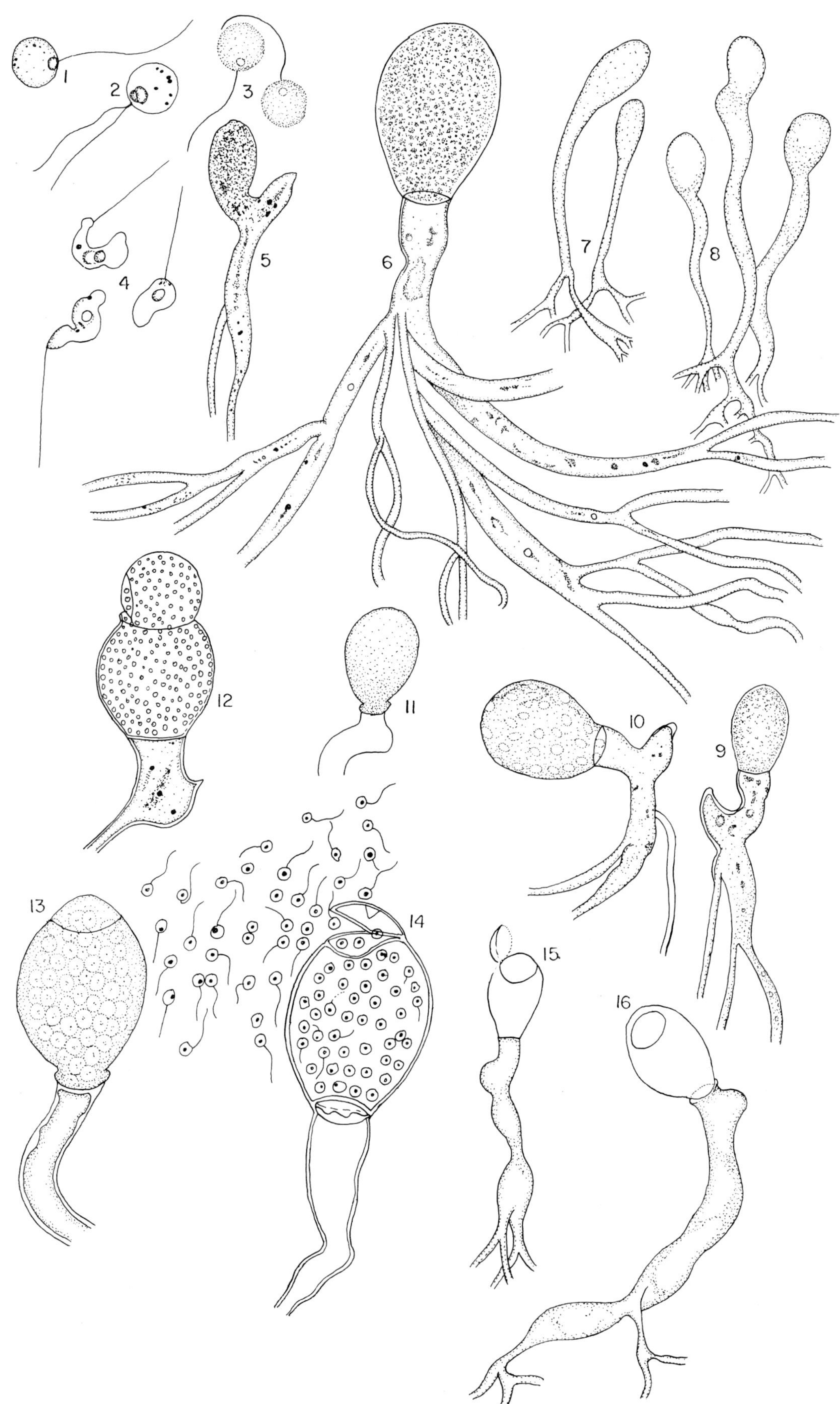

Macrochytrium

MACROCHYTRIUM Minden

Centralbl. Bakt. Parasitenk. Infektionskr. abt. 2, 8:824, 1902; Falck, Mykol. Untersuch Ber. 2:249, 1916

Plate 98

This monocentric operculate genus with one species, *M. botrydioides* Minden, occurs as a saprophyte on twigs and decaying fruits in pools of stagnant water and has been found in Germany, Denmark, Michigan and Minnesota, U. S. A. It is the largest known chytrid and may be visible to the naked eye. Its extramatrical zoosporangia may be 300 to 800 μ long by 200 to 650 μ broad with an intramatrical rhizoidal axis 400 to 450 μ long by 60–90 μ in diameter.

The early developmental stages have not been observed, but in germination the planospore probably develops a long broad germ tube which increases in diameter equal to that of the planospore and becomes the rhizoidal axis. It branches in the substratum and develops into the coarse rhizoidal system (fig. 6). Minden's figures suggest that the sporangial rudiment buds out of the side of the germ tube or axis near the apex and then grows upward (fig. 5), but Johnson's (1968) drawings of young thalli (figs. 7, 8) suggest that the planospore develops directly into a zoosporangium. In the early stages the thallus is unicellular and filled with dark brownish protoplasm which later accumulates in the distal enlarged portion. As the incipient zoosporangium increases in size the apex of the thallus is pushed aside and appears as a blunt or angular appendage on the thallus (figs. 5, 9, 10). Such appendages, however, are not clearly evident in Johnson's drawings, although figures 15 and 16 show bulges beneath the zoosporangia. The latter are delimited from the rhizoidal axis by a concave septum, and at maturity the content cleaves into planospore initials (fig. 13).

At dehiscence a smooth-walled broad convex operculum with a peg-like projection (fig. 14) is lifted up, and the emerging planospores are surrounded by a delicate vesicle (fig. 12) which ruptures by the time it is half the diameter of the zoosporangium. Johnson did not observe an enveloping vesicle in his specimens. The freed planospores are amoeboid (fig. 4) at first but later round up and swim away (fig. 14).

Observations on the presence of resting spores are doubtful. Minden was not certain that he saw them, but he illustrated a fairly thick-walled sporangium-like body (fig. 9) which he believed might be a resting spore or resistant zoosporangium. Johnson, also, failed to find resting spores or resistant sporangia.

The relationships and classification of *Macrochytrium* are uncertain at present. The organization and development of its thallis are somewhat similar to those of some species of *Blastocladiella*, but the presence of an operculum excludes it from the Blastocladiales. Minden (1911) classified it in the Hyphochytriaceae as he interpreted this family; Fitzpatrick listed it among the doubtful Chytridiales, and Sparrow (1943) erected the subfamily Macrochytrioideae in the family Chytridiaceae for it. Later (1960) he placed it in the subfamily Chytridioideae next to *Karlingiomyces*. He identified the fungus which Crasemann (1954) had studied physiologically as a probable spe-

PLATE 99

Figs. 1-27. *Caulochytrium gloeosporii* Voos and Olive. Sketches and diagrams based on photographs by Voos and Olive, 1968; and Voos, 1969.

Figs. 1, 2. Elongate, 3.5 × 8.5 μ, planospores with a flagellum attached slightly anterior to a basal knob, with one or more hyaline refractive globules.

Fig. 3. Amoeboid planospores.

Fig. 4. Encysted planospore.

Fig. 5. Encysted planospore germinating to produce a secondary planospore.

Fig. 6. Empty planospore cyst after germination.

Fig. 7. Encysted planospore on a conidium of *Gloeosporium* sp.

Figs. 8, 9. Germinated planospores on the mycelium of the host with endobiotic haustoria.

Fig. 10. Germinated planospore at a distance from the host mycelium with an endobiotic haustorium and epibiotic rhizoids.

Fig. 11. Sessile zoosporangium with several exit papillae.

Fig. 12. Diagram of a sessile zoosporangium parasitic on the host conidia, discharging planospores.

Fig. 13. Sessile zoosporangium surrounded by discharged amoeboid planospores.

Fig. 14. Hypothetical sketch of early fusion stage of gametes based on Voos' statement that "one of two zoospores sticks to the second one usually by the peculiar posterior protrusion"...

Figs. 15, 16. Biflagellate zygotes shortly after fusion.

Fig. 17. Biflagellate zygote with a large contractile vacuole near the posterior end.

Fig. 18. Binucleate encysted zygote with a large contractile vacuole.

Figs. 19-21. Hypothetical sketches based on Voos' description.

Fig. 19. Encysted diploid zygote after karyogamy.

Fig. 20. Germinating diploid zygote with a rhizoid, forming a sporangiophore or stalk.

Fig. 21. Later stage of stalk formation.

Fig. 22. Aerial zoosporangium formed at the tip of the sporangiophore or stalk.

Fig. 23. Eight-nucleate aerial zoosporangium, stained in haematoxylin.

Fig. 24. Amoeboid planospores discharged around an aerial zoosporangium.

Fig. 25. Planospore from an aerial zoosporangium.

Fig. 26. Collapsed germinated zygote with extramatrical branched rhizoids.

Fig. 27. Diagram illustrating the probable life cycle of *Caulochytrium gloeosporii.*

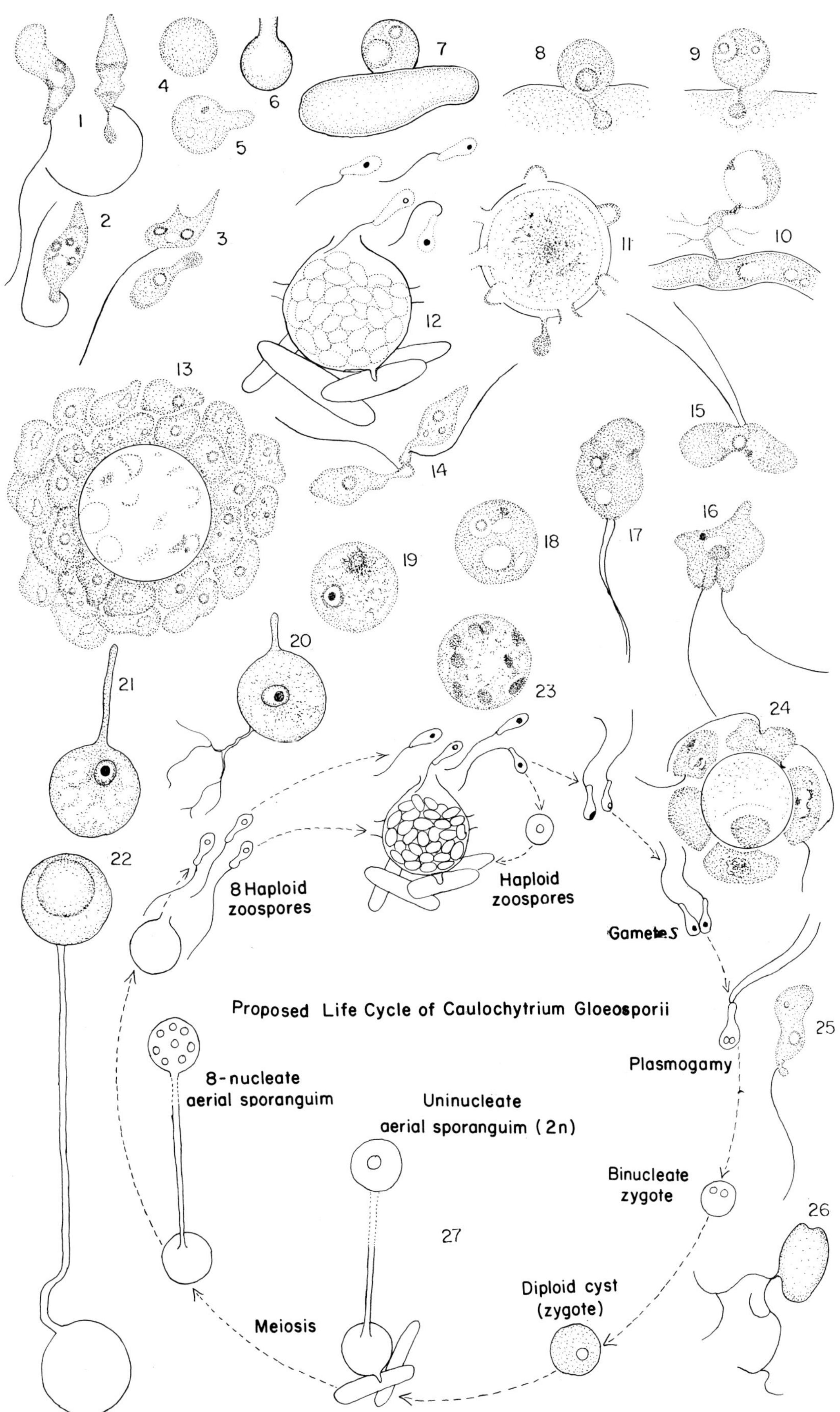
1
2
3
4
5
6
7
8
9
10
11
12
13
14
15
16
17
18
19
20
21
22
23
24
25
26
27
8 Haploid
zoospores
Haploid
zoospores
Gametes
Proposed Life Cycle of Caulochytrium Gloeosporii
Plasmogamy
8-nucleate
aerial sporanguim
Uninucleate
aerial sporanguim (2n)
Binucleate
zygote
Diploid cyst
(zygote)
Meiosis

Caulochytrium

cies of *Cylindrochytridium*. Whiffen (1944) omitted *Macrochytrium* from her key to the monocentric chytrids on the grounds that it is impossible to determine its type of development from the descriptions given. If, as Minden's figures suggest, the zoosporangia are formed as a bud from the germ tube or rhizoidal axis it might be included in the Entophlyctaceae, but this is hypothetical. In view of the uncertainty about its type of development and relationships it is grouped here among the monocentric eucarpic genera of doubtful affinity for the time being.

CAULOCHYTRIUM Voos and Olive

Mycologia 60:731, 1968

Plate 99

This monotypic inoperculate genus is unique among the eucarpic monocentric chytrids in the development of aerial zoosporangia from the germinated zygote in addition to sessile zoosporangia from planospores, and the fusion of isogametes while they are actively motile. *Caulochytrium gloeosporii* Voos and Olive is the only known genus among such chytrids which develops such aerial zoosporangia, and this makes it difficult to classify the species in any of the known families. Accordingly, it is listed among the genera of doubtful affinity in relation to the Rhizidiaceae. Possibly, on the basis of the organization and development of sessile sporangial thalli it might be placed arbitrarily in this family for the time being. However, it and others like it which may be discovered in the future may prove to be representative of a separate family. It is an obligate parasite, principally of the conidia of *Gloeosporium* sp. and one of its mutants, and will not grown on synthetic agar media or other fungi or algae. However, it does not kill its host, so that the latter with the parasite may be maintained on agar media.

Unfortunately for the present purpose Voos and Olive did not illustrate this species by drawings, and some of their photographs are out of focus and not clear as to details. Accordingly, the sketches shown in plate *99* of some of these photographs are inadequate.

The planospores of *C. gloeosporii* are elongate (figs. 1, 2) with one or more hyaline refractive globules, a contractile vacuole in the amoeboid stage, and an elastic basal protrusion above which the flagellum is inserted. No nuclear cap has been seen. After discharge they may swim directly away or become amoeboid (fig. 3) and creep around the periphery of the zoosporangium (fig. 13). Usually, they come to rest and encyst on the host (figs. 7–9), but occasionally they form an exit papilla (fig. 5) and release a secondary planospore, leaving an empty cystospore behind (fig. 6). Apparently, the secondary planospore is similar in size and structure to the primary one. In germinating on and infecting the host the encysted planospores forms a germ tube which penetrates the host wall and forms a knob-like haustorium at its tip (figs. 8, 9). If germination occurs at some distance from the host extramatrical rhizoids may develop from the germ tube (fig. 10) in addition to the endobiotic haustorium. The planospore body enlarges and eventually develops into a sessile zoosporangium which may bear one to several exit papillae (fig. 11). Occasionally, the exit papillae may be elongated into short necks. The tips of the papillae deliquesce, and as the planospores emerge they may swim directly away (fig. 12) or become amoeboid around the zoosporangium (fig. 13), as noted earlier.

Such planospores are facultative. They may develop into sessile zoosporangia or fuse in pairs as gametes while actively motile. Sometimes 3 to 5 gametes may fuse simultaneously. In the process of fusion

PLATE 100

Figs. 1-25. *Canteria apophysata* (Canter) Karling. (Canter, 1947, 1961.)

Fig. 1. Ovoid and spherical, 3 μ diam., planospores with a hyaline refractive globule.

Fig. 2. Germination of a planospore on *Mougeotia*.

Fig. 3. Development of a broad germ tube from which the apophysis and tubular haustorium or rhizoid are forming.

Figs. 4-6. Later developmental stages; apophysis formed immediately underneath the planospore cyst; rhizoid or haustorium expanding to become ovoid, somewhat irregular, and oblong; no increase in size of planospore cyst; growth endogenous

Figs. 7-9. Exogenous growth, protoplasm has begun to move outward and is enlarging the zoospore cyst into an incipient zoosporangium.

Figs. 10, 11. Large and small mature zoosporangia with the definitive globules of planospore.

Fig. 12. Early stage of dehiscence, apex of the zoosporangium has deliquesced, protoplasm moving out and surrounded by a thin membrane.

Fig. 13. Later stage, emerged protoplasm surrounded by a vesicular membrane.

Fig. 14. Discharge of planospores; vesicular membrane has ruptured at the apex.

Fig. 15. Empty zoosporangium showing ruptured vesicular membrane at apex.

Fig. 16. Early stage of endogenous resting spore or zygospore development; conjugation tubes from each thallus have fused to form a swelling or rudiment of the zygospore at Z.

Figs. 17, 18. Later stages showing enlargement of intercalary zygospore rudiment.

Figs. 19, 20. Mature intercalary zygospores with attached conjugation tubes and thalli.

Figs. 21-24. Variations in sizes and shapes of smooth-walled, hyaline zygospores containing a large refractive globule.

Fig. 25. Conjugation tube of a thallus growing towards another thallus which is forming a sporangial rudiment.

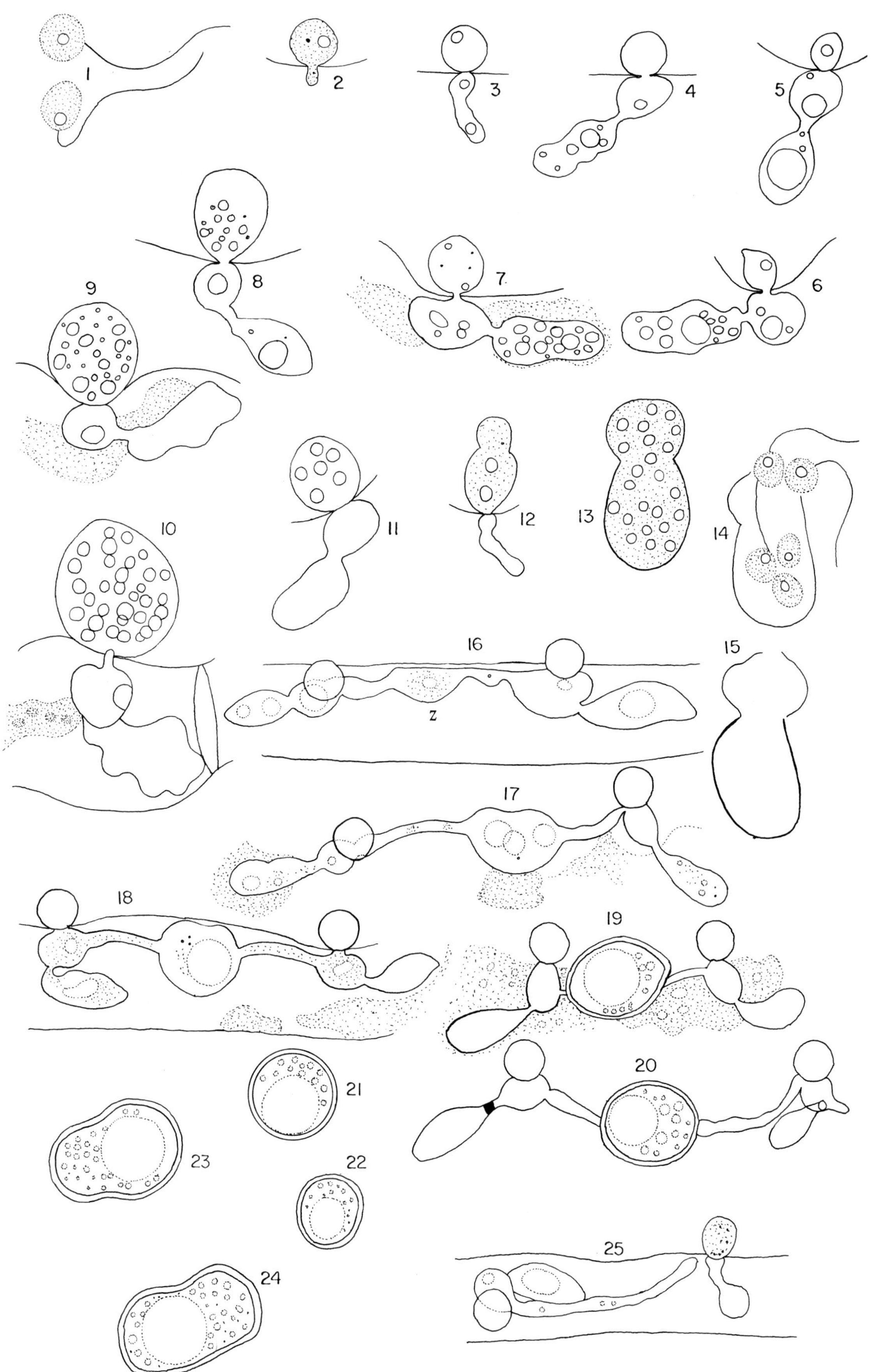

Canteria

"one of the zoospores sticks to the second one usually by the peculiar posterior protrusion." (fig. 14). Fusion is completed in a few seconds, and after a brief tug of war the gametes unite to form a biflagellate zygote which swims away and eventually becomes amoeboid (figs. 15–17). Karyogamy does not immediately follow plasmogamy, and the zygote remains binucleate (fig. 18) with a conspicuous contractile vacuole. Encystment soon follows, and karyogamy presumably occurs at this stage, although this has not been observed.

The encysted zygote remains dormant for 12 to 18 hours after which it develops a sporangiophore or stalk (figs. 20, 21) which may become 62 to 92 μ in length within 5 to 6 hours. Once this stalk has attained its full length an aerial zoosporangium develops at its tip in 10 to 15 minutes as the content of the zygote and stalk migrate into it. At this stage the masses of aerial zoosporangia gives the whole colony the appearance of a pincushion. Some of these young zoosporangia have been found to be uninucleate with a large nucleus which suggests that karyogamy occurred in the encysted zygote, but it is not improbable that it might occur in the incipient aerial zoosporangium. Apparently, 3 mitoses follow during which meiosis is believed to occur, and the zoosporangium commonly contains 8 nuclei at this stage (fig. 23).

The aerial zoosporangia are not dislodged by air currents, but they adhere readily when touched with a glass needle. When placed in water or on agar they develop 1 or 2 exit tubes and discharge, usually 8 planospores (fig. 24) which are similar (fig. 25) to those formed in the sessile zoosporangia. However, they do not behave as gametes but encyst and develop into sessile zoosporangia. Inasmuch as single planospores from either type of zoosporangium can complete the entire life cycle *C. gloeosporii* is homothallic.

CANTERIA Karling

Arch. f. Mikrobiol. 76:129, 1971

Plate 100

This monotypic inoperculate genus includes a parasite of *Mougeotia* which Canter (1947) described as *Phlyctidium apophysatum*. Her subsequent (1961) discovery that the resting spores are formed endobiotically by the conjugation of two thalli showed that this species does not belong in the invalid genus *Phlyctidium* as it is still interpreted by many chytridiologists. Accordingly, a new genus was created for it and named in honor of Hilda M. Canter who has contributed so much to our knowledge of the chytrid parasites of planktonic and other algae.

Canteria apophysata (Canter) Karling is characterized by epibiotic inoperculate zoosporangia and endobiotic sexually formed resting spores or zygospores. The encysted planospore infects the host by a broad germ tube (fig. 2) which elongates and increases in diameter. Soon, a swelling develops in it underneath the host wall (fig. 3) and enlarges to become the apophysis (fig. 4). The remainder of the germ tube elongates further, enlarges and becomes the tubular unbranched rhizoid or haustorium which may be ovoid (fig. 5), oblong (figs. 6, 7) or slightly irregular in outline (figs. 8, 10) at maturity. Meanwhile, the plano-

Plate 101

Figs. 1-17. *Cladochytrium tenue* Nowakowski. Fig. 1 after Nowakowski, 1876; figs. 2, 3, 7, 8 drawn from living material; figs. 4-6, 11-17, after Karling, 1945.

Fig. 1. Portion of a rhizomycelium with tenuous branched filaments, rhizoids, intercalary swellings, and incipient zoosporangia.

Fig. 2. Similar portion of a rhizomycelium with a mature terminal and an incipient intercalary zoosporangium.

Fig. 3. Spindle organs or intercalary swellings.

Figs. 4-6. Proliferating spindle organs; incipient zoosporangium forming from one of the cells in fig. 5.

Fig. 7. Discharge of planospores.

Fig. 8. Spherical, 4.5-5.5 μ diam., planospores with a hyaline refractive globule.

Fig. 9. Early germination stage.

Fig. 10. Later germination state showing development of the thallus.

Fig. 11. Young resting spore filled with refractive globules.

Fig. 12. Later stage of development; wall has thickened and globules are coalescing.

Figs. 13-15. Resting spores formed in the intercalary swellings.

Fig. 16. Germination of resting spore.

Fig. 17. Germinated resting spore with a stalked zoosporangium.

Figs. 18-21. *Cladochytrium polystomum* Zopf. (Zopf, 1884.)

Fig. 18. Germinated planospores with a brownish-orange refractive globule.

Fig. 19. Portion of a rhizomycelium with a spindle organ and rhizoids.

Fig. 20. Zoosporangium with 2 long and 3 short exit canals.

Fig. 21. Intercalary zoosporangium with large and small planospores.

Figs. 22-27. *Cladochytrium setigerum* Karling. (Karling, 1951.)

Fig. 22. Spherical, 3-3.4 μ diam., planospores with a minute hyaline refractive globule.

Fig. 23. Zoosporangium discharging planospores; setae omitted.

Fig. 24. Germinated planospore and early development of the thallus.

Fig. 25. Portion of the delicate rhizomycelium and 2 setigerous zoosporangia.

Fig. 26. Enlarged portion of the rhizomycelium showing setae on the tenuous portion; setae omitted on the zoosporangium.

Fig. 27. Zoosporangium with branched and unbranched setae.

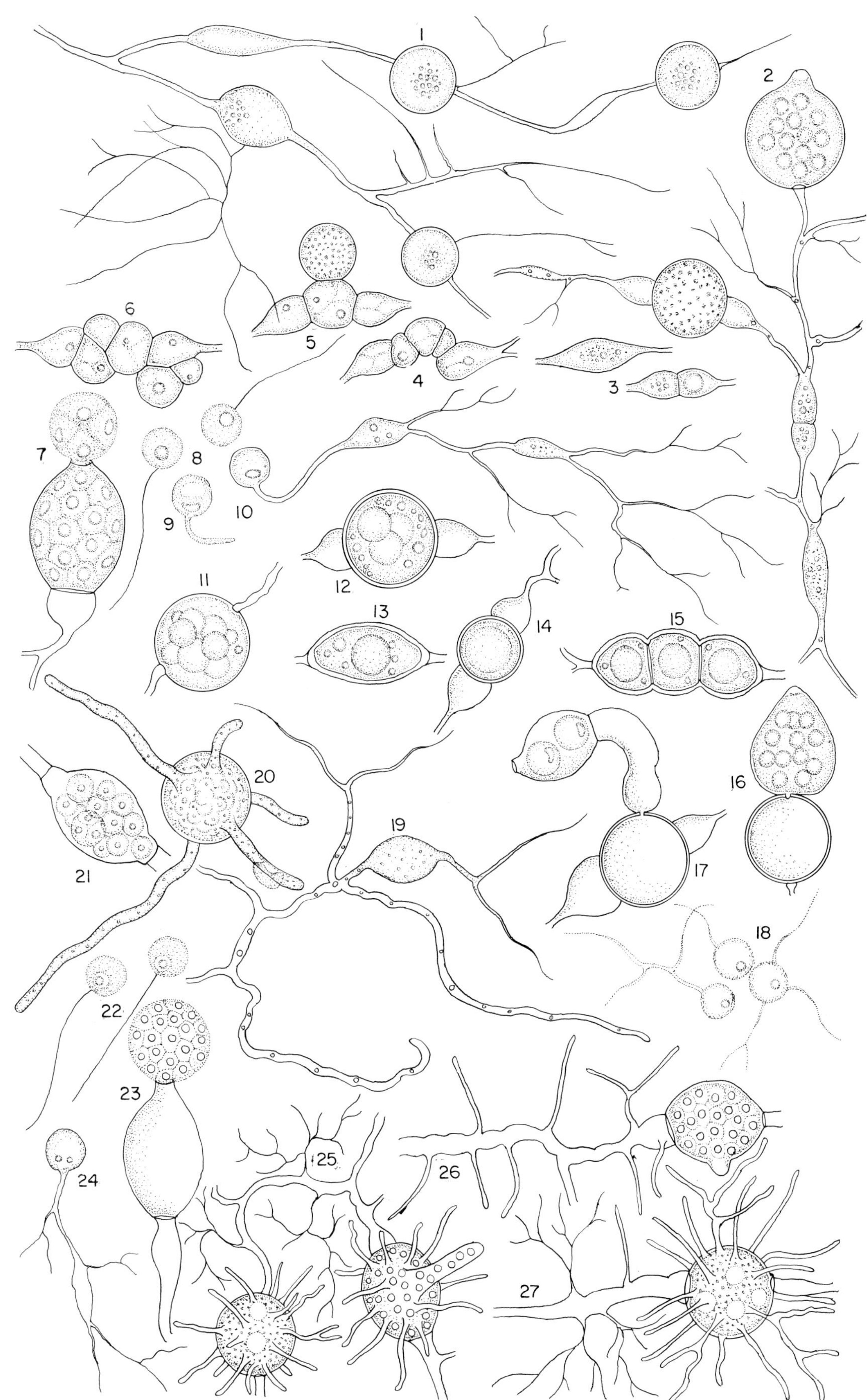

Cladochytrium

spore cyst remains unchanged in size (figs. 2–5) until the endobiotic absorbing system is fully developed, and up to this point growth has been endogenous. Then, movement of absorbed nutriments is outward, and exogenous development apparently occurs. The encysted planospore gradually increases in size (fig. 7) and grows into a zoosporangium (figs. 8, 9). At maturity of the latter the apophysis and haustorium become empty (figs. 10, 11).

In dehiscing, the apex of the zoosporangium deliquesces, and a portion of the undifferentiated sporeplasm flows out and is enveloped by a vesicular membrane- possibly the inner wall of the zoosporangium or an endosporangium. After the individual planospores are differentiated one ruptures the membrane, which initiates movement, motility, and dispersal (fig. 14). In empty zoosporangia (fig. 15) the thin ruptured vesicular membrane is still visible as a discrete entity.

In sexual reproduction two thalli in a host cell develop conjugation tubes which grow towards each other and apparently fuse at their ends, although such fusions have not been observed. A swelling develops at the point of fusion (fig. 16) which is recognizable as the zygospore or resting spore rudiment. It enlarges (figs. 17, 18) and eventually matures and becomes thick-walled (figs. 19–24). The zygospore may lie equidistant or nearly so between the conjugating thalli (figs. 17, 18, 20) or it may be almost adnate (fig. 19) to one of them. According to Canter (1961c) "there does not appear to be any morphological distinction between the conjugation thalli and no consistency in size. Sometimes both are more or less equal of size; at other times unequal." Figure 25 which shows one thallus with a conjugation tube growing towards an incipient sporangial thallus suggests that the planospores of this species are facultative; they may develop into either gametic or sporangial thalli.

Whether or not the marine fungus *Aplanochytrium kerguelensis* Bahnweg and Sparrow (1972) is to be regarded as a chytrid is uncertain. Its monocentric eucarpic thallus with radiating branched rhizoids occurs as a saprophyte on pine pollen and produces aplanospores which are released by the rupture of the sporangium wall. Resting spores are unknown. Tests for cellulose and chitin in the cell walls have not been made to determine if this genus mighf be chyrid-like or saprolegniaceous in this respect.

Plate 102

Figs. 28-47, 51-53. *Cladochytrium replicatum* Karling. Figs. 41, 42, 45, 47 drawn from living material; figs. 28-40, 43, 44, 46 after Karling, 1937.

Fig. 28. Living spherical, 4.5 μ diam., planospore with a golden-orange refractive globule.

Fig. 29. Enlarged, fixed and stained planospore with a nuclear cap, nucleus, nucleolus, and rhizoplast.

Fig. 30. Fixed and stained planospore with its nucleus lying on the substratum.

Fig. 31. Infection of the substratum and development of the germ tube.

Fig. 32. Planospore nucleus migrating down into germ tube.

Fig. 33. Elongate planospore nucleus has arrived in the primary spindle organ.

Fig. 34. Intranuclear mitosis of nucleus of a primary spindle organ.

Fig. 35. Appearance of a migrating nucleus in a tenuous portion of the rhizomycelium.

Fig. 36. Portion of a rhizomycelium showing a nucleus in the upper cell of a septate spindle organ; the nucleus of the lower cell has migrated down into another spindle organ.

Fig. 37. Simultaneous intranuclear mitosis in an incipient zoosporangium.

Fig. 38. A tetranucleate elongate spindle organ.

Fig. 39. Median section of a zoosporangium; cleavage furrows beginning at the periphery of the protoplasm.

Fig. 40. Cleavage segment with dark-staining bodies around the nucleus.

Fig. 41. Mature terminal zoosporangium shortly before cleavage, with a tapering exit canal.

Fig. 42. Discharge of planospores.

Fig. 43. Intercalary immature resting spore filled with refractive globules.

Fig. 44. Similar stage of a fixed and stained uninucleate, resting spore.

Fig. 45. Mature resting spore; most of refractive globules have coalesced to form a large central one.

Fig. 46. Fixed and stained uninucleate resting spores formed in a septate spindle organ.

Fig. 47. Germination of a spiny resting spore.

Figs. 48-50, 52. *Cladochytrium aurantiacum* Richards. (Richards, 1956.)

Fig. 48. Spherical, 6-7 μ diam., planospore with a large orange, 3.7-4.2 μ diam., refractive globule.

Fig. 49. Portion of the rhizomycelium.

Figs. 50-53. Variations in sizes and shapes of spindle organs in comparison with figs. 51 and 53 of *C. replicatum.*

Figs. 54, 55. *Cladochytrium tainum* Shen and Siang. (Shen and Siang, 1948.)

Fig. 54. Spherical, 11 μ diam., planospores with a hyaline, 7.2 μ diam., refractive globule; drawn to scale with figs. 28 and 48 to show comparative sizes.

Fig. 55. Portions of the rhizomycelium showing spindle organs and 2 resting spores.

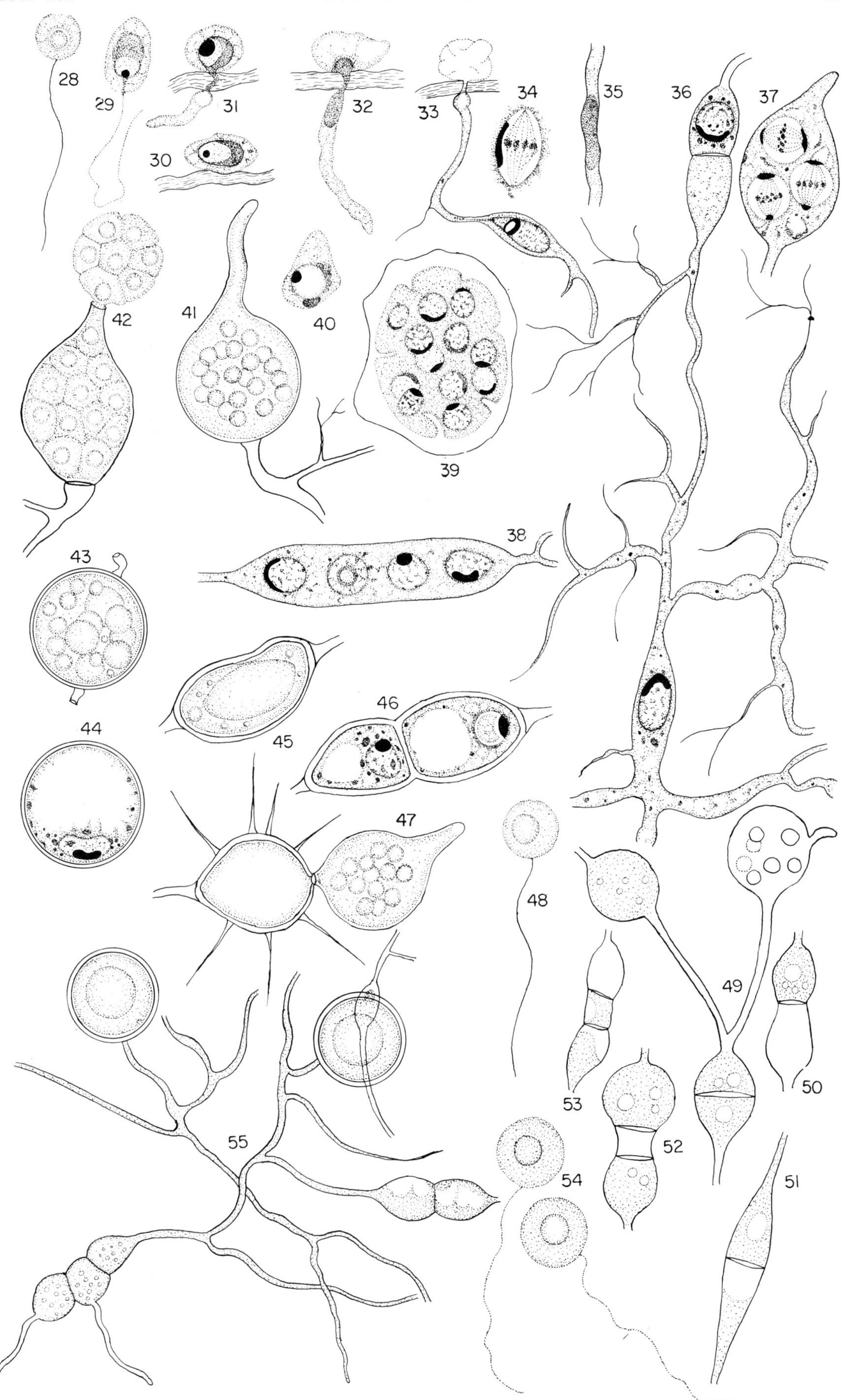

Cladochytrium

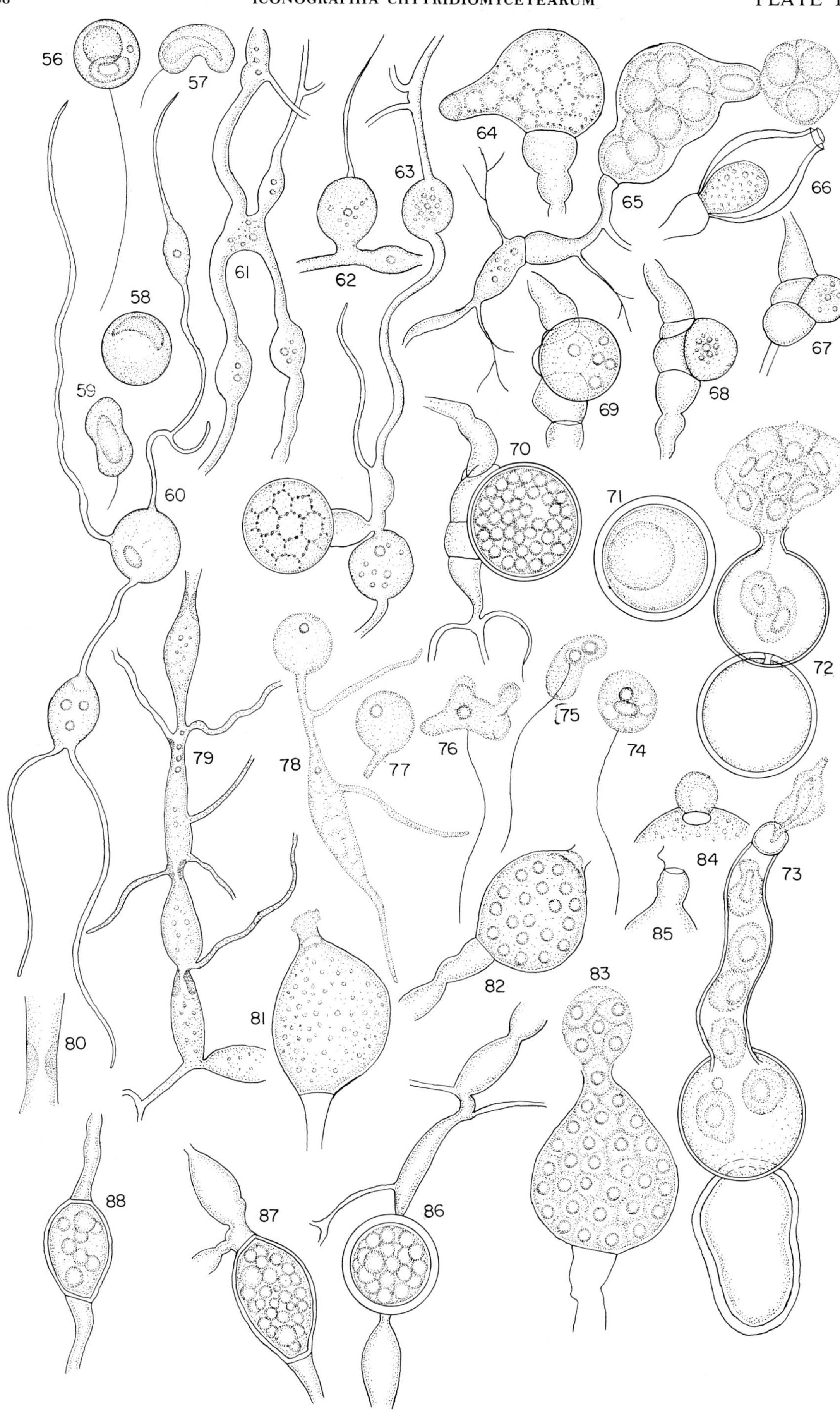

Cladochytrium

Chapter VII

CLADOCHYTRIACEAE

As interpreted here this family includes 5 inoperculate and 3 operculate genera and approximately 29 fairly well known species. In addition, several unnamed species have been assigned to this family. The classification of the operculate genera in a separate family, Megachytriaceae, is not warranted, according to Whiffen (1944) and Karling (1966) on the grounds that several of the genera are so strikingly similar in development, organization, structure and reproduction that it is almost impossible to distinguish them except for the presence of an operculum in some of the genera. Accordingly, it is merged with the Cladochytriaceae.

The thallus of this family is polycentric in that several zoosporangia and resting spores develop terminally or intercalarly on tenuous and somewhat mycelium-like filaments which also may bear rhizoids and intercalary swellings. Because the thallus is polycentric, extensive, somewhat mycelium-like and bears rhizoids in all but 2 genera Karling (1932) proposed the purely descriptive name, rhizomycelium, for it without implying any phylogenetic significance, and this name has been generally accepted. The thallus of *Coenomyces* and *Megachytrium* is a typical septate mycelium without rhizoids and well-defined intercalary swellings, and it remains to be seen whether or not they will continue to be classified in the Cladochytriaceae. In this connection it may be noted that Karling (1939) suggested the creation of a separate family, Myceliochytriaceae, for these non-rhizoidal mycelioid genera with posteriorly uniflagellate planospores. However, this suggestion no longer seems appropriate because it has become obvious that the presence or absence of rhizoids on the thallus is not a family characteristic. *Catenaria*, *Catenomyces*, and *Catenophlyctis*, for example, may have a cladochytriaceous rhizomycelium with abundant rhizoids, but they are placed in a separate family from the true chytrids as they are interpreted at present.

The tenuous filaments of the rhizomycelium may be delicate, 1.5 μ diam., or coarse and up to 20 μ in diameter. The intercalary swellings are usually conspicuous, narrowly to broadly fusiform or irregular, septate or non-septate. They were earlier described as "sammelzellen" or "collecting cells" and regarded as the portions in which the protoplasm accumulates, but since that time they have been referred to as spindle organs or mere intercalary swellings. They may develop directly into zoosporangia or resting spores or divide extensively after which one or many of the cells forms resting spores.

Anastomosis of fusion of the tenuous filaments has been reported in several species, and such and other cell fusions might be indicative that a parasexual cycle operates in some of the species, according to Alexopoulos (1962). Otherwise, no conclusive evidence of sexuality has been reported in this family. The fragmentary and inconclusive evidence reported so far will be discussed in more detail in the description of the various genera.

PLATE 103

Figs. 56-73. *Cladochytrium hyalinum* Berdan. Figs. 56-64, 67-73 after Berdan, 1941; figs. 65, 66 drawn from living New Zealand material.

Fig. 56. Spherical, 8-10 μ diam., planospore with a 4-7 μ diam., hyaline refractive globule, and a 40-50 μ long flagellum.

Figs. 57-59. Changes in shape of the refractive globule.

Fig. 60. Germinated planospore with 3 germ tubes and 2 intercalary spindle organs.

Fig. 61. Anastomosis of tenuous filaments of rhizomycelium; possibly representative of parasexuality.

Fig. 62. Incipient zoosporangium developing directly from the tenuous portion.

Fig. 63. Incipient terminal zoosporangium.

Fig. 64. Zoosporangium with alveolar protoplasm.

Fig. 65. Discharge of planospores from an irregular zoosporangium.

Fig. 66. Common proliferation of a zoosporangium.

Figs. 67-70. Stages in the formation of a resting spore from a cell of a multicellular spindle organ.

Fig. 71. Mature, hyaline, smooth-walled resting spore.

Figs. 72, 73. Germination of resting spores by the development of a zoosporangium on their surface; zoosporangium in fig. 73 with a long neck.

Figs. 74-88. *Cladochytrium crassum* Hillegas. (Hillegas, 1941.)

Fig. 74. Spherical, 5-6 μ diam; planospores with a hyaline refractive globule and a 25-35 μ long flagellum.

Figs. 75, 76. Changes in shapes of planospores.

Fig. 77. Early germination stage.

Fig. 78. Later germination stage with the primary spindle organ.

Fig. 79. Portion of the coarse rhizomycelium with trabeculae.

Fig. 80. Enlarged portion of the rhizomycelium with trabeculae.

Fig. 81. Fully grown zoosporangium with a plug of opaque material in the exit orifice.

Fig. 82. Later stage; refractive material has coalesced into definitive globules.

Fig. 83. Discharge of planospores.

Fig. 84. Vesicle at the mouth of the exit pore.

Fig. 85. Exit orifice with attached flap.

Figs. 86-88. Intercalary, resting spores with a smooth, light-brown wall; filled with refractive globules.

CLADOCHYTRIUM Nowakowski, pro parte.

In, Cohn, Beitr. Biol. Pflanzen 2:92, 1876. (Sensu recent. Schroeter in Engler u. Prantl, Naturl. Pflanzenf. 1:81, 1893)

Pyroctonum Prunet, pro parte, Comp. Rendu, Acad. Sci., Paris 119, pt. 2, 108, 1894.

Plates 101-104

At the present time this inoperculate genus includes 5 well known and several imperfectly known and questionable species. Of the latter *C. aureum* Karling has been merged with *C. replicatum* Karling (see Karling, 1967); *C. irregulare* de Wildemann is probably another chytrid similar in some respects to *Mitochytridium; C. polystomum* Zopf may prove to be identical with *C. replicatum*; and *C. cornutum* de Wildemann is known only by its rhizomycelium, toothed zoosporangia, and planospores. Other species have been assigned to this genus from time to time, but these have proven to be either invalid or species of *Physoderma*.

Most of the species are almost worldwide in distribution and occur as saprophytes in submerged decayed vegetation, soil, and water from which they may be trapped readily on cellulosic substrata such as cellophane, boiled grass leaves, and moribund or dead algae. Several species have been isolated from such substrata and grown on various agar media. *Cladochytrium aneurae* Thirumalachar is reported to be nonaquatic and parasitic in the liverwort *Aneura*. The most common species is *C. replicatum* which may be encountered in almost any submerged dead or decaying vegetation.

The thallus of *Cladochytrium* is a typical rhizomycelium consisting of extensive, branched tenuous filaments which bear rhizoids, intercalary septate or continuous fusiform swelling, intercalary or terminal zoosporangia and resting spores (figs. 1, 2, 55, 65, 96). The tenuous portions may be fine, 1.5 u diam, (fig. 1) or very coarse (fig. 79). The intercalary swellings are usually broadly fusiform or spindle-shaped and may be continuous or septate (figs. 3, 19, 50, 53). Sometimes they may divide several times and produce a pseudoparenchyma-like group of cells (figs. 4–6, 67–70) from one of which a zoosporangium or resting spore may develop as in some species of *Nowakowskiella*. The intercalary and terminal zoosporangia may be apophysate (figs. 7, 64) or non-apophysate, and at maturity discharge the planospores through a pore (fig. 84), papilla (figs. 7, 23, 83), or a short or long tube (fig. 41). In *C. polystomum* 1 to 6 long necks (fig. 20) are formed. Proliferation of the zoosporangia (fig. 66) occurs commonly throughout the genus as in *Nowakowskiella*. The planospores usually emerge to form a globular mass at the exit orifice (figs. 7, 23, 42, 83), and after remaining quiescent for a while they separate and swim away. Under certain conditions, however, they may swarm in a vesicle for a few minutes before dispersing. They are predominantly spherical and vary from 3 μ in *C. setigerum* Karling (fig. 22) to 8–10 μ in *C. hyalinum* Berdan (fig. 56) and 11 μ in diameter in *C. tianum* Shen and Siang (fig. 54). In most species they contain a single small or a very large plastic hyaline refractive globule which may undergo marked changes in shape (figs. 57–59) as the planospores creep about. In *C. replicatum* and *C. aurantiacum* Richards the globule is brilliantly golden to golden-red in color and very conspicuous. In fixed and stained zoospores of *C. replicatum* a conspicuous nuclear cap (fig. 29), nucleus, nucleolus, and rhizoplast are visible.

Cytological studies on this species indicate that the organization and replication of the rhizomycelium is associated with the distribution of the nuclei in the thallus, and this association probably occurs in other species, also. So far nuclei have been found only in spindle organs and incipient zoosporangia or resting

PLATE 104

Figs. 89-94. *Cladochytrium cornutum* de Wildemann. (de Wildemann, 1896.)

Fig. 89. Planospores with a small hyaline globule.

Fig. 90. Early stage of zoosporangial development.

Figs. 91, 92. Young zoosporangia with 3 and 2 apical teeth, resp.

Fig. 93. Fully grown zoosporangium with 5 teeth and a large central vacuole.

Fig. 94. Parasitized zoosporangium with 3 collapsed vesicles and 2 exit papillae.

Figs. 95-99. *Cladochytrium aneurae* Thirumalacher in *Aneura*. (Thirumalacher, 1947.)

Fig. 95. Two non-septate spindle organs.

Fig. 96. Portion of a rhizomycelium with continuous and septate spindle organs and incipient zoosporangia.

Figs. 97, 98. Zoosporangia filling the host cells and conforming to their sizes and shapes.

Fig. 99. Resting spores.

Figs. 100, 101. *Cladochytrium irregulare* de Wildemann in aquatic grasses; Tubular zoosporangia; possibly a species of *Mitochytridium*. (de Wildemann, 1895.)

Figs. 102-108. *Physocladia obscura* Sparrow. (Sparrow, 1931.)

Fig. 102. Portion of a rhizomycelium growing among pine pollen grains.

Fig. 103. Habit of a reduced rhizomycelium with an empty apical, proliferating dark, thick-walled zoosporangium, a 2nd immature zoosporangium, a spindle organ, and rhizoids.

Fig. 104. Discharging zoosporangium in optical section: planospores swarming in a vesicle.

Fig. 105. Spherical, 4.2 μ diam., planospore with a hyaline refractive globule.

Figs. 106, 107. Germination stages of planospores.

Fig. 108. Rough, dark-and brown-walled resting spore with vacuolate content.

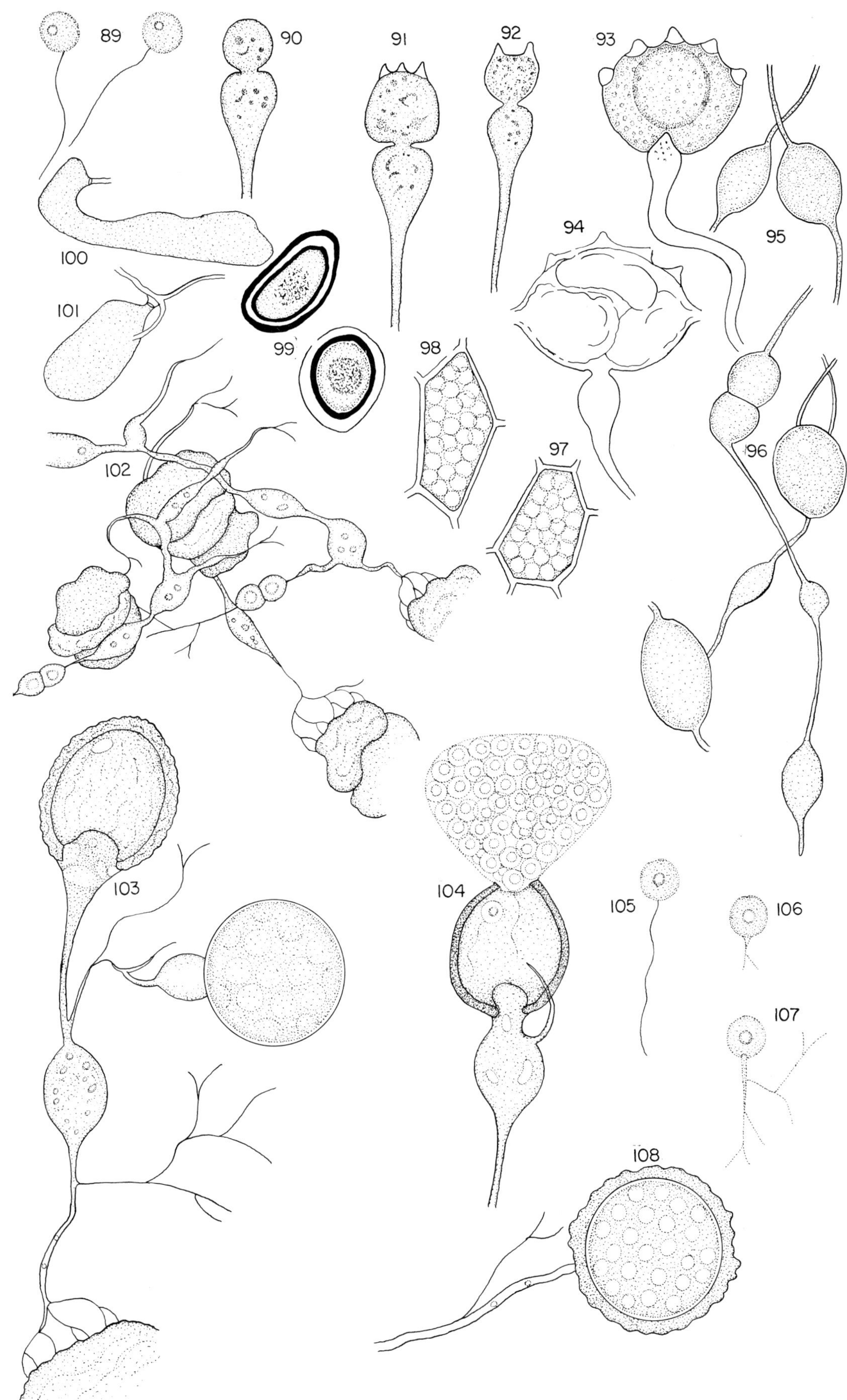

Cladochytrium, Physocladia

spores and not in the tenuous portions except in cases of nuclear migration (fig. 35). The sites occupied by the nuclei, thus, appear to be the center of growth and reproduction. A fairly broad germ tube is formed during germination of the planospore (fig. 31), and the nucleus moves down into it (fig. 32). A swelling develops in the tube where the nucleus comes to rest, and this enlargement soon becomes recognizable as the primary spindle organ (fig. 33). The nucleus divides with an intranuclear division spindle (fig. 34), and the fusiform swelling becomes bi- and sometimes tetranucleate (fig. 38) before becoming septate. The nucleus in the primary swelling may migrate into the tenuous filament as the rhizomycelius elongates, and where it comes to rest a second spindle organ develops (fig. 36). In this manner, then, the thallus is continually replicated as it spreads in the substratum.

The resting spores are borne terminally (fig. 55) or intercalary (figs. 12–15, 86–88) like the zoosporangia, and in *C. replicatum* (figs. 44, 46) they have been found to be uninucleate. In some cases the entire spindle organ (figs. 13–15, 45) or its cells (figs. 15, 46) become thick-walled and transformed into resting spores. In *C. hyalinum* Berdan (figs. 67–70) and also in *C. tenue* Nowakowski (fig. 5), the resting spore may be formed as a bud from one of the cells of the spindle organ. So far no evidence of sexuality has been reported in relation to their development. However, anastomosis of the tenuous filaments of the rhizomycelium occurs commonly (fig. 61), and it has been suggested that this might indicate the occurrence of parasexuality in the Cladochytriaceae. However, formed, the resting spores of 3 species have been shown to function as prosporangia when they germinate (figs. 16, 17, 47, 72, 73).

PHYSOCLADIA Sparrow

Mycologia 24:285, 1932.

Plate 104, Figs. 102-108

This monotypic inoperculate genus was created for a saprophyte, *P. obscura* Sparrow, which developed on staminate cones of pine in a water culture containing *Sphagnum*. Its thallus consists of a rhizomycelium (fig. 102) with tenuous portions, septate spindle organs and continuous swellings and rhizoids. The zoosporangia are terminal (fig. 103) with a dark thick wall, proliferate, and discharge planospores through a pore (fig. 104). These swarm in a vesicle for several minutes before dispersing. The rough, dark, thick-walled resting spores (fig. 108) are borne terminally like the zoosporangia.

The organization, morphology, and development of the thallus are so similar to those of *Cladochytrium* that *P. obscura* will doubtless prove to be a species of this genus. In the present writer's judgment, dark, thick-walled proliferating zoosporangia with a discharge pore, the swarming of planospores in a vesicle, and the formation of thick-walled dark resting spores are specific rather generic characters. Occasional, swarming of planospores in a vesicle occurs in *Cladochytrium* and *Nowakowskiella* under certain conditions, also, and this behavior cannot be regarded as a generic character.

PLATE 105

Figs. 1-8. *Nowakowskiella elegans* Nowakowski. Fig. 1 after Nowakowski, 1876; figs. 2-4 after Sparrow, 1933; figs. 5-8 drawn from New Zealand specimens.

Fig. 1. Terminal portion of a rhizomycelium with an empty proliferated sporangium and another one about to dehisce; drawing shortened.

Fig. 2. Portion of a rhizomycelium with rhizoids, intercalary swellings, and zoosporangia.

Figs. 3, 4. Stages in the discharge of planospores.

Fig. 5. Spherical, 5-7.5 μ diam. planospores with a large hyaline refractive globule.

Fig. 6. Amoeboid planospores.

Fig. 7. Intercalary, smooth, hyaline resting spore with a large central globule.

Fig. 8. Germination of a resting spore which has functioned as a prosporangium.

Figs. 9-25. *Nowakowskiella ramosa* Butler. Figs. 17-19 after Butler, 1907; figs. 9-16, 18, 21-23, after Karling, 1944; figs. 24, 25 drawn from New Zealand specimens.

Fig. 9. Enlarged portion of a rhizomycelium with rhizoids, anastomosis of tenuous filaments, and a group of 4 resting spores subtended by a group of parenchymatous cells.

Fig. 10. Discharge of planospores.

Fig. 11. Proliferating zoosporangium with a long neck.

Fig. 12. Mature intercalary zoosporangium with an operculate dehiscence pore.

Fig. 13. Amoeboid planospores with a 36-40 μ long flagellum.

Fig. 14. Spherical, 6.6-8.8 μ diam., planospores with a large hyaline globule.

Figs. 15, 16. Germination stages of planospores.

Fig. 17. Initial stage of resting spore development from a short lateral branch.

Fig. 18. Later stage; apical cell has divided and the initials of the spores are budding out.

Fig. 19. Division of cells underneath an incipient spore which is forming at the tip of a single terminal branch.

Fig. 20. Mature smooth, hyaline resting spore formed at the apex of a mass of pseudoparenchymatous cells which have anastomosed with another inflated branch.

Figs. 21-23. Division and proliferation of the cells of an intercalary swelling and the initial stage of resting spore formation as a bud from one of the cells.

Fig. 24. Germination of a resting spore which has functioned as a prosporangium.

Fig. 25. Germination of a resting spore which has functioned partly as a sporangium and partly as a prosporangium.

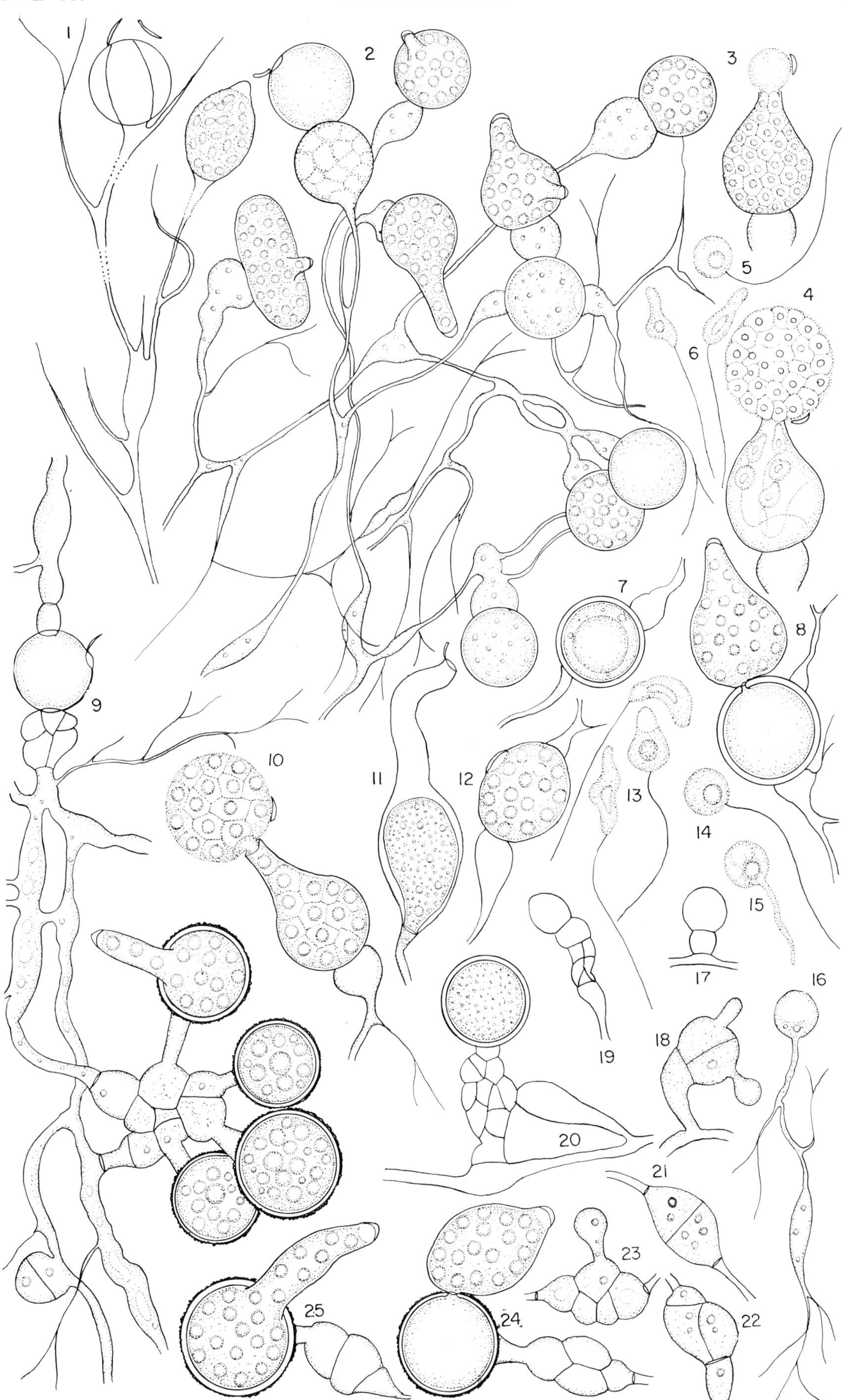

Nowakowskiella

Nevertheless, *Physocladia* is illustrated and described separately here so that readers may draw their own conclusions about its validity as a separate genus.

NOWAKOWSKIELLA Schroeter

Engler und Prantl, Naturl. Pflanzenf 1:82, 1893.

Cladochytrium Nowakowski, pro parte. *In* Cohn, Beitr. Biol. Pflanzen 2:92, 1876.

Plates 105-108

This genus is the operculate counterpart of *Cladochytrium* and includes 12 species and several unidentified specimens (Willoughby, 1961, 1965; Karling, 1968) which occurs as saprophytes in decayed vegetation in soil and water. They are worldwide in distribution, and most of them have been trapped on cellulosic substrata and readily grown on agar media. *Nowakowskiella pitcairnensis* Karling, however, has a predilection for fatty substrata such as bits of boiled hemp seeds and grows sparingly, if at all, on cellulosic substrata.

The development of the rhizomycelium is so similar to that of *Cladochytrium* (figs. 16, 41, 69) that it need not be described again. At maturity the thallus or rhizomycelium consists of frequently branched tenuous filaments with branching rhizoids at irregular intervals, spindle organs or intercalary enlargements

PLATE 106

Figs. 26-38. *Nowakowskiella sculptura* Karling. (Karling, 1961.)

Figs. 26, 27. Spherical, 3-3.8 μ diam., and amoeboid planospores with a small hyaline refractive globule and a 16-18 μ long flagellum.

Fig. 28. Portion of the rhizomycelium showing the abundance of intercalary swellings or spindle organs.

Fig. 29. Terminal endooperculate zoosporangium shortly before dehiscence.

Fig. 30. Discharge of planospores.

Fig. 31. Division or proliferation of an intercalary spindle organ.

Fig. 32. A later stage of proliferation of the spindle organ cells to form a pseudoparenchyma.

Fig. 33. A spindle organ that has divided into 13 cells, each of which bears a bud or potential resting spore.

Fig. 34. Proliferated spindle organ bearing 6 immature resting spores on stalks.

Fig. 35. Proliferated spindle organ bearing a smooth, a sculptured, and a warty mature hyaline resting spore.

Fig. 36. A single sculptured resting spore borne at the apex of a septate spindle organ.

Fig. 37. Early germination stage of a sculptured resting spore.

Fig. 38. Germinated smooth-walled resting spore which has formed an endooperculate zoosporangium.

Figs. 39-52. *Nowakowskiella hemisphaerospora* Shanor. Figs. 39-47 after Shanor, 1942; figs. 48-50, 52 after Karling, 1967; fig. 51 after Dogma, 1969.

Fig. 39. Zoosporangium discharging planospores.

Fig. 40. Spherical, 4.4-6.3 μ diam., planospores with a hyaline globule and a 35-40 μ long flagellum.

Fig. 41. Germinated planospore with rhizoids and the primary spindle organ.

Fig. 42. Terminal thick-walled cell with dense coarsely granular protoplasm in which the resting spore will develop.

Figs. 43-47. Stages in the aggregation of the protoplasm into the apex of the cell and its investment by a wall to form resting spores.

Fig. 48. Large terminal sporangium-like vesicle whose protoplasm is contracting and cleaving into 7 segments.

Fig. 49. The same vesicle with 7 resting spores.

Figs. 50, 52. Germinated resting spore which has functioned as a prosporangium.

Fig. 51. Unusual germinated resting spores in which the zoosporangium lies partly within and outside of the spore.

Figs. 53-59. *Nowakowskiella granulata* Karling. (Karling, 1944.)

Figs. 53, 54. Spherical, 5-6 μ diameter planospores with golden-brown granules and a 35 μ long flagellum.

Fig. 55. Mature endooperculate zoosporangium with a plug of translucent matrix in its apex, and a portion of the rhizomycelium.

Fig. 56. Upper part of a mature zoosporangium; plug of translucent matrix fills the orifice and short neck above the operculum.

Fig. 57. Some variations in the shapes of the opercula.

Fig. 58. Discharge of planospores.

Fig. 59. Intercalary, smooth, hyaline resting spore with a large hyaline globule and several smaller ones.

Figs. 60-67. *Nowakowskiella elongata* Karling. (Karling, 1944.)

Fig. 60. Spherical, 5-6 μ diam., planospore with a large hyaline globule.

Fig. 61. Small elongate, septate zoosporangium discharging planospores.

Figs. 62-64. Some variations in the shapes and sizes of elongate zoosporangia.

Fig. 65. Young and mature, hyaline, smooth walled intercalary resting spores.

Fig. 66. Ovoid, mature resting spore with a large central globule surrounded by smaller ones.

Fig. 67. Germinated resting spore which has functioned as a prosporangium.

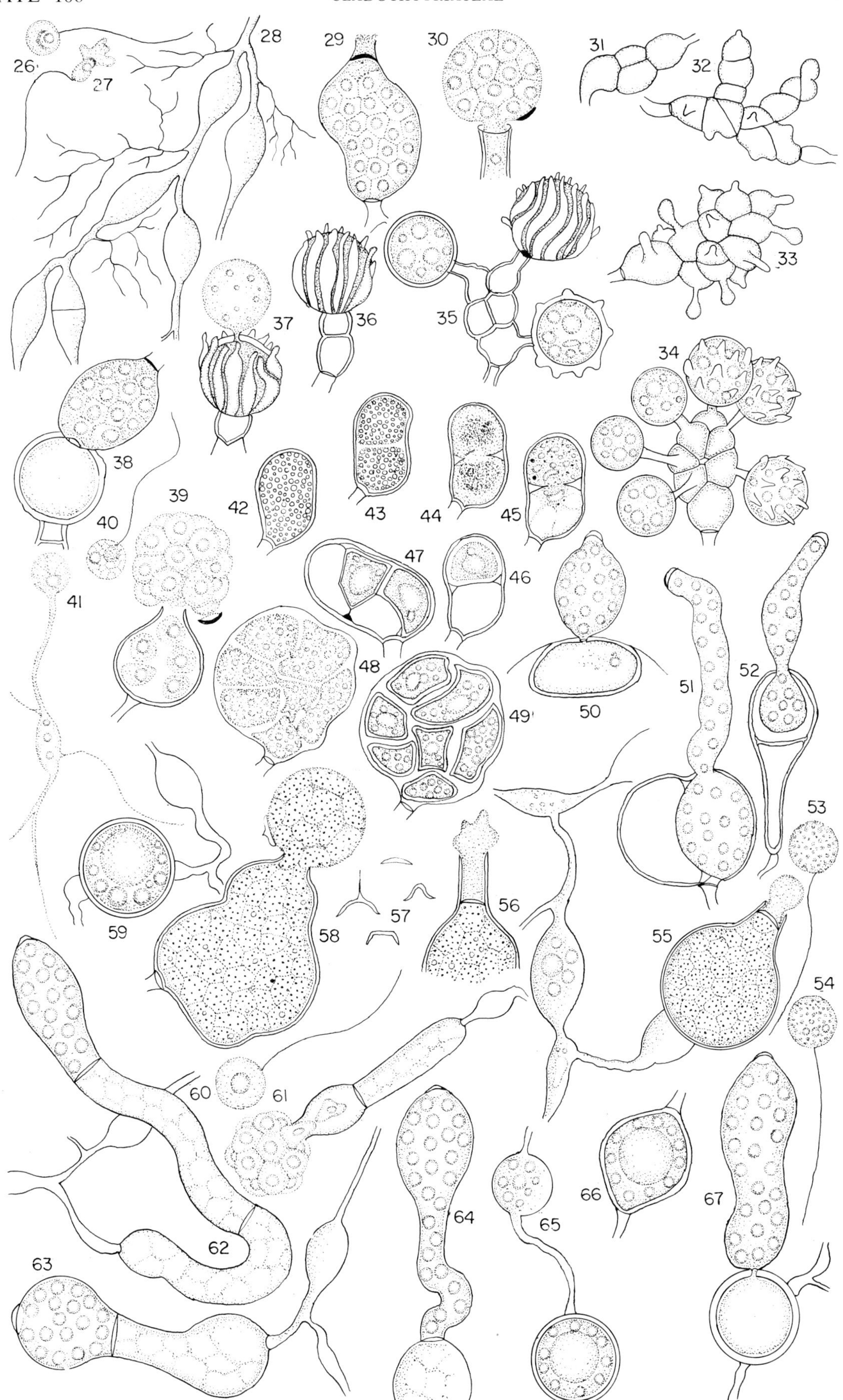

Nowakowskiella

which may be septate, non-septate and catenulate, terminal or intercalary zoosporangia, and resting spores. In some species, *N. elegans* (Nowakowski) Schroeter (figs. 2). *N. delica* Whiffen (fig. 86) and *N. pitcairnensis* (fig. 110), for example, the tenuous filaments may be fine and delicate, as in some species of *Cladochytrium*, but in *N. ramosa* Butler (fig. 9), *N. profusa* Karling (fig. 70) and *N. crassa* Karling (fig. 88) they may be quite coarse. In *N. crassa* specimens collected in New Zealand they were often as much as 20 μ in diameter. Anastomosis of the filaments (figs. 2, 9) occurs fairly commonly in several species. The intercalary swellings or spindle organs vary markedly in size, shape, and abundance. In *N. profusa* well-defined spindle organs are rare or lacking (fig. 70), but in *N. sculptura* Karling (fig. 28) and *N. crassa* (fig. 88) they are usually very abundant. In most species they are usually narrowly (figs. 2, 86, 110) to broadly fusiform (figs. 21, 28, 55, 63, 88, 96, 101) and non-septate or septate, and in *N. pitcairnensis* they may often be constricted and have a catenulate appearance (figs. 110, 112). In some species they may divide, particularly in relation to resting spore development, in several planes, become dictyosporus in appearance (figs. 9, 20, 23, 32, 33) and form a pseudoparenchyma. The rhizoids arise at irregular intervals along the length of the tenuous filaments and sometimes from the apex, side, or base of the zoosporangia and resting spores. They usually branch extensively and run out to fine points at their extremities.

The operculate, terminal and intercalary zoosporangia may be apophysate or non-apophysate and vary markedly in size and shape, but the majority of them are usually broadly or narrowly pyriform with smooth hyaline walls. In *N. elongata* Karling (figs. 60–64) they are usually elongate, coiled or curved and septate, and in *N. atkinsii* Sparrow (figs. 96, 99) they may bear setae or pointed pegs. Occasionally, the setae may branch at their tips (fig. 98) and become rhizoids. At maturity zoosporangia develop relatively long or short necks or narrow to broad papillae through which the zoospores are discharged. Most species are exooperculate, but *N. sculptura* (figs. 29, 38) *N. macrospora* Karling (figs. 79, 83), *N. granulata* Karling (figs. 55, 56) and *Nowakowskiella* sp. Karling (1968) are endooperculate. In *N. multispora* Karling (figs. 101, 102) and *N. profusa* Karling both endo- and exooperculate zoosporangia occur. In the latter species Willoughby (1961), and Chambers, Markus and Willoughby (1967) reported that dehiscence may occasionally be inoperculate as well as exo- and endooperculate.

In the discharge of the planospores the operculum is pushed off or out or remains attached at the edge of the exit orifice as the planospores emerge (figs. 3, 4, 10, 30, 39, 58). The latter emerge slowly and form a globular mass at the exit orifice that is usually

PLATE 107

Figs. 68-76. *Nowakowskiella profusa* Karling. (Karling, 1944.)

Fig. 68. Spherical, 4-5.5 μ diam., planospore with a small hyaline refractive globule.

Fig. 69. Germinated planospore and portion of a young thallus with a rare well-defined primary spindle organ.

Fig. 70. Portion of the rhizomycelium with intercalary and terminal young, proliferating, and dehiscing zoosporangia; well-defined intercalary spindle organs lacking.

Figs. 71-73. Elongate, ovoid and spherical resting spores with numerous refractive globules and smooth yellowish brown walls.

Fig. 74. Germinated resting spore with a septate zoosprangium.

Fig. 75. Germinated resting spore with a sessile zoosporangium.

Fig. 76. Germinated resting spore which has functioned directly as a sporangium.

Figs. 77-83. *Nowakowskiella macrospora* Karling. (Karling, 1945.)

Fig. 77. Spherical, 10-12 μ diam., planospore with a 3-5 μ diam., hyaline refractive globule and several smaller ones at the posterior end.

Fig. 78. Amoeboid planospore.

Fig. 79. Endooperculate, apophysate zoosporangium and a portion of the rhizomycelium.

Fig. 80. Variations in the shapes of endoopercula.

Figs. 81, 82. Ovoid and spherical resting spores with large and small refractive globules and faintly yellowish brown walls.

Fig. 83. Germinated resting spore.

Figs. 84-87. *Nowakowskiella delica* Whiffen. (Whiffen, 1943.)

Fig. 84. Spherical, 5.7-7.5 μ diam., and amoeboid planospores with a minute hyaline refractive globule and a 30 μ long flagellum.

Fig. 85. Mature apophysate zoosporangium.

Fig. 86. Portion of an intramatrical rhizomycelium; host tissue omitted.

Fig. 87. Mature resting spore with numerous refractive globules and a smooth hyaline wall.

Figs. 88-95. *Nowakowskiella crassa* Karling. (Karling, 1949.)

Fig. 88. Portion of the very coarse rhizomycelium with large non-septate intercalary spindle organs.

Figs. 89, 91. Amoeboid planospores.

Fig. 90. Spherical, 4.5-5 μ diam., planospore with a small hyaline refractive globule and a 24-26 μ long flagellum.

Fig. 92. Mature terminal zoosporangium.

Fig. 93. Discharge of planospores from an intercalary zoosporangium.

Figs. 94, 95. Ovoid and spherical resting spores with hyaline, smooth walls and filled with small hyaline refractive globules.

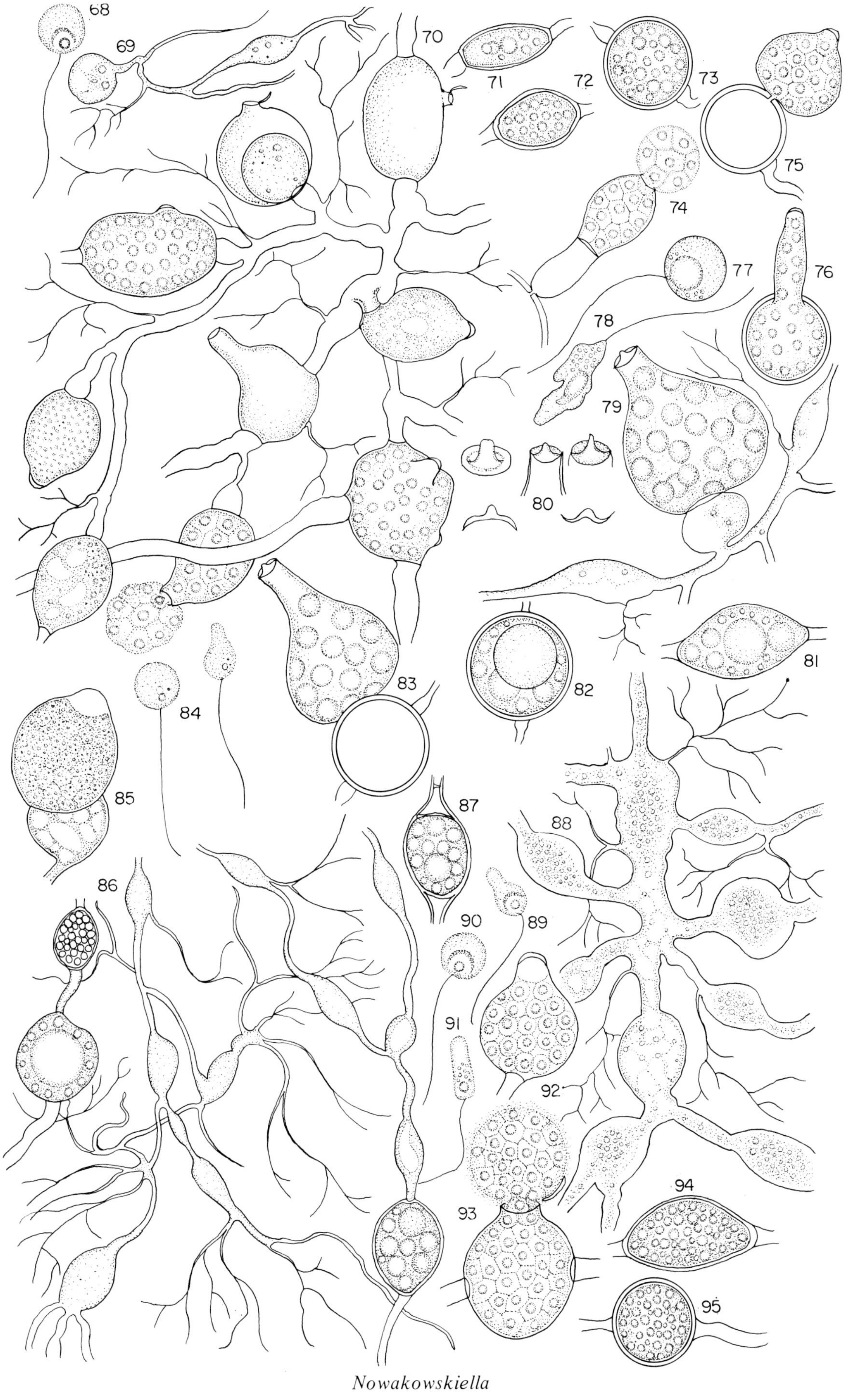

Nowakowskiella

enveloped by a layer of matrix, and after a short while the zoospores become motile and dart away. Very often they become amoeboid (figs. 6, 13, 78, 89, 108) and creep about. Occasionally under certain conditions they may swarm actively for a short while in an enveloping vesicle before dispersing. The planospores in most species are spherical in shape and vary in diameter from 3–3.8 μ in *N. sculptura* (fig. 26) to 10–12 μ in *N. macrospora* (fig. 77). In all known species except *N. granulata* they contain a hyaline refractive globule which may be as small as 0.8–1.2 μ in diameter in *N. profusa* and *N. sculptura* and as large as 3–5 μ in *N. macrospora*. In *N. granulata* a large refractive globule is lacking and is replaced by numerous golden-brown granules of fairly uniform size (figs. 53, 54). The flagellum is posteriorly attached and varies in length from 40 μ to 42 in *N. ramosa* and *N. macrospora*, respectively, to 12–14 μ in *N. multispora*. After discharge of the planospores the zoosporangia may proliferate once to several times (figs. 1, 11).

The ultrastructure of mature zoosporangia and incipient planospores was studied by Chambers *et al*, as noted in the Introduction, and they found in addition to the other structures shown in pl. 1, fig. 40 a conspicuous fibrous body immediately adjacent to the refractive globule. They interpreted this body as possibly representing a primitive photoreceptor organelle, and Fuller and Reichel regarded it as the same as the "rumposome" which they described in planospores of *Monoblepharella*.

Resting spores are known in all species except *N. atkinsii*, and the wall is yellowish to dark brown. In *N. sculptura*, the wall is usually sculptured or ridged (figs. 35–37), but sometimes it is smooth or bears blunt pegs (fig. 34). In the majority of species the spores usually contain a large hyaline refractive globule surrounded by smaller ones (figs. 7, 59, 65, 66), but in *N. delica* (fig. 86), *N. macrospora* (figs. 94, 95) and *N. pitcairensis* (figs. 111–113) the content is coarsely globular in appearance. In germinating the resting spores may function as a sporangium with a broad neck (fig. 25) or as a prosporangium (figs. 8, 24, 67, 75, 83, 114). In *N. hemisphaerospora* Shanor the resulting zoosporangium may sometimes occur partly within and partly on the outside of the germinated spore (fig. 52).

No conclusive evidence of sexuality has been observed in *Nowakowskiella*. Butler (1907) observed fusion or anastomosis of a branch with a dictyosporus terminal swelling bearing a resting spore (fig. 20) in *N. ramosa* and stated that such branches simulated antheridia. He did not, however, believe that the presence of such structures are indicative of sexual reproduction. Shanor (1942) described the separation of the contents of incipient resting spores (figs. 42, 43) into 2 or more parts (fig. 47) after which a cross septum begins to develop between them (fig. 44). One portion of the content fuses or migrates into the other (fig. 45) after which the septum is completed. The fused content becomes enveloped by a wall and develops into a resting spore (fig. 46) which is accompanied by an empty cell. Shanor believed that this migrating or fusion represents sexual fusion of the contents, and he interpreted the empty cell as a male. However, Karling (1967) found that a large number of spores without accompanying empty cells (figs. 48, 49) may develop in so-called container cells. Dogma (1969) also found as many as 13 resting spores in a container cell. It is accordingly questionable that Shanor's observations relate to sexual reproduction, and the fusions reported by him are interpreted here as contraction of the content of incipient resting-spore container cell and its investment by a thick wall. Such types of resting spore development occur commonly in *Neprochytrium stellatum*, *Sparrowia parasitica*, and often in *Karlingia rosea*. Whether or not the anastomosis of the tenuous filaments (figs. 2, 9) has any sexual significance is not evident at present, but Alexopoulos (1962) suggested that the fusion of cells "may indicate that at least a parasexual cycle is in operation."

Plate 108

Figs. 96-100. *Nowakowskiella atkinsii* Sparrow. Figs. 96, 97 after Sparrow, 1950; figs. 98-100 after Karling, 1967.

Fig. 96. Portion of the rhizomycelium with 2 zoosporangia bearing setae.

Fig. 97. Ovoid, 3 × 5 μ, planospore with a single hyaline refractive globule.

Fig. 98. Spindle organ bearing setae, 2 of which have branched at the tips to become rhizoids.

Fig. 99. Zoosporangium bearing sharp pegs.

Fig. 100. Occasional, smooth zoosporangium.

Figs. 101-106. *Nowakowskiella multispora* Karling. (Karling, 1964.)

Figs. 101. Portion of a dense rhizomycelium growing at the edge of and within a piece of cellophane showing the great abundance of resting spores.

Fig. 102. Endooperculate zoosporangium with a plug of matrix above the operculum.

Fig. 103. Discharge of planospores.

Fig. 104. Spherical, 3-3.9 μ diam., and amoeboid planospores with a small hyaline globule and a 12-14 μ long flagellum.

Fig. 105. Germinated resting spore which has functioned as a prosporangium.

Fig. 106. Portion of germinated resting spore with a septate zoosporangium.

Figs. 107-114. *Nowakowskiella pitcairnensis* Karling. (Karling, 1968.)

Figs. 107, 108. Spherical, 3-3.2 μ diam., and amoeboid planospores with a hyaline refractive globule and a 22-25 μ long flagellum.

Fig. 109. Germination of a planospore.

Fig. 110. Portion of a *Cladochytrium*-like delicate rhizomycelium with 4 zoosporangia growing at the edge of boiled hemp seed bits; intercalary enlargements septate or continues, often catenulate.

Figs. 11-113. Mature constricted, spherical and ovoid resting spores with smooth, light-brown walls, and filled with coarse granules.

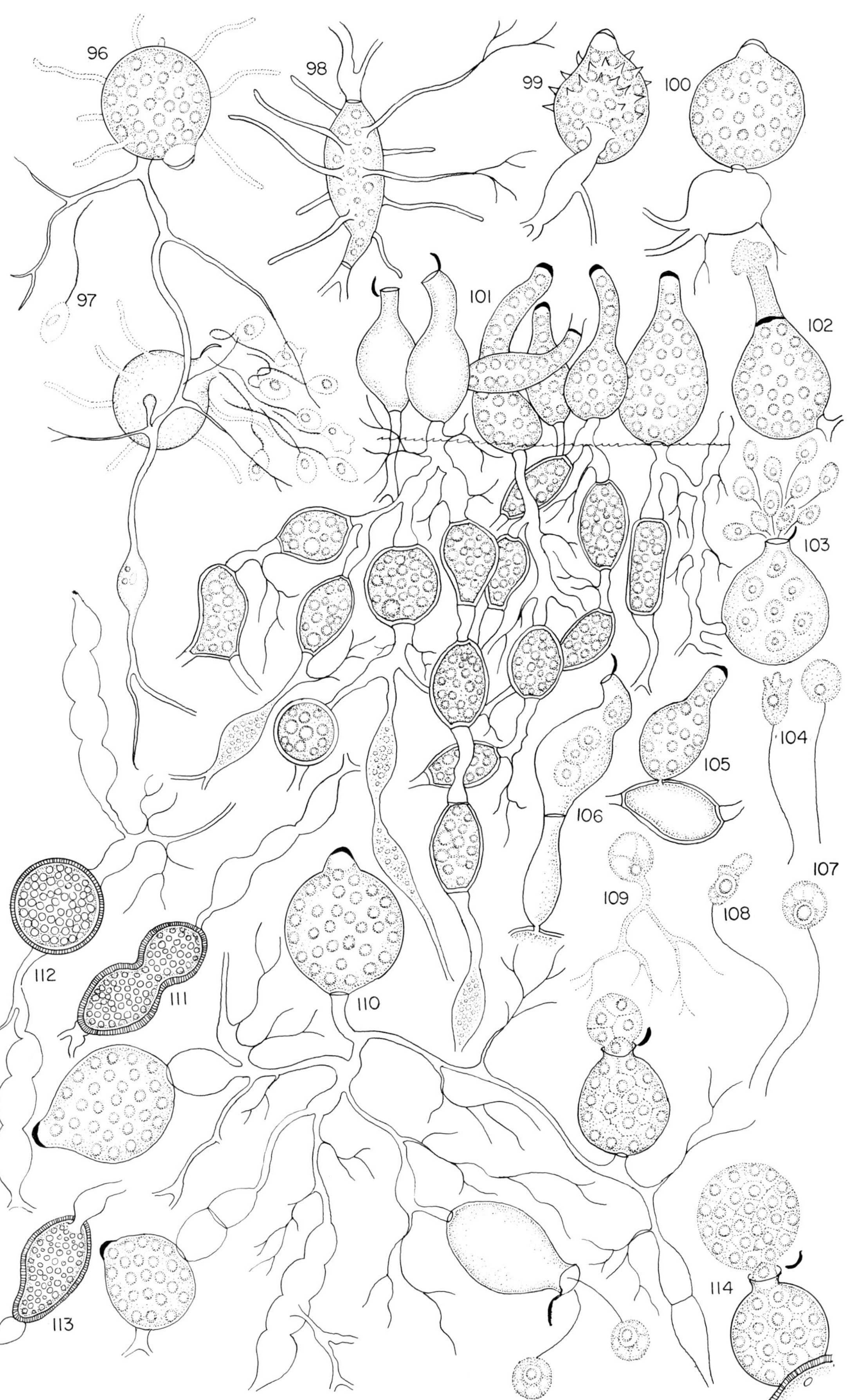

Nowakowskiella

AMOEBOCHYTRIUM Zopf

Nova Acta Acad. Leop.-Carol. 47:181, 1884.

Plate 109

This polycentric genus includes one species, *Amoebochytrium rhizidioides*, which Zopf found to be growing as a saprophyte in the gelatinous sheath of *Chaetophora elegans*, and later Harder (1948) reported it from soil. So far as it is known it is characterized by aflagellate amoebospores, and a rhizomycelium consisting of branched tenuous portions, intercalary spindle-shaped swellings and terminal or intercalary zoosporangia. Resting spores are unknown.

The amoebospores contain an unusually large yellowish refractive globule, and undergo marked changes in shape as they creep around in the slime around *Chaetophora* (figs. 1–4). After a while they round up, encyst, and germinate by the formation of one or two germ tubes (figs. 6–8). This tube elongates, branches, and develops into the rhizomycelium (fig. 15). Germination apparently occurs commonly *in situ* (figs. 14, 15). The zoosporangia usually develop from the intercalary swellings as the protoplasm accumulates in the latter, but sometimes they develop directly by enlargement of the amoebospore body. The exit canal of the mature intercalary zoosporangium appears to develop by increase in diameter of the tenuous portion of the rhizomycelium at the apex of the zoosporangium (fig. 9), and after disarticulation a cross septum (endooperculum?) becomes visible in the canal (figs. 10–13). At dehiscence a pore is formed in this septum, and the amoebospores creep out. Zopf believed it possible that there might be flagella, but after repeated observations he failed to observe them. In the event a posteriorly directly flagellum is present and the cross septum is pushed out as an endooperculum *A. rhizidioides* should be classified as a species of *Nowakowskiella*.

Zopf regarded it as intermediate between *Rhizidium* and *Cladochytrium*, but subsequent workers, including Fischer (1892), Schroeter (1893), Sparrow (1943, 1960) and Whiffen (1944), have classified it in the family *Cladochytriaceae*.

POLYCHYTRIUM Ajello

Mycologia 34:442, 1942.
Plate 110

At present this inoperculate genus includes the type species, *P. aggregatum* and another questionable unidentified one, *Polychytrium* sp. The type species may occur on decayed vegetation in nature, but it has a predilection for chitin and can be cultured on a vitamin free synthetic medium. It is similar to species of *Cladochytrium* by its inoperculate zoosporangia but differs by its fairly uniform and almost isodiametric rhizomycelium which lacks intercalary spindle-shaped swellings and bears tuberculate and smooth zoosporangia in clusters (fig. 1). Also, its nuclei are fairly evenly distributed in the rhizomycelium (fig. 2) and not confined to intercalary swellings as in *Cladochytrium replicatum*. The living planospores lack a refractive globule but may contain a lunate body (fig. 3), and in fixed and stained planospores (fig. 4) a nuclear cap is quite conspicuous and envelopes the upper half of the nucleus. In germination they develop a broad blunt germ tube (fig. 5), and the larger portion of the planospore body becomes thick-walled and persists as an appendage (fig. 6) on the developing thallus. In the meantime, the germ tube broadens markedly (fig. 6, 7), and branches dichotomously. Then by continued elongation of the branches and more or less regularly repeated dichotomy (fig. 11) an intricate rhizomycelial network of variously sized vegetative branches are formed which bear rhizoids and terminal or intercalary zoosporangia (figs. 1, 9). Lateral, intercalary and terminal resting spores (fig. 12) were found in a chitinophilic polycentric species by Karling (1949) which he thought might belong to *P. aggregatum*, but inasmuch as no zoosporangia were present

PLATE 109

Figs. 1-18. *Amoebochytrium rhizidioides* Zopf from the gelatinous matrix of *Chaetophora elegans*. (Zopf, 1884.)

Figs. 1-4. Changes in shape of an aflagellate amoebospore and its large yellowish refractive body in the matrix surrounding the algae.

Figs. 5-8. Encysted amoebospore and successive germination stages.

Fig. 9. Intercalary zoosporangium; refractive material coalescing to form the definitive globules of the amoebospores.

Fig. 10. Zoosporangium with 3 amoebospores as indicated by the refractive globules; cross septum (?) in exit canal.

Fig. 11. Similar but apophysate zoosporangium with a short neck and 3 amoebospores; empty fusiform swelling in the attached filament.

Fig. 12. Zoosporangium with approximately 18 amoebospores, as indicated by refractive globules.

Fig. 13. Mature zoosporangium with 7 amoebospores and a long exit canal; septum still present in the canal but with a plug beneath it.

Fig. 14. Zoosporangium with 2 amoebospores, one of which has germinated *in situ*.

Fig. 15. Similar zoosporangium with several *in situ* germinated amoebospores; only two of the germ tubes are drawn to their full lengths to show the branching and presence of intercalary swellings.

Figs. 16-18. Empty zoosporangia with exit canals of various lengths.

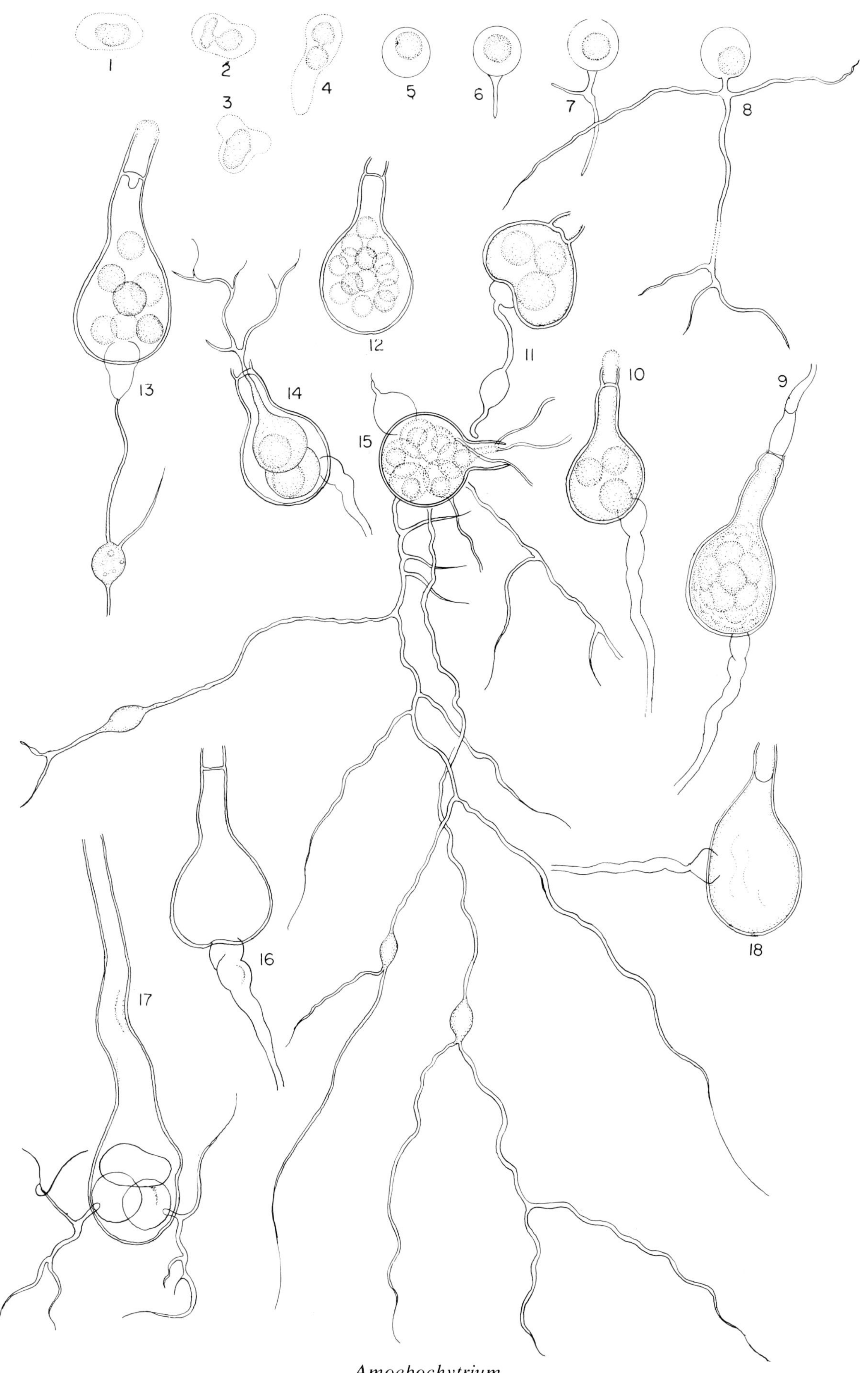

Amoebochytrium

and the thallus possessed intercalary swellings as in *Cladochytrium* it is not certain that it relates to Ajello's species.

In *Polychytrium* sp. Sparrow (1965), which develops on cellophane and vegetable debris, the rhizomycelium is isodiametric (fig. 13) without intercalary swellings and devoid of rhizoids save near the base in the substratum. At irregular intervals stout septa of refractive material occur in the rhizomycelium, and the planospores possess a conspicuous colorless refractive globule (figs. 14, 15). No resting spores have been observed in this species. It is a doubtful species which probably will be found to belong to another or new genus.

Polychytrium has several characteristics in common with species of the Blastocladiales, but Ajello (1948) concluded that the monopolar type of germination and the absence of a side body in the planospores indicate a closer affinity with the Chytridiales.

SEPTOCHYTRIUM Berdan

Amer. J. Bot. 26:461, 1939.

Plates 111, 112

This operculate genus is quite similar to *Nowakowskiella*, and so far as it is known differs only by the presence of septa, trabeculae, plugs, and constrictions in the tenuous filaments of most species. However, it remains to be seen whether or not the presence or absence of such structures are generic characters or distinctions, because trabeculae have been observed in *Cladochytrium crassum* (Hillegas, 1941) and *Nowakowskiella hemisphaerospora* (Dogma, 1969), also. *Septochytrium* is reported to include 4 saprophytic species which occur in soil and water and may be isolated readily on cellulosic substrata. However, *S. macrosporum* Karling was assigned temporarily to this genus and may prove to be a species of *Nowakowskiella* because it lacks septa or trabeculae. Also, the fungus described by Willoughby (1964) as *S. marylandicum* Karling may be a different species or variety because its planospores are much larger than those described by Karling (1951).

The early stages of development of the rhizomycelium (figs. 3, 4, 23, 24, 31) are similar to those of *Cladochytrium* and *Nowakowskiella* in most species, but in *S. marylandicum* Willoughby reported that a part of the planospores develops into a primary zoosporangium and rhizoidal axes, while the remainder of it becomes thick-walled and persists as a cyst (figs. 32–34). Tenuous filaments with broadly fusiform swellings then develop from the primary zoosporangium (fig. 33) which normally does not dehisce but serves as a "vegetative center for the developing thallus." Occasionally in *S. macrosporum* the planospores gives rise to a monocentric thallus and develops directly into a zoosporangium.

At maturity the rhizomycelium is usually coarse and very extensive, and parts of it usually grow out into the surrounding water where the tenuous filaments may become curved and coiled (fig. 20). In *S. plurilobulum* Johanson they may be irregular in contour (fig. 13), while in *S. macrosporum* they are more even, and attain a diameter of 5 to 15 μ. In *S. marylandicum* anastomosis of the filaments may occur (fig. 35A). The septa may extend partly (fig. 37) or fully across the filaments or be trabeculate and plug-like. The intercalary swellings or spindle organs may be broadly or narrowly fusiform (figs. 5, 19, 24, 31), or irregular in outline (fig. 35) and non-septate. Rhizoids occur at irregular intervals along the length of the filaments, at their extremities, or from the periphery of the zoosporangia and spindle organs.

As in *Cladochytrium* and *Nowakowskiella* the zoosporangia develop terminally or intercalarly (figs. 13, 35) and vary markedly in size and shape, particularly in *S. variabile* Berdan. In *S. macrosporum* the spherical zoosporangia may attain a diameter of 280 μ, and in *S. variabile* (fig. 5) and *S. plurilobulum* (fig. 13) the primary zoosporangium may be larger than the secondary ones. At maturity they develop a low or prominent exit papilla (figs. 6, 7, 19) in most species, but in *S. marylandicum* long exit canals are formed (figs. 35, 42) for the discharge of the planospores.

PLATE 110

Figs. 1-10. *Polychytrium aggregatum* Ajello. Fig. 12 *P. aggregatum*?; figs. 1, 8, 12 after Karling, 1949; figs. 2-7 after Ajello, 1948.

Fig. 1. Portion of rhizomycelium with rhizoids, tuberculate, and smooth walled zoosporangia; proliferating sporangium at A; discharge of planospores at B. Drawn from living material.

Fig. 2. Portion of a filament showing distribution of nuclei.

Fig. 3. Living planospore with opaque lunate body.

Fig. 4. Fixed and stained planospore with conspicuous nuclear cap.

Fig. 5. Germination of planospore on chitin with a broad germ tube.

Fig. 6. Portion of young thallus branching dichotomously with persistent thick-walled planospore cyst.

Figs. 7, 11. *In toto* stained later stages in development of thallus on chitin showing distribution of nuclei and dichotomous branching of rhizomycelium.

Fig. 8. Intranuclear mitosis with 5 chromosomes.

Fig. 9. Young lateral and terminal zoosporangia.

Fig. 10. Cleavage segment in zoosporangium showing dense nuclear cap.

Fig. 12. Possibly, resting spores of *Polychytrium aggregatum*.

Figs. 13-15. *Polychytrium* sp. Sparrow. (Sparrow, 1965.)

Fig. 13. Portion of thallus with zoosporangia; rhizoids lacking.

Figs. 14, 15. Discharge of planospores.

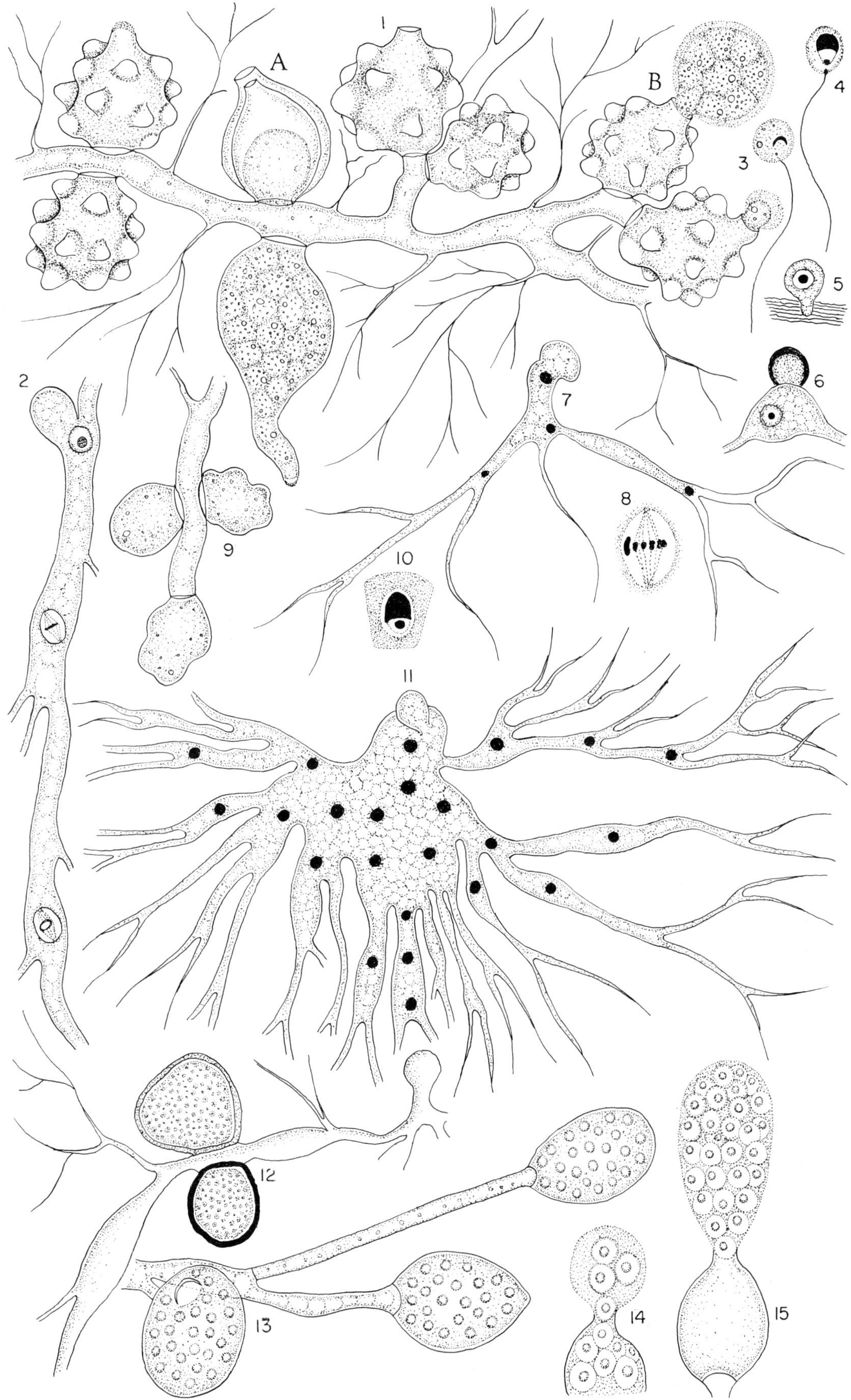

Polychytrium

These canals may become curved and slightly irregular in contour, branch once to several times, and bear an endooperculum (figs. 35, 38, 41, 42). All other species are exooperculate. The operculum is pushed off or out at dehiscence, and the first planospores emerge in a globular mass (figs. 7, 35B) which is usually enveloped by a thin layer of matrix. These disperse (fig. 40) after a short while and the remaining planospores emerge singly. In *S. variabile* they contain a single hyaline refractive globule (fig. 1), but in *S. macrosporum* several minute ones are present in addition to a larger one (figs. 18, 19, 21). In *S. plurilobulum* (figs. 11) and *S. marylandicum* (figs. 28–30) the large globule is replaced by numerous hyaline refractive granules which gives the planospores a greyish-granular appearance. In *S. macrosporum* the planospores are among the largest of all known chytrids and may vary from 11–13 μ in diameter, while in *S. marylandicum* they may be only 3.8 to 4.8 μ.

Resting spores have been reported in all species, and these develop at the ends of branches (figs. 13, 14) or from intercalary swellings (figs. 25, 26). Occasionally, the contents of a large swelling contracts and becomes invested with a wall so that the spore fills only a part of the swelling (fig. 26). In *S. plurilobulum* they are deeply lobed (figs. 13, 15), while in *S. macrosporum* they may be smooth-walled (fig. 25), slightly encrusted (fig. 26), or bear numerous pegs or blunt hypha-like branches (fig. 27). In this species they may reach a diameter of 115 μ with a wall 4 to 6 μ thick. In the process of germination the spores function as prosporangia in *S. plurilobulum* (figs. 16, 17) and *S. variabile* (figs. 9, 10). In the latter species the incipient zoosporangium may become greatly elongate, tubular, cylindrical, oiled and curved (fig. 10) with a septum delimiting the zoosporic portion at the tip.

So far no evidence of sexuality in the development of the resting spores has been reported in *Septochytrium*.

MEGACHYTRIUM Sparrow

Occ. Papers Boston Soc. Nat. History 8:9, 1931;
Amer. J. Bot. 20:73, 1933.

Plate 113

This monotypic operculate genus is characterized by an intra-extramatrical coarse, 5–7 μ diam., extensive and profusely branched, occasionally septate, nonrhizoidal undulating mycelium which bears intercalary and terminal zoosporangia and resting spores. The type species, *M. westonii* Sparrow, parasitizes *Elodea canadensis* and causes pronounced discoloration and disintegration of the leaves. Its thallus, like that of *Coenomyces*, is typically mycelial (figs. 3–6) and does not bear rhizoids at irregular intervals along its surface as do other genera of the Cladochytriaceae.

In germinating the spherical planospore forms a blunt-ended germ tube (fig. 2) which penetrates the host or germinates on its surface and ultimately develops into a profusely branched, undulating and extensive mycelium which becomes intra- and extramatrical. Its terminal branches are highly refractive and often anastomose (fig. 3). Terminal and intercalary swellings develop at intervals along the length of the mycelium, and these eventually become zoosporangia or resting spores (figs. 5, 6). The former may be spherical or clavate but more often irregular in shape (fig. 1), occasionally apophysate, and develop an operculate papilla or short neck for the discharge of the planospores. The resting spores are smooth, hyaline, ovoid with truncated ends, germinate after a short or no resting period, and function as prosporangia (fig. 6) in this process.

Plate 111

Figs. 1-10. *Septochytrium variabile* Berdan. Figs. 1-5, 8-10 after Berdan, 1939, 1942; figs. 6, 7 drawn from living material.

Fig. 1. Spherical, 4-6 μ diam., planospore with a hyaline refractive globule; flagellum 30-45 μ long.

Figs. 2, 3. Early germination stages of planospores.

Fig. 4. Later stage, 2 swellings developing in tenuous filaments.

Fig. 5. Portion of a rhizomycelium with a primary and 3 secondary zoosporangia, broadly fusiform non-septate swellings on the septate or trabeculate filaments and rhizoids.

Fig. 6. Small, mature zoosporangium with a low exit papilla under which is a mass of translucent matrix.

Fig. 7. Discharge of planospores which are enveloped by a thin layer of matrix.

Fig. 8. Mature pale-amber to dark-brown resting spore with a layered wall, and a large brownish globule surrounded by smaller ones.

Fig. 9. Germinated resting spore with a sessile zoosporangium.

Fig. 10. Germinated resting spore with a stalked, curved and septate zoosporangium; only part of resting spore shown

Figs. 11-17. *Septochytrium plurilobulum* Johanson. (Johanson, 1943.)

Fig. 11. Ovoid, 7-8 μ diam., planospores with several refractive globules.

Fig. 12. Germinated planospore.

Fig. 13. Portion of a septate rhizomycelium bearing zoosporangia, irregular swellings, rhizoids, and deeply lobed resting spores.

Fig. 14. Early stage in resting spore development at the ends of branches.

Fig. 15. Large, deeply lobed, greenish grey resting spore with a dark smooth wall and containing several large refractive globules.

Figs. 16, 17. Germination of resting spores.

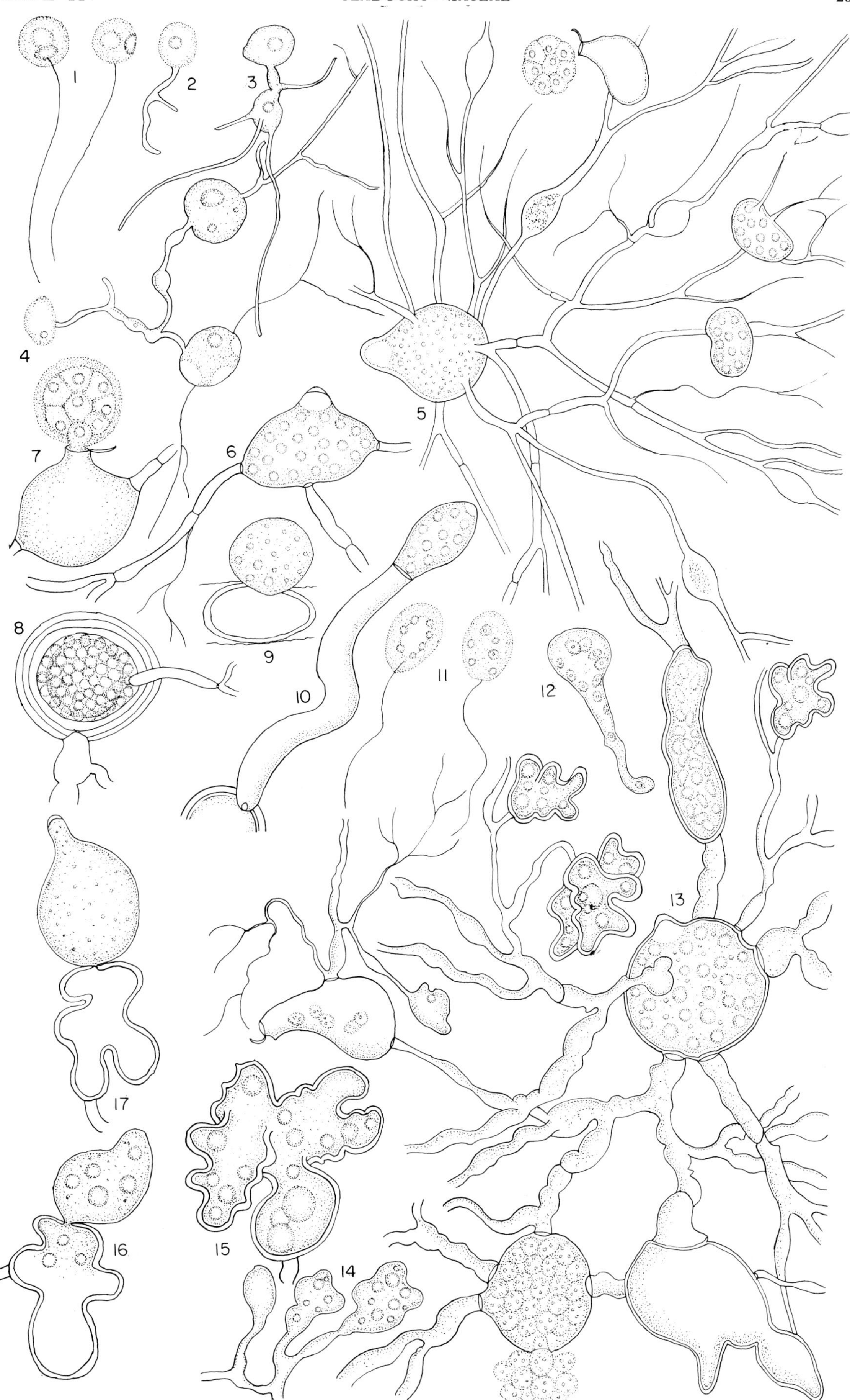

Septochytrium

Sparrow (1943, 1960) placed this genus in the operculate series and the family Megachytriaceae, but Whiffen (1944) classified it in the Cladochytriaceae, a classification which is followed tentatively here.

COENOMYCES Deckenbach

Scripta Bot. Horti. Univ. Imper. Petropol. 19:115, 1902-1903; Flora 92:265, 1903.

This inoperculate monotypic genus is characterized by an intra- and extramatrical slender, septate mycelium without rhizoids, intercalary varicose swellings, zoosporangia, and posteriorly uniflagellate planospores. Accordingly, its mycelium is like that of the higher fungi, but is zoosporangia and planospores are like those of the Chytridiales or Blastocladiales. *Coenomyces consuens* Deckenback grows in the gelatinous surrounding envelope and beneath the sheath of filaments of the blue green algae *Calothrix, Rivularia* and *Tildenia* (fig. 1), and since its discovery by Deckenbach it has been reported in Denmark by Petersen (1906), the U. S. A. by Sparrow (1943), and Albania by Komárek (1958).

After encysting the planospores (figs. 2, 3) germinate in a mono- or bipolar manner (figs. 4, 5), and the germ tubes branch to form a delicate septate mycelium which grows extramatricully in the surrounding gelatinous envelope or penetrates the sheath of the filaments and develops intramatrically (fig. 1). Varicose intercalary swellings develop at irregular intervals along its length as the mycelium spreads, so that several algal filaments may be held together in meshwork of mycelium (fig. 1). As the zoosporangia mature their content becomes golden orange in color. Usually, the zoosporangia develop very long narrow exit canals (figs. 10, 12, 13) through which the planospores emerge after the tips have deliquesced. The species observed by Komárek is similar to that described by Deckenbach except for its larger zoosporangia. Resting spores have not been observed.

The relationships and classification of *Coenomyces* are uncertain because its life cycle is not fully known, and it appears to combine characteristics of the higher fungi as well as the Chytridiales or Balstocladiales. Deckenback, accordingly, established a new class, Coenomycetes, intermediate between the Phycomycetes and Eumycetes to accommodate it and Sorokin's *Aphanistis*. Sparrow (1943, 1960) classified it in the Cladochytriaceae next to *Polychytrium*, but Cejp (1957) and Komárek included it in the family Catenariaceae of the Blastocladiales. It is classified here only provisionally in relation to the Cladochytriaceae.

REFERENCES TO THE CLADOCHYTRIACEAE

Ajello, L. 1942. *Polychytrium*, a new cladochytriaceous genus. Mycologia 34:442–451, 16 figs.

———. 1948a. A cytological and nutritional study of *Polychytrium aggregatum* I. Cytology. Amer. J. Bot. 3511-2, 49 figs.

PLATE 112

Figs. 18-27. *Septochytrium macrosporum* Karling. (Karling, 1942.)

Figs. 18, 19. Portions of the rhizomycelium with intercalary and lateral zoosporangia, coarse tenuous filaments, fusiform intercalary swellings, and rhizoids.

Fig. 20. Coiled portion of the rhizomycelium growing in water at the edge of a piece of cellophane.

Fig. 21. Spherical, 11-13 μ diam., planospore with one larger and 5 minute refractive, hyaline globules.

Fig. 22. Fixed and stained planospore with a nuclear cap (?), nucleus, and nucleolus.

Fig. 23. Germination of a planospore with 1 germ tube.

Fig. 24. Germinated planospore with 3 germ tubes; intercalary swellings have developed in 2 of the tubes.

Fig. 25. Spherical, smooth, light-brown resting spore with numerous refractive globules.

Fig. 26. Similar amber to yellow-brown resting spore which has developed in the upper portion of an intercalary swelling; wall 6 μ thick with the outer surface slightly encrusted.

Fig. 27. Resting spore bearing blunt pegs and short, coarse, hypha-like branches.

Figs. 28-45. *Septochytrium marylandicum* Karling. Figs. 29-31, 35-40, 42 after Karling, 1951; figs. 28, 32-34, 41, 43-45 after Willoughby, 1964.

Fig. 28. Spherical, 6 μ diam., planospore with several minute refractive globules and a 32 μ long flagellum.

Fig. 29. Spherical, 3.8-4.7 μ diam., planospore with numerous refractive globules and a 24-27 μ long flagellum.

Fig. 30. Amoeboid oblong planospore.

Fig. 31. Germinated planospore with 3 intercalary swellings in the branched germ tube.

Fig. 32. "Young thallus with only the primary sporangium developed and showing a thickened persistent planospore cyst."

Fig. 33. "Extensive growth of the primary sporangium, but with the latter still recognizable."

Fig. 34. Primary zoosporangium with thick-walled planospore cyst at the apex.

Fig. 35. Portion of the rhizomycelium with 3 intercalary endooperculate zoosporangia, tenuous filaments, rhizoids, and irregular and fusiform intercalary swellings; anastomosis of filaments at A.

Fig. 36. Portion of a filament with a septum.

Fig. 37. Intercalary swelling with a partial septum near one end.

Fig. 38. Apical portion of exit canal with an endooperculum.

Fig. 39. Variations in shapes of endoopercula.

Fig. 40. Discharge and dispersal of planospores.

Fig. 41. Zoosporangium with a short neck.

Fig. 42. Intercalary zoosporangium with a long curved neck.

Figs. 42-45. Resting spores with a thick yellow wall.

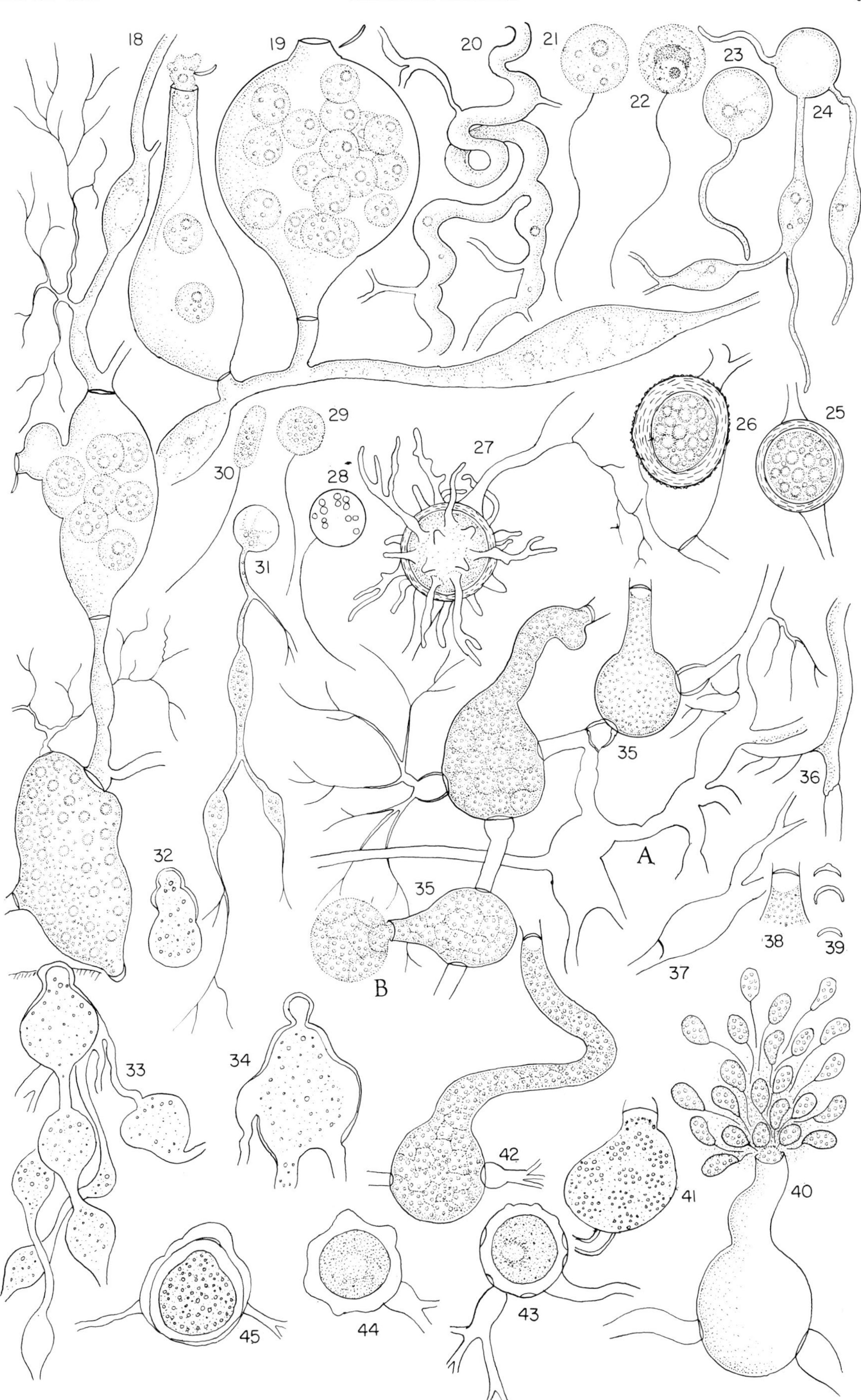

Septochytrium

_______. 1948b. A cytological and nutritional study of *Polychytrium aggregatum*. II. Nutrition. Ibid. 35:135-140, 1 fig.

Alexoupolos, C. J. 1962. Introductory mycology. 2nd. ed., IX-613, 194 figs. Wiley and Sons, Inc., New York.

Berdan, H. 1939. Two new genera of operculate chytrids. Amer., J. Bot. 26:459-463, 2 figs.

_______. 1941. A developmental study of three saprophytic chytrids. I. *Cladochytrium hyalinum* sp. nov. Ibid. 28:422-438, 84 figs.

_______. 1942. A developmental study of three saprophytic chytrids. III. *Septochytrium variabile* Berdan. Ibid. 29:260-270, 52 figs.

Butler, E. J. 1907. An account of the genus *Pythium* and some Chytridiaceae. Mem. Dept. Agric. India, Bot. ser. 1:1-160, 10 pls.

Cejp, K. 1957. Houby 1:87-89, Cesk. Akad. Ved.

Deckenbach, C. von 1902-1903. *Coenomyces consuens*, n. g., n. sp. Scripta Bot. Horti. Univ. Petropol. 19:1-42, 2 pls.

_______. 1903. *Coenomyces consuens*, n.g., n. sp. Flora 92:253-283, pls. 6, 7.

Dogma, I. J. 1969. Observations on some cellulosic chytridiaceous fungi. Arch. f. Mikrobiol. 66: 203-219, 70 figs.

Harder, R. 1948. Uber das Vorkommen niederer Phycomyceten in duetschen Boden. Nachr. Akad. Wiss. Göttingen. Math. - Physik. Kl. 1048:5-7.

Hillegas, A. B. 1941. Observations on a new species of *Cladochytrium*: Mycologia 33: 618-632, 40 figs.

Jaczewaski, A. A. 1931. Operedelitel gribov. Sovershennyi griby (Diploidyne stadii) Determination of fungi. Perfect fungi, diploid species). Moskova Leningrad, Gosudarstvennoe izdatel'stovo sel' skokloziaiste-vennoi i kolkhoznokoopera-tivnoi literaury. I. Fikomitsety (Phycomycetes). 294 ppl, 329 figs.

Johanson, A. E. 1943. *Septochytrium plurilobulum* sp. Nov. Amer. J. Bot. 30:619-622, 1 fig.

Karling, J. S. 1931. Studies in the Chytridiales. VI. The occurrence and life history of a new species of *Cladochytrium* in cells of *Eriocaulon septangulare*. Amer. J. Bot. 18:526-557, pls. 42-44.

_______. 1932. Studies in the Chytridiales. VII. The organization of the chytrid thallus. Ibid. 41-74, 138 figs.

_______. 1935. A further study of *Cladochytrium replicatum* with reference to its distribution, host range, and culture on artificial media. Ibid. 22:439-452, 29 figs.

_______. 1937. The cytology of the Chytridiales with special reference to *Cladochytrium replicatum*. Mem. Torrey Bot. Club 19:3-92, 2 text-figs., 6 ls.

_______. 1939. A new fungus with anteriorly uniciliate zoospores: *Hyphochytrium catenoides*. Amer. J. Bot. 26:512-519, 18 figs.

_______. 1942. A new chytrid with giant zoospores: *Septochytrium macrosporum* sp. nov. Ibid. 29: 616-622, 15 figs.

_______. 1944. Brazilian chytrids. I. Species of *Nowakowskiella*. Bull. Torrey Bot. Club 71:374-389, 69 figs.

_______. 1945. Brazilian chytrids. V. *Nowakowskiella macrospora*. n. sp., and other polycentric species. Amer. J. Bot. 32:29-35, 51 figs.

_______. 1949. *Nowakowskiella crassa* sp. nov., *Cladochytrium aureum* sp. nov., and other polycentric chytrids from Maryland. Bull. Torrey Bot. Club. 76:294-301, 17 figs.

_______. 1951. *Cladochytrium setigerium* sp. nov. and *Septochytrium marylandicum* sp. nov. from Maryland. Ibid. 78:38-43, 30 figs.

_______. 1961. *Nowakowskiella sculptura* sp. nov. Trans. Brit. Mycol. Soc., 44:453-457, 23 figs.

_______. 1964. Indian chytrids. IV. *Nowakowskiella multispora* sp. nov. and other polycentric species. Sydowia 17:314-319, 8 figs.

_______. 1967. Some zoosporic fungi of New Zealand. VIII. Cladochytriaceae and Physodermataceae. Ibid. 20:129-136, pl. 25.

_______. 1968. Zoosporic fungi of Oceania. V. Cladochytriaceae, Catenariaceae and Blastocladiaceae. Nova Hedwigia 25:91-201, pls. 21-24.

Komárek, J. 1958. *Coenomyces consuens* Deckenb. V. Albánii Ceská Mykol. 12:110-113, 5 figs.

Nowakowski, L. 1876. Beitrag zur Kenntniss der Chytridiaceen. *In* Cohn Beitr. Biol. Pfanzen 2:73-100, pls. 4-6.

Petersen, H. E. 1906. Om Forekomsten of *Coenomyces consuens* i Danmark. Bot Tidsskrift. 27:XXII-XXIII.

Richards, M. 1956. Some inoperculate chytrids from South Wales. Trans. Brit. Mycol. Soc. 39:261-266, 5 figs., pls. 6, 7.

Plate 113

Figs. 1-6. *Megachytrium westonii* Sparrow in and on *Elodea canadensis*, host tissue omitted. (Sparrow, 1933.)

Fig. 1. Intercalary zoosporangium discharging spherical, 5 μ diam., planospores with a hyaline refractive globule; operculum lying free.

Fig. 2. Germination of a planospore with a branched blunt-ended germ tube.

Fig. 3. Enlarged portion of a thallus showing anastomosis of terminal refractive branches.

Fig. 4. Portion of a thallus with a terminal zoosporangium at the end of a short branch; rhizoids lacking.

Fig. 5. Portion of a profuse mycelial growth with intercalary zoosporangia, swellings, and resting spore; rhizoids lacking.

Fig. 6. Portion of a mycelium with a smooth, hyaline resting spore, 2 zoosporangia, and 2 germinated resting spores which have functioned as prosporangia.

Megachytrium

Roberts, J. M. 1948. Developmental studies of two species of *Nowakowskiella* Schroeter: *N. ramosa* Butler and *N. profusa* Karling. Mycologia 40: 127-157., 2 figs.

Schroeter, J. 1893. Phycomycetes. *In* Engler u. Prantl. Natürl. Pflanzenf. 1:63-141.

Shanor, L. 1942. A new fungus belonging to the Cladochytriaceae. Amer. J. Bot. 29:174-179, 38 figs.

Shen, S. C., and W. N. Siang. 1948. Studies in the aquatic Phycomycetes of China. Sci. Repts. Nat. Tsing University, ser. B:Biol. and Psychol. Sci. 3: 179-203, 13 figs.

Sparrow, F. K. Jr., 1931. Two new chytridiaceous fungi from Cold Spring Harbor. Amer. J. Bot. 17:615-623, pl. 45.

———. 1931. A note on a new chytridiaceous fungus parasitic in *Elodea*. Occ. Papers Boston Soc. Nat. Hist. 8:9-10.

———. 1932. Observations on the aquatic fungi of Cold Spring Harbor. Mycologia 24:268-303, 4 figs., pls. 7, 8.

———. 1933. Observations on operculate chytridiaceous fungi collected in the vicinity of Ithaca, New York. Amer. J. Bot. 20:63-77, pls. 2, 3.

———. 1943. Aquatic Phycomycetes. XV. 785 pp., 634 figs. Univ. of Michigan Press. Ann Arbor.

———. 1950. Some Cuban Phycomycetes. J. Wash. Acad. Sci. 40:50-55, 30 figs.

———. 1965. The occurrence of *Physoderma* in Hawaii, with notes on other Hawaiian Phycomycetes. Mycopath. et Mycol. Appl. 25:119-143, pls. 1-7.

Thirumalacher, M. J. 1947. Some fungal diseases of Bryophytes in Mysore Trans. Brit Mycol. Soc. 31:7-12, 8 figs.

Whiffen, A. J. 1943. New species of *Nowakowskiella* and *Blastocladia*. J. Elisha Mitchell Sci. Soc. 59: 37-43, pls. 2-4.

Wildemann, E. de. 1895. Notes mycologiques IV. Ann. Soc. Belge Micro. (Mém.) 19:85-114, 3 figs., pls. 3, 4.

———. 1896. Notes mycologiques. VII. Ibid. 20:21-64, 1 fig., pls. 1-3.

Willoughby, L. G. 1961. The ecology of some lower fungi at Esthwaite water. Trans. Brit. Mycol. Soc. 44:305-332, 17 test-figs., pls. 22, 23.

———. 1964. A study of the distribution of some lower fungi in soil. Nova Hedwigia 7:133-150, pls. 17 (1)-26 (10).

———. 1965. A study of Chytridiales from Victoria and other Australian soils. Arch. f. Mikrobiol. 52: 101-131, 12 figs.

Zopf, W. 1884. Zur Kenntniss der Phycomyceten. I. Zur Morphologie und Biologie der Ancylisteen und Chytridiaceen. Nova Acta Acad. Leop-Carol. 47:143-236, pls. 12-21.

Plate 114

Figs. 1-16. *Coenomyces consuens* Deckenbach. Figs. 1-9, 12-14, 16 after Deckenbach; figs. 10, 11, 15 after Komárek, 1958.

Fig. 1. Septate mycelium with intercalary varicose swellings on and in filaments of *Calothrix*.

Figs. 2, 3. Planospores.

Figs. 4, 5. Mono- and bipolar germination of planospores, resp.

Fig. 6. Rare infection of *Nemalion* cell.

Fig. 7. Incipient zoosporangium growing out from the intramatrical mycelium.

Fig. 8. Incipient zoosporangium developing on the epibiotic mycelium.

Fig. 9. Well developed epibiotic, septate mycelium bearing an almost mature zoosporangium.

Fig. 10. Obclavate, proliferating (?) zoosporangium.

Fig. 11. Mature zoosporangium and 2 sporangial rudiments developing on the mycelium.

Fig. 12. Injured or slightly crushed zoosporangium with planospores and a long neck.

Figs. 13, 14, 16. Empty zoosporangia with short and long exit canals, resp.

Fig. 15. Portion of the septate mycelium with an irregular intercalary swelling

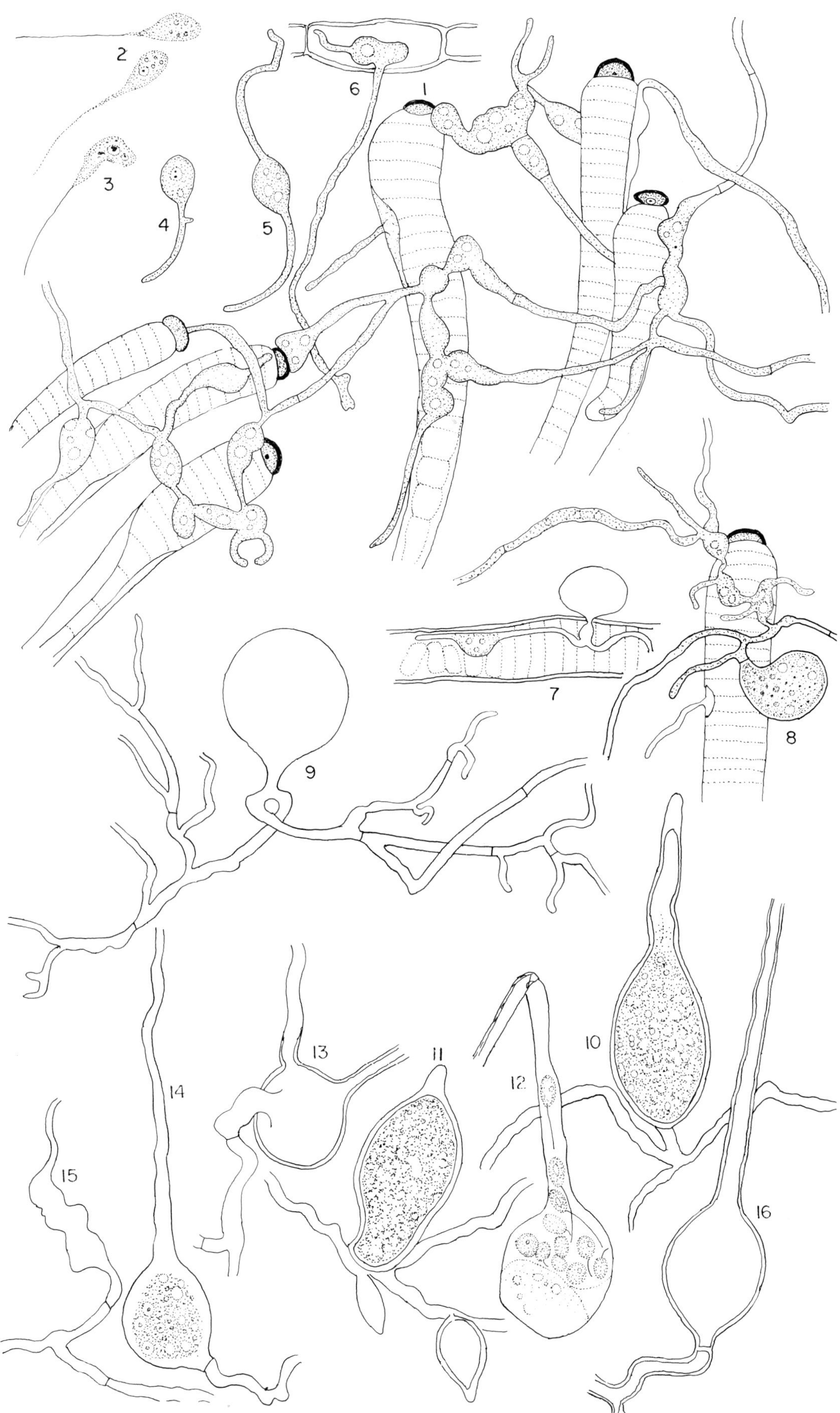

Coenomyces

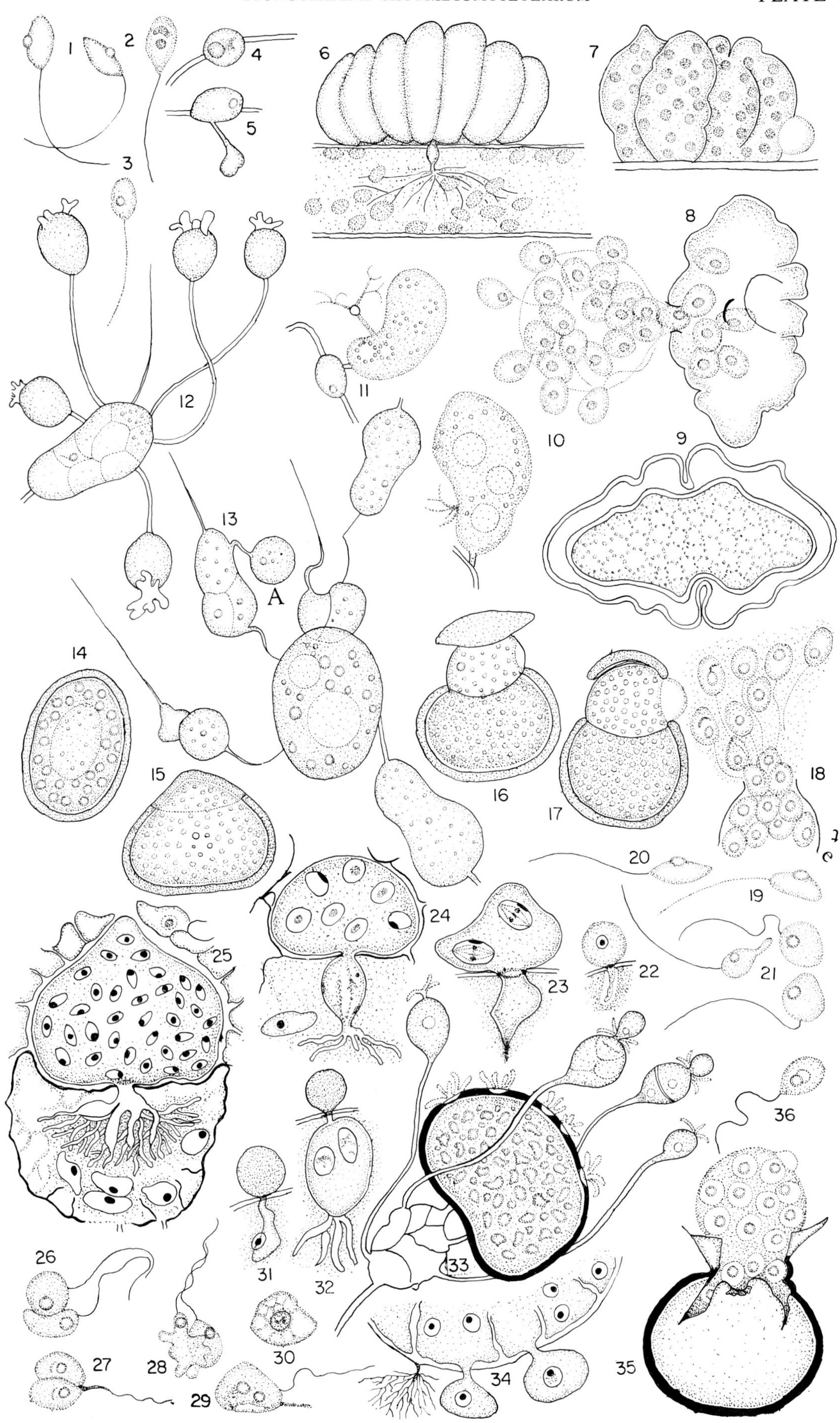
1
2
3
4
5
6
7
8
9
10
11
12
13
A
14
15
16
17
18
19
20
21
22
23
24
25
26
27
28
29
30
31
32
33
34
35
36

Physoderma

Chapter VIII

PHYSODERMATACEAE

The life cycle of the fully-known members of this family differs from that of the species of the Cladochytriaceae by the development of a monocentric epi-endobiotic eucarpic rhizidiaceous thallus in addition to an extensive polycentric endobiotic rhizomycelium. The monocentric thallus consists of an ephermeral epibiotic zoosporangium that produces planospores and an endobiotic absorbing haustorial or rhizoidal system which may or not be oriented on an apophysis. Accordingly, this thallus may be strikingly similar in development, structure and organization to that of the sporangial stages of *Rhizophydium* and *Phlyctochytrium* of the family Rhizidiaceae.

The endobiotic thallus, on the other hand, is polycentric and extensive in the host tissues with tenuous filaments or rhizoids which bear terminal and intercalary turbinate and spindle-like enlargements, and resting sporangia. The latter function as sporangia during germination by developing a protruding endosporangium and producing planospores which may give

PLATE 115

Figs. 1-20. *Physoderma maculare* Wallroth on *Alisma Plantago-aquatica*. Figs. 1, 4-6, 9, 20 after Clinton, 1902; figs. 2, 3, 7, 8, 10-19 after Sparrow, 1964.

Fig. 1. Ellipsoidal planospores with a lateral refractive globule from an ephemeral zoosporangium.

Fig. 2. Ovoid planospores, 5 × 8 μ, with a colored globule and a nuclear cap from an ephemeral zoosporangium.

Fig. 3. Similar smaller, 3 × 5 μ, planospore with a colorless globule from an ephemeral zoosporangium.

Figs. 4, 5. Encysted planospore on the host cell and its germination, resp.

Fig. 6. Deeply-lobed, epibiotic ephemeral zoosporangium with a small endobiotic apophysis and branching rhizoids.

Fig. 7. Similar mature ephemeral zoosporangium with colored globules and a basal exit papilla.

Fig. 8. Discharge of planospores with a colored refractive globule.

Fig. 9. Proliferated zoosporangium as seen in surface view.

Fig. 10. Primary turbinate organ with a tuft of haustoria or rhizoids.

Fig. 11. Similar young turbinate organ before septation with a tuft of rhizoids, germ tube, and planospore cyst.

Fig. 12. Multiseptate primary turbinate organ with 5 elongate filaments bearing secondary turbinate organs with digitate haustoria at their apex.

Fig. 13. Portion of thallus with turbinate organs and an almost mature resting sporangium; incipient resting sporangium developing at A.

Fig. 14. Mature resting sporangium with an ovoid central globule surrounded by numerous smaller one, and an amber-colored wall.

Fig. 15. Early stage in germination of resting sporangium showing line of dehiscence of the cap.

Figs. 16, 17. Later stages of germination with an endosporangium which has pushed up the circumscissile lid or cap; exit papilla lateral.

Fig. 18. Discharge of planospores from an endosporangium.

Figs. 19, 20. Ellipsoidal planospores from a resting spore with colorless refractive globules.

Figs. 21-36. *Physoderma pulposum* Wallroth on *Chenopodium* and *Atriplex*. (Y. Lingappa, 1959 a-c.)

Fig. 21. Ovoid, 2.5 × 3.7 μ diam., and amoeboid planospores with a golden-red refractive globule.

Fig. 22. Infection by the uninucleate planospore.

Fig. 23. Young apophysate ephemeral zoosporangium with 2 nuclei dividing by intranuclear division spindles.

Fig. 24. Young multinucleate ephemeral apophysate zoosporangium with haustoria or rhizoids in the enlarged basal host cell; only outlines of enveloping gall cells indicated.

Fig. 25. Mature multinucleate ephemeral zoosporangium lying in a gall; dense bushy haustoria or rhizoids lying in a multinucleate symplast formed by the lysis of several cell walls.

Fig. 26. Pair of active isoplanogametes.

Fig. 27. Quiescent pair of gametes.

Fig. 28. Pair of gametes becoming amoeboid shortly before plasmogamy.

Fig. 29. Biflagellate zygote with one of the flagella being absorbed.

Fig. 30. Sedentary zygote with a fusion nucleus.

Fig. 31. Germination of a zygote on the host and development of the rudiment of the primary turbinate organ.

Fig. 32. Binucleate primary turbinate organ with rhizoids at its base.

Fig. 33. Septate turbinate organ bearing a mature resting spore and 4 filaments which are terminated by incipient turbinate organs and haustoria; mature resting spore with pits and haustoria.

Fig. 34. Portion of a large turbinate organ whose content has undergone cleavage into numerous segments, 2 of which have formed uninucleate buds.

Fig. 35. Germination of a resting sporangium; endosporangium has cracked open the wall and grown partly out of the spore.

Fig. 36. Planospores from a germinated resting sporangium.

rise to both the monocentric and polycentric phases in some species. Possibly, some species may prove to be short-cycled or micro-cylic and do not develop the eucarpic monocentric phase. This is suggested, at least, by Sparrow and Griffins (1961) studies on *Physoderma australasica* (McAlp.) Walker var. *sparrowii* Saville and Parmlee.

Sexual reproduction is reported to occur by fusion of isoplanogametes from the epibiotic ephemeral zoosporangia, and the fusion product or zygote is believed to develop into the endobiotic polycentric thallus. According to these reports the monocentric eucarpic thallus is haploid, and the polycentric thallus is diploid with meiosis occurring possibly before or during germination of the resting sporangia. The planospores produced by the ephemeral zoosporangia of some species are reported to be facultative. They may function as zoospores and give rise to additional monocentric thalli and epibiotic zoosporangia, or fuse in pairs as gametes. In the event some species are found to be microcyclic with only an endobiotic phase, "sexuality might be lacking altogether, or planospores from germinating resting sporangia might function as gametes," according to Y. Lingappa (1959).

This family is interpreted here to include a single genus *Physoderma*; but it is quite likely that its so-called synonym, *Urophlyctis*, may prove to be different and valid.

PHYSODERMA Wallroth

Flora Crypt. Germ. 2:192, 1833.

Urophlyctis Schroeter, Cohn's Krypt Fl. Schlesiens 3:196, 1889.

Plates 115-120

Physoderma was the first of the chytrid genera to be named, 17 years before the chytrids were recognized as a distinct group of fungi by Braun in 1850, but it was not included in the Chytridiales until much later. For several decades it was confused with *Protomyces* and several genera of the Ustilaginales and Uredinales, probably because of the similarity of the symptoms produced in the hosts in many cases; but the discovery and establishment of the genus *Cladochytrium* by Nowakowski (1878) soon led to the recognition that *Physoderma* was similar to it in many respects and belonged among the chytrids instead of the smuts. Later, several of its species were confused with *Synchytrium* (Karling, 1956). In 1882 Schroeter discovered the eucarpic, monocentric zoosporangial stage and uniflagellate planospores in *P. pulposum* Wall., and it became obvious that Wallroth's genus was a taxon of chytrids.

Physoderma was monographed by Karling (1950), and the genus *Urophlyctis* was merged with it. At

PLATE 116

Figs. 37-49. *Physoderma alfalfae* Karling on *Medicago*. Figs. 37-42, 47-49 after Jones and Drechsler, 1920; figs. 43, 44 drawn from sectioned and stained preparations.

Fig. 37. Section of the host epidermal region showing young primary turbinate organs, *ta-tg*, in enlarging host cells.

Fig. 38. Primary multiseptate turbinate organ, *ta*, from whose cells, filaments, and secondary turbinate organs, *tb-te*, have developed; host cell enlarged and the walls of cells beneath are partially lysed.

Fig. 39. Peripheral portion of an actively growing endobiotic thallus dissected from a living host cell with turbinate organs, *ta-te*, in successive order; mature and immature resting sporangia, *rb*, encircled at the apex by digitale haustoria.

Fig. 40. Abnormally thick-walled filaments or tenuous portions and turbinate organs.

Fig. 41. Section of a turbinate organ with 3 evacuated cells, and a 4th one forming an incipient resting sporangium.

Fig. 42. Section of a resting sporangium with 8 nuclei and a central vacuole.

Figs. 43, 44. Metaphase and anaphase stages of mitosis in a young resting sporangium; division spindle intranuclear.

Fig. 45. Section through a resting sporangium with 5 nuclei and numerous red-staining granules.

Fig. 46. Section through a resting sporangium with 11 normal sized and 4 enlarged nuclei in the center.

Fig. 47. Nearly mature resting sporangium with 11 digitate haustoria in a zonate arrangement.

Fig. 48. Mature resting sporangium viewed from above with 13 pits which mark the former location of the haustoria.

Fig. 49. Mature resting sporangium in profile showing a basal concavity.

Figs. 50-56. *Physoderma pluriannulata* Karling on *Sanicula*. Figs. 50-52 after Jones and Drechsler, 1920; figs. 53-56 after Sparrow, 1957.

Fig. 50. Development of the endobiotic thallus in a gall with a primary, *tá*, secondary, *tb*, and tertiary, *tc*, turbinate organs formed at the ends of filaments, and primary, *ra*, and secondary, *rb*, resting sporangia; *hn*, host nuclei; wall of host cells partly lysed and resulting in a multinucleate symplast which envelopes the parasite.

Fig. 51. Peripheral portion of an actively growing endobiotic thallus showing 8 turbinate organs of the second order, of which 7 have produced turbinate organs of the tertiary order as well as resting sporangia.

Fig. 52. Nearly mature resting sporangium showing 22 digitate haustoria in a zonate arrangement.

Fig. 53. Mature resting sporangium with a dark-brown wall.

Fig. 54. Germination of resting sporangium; the endosporangium has ruptured the spore wall and protrudes out; sporeplasm enveloped by a gelatinous matrix.

Fig. 55. Discharge of planospores through a fissure in the gelatinous matrix.

Fig. 56. Ovoid, 7×5 μ, planospore from a germinated resting sporangium with a hyaline refractive globule, nuclear cap and anterior globules, flagellum 25-27 μ long.

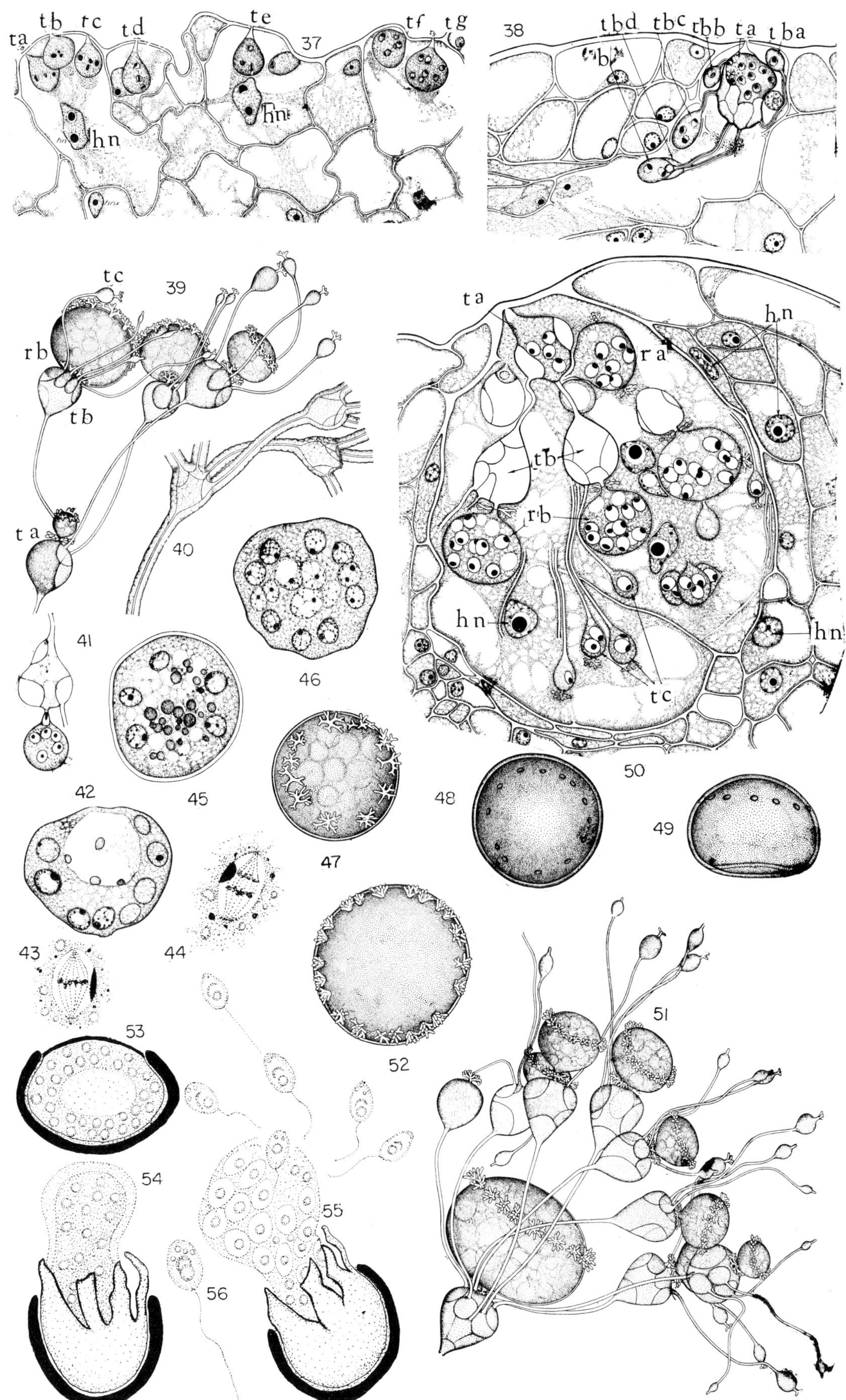
ta
tb
tc
td
te
37
tf
tg
hn
hn
38
tbd
tbc
tbb
ta
tba
tc
39
rb
tb
ta
40
ta
hn
ra
tb
rb
hn
hn
tc
41
46
45
42
47
48
50
49
43
44
53
52
51
54
55
56

Physoderma

PLATE 117

Figs. 57-66. *Physoderma menyanthes* DeBary on *Menyanthis trifoliata*. Figs. 57-62, 65, 66 after Sparrow, 1946; figs. 63, 64 after Clinton, 1902.

Fig. 57. Planospores, 5 × 8-9 μ, from a germinated resting sporangium, with a hyaline refractive globule and a nuclear cap.

Fig. 58. Germinated resting sporangium with a protruding endosporangium; lid or operculum on the left side.

Fig. 59. Ephemeral slipper-shaped apophysate zoosporangium with a tuft or rhizoids.

Fig. 60. Discharge of planospores from the end of an elongate zoosporangium; unexpanded portion of planospore cyst at apex.

Fig. 61. Planospores, 3-5 × 5-7 μ, from an ephemeral zoosporangium; note difference in size compared with those in fig. 57.

Fig. 62. Infection of the host cell; primary turbinate organ 2-celled; planospore cyst persistent.

Fig. 63. Primary 2-celled turbinate organ bearing 4 filaments; planospore cyst persistent.

Fig. 64. Septate primary turbinate organ bearing a resting sporangium and 3 filaments; secondary turbinate organ at A.

Fig. 65. Two-celled spindle or turbinate organ similar to those developed in some species of *Cladochytrium*.

Fig. 66. Resting sporangium attached to a turbinate organ; wall pale- to deep-amber in color.

Figs. 67-72. *Physoderma butomi* Büsgen on *Butomus umbellatus*. (Büsgen, 1877.)

Fig. 67. Germinated resting sporangium with a protruding endosporangium; lid attahed at the left side.

Fig. 68. Planospores, 7 μ long, with a hyaline eccentric protruding refractive globule.

Fig. 69. Infection of host cell and a 3-celled primary turbinate organ bearing 2 filaments.

Fig. 70. Ephemeral zoosporangium with a thick-walled apex, apparently unexpanded portion of zoospore cyst.

Fig. 71. Ephemeral zoosporangium with planospores.

Fig. 72. Tuft of detached rhizoids from an ephemeral zoosporangium.

Fig. 73. *Physoderma flammulae* de Wildemann on *Ranunculus flammula*. (Büsgen, 1877.)

Fig. 73. Mature resting sporangium with a tuft of haustoria and attached by a tubular projection to a 4-celled turbinate organ.

Figs. 74-88. *Physoderma maydis* Miyabe on *Zea mays*. Figs. 74, 78, 83-85 after Tisdale, 1919; figs. 76, 77, 79, 80-82, 88 after Sparrow, 1947; fig. 75 after Couch, 1953; figs. 86, 87 drawn from living material.

Fig. 74. Ellipsoidal, 3-4 × 5-7 μ, planospores from a germinated resting sporangium with a hyaline refractive globule.

Fig. 75. Similar planospore, fixed and stained.

Fig. 76. Living ellipsoidal planospore, 5 × 7 μ, from a germinated resting sporangium with a lateral refractive globule.

Fig. 77. Fixed and stained planospore with a nuclear cap from a germinated resting sporangium.

Fig. 78. Germinating resting sporangium; lid at the apex of the protruding endosporangium.

Fig. 79. Ephemeral slipper-shaped apophysate zoosporangium with a tuft of rhizoids.

Fig. 80. Empty ephemeral zoosporangium with the exit orifice at one end.

Figs. 81, 82. Living ellipsoidal, 3 × 5 μ and fixed and stained planospores, resp., from an ephemeral zoosporangium; note smaller size compared with those in figs. 76 and 77.

Figs. 83, 84. Germinating planospores in water.

Fig. 85. Germinated planospore on the host with the primary spindle or turbinate organ within.

Figs. 86, 87. Portions of the endobiotic thallus, growing in young seedling leaves of albino *Zea mays*; spindle-like enlargements or organs septate; incipient resting sporangia developing on short stalks at B.

Fig. 88. Mature resting sporangium in side view.

Figs. 89-93. *Physoderma potteri* Karling on *Lotus corniculatus*. Figs. 89-92 after Bartlett, 1926; fig. 93 after Karling, 1967.

Fig. 89. Young resting sporangium filled with globules and bearing a tuft of apical haustoria, borne at the apex of a turbinate organ.

Fig. 90. Similar more mature resting sporangium at the apex of a turbinate organ which bears 3 additional filaments and secondary turbinate organs.

Fig. 91. Fixed and stained section of a young multinucleate resting sporangium.

Fig. 92. Side view of a mature brown resting sporangium with a pronounced ornamentation of its surface.

Fig. 93. Probably an empty ephemeral zoosporangium of *P. potteri*; found on submerged leaves of a plant heavily infected with the endobiotic thallus and resting sporangia in New Zealand.

Figs. 94-97. *Physoderma corchori* B. T. Lingappa on *Corchoria* spp. (B. T. Lingappa, 1955.)

Fig. 94. Spindle-shaped septate and continuous intercalary swellings or organs in the rhizomycelium.

Fig. 95. Mature resting sporangium with a brown wall.

Fig. 96. Germinated resting sporangium; protruding endosporangium with a subapical orifice has pushed off a lid and is discharging planospores.

Fig. 97. Ellipsoidal planospore 3.5 × 6 μ, with an eccentric hyaline refractive globule; flagellum 18 μ long.

Figs. 98, 99. *Physoderma commelinae* B. T. Lingappa on *Commelina nudiflora*. (B. T. Lingappa, 1955.)

Fig. 98. Germinated resting sporangium with a protruding endosporangium which has pushed off the lid.

Fig. 99. Elongate, 5 × 7 μ, planospore with an eccentric refractive globule; flagellum 20 μ long.

Figs. 100-106. *Physoderma australasica* (McAlp.) Walker var. *sparrowii* Saville and Parmlee on *Claytonia virginica*. (Sparrow, 1961.)

Figs. 100, 101. Germination of a resting sporangium and discharge of planospores, resp.

Fig. 102. Enlarged view of ovoid, 5 μ diam., planospore with an eccentric hyaline refractive globule.

Fig. 103. Five primary turbinate organs in a host cell which have developed from planospores of germinated resting sporangia.

Fig. 104. Eight-day-old thallus after exposing host to R. S. planospores.

Figs. 105, 106. Variations in shapes and sizes of the intercalary spindle or turbinate organs.

Fig. 107. *Physoderma australasica* (McAlp.) Walker; resting sporangium attached to a multiseptate turbinate organ in a cell of *Claytonia virginica*. (Sparrow, 1947.)

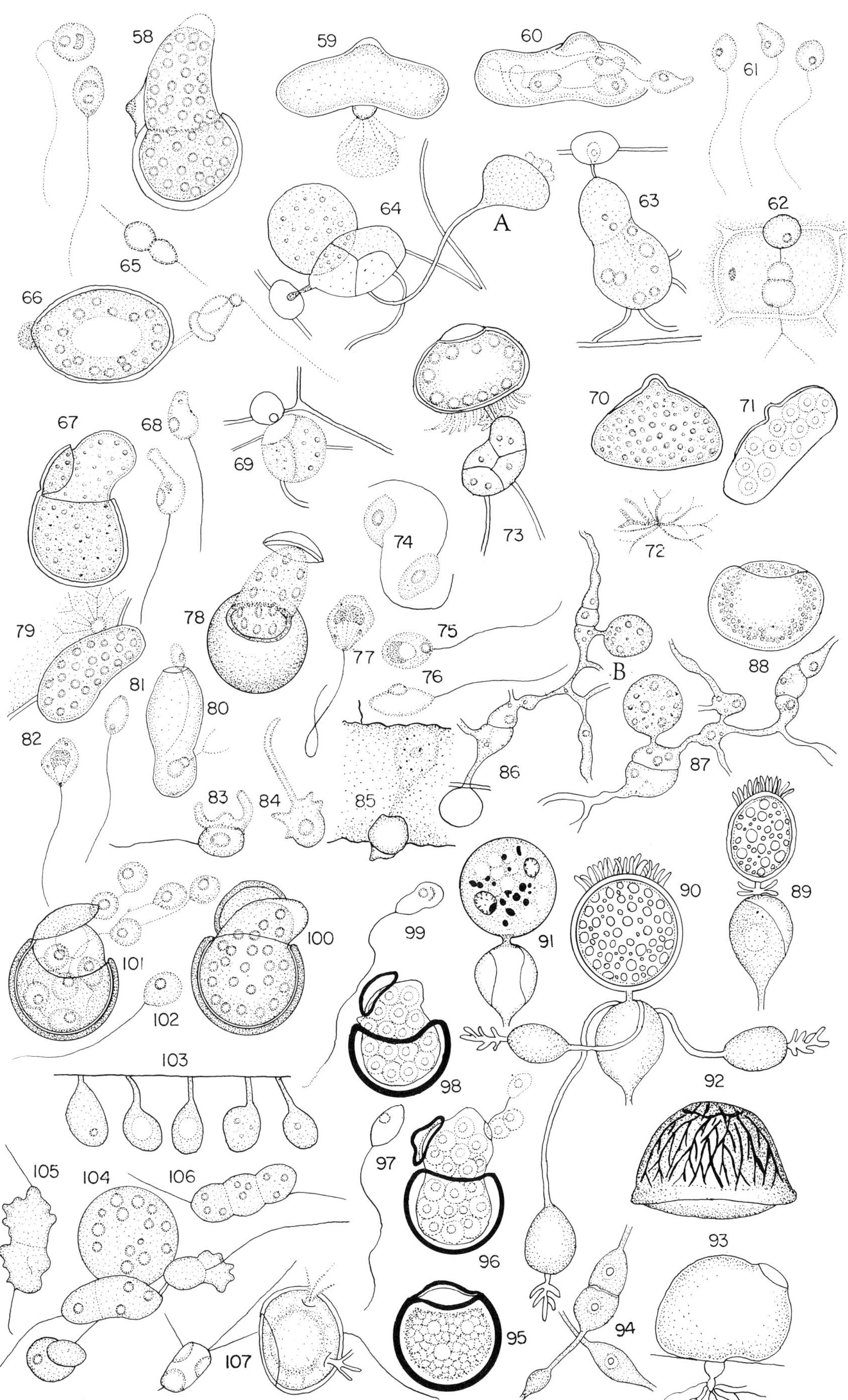
58
59
60
61
62
63
64
A
65
66
67
68
69
70
71
72
73
74
75
76
77
78
79
80
81
82
83
84
85
86
B
87
88
89
90
91
92
93
94
95
96
97
98
99
100
101
102
103
104
105
106
107

Physoderma

PLATE 118

Figs. 108-118. *Physoderma lycopi* Sparrow on *Lycopus americanus*. (Sparrow, 1957.)

Fig. 108. Mature slightly flattened brown resting sporangium in a side view without pits but with a large central vacuole and lid at the apex.

Fig. 109. Germinated resting sporangium discharging planospores.

Fig. 110. Spherical, 5.5-6.8 μ diam., and ovoid, 4-5 × 6, 8-7 μ, planospores from a germinated resting sporangium with a hyaline eccentric refractive globule, 3 apical granules, and a nuclear cap.

Fig. 111. Young ephemeral thallus with a tuft of rhizoids.

Figs. 112, 113. Mature ephemeral zoosporangium with hyaline refractive globules, and discharge of planospores, resp.; portion of planospores cyst persistent on the zoosporangia.

Fig. 114. Small, 3.9-4.7 μ diam., planospore from an ephemeral zoosporangium with a hyaline refractive globule; anterior granules lacking; note difference in size compared with fig. 110.

Figs. 115, 116. Developmental stages of ephemeral zoosporangia with orange-colored globules.

Fig. 117. Mature ephemeral zoosporangium with orange-colored globules.

Fig. 118. Spherical, 3.9-4.7 μ diam., planospore with an orange-colored globule.

Figs. 119-122. *Physoderma majus* Schroeter on *Rumex orbiculatus*. (Sparrow and Y. Lingappa, 1960.)

Fig. 119. Distal portion of a thallus dissected out of a gall and mounted in lactophenol and cotton blue; mature resting sporangium with digitate haustoria and 4 filaments developed from peripheral segments of a turbinate organ; filaments bearing incipient turbinate organs at their tips.

Fig. 120. Possibly, the remains of an ephemeral zoosporangium surrounded by gall cells with a lysigenous cavity at the base.

Fig. 121. Multinucleate primary turbinate organ before septation.

Fig. 122. Primary turbinate organ with peripheral segments and an incipient resting sporangium.

Figs. 123-135. *Physoderma calamii* Krieger on *Acorus calamus*. (Sparrow, 1964.)

Figs. 123, 124. Germinated resting sporangia; prominent exit papilla shown on the endosporangium in fig. 124.

Fig. 125. Swelling of the discharge papilla.

Fig. 126. Spherical, 5-6 μ diam., and ovoid planospores from a germinated resting sporangium with a hyaline refractive globule and nuclear cap.

Fig. 127. Encysted R. S. planospore.

Fig. 128. Unilateral germination of a R. S. planospore; part of cyst wall unexpanded.

Fig. 129. Elongate slightly curved ephemeral zoosporangium with a tuft of rhizoids; unexpanded part of planospore cyst on the upper periphery.

Fig. 130. Elongate tubular ephemeral zoosporangium with its content in the ring stage as seen from above.

Fig. 131. Ephemeral zoosporangium about to discharge planospores from both ends.

Fig. 132. Spherical, 4 μ, and ovoid planospores from an ephemeral zoosporangium; note difference in size compared with those in fig. 126.

Fig. 133. Germinated planospores from ephemeral zoosporangium which have formed a short germ tube and a septate spindle organ.

Fig. 134. Young resting sporangium developing from 1 cell of a septate spindle organ.

Fig. 135. Older resting sporangium with digitate haustoria at its base.

Figs. 136-149. *Physoderma* sp. on *Asclepias incarnata*. (Sparrow and Johns, 1970.)

Fig. 136. Discharge of planospores from a germinated resting sporangium.

Fig. 137. Encysted R. S. planospores with a large hyaline refractive globule.

Fig. 138. Ephemeral zoosporangium with hyaline refractive globules.

Fig. 139. Discharge of planospores, 3 × 5-7 μ, with a hyaline refractive globule; note smaller size than those in fig. 137.

Fig. 140. Ephemeral zoosporangium with ochraceous salmon-colored refractive globules which are stippled uniformly.

Fig. 141. Discharge of fusiform, 3 × 5-7 μ, planospores with an ochraceous salmon-colored refractive globule.

Fig. 142. Pair of planospores (gametes?), one with a hyaline and the other with a salmon-colored refractive globule.

Fig. 143. Early stage of fusion

Figs. 144, 145. Completion of fusion.

Fig. 146. Motile fusion product or zygote with 2 parallel flagella which will soon appear as one.

Fig. 147. Types of intercalary swellings or turbinate organs.

Fig. 148. Young resting sporangium with a tuft of haustoria.

Fig. 149. Mature resting sporangium with a thin, 1.5-2 μ, pale-brown wall.

Figs. 150-157. *Physoderma* sp. Sparrow on *Juncus pelocarpus*. (Sparrow, 1970.)

Fig. 150. Germinated resting sporangium and discharge of planospores.

Fig. 151. Enlarged slightly ellipsoidal R. S. planospore, 5 × 7 μ, with an eccentric refractive globule.

Fig. 152. Portion of endobiotic thallus with 2 incipient resting sporangia formed in relation to intercalary swellings or turbinate organs.

Figs. 153, 154. Multiseptate turbinate organ and young turbinate organs bearing apical haustoria.

Fig. 155. Portion of an endobiotic thallus with 2 septate intercalary swellings and a young resting sporangium with haustoria at the apex.

Fig. 156. Young resting sporangium with a tuft of haustoria.

Fig. 157. Mature resting sporangium with a relatively thin, 1.5-2 μ, light-brown wall, a large central vacuole, and numerous refractive globules.

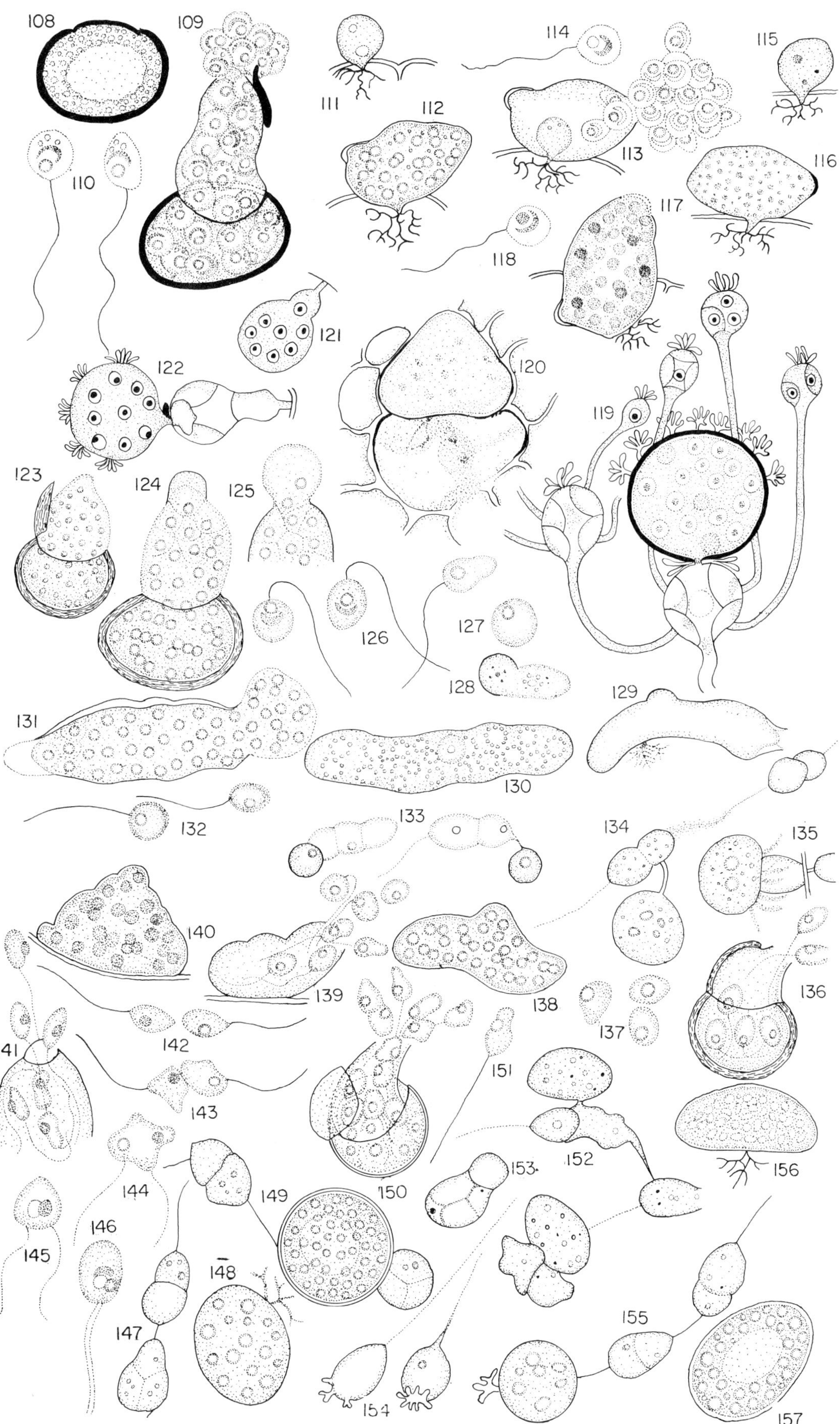
108
109
110
111
112
113
114
115
116
117
118
119
120
121
122
123
124
125
126
127
128
129
130
131
132
133
134
135
136
137
138
139
140
141
142
143
144
145
146
147
148
149
150
151
152
153
154
155
156
157

Physoderma

that time approximately 50 species were recognized, but in the last two decades 30 or more additional ones as well as several unnamed species have been added to the genus. Most of the older known and recently added species have been identified on the basis of their hosts, host reaction to infection, and sizes, shapes, and color of the resting sporangia, but their full life cycles are unknown in many cases. Quite likely, several of them will prove to be identical, particularly among the numerous species added by Thirumalachar and his associates (1948–1964) from India. On the other hand, many more valid species will likely be discovered so that *Physoderma* may possibly become as large as *Synchytrium*. The excellent life history and host range studies by Sparrow and his students (1940–1974) have greatly extended our knowledge of the genus, and it is hoped that they will continue until all species are fully known.

Physoderm species are worldwide in distribution and parasitize plants which range taxonomically from the water ferns and sedges to the composites. Some species appear to have a limited host range while others will infect a large number of plants. A species discovered by Sparrow, Griffin, and Johns (1961) on *Agropyron repens*, for example, was subsequently (Sparrow and Griffin, 1964) found to infect many genera and species of the Festuceae, Hordeae, Agrostideae, and Phalarideae of the Gramineae, *Andropogon* of the Andropoganaceae, as well as *Portentilla anserina* of the Rosaceae under controlled experimental conditions. Particulary significant was the discovery that the eucarpic monocentric zoosporangial stage was not as particular about its hosts but occurred on a large number of plants in which the endobiotic stage failed to develop. Some species such as *P. alfalfae* (Pat. and Lagerh.) Karling, *P. pluriannulata* (Berk, and Curtis) Karling and others cause large galls, excrescences, and malformations on their hosts, while others cause only discolorations, streaks and pustules on the leaves. *Physoderma dulichii* Johns (1957, 1966), for example, is confined to the upper epidermal cells of its host and causes only brownish discolored, irregular areas.

As to the life cycle of the fully known species, it includes a eucarpic, epi-endobiotic monocentric rhizidiaceous and an extensive endobiotic polycentric

PLATE 119

Figs. 158-179. *Physoderma* sp. on *Agropyron repens*. (Sparrow, Griffin and Johns, 1961.)

Fig. 158. Germination of resting sporangium: protoplasm of endosporangium in the "ring stage."

Fig. 159. Discharge of R. S. planospores.

Fig. 160. Spherical, 7 μ diam., R. S. planospore shortly after discharge with a large hyaline refractive globule.

Fig. 161. Planospore during motile stage.

Figs. 162-165. Germination of R. S. planospore, infection, and development of primary turbinate organ.

Fig. 166. Portion of the endobiotic thallus with intercalary and terminal septate turbinate organs, delicate tenuous filaments, and 2 young resting sporangia at A.

Fig. 167. Development of the resting sporangium at the tip of a tubular projection from a turbinate organ.

Figs. 168, 169. Young, and pale-amber mature resting sporangia.

Fig. 170. Early developmental stage of eucarpic monocentric thallus.

Fig. 171. Later stage, part of planospore cyst at right.

Fig. 172. Asymmetrical elongation of young zoosporangium; part of thick planospore cyst wall at right.

Fig. 173. Elongate zoosporangium with its protoplasm in the "ring stage."

Fig. 174. Zoosporangium with hyaline refractive globules.

Fig. 175. Spherical, 4-5 μ diam. and elongate 4-5 × 6-7 μ planospores with a hyaline refractive globule.

Fig. 176. Elongate zoosporangium with orange refractive globules.

Figs. 177, 178. Discharge of planospores with orange refractive globules.

Fig. 179. Spherical, 4-5 μ diam. and elongate 4-5 × 6-7 μ planospores with orange refractive globules.

Figs. 180, 181. *Physoderma* sp. on *Mentha arvensis*, possibly *P. menthae* Schroeter. (Sparrow and Johns, 1959.)

Fig. 180. Young epibiotic zoosporangium on a glandular trichome of the host with an apophysis and bushy rhizoids.

Fig. 181. Empty proliferating zoosporangium with exit orifice at the opposite end from the thickened portion of planospore cyst.

Figs. 182, 183. *Physoderma* sp. on *Calthia palustris*, possibly *P. calthae* Bucholtz. (Sparrow and Johns, 1959.)

Fig. 182. Side view of a slightly gibbose epibiotic empty zoosporangium with bushy rhizoids.

Fig. 183. Mature epibiotic zoosporangium with a broad discharge papilla at the top and the planospore cyst at the opposite end.

Figs. 184-186. *Physoderma* sp. on underwater leaves of *Sium suave*, possibly *P. vagans* Schroeter. (Sparrow and Johns, 1959.)

Fig. 184. Nearly mature epibiotic zoosporangium with a broad lateral exit papilla and an apical planospore cyst.

Fig. 185. Similar zoosporangium discharging planospores and proliferating.

Fig. 186. Empty proliferating zoosporangium with bushy rhizoids.

Figs. 187-189. *Physoderma* sp. on plants of *Sanicula marylandica* which were heavily infected with *P. pluriannulata*; possibly *P. pluriannulata*. (Sparrow and Johns, 1959.)

Fig. 187. Epibiotic zoosporangium with an apical planospore cyst and broad stubby digitate haustoria or rhizoids.

Fig. 188. Proliferating zoosporangium with an apical planospore cyst.

Fig. 189. Zoosporangium with a broad digitate haustorium.

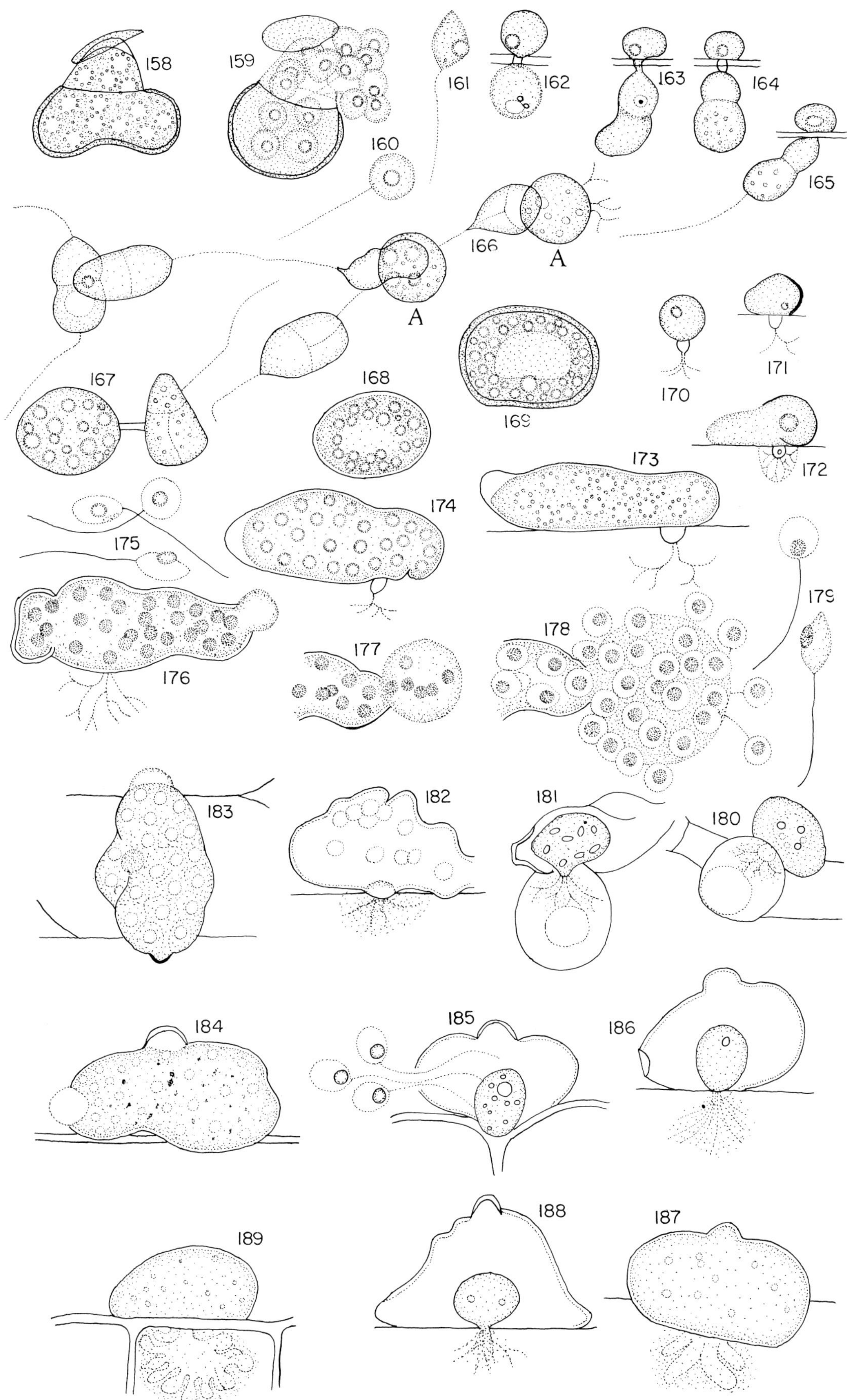
158
159
160
161
162
163
164
165
166
A
A
167
168
169
170
171
172
173
174
175
176
177
178
179
180
181
182
183
184
185
186
187
188
189

Physoderma

phase. However, the eucarpic monocentric stage has been found only in 15 named and 6 unnamed species so far. Planospores from germinated resting sporangia, hereafter referred to as R. S. planospores, give rise to the monocentric eucarpic thallus or the polycentric endobiotic rhizomycelium. In *P. pulposum, P. lycopi* Sparrow, and *P. maydis* Miyabe they are reported to develop into the former thallus, while in *Physoderma* sp. on *Agropyron repens, P. maculare* Wall., *P. calami* Sparrow, *P. menyanthis* De Bary, and *P. dulichii* they may give rise to either or both stages of the life cycle. In species in which no eucarpic monocentric stage is known, i.e., *P. australasica* var. *sparrowii, P. corchori* B. T. Lingappa, *P. pluriannulata*, etc., it is presumed that the R. S. planospores give rise only to the endobiotic polycentric stage. Particularly noteworthy is that in *P. maculare*, (fig. 7), *P. lycopi* (figs. 112, 117), *Physoderma* sp. on *Asclepias* (figs. 138, 140), and *Physoderma* sp. on *Agropyron* (figs. 174, 177), they give rise to 2 types of epibiotic zoosporangia; one with hyaline and another with colored refractive globules, and it is likely that this will be found to occur in other species, also.

The R.S. planospores (figs. 19, 20, 36, 56, 57, 74, 77, 110, 126) in different species are ellipsoidal, ovoid, or subspherical in shape with a conspicuous colorless, or red, orange, yellow or salmon-colored refractive globule. After infecting the host with a short germ tube (fig. 22) the planospore enlarges to become an incipient zoosporangium, but the enlargement is frequently asymmetrical. Only a portion of the planospore cyst expands (figs. 128, 171, 172), and the remainder persists as a thickened part or appendage of the zoosporangial wall (figs. 70, 71, 112, 113, 129, 176, 181, 184, 188). The nucleus (fig. 22) remains in the planospore body and divides with an intranuclear division spindle (fig. 23). In the meantime, the germ tube may enlarge locally into an apophysis (fig. 23) and later branches at its tip to form the bushy rhizoidal or haustorial absorbing system (figs. 25, 34, 79, 113, 129). In *P. pulposum* the walls of cells adjacent to the infected one usually lyse so that the rhizoids are eventually enveloped by a multinucleate symplast (fig. 25). At the same time adjacent epidermal and other cells are stimulated to divide and proliferate until the mature zoosporangium becomes surrounded by a gall.

Thus, the development of the eucarpic monocentric thallus is essentially identical with that of the zoosporangial thalli of *Rhizophydium* and *Phlyctochytrium*. The so-called ephemeral epibiotic zoosporangia vary markedly in size and shape. In *P. pulposum* they are predominantly pyriform and may be up to 350 μ in greatest diameter, and in other species they may be almost hemispherical (figs. 70, 140), or greatly elongate, slightly gibbose, and slipper-shaped with the long axis parallel to the surface of the host (figs. 59, 60, 79, 129–131, 182). In *P. maculare* they are usually deeply lobed in a vertical plane (figs. 6, 7) and resemble somewhat the appearance of a peeled tangerine. In *P. calamii* they are predominantly tubular (figs. 129–131) or broadly slipper-shaped and may be up to 90 μ long by 15 μ in diameter with discharge papillae at both ends (fig. 131). The zoosporangia dehisce by the deliquescence of 1 or 2 lateral (figs. 7, 8, 60, 113) or apical (figs. 25, 141) papillae, and the initial group of planospores emerge to form a globular mass at the exit orifice before dispersing. Proliferation of dehisced zoosporangia (figs. 9, 181, 185, 186, 188), occurs commonly in all species in which the eucarpic monocentric stage is known.

According to Y. Lingappa the endobiotic polycentric thallus or rhizomycelium of *P. pulposum* originates from an infecting zygote (fig. 30), but this has not been demonstrated in other species. Planospores from ephemeral zoosporangia as well as from germinated resting sporangia are reported to originate the endobiotic stage in some species. However, in *Physoderma* sp. on *Asclepias* Sparrow (1970b) observed that planospores from ephemeral zoosporangia consistently failed to repeat the monocentric stage and frequently fused in pairs (figs. 142–146) which suggests that the endobiotic stage arises from a fusion product in this species, also. Whatever the origin of the infecting cell, its germ tube enlarges within the host cells to become the primary turbinate (figs. 11, 31, 32, 63, 103 121, 163–165) or spindle-

PLATE 120

Life cycle diagram of *Physoderma pulposum* indicating sexuality and alternation of haplo- and diplophases. After Y. Lingappa, 1959 (slightly modified.)

A. Planospores (zoospores, gametes.)

B. Sedentary zoospore on the host cell.

C. Infection by zoospore.

D. Incipient ephmeral zoosporangium enveloped by gall cells; nuclei dividing with intranuclear division spindles.

E. Incipient multinucleate ephemeral zoosporangium with bushy rhizoids; nuclei of enveloping gall cells are dividing.

F. Gall enveloping a mature ephemeral zoosporangium.

G. Ephemeral zoosporangium discharging facultative planospores which may function as gametes or zoospores.

H. Gamete.

I. Cluster of gametes.

J. Plasmogamy of gametes.

K. Motile zygote.

L. Karogamy.

M. Infecting zygote.

N. Simultaneous mitosis in incipient primary turbinate organ.

O. Occasional monocentric development of primary turbinate organ into a resting sporangium.

P. Septate primary organ from which filaments and secondary incipient turbinate organs have developed.

Q. Resting sporangium.

R. Germinated resting sporangium discharging zoospores.

S. Zoospore from germinated resting sporangium.

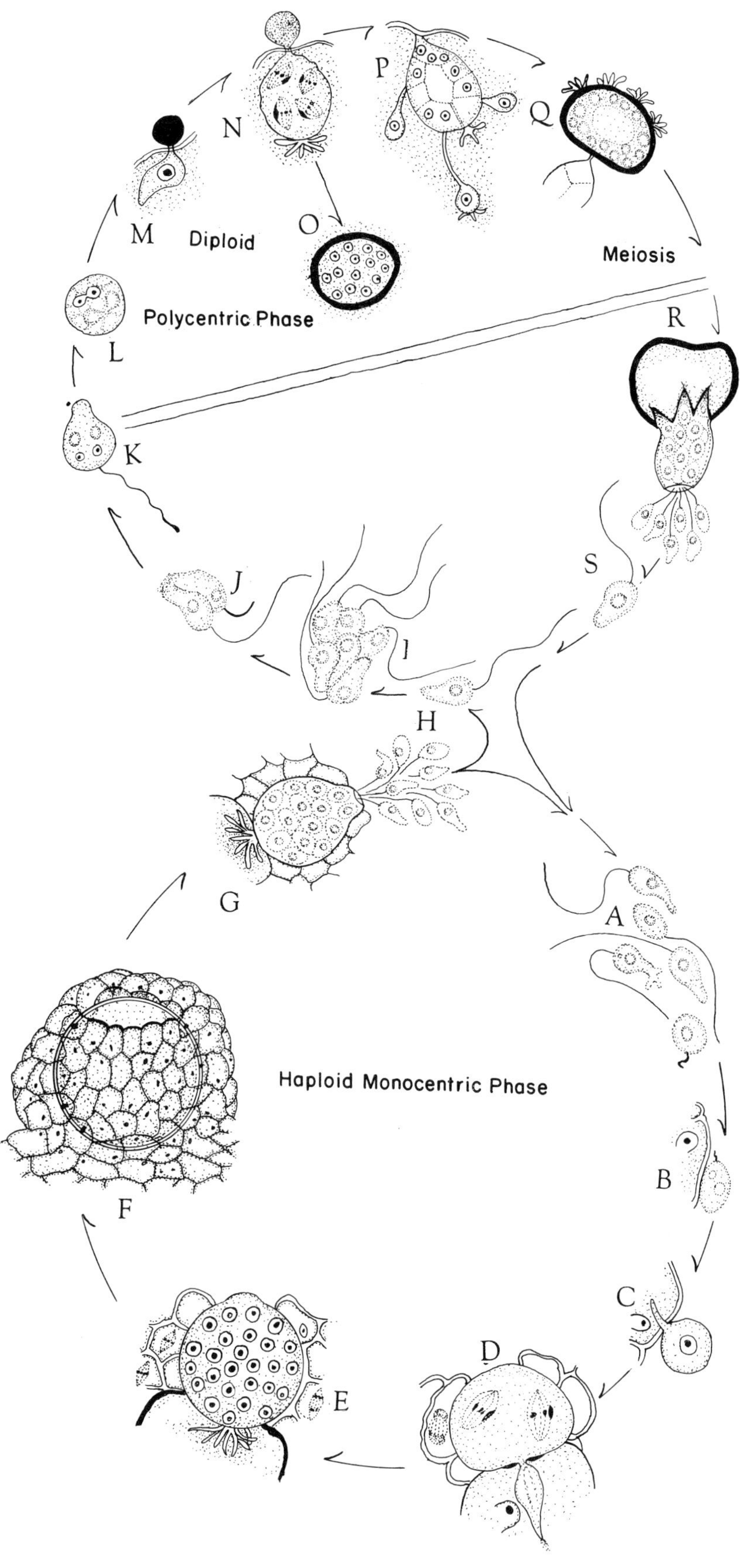
P
N
Q
M
Diploid
O
Meiosis
Polycentric Phase
R
L
K
S
J
I
H
G
A
Haploid Monocentric Phase
B
F
C
D
E

Physoderma

shaped organ (figs. 62, 133). It usually develops a tuft of haustoria or rhizoids at its base (figs. 10, 11, 32), and in the meantime the planospore or zygote nucleus moves down into it (fig. 31) where it and its derivatives divide several times until the organ becomes multinucleate (figs. 32, 37, 50, 121, 122).

In species with distinctly top-shaped turbinate organs these enlargements are reported to divide tangentially and become multiseptate and multicellular, but in *P. pulposum* and *P. pluriannulata* Y. Lingappa (1959b) and Sparrow and Y. Lingappa (1960) reported that the multinucleate content of the turbinate organs undergoes centripetal and tangential cleavage into numerous segments (fig. 34). Some of these segments bud out and form filaments and secondary turbinate organs with haustoria at their tips. The filaments or tenuous portions of the rhizomycelium or thallus in such species are fairly coarse and ribbon-shaped and may become thick-walled with a narrow lumen (fig. 40). The primary resting sporangium usually develops at the end of a short projection or filament at the apex of the turbinate organ (figs. 33, 39, 51, 90, 119, 122). During the development and growth of the filaments the nucleus in the segments of the primary turbinate organ moves into them, and where they come to rest, usually at the tips, secondary swellings or turbinate organs develop. These in turn become multinucleate and multiseptate, and from their cells develop a secondary resting sporangium and secondary filaments which bear tertiary turbinate organs at their tips. In this manner tertiary, quarternary, etc., turbinate organs, filaments and resting sporangia are formed in succession so that the structure and organization of the endobiotic thallus are continually duplicated as it spreads in the host tissue (figs. 32, 37, 50, 51). As this occurs the walls of the host cells undergo lysis with the result that large lysigenous cavities are formed in the host (figs. 38, 50). Sometimes, the primary turbinate organ may become thick-walled and is transformed directly into a resting sporangium (Pl. 120, fig. 0) without further development.

In numerous other species the primary secondary and tertiary enlargements in the rhizomycelium are not typically top-shaped but elongate or irregular, transversely (figs. 65, 106, 133) or otherwise septate and similar in many respects to the spindle organs of *Cladochytrium*. In such species the tenuous filaments are delicate and rhizoid-like (figs. 134, 147, 155, 166), and run from cell to cell without causing lysis of the host cell walls. The resting sporangia in such species are usually formed laterally at the tip of a tubular projection from the intercalary or terminal spindle-like or irregular enlargements (figs. 13A, 86, 87, 104, 152, 155, 167).

However formed, the resting sporangia are multinucleate (figs. 42–46, 91, 119) and at maturity they are usually flattened slightly or considerably on one surface (figs. 15, 33, 49), slightly concave, or almost hemispherical (fig. 92). They vary in size, 12–40 × 18–60 μ in the various species, but in *P. punctiformis* (Speg.) Karling they are reported to measure 50 × 75–100 μ. They have a reddish-brown, pale chestnut-brown, golden-brown, amber, pale yellow, or straw-colored, 1.5–4.5 μ thick, smooth, undulate, verrucose or ridged epispore and a thinner hyaline endospore. In some species the 2-layered wall may be only 1 or 2 μ thick. In living material their contents appear as a large central globule (figs. 14, 53, 108, 149, 157) surrounded by smaller ones which may be hyaline, deep-orange, and brown as seen through the sporangium wall. In most species the resting sporangia have been found to bear digitate haustoria in a zonate arrangement around the upper portion or near the equator (figs. 33, 47, 52, 119). These disappear as the sporangia mature, but their former positions and numbers are indicated by pits in the wall. After examining the resting sporangia in a fairly large number of species Jones and Drechsler (1920) reported "that a certain range in number of pits was found to be characteristic of species."

The resting sporangia in some species may remain dormant for several years before germinating, but germination may be induced by various artificial means such as aging and low temperatures. In *P. hydrocotylidis* Viegas and Teixeira, Sparrow (1968) found that they will germinate soon after they reach maturity. So far, germination has been observed in 25 named species and 3 unnamed species, and among these 18 named and 2 unnamed species a circumcissile lid or operculum (figs. 16, 17, 58, 95, 96, 100, 101, 109, 123, 136, 150) is pushed off or lifted up in the process. In the other 6 named and 1 unnamed species the wall of the sporangium is cracked open during germination (figs. 35, 54, 55). In *P. menyanthis*, however, Clinton (1902) and Sparrow (1946) observed that a lid may be pushed up, or the wall is cracked open. In the process of germination the refractive material in the sporangium becomes progressively dispersed so that the protoplasm becomes more optically homogeneous. As it expands a thin-walled, stubby, sac-like vesicle or endosporangium is formed which cracks the wall, or pushes up or off a lid or operculum. Soon, the dispersed refractive material takes on a "ring stage" appearance (fig. 158) and coalesces to form the definitive globules of the planospores as an apical (figs. 54, 123–125), subapical or lateral (figs. 16, 17) exit papilla develops. As its tip deliquesces and expands (figs. 123–125) the planospores emerge (fig. 109) and disperse (figs. 18, 35, 150). These planospores are slightly larger than those produced by ephemeral zoosporangia.

Sexual reproduction in *Physoderma* has been reported to occur by fusion of isoplanogametes from ephemeral zoosporangia and resting sporangia as well as by copulation of cells of the turbinate organs, but the latter reports have been refuted. Historically, it is interesting to note that very early Schroeter (1882) described the copulation of two cells in what is now recognized as the turbinate organ in *P. pulposum*.

As a result one cell becomes empty while the other buds out and enlarges to become the resting sporangium. However, in 1892 Fischer refuted Schroeter's report of sexuality and pointed out that what appeared to be the formation of a zygote, was nothing more than the division of the incipient "sammelzellen" or turbinate enlargements and subsequent development of the daughter cells. One of the daughter cells budded out to form a resting sporangium as the protoplasm moved into it, leaving the empty cell adhering to the resting sporangium as a vesicle. Wilson (1915, 1920) reported heterogamous fusion between small uniflagellate and large biflagellate heterocont planospores from germinated resting sporangia in *P. alfalfae*, but his account of the morphology and development of this species is so confusing that it throws serious doubt on his observations. Thind and Sharma (1959) also reported anteriorly biflagellate heterocont planospores in a *Physoderma* sp. on *Trifolium alexandrianum* and their fusion to form "zygospores."

Sparrow (1940), on the other hand, suggested that the epibiotic zoosporangia and their planospores might be gametangia and gametes, respectively, and that the gametes fuse to form a zygote which develops into the endobiotic polycentric thallus. However, he was unable to confirm this hypothesis in later (1946, 1947) studies on *P. menyanthis* and *P. maydis*, although he observed that the planospores from the epibiotic zoosporangia were smaller than those produced by the resting sporangia. Later, he (1957) and his associates (1961) found support for this hypothesis in *P. lycopi* and *Physoderma* sp. on *Agropyron repens* in which 2 types of zoosporangia occur - one with hyaline and another with orange refractive globules (figs. 112, 117, 140, 174, 176). A few fusions between the different planospores were observed in *P. lycopi* but these did not occur in "sufficient numbers to prove conclusive evidence of sexuality."

In 1959a, b, Y. Lingappa reported that the planospores from epibiotic zoosporangia in *P. pulposum* are facultative and may give rise to additional ephemeral zoosporangia or may fuse as gametes to form a zygote (figs. 26–30) which develops into the endobiotic polycentric thallus (fig. 31). She found that fusions occurred only between gametes from 2 or more zoosporangia and not between planospores from germinated resting sporangia, and that gametes often fused with zygotes. Although she did not observe meiosis, she believed that it might occur during germination or the resting sporangia. According to her, the endobiotic polycentric phase of *Physoderma* is diploid, and the eucarpic monocentric phase is haploid, as shown in plate 120. More recently Sparrow and Johns (1970) reported that a species on *Asclepias*, as noted before, produces 2 types of epibiotic zoosporangia and planospores, one with hyaline and another with ochraceous salmon colored refractive globules (figs. 138, 140), and that numerous fusions occurred between the different planospores (figs. 142–146). Although they did not observe the development of the fusion product or zygote into the endobiotic polycentric thallus it is significant that, unlike in most other species, the planospores from epibiotic zoosporangia were unable to repeat the zoosporangial phase. This suggests strongly that in this species they are not facultative and can only function as gametes.

Whether *Urophlyctis* is to be maintained as a separate genus or merged with *Physoderma* remains to be seen. It was established by Schroeter (1892–1897) for species which develop epibiotic zoosporangia and form resting sporangia by copulation of the contents of 2 turbinate organ cells, but these distinctions are no longer valid. Magnus (1897, 1901), nonetheless, supported his distinctions of the genus and added other reasons for separating it from *Physoderma*. He maintained that the resting sporangia of *Physoderma* are globose and ellipsoidal while those of *Urophlyctis* are flattened on one surface and almost hemispherical in shape. Furthermore, *Urophlyctis* species cause marked hypertrophy and malformation of the infected host tissues while *Physoderma* species cause only discoloration and slight malformation. Although host reactions and symptoms are convenient means of superficial identification, they are not always associated with distinct morphological difference in the causal agents. *Physoderma flammulae* (Büsgen) de Wildemann, for instance, has flattened *Urophlyctis*-like resting sporangia (fig. 73) with digitate haustoria, but according to Büsgen it causes no malformation of the host. Also, the resting sporangia of *P. maydis* (fig. 88) are flattened on one surface, but this species causes only streaks on the leaves and sheaths and slight enlargement of infected cells.

In the previous decade Sparrow and Y. Lingappa (1960) revived the above concept of host reaction differences between the genera and coupled them with structural differences exhibited by the endobiotic thallus and epibiotic zoosporangia. "Our findings as to the nature of the endobiotic thallus of *Urophlyctis majus*, add another species to the group composed of *U. pulposa, U. pluriannulatus, U. alfalfae*, etc., which form top-shaped turbinate organs, ribbon-like rhizoids, resting spores formed apically on the turbinate organs, and which produce lysigenous cavities within a host gall. Distinct from this group is a galaxy of species with irregular or elongate, transversely or otherwise septate, turbinate organs, delicate rhizoids which wander from cell to cell as Magnus (1897) puts it, rather than producing lysis of the host cells, and which form resting spores for the most part laterally at the tip of a tubular prolongation from the turbinate organ. The latter group, considered to be true species of *Physoderma*, have epibiotic sporangial characters which are also distinctive." In 1962 Sparrow emphasized again the differences between the epibiotic zoosporangia of *Physoderma* and those of so-called *Urophlyctis* species, but it is to be noted that the zoosporangia found on *Sanicula marylandica* (figs. 187–189) in association with the endobiotic phase of *P.* (*Urophlyctis*) *pluriannulata* "are distinctly *Physoderma*-like." He, also, stressed again the differences in the effects produced in the hosts as well as the character of the tenuous filaments of the rhizomycelium and the manner in which the resting sporangia are developed and borne.

In this connection it should be noted again that in 5 species which cause marked hypertrophy and lysigeny of the infected host cells the resting sporangial wall is cracked open (figs. 35, 54, 55) by the endosporangium during germination. However, this occurs, also, in *P. dulichii* which does not cause hypertrophy. In the remaining species where germination has been observed a circumscissile lid is lifted off or up. Karling (1950) suggested that if this lid is regarded as an operculum *Physoderma* might be distinguished as an operculate genus and *Urophlyctis* as inoperculate. However, Clinton's (1902) and Sparrow's (1946) reports that a lid may be present or lacking in *P. menyanthis* raises some doubts about the generic value of this character. Furthermore, germination of the resting sporangia has been observed in so few of the 80 or more species that it is difficult to draw general conclusions in this respect at the present time.

REFERENCES TO THE PHYSODERMATACEAE

Bartlett, A. W. 1926. On a new species of *Urophlyctis* producing galls on *Lotus corniculatus* Linn. Trans. Brit. Mycol. Soc. 11:266-280, 26 figs., pls. 11-14.

Braun, A. 1850. Betrachtungen über die Erscheinung der Pflanze. 363 pp. Leipzig.

Büsgen, M. 1887. Beitrag zur Kenntniss der Cladochytrien. *In* Cohn's Beitr. Biol. Pflanzen 4:469-483, pl. 15.

Clinton, G. P. 1902. *Cladochytrium alismatis*. Bot. Gaz. 33:49-61, pls. 2-4.

Couch, J. N. 1953. The occurrence of thin-walled sporangia in *Physoderma zeae-maydis* on corn in the field. J. Elisha Mitchell Sci. Soc. 63:182-184, 8 figs.

Fischer, A. 1892. Phycomycetes. Die Pilze Deutschlands, Oesterreichs und der Schweiz. Rabenhorst Kryptogamen-Fl. 1:1-490. Leipzig.

Johns, R. M. 1957. A new species of *Physoderma* on *Dulichium*. Mycologia 48:433-438, 12 figs.

————. 1966. Morphological and ecological study of *Physoderma dulichii*. Amer. J. Bot. 53:34-45, 26 figs.

Jones, F. R. and C. Drechsler. 1920. Crownwart of alfalfa caused by *Urophlyctis alfalfae*. J. Agr. Res. 20:295-331, pls. 47-56.

Karling, J. S. 1950. The genus *Physoderma* (Chytridiales). Lloydia 13:20-71.

————. 1956. Unrecorded hosts and species of *Physoderma*. Bull. Torrey Bot. Club 83:292-299.

————. 1967. Some zoosporic fungi of New Zealand. VIII. Cladochytriaceae and Physodermataceae. Sydowia 20:129-136, pl. 25.

Lingappa, B. T. 1955. The two new species of *Physoderma* from India. Mycologia 47:109-121, 30 figs.

Lingappa, Y. 1959a. Sexuality in *Physoderma pulposa* Wallroth. Ibid. 51:151-158, 55 figs.

————. 1959b. The development and cytology of the epibiotic phase of *Physoderma pulposum*. Amer. J. Bot. 46:145-150, 14 figs.

————. 1959c. Development and cytology of the endobiotic phase of *Physoderma pulposum*. Ibid. 46:233-240, 31 figs.

Magnus, P. 1897. On some species of the genus *Urophlyctis*. Ann. Bot. 11:87-96, pls. 7, 8.

————. 1901. Ueber eine neue unterirdisch lebende Art der Gattung *Urophlyctis*. Ber. deut. Bot. Gesell. 19:(149)-(153), pl. 27.

McAlpine, D. 1896. Australian fungi. Agr. Gaz. N. S. W. 7:147-156.

Nowakowski, L. 1876. Beitrag zur Kenntniss der Chytridiaceen. *In* Cohn's Beitr. Biol. Pflanzen 2: 73-100, pls. 4-6.

Pavgi, M. S., and M. J. Thirumalachar. 1954. Some new or interesting *Physoderma* species from India. Sydowia 8:90-95, pls. 4, 5.

Schroeter, J. 1882. Untersuchungen der Pilzgattung *Physoderma*. Jahresb. Schles. Gesell. Vaterländ. Cultur 60:198-200.

————. 1885-1889. Die Pilze Schlesiens. Cohn's Kryptogamenfl. Schlesiens 3:1-814.

————. 1893-1897. Phycomycetes. In Engler und Prantl, Naturl. Pflanzenf. 1(1):63-141.

Sparrow, F. K., Jr. 1940. Chytridiaceous fungi in relation to disease of flowering plants, with special reference to *Physoderma*. Rept. Proc. 3rd Internat. Cong. for Microbiology. New York, VIII, 833 pp.

————. 1946. Observations on chytridiaceous parasites of phanerogams. *Physoderma menyanthes* de Bary. Amer. J. Bot. 33:112-118, 41 figs.

————. 1947a. Observations on chytridiaceous parasites of phanerogams. II. A preliminary study of the occurrence of ephemeral sporangia in the *Physoderma* disease of maize. Ibid. 34:94-97, 17 figs.

————. 1947b. Observations on chytridiaceous parasites of phanerogams. III. *Physoderma claytoniana* and an associated parasite. Ibid. 34:325-329, 17 figs.

————. 1953. Observations on chytridiaceous parasites of phanerogams. IV. *Physoderma aponogetonis* sp. n. parasitic on *Aponogeton*. Trans. Brit. Mycol. Soc. 36:347-348, 1 fig.

————. 1957 Observations on chytridiaceous parasites of phanerogams. VII. A *Physoderma* on *Lycopus americanus*. Amer. J. Bot. 44:661-665, 26 figs.

————. 1962. *Urophlyctis* and *Physoderma*. Trans.

Mycol. Soc. of Japan 3:15-16.

———. 1964a. Observations on chytridiaceous parasites of phanerogams. XIV. *Physoderma calamii.* Amer. J. Bot. 51:958-963, 27 figs.

———. 1964b. Observations on chytridiaceous parasites of phanerogams. XIII. *Physoderma maculare* Wallroth. Arch. f. Mikrobiol. 48:136-149, 31 figs.

———. 1965. Concerning *Physoderma graminis.* Mycologia 57:624-627, 1 fig.

———. 1968. *Physoderma hydrocotylidis* and other interesting Phycomycetes from California. J. Elisha Mitchell Sci. Soc. 84:62-68, 34 figs.

———. 1970. Observation on chytridiaceous parasite of phanerogams. XVIII. A *Physoderma* on *Juncus pelocorpus* Mey. Arch. f. Mikrobiol. 70: 104-109, 17 figs.

———. 1974a. Observations on chytridiaceous parasites of phanerogams. XIX. A *Physoderma* on Eurasian milfoil (*Myriophyllum spicatum*). Amer. J. Bot. 61:174-180, 11 figs.

———. 1974b. Observations on chytridiaceous parasites of phanerogams. XX. Resting spore germination and epibiotic stage of *Physoderma butomi* Schroeter. Ibid 61:203-208, 25 figs.

———, and J. E. Griffin. 1961. Observation on chytridiaceous parasites of phanerogams. XII, Further studies of *Physoderma claytonianum* var. *sparrowii.* Arch. f. Mikrobiol. 40:275-282, 18 figs.

———, and J. E. Griffin. 1964. Observations on chytridiaceous parasites of phanerogams. XV. Host range and species concepts studies in *Physoderma.* Ibid. 49: 103-111.

———, J. E. Griffin and R. M. Johns. 1961. Observations on chytridiaceous parasites of phanerogams. XI. A *Physoderma* on *Agropyron repens.* Amer. J. Bot. 48:850-858, 53 figs.

———, and R. M. Johns. 1959. Observations on chytridiaceous parasites of phanerogams. IX. Epibiotic sporangial stages of *Physoderma* collected in the field. Arch. f. Mikrobiol. 34:92-102, 29 figs.

———, and R. M. Johns. 1970. Observations on chytridiaceous parasites of phanerogams. XVII. Notes on a *Physoderma* parasitic on *Asclepias incarnata.* Ibid. 70:72-81, 20 figs.

———, and Y. Lingappa. 1960. Observations on chytridiaceous parasites of phanerogams. VII. *Urophlyctis (Physoderma pluriannulatus* and *U. majus).* Amer. J. Bot. 47:202-209, 18 figs.

———, and M. S. Saunders. 1974. Observations of chytridiaceous parasites of phanerogams. XXII. A *Physoderma* on *Ranunculus septentrionales.* Arch. f. Mikrobiol. 100:41-40, figs. 1-23.

Thind, K. S., and S. R. Sharma. 1959. *Physoderma* on *Trifolium alexandrianum* Linn. Indian Phytopath. 12:122-130, 18 figs.

Thirumalachar, M. J. 1948. A chytridiaceous parasite of *Limnanthemum indicum.* Indian Phytopath. 2:127-131, 9 figs.

———, and M. D. Whitehead. 1951. An undescribed species of *Physoderma* on *Aeschymonene indica.* Mycologia 43:430-436, 17 figs.

———, and M. S. Pavgi. 1954. Some new or interesting *Physoderma* species from Inda. Bull. Torrey Bot. Club 81:149-154, 6 figs.

———, and M. S. Pavgi. 1956. Some new or interesting *Physoderma* species from India. III. Sydowia 10:112-117, pls. 1, 2.

———, and M. S. Pavgi. 1964. Some new and interesting species of *Physoderma* from India—IV. Ibid. 17:28-32, pls. 8, 9.

Tisdale, W. H. 1919. *Physoderma* disease of corn. J. Agric. Res. 16:137-154, pls. A, B. 10-17.

Voorhees, R. K. 1933. Effect of certain environmental factors on the germination of the sporangia of *Physoderma zeae-maydis.* J. Agric. Res. 47: 609-615.

Walker, J. 1962. Notes on plant parasitic fungi. I. Proc. Linn. Soc. of New South Wales 87:162-176, pl. 3.

Wallroth, F. G. 1833. Flora cryptogamia Germaniae 2:1-923. Nüremburg.

Wilson, O. T. 1915. The crown gall of alfalfa. Science II, 41:797.

———. 1920. Crown gall of alfalfa. Bot. Gaz. 70: 51-68, pls. 7-10.

POLYCENTRIC GENERA OF DOUBTFUL AFFINITY

Under this heading are included two operculate, and 1 inoperculate genera, and an unidentified inoperculate taxon of polycentric fungi which produce posteriorly uniflagellate chytrid-like planospores in zoosporangia. Three of these species have not been seen since Sorokin described them almost a hundred years ago, but the unidentified taxon has recently been described by Willoughby (1966). Mycologists have doubted the existence of Sorokin's genera because they have not been rediscovered and the fact that one of them combines a chytrideaceous type of asexual reproduction with a zygomycetous type of sexuality. The present writer does not concur with this doubt and believes that separate families will probably have to be erected to accommodate them and other similar species which might be discovered in the future.

The thallus of *Zygochytrium* and *Tetrachytrium* consists of a basal holdfast and a central axis with branches that bear zoosporangia, while in *Saccopodium* and the unidentified taxon it is extensive and more mycelioid in appearance. *Zygochytrium* has a typical zygomycetous type of sexual reproduction, and in *Tetrachytrium* fusion of pairs of isomorphic planogametes occurs. Sexual reproduction is unknown in the other genera. A more detailed description of the structure, development and reproduction is presented below for each genus.

ZYGOCHYTRIUM Sorokin

Bot. Zeit. 32:308, 1874; Bull. Soc. Nat. Kazan 4:12, 1874.

Plate 121, Figs. 1-20

This operculate polycentric genus includes one saprophytic species, *Z. aurantiacum* Sorokin, which was found on submerged cadavers of insects in Russia 102 years ago. It is characterized by an orange red hyphal thallus with a lobed holdfast, a dichotomously branched central axis and 2 similarly branched secondary axes, one of which bears an operculate zoosporangium while the other is sterile (fig. 1). At maturity the operculum is pushed off and aside as the protoplasm flows out of the zoosporangium (figs. 2–4). The protoplasm is naked at first, but in a short while it rounds up and becomes enveloped by a membrane as the vermilion granules aggregate in the center (figs. 5–7). It soon cleaves into planospores (figs. 8–9) which later swarm within the membrane and are eventually liberated as it ruptures (fig. 10). These germinate by first forming the rudiment of the foot (fig. 12) and then developing the central axis and branches at the opposite pole (figs. 13–14).

Sexual reproduction is isogamous and occurs by the development of progametangia between two branches of the same thallus (fig. 15) from which 2 gametangia are delimited by cross walls after they have come into contact (fig. 16). The wall between the gametangia dissolves, and their red contents fuse (fig. 17) to produce an incipient zygospore (fig. 18) between them. At maturity the zygospore has a thick, blood red warty or verrucose exospore, a hyaline endospore, and red, granular content (fig. 19). It germinates readily as the endospore and expanding content crack the exospore and elongates into a tube (fig. 20). The fate of this tube is not known; possibly it may develop a sporangium directly at its apex as in the Mucorales, or give rise to dichotomously branched thallus.

As described and illustrated by Sorokin, *Z. aurantiacum* is homothallic, and its type of sexual reproduction is clearly zygomycetous. It is a unique species in that it combines this type sexuality with a chytridiomycetous type of asexual reproduction, and for this reason it is difficult to classify and place it in any of the known familial taxa. Since its discovery a century ago it has not been found since, and this fact together with its unusual life history had led several mycologists to reject it as a valid genus. Others have placed it in various groups. Minden (1911) placed it in the Hyphochytriaceae, and Schroeter (1892) listed it as an "anhang" to his family Oochytriaceae. Fitzpatrick (1930) listed it as a doubtful chytrid; Bessey (1900) noted it as an appendage to the operculate Cladochytriaceae; and Sparrow (1960) regarded it as of doubtful affinity. The present writer has no doubts about its validity because of Sorokin's excellent figures and description of it, and also the fact that his reports on and descriptions of other chytrids have been proven to be correct to a large degree by subsequent workers. The presence of two diverse types of reproduction is not sufficient reason for rejecting it. The chytrids exhibit such variations in sexual reproduction that it is quite likely that species like *Zygochytrium* exist. In fact, members of the subfamily Polyphagoideae, an undoubted group of chytrids, have a fairly similar type of sexual reproduction.

PLATE 121

Figs. 1-20. *Zygochytrium aurantiacum* Sorokin. (Sorokin, 1874.)

Fig. 1. Branched thallus with a foot or holdfast, 2 operculate zoosporangia, and 2 sterile branches.

Fig. 2. Initial stage in dehiscence.

Figs. 3, 4. Protoplasm emerging through an opening; operculum pushed aside.

Fig. 5. Emerged protoplasm from an empty zoosporangium with the red granules aggregating in the center.

Fig. 6. Emerged protoplasm enveloped by a distinct membrane.

Figs. 7-9. Successive stages in sporogenesis.

Fig. 10. Ruptured and collapsed membrane with discharged planospores.

Fig. 11. Enlarged amoeboid planospores.

Figs. 12-14. Germination stages of encysted planospores.

Fig. 15. Progametangia developed between branches of the same thallus.

Fig. 16. Gametangia delimited by cross walls.

Fig. 17. Breakdown of wall between gametangia and dusion of protoplasm.

Fig. 18. Young zygospore.

Fig. 19. Median section through a mature zygospore with a thick blood red, warty, or verrucose exospore and a hyaline endospore.

Fig. 20. Germination of zygospore by the elongation of the endospore into a long tube.

Figs. 21-40. *Tetrachytrium triceps* Sorokin. (Sorokin, 1874, 1883.)

Figs. 21, 22. Two thalli with 3 branches bearing zoosporangia, a sterile curved branch, a main axis, and a foot.

Figs. 23, 24. Protoplasm emerging from an operculate zoosporangium and rounding up.

Figs. 25-27. Emerged protoplasm enveloped by a membrane and cleaving into 4 planospores.

Fig. 28. Rupture and collapse of membrane liberating 4 planospores.

Fig. 29. Motile planospores.

Figs. 30-33. Fusion of isogametes (figs. 31-33 after Sorokin, 1883.)

Fig. 34. Fusion cell or zygote.

Figs. 35, 36. Germination of zygote.

Figs. 37-39. Coiling of sterile branches. (Sorokin, 1883.)

Fig. 40. Coiling, tangled, wefted sterile branches between 2 thalli.

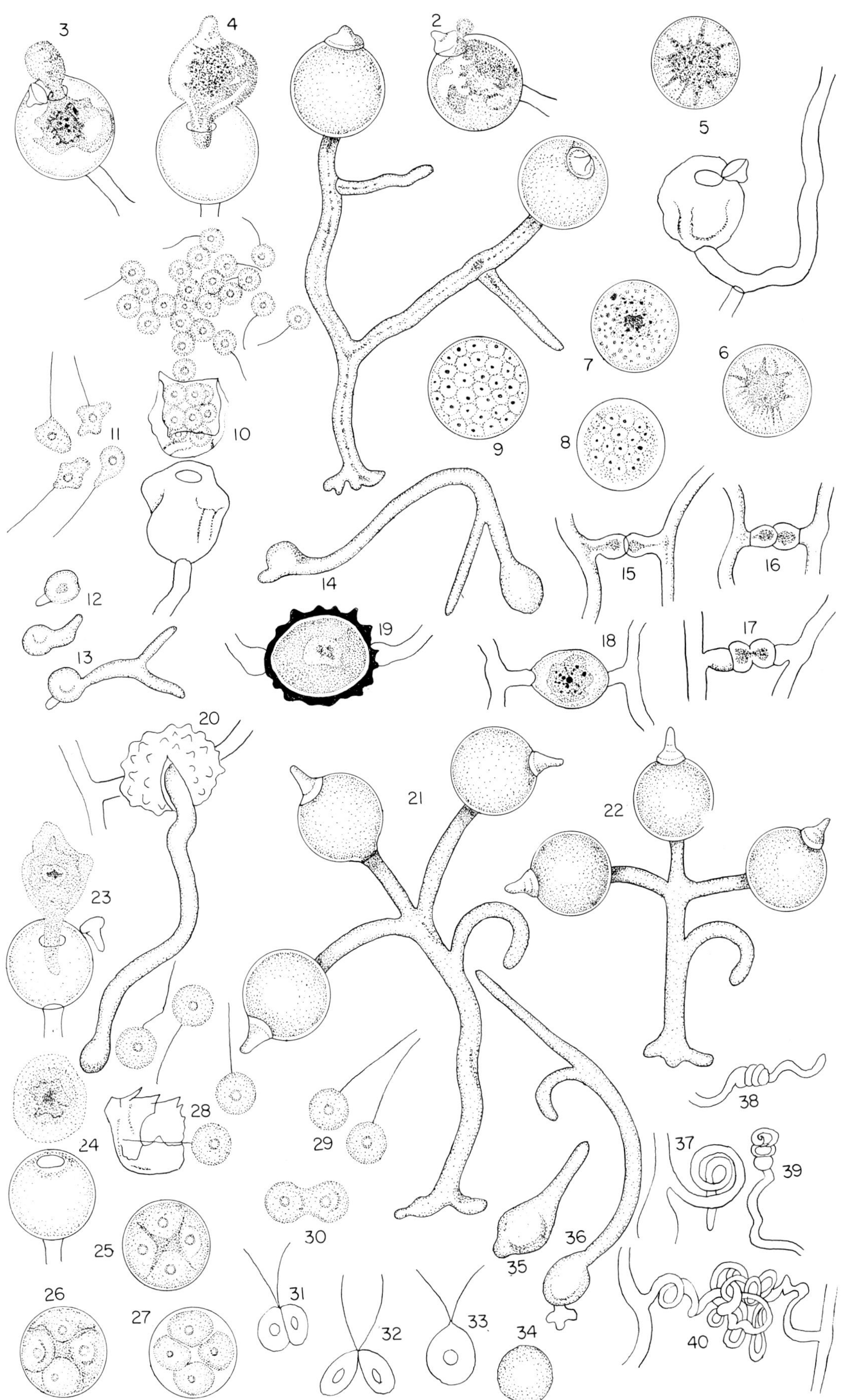

Zygochytrium, Tetrachytrium

TETRACHYTRIUM Sorokin

Bot. Zeit. 32:3, 1874; Bull. Soc. Nat. Kazan 4:15, 1874

Plate 121, Figs. 21-40

Tetrachytrium is an operculate polycentric genus of one species, *T. triceps* Sorokin, with bluish colored protoplasm which occurs as a saprophyte on wood, grass stems, and rarely on cadavers of coleoptera in southern Russia. It resembles *Zygochytrium* by its hypha-like thallus with a lobed holdfast and a central axis bearing 4 branches, 3 of which are terminated by operculate zoosporangia while the 4th one is sterile and curved (figs. 21, 22). As in *Zygochytrium* the protoplasm pushes off the pointed operculum as it emerges (fig. 23) and floats away as a naked mass after which it rounds up (fig. 24) and becomes enveloped by a thin membrane. It soon cleaves into 4 large blue-grey planospores (figs. 25–27), which are liberated as the membrane bursts (fig. 28). Some of these fuse in pairs (figs. 30–32) to produce an ovoid to spherical zygote (?) (figs. 33, 34) which germinates and developes into a sporangial thallus (figs. 35, 36). Apparently, no resting spores are formed from the fusion of these isogametes. The planospores or gametes which do not pair and fuse degenerate, and fusion, apparently, is a prerequisite for perpetuation of the sporangial thallus. Sorokin found *T. triceps* again in 1883 and reported that the sterile hypha may become tightly coiled (figs. 37–39). Occasionally, such hyphae from separate thalli become tangled or wefted together (fig. 40), suggesting the possible early stages of zygospore development, but Zorokin rejected this probability.

The affinities and taxonomic position of this species are obscure and uncertain. Sorokin regarded it as a chytrid and included it (1889) among his so-called Syphomycetes. Schroeter (1893) and Minden (1911) placed it in the Hyphochytriacene, but Fitzpatrick (1930) and others regarded it as a doubtful chytrid.

SACCOPODIUM Sorokin

Hedwigia 16:88, 1877.

Plate 122, Figs. 1-3

This monotypic genus is characterized by a polycentric tubular, branched, non-septate intramatrical thallus which produces 6–12 (4–5 μ diam.) zoosporangia in clusters at the tips of unbranched extramatrical sporangiophores (figs. 1–3). The zoosporangia dehisce at their apices and discharge oblong, minute, 1–1.5 μ, planospores (fig. 2). Neither flagella on the planospores, nor resting spores have been observed. Because it is incompletely known and has not been found since Sorokin's time chytridologists have questioned the validity of this genus as a chytrid. However, Sparrow's (1943) incomplete observations on a somewhat similar fungus in *Cladophora* has convinced him that *Saccopodium* is a valid chytrid genus.

This viewpoint has been strengthened by the more recent discovery by Willoughby (1966) of an extramatrical chytrid on leaf litter which bears primary, secondary and tertiary zoosporangia in succession on sporangiophores (figs. 4–24) in much the same manner as in some species of *Phytophthora* and *Pythium*. Its planospores are typically chytridiaceous (fig. 4), and during germination produce a long germ tube which becomes attached to the substratum (figs. 7–9) and functions as the primary isodiometric sporangiophore. Only rarely did Willoughby find a branching of the germ tube and the development of a tubular rhizoidal system (figs. 10, 11). Meanwhile, the planospore body enlarges to become the primary zoosporangium (figs. 12–16), and after it has discharged planospores a branch develops beneath it and becomes the secondary sporangiophore (figs. 17–19). This branch bears the secondary zoosporangium (figs. 20, 21),

PLATE 122

Figs. 1-3. *Saccopodium gracile* Sorokin. (Sorokin, 1877, 1883.)

Fig. 1. Portion of a thallus in *Cladophora* sp. with 2 sporangiophores bearing clusters of zoosporangia.

Fig. 2. Discharged planospores, 1-1.5 μ diam., from a cluster of zoosporangia.

Fig. 3. Cluster of empty zoosporangia.

Figs. 4-24. Leaf-litter Chytrid sp. on leaves of alder and oak. (Willoughby, 1966.)

Fig. 4. Spherical, 4.7-5.9 μ diam., planospore with an orange refractive globule and a 33 μ long flagellum.

Figs. 5, 6. Planospores showing profile and surface views, resp., of refractive globule.

Figs. 7-9. Germination stages of a planospore; germ tube becoming a primary sporangiophore, and planospore body enlarging into primary zoosporangium, resp.

Figs. 10, 11. Young thalli with branched rhizoids in detritus associated with leaf material.

Fig. 12. Fully grown primary zoosporangium with pigmented protoplasm.

Figs. 13-15. Mature primary zoosporangia with definitive orange refractive globules and sporangiophores of variable lengths.

Fig. 16. Cleavage of protoplasm into zoospore initials.

Figs. 17-19. Stages in the development of the secondary sporangiophore beneath, following the dehiscence of primary zoosporangium.

Fig. 20. Young secondary zoosporangium and sporangiophore.

Fig. 21. Thallus with empty primary and secondary zoosporangia.

Fig. 22. Development of tertiary sporangiophore as a branch beneath empty zoosporangia.

Fig. 23. Thallus with empty primary and secondary zoosporangia and a fully grown tertiary zoosporangium.

Fig. 24. Thallus with empty primary, secondary and tertiary zoosporangia.

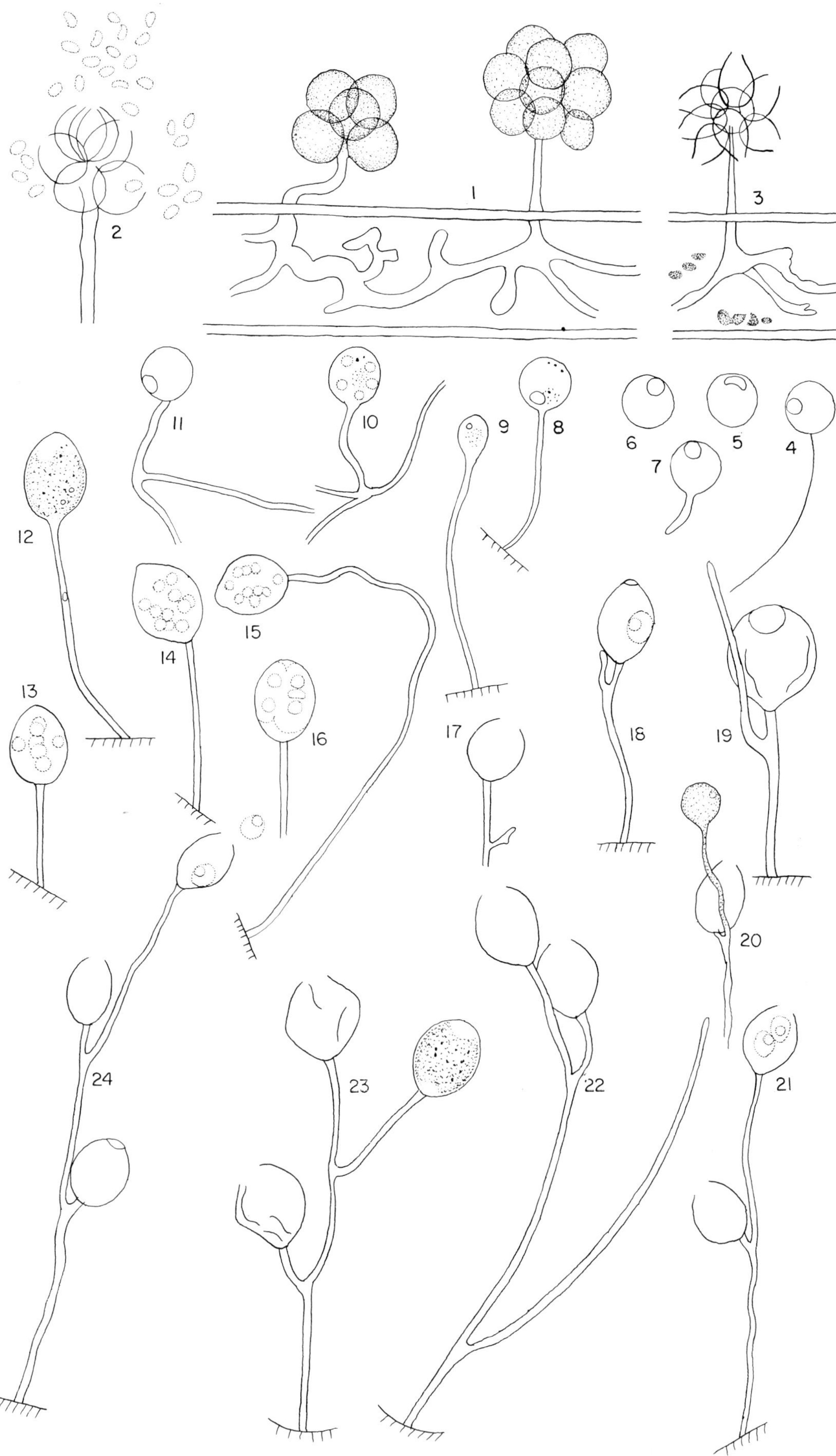

Saccopodium, *Chytrid* sp.

and after its dehiscence the tertiary sporangiophore and zoosporangium develop (figs. 22–24).

Willoughby did not observe resting spores, and our knowledge of this chytrid is incomplete. However, his observations lend credence to the existence of *Saccopodium, Tetrachytrium, Zygochytrium*, and other chytrids which bear zoosporangia on extramatrical sporangiophores.

IMPERFECTLY KNOWN, DOUBTFUL, AND EXCLUDED GENERA

RHIZIDIOCYSTIS Sideris

Phytopath. 19:376, 1929.

Nephromyces Sideris (non *Nephromyces* Giard, 1888). Pineapple News 11:102-106, 1927.

This genus was created for a parasite, *R. ananasi* Sideris, which kills the root hairs of pineapples in Hawaii. Although it does not produce planospores Sideris regarded it as a chytrid and classified it in the family Cladochytriaceae. Its vegetative thallus is an abundant white mycelium which is packed in the air spaces between the root hairs and consists of thin non-rhizoidal, filamentous arachnoid hyphae, about 1 μ in diameter, which bear intercalary spherical, oblong, triangular and variously-shaped swellings, 3–6 μ in diameter. Sideris described these as "sammelzellen" or turbinate cells. These give rise to 1–8 branches which produce kidney-shaped "sporangia" at their tips. Such "sporangia" do not produce planospores but develop on their concave side an emmission collar, 1–2 μ diam., by 5–15 μ long, that penetrates the root hair and into which it discharges the protoplasmic content of the "sporangia." This content grows at the expense of the root hair, killing and assimilating the protoplasm of the host, but Sideris did not indicate whether the discharged parasite's content is naked in the host or occurs as a broad mycelium.

Other "sporangia" produce resting spores or hypnospores on their convex side. These begin as a tube which enlarges at its tip and eventually become spherical with a rough to echinate surface as they mature and age. Occasionally, more than one hypnospore develops from a "sporangium", either in chains or from different tubes, but not more than one reaches maturity. These do not produce planospores during germination, but apparently form a germ tube or hypha, although Sideris was not certain about this.

The vegetative thallus of this fungus is somewhat similar in appearance to that of some members of the Cladochytriaceae, according to Sideris' photographs (fig. 6), and the production of the resting spores on short branches of intercalary swellings is slightly reminiscent of resting spore development in *Nowakowskiella ramosa, N. sculptura*, and some species of *Physoderma*; but the lack of planospore production in *Rhizidiocystis* precludes it from the chytrids and the family Cladochytriaceae as they are presently recognized. Sideris is of the opinion that the non-production of planospores by the so-called zoosporangia "is due to a change of habitat, namely, changing from an aquatic to a terrestrial mode of life, a condition forced upon it by its parasitic behaviour."

MYCELIOCHYTRIUM Johanson

Torreya 45:104, 1945.

The fungus for which this genus was created was first observed by the present writer in 1938 and described in connection with a paper on Brazilian chytrids before the annual meeting of the Mycological Society of America at St. Louis, Missouri, in 1946. In the meantime Johanson had rediscovered it in 1944 and given it the name *M. fulgens* in 1945. Since that time it and many other similar species have been found by Couch (1949, 1950, 1955), Karling (1954), Rothwell (1957) and numerous other workers to be members of the Actinoplanales. Accordingly, it excluded from the Chytridiales.

NEPHROMYCES Giard

C. R. Acad. Sci., Paris 106:1180, 1888

This genus was established by Giard for three parasites which occur in the kidneys of the ascidean genera *Mogula, Lithonephrya* and *Aneurella* in France, but he did not illustrate them. The thallus of the parasites consists of a delicate entangled, non-septate mycelium whose free ends are terminated by spheroid swellings. The delicate mycelium also produces a large number of thicker, irregularly cylindrical tubes which give rise to terminal variously-shaped zoosporangia, which in the parasites of *Mogula* may be bifurcate at the tips. The parasite in *Lithonephrya* is similar to the former, but is characterized by regularly pyriform zoosporangia. These produce minute spherical planospores with a basal granule and a long delicate flagellum, but Giard did not state whether the attachment of the flagellum is posterior, anterior, or lateral.

In the autumn months numerous zygospores are formed in the kidneys of the hosts. These are much larger than the planospores and are borne in an isolated manner where several filaments of the mycelium conjugate. They have a finely granulate and even a slightly echinulate envelope, and in the beginning of February they germinate by emitting two equal filaments or germ tubes.

The species reported by Harant (1931) from the kidneys of *Ctemicella appendiculata* and grown in cultures exhibits a wide diversity of form from that of spirochaetes, yeasts, bacilli up to 80 μ long by 5–6

μ in diameter, spores of dermatophytes, or fine mycelial filaments with intercalary enlargements, which apparently bear the zoosporangia. These produce from 5 to 20 planospores with an apical (anterior?) long flagellum which fuse in pairs to form biflagellate motile zygotes. These germinate in a few hours to produce a mycelium. Harant did not observe any conjugation of hyphae and the production of zygospores. He regarded it as an "incontestable" chytrid, and because of the intercalarly enlargements placed it close to *Catenaria* and in the family Cladochytriaceae. Unfortunately, his drawings do not indicate any distinct chytridiaceous characteristics.

Giard, also, regarded *Nephromyces* as a chytrid allied to *Catenaria*, but its affinity and relationships will remain obscure until it has been studied more intensively and is fully known. Nonetheless, the development of zygospores by conjugation of mycelial filaments and the germination of the spores by germ tubes, as reported by Giard, brings to mind *Zygochytrium*.

REFERENCES TO POLYCENTRIC GENERA OF QUESTIONABLE AFFINITY

Bessey, E. A. 1950. Morphology and taxonomy of fungi. IX-791 pp., 210 figs., Blakeston, Philadelphia.

Couch, J. N. 1949. A new group of organisms related to *Actinomyces*. J. Elisha Mitchell Sci. Soc. 65: 315-318.

———. 1950. *Actinoplanes*. A new group of the Actinomyceteles. Ibid. 66:87-92.

———. 1955. A new genus and family of the Actinomyceteles, with a revision of the genus *Actinoplanes*. Ibid. 71:148-155.

Fitzpatrick, H. M. 1930. The lower fungi – Phycomycetes. XI 331 pp. McGraw Hill, New York.

Giard, A. 1888. Sur les *Nephromyces*, genere nouveau, de Champignons parasites du rein des Molgulidees. C. R. Acad. Sci. Paris 106:1180-1182.

Harant, H. 1931. Contribution a l'histoire naturelle des ascides et de leurs parasites. Ann. Inst. Oceanograph. (Monaco) 8:231-389, 61 figs.

Johanson, A. E. 1945. A new mycelioid chytrid: *Myceliochytrium fulgens* gen. nov. et sp. nov. Torreya 45:104105.

Karling, J. S. 1954. An unusual keratinophilic microorganism. Proc. Indiana Acad. Sci. 63:83-86, 10 figs.

Minden, M. von. 1911. Chytridineae. Kryptogamenfl. Mark Brandenburg 5:209-422.

Rothwell, E. M. 1957. A further study of Karling's keratinophilic organism. Mycologia 49:68-72, 2 figs.

Schroeter, J. 1893. Phycomycetes. In Engler u. Prantl, Naturl. Pflanzenf. 1:63-141.

Sideris, C. P. 1927. A root-hair parasite of pineapples. Pineapple News 11:102-106.

———. 1929. *Rhizidiocystis ananasi* Sideris, nov. gen. et sp., a root hair parasite of pineapples. Phytopath. 19:367-382, 9 figs.

Sorokin, N. W. 1874a. Einige neue Wasserpilze (*Zygochytrium aurantiacum, Tetrachytrium triceps*). Bot. Zeit. 32:305-315, pl. 6.

———. 1874b. Aperçu systématique du groupe Siphomycètes. Bull. Soc. Nat. Kazan 4:1-26, pls. 1, 2.

———. 1877. Vorläufige Mittheilungen über zwei mikroskopische Pilze-*Pophyrtroma tubularis* und *Saccopodium gracile*. Hedwigia 16:87-88, 2 figs.

———. 1883. Aperçu systématique des Chytridiacées recoltées en Russia et dans l'Asie Central. Arch. Bot. Nord. France 2:1-44, 54 figs.

———. 1889 Materiaux pore la flore cryptogamique de l'Asie Central. Rev. Mycol. 11:69-85, 136-151, 207-208, 1890 Ibid. 12:3-16, 49-61, pls. 76-11.

Sparrow, F. K. 1943. The aquatic Phycomycetes . . . XIX-785 pp., 634 figs., Univ. Michigan Press, Ann Arbor.

———. 1960. Aquatic Phycomycetes..., 2nd ed., XXI-1187 pp., 91 figs., Univ. Michigan Press, Ann Arbor.

Willoughby, L. G. 1966. An unusual chytrid from incubated leaf litter. Trans. Brit. Mycol. Soc. 49: 451-455, 27 figs.

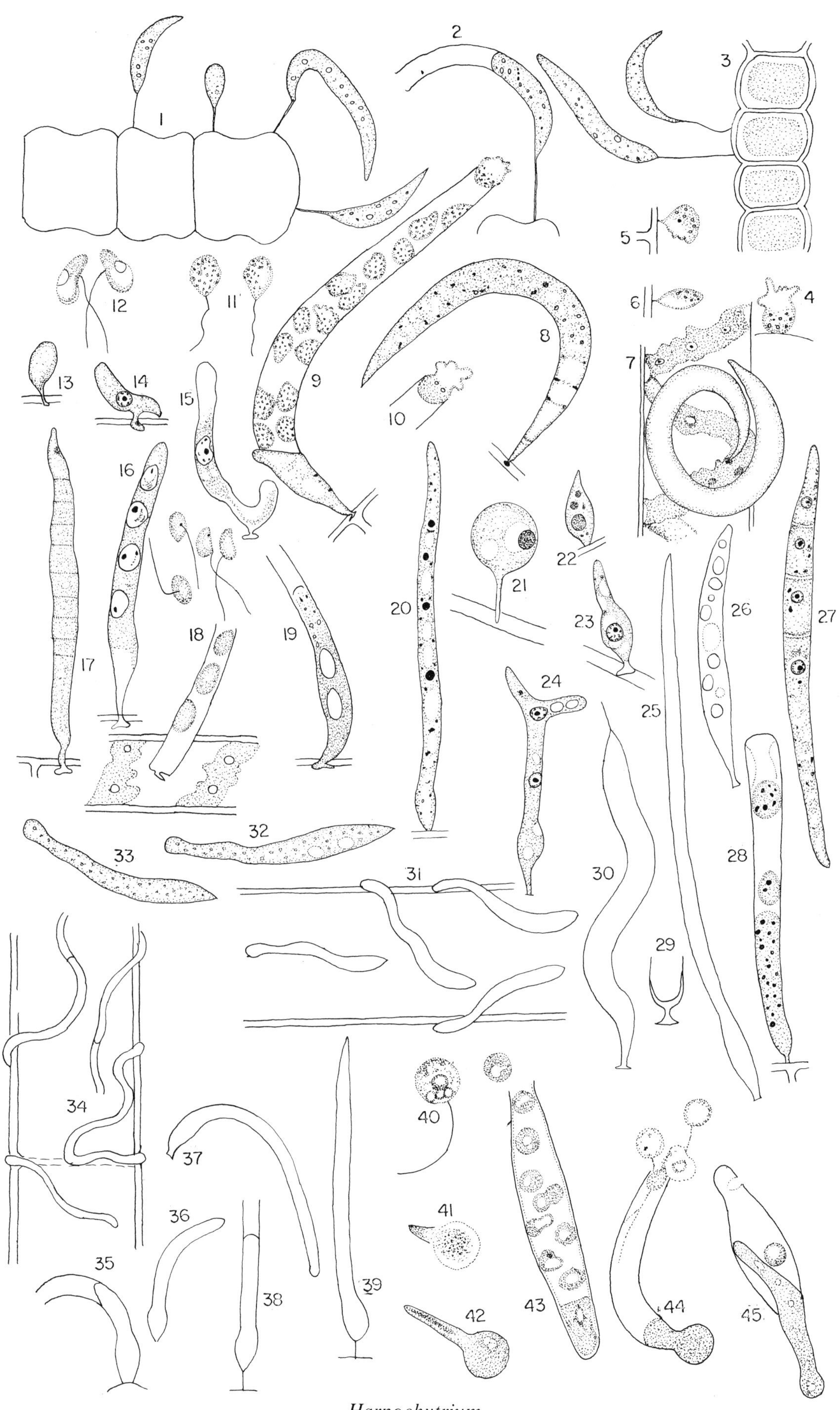

Harpochytrium

Chapter IX

HARPOCHYTRIALES

Emerson and Whisler, Arch. f. Mikrobiol.

61:208, 1968

HARPOCHYTRIACEAE Wille

(sensu recent. Emerson and Whisler,

Arch. f. Mikrobiol. 61:210, 1968)

Emerson and Whisler emended Wille's family Harpochytriaceae and established a new order for it among the Chytridiomycetes on the basis of their studies on *Harpochytrium* and *Oedogoniomyces* which they had isolated from various substrata and grown in axenic cultures on and in synthetic media. According to their diagnosis, this order and family are characterized by a eucarpic, monoaxial, coenocytic, epibiotic thallus with an extracellular, basal, disc-like holfast and posteriorly uniflagellate planospores borne in zoosporangia which proliferate in *Harpochytrium* or are delimited and mature in basipetal succession in *Oedogoniomyces*. These dehisce by a circumscissile rupture of the wall at or near the apex which delimits a cap-like portion that may become detached or remain attached at one side of the zoosporangium. Neither exit papillae nor tubes are developed for the discharge of the planospores. As in the

PLATE 123

Figs. 1-3. *Harpochytrium hyalothecae* Lagerheim on *Hyalotheca dissiliens* and other algae. Figs. 1, 2 after Lagerheim, 1890; fig. 3 after Gobi, 1899.

Fig. 1. Stalked, curved thalli.

Fig. 2. Proliferation of the zoosporangium.

Fig. 3. Stalked thalli on *Sphaerozosma vertebratum*.

Figs. 4-11. *Harpochytrium hedinii* Wille on *Spirogyra* and *Zygnema*. (Atkinson, 1903.)

Fig. 4. Amoeboid planospore shortly after escaping from a zoosporangium.

Figs. 5, 6. Attachment of planospores to the alga and elongation, resp.

Fig. 7. Coiled thallus on *Spirogyra*.

Fig. 8. Curved U-shaped thallus.

Fig. 9. Discharge of planospores; zoosporangium beginning to proliferate at the base.

Fig. 10. Amoeboid planospore in the exit orifice.

Fig. 11. Ovoid planospores, $1 \times 5\ \mu$, (?) with granular content.

Figs. 12-19. *Harpochytrium hedinii* (*Rhabdium acutum*) on *Spirogyra*. (Dangeard, 1903.)

Fig. 12. Actively motile reniform planospores with a nucleus.

Fig. 13. Attachment and elongation of planospore.

Figs. 14-16. Uni- and tetra-nucleate thalli, resp.

Fig. 17. Segmentation of the protoplasm into planospore initials.

Fig. 18. Discharge of planospores.

Fig. 19. Proliferating zoosporangium.

Figs. 20-29. *Harpochytrium tenuissimum* Korsch. emend. Jane, on various algae. Fig. 20 after Korschikoff, 1931; figs. 21-29 after Jane, 1942, 1946.

Fig. 20. Erect, elongate 4-nucleate thallus.

Fig. 21. Enlarged view of vital stained germinating encysted planospore.

Fig. 22. Young sporeling, fixed in formol.

Fig. 23. Uninucleate thallus with a swollen base.

Fig. 24. Binucleate branched thallus.

Fig. 25. Greatly elongated thallus (Jane, 1946.)

Fig. 26. Thallus with oil globules which showed no change after 9 weeks; apparently in a resting condition.

Fig. 27. Segmentation of upper portion of protoplasm into uninucleate planospore initials.

Fig. 28. Zoosporangium with 2 planospores in the upper half.

Fig. 29. Base of a thallus with thickened wall.

Fig. 30. *Harpochytrium apiculatum* Pascher; thallus with a fine apical bristle. (Pascher, 1938.)

Figs. 31-33. *Harpochytrium adpressum* Scherffel forma *hedinii* Jane. Fig. 31 after Pascher, 1938; figs. 32, 33 after Scherffel, 1926.

Fig. 31. Four thalli lying on an algal filament.

Figs. 32, 33. Shapes of thalli.

Fig. 34. *Harpochytrium* (?) *vermiculare* Pascher; thalli on *Spirogyra* (Pascher, 1938.)

Figs. 35-37. *Harpochytrium botryococci* Jane; thalli on *Botryococcus*. (Jane, 1946.)

Figs. 38, 39. *Harpochytrium monae* Jane; thalli on *Zygnema* and *Zygogonium*. (Jane, 1946.)

Figs. 40-45. *Harpochytrium hedinii* from axenic cultures; sketched from photographs by Emerson and Whisler, 1968.

Fig. 40. Spherical planospore with a large central globular body and smaller globules.

Figs. 41, 42. Germination of the planospores.

Fig. 43. Zoosporangium with planospores and residual protoplasm in the base.

Fig. 44. Final stage of planospore release.

Fig. 45. Proliferation of a zoosporangium; apical cap or operculum attached at the apex.

Chytridiales, the wall of the thallus is composed largely of chitin, and intensive histochemical texts have not indicated the presence of cellulose.

Emerson and Whisler did not indicate where they would place the Harpochytriales among the Chytridiomycetes, and it remains to be seen whether or not their order will stand as distinct from the Chytridiales and Blastocladiales. Nonetheless, they are to be highly commended for their excellent study of these fungi. In our present state of knowledge of the chytrids as well as of these fungi, the Harpochytriaceae might be accommodated better perhaps in the Chytridiales as a separate family of eucarpic chytrids, possibly after the Rhizidiaceae. Certainly, the manner of dehiscence of all but the apical zoosporangium in *Oedogoniomyces* is unique, but the delimitation of an apical cap by a circumscissile weakening and rupture of the sporangium wall in *Harpochytrium* does not appear to be distinctly different from the delimitation of an operculum in certain chytrids. In *Karlingia dubia* and *Rhizophlyctis oceanis*, for example (see Karling, 1949, figs. 36 - 39; 1969, figs. 21, 22, 52, 53; Willoughby, 1957, fig. 3), a circumscissile ring or weakening of the wall occurs whereby the operculum is eventually delimited. Also, the apical cap on the zoosporangium of *Harpochytrium* and on the terminal one in *Oedogoniomyces* may be strikingly similar in appearance to the operculum of *Chytridium microcystidis* Skuja, for instance (see pl. 67 fig. 51). Furthermore, it may be mentioned that the stalked thalli of some species of *Harpochytrium* are not unlike in appearance, at least, those of some of the eucarpic chytrids with an unbranched rhizoid which were included formerly in *Phlyctidium*. While such chytrids are reported to be parasitic they might be only epiphytic, and it is not definitely known nor has it been proven that the so-called rhizoid draws nourishment from the host. Possibly, axenic cultures of such chytrids in liquid media may show that the thallus can draw its nourishment directly from the substrate as it does in *Harpochytrium* and *Oedogoniomyces.*

Proliferation of zoosporangia from an uncleaved portion of protoplasm in the base of the thallus occurs in eucarpic, monocentric, and polycentric chytrids *(Physoderma)* and other zoosporic fungi (Traustochytriaceae, for example) also, and of itself is not a distinctive generic, familial, or ordinal characteristic. Also, the planospores, particularly of *Harpochytrium*, are chytrid-like, and the presence or absence of a distinct nuclear cap is neither exclusive nor inclusive because nuclear caps have been reported to be present in the planospores of chytrids (pl. 1, figs. 31-39). The planospores of *Oedogoniomyces* sp., isolate CR84, on the other hand, have some of the characteristics of those of the Monoblepharidales, as Emerson and Whisler, and Travland and Whisler (1971) pointed out.

In structure, size, and shape the thalli of some species of *Harpochytrium* and those of *Oedogoniomyces*, respectively, are similar to those of *Amoebidium* and the monoaxial Eccrinales such as *Enterobryus* and *Eccrina* of the Trichomycetes, as Emerson and Whisler emphasized. However, these genera do not produce flagellate planospores and the walls of their thalli react *positively* to tests for cellulose but not for chitin. Accordingly, Emerson and Whisler believe there is greater justification for including the Harpochytriales in the Chytridiomycetes than for associating them with the Trichomycetes.

HARPOCHYTRIUM Lagerheim

Hedwigia 29:143, 1890

Fulminaria Gobi. 1899-1900. Scripta Bot. Horti Univ. Imp. Petropol Fasc. 15:283.

Plate 123

This genus includes 5 species and 6 forms, according to Jane (1946), who excluded the pigmented species which has been added from time to time. Also, he reduced several other species which had been named and placed in this genus to forms of the more characteristic species, and his classification will be followed here primarily as a convenience. *Harpochytrium* is characterized by posteriorly uniflagellate planospores and a tapering cylindrical, up to 200 μ long by 8–10 μ in diameter, coenocytic, but rarely septate, straight, curved, falcate or coiled, erect or adpressed, rarely branched, non-rhizoidal, epiphytic thallus which is attached to algae, debris and various substrata by a stalk, foot or holdfast. At maturity the protoplasm of the upper half or more of the thallus segments into planospore initials while that of the lower portion remains unsegmented and gives rise later to a secondary zoosporangium within the primary one. No resting spores have been observed, but Jane (1942) reported that in lens paper cultures thalli may go into a resting condition and remain unchanged for 9 weeks (fig. 26), or aplanospores containing numerous fat globules and a thick wall may be formed in culture.

The developmental sequence of *Harpochytrium* as seen through the light microscope is relatively simple and involves the attachment of the planospore to the substratum by a stalk or foot (figs. 4–6, 13, 14, 21–23) and its elongation into a straight, curved, coiled, erect or adpressed, epiphytic thallus (figs. 15–17). In this process the nucleus of the planospore (fig. 21) and young thalli (figs. 14, 15, 23) apparently divides whereby the elongate thallus becomes multinucleate (figs. 16, 20). At maturity, as noted earlier, the protoplasm in the upper half or more of the thallus cleaves into a single or rarely more than one row of uninucleate planospore initials (fig. 27) while that in the lower portion remains unsegmented. In preparation for dehiscence a circumscissile weakening and rupture of the wall occurs just below the apex resulting in cap-like terminal portion (fig. 45) which may be displaced to 1 side or removed completely as suggested by Atkinson (1903) and confirmed by Emerson and Whisler (1968). According to the latter authors, the planospores, 5–15 or more in number, are spherical (fig. 40) with a whiplash posteriorly attached flagellum, but Dangeard reported and illustrated them to be ovoid to reniform with the flagellum attached near the middle or laterally (figs. 12, 18). Atkinson showed them to be ovoid with numerous refractive granules (figs. 9, 11), and reported that after emerging they become amoeboid and creep about for a while, after which they round up and dart away. After swimming about for 10 to 15 minutes they come to rest on the substratum (fig. 4) and become amoeboid again. They become attached to the substratum by an elongate pseudopod (fig. 5), according to Atkinson, but Gobi (1899) and Jane (1946) reported that the flagellum serves as the attachment organ. In liquid broth axenic cultures Emerson and Whisler found that the thalli do not develop a holdfast and become attached, but remain freely suspended.

The unsegmented protoplasm in the lower part of the thallus or zoosporangium grows into a secondary incipient zoosporangium (figs. 19, 28, 45) and undergoes the same maturation phases as the primary one. A few tertiary zoosporangia have been reported. The secondary and tertiary zoosporangia are usually shorter than the primary one and do not fill the thallus completely. Proliferation of the zoosporangium is basically similar to that which occurs in several genera and species of the Chytridiales.

As noted previously the planospores, development of the thallus, sporogenesis and mitosis of *H. hedeni* have been studied ultrastructurally by Travland and Whisler (1971, 1973). During encystment of the planospore the axoneme persists; the nuclear cap membrane undergoes blebbing and gradually disappears as such as the endoplasmic reticulum ramifies the ribosomal complex; the ribosome become dispersed in the cytoplasm; the striated rootlet (disk) material and rumposome disappear, and the mitochondria elongate. In the process of germination the 2 centrioles maintain a close association with the nucleus but are isolated from the plasma membrane. Together with a small dictyosome they form a centrosomal complex. In preparation for sporogenesis the process of organelle dispersal is reversed. The axoneme, striated rootlet (disk) and rumposome are reformed, and the ribosomes aggregate into a nuclear cap. At the same time the mitochondria round up and change in staining capacity. Finally, cleavage furrows delimit the planospore initials, and towards the end of cleavage one of the centrioles elongates to form the functional kinetosome or flagellar base.

Since the time of its discovery *Harpochytrium* has been classified among the Chytridiales and algae. Lagerheim regarded it as a chytrid, a view which Dangeard, Atkinson, Fischer (1892), Schroeter (1893), Minden (1911–1915), and Fitzpatrick (1930) accepted. Gobi, on the other hand, did not regard it as a plant. Wille (1900) placed it among the algae as a reduced form which has lost its chlorophyll, and erected the family Harpochytriaceae for its reception. This view received support from Scherffel's (1926) inclusion of a species with a green chromatophore in *Harpochytrium*, thereby implying that the colorless species were algae without pigment. Several algologists, including Pascher (1938) and Fritsch (1956), as well as some mycologists (Gwynne-Vaughan and Barnes, 1937), followed Scherffel's view and classified all species as members of the Xanthophyceae. Jane (1942, 1944, 1946) made an intensive study of both the colorless and pigmented species and segregated the green species in the algal genus *Chytridiochloris*. He interpreted the colorless species to be fungi, but he did not classify them in any of the existing orders of the Phycomycetes. His view that they are fungi has been fully confirmed by Emerson and Whisler and Travland and Whisler who showed that they are typically chytridiomycetous insofar as they are known.

OEDOGONIOMYCES Kobayashi and Okubo

Bull. Nat. Sci. Mus. (Tokyo) 1:62, 1954 (Sensu recent. Emerson and Whisler, Arch. f. Mikrobiol. 61:195-211, 1968)

Plate 124

This genus of one, *O. lymnaeae* Kobayashi and Okubo, or possibly more species or varieties, was erected for a fungus which occurred on the shells of pond snails and grass leaves in Japan and was so named because it exhibits some of the characteristics of *Oedogonium* in the dehiscence of its zoosporangia. Four isolates of *Oedogoniomyces* were subsequently made by Emerson and Whisler (1959, 1968) from various substrata and in soil in Costa Rica and California, U.S.A., and grown in axenic cultures on and in various media. These were designated by numbers but not definitely identified as *O. lymnaeae*.

Oedogoniomyces is characterized by large, 6–9 μ diam., monoblepharid-like, posteriorly uniflagellate planospores (fig. 20) and an elongate, up to 0.5–1 cm long by 20–37 μ in diameter, filamentous, cylindrical, straight or curved (fig. 1), coiled or spiral (fig. 3), apically tapering (fig. 5) thallus which is attached to the substratum by a disc-like holdfast or appressorium (fig. 3). At maturity most of the thallus becomes septate in basipetal succession (fig. 3), producing relatively short cylindrical segments which successively become zoosporangia (figs. 6, 7). These dehisce by a circumcissile rupture of the wall near the apex of each segment (fig. 25), and as each zoosporangium opens a cap-like piece may be pushed aside as in the dehiscence of the sporangium of *Oedogonium*. However, in the dehiscence of the apical zoosporangium of a filament a conical cap (fig. 23) is pushed off as in

Harpochytrium. Approximately 1/4 to 1/2 of the thallus remains permanently vegetative in strain CR22, according to Emerson and Whisler, and apparently is not segmented into zoosporangia. In their isolates CR84 and CR90, however, "entire thalli become septate almost to the very base."

The segments of the thallus may, also, become thick-walled gemmae in *O. lymnaeae* (figs. 17, 18), which later separate and become dormant. In germination they form planospores (fig. 26) which may germinate within the segments (fig. 19). Apparently, the gemmae are to be regarded as dormant zoosporangia, and in one isolate they resisted drying for as long as 7 years. These represent the only known resting stage, although Kobayashi and Okubo described and illustrated chlamydocysts which Emerson and Whisler interpreted to be resting sporangia of *Allomyces*.

The development of the vegetative thallus is similar to that of *Harpochytrium*. The encysted planospore, (fig. 21) form a germ tube which elongates (figs. 12, 13, 22) into the thallus and becomes attached to the substratum by a holdfast. Kobayashi and Okubo's figs. 12 and 13 suggest that the holdfast might originate from the planospore cyst, but Emerson and Whisler reported that it is non-cellular and excreted by the thallus.

Reichle's (1972) study of the fine structure of *Oedogoniomyces* planospores showed that ultrastructurally they are very similar to those of *Monoblepharella*. "Both have rumposomes, a striated disk, and an accessory kinetosome which had previously been thought absent in *Monoblepharella*."

Kobayashi and Okubo placed *Oeodogoniomyces* in the order Blastocladiales and the family Blastocladiaceae on the basis of the planospore structure and presence of thick-walled chlamydocysts, which, as noted above, have been interpreted subsequently to be resting sporangia of *Allomyces*.

REFERENCES TO THE HARPOCHYTRIALES

Atkinson, G. F. 1903. The genus *Harpochytrium* in the United States. Ann. Mycologici 1:479-502, pl. 10.

———. 1904. Note on the genus *Harpochytrium*. J. Mycol. 10:3-8, figs. 24-26, pl. 72.

Dangeard, P. A. 1903. Un noveau genre de Chytridiacées; le *Rhabdium acutum*. Ann. Mycologici 1:61-64, pl. 2.

Emerson, R., and H. C. Whisler. 1959. The nature and relationships of *Oedogoniomyces*. Proc. IX Intern. Bot. Congr. 2:103-104.

———. 1968. Cultural studies of *Oedogoniomyces* and *Harpochytrium*, and a proposal to place them in a new order of the aquatic Phycomycetes. Arch. f. Mikrobiol. 61:195-211, 19 figs.

Fischer, E. 1892. Phycomycetes. Die Pilze Deutschlands, Oesterreichs und der Schweiz. Rabenh. Kryptogamen.- Fl. 1:1-490. Leipzig.

Fitzpatrick, H. M. 1930. The lower fungi—Phycomycetes. McGraw-Hill, New York.

Fritsch, F. E. 1956. The structure and reproduction of the algae. Vol. 1. Cambridge Univ. Press.

Gobi, C. 1899. *Fulminaria mucophila* nov. gen. et spec. Script. Bo. Horti Univ. Imp. Petropol. Fasc. 15:283-292, pl. 7, figs. 1, 2.

Gwynne-Vaughan, H. C. I., and B. Barnes. 1937. The structure and development of the fungi. Cambridge Univ. Press.

Jane, F. W. 1942. *Harpochytrium tenuissimum* Korsch.

Plate 124

Figs. 1-26. *Oedogoniomyces lymnaeae* Kobayashi and Okubo and *Oedogoniomyces* spp. on various substrata. Figs. 1-5, 15, 20-24, 26 after Emerson and Whisler, 1968; figs. 6-14, 16-19, 25 after Kobayashi and Okubo, 1954. Figs. 15, 20-24, 26 sketched from photographs.

Figs. 1-3. Variations in sizes and shapes of thalli.

Fig. 4. Portion of a septate thallus of zoosporangia with discharged planospores.

Fig. 5. Tapering apical portion of a thallus of isolate CR84.

Fig. 6. Similar portion greatly enlarged with planospores in the apical zoosporangium.

Fig. 7. Zoosporangium about to dehisce through a circumscissile thin band at the apex.

Fig. 8. Separation of partly empty short cylindrical zoosporangia.

Fig. 9. Narrow zoosporangium with 2 rows of planospores.

Figs. 10, 11. Greatly enlarged, 6-9 μ diam., planospores with nuclear cap and small granules.

Figs. 12, 13. Sporelings.

Fig. 14. Base of an empty thallus.

Fig. 15. Base of a thallus with a holdfast in isolate CR90, attached to a corn grain.

Fig. 16. Fused bases or holdfasts of 8 fasciculate thalli.

Fig. 17. Row of immature gemmae.

Fig. 18. Mature, thick-walled, separating gemmae.

Fig. 19. Germinated gemmae; 2 with sporelings.

Fig. 20. Monobleparid-like planospore of isolate CR84 with a lunate body like a nuclear cap, anterior granules, and a grouping of granular material at the posterior end.

Fig. 21. Encysted planospore of isolate CR84.

Fig. 22. Enlarged view of a sporeling os isolate CR90.

Fig. 23. Starved thallus of isolate CR90 with a single apical dehisced zoosporangium; cap persistent at the apex.

Fig. 24. Three empty attached zoosporangia of isolate CR84 showing the circumscissile rupture of the wall at the apical end.

Fig. 25. Enlarged view of an initial stage of the circumscissile rupture of the wall at the apical end of a zoosporangium.

Fig. 26. Row of gemmae of isolate CR90 with some planospores and sporelings at the upper end.

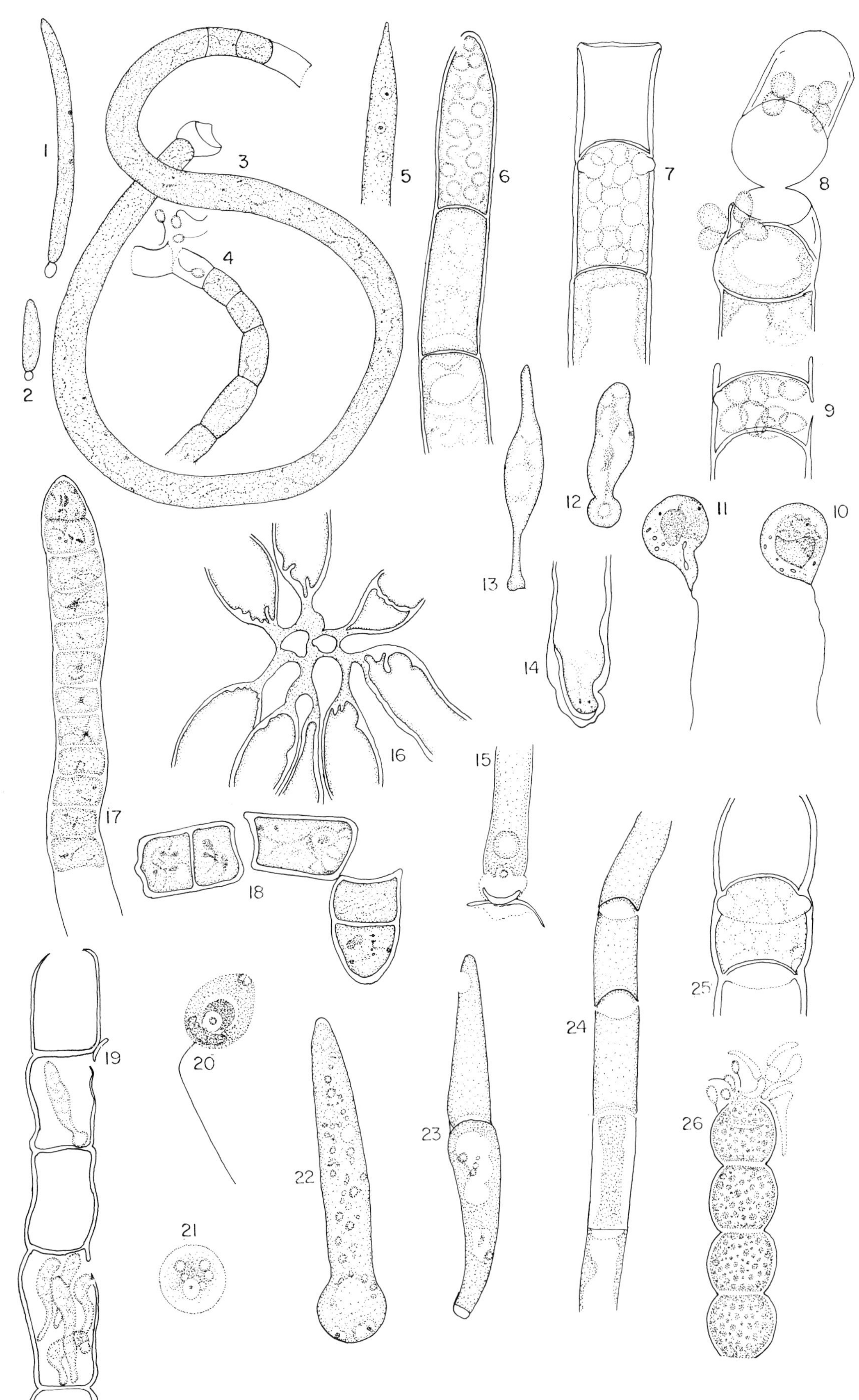

Oedogoniomyces

New Phytol. 41:91-100, 37 figs.
———. 1944. The genus *Chytridiochloris*. Ibid. 43:154-163, 54 figs.
———. 1946. A revision of the genus *Harpochytrium*. J. Linn. Soc., London, Bot. 53:28-40, 28 figs.
Karling, J. S. 1949. New monocentric eucarpic operculate chytrids from Maryland. Mycologia 41: 505-522, 78 figs.
———. 1969. Zoosporic fungi of Oceania. VII. Fusions in *Rhizophlyctis*. Amer. J. Bot. 56:211-221, 108 figs.
Kobayashi, Y., and M. Okubo. 1954. On a new genus *Oedogoniomyces* of the Blastocladiales. Bull. Nat. Sci. Mus. (Tokyo) n.s. 1:59-66, 6 figs., pls. 21, 22.
Korschikoff, A. A. 1931. Notizen über einige neue apochloritische Algen. Arch. f. Protistenk. 74: 249-258, 22 figs.
Lagerheim, G. 1890. *Harpochytrium* und *Achlyella*, zwei neue Chytridiaceen Gattungen. Hedwigia 29:142-145, pl. 2.
Minden, M. von. 1915. Chytridiineae, Ancylistineae, Monoblepharidineae, Saprolegniineae. Kryptgamenfl. Mark Brandenburg 5:193-630.
Pascher, A. 1938. Heterokonten. Rabenhorst's Kryptogamen—Fl. 11:641-832.
Reichle, R. E. 1972. Fine structure of *Oedogoniomyces* zoospores, with comparative observations of *Monoblepharella* zoospores. Canad. J. Bot. 50: 819-824, pls. 1-7.
Scherffel, A. 1926. Beitrage zur Kenntnis der Chytridineen. Teil II. Arch. f. Protistenk. 54:510-528, pl. 28.
Schroeter, J. 1893-1897. Chytridineae. *In* Engler u. Prantl, Natürl. Pflanzenf. 1:64-87.
Travland, L. B., and H. C. Whisler. 1971. Ultrastructure of *Harpochytrium hedenii*. Mycologia 63: 767-789, 18 figs.
Whisler, H. C., and L. B. Travland. 1973. Mitosis in *Harpochytrium*. Arch. f. Protistenk. 115:69-74, pls. 2-6.
Wille, N. 1900. Algen aus dem nordlichen Tibet von Dr. S. Hedin im Jahre 1896 gesammelt. Petermann's Georg. Ergänzungsb. 28:370-371.
Willoughby, L. G. 1957. Studies on soil chytrids. II. On *Karlingia dubia* Karling. Trans. Brit. Mycol. Soc. 40:9-16, 5 figs.

Chapter X

BLASTOCLADIALES

As interpreted by most present-day chytridiomycetologists this order includes 3 families, Catenariaceae, Coelomomycetaceae, and Blastocladiaceae, 8 or possibly 13 genera, approximately 50 species, and several varieties, which are either parasites of aquatic larvae of insects, eggs of liver flukes, adult nematodes and aquatic fungi, or saprophytes in water and soil. The terricolous species have been trapped and isolated on various substrata floated on watered soil samples, but in most cases it is not known what their natural substrata are. The aquatic saprophytes occur commonly on rosaceous fruits, twigs of ash, birch, oak and horse chestnut, or on bits of animal debris. So far as they are known species of this order may vary markedly in size, organization, and type of development in the various genera and families, but they have several common characteristics. Among these are the production of posteriorly uniflagellate planospores or gametes with a conspicuous nuclear cap in many species and the method of germination of the resistant sporangia which involves swelling of the contents, cracking of the thick outer wall, and the protrusion of an endosporangium with planospores as in some species of *Physoderma*. This type of germination has been observed in all families and in all but one genus. Bipolar germination of the planospores has been cited as another common characteristic, but in some species germination may be monopolar.

The thallus of some species may be sessile small, globular and monocentric as in the rhizidiaceous chytrids. In others it may be greatly elongate, polycentric and hypha-like with rhizoids along its length, and bears a succession of zoosporangia and resistant sporangia as in the Cladochytriaceae. In the parasitic species of aquatic larvae of insects, rhizoids and thin-walled zoosporangia appear to be lacking or unknown, and the hyphae are usually coarse, cylindrical, or blunt, and lack walls. In the larger saprophytic species the thallus usually consists of a basal cell or axis which is anchored to the substratum by branched rhizoids, branches dichotomously, subdichotomously, or sympodially at its apex, and bears zoosporangia, resistant sporangia, or gametangia at the ends of the branches. Thus, the form of the thallus in this order may parallel that of the chytrids, Leptomitales, and the true mycelial fungi. It may be non-septate, pseudo-septate, or septate, and the walls are composed largely of chitin-glucan and do not give a positive reaction to tests for cellulose.

Asexual reproduction in this order occurs by planospores from thin-walled zoosporangia and resistant sporangia, although the former structure is unknown or lacking in the Coelomomycetaceae. Details of this process of reproduction will be given later for the separate families and genera.

Sexual reproduction is unknown in most members, but in 10 species among *Catenaria*, *Blastocladiella* and *Allomyces* it occurs by the fusion of iso- or aniso-planogametes. In species with anisogamous reproduction, an alternation of isomorphic gametophytic and sporophytic generations occurs. But in some species with isogamous reproduction the gametophyte may resemble the sporophyte, or be represented in other species by encysted haploid planospores from germinated resistant sporangia which function holocarpically as gametangia and produce 4 gametes each. Male and female gametangia develop on the same gametophyte in *Allomyces*, and such thalli are apparently homothallic. In long-cycled genera the gametophytic thallus is reported to bear either a + or − gametangium. In species which have been studied cytologically the sporphyte is diploid from its inception as a zygote to mature development, and meiosis occurs during germination of the resistant sporangia. These produce haploid planospores which germinate and grow into haploid gametophytes. Further details of the type of sexual reproduction and life cycles will be given with the descriptions and illustrations of the various genera.

So far as is known the Blastocladiales appears to occupy a niche between the Chytridiales and Monoblepharidales, but the distinctions between them and the Chytridiales are becoming less and less sharp. Formerly, the reported presence of a conspicuous nuclear cap and side body in the planospores and a bipolar mode of germination were regarded as distinctive characteristics of the Blastocladales, but as noted in the Introduction ultrastructural studies of chytrid planospores have revealed the same basic structure with the possible exception of a membrane-enclosed nuclear cap. Furthermore, the thallus of some members of this order (*Blastocladiella stübenii, Catenophlyctis variabilis, Catenaria* species, etc. for example) may be strikingly similar in structure, development, and organization to those of monocentric rhizidiaceous or of polycentric cladochytriaceous chytrids, and germination of the resistant sporangia is basically the same as in some species of *Physoderma* of the Physodermataceae. Thus, the characteristics of some species merge with those of members of the Chytridiales, and sometimes, as in the case of *Catenophlyctis variabilis* for example, it is difficult to determine with certainty the order to which they belong. Certainly the affinities of the Blastocladiales with the Chytridiales are pronounced as many chytridimycetologists have indicated in their studies. Cantino (1950, 1955) in particular has emphasized this relationship on the basis of the synthetic capacities of the two orders.

As to the relations of the 3 families and their genera to each other, Sörgel (1952) postulated, as shown in Plate 165, that the families Catenariaceae and Coelomomycetaceae might be sideline off-shoots of but overlap to some degree with the family Blastocladiaceae, and that the last-named family may have given rise in a direct line to the Monoblepharidiales. Emerson (1950) however, believed it unlikely that a haplontic (?) genus like *Monoblepharis* evolved from species with an isomorphic or diplontic life cycle as in members of the subgenera *Allomyces* and *Cystogenes*. Included in Sörgel's scheme are listed the genera *Brevicladia, Ramocladia, Leptocladia,* and *Allocladia* which Sörgel named and described briefly without diagnoses before his death in a glider accident. He retained *Sphaerocladia* as a genus and placed it in the Catenariaceae because its thallus may be elongate, branched and similar to those of the polycentric specimens of *Catenophlyctis* and *Catenaria*.

The similarity of the life cycles of the Blastocladiales to those of some algae has been regarded by some investigators as indicating a relationship of this order with algae. As pointed out by C. W. Wilson (1952) *Ulva* and *Cladophora*, for example, have cycles similar to that of *Allomyces (Euallomyces)*. Also, the reduction of the gametophyte in the Cyclosporales of the brown algae is similar to that found in *Cystocladiella* and *Cystogenes*, and the lack of a haploid stage under certain conditions in *Ectocarpus* is comparable to the omission of the sexual phase in the subgenera *Blastocladiella* and *Brachyallomyces*. In summarizing the research on the existence of diploid plants in green algae Schussnig (1930) pointed out that where diploid and haploid plants occur the zygote, as in *Allomyces*, does not become dormant. Also, in species where equal and separate haploid and diploid plants occur a delay of gametogenesis after chromosome reduction as well as of meiosis following gametic karyogamy is necessary. These similarities in life cycles are probably the result of convergent evolution or are parallelisms of development and may not indicate relationships. Certainly, present-day evidence indicates strongly that the Blastocladiales are closely related to the Chytridiales.

CATENARIACEAE

This is a relatively small family of 3 genera, *Catenophlyctis, Catenomyces* and *Catenaria*, and approximately 10 species. The first of the genera was formerly regarded as a possible member of the Rhizidiaceae while the latter two were included in the Cladochytriaceae. As interpreted here, the Catenariaceae is characterized by a predominantly polycentric cladochytriaceous type of thallus or rhizomycelium which consists of catenulate zoosporangia and resting spores connected by short or long, septate or non-septate isthmuses, and branched rhizoids borne at the ends of the rhizomycelium and along the isthmuses as well as from the surface of the zoosporangia and resting sporangia. Occasional monocentric, holocarpic, olpidioid thalli develop in *Catenaria* growing in pollen grains, and in this genus as well as in *Catenomyces* monocentric, eucarpic, rhizidiaceous thalli may develop, also. In *Catenophlyctis* the latter type of thallus occurs frequently and abundantly, and this genus is both monocentric and polycentric. Accordingly, so far as present knowledge goes the Catenariaceae appears to have links with the Rhizidiaceae and Cladochytriaceae. However, it remains to be shown that the Catenariaceae is a coherent group of genera. Nolan (1970) found a difference in the nutritional

PLATE 125

Figs. 1-17. *Catenophlyctis variabilis* Karling. (Karling, 1947, 1965.)

Fig. 1. Thirteen monocentric thalli in a fragment of human hair; intramatrical rhizoids not visible.

Figs. 2, 3. Ovoid, 2-2.5 × 3.8-4.5 μ, and amoeboid planospores with finely granular protoplasm.

Fig. 4. Fixed and stained planospore with a conspicuous nuclear cap, nucleus, nucleolus, side bodies and whiplash flagellum.

Fig. 5. Encysted, stained, uninucleate planospore.

Fig. 6. Germination of planospore on snake skin; nucleus has migrated down into the swelling of the germ tube.

Fig. 7. A later but early developmental stage of a uninucleate monocentric thallus; planospore cyst persistent.

Figs. 8, 9. *Rhizophlyctis*-like germination of planospores and their direct development into incipient zoosporangia with rhizoids, fully and partly extramatrical.

Fig. 10. Early development of an intramatrical monocentric thallus from an enlargement of the germ tube; planospore cyst persistent.

Fig. 11. A *Rhizidium*-like monocentric thallus with one central rhizoidal axis.

Fig. 12. Portion of a *Rhizophlyctis*-like thallus with a long exit canal and delimited planospores.

Fig. 13. Initial stage of planospore discharge.

Fig. 14. Later stage of planospore discharge; planospores swarming within the zoosporangium.

Fig. 15. "Germination" of a thick-walled dormant or resistant (?) sporangium; endosporangium has cracked the outer wall and is protruding as a broad papilla.

Fig. 16. Median view of a "germinated" resistant sporangium with a long neck.

Fig. 17. A small portion of a polycentric thallus growing in snake skin with zoosporangia and thick-walled dark brown resistant (?) sporangia connected by tubular, infrequently septate isthmuses.

Catenophlyctis

requirements of *Catenaria* and *Catenophlyctis* and suggested that "it may be necessary to erect a transitional family in the Order Chytridiales or Order Blastocladiales to accommodate organisms whose morphological, nutritional and biochemical aspects indicate affinities with members of both of these orders."

Sexual reproduction has been reported in only one species, *Catenaria allomycis* Couch, in which the planospores from germinated resting spores encyst and function as holocarpic gametangia, giving rise to 4 isoplanogametes. These fuse in pairs and form motile biflagellate zygotes which later infect the host. Thus, the life cycle and type of sexual reproduction are similar to those of the subgenera *Cystocladiella* of *Blastocladiella* and *Cystogenes* of *Allomyces*. On these grounds *Catenaria allomycis* might possibly be accommodated better as a member of the Blastocladiaceae.

CATENOPHLYCTIS Karling

Amer. J. Bot. 52:133, 1965

Perirhiza Karling, 1946. Amer. J. Bot. 33 (suppl. 3):219

Plates 125, 126

This inoperculate genus appears to be unique at present in that it develops both monocentric and polycentric thalli which vary to a high degree. The monocentric thalli may exhibit an *Entophlyctis* and *Rhizophlyctis* type of development, and in some instances they may resemble the thallus of *Rhizidium* except for their intramatrical position. The polycentric thalli may be very similar to those of *Catenaria anguillulae* with the exception that the resting or dormant sporangia do not lie in a vesicle. Also, at the edges of bits of fibrin film the largely extramatrical thallus may be tree-like and somewhat similar in general appearance to the abnormally growing thalli of *Blastocladiella variabilis* as illustrated by Stûben (1939, figs. 11, 12). So far no other genus has been found which exhibits such a wide and frequent range in organization, and *C. variabilis* Karling may be both rhizidiaceous and catenariaceous in structure and appearance. Both monocentric and polycentric thalli appear to be totipotent for the two types of organization, according to Karling (1951). However, monozoosporic isolates of monocentric thalli yielded a preponderance of monocentric offspring in the course of 6 generations, while a similar number of generations of polycentric monozoosporic isolates gave rise to a relatively larger number of polycentric thalli.

Apparently, *Catenophlyctis* is a monotypic genus at present because Sparrow's (1950) *Phlyctorhiza peltata* (fig. 20) appears to be identical with *C. variabilis*. The latter species was first named *Perirhiza endogena* and later described as *Phlyctorhiza variabilis* on the grounds that its thalli seemed to be predominantly monocentric and somewhat similar to those of *P. endogena*. Later, as it was found to develop extensive polycentric thalli it was made the representative of a new genus, *Catenophlyctis*, and transferred to the family Catenariaceae. This species is a common soil inhabitant throughout the world and has a predilection for keratinic substrata, although it can be trapped on fibrin film and cellulosic substrata such as boiled grass leaves, also. It grows readily on synthetic media and has been shown by Rothwell (1956) to utilize inorganic sulfur and requires biotin and nicotinamide for growth.

In living material the planospores are ovoid to elongate with numerous refractive granules (fig. 2, 3), but in fixed and stained preparations a conspicuous nuclear cap, nucleus, and side bodies (fig. 4) are visible. Germination is predominantly monopolar, and the intramatrical enlargement or incipient zoosporangium in monocentric thalli begins as a swelling in the short germ tube (fig. 6) as in *Entophlyctis* or *Endochytrium*. Meanwhile, the planospore cyst may persist and later appear as an appendage on the developing zoosporangium (figs. 7, 10). As the latter enlarges it usually becomes angular in median outline as the rhizoids extend out at its extremities. In other cases the encysted planospore may develop directly into the incipient zoosporangium on the surface of the substratum or partly within it (figs. 8, 9) with radiating rhizoids as in *Rhizophlyctis* and *Karlingia*. In instances when the germ tube is long and the incipient zoosporangium develops considerably above its branches, a thallus like that of *Rhizidium* (fig. 11) may develop with a single basal rhizoidal axis.

For dehiscence, one or more long, straight, curved or coiled, or short exit tubes, or prominent papillae develop on the thin-walled zoosporangia and penetrate to the outside of the substratum (figs. 12, 14). If the sporangial wall is thickened the exit tubes rupture the outer layer of the wall and protrude out as an endosporangium (figs. 15–17). The tip of the tubes or papilla deliquesces, and the initial emerging planospores form a globular mass at the exit orifice (fig. 13). As these disperse the planospores within the zoosporangium begin to swarm and emerge in a stream from the exit tube (fig. 14).

Plate 126

Figs. 18, 19. *Catenophlyctis variabilis*. (Karling, 1965.)

Fig. 18. Portion of an unusually compact thallus with mostly spherical thin-walled zoosporangia connected by isthmuses and bearing long spine-like rhizoids; growing in a bleached corn leaf.

Fig. 19. Portion of a tree-like thallus growing at the edge of a piece of fibrin film bearing light to dark brown resting (?) or dormant sporangia and incipient ovoid to broadly fusiform thin-walled zoosporangia.

Fig. 20. *Phylctorhiza peltata* Sparrow; monocentric thallus discharging planospores. (Sparrow, 1950.)

Catenophlyctis

The development of polycentric thalli begins like that of the monocentric ones, either from a swelling in the germ tube or by direct enlargement of the planospore body. As the rhizoids of such thalli elongate swellings occur in them and become incipient secondary zoosporangia. Meanwhile the initial swelling develops as the primary zoosporangium, and is usually recognizable by the presence of the attached planospore cyst (fig. 17). Tertiary incipient zoosporangia develop from swellings in the rhizoids of the secondary zoosporangia, and this process is repeated until an extensive thallus with tertiary, quartenary, etc., zoosporangia is formed. Such zoosporangia are connected by long or short isthmuses which may be continuous or occasionally septate with rhizoids arising along their lengths (fig. 17) as well as from the periphery of the zoosporangia. In instances when replication occurs largely in one direction the thallus looks strikingly like that of *Catenaria* species. When it occurs almost radically, a fairly compact mass of zoosporangia, isthmuses, and rhizoids may develop (fig. 18). The rhizoids arising from the periphery of the zoosporangia may be almost elongately spine-like, single, or branched at their tips.

At the edges of human fibrin film the thallus may look so different that it may be mistaken for that of another fungus unless its source and origin are known. On this substratum the thalli may become quite coarse, large, extensively branched and often tree-like (fig. 19) with terminal and subterminal thin-walled zoosporangia and yellowish-amber to light- and dark-brown dormant sporangia or resistant sporangia (?), and terminal rhizoids which may extend for distances of 2.5–5 mm in the surrounding water, and become markedly curved and coiled. The main axis of such thalli may attain a diameter up to 74 μ and bear numerous rhizoids as well as branches.

Whether or not the yellowish-amber to light- and dark-brown thick-walled dormant sporangia borne on monocentric and polycentric thalli are true resistant sporangia is questionable. They are not formed by the contraction of the contents of sporangium-like bodies as in some species of *Catenaria* but are formed apparently in the same manner as those of *Catenaria sphaerocarpa* and *C. indica*. However, they "germinate" in a manner similar to that of the resting spores of *Catenaria anguillulae, C. allomycis*, and *C. vermicola*. The outer wall is cracked open locally (figs. 15, 16, 19) as a broad papilla or tube protrudes for the emission of planospores. The fact that some thinner-walled zoosporangia may "germinate" in a similar manner (fig. 17) suggests that these thick-walled bodies might be only encysted dormant zoosporangia.

CATENOMYCES Hanson

Torreya 44:30: 1944; Amer. J. Bot. 32:431–438, 52 figs. 1945

Plate 127

This monotypic endooperculate genus occurs as a saprophyte in soil and water and has been isolated in monozoosporic cultures (Hanson, 1945) on cello-

PLATE 127

Figs. 1-24. *Catenomyces persicinus* Hanson in bits of cellophane. Figs. 1-21, 27, 28 after Hanson, 1945; figs. 22-26 after Sparrow, 1965.

Fig. 1. A small portion of an intricate rhizomycelium with 3 zoosporangia, connecting hyphae or filaments, and rhizoids.

Figs. 2, 3. Spherical, 3.7-4.5 μ diam., and amoeboid planospores with several golden refractive globules; flagellum about 30 μ long.

Fig. 4. Fixed and stained planospore with a conspicuous nuclear cap, nucleus and nucleolus.

Fig. 5. Monopolar germination of a planospore.

Figs. 6, 7. Bipolar (?) germination stages of a planospore with sterile filaments arising on the upper periphery.

Figs. 8-10. Early stages in the development of the thallus; planospore cyst persistent.

Fig. 11, 12. Deliquescence of the tip of an exit tube and initial stages of the formation of a plug of matrix in the exit orifice.

Figs. 13, 14. Later stages; sporeplasm receding in the canal prior to the development of the endooperculum.

Fig. 15. Later stage; operculum fully formed and invaginated.

Fig. 16. Evagination of the operculum and its outward elevation by the planospores; viscid matrix has dispersed.

Fig. 17. Rupture of operculum at one edge and the emergence of planospores one by one.

Figs. 18-21. Various appearances and structure of the endoopercula.

Fig. 22. Portion of a thallus of *C. persicinus* (?) with coarse filaments, rhizoids, and a zoosporangium with 2 septate exit canals.

Fig. 23. Plug of matrix in the exit orifice.

Fig. 24. Enlarged view of ovoid planospore.

Fig. 25. Discharge of planospores through a pore in the basal membrane or septum in the exit canal.

Fig. 26. Empty zoosporangium showing the exit pore in the septum.

Figs. 27, 28. Hyaline thick-walled intercalary bodies with numerous small and large golden refractive globules; probably resting spores.

Catenomyces

phane and other cellulosic substrata and subsequently reported from other parts of the world (Karling 1949, 1967; Sparrow, 1965; Johnson, 1973). It is characterized by a profuse, fairly coarse, polycentric nonseptate rhizomycelium, somewhat similar to that of *Nowakowskiella* and *Catenaria* with occasional intercalary enlargements, zoosporangia, connecting filaments, rhizoids, and possibly resting spores (fig. 1, 28). Occasionally, monocentric thalli develop. *Catenomyces persicinus* Hanson is readily recognizeable in and on substrata by the peach, apricot or golden color of the globules in the hyphae, zoosporangia, planospores, and resting bodies. Another similar unidentified *Catenomyces* sp. has been found in Iceland, and Johnson (1973) reported it to be inoperculate.

In living material the planospores contain several golden globules (figs. 2, 3) but when fixed and stained a conspicuous nuclear cap, nucleus, and nucleolus become visible (fig. 4). The flagellum is about 30 μ long with a fine lash at its tip, and as the planospore comes to rest the flagellum is gradually absorbed. Germination is usually monopolar (fig. 5) but sterile filaments may develop from the opposite periphery of the planospore (figs. 6–9) and extend into the surrounding water. The germ tube is usually coarse and blunt (figs. 6, 7), and as it elongates and branches an extensive rhizomycelium is formed.

The zoosporangia usually develop from enlargements in the rhizomycelium, but sometimes the primary one may develop as an enlargement of the germ tube (fig. 10). They vary markedly in sizes and shapes and may develop 1 to 9 exit canals or tubes which frequently branch. However, only one of these may be functional. At maturity the tip of the functional one deliquesces (fig. 11) and a globular mass of viscid matrix begins to exude (figs. 12, 13). This continues as the orifice at the tip dilates slightly until it is filled with a plug of this matrix. Also, this material may extend downward and fill a considerable portion of the canal (figs. 15, 18). In the meantime, the granular protoplasm recedes in the canal (figs. 13, 14), and its upper convex surface becomes denser. Apparently, the endooperculum is formed by condensation of protoplasm at this surface, and when fully formed it is at first convex. Later, it invaginates (fig. 15). After a few hours it evaginates and becomes convex again about the time the viscid matrix above disperses (fig. 16). The first planospores are discharged under such pressure or with such force that the operculum is torn loose and carried out with the spores. The remaining planospores emerge one by one, and if the operculum is not ripped out it is raised up by each planospore as it passes out (fig. 17). These spores usually form a temporary but naked group or globular mass at the exit orifice, but in a few instances they were observed to be enveloped by a distinct hyaline matrix which soon dispersed. The endoopercula may be distinctly apiculate and bear a tenuous hyaline spine (figs. 18–21).

Resting spores have not been conclusively demonstrated in *C. persicimus*, but hyaline, thick-walled, ovoid, elliptical and almost triangular bodies with large and small golden refractive globules were found some-

PLATE 128

Figs. 1-28. *Catenaria anguillulae* Sorokin. Figs. 2, 28 after Villot, 1874; fig. 3 after Sorokin, 1876; figs. 4, 11, 14, 15-17, 19-24, 27 after Karling, 1934, 1968; fig. 5 after Butler, 1928; fig. 6 after Butler and Buckley, 1928; figs. 7, 8, 18, 25, 26, 29 after Couch, 1945; fig. 9 after Butler and Humphries, 1932; fig. 10 after Gaertner 1962; fig. 13 after Dangeard, 1885.

Fig. 1. Portion of a rhizomycelium with catenulate zoosporangia, isthmuses and rhizoids; drawn from living material by Karling.

Fig. 2. Spherical, 2 μ diam., non-motile (encysted?) planospores with a hyaline refractive globule.

Fig. 3. Flagellate planospores, 1.5-2 μ diam., with a hyaline refractive globule.

Fig. 4. Ovoid, 4-5 × 6-7.5 μ, and elongate planospores with a nuclear cap, nucleus, and granules.

Fig. 5. Ovoid 4-4.5 × 6.5-7.5 μ, and subspherical planospores.

Fig. 6. Beaked, 4-5 × 6-7 μ, planospores.

Figs. 7, 8. Ovoid, 3.8-5.4 × 6.7-8 μ, fixed and stained planospores with a nuclear cap, nucleus, side body and granules.

Fig. 9. Monopolar germination of planospores.

Figs. 10, 11. Primary swellings or sporangial rudiment forming in the germ tube; planospore cyst adherent.

Figs. 12, 13. Bipolar germination of planospores; planospore body developing directly into the primary swelling.

Fig. 14. Primary swelling formed in the germ tube and transversely to it.

Fig. 15. Monocentric eucarpic thallus developed in an agar film hanging drop.

Fig. 16. Monocentric, holocarpic *Olpidium*-like thallus in a pollen grain.

Fig. 17. Beginning of planospore discharge through a long exit canal.

Fig. 18. Later stage, initially emerged planospores forming a globular mass at the exit orifice.

Figs. 19-22. Stages in formation of resting spores by the contraction of the protoplasm and its investment by a hyaline to yellowish wall, leaving a "pip" or projection at the apex.

Fig. 23. Intercalary swelling which has been transformed completely into a resting spore without contraction of its protoplasm.

Figs. 24, 25. Germination of resting spores; wall of the spore has been cracked open by a long exit canal.

Figs. 26. Planospore from a germinated resting spore.

Fig. 27. Germinated resting spore; the exit canal developed rhizoids and elongated into an extensive rhizomycelium without the production of planospores.

Fig. 28. The first illustration of a resting spore in a sporangium-like structure.

Fig. 29. Enlarged view of perforated septum.

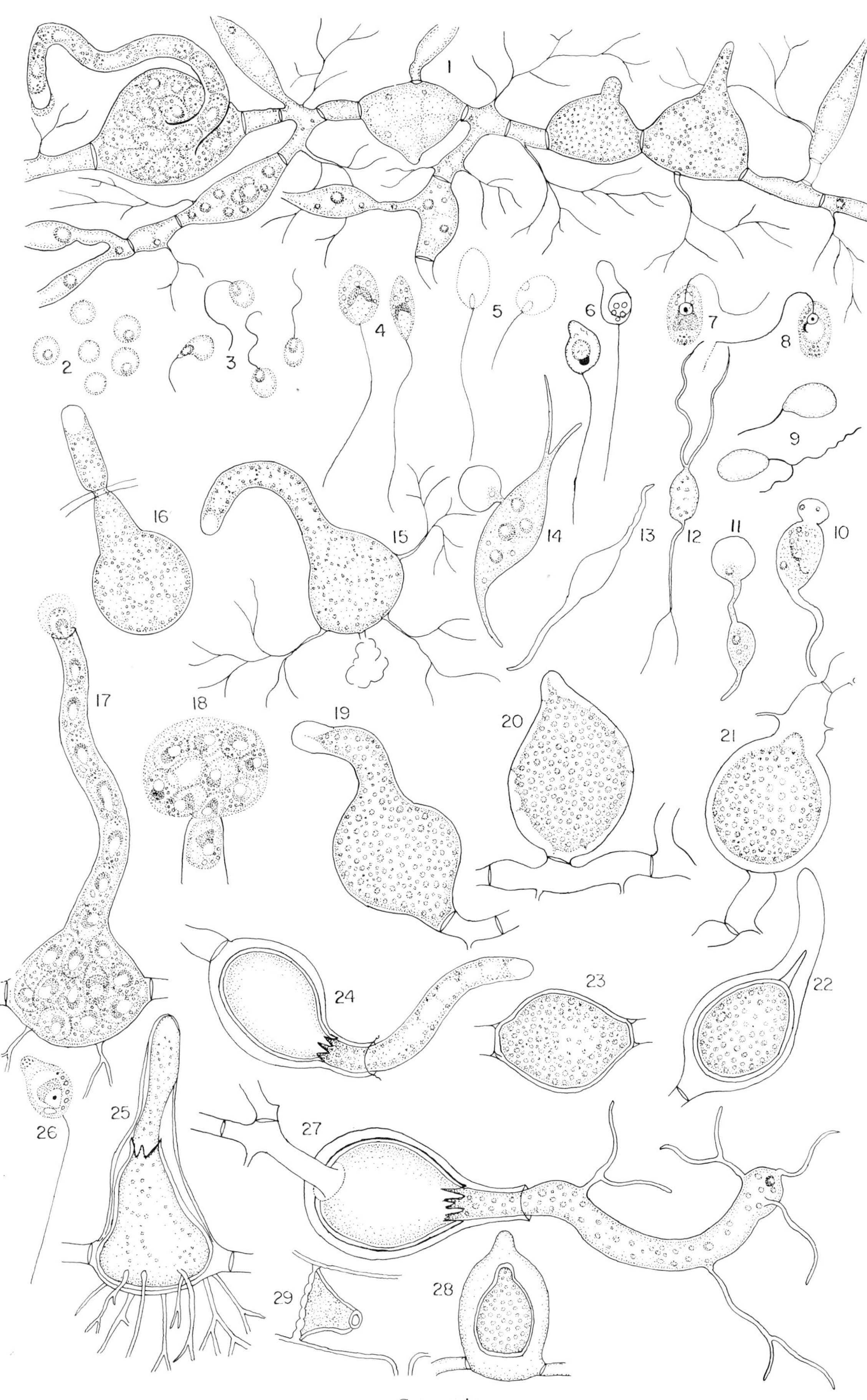

Catenaria

times on the rhizomycelium (figs. 27, 28). These have not been germinated, and their significance in the life cycle of this species is uncertain.

It is not certain that the salmon-pink species described by Sparrow (1965) in cellophane from Hawaii is identical with Hanson's species although he reported it as such with some doubts. Its coarse rhizomycelium with hyphal-like filaments, which may excede 8 μ in diameter (fig. 22), bears rhizoids and zoosporangia singly or in clusters at the tips of the extramatrical branches. The zoosporangia bear 1 to 3 exit tubes which are septate at the base or have a basal membrane (figs. 23, 25, 26), and usually develop a plug of matrix in the exit orifice (fig. 23). During dehiscence a pore is formed in the basal membrane through which the planospores emerge (figs. 25, 26). No endoopercula nor resting spores were observed.

Possibly the polycentric orange colored endooperculate species which Willoughby (1962) described as *Rhizophlyctis rosea* may be *Catenomyces persicinus* or another species of this genus, but he did not refer it to Hanson species because of the lack of apiculate endoopercula. Large sterile swellings developed consistently on the rhizomycelium, and the orifices of the exit tubes were filled with a plug of gelatinous matrix. The planospores were slightly larger than those of *C. persicinus* with granular content.

Hanson (1945) included *Catenomyces* temporarily in the Cladochytriaceae on the basis of the structure and organization of the thallus, but Sparrow (1960) transferred it to the Catenariaceae. Knox (1970), on the other hand, believed that it should be merged with *Catenaria*.

CATENARIA Sorokin

Ann. Sci. Nat. Bot. VI, 4:67, 1876

Plates 128-130

Catenaria is characterized by a coarse cladochytriaceous type of rhizomycelium with terminal and intercalary zoosporangia and resting spores connected by long or short, septate or non-septate tubular isthmuses which may or may not bear rhizoids. Such rhizoids may be borne, also on the periphery of the zoosporangia and resting spores. Occasionally, the thallus may be monocentric and holocarpic as in *Olpidium* or monocentric and eucarpic as in species of *Entophlyctis*. Eight species* are included in this genus at present, but it is not certain that all of them belong here. *Catenaria sphaerocarpa* Karling, for example, has been questioned as a valid member because of the chytridiaceous nature of its planospores, but similar planospores have been reported by Villot (1874) and Sorokin in *C. anguillulae* Sorokin and by Singh and Pavgi (1970) in *C. indica* Singh and Pavgi. Probably more than one species is lumped under the binomial *C. anguillulae*; at least, this is suggested by the differences in size and structure of the planospores reported by several workers. The genus is almost worldwide in distribution, and some of the species occur as saprophytes in soil, vegetable debris and moribund freshwater algae from which they may be isolated readily on cellulosic and keratinic substrata. Others are parasitic or weakly so in eggs of microscopic animals and liver flukes, mites, adult rotifers, midge eggs, numerous species of nematodes, and *Daphnia, Allomyces,* and *Blastocladiella*. Three species, *C. anguillulae, C. allomycis* Couch, and *C. vermicola* Birchfield, have been grown and maintained on various types of synthetic media, and the nutritional requirements of *C. anguillulae* have been studied intensively by Nolan (1970).

The planospores of the type species, *C. anguillulae*, have been reported to be of different sizes and structure. Villot, Sorokin, and Sparrow (1932) reported them to be spherical, 2, 1.5–2, and 2 μ diam., respectively, with 1 or 2 refractive globules (figs. 2,3), but subsequent workers described them as ovoid, 3.8–5.4 × 6.7–8 μ, with a tapering anterior end (figs. 4–8), a distinct nuclear cap and nucleus, side body, several small refractive bodies, and a whiplash flagellum. On the basis of

*Martin (personal communication) reported two additional species in midge eggs which differ from *C. spinosa* in dimension and host range and in the shapes of the appendages on the resting sporangial wall. In one the appendages are branched and in the other they are simple.

PLATE 129

Figs. 30-39. *Catenaria sphaerocarpa* Karling saprophytic in numerous algae. (Karling, 1938.)

Fig. 30. Spherical 4-4.8 μ diam., planospore with a conspicuous hyaline refractive globule and nuclear cap.

Fig. 31. Germination of a planospore in water.

Fig. 32. Young thallus with a primary sporangial rudiment and elongate swellings developing in the branched germ tube and rhizoids.

Fig. 33. Portion of a thallus with 6 zoosporangia, isthmuses, and rhizoids.

Figs. 34-36. Stages in the deliquescence of the exit canal tip and initial phases of planospore discharge.

Figs. 37-39. Intercalary, elongate, ovoid, and spherical resting spores with a thick dark-brown smooth wall and evenly granular content.

Figs. 40-43. *Catenaria vermicola* Birchfield parasitic in numerous nematodes. (Birchfield, 1960.)

Fig. 40. Portion of a nematode with filaments, zoosporangia and resting spores; rhizoids lacking.

Fig. 41. Discharge of planospores.

Fig. 42. Flagellate, ovoid, 1.3-3.1 × 2.6-4.2 μ, and nonflagellate planospores.

Fig. 43. Germination of a resting spore.

Fig. 44. *Catenaria verrucosa* Karling saprophytic in snake skin. (Karling, 1967.)

Fig. 44. Portion of a thallus with 3 empty zoosporangia and another one with 6 quiescent planospores; verrucose resting spores with a brown wall lying in intercalary and terminal enlargements.

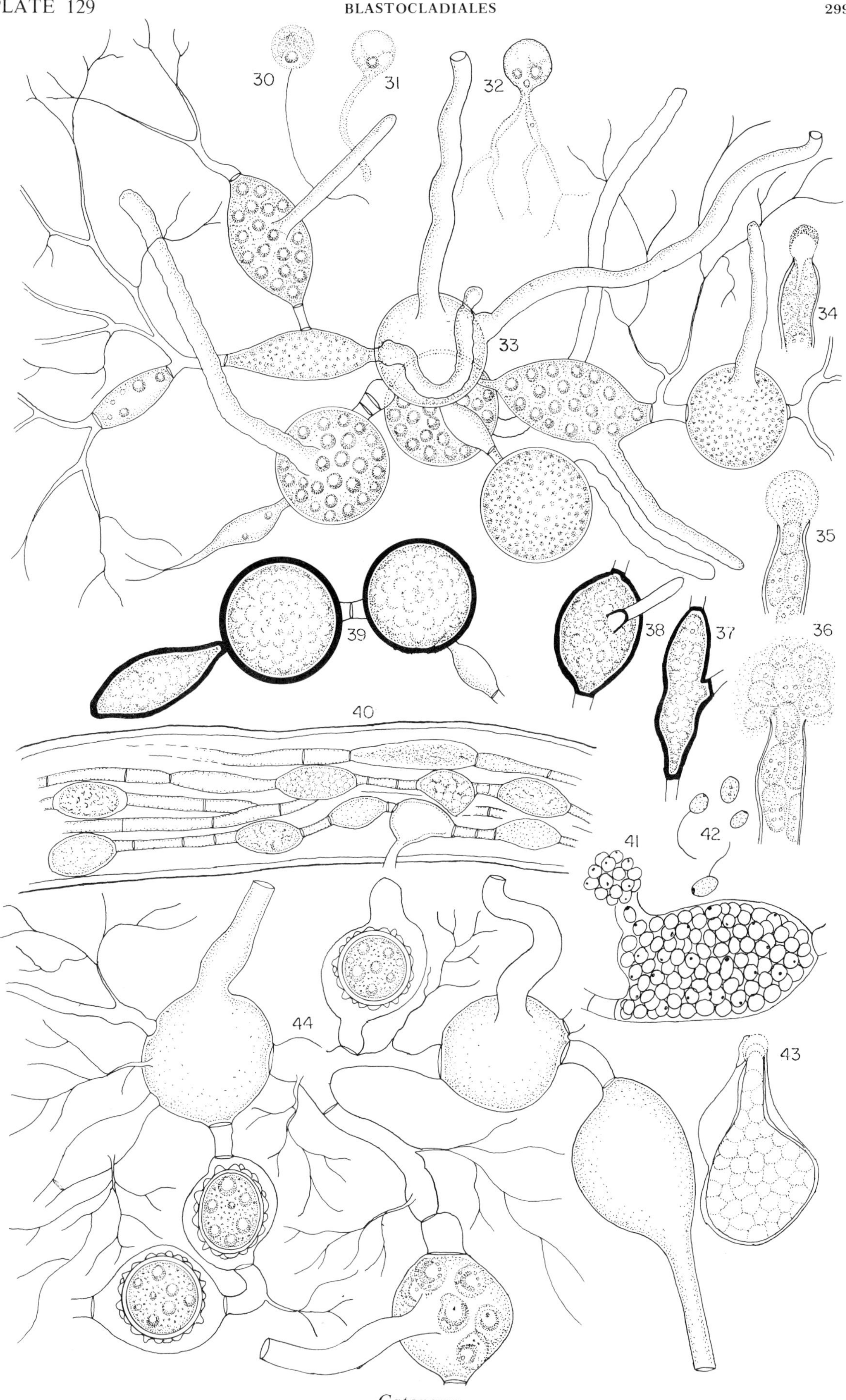

Catenaria

these reported differences it is possible that Villot's, Sorokins's and Sparrow's species might relate to *C. vermicola* in which the planospores are 1.3–3.1 × 2.6–4.2 μ. In *C. sphaerocarpa* they are spherical, 4–4.8 μ in diam., with a conspicuous hyaline refractive globule and nuclear cap (fig. 30), and in *C. indica* they are ovoid, 4.4–6 μ, with a small globule (fig. 69). In *C. spinosa* the elliptic, 4.5–7.5 × 7.0–11.0 μ, planospores are characterized by numerous anterior filose pseudopodia by which they move in the gelatinous egg matrix (fig. 75).

So far two ultrastructural studies have been made of the planospores by Fuller (1966, 1968) and Chong and Barr (1974). Like those of other members of the Blastocladiales and several chytrids the planospores contain the usual organelles (nuclear cap, mitochondrion, lateral lipoid bodies, microtubules, gamma bodies, etc.) and the kinetosome consists of 9 sets of 3 microtubules. These run forward in the planospores, ensheath the nucleus and nuclear cap, and terminate at the anterior of the cells.

Germination of the planospores may be monopolar (figs. 9–11, 31) or bipolar (figs. 12, 13), and the planospore cyst may persist until after the incipient endobiotic thallus has become established (figs. 10, 11, 14, 15, 58). In *C. sphaerocarpa* (fig. 32) the cyst may be functional and develop into the primary zoosporangium, but in other species the primary swelling or incipient zoosporangium develops in the endobiotic germ tube (figs. 11, 14, 57, 58). As noted, this swelling may develop directly into a zoosporangium in the cases of monocentric eucarpic (figs. 14, 61) or holocarpic thalli (fig. 16). Usually, the germ tube branches and develops a succession of swellings connected by isthmuses until an extensive rhizomycelium is formed (figs. 1, 33, 44, 62). Rhizoids usually develop from the isthmuses, swellings, and the priphery of the zoosporangia, but in *C. indica* they are lacking on the isthmuses (fig. 68). The rhizoids may be richly branched, extensive, and run out to fine filaments, in some species, but in *C. allomycis* and *C. spinosa* they end bluntly (figs. 45, 62), are sparse, and almost vestigial in the former species. Also, they are lacking on the immature thalli (fig. 60). In *C. vermicola* no rhizoids were present on thalli parasitizing numerous nematodes (fig. 40) but developed on thalli growing on artificial media. As the thallus matures septa are usually formed across the isthmuses (figs. 1, 33, 44, 62), and in *C. anguillulae* and *C. allomyces* these are pictured as being perforated by openings (figs. 29, 67). Apparently, in these species they should be regarded as pseudosepta. The incipient zoosporangia are delimited from the remainder of the thallus by septa or pseu-

PLATE 130

Figs. 45-67. *Catenaria allomycis* Couch parasitic in *Allomyces*. (Couch, 1945.)

Fig. 45. Two resting spores lying in intercalary swellings of the host with blunt-ended rhizoids.

Fig. 46. Early germination stage of a resting spore; wall is cracked by the protruding endosporangium.

Fig. 47. Slightly enlarged view of a germinating resting spore with emerging planospores which are sluggishly motile and encyst near the exit orifice.

Fig. 48. A cluster of encysted planospores 5.4-6.4 μ, or potential gametangia near the exit orifice; several have developed a papilla.

Fig. 49. Enlarged view of incipient gametangium.

Fig. 50. Content of gametangium has cleaved into 4 gametes.

Fig. 51. Emergence of isoplanogametes.

Fig. 52. Three ovoid, 3 × 4.4-4.7 μ, isoplanogametes.

Fig. 53. Fusion of 2 isoplanogametes.

Fig. 54. Biflagellate, elongately ovoid, 3.5-7.6 μ, zygote.

Fig. 55. Closely appressed flagella which frequently appear as one on the zygote.

Figs. 56-58. Germination stages of the zygote to form a sporangial thallus.

Fig. 59. Germ tube of zygote growing parallel to the host wall.

Fig. 60. Portion of a young polycentric thallus lacking rhizoids.

Fig. 61. Monocentric eucarpic thallus.

Fig. 62. Portion of a pseudoseptate polycentric thallus with numerous blunt rhizoids.

Fig. 63. Exit tube beginning to develop on a zoosporangium.

Fig. 64. Same zoosporangium 4 hours later; exit canal has penetrated the host wall and extended considerably beyond it.

Fig. 65. Discharge of the initial mass of planospore, surrounded by a gelatinous layer.

Fig. 66. Ovoid, 5-6.3 × 6.3-7 μ, planospores with nuclear cap, nucleus, granules, and a whiplash at the end of the flagellum.

Fig. 67. Enlarged view of perforated cross septum of the thallus.

Fig. 68-73. *Catenaria indica* Singh and Pavgi saprophytic in keratinic substrata. (Singh and Pavgi, 1970.)

Fig. 68. Portion of a thallus with terminal and intercalary zoosporangia connected by long isthmuses; rhizoids arising from the zoosporangia but not arising along the isthmuses.

Fig. 69. Ovoid, 4.4-6 μ, planospore with a small hyaline globule.

Fig. 70. Zoosporangium shortly before dehiscence.

Fig. 71. Apex of exit canal.

Fig. 72. Discharge of planospores, enveloped in a thin membrane(?).

Fig. 73. Resting spore with a thick, warty, dark brown wall "formed within the lobate sporangial segments."

Fig. 74-75. *C. spinosa* Martin (1975).

Fig. 74. Isolated spiny resting sporangium filling container cell; spines hollow except near the base, derived from rhizoids.

Fig. 75. Planospore with anterior filose pseudopodia, a nuclear cap, cluster of lipid globules, and a side body.

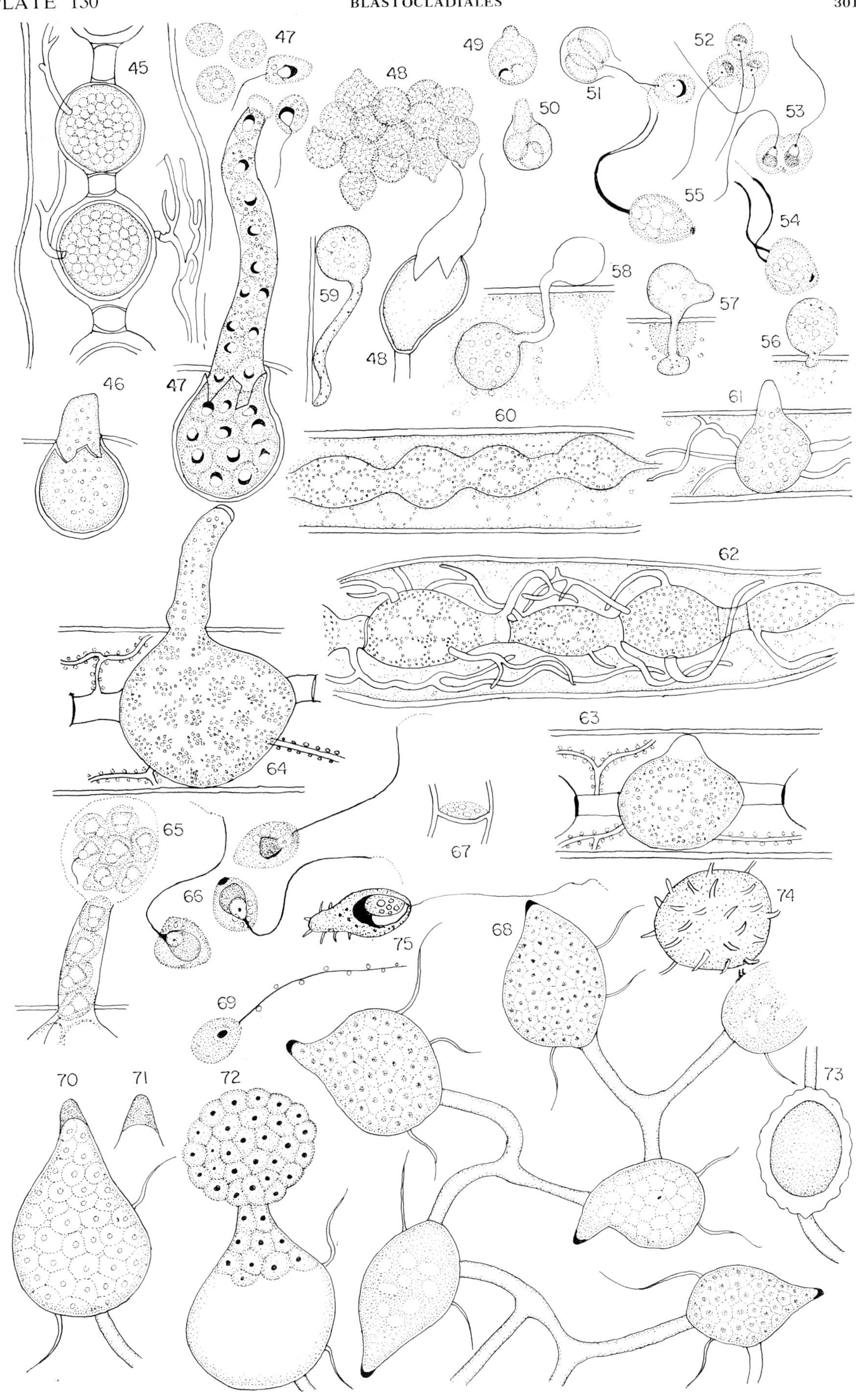
45
46
47
47
48
48
49
50
51
52
53
54
55
56
57
58
59
60
61
62
63
64
65
66
67
68
69
70
71
72
73
74
75

Catenaria

dosepta and at maturity they usually develop long, straight, curved or coiled necks for discharge of the planospores (figs. 1, 17, 33). As the tip of the neck deliquesces a globule of matrix exudes (fig. 17), and the first mass of planospores emerges into and enlarges it (figs. 18, 65) until it disperses in the surrounding water. The initial mass of planospores soon disperses, and those in the zoosporangium begin to swarm and emerge singly in succession.

The resting spores or sporangia in all species but *C. sphaerocarpa* and *C. spinosa* are usually formed by the contraction of the content of swellings or sporangium-like structures and its investment (figs. 19–22) by a relatively thick, smooth, verrucose, or warty wall. As a consequence they lie in and only partly fill a vesicle (figs. 22, 25, 28, 44, 45). In *C. sphaerocarpa* (figs. 37–39) and *C. spinosa* (fig. 74) however, the entire swelling or sporangium-like structure is transformed into a dark thick-walled or spiny spore without contraction of the content, and this may occur occasionally in *C. anguillulae*, (fig. 23), also. In germinating the spore functions as a sporangium and develops an endosporangium (?) which cracks the wall and grows into a short or greatly elongate, straight, curved, or coiled tube (figs. 24, 25, 47, 48) like that on the zoosporangium. Sometimes, the tube continues to elongate (fig. 27) and develop into an extensive rhizomycelium with swellings, isthmuses and rhizoids without the development of planospores.

Sexual reproduction is known only in *C. allomycis*, although Karling (1934) reported occasional fusions of planospores in *C. anguillulae* which separated subsequently. In *C. allomycis* the planospore produced by the germinated resting sporangia are sluggishly motile (fig. 47) and soon encyst near the exit orifice. These enlarge slightly, develop a papilla (fig. 49), and their content cleaves into 4 segments (fig. 50) which escape through the deliquesced papilla and become isoplanogametes (figs. 51, 52). These fuse in pairs (fig. 53) to produce a motile biflagellate ovoid zygote (figs. 54, 55). At first the two flagella are distinct as such, but later they become so closely associated that they may appear as one except at the tips (fig. 55). Accordingly, sexual reproduction and the life cycle of *C. allomycis* is identical with those of the subgenera *Cystocladiella* and *Cytogenes* of the Blastocladiaceae. Presumably, meiosis occurs during germination of the resting sporangium, and in that event the planospores and cystospores produced represent the short gametophytic generation. On the other hand, the sporophytic generation would be comparatively long, embracing the zygote and mature rhizomycelium, zoosporangia, and resting sporangia.

The zygote comes to rest on the host, encysts, and germinates (figs. 56, 57) by a broad germ tube at whose tip an endobiotic thallus forms (fig. 60). It may develop into a monocentric, eucarpic thallus (fig. 61) or an extensive rhizomycelium. The first formed thalli give rise to zoosporangia; later ones develop both zoosporangia and resting sporangia, and the last thalli form only resting sporangia, according to Couch.

REFERENCES TO THE CATENARIACEAE

Birchfield, W. 1960. A new species of *Catenaria* parasitic on nematodes of sugarcane. Mycopath. et Mycol. Appl. 60:331-338, 3 figs.

Buckley, J. J. C. and P. A. Clapham. 1929. The invasions of helminth eggs by chytridiacean fungi. J. Helminth. 7:1-14, 19 figs.

Butler, E. J. 1928. Morphology of the chytridiacean fungus, *Catenaria anguillulae*, in live-fluke eggs. Ann. Bot. 42:813-821, 19 figs.

———, and J. J. C. Buckley, 1928. *Catenaria anguillulae* as a parasite of the ova of *Fasciola hepatica*. Sci. Proc. Roy. Dublin Soc., n.s., 18: 497-512, pls. 23-26.

———, and A. Humphries. 1932. On the cultivation in artificial media of *Catenaria anguillulae*, a chytridiacean parasite of the ova of the liver fluke, *Fasciola hepatica*. Ibid. 20 (25):301-324, pls. 13-18.

Chong, J., and D. J. S. Barr, 1974. Ultrastructure of the zoospores of *Entophlyctis confervae-glomeratae, Rhizophydium patellarium*, and *Catenaria anguillulae*. Can. J. Bot. 52:1197-1204, 38 figs.

Constantineanu, M. J. C. 1901. Contributions a la flore mycologique de la Roumanie. Rev. Gen. Bot. 13:369-389, 9 figs.

Couch, J. N. 1945. Observations on the genus *Catenaria*. Mycologia 37:163-193, 73 figs.

Dangeard, P. A. 1885. Note sur le *Catenaria anguillulae* Sorok. Bull. Soc. Linn. Normandie 3 ser. 9:126-135, 12 figs.

Fuller, M. S. 1966. Structure of the uniflagellate zoospores of aquatic Phycomycetes. Colston Papers 18:67-84, 26 figs.

———. 1968. Microtuble-kinetosome relationships in the motile cells of the Blastocladiales. Zeitschrift f. Zellforsch. 87:526-533, 6 figs.

Gaertner, A. 1962. *Catenaria anguillulae* Sorokin als Parasit in den Embryonen von *Daphnia magna* Strauss nebst Beobachtungen zur Entwickelung, zur Morphologie und zum Substratverhalten des Pilzes. Arch. Mikrobiol. 43:280-289, 15 figs.

Hanson, A. M. 1944. Three new saprophytic chytrids. Torreya 44:30-33.

———. 1945. A morphological, developmental, and cytological study of four saprophytic chytrids. I. *Catenomyces persicinus* Hanson. Amer. J. Bot. 32:431-438, 52 figs.

Ichida, A. A., and M. S. Fuller. 1968. Ultrastructure of mitosis in the aquatic fungus *Catenaria anguillulae*. Mycologia, 60:141-155, 13 figs.

Johnson, T. W. Jr., 1973. Aquatic fungi of Iceland: Some polycentric species. Ibid. 65:1337-1355, 42 figs.

Karling, J. S. 1934. A saprophytic species of *Catenaria* isolated from roots of *Panicum variegatum*. Mycologia 26:528-543, 3 figs., pls. 57, 58.

———. 1938. A further study of *Catenaria*. Amer. J. Bot. 25:328-335, 34 figs.

———. 1946. Keratinophilic chytrids. Amer. J. ot. 33, Suppl. no. 3, p. 5s.

———. 1947. Keratinophilic chytrids. II. *Phlyctorhiza variabilis* n. sp. Ibid. 34:27-32, 48 figs.

———. 1949. *Nowakowskiella crassa* sp. nov., *Cladochytrium aureum* sp. nov., and other polycentric chytrids from Maryland. Bull. Torrey Bot. Club 76:294-301, 17 figs.

———. 1951. Polycentric strains of *Phlyctorhiza variabilis*. Amer. J. Bot. 38:772-777, 3 figs.

———. 1965. *Catenophlyctis*, a new genus of the Catenariaceae. Ibid. 52:133-138, 12 figs.

———. 1967. Some zoosporic fungi of New Zealand. X. Blastocladiales. Sydowia 20:144-150, pls. 29-31.

———. 1968. Zoosporic fungi of Oceania. V. Cladochytriaceae, Catenariaceae, and Blastocladiaceae. Nova Hedwigia 15:191-201, pls. 21-23.

Knox, J. S. 1970. Biosystemic studies of aquatic Chytridiales and Blastocladiales. Ph.D. thesis, Virginia Poly. Inst., State Univ.

Martin, W. W. 1975. A new species of *Catenaria* parasitic in midge eggs. Mycologia 67:264-272, 15 figs.

Nolan, R. A. 1970a. The phycomycete *Catenaria anguillulae*: growth requirements. J. Gen. Microbiol. 60:167-180.

———. 1970b. Sulfur source and vitamin requirements of *Phlyctorhiza variabilis*. Amer. J. Bot. 43:28-32.

Schussnig, B. 1930. Der Generation-und Phasenwechsel bei den Chlorophyceen. Oestrr. Bot. Zeitschr. 79:58-77.

Singh, U. P., and M. S. Pavgi. 1970. Two noteworthy members of the Catenariaceae from India. Mycologia 62:587-590, 6 figs.

Sorokin, N. 1876. Note sur les Vegetaux parasites des Anguillulae. Ann. Sci. Nat. 6 ser., 4:62-71, pl., 3.

Sparrow, F. K., Jr. 1932. Observations on the aquatic fungi of Cold Spring Harbor. Mycologia 24: 268-303, 4 figs., pls. 7, 8.

———. 1950. Some Cuban Phycomycetes. J. Wash. Acad. Sci. 40:50-55, 30 figs.

———. 1965 The occurrence of *Physoderma* in Hawaii, with notes on other Hawaiian Phycomycetes. Mycopath. et Mycol. Appl. 25:119-143, pls. 1-7

———. 1960. Aquatic Phycomycetes. 2nd. ed. XXV. 1187 pp. Univ. Michigan Press, Ann Arbor.

Stüben, H. 1939. Uber Entwickelungsgeschichte und Ernärungsphysiologie eines neues Phycomyceten mit Generationswechsel. Planta, Arch. Wiss. Bot. 30:353-383, 17 figs.

Villot, A. 1874. Monograph des Dragonneaux. Arch. Zool. Exp. Gen. 3:181-238, pls. 4-9.

Willoughby, L. G. 1961. The ecology of some lower fungi at Esthwaite Water. Trans. Brit. Mycol. Soc. 44:305-332, 17 figs., pls. 22, 23.

Wilson, C. W. 1952. Meiosis in *Allomyces*. Bull. Torrey Bot. Club. 79:139-160, 23 figs.

COELOMOMYCETACEAE

This small family includes one genus, *Coelomomyces*, and several species which are internal obligate parasites of numerous genera and species of mosquitos and a few other insects. Its members have been reported from all continents and several tropical islands, and they will doubtless prove to be even more more widely distributed wherever mosquitos occur. These fungi have attracted a great deal of attention and interest among parasitologists, medical entomologists, public health workers, and mycologists as possible means of controlling mosquitos and malaria biologically, and a large number of papers have been published on their distribution, hosts, pathogenicity, etiology, and ecology since Keilin (1921) described the first species. At first the affinities of *Coelomomyces* were unknown or thought to be with the Sporomyxa and Chytridiales, but in 1945 Couch created a new family for it among the Blastocladiales on the basis of the manner of resistant sporangial germination and the structure of the planospores.

As interpreted by Couch, this family is characterized by a sparse or abundant, slender or coarse, multinucleate, non-septate or coenocytic, dichotomously or irregularly branched mycelium which *lacks a wall* and possibly rhizoids, and thin- or thick-walled resting sporangia borne at the ends of hyphal branches. The lack of a wall on the mycelium is unique among filamentous fungi, and it is possibly the result of the unusual position and obligate parasitism in the coelom of its hosts. Although the mycelium lacks a wall, the species have, nevertheless, retained the ability of wall formation in the development of resistant sporangia. Ultrastructural studies have shown that the wall-less mycelium and hyphal bodies (*hyphagens*) are bounded by a single plasma membrane whose surface is convoluted into protuberances or microvilli. The abundance of the latter structures suggests that they function to increase the absorptive area, or possibly they may serve as microhaustoria.

Although the resistant sporangia, their method of germination, the production of uniflagellate planospores and their structure in this family are similar to those of members of the Blastocladiaceae the relationships of the Coelomomycetaceae and its position in the Blastocladiales were obscure until recently Whisler *et al* (1974, 1975) discovered that *C. psorophorae* produces isogametes. These are derived from heterothallic, wall-less gametangia in the alternate host, *Cyclops vernalis*, fuse to produce a diploid zygote that subsequently infects the mosquito host, and gives rise to a diploid coenocytic thallus. In view of these brilliant discoveries Whisler *et al* (1975) suggest that *C. psorophorae* has an *Euallomyces* type of life cycle and displays a number of similarities to the life history of *Blastocladiella variabilis*. However, unlike some species of the Catenariaceae and Blastocladiaceae the thallus of the Coelomycetaceae is not chytrid-like, apparently lacks rhizoids, and so far as is known is unique by its wall-less mycelium that seemingly derives its nourishment from the hosts through the plasma membrane.

Whether or not *Myiophagus* (see pl. 23) belongs in this family is open to question. Its sporangia fill the body of its hosts as in *Coelomomyces*, but its mycelium is septate and has a well-defined wall. If it is included here the Coelomomycetaceae must be emended to include it.

COELOMOMYCES Keilin, emed Couch

J. Elisha Mitchell Sci. Soc. 61:128. 1945.

Coelomomyces Keilin, Parasitology 13:226. 1921.
Zographia Bogoyavlensky, Russkii arkhiv protistologii 1:113, 1922.

Plates 131, 132, 132A

At the present time this genus includes approximately 40 varieties and species, and some unidentified species which parasitize many genera and species of mosquitos, nymphs, the black swimming water bug *Notonecta* sp., ova of the sand fly *Phleobotomus* sp., and larvae of *Chironomous, Orthocladius* and *Cricotopus*. Doubtless, many more species will be found as the aquatic larval stages of other insects are studied for parasites. Presently, so far as they are known species distinctions are based primarily on the character, structure and ornamentation of the outer wall, size, and shape of the resistant sporangia. Also, the relative abundance and diameter of the hyphae and mycelium are used as supplementary criteria of identification. As seen by the light microscope the outer wall of the resistant sporangia may be smooth, striated, crinkled, cribose, banded or ridged, punctate or pitted, and spiculate, and the ridges may anastomose to form a meshlike network of polygons on the outer surface in the different species, and these characters have been confirmed by scanning electron microscopy (Anthony, Chasman, and Hazard (1971), and Bland and Couch (1973). The resistant sporangia vary from subspherical to ovoid with one surface slightly or markedly flattened or concave, and range from 12×21 to $46 \times 100\ \mu$ in size. At maturity they may be hyaline, pale to bright yellow, dark-brown and maroon in color. Additional distinctions are made on the size, shape, depth, and distribution of pits in the wall, the number of bands or ridges, whether they are high, low, or equal in height, or oblong or circular, and anastomose. These characters may vary considerably in the same species, and it is not unlikely that some of the species may prove to be identical. Nevertheless, Couch and Umphlett (1963) reported that "the species of *Coelomomyces* can be arranged in about 10 natural groups according to shape and size of the resting sporangia and particularly the structure of the wall." Bland and Couch, however, report that the known species fall into 8 recognizable groups on the basis of ornamentation as revealed by scanning microscopy.

Infection of specific mosquito host larvae with a single species under controlled laboratory conditions have proven to be difficult although Walker (1938), Muspratt (1946), Laird (1959), Madelin (1968), and

PLATE 131

Figs. 1, 2 after Keilin, 1921; figs. 3, 4 after Iyengar, 1935; figs. 5-13, 17 after Umphlett, 1962; figs. 14, 15 after Keilin *et al.*, 1963; figs. 18-22 after Martin, 1969.

Fig. 1. Three gills, G, and the respiratory siphon, S, of *Stegomyia scutellaris* filled with resistant sporangia of *Coelomonyces stegomyiae* Keilin.

Fig. 2. Transverse section of the midgut of the same host showing intestinal epithelium surrounded by the multinucleate mycelium of *C. steomyiae*; immature resting sporangium at the right.

Figs. 3, 4. Young thalli of *C. indiana* Iyengar in the body of *Anopheles* with hyphae (?) at the base.

Fig. 5. "Young mycelium of *C. dodgei* Couch in larvae of *Anopheles crucians* showing branching of a rhizoid-like hypha; one of the branch tips is flared."

Fig. 6. Portion of the mycelium of *C. pentangulatus* Couch in larvae of *Culex erraticus* showing the method of branching.

Figs. 7, 8. Types of mycelial branching in *C. dodgei.*

Fig. 9. "Hyphal body formation in hypha of *C. dodgei.*"

Figs. 10-13. "Hyphal bodies of *C. pentangulatus.*"

Figs. 14, 15. Simple and deeply-lobed thalli of *C. psorophorae* Couch from an adult female of *Aedes melanimon*; inadequately sketched from photographs.

Fig. 16. "Free floating hyphal segment of *C. pentangulatus,* probably developed from a hyphal body."

Fig. 17. Intranuclear division in *C. dodgei.*

Fig. 18. Subspherical hyphagen or hyphal body of *C. punctatus* Couch and Dodge in larvae of *Anopheles quadrimaculatus.*

Fig. 19. Electron micrograph of the surface of a hyphagen of *C. punctatus* showing convoluted border and projecting microvilli; sketched from a photograph.

Fig. 20-22. Stages in the development of the mycelium from a hyphagen or hyphal body of *C. punctatus*; sketched from photographs.

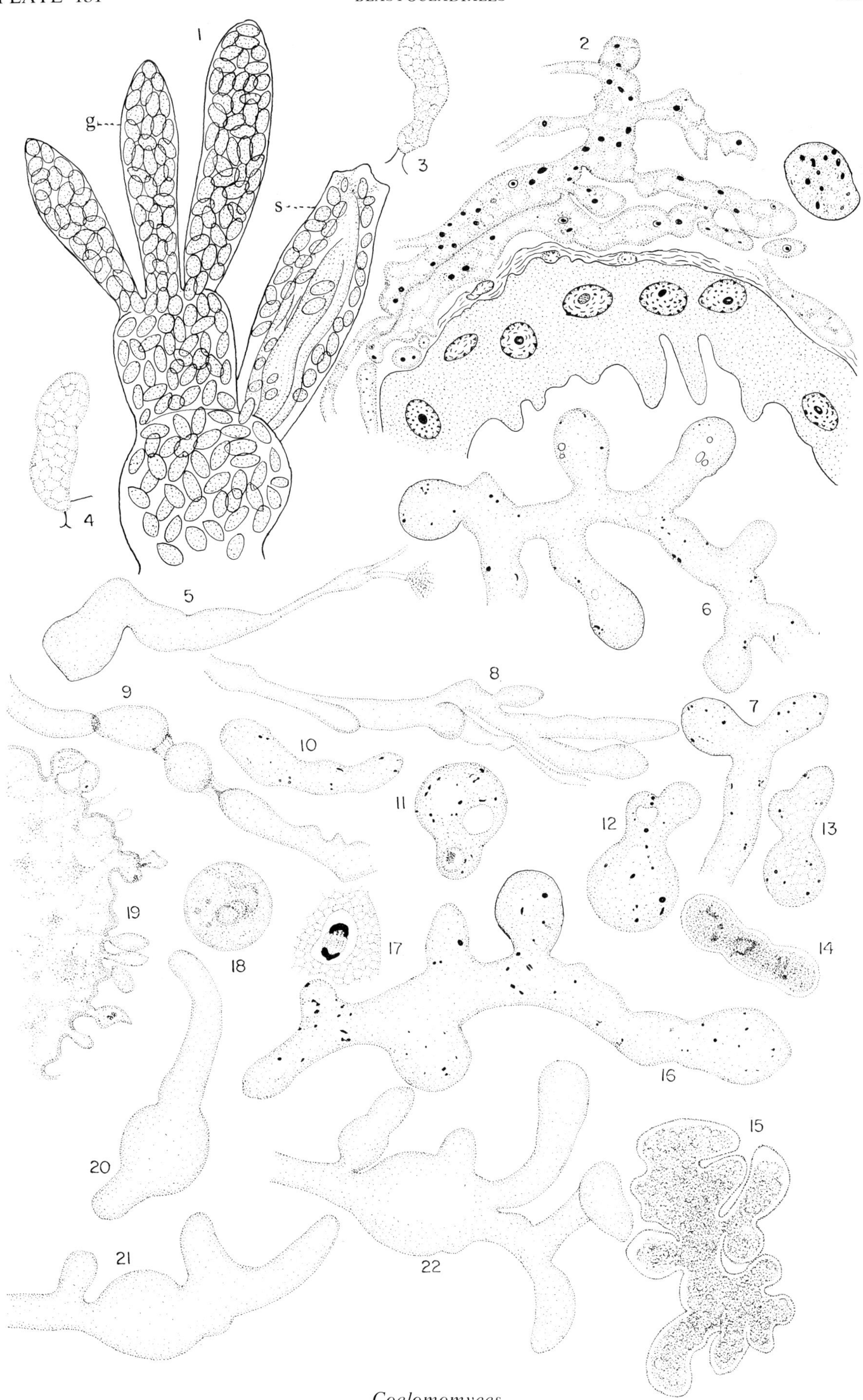

Coelomomyces

Couch (1968, 1973) secured a few to a large percentage of infections under certain conditions. However, but few successful cross inoculations have been made, and it is not certain at present how specific most *Coelomomyces* species are as to their mosquito hosts, although Couch (1968) stated that "as a rule each species has its own particular mosquito host...." In 1963 he and Umphlett noted that "thirteen species have been reported on one host species with the exception noted below. Six of these are of no significance in indicating host range or host specificity since each of these is known from only one host larva. A seventh species has been reported once as abundant on nymphs of notonecta sp. The remaining six species range from fairly abundant to rare in Georgia and do not afford significant information on host range. Other species have been reported attacking several different mosquito hosts." Among these are species which appear in various forms on several hosts. Also, it is not unlikely that biological races might be found among species which have been reported on more than one host.

Except in *C. psorophorae* the method and agent of infection has not been observed so far, but it probably occurs by planospores or zygotes which attack the thorax and head of first instar larvae, although Couch and Dodge (1947) and Umphlett (1961) failed to secure infection of larvae by planospores in petri dishes. Although infection usually occurs in the 1st instar of the larvae, Couch (1968) found that any of the 4 instars may become infected when germinating zoosporangia are added to them. The fungus first becomes visible as "specks" in the blood fluid of first instar larvae, according to Muspratt (1946a). Possibly, these specks are hyphal bodies derived from infecting planospores. These are abundant in second instar larvae as sphaeroid bodies (fig. 18) which Martin (1969) appropriately named *hyphagens* because they proliferate and give rise to hyphae (figs. 20–22). However, hyphal bodies or hyphagens of various sizes and shapes (figs. 10–13) may be formed by constriction and segmentation of the mycelium (fig. 9). These as well as elongate, branched hyphal segments (fig. 16) and resting sporangia are carried along with the circulation of the hemolymph and can be clearly seen passing through the heart of the insect. The surface or plasma membrane of the hyphagens and mycelium is not smooth, and ultrastructural studies of *C. punctatus* have shown that it is convoluted with numerous protuberances or microvilli (fig. 19), according to Martin, which apparently increase the absorptive area.

The hyphagens, as noted above, give rise to the mycelium which may be sparse or abundant and varies from 3 to 16 μ in diam. in different species. The branching may be dichotomous, subdichotomous, or irregular (figs. 6–8, 22) with fairly long or short blunt branches. Sometimes the branching may be such that coralloid, deeply-lobed thalli are formed (fig. 15). After the mycelium has become established in the hemocoel it may fragment additionally into segments which circu-

PLATE 132

Figs. 23, 26, 27, 30, 31 after Umphlett, 1964; figs. 24, 25, 28, 29, 35 after Martin, 1969; figs. 32, 41 after Iyengar, 1935; figs. 33, 34 after Keilin, 1921; fig. 36 after De Meillon and Muspratt, 1943; figs. 37-40 after Couch and Dodge, 1947; figs. 42, 43 after Laird, 1959; fig. 44 after Weiser and Vavra, 1964; fig. 45 after Laird, 1956; figs. 46, 47 after Manier *et al.*, 1970.

Fig. 23. Portion of a mycelium of *C. pentangulatus* showing aggregation of nuclei in hyphal swellings in preparation for sporangial development.

Fig. 24. Resistant sporangium of *C. punctatus* developing at the end of a short hypha derived from a hyphagen.

Fig. 25. Resistant sporangium of *C. punctatus* developing from the surface of a hyphagen.

Fig. 26. Young resistant sporangium of *C. pentangulatus* separating from a hypha.

Fig. 27. Similar resistant sporangium of *C. pentangulatus* showing distribution of nuclei.

Fig. 28. "Incipient resistant sporangium within the expanded hyphal sheath prior to wall formation"; *C. punctatus*; sketched from a photograph.

Fig. 29. Severed hyphal sheath or membrane after liberation of a resistant sporangium of *C. punctatus*.

Fig. 30. Resistant sporangium of *C. pentangulatus* after liberation.

Fig. 31. Post-liberated resistant sporangium of *C. pentangulatus* showing germination slit.

Fig. 32. Section through a resistant sporangium of *C. indiana* showing ridged wall and distribution of nuclei.

Fig. 33. Section of an immature resistant sporangium of *C. stegomyiae* showing pitted outer wall.

Fig. 34. Enlarged drawing of the outer wall of *C. stegomyiae*.

Fig. 35. Ultrastructure of the wall of a resistant sporangium of *C. punctatus* in the region of a pit, P.; tl = tectum layer, ol = outer layer, fl = foot layer, ml. = middle layer, il = inner layer, c = columellae. Sketched from an electron micrograph.

Fig. 36. First illustration of a germinating resistant sporangium of *Coelomomyces*.

Figs. 37, 38. Germination stage and discharge of planospores in *C. lativittatus*.

Figs. 39, 40. Elongate, 2.6-3.8 × 5.2-6.3 μ, tapered and subspherical, 4-5 μ diam., planospores of *C. lativittatus* Couch and Dodge with nuclear caps, nucleus, side body, anterior lipoidal bodies, and whiplash flagellum.

Fig. 41. Side view of the ridged resistant sporangium of *C. anophelesica* Iyengar.

Figs. 42, 43. Resistant sporangia of *C. finlayae* Laird and *C. macleayae* Laird resp., showing sculpturing of the wall. Copied by permission of the National Research Council of Canada.

Fig. 44. Resistant sporangium of *C. chironomi* Rasin; ridges have formed a series of 5 to 6 sided polyons.

Fig. 45. Resistant sporangium of *C. solomonis* Laird with protuberance on the wall. Copied by permission of the National Research Council of Canada.

Figs. 46, 47. Resistant, smooth-walled hyaline resistant sporangia of *C. ruzetae* Manier *et al.*

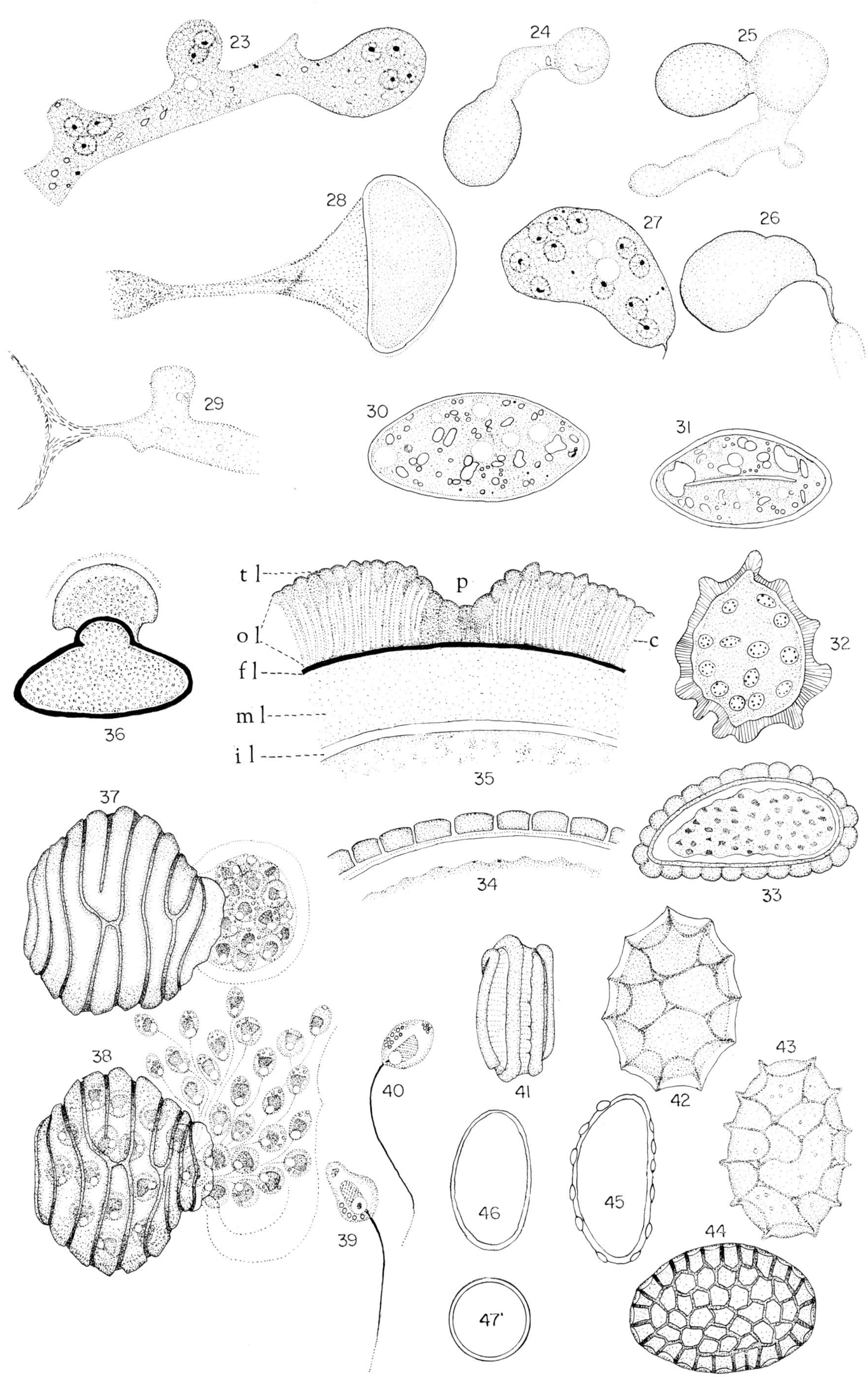

Coelomomyces

late and thus provides a wider distribution in the body cavity. Anastomoses of hyphae have been reported in *C. stigomyiae, C. notonectae* and *C. dodgei* by Keilin (1921), Bogoyavlensky (1922) and Couch (1945), respectively, but their significance in the life cycles of these fungi is not apparent.

The attachment of the hyphae and mycelium and their exact relation to the host tissues is not altogether clear. Keilin (1921) reported that in *C. stegomyiae* "it is well developed around the viscera, especially the midgut (fig. 2), and the five anterior intestinal coeca, forming two to three concentric layers, so closely attached to the organs, that in some places it is difficult to separate the tissues of the host from the surrounding mycelium". Umphlett (1962), also, noted that the mycelium of *C. dodgei* remains in contact with the host tissues for some distance before the hyphal tips turn away and grow out into the body cavity. Iyengar (1931) and Umphlett (1962) illustrated rhizoid-like filaments at the ends of hyphae (figs. 3–5), but their true nature and function are not known. Iyengar believed that they penetrate and fix the mycelium to the fat body of the host, but Umphlett was not certain about whether they are true rhizoids or empty, wrinkled hyphal membranes which become drawn out into filaments during dissection. Apparently, the adipose tissue is the chief sources of nutriment of the fungus in larvae. At least, the fat body is reported to lose its characteristic appearance, shrink, collapse and degenerate as the result of infection and spread of the parasite. Another effect of infection may be the suppression of the imaginal buds so that the wing and leg rudiments fail to develop fully. Usually, infected larvae are killed before pupating.

The resistant sporangia are formed from the ends of the mycelium and its lateral branches within a few days after the mycelium has become established. Iyengar (1935) estimated that if infection occurs during the first instar of larvae, fully developed resting sporangia may be formed within 6 days after infection. Sporangia may develop, also, from the surface of a hyphagen (fig. 25), or a short branch of it (fig. 24), or from buds on the mycelium (fig. 23) in which the nuclei accumulate. Such buds and ends of hyphae increase in diameter, become subspherical and ovoid and invested with a well-defined membrane and eventually separate from the mycelium as multinucleate sporangia (fig. 27). The process of separation is somewhat complicated and involves vacuolation of the protoplasm and pinching off from the hyphae and hyphal membrane at or near the base of the incipient sporangium. This continues until the sporangium is attached only by a thread (fig. 26), the remains of the hyphal membrane. The thread is eventually severed (fig. 27), and the sporangium is released and becomes free-floating in the coelom. There it matures and develops a thick wall with a germination slip (fig. 31), according to Umphlett (1964). However, in *C. punctatus*, according to Martin, most sporangia complete their development while still enveloped by and attached to the hyphal sheath (fig. 28). When the outer wall has attained its final thickness and become ornamented and the germination slit has developed, the sporangium separates from the hyphal sheath, leaving a ruptured crescentric membrane behind (fig. 29).

Thin-walled sporangia have been reported in *C. stegomyiae, C. indiana, C. anophelisca* and *Coelomomyces* sp., type C. by Keilin (1921), Inyengar (1935), and Muspratt (1946), but their signficance in the life cycle of these species is not known. Probably, they are only ordinary resistant sporangia whose walls have not thickened perceptibly. At least, Couch and Umphlett (1962) reported that in a species from Florida these sporangia grade imperceptibly into the thick-walled ones and are not sharply distinctive. Also, they are reported to germinate in the same manner as the thick-wall sporangia, and this has been confirmed by Madelin and Beckett (1972) in *C. indicus*. As the wall of the latter type thickens and the ornamentations develop on the outer layer it becomes pale yellow, bright yellow, orange to deep-brown in the different species. Most workers have reported the wall to be two-layered (figs. 33, 34), but Martin's ultrastructural study of *C. punctatus* showed that it is 3-layered and rather complex in structure (fig. 35).

Germination of the resting sporangia was first observed by Rasin (1929) and De Meillon and Muspratt in 1943 (fig. 36) and confirmed by subsequent workers (Madelin and Beckett, 1972; Whisler, *et al* (1972). The process first becomes visible by the opening of the germination slit and a slight protrusion of the sporangial content which is bounded by the inner sporangial wall. The latter is single but double contoured, according to Couch and Umphlett (1962), but De Meillon and Muspratt reported it to be two-layered; an interpretation which is supported to some extent by Martin's demonstration of an inner wall in the sporangium. After cleavage into planospores has been completed, the spore mass pushes out slowly at first but with gradually increasing speed. The cover-

PLATE 132A

Fig. 1. Life cycle of *Coelomomyces psorophorae*. (Whisler *et al.*, 1975).

A = zygote which infects larvae of *Culiseta inornata*.
B = development of hyphal bodies, mycelium, and thick-walled resistant sporangia.
C = release of zoospores.
D = zoospores of opposite mating type.
E = infection of alternate host, *Cyclops vernalis*; each zoospore develops into a thallus and eventually gametangia.
F = gametes of opposite mating type fuse either in or outside of *Cyclops* to form the mosquito-infecting zygote.

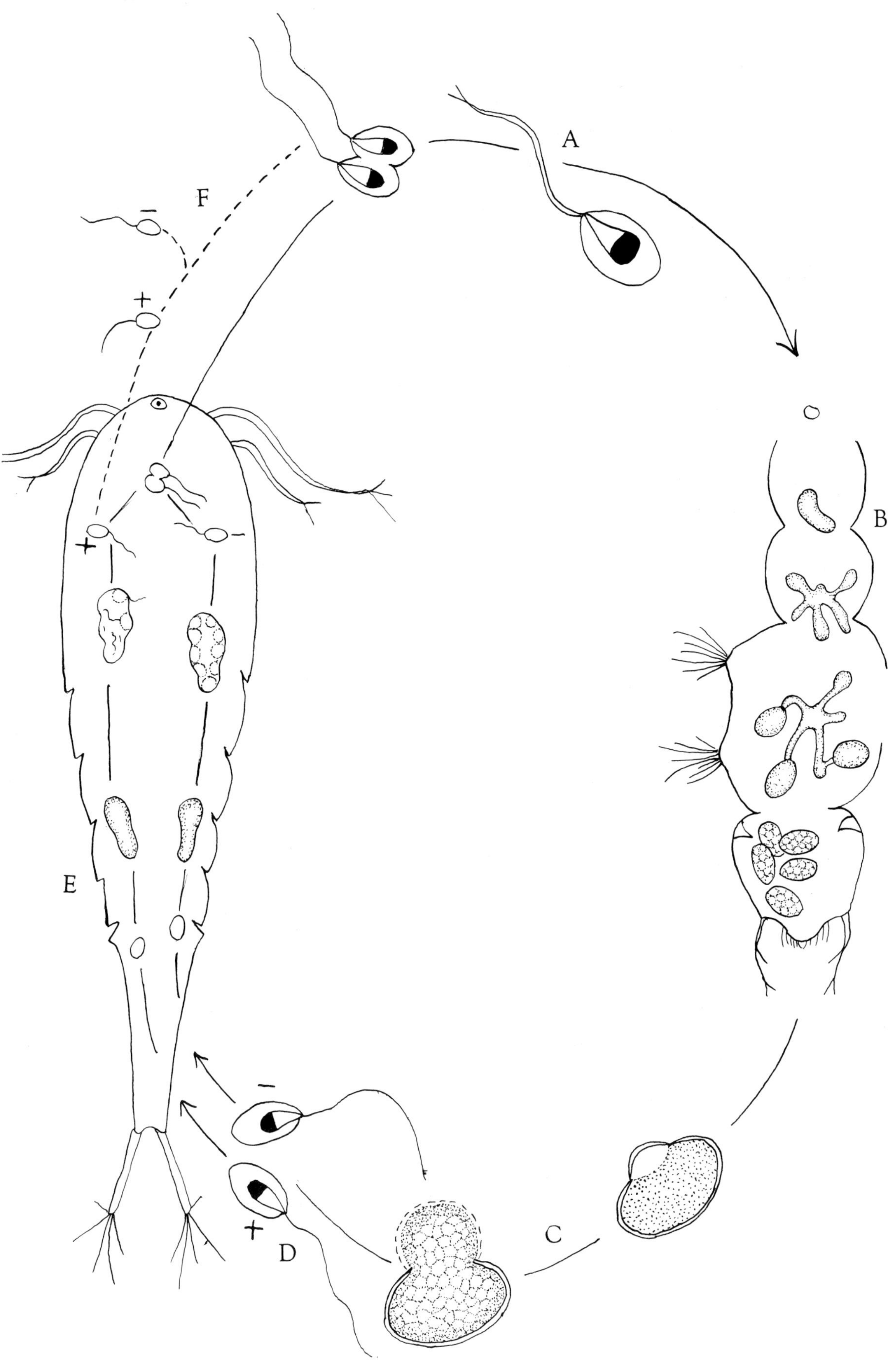

Coelomomyces psorophora

ing membrane wall is stretched thereby so that the enlarging mass becomes almost hemispherical (fig. 36) to ovoid in shape (fig. 37). The outer part of the enveloping wall then fragments and moves away, and the planospores are retained by a thin membrane. The spores begin to rock back and forth, become increasingly active and eventually seething in motion just before the membrane bursts to release them (fig. 38). The planospores (figs. 39, 40) are ovoid or tapering at the anterior end, and in fixed and stained preparations the nucleus, nuclear cap, side and anterior lipoid granules, and a whiplash flagellum are visible and conspicuous. In light and electron microscopy studies on *C. psorophorae* Whisler *et al* (1971) found that in the planospore initials in a germinating resting sporangium each nucleus carries a lipid crown, and at release the planospores display a nuclear cap and a single basal mitochondrion similar to that of *Blastocladiella emersonii*. The outer typically alveolar wall as well as the inner wall and discharge plug possess a number of sublayers. In *C. punctatus* Martin (1971) found that the ultrastructure of the planospores is basically similar, with some exceptions, to that of other blastocladiaceous species. The axial components (i.e. axoneme, kinetosome with its distal doublet to triple arrangement of microfibrils, nucleus with a porous membrane, and double-membraned nuclear cap) have a linear arrangement, and the last 2 named organelles are of the shape of an inverted cone. The nuclear cap covers only the upper portion of the nucleus, and the larger part of the nucleus occurs as a hatchet-shaped spur and extends laterally. The large single mitochondrion extends anteriorly from a level below the kinetosome to the nuclear cap summit. The lipid sac is strikingly similar to that described by Cantino and Truesdell in *B. emersonii*. However, no second or vestigal kinetosome, striated rootlet, and a basal plate at the distal end of the kinetosome were found by Martin. Particularly noteworthy is the presence of a paracrystalline rod-shaped body composed of an array of solid fibers at one side of the nucleus and nuclear cap, a body similar to that which occurs in *Callimastix*. Whisler *et al* (1972) and Madelin and Beckett (1972) have confirmed Martin's observations on the presence of this body *C. psorophorae* and *C. indicus*, respectively.

Whisler, *et al* (1974) discovered that *C. psorophorae* has an alternate host, *Cyclops vernalis*. The uniflagellate planospores from resistant sporangia in *Culiseta inornata* infect *Cyclops* where they give rise to heterothallic wall-less gametangia which produce isogametes. These fuse to produce diploid zygotes which are freed, infect mosquito larva, and produce a coenocytic thallus from which the resistant sporangia are subsequently formed.

REFERENCES TO THE COELOMOMYCETACEAE

Anthony, D. W., H. C. Chapman, and E. I. Hazard. 1971. Scanning electron microscopy of the sporangia of species of *Coelomomyces* (Blastocladiales: Coelomomycetaceae). J. Invertbr. Path. 17:395-430, 20 figs.

Area, Leao, A. E., and C. Pderoso. 1964. *Coelomomyces differit* n. p. parasito de ovas de *Phlebotomus*. Atas Soc. Biol. Rio de Janeiro 8:55-56.

Bland, C. E., and J. N. Couch. 1973. Scanning electron microscopy of sporangia of *Coelomomyces*. Canad. J. Bot. 51:1325-1330, pls. I-VII.

Bogoyavlensky, N. 1922, *Zografia notonectae* n.g., n. sp., Ark. Pro Russ. Arkh. Protistol. 1:113-119, pl. 10.

Chapman, C. and D. B. Woodward. 1966. *Coelomomyces* (Blastocladiales: Coelomomycetaceae) infections in Louisiana mosquitos. Mosquito News 26:121-123.

Coluzzi, M. and J. A. Rioux. 1962. Primo reperto in Italia di larve di *Anopheles* parassitate da funghi del genere *Coelomomyces* Keilin, Descrizioni di *Coelomomyces raffaeli* n. sp. (Blastocladiales, Coelomomycetacae). Riv. Malariol. 41:1-

Couch, J. N. 1945. Revision of the genus *Coelomomyces* parasitic in insect larvae. J. Elisha Mitchell Sci. Soc. 61:124-136, pls. 1, 2.

———. 1960. Some fungal parasites of mosquitos. *In* Biological Control of Insects of Medical Importance. A1BS Tech. Rept.

———. 1968. Sporangial germination of *Coelomomyces punctatus* and the conditions favoring the infection of *Anopheles quadrimaculatus* under laboratory conditions. *In* Proc. Joint U.S.-Japan Seminar on Microbial Control of Insect Pests. K. Aizawa, Ed. Fukuoka, Japan, pp. 93-105, 13 figs.

———. 1973. Mass production of *Coelomomyces*, a fungus that kills mosquitos. Proc. Nat. Acad. Sci. 69:2043-2047, 2 figs.

———, and H. R. Dodge. 1947. Further observations on *Coelomomyces* parasitic in mosquito larvae. J. Elisha Mitchell Sci. Soc. 63:69-79, pls. 15-20.

———, and C. J. Umphlett. 1963. *Coelomomyces* Infections. *In* Steinhaus, Insect Pathology 2: 149-188, 15 figs.

De Meillon, B., and J. Muspratt. 1943. Germination of the sporangia of *Coelomomyces* Keilin. Nature 152:507, figs. a-e.

Iyengar, M. O. T. 1935. Two new fungi of the genus *Coelomomyces* parasitic in larvae of *Anopheles*. Parasitology 27:440-449, 5 figs.

Karling, J. S. 1967. Some zoosporic fungi of New Zealand. X. Blastocladiales. Sydowia 20:144-150, pls. 29-31.

Keilin, D. 1921. On a new type of fungus: *Coelomomyces stegomyiae*, n.g., s. sp., parasitic in the body cavity of the larvae of *Stegomyia scutellaris* Walker (Diptera, Nematocera, Culicidae). Parasitology 13:225-234, 7 figs.

———. 1927. On *Coelomomyces stegomyiae* and *Zografia notonectae*, fungi parasitic in insects.

Ibid. 19:365-367.

Keilin, W. R., T. B. Clark, and J. E. Lindegren. 1963. A new host record for *Coelomomyces psorophorae* Couch in California (Blastocladiales: Coelomomycetaceae). J. Insect Path. 5:167-173, 8 figs.

Khaliulin, G. L., and S. K. Ivanov. 1973. Paraziticheskii grib *Coelomomyces* sp. Tichinok komarov v Maüskoi ASSR. Med. Parazitol Parazit Bolezn 42:487.

Laird, M. 1956a. A new species of *Coelomomyces* (fungi) from Tasmanian mosquito larvae. J. Parasitol. 42:53-55.

________. 1959a. Parasites of Singapore mosquitos with particular reference to the significance of larval epibionts as an index of habit pollution. Ecology 40:206-221.

________. 1959b. Fungal parasites of mosquito larvae from the oriental and Australian regions, with a key to the genus *Coelomomyces* (Blastocladiales: Coelomomycetaceae). Canad. J. Zool. 37:781-791.

________. 1960. *Coelomomyces*, and the biological control of mosquitos. *In* Control of Insects of Medical Importance. AIBS Tech. Rept.

________. 1961. New American locality records for four species of *Coelomomyces* (Blastocladiales: Coelomomycetaceae). J. Insect Path. 3:249-253.

________. 1966. Integrated control and *Aedes polynesiensis*: An outline of the Tokelau project, and its results. World Health organization (EBL 66) 69:1-9.

Lavits'ka Z. H. and D. H. Colles. 1960. A field experiment with a fungal pathogen of mosquitos in the Tokelau Island. Verh. XI. Int. Kongress f. Entomologie II. :867.

________., H. I. O. Dudka, and D. B. Tsarychkova. 1967. The fungus *Coelomomyces quadrangulatus* Couch, parasite of mosquito larvae. Dapov Akad. Nauk, Ukr. rss. 29:116-118.

Lum, P. T. M. 1963. The infection of *Aedes taeniorhynchus* (Wiedemann) and *Psophora howardii* Coquillett by the fungus *Coelomomyces*. J. Insect pat. 5:157-166, 6 figs.

Madelin, M. F., 1968. Studies on the infection by *Coelomomyces indicus* of *Anopheles gambiae*, J. Elisha Mitchell Sci. Soc. 84:115-124.

________, and A. Beckett 1972. The production of planouts by thin-walled sporangia of the fungus *Coelomomyces indicus* a parasite of mosquitos. J. Gen. Microbiol. 72:185-200, 27 figs.

Manier, J. F., J. A. Rioux, F. Coste, and J. Maurand. 1970. *Coelomomyces tuzetae* (Blastocladiales-Coelomomycetaceae), parasite des larves de chironomes (Diptera, Chironomidae). Ann. Parasit. Hum, Comp. 45:119-128, 6 figs.

Martin, W. W. 1969. A morphological and cytological study of development in *Coelomomyces punctatus* parasitic in *Anopheles quadrimaculatus*. J. Elisha Mitchell Sci. Soc. 85:59-72, 44 figs.

________. 1971. The ultrastructure of *Coelomomyces punctatus* zoospores. Ibid. 87:209-221, 20 figs.

Muspratt, J. 1946a. On *Coelomomyces* fungi causing high mortality of *Anopheles gambiae* larva in Rhodesia. Ann. Trop. Med. Parasit. 40:10-17, pl. 1.

________. 1946b. Experimental infection of larvae of *Anopheles gambiae* (Dept., Culcidae) with a *Coelomomyces* fungus. Nature 158:202.

________. 1963. Destruction of the larvae of *Anopheles gambiae* Giles by a *Coelomomyces* fungus. Bull. World Health Organization 29:81-86.

Rajapaksa, N. 1964. Survey for *Coelomomyces* infections of mosquito larvae in the southwest coastal regions of Ceylon. Ibid. 30:149-151.

Rasin, K. 1929. *Coelomomyces chironomi* n. sp., houba cizopasici v dutine telni larev chironoma. Biol. Spisy Vys. skoly Zverolek 8:1-13.

Rioux, J. A., and J. Peck, 1960. *Coelomomyces grassei* n. sp., parasite d'*Anopheles gambiae* (note preliminaire). Acta Trop. 17:179-182.

Rodhain, F. 1968. Elements d'une revision des champignons du genre *Coelomomyces*, parasites des Moustiques. These Medicine, Paris.

________. 1969. Sur la presence d'une champignons du genre *Coelomomyces* en république de Haute-volta. Ann. Parasitol. Hum. Comp. 44:262-264, 3 figs.

________, and P. Gayral. 1971. Nouveaux cas de parasitisme de larves d'*Anopheles* par des Champignous du genre *Coelomomyces* en république de Haute-Volta. Ibid. 44:295-300.

________, and J. Brengues. 1974. Présence de Champignous du genre *Coelomomyces* chez des *Anopheles* en Haute-Volta. Ibid. 49:241-246, 2 figs.

Shemanchuk, J. A. 1959. Note on *Coelomomyces psorophorae* Couch, a fungus parasitic on mosquito larvae. Canad. Entomol. 91:743-744.

Steinhaus, E. A. 1963. Insect pathology, an advanced treatise. Vol. 2, 689 pp. Academic Press, New York.

Umphlett, C. J. 1961. Comparative studies in the genus *Coelomomyces* Keilin, Ph.D. Thesis, Univ. N. Carolina, Chapel Hill, N. C.

________. 1962. Morphological and cytological observations on the mycelium of *Coelomomyces*. Mycologia 54:540-554, 42 figs.

________. 1964. Development of the resting sporangia of two species of *Coelomomyces*. Ibid. 56:488-497, 35 figs.

________. 1968. Ecology of *Coelomomyces* infections of mosquito larvae. J. Elisha Mitchell Sci. Soc. 84:108-114, 6 figs.

Van Thiel, P. H. 1954. Trematode, Gregarine, and fungus parasites of *Anopheles* mosquitos, J. Parasitol. 40:271-279.

Walker, A. J. 1938. Fungal infections of mosquitos, especially of *Anopheles costalis*. Ann. Trop. Med. Parasitol. 32:231-244.

Weiser, J., and J. Vavra. 1964. Zur Verbreitung der *Coelomomyces*-Pilze in europischen Insekten. Zeitschr. f. Tropenmed. u. Parasitol. 15:38-42, 1 fig.

———, and V. J. E. McCauley, 1971. Two *Coelomomyces* infections of Chironomidae (Diptera) larvae in Marion Lake, British Columbia, Canad. J. Zool. 49:65-68, 3 figs.

Whisler, H. C., J. A. Shemanchuk, and L. B. Travland. 1971. Cytology of *Coelomomyces psorophorae* from Mosquito larvae collected in Southern Alberta. Amer. J. Bot. 58:475.

———. 1972a. Germination of the resistant sporangia of *Coelomomyces psorophorae*, J. Inver. Path. 19:139-147, 22 figs.

———, S. L. Zebold, and J. A. Shemanchuk. 1974. Alternate host for mosquito parasite *Coelomomyces*. Nature 251:715-716, 1 fig.

———. 1975. Life history of *Coelomomyces psorophorae*. Proc. Nat. Acad. Sci. 72:693-696, 3 figs.

BLASTOCLADIACEAE

This is the largest family of the Blastocladiales with approximately 35 species distributed among the genera *Blastocladiella, Blastocladiopsis, Blastocladia, Microallomyces,* and *Allomyces*. Possibly, *Ramocladia, Allocladia, Leptocladia*, and *Brevicladia* with 5 species will be included in this family if they prove to be valid genera. Sörgel (1952) described and illustrated these genera fragmentarily without diagnosis, but Emerson and Robertson (1974) have suggested that *Allocladia* may be *Allomyces reticulatus*, and Emerson (personal communication) believes that *Ramocladia reticulata* Sörgel is a species of *Blastocladiopsis* (to be described shortly by Robertson.)* Another genus, *Callimastix*, with one parasitic species on copeopods is regarded by some investigators as a member of this family, but its classification here is highly questionable.

Species of this family are worldwide in distribution and except for *Blastocladiella anabaenae* Canter and Willoughby occur as saprophytes on a variety of plant and animal debris in soil and water, from which they may be isolated and grown readily on synthetic media. Their thalli vary markedly in size and extent. Some species are monocentric and chytrid-like with a non-septate sessile or septate stalked reproductive structure and radiating or basal rhizoids as in some members of *Blastocladiella*, while others are larger, more extended and branched as in *Allomyces* and *Blastocladia*. In these larger species the pseudoseptate or non-septate thalli usually consist of a basal cell or axis which is anchored to the substratum by branched rhizoids and branches dichotomously, subdichotomously, or sympodially at the apex and bears an indeterminate number of zoosporangia and resistant sporangia or gametangia at the ends of the branches. Some species of *Blastocladia* bear fine setae in addition to the two former reproductive structures. Planospores are produced in thin-walled zoosporangia and germinate with papillae or short tubes. The resistant sporangia are borne in thin-walled container cells which they may fill completely or partly and have a thick, smooth or variously ornamented wall. In germination this wall cracks open and releases planospores through papillae or short tubes.

In some genera separate, equal or unequal gametophytic and sporophytic generations occur. In some such species the gametophyte consists only of a cyst which functions holocarpically as a gametangium while the sporophyte is large and dominant in the life cycle. In other species the gametophyte is equal to the sporophyte and bears male and female gametangia. Sexuality is unknown or lacking in some species, and is isogamous or anisogamous in others. Fusion occurs be-

*Canad. J. Bot. 54:11, 1976.

PLATE 133

Figs. 1-14. *Blastocladiella simplex* Matthews on dead flies and agar. (Matthews, 1937.)

Figs. 1, 2. Fixed and stained ovoid, 3-4 × 5.5-7 μ, planospores from a zoosporangium and a germinated resting sporangium, resp. showing nucleus, nucleoleus, nuclear cap and lipoid globules.

Figs. 3, 4. Germinating planospores.

Figs. 5, 6. Sessile and stalked thalli grown on agar.

Fig. 7. Portion of a stalked thallus with the protoplasm accumulating in the apex.

Fig. 8. Zoosporangium delimited by a cross wall with an apical exit papilla.

Fig. 9. Discharge of planospores; initially emerging planospores enveloped by a thin vesicular membrane.

Figs. 10, 11. Young and mature resting sporangia, resp., showing sculptured outer wall in fig. 11.

Fig. 12. Germinating resting sporangium; outer wall has cracked and an exit tube has developed.

Fig. 13. Discharge of planospores from a resting sporangium.

Fig. 14. Germinated resting sporangium with a long exit canal.

Figs. 15-27. *Blastocladiella stomatophilum* Couch and Whiffen on boiled corn and grass leaves. (Cox, 1939.)

Figs. 15, 16. Living and fixed and stained ovoid, 3.5 × 6.5 μ, planospores, resp., with nucleus, nuclear cap, and several small refracfive globules.

Figs. 17, 18. Germination of planospores and the formation of rhizoids.

Figs. 19-22. Developmental stages of the clavate thallus on the surface of the substratum; endobiotic rhizoid not shown.

Fig. 23. Delimitation of apical zoosporangium by a septum.

Fig. 24. Apical zoosporangium shortly before discharge of planospores.

Fig. 25. Largely empty zoosporangium.

Fig. 26. Sessile zoosporangium with a few planospores aggregated at the exit orifice.

Fig. 27. Stalked zoosporangium showing the basal rhizoidal axis and its relation to the stomata of the substratum.

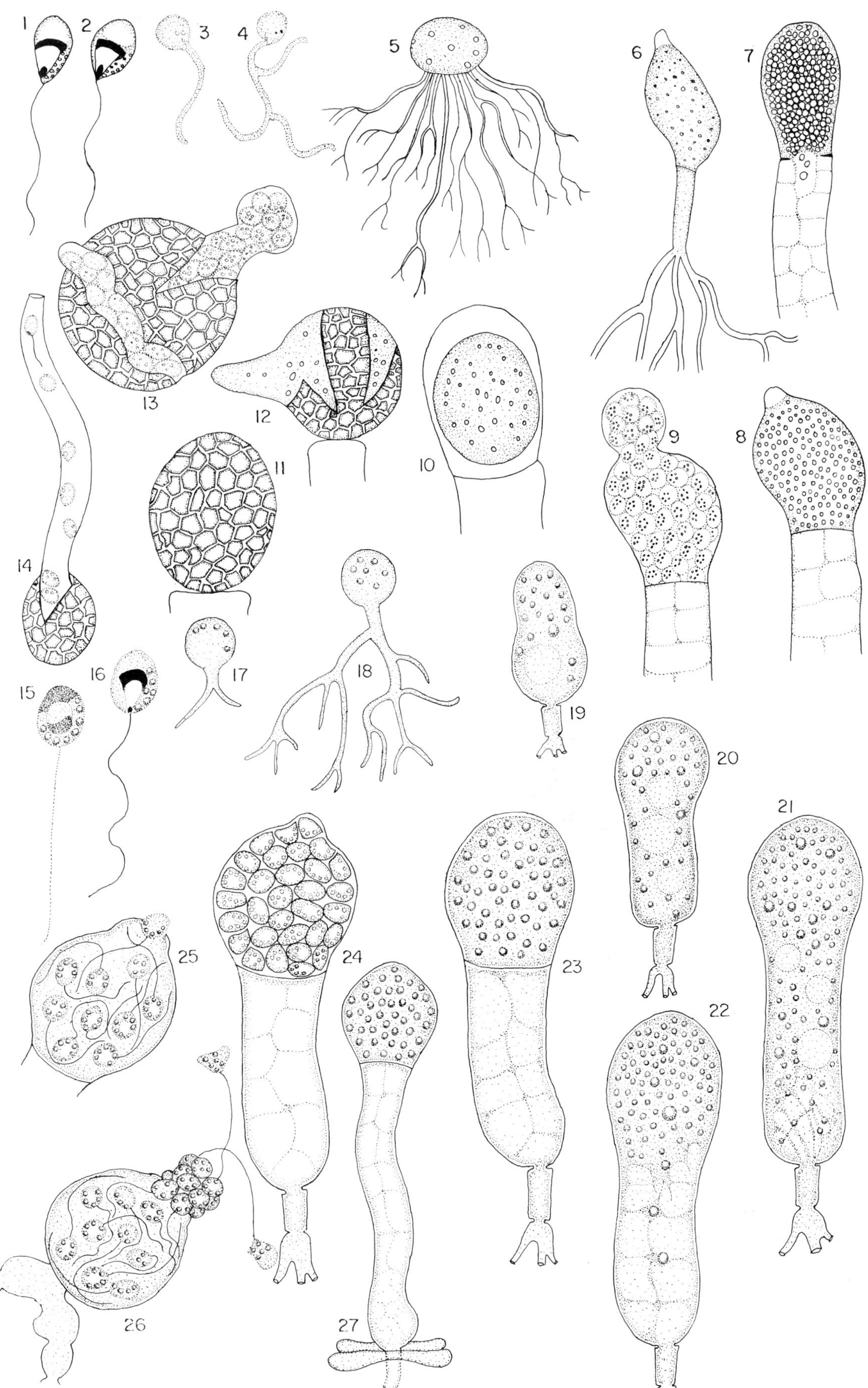

Subgenus *Blastocladiella*

tween pairs of iso- or anisoplanogametes to produce a biflagellate zygote which gives rise to the sporophyte. In such species which have been studied cytologically it is known that the sporophyte is diploid throughout its development and that meiosis occurs during the germination of resistant sporangia. Accordingly, the R.S. planospores are haploid and give rise to the haploid gametophyte.

BLASTOCLADIELLA Matthews

J. Elisha Mitchell Sci. Soc. 53:194, 1937

Rhopalomyces Harder and Sörgel 1938, Nachtr. Gesell. Wis. Göttingen, Math.-Physik. Kl., Fachgruppe VI (Biol.) n.f. 3:123. non Corda *et al.*
Clavochytrium Cox. 1939. J. Elisha Mitchell Sci. Soc. 55:389.
Sphaerocladia Stüben, 1939. "Planta," Arch. Wiss. Bot. 30:364.

Plates 133-140

This genus is reported to include 12 species at present, 11 of which occur as saprophytes in soil and water and may be trapped on animal and plant substrata and readily grown on and in various synthetic media. A few mutants or strains have been isolated and cultured in one of these species, *B. emersonii*. The 12th species, *B. anabaenae*, parasitizes species of *Anabaena* and has not been cultured apart from its hosts. Because it can be grown so readily in synchronized culture and manipulated experimentally, *B. emersonii* Cantino and Hyatt, next to species of *Allomyces*, has been studied more intensively than any other single species of the Blastocladiaceae, and the literature on its metabolism, pathways of morphogenesis, development, and biosynthesis of various compounds, molecular components, ultrastructure of its organelles, and successive developmental stages, sex determination, and other aspects is very extensive.

The thallus of *Blastocladiella* consists either of an extramatrical, sessile, globular enlargement with basal (figs. 5, 116, 117) or radially oriented rhizoids (figs. 64, 93, 94, 168), or a more or less elongate, clavate, septate or non-septate body which usually bears a single reproductive organ at the apex and branched rhizoids or a holdfast at the base (fig. 6, 23, 24, 27, 121, 122, 172, 181–183). However, under certain conditions abnormal thalli may develop which branch (fig. 182) and bear several reproductive organs at the apex (fig. 171). In such thalli the rhizoids may arise along the length of the central stalk and branch several times. Also, the basal cell or stalk may become multiseptate (fig. 122), and the septa extend fully across the cell and are not pseudosepta as in *Allomyces*. In overall size the thalli may be quite large, but this varies considerably under certain conditions and on different substrata or synthetic media. In stalked specimens of *B. simplex* the thallus may be 30–1005 μ high by 8–40 μ in diameter with zoosporangia 15–180 μ and resting sporangia 15–180 μ in diameter.

The successive developmental stages of *Blastocladiella* species such as *B. simplex, B. laevisperma, B. asperosperma*, etc. of the subgenus *Blastocladiella* are well known from light microscope studies of living and fixed and stained preparations, and they may serve in general to illustrate the genus as a whole. The planospores (figs. 1, 28, 44, 45) with its conspicuous nuclear cap, and "side body" (composed of the mitochondrion and lipid sac) absorbs its flagellum and

PLATE 134

Figs. 28-41. *Blastocladiella laevisperma* Couch and Whiffen on grass leaves. (Couch and Whiffen, 1942.)

Fig. 28. Swimming, ovoid, 3.8-4.6 × 6-6.5 μ, planospore with refractive globules, nucleus and nuclear cap; side body not visible.

Figs. 29-31. Germination stages of planospores.

Figs. 32-35. Developmental stages of stalked zoosporangium and discharge of planospores.

Fig. 36. Empty, stalked, septate sporangial thallus showing successive rings in the stalk.

Fig. 37. Stalked, septate resting sporangial thallus with a reddish-brown smooth resting sporangium in the upper portion; successive rings in the stalk.

Fig. 38. Almost sessile resting sporangial thallus with a spherical resting sporangium.

Fig. 39. Sessile non-stalked resting sporangial thallus with the resting sporangium filling the thallus almost completely; ridges on the wall of the sporangium.

Fig. 40. Germination of a resting sporangium; outer wall cracked.

Fig. 41. A large thallus parasitized by several smaller ones; cannibalism.

Figs. 42-53. *Blastocladiella asperosperma* Couch and Whiffen on grass leaves. Figs. 42, 45-53 after Couch and Whiffen, 1942; figs. 43, 44 drawn from Fiji material by Karling.

Fig. 42. Stalked, septate sporangial thallus on grass leaf shortly before discharge of planospores from 1 apical papilla; sessile zoosporangium at the right with 3 exit papillae.

Fig. 43. Discharge of planospores from 2 papillae.

Fig. 44. Ovoid, 3.6-4.6 × 6-7 μ, swimming planospore.

Fig. 45. Fixed and stained planospore with a whiplash flagellum, nucleus, nuclear cap, refractive globules, and side body.

Fig. 46. Empty, stalked, septate sporangial thallus with successive rings in the stalk.

Fig. 47. Sessile resting sporangial thallus with a yellowish-brown warted resting sporangium partly filling the thallus.

Fig. 48. Similar resting sporangium in a stalked septate thallus.

Fig. 49. Section of the wall of a resting sporangium.

Figs. 50-53. Stages in the germination of the resting sporangium.

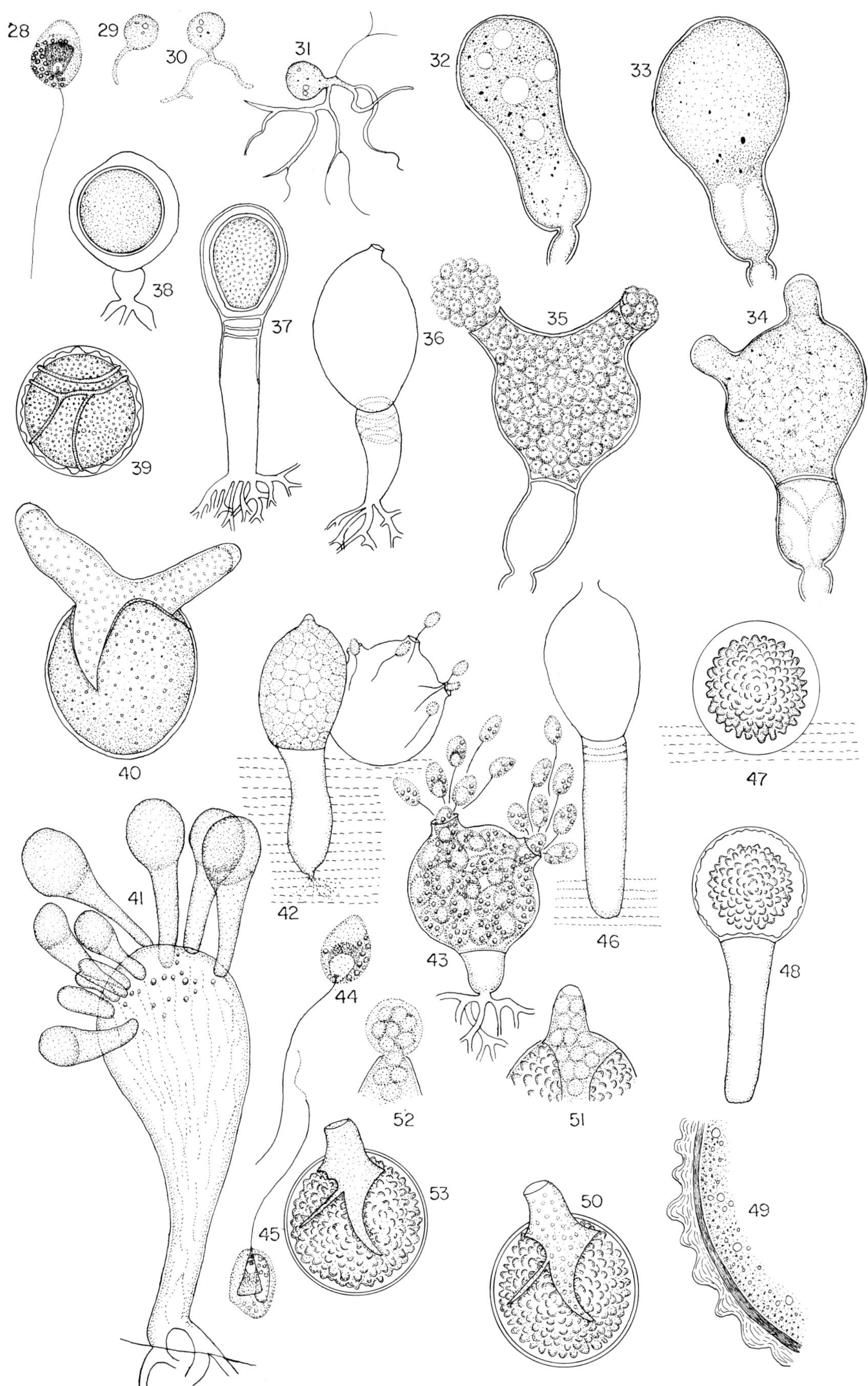

Subgenus *Blastocladiella*

germinates with a broad germ tube (figs. 3, 4) which branches and eventually becomes the rhizoidal system. The planospore body may enlarge directly and become a globular incipient zoosporangium (fig. 5), or it may elongate to form a clavate, subcylindrical elongate structure (figs. 6, 32–35) with rhizoids at its base. As such thalli mature the protoplasm gradually accumulates in the apex (figs. 6, 8, 32, 33) and is delimited by a cross septum (figs. 8, 34). This apical portion becomes the thin-walled zoosporangium which develops one (fig. 8) or more (fig. 42) papillae or short necks (figs. 34, 35) through which the planospores emerge (figs. 35, 42, 43). The initially emerging planospores may push out the thin papilla wall and be enveloped by it for a few seconds before dispersing in *B. simplex* (fig. 9), but in other species the tip deliquesces and the initial mass of planospores is enveloped by a thin layer of matrix (figs. 95, 96). The remaining planospores begin to swirl in the zoosporangium and emerge one by one. Such planospores apparently give rise to additional generations of thin-walled zoosporangial or to resistant sporangial thalli. Both types may occur together, but the latter are reported to become more numerous as the cultures age or fresh water is withheld from aquatic cultures. However, in *B. britannica*, according to Hornstein and Cantino (1961), formation of resting sporangia is affected by several environmental factors but most strikingly by white light. In this species Hoerner and Zetsche (1974), also, found that the development of thin-walled zoosporangia and resistant sporangia is regulated by light. Blue light induces the formation of the former organs, "whereas in red light differentiation to thick-walled sporangia occurs as it does in the dark."

The development of the resistant sporangial thalli is similar to that of the zoosporangial thalli. In the early stages they are usually indistinguishable, but after the former have matured the protoplasm retracts from the wall of the upper delimited portion of the thallus in *B. laevisperma* and *B. asperosperma* or from the wall of the globular portion in *B. novaze-ylandiae* (figs. 99, 100). It soon becomes invested by a wall which may be smooth (figs. 37, 38, 102–105) or warted (figs. 47, 48) or irregularly reticulated, hyaline, pale-yellow, brown or reddish in various species. These resistant sporangia germinate fairly readily, and in the process the outer wall is cracked open by one or more thin-walled papillae (figs. 12, 13, 50–53, 130) or long exit canals (fig. 14) through which the planospores emerge. These are capable of developing into either zoosporangial or resting sporangial thalli, so far as is known.

The developmental stages of *B. novae-zeylandiae* and *B. anabaenae* are basically the same as described above, but the germinating planospore does not expand equally in all directions. A portion of it remains unexpanded, becomes invested with a thick wall (figs. 89, 90), and persists as an appendage on the zoosporangial (figs. 81–84, 91, 92, 95) and resting sporangial thalli (figs. 100, 103, 104). Furthermore, the germinating planospore may develop several germ tubes (fig. 90) that develop into rhizoidal axes on the incipient thalli. As a result they become polyrhizoidal and *Rhizophlyctis*-like (figs. 64, 78, 79, 83). Also, in *B. novae-zeylandiae* they may be apophysate (figs. 91, 95, 103).

The structure of the planospores and developmental stages of *B. emersonii* have been studied intensively by electron microscopy and for these reasons they are described separately here. Such studies by Lovett (1963), Cantino *et al* (1963, 1968, 1969, 1970), Fuller (1966), Reichle and Fuller (1967) and Lessie and Lovett (1968) have shown that the planospores have a complex structure (figs. 108, 109). The most conspicuous internal structure is the nuclear cap enveloping the upper portion of the nucleus (*nc*). Lovett succeeded in isolating the caps (fig. 111) from the nuclei, purified them by differential centrifugation, and analysed their content for protein, RNA, DNA, nitrogen, and phosphorus.

PLATE 135

Figs. 54-65. *Blastocladiella britannica* Willoughby on various organic substrata. (Willoughby, 1959.)

Fig. 54. Ovoid, 4.5-6 × 7.5-11 μ, planospores with a conspicuous nuclear cap and refractive globules.

Figs. 55, 56. Germination of planospore and a young thallus with 1 rhizoidal axis on agar, resp.

Fig. 57. Young thallus with 2 rhizoidal axes.

Fig. 58. Mature zoosporangium.

Fig. 59. Mature stalked septate thallus on hemp seed; planospore outlines in upper portion.

Figs. 60-63. Variations in shapes and sizes of empty zoosporangia.

Fig. 64. Resting sporangium in optical section filling only a part of the spherical portion of a multirhizoidal thallus.

Fig. 65. Similar resting sporangium in surface view; black dots represent pits in the outer wall.

Figs. 66-86. *Blastocladiella anabaenae* Canter and Willoughby parasitizing *Anabaena flos-aquae* and *A. circinali*. (Canter and Willoughby, 1964.)

Figs. 66-68. Ovoid, 4-6.3 × 8-9.5 μ, planospores with a conspicuous nuclear cap and refractive globules; basal pseudopodia in figs. 67 and 68.

Figs. 69, 70. Germination of planospores.

Fig. 71. Germination on a rotifer egg.

Figs. 72-74. Contact and infection of *Anabaena* cells by the rhizoids.

Figs. 75-78. Similar stages; planospore cyst only partly expanded and persistent at the apex of the incipient zoosporangia.

Fig. 79. Portion of a thallus approaching maturity of the zoosporangium.

Fig. 80. Mature zoosporangium shortly before dehiscence; only nuclear caps and refractive granules of planospores shown.

Figs. 81-84. Variations in shapes and sizes of empty zoosporangia; persistent part of planospore cyst present.

Figs. 85, 86. Probable ovoid and subspherical resting sporangia (?) with a smooth thick wall; sketched from photographs.

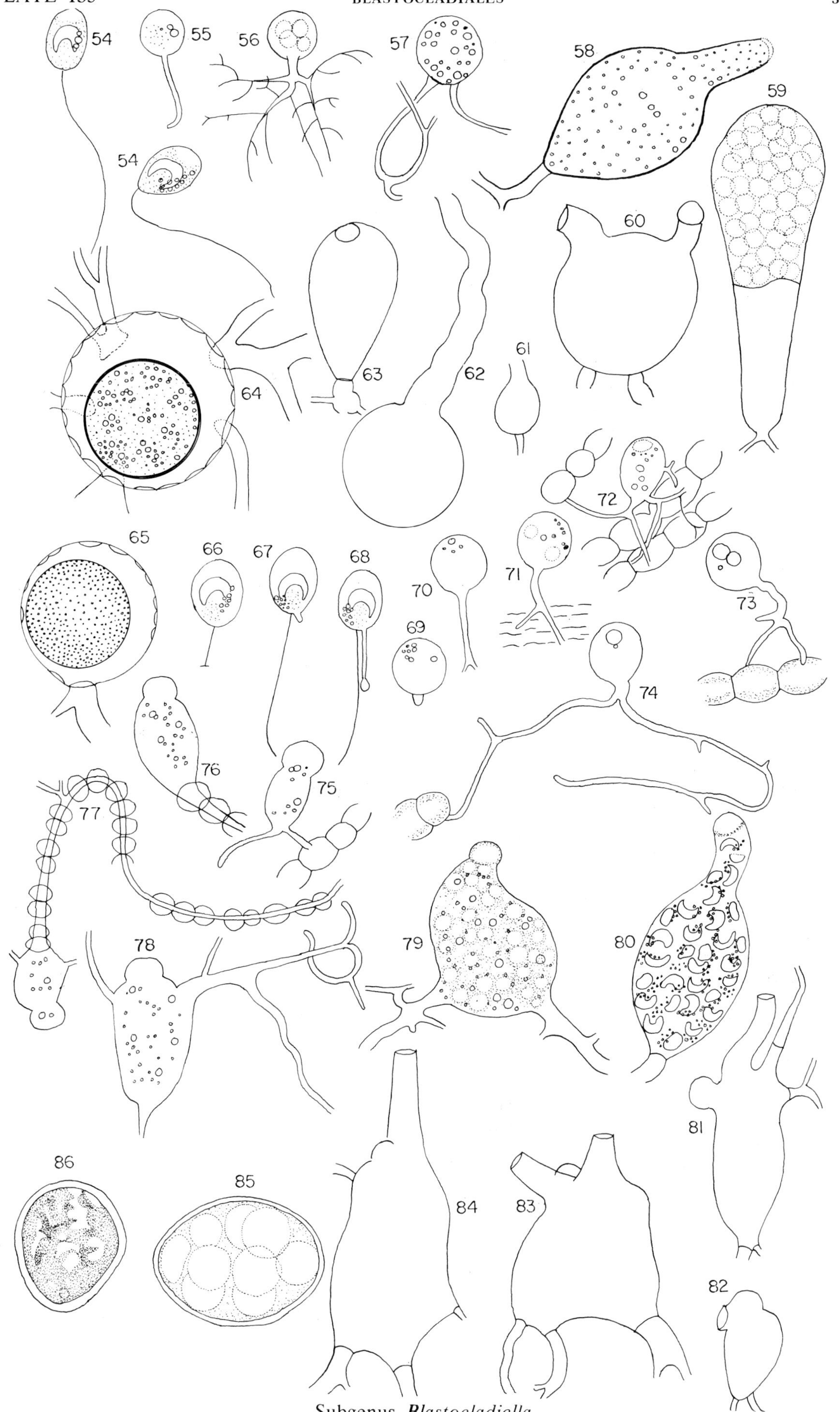

Subgenus *Blastocladiella*

He found that the cap is strongly basophilic and composed of 60% protein and 40% RNA. It represents 18% of the dry weight and 69% of the total RNA of the planospore, and the purified cap particles contain 37% protein and 63% RNA. Lovett concluded that the nuclear cap is an unusual intracellular "packet" of ribosomes. His study, thus, confirmed by more precise methods and results the earlier studies by Turian (1955, 1956) and Turian and Kellenberger (1956) of the nuclear caps of *Allomyces* gametes. In addition to the ribosomal nuclear cap Cantino *et al* (1969) found 2 satellite ribosomal packages or visicle-like organelles with a double membrane. One of these vesicles occurs near the nuclear cap, and the other is almost wholly enclosed within the mitochondrial chamber. Other organelles revealed by these studies are a single basal mitochondrion (*m*) which is perforated and passed through by the kinetosome and its closely associated lipid sac (*l*) with globles, a banded rootlet (*r*), a basal structure (*b*), a posterior whiplash flagellum (*f*) with a sheath and 9 outer double and 2 inner single fibers (fig. 110), and gama particles (*g*) at the anterior end. The latter are complicated 3-dimensional organelles (fig. 109a) which appear during sporogenesis and disappear as the planospore germinates. In *B. britannica* they differ from those of *B. emersonii* in that they do not contain the well defined 3 hole matrix, according to Cantino and Truesdell (1971). The lipid sac in *B. emersonii* is a complicated structure and is made up of a lipid sac matrix, a spoon bowl-shaped organelle, and a continuous unit membrane (Cantino and Truesdell, 1970).

As to the changes which these organelles undergo, Lovett (1968), Soll *et al* (1969), and Truesdell and Cantino (1970) showed that as the planospores germinate dedifferentiation occurs. The nuclear cap breaks down as such, and its ribosomes become dispersed in the cytoplasm (fig. 112). At the same time the single mitochondrion (figs. 107–109) fragments so that portions are dispersed in the planospore body, germ tube, and rhizoids (figs. 112, 113). The gamma particles give off vesicles which fuse with the plasma membrane, or adhere to the nuclear cap, and in the final stages membranes of different particles fuse to produce large vacuoles containing electron dense matrices (Truesdell and Cantino, 1970). Further developmental ultrastructural studies were made by Lessie (1967) and by Lessie & Lovett (1968) from the log phase to the completion of planospore differentiation. They found that "after induction zoosporangium differentiation requires a 2½ hr. period in which the nuclei divide, a cross wall forms to separate the basal rhizoidal region (figs. 119, 120), and an apical papilla is produced." The latter (fig. 124) is formed by a localized breakdown of the wall and the deposition of papilla material by secretory granules. In addition to other ultrastructural details of mitosis, and differentiation, they found that the cleavage planes are formed by fusion of vesicles in the cytoplasm. Probable fusion of mitochondria occurs to form the single mitochondrion of the planospores, and the ribosomes aggregate around the nucleus (fig. 125) to form the nuclear cap whose membrane develops by the fusion of many small vesicles.

The light-yellow to deep-brown resistant sporangia (fig. 129) possess a thick pitted or unpitted wall impregnated with melanin and enclose globules and small quantities of Y-carotene. Resistant sporangia occur sporadically in gross cultures, but Cantino (1950) reported that their formation can be controlled at will by the addition of 10^{-2} sodium bicarbonate on agar media. In the absence of added bicarbonate only thin-walled zoosporangial thalli developed. In more intensive studies Cantino and Hyatt (1953) found that R.S. planospores yielded populations of less than 2% orange (O), 98% of ordinary colorless (OC), and late colorless (LC) thalli which were distinguishable from another by color, rate of growth, method and duration of motility, viability and size of the planospores, and potentialities of the resistant sporangia developed from them. The planospores of the orange thalli were smaller, $4 \times 6\ \mu$, but were incapable of fusing with any

PLATE 136

Figs. 87-105. *Blastocladiella novae-zeylandiae* Karling on boiled corn leaves and snake skin. (Karling, 1967.)

Figs. 87, 88. Ovoid, $5.5\text{-}6.6 \times 8.5\text{-}9.9\ \mu$, and amoeboid planospores with a conspicuous nuclear cap and yellowish-orange refractive granules.

Fig. 89. Young monorhizoidal thallus, part of the planospore cyst unexpanded.

Fig. 90. Young thallus with 4 germ tubes arising from the periphery and which will develop into rhizodal axes.

Fig. 91. Apophysate, *Diplophylctis*-like thallus with part of the planospore cyst as an appendage.

Fig. 92. Largely extramatrical almost mature thallus with 2 rhizoidal axes.

Fig. 93. Spherical polyrhizoidal, *Rhizophlyctis*-like thallus with 8 incipient exit canals.

Fig. 94. Same thallus 5 hours later; 4 of the tubes have elongated into exit canals; refractive granules in the ring stage.

Fig. 95. Subspherical zoosporangium shortly before discharge of planospores.

Fig. 96. Initial mass of emerged planospores enveloped by a thin layer of matrix.

Fig. 97. Fairly thick-walled dormant zoosporangium.

Fig. 98. "Germination" of a thick-walled dormant zoosporangium.

Figs. 99-101. Stages in contraction of the protoplasm in the formation of the resting sporangium.

Fig. 102. Resting sporangium which fills only part of the spherical portion of a polyrhizoidal thallus; refractive globules yellowish-orange or golden; wall smooth, non-punctate and golden-brown.

Figs. 103, 104. Resting sporangia which fill the spherical part of the thallus almost completely; unexpanded portion of the planospore cyst persistent.

Fig. 105. Small resting sporangium.

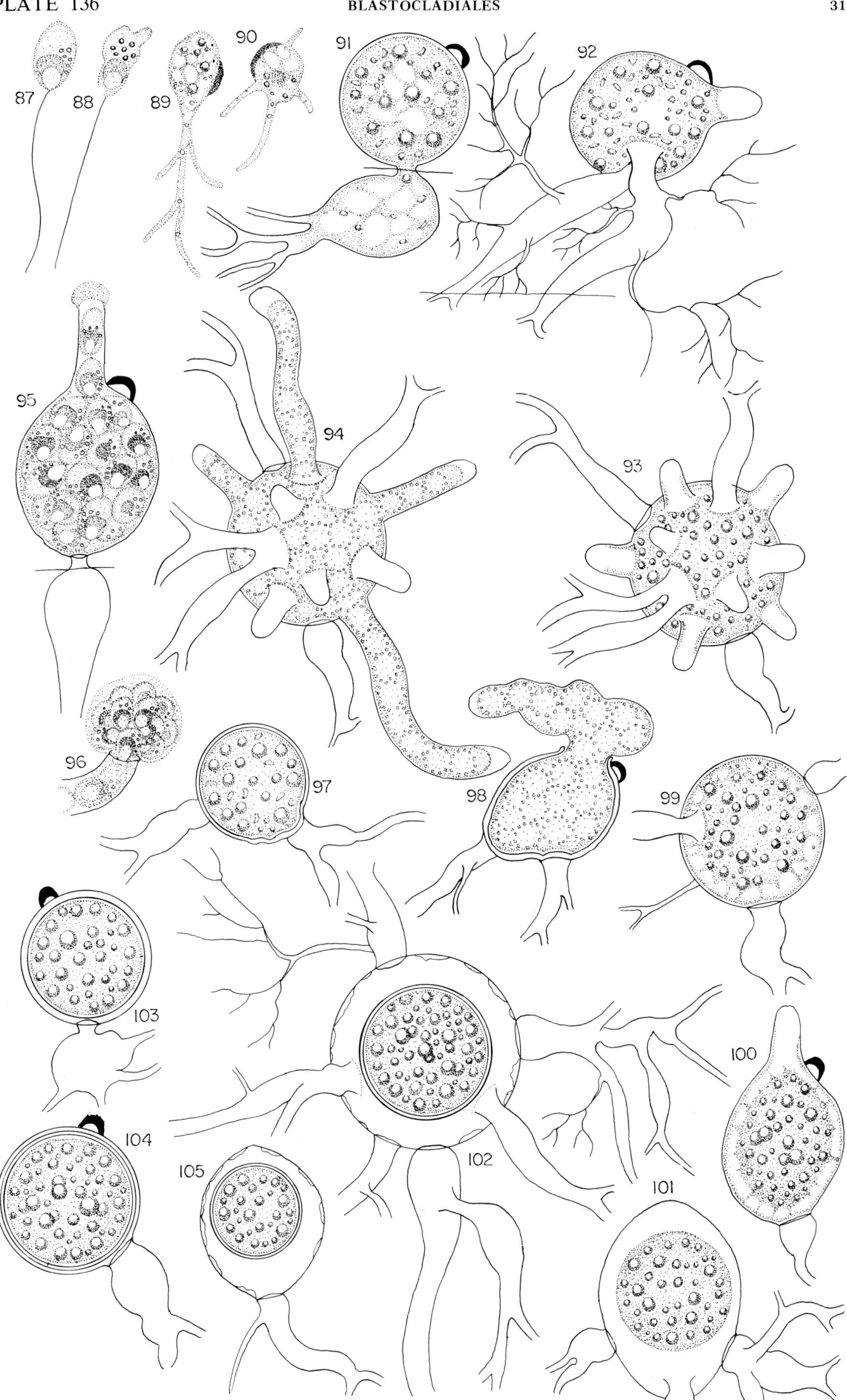

Subgenus *Blastocladiella*

other planospores. Later, Cantino and Hornstein (1954) isolated an orange mutant (BEM) from cultures of *B. emersonii* whose planospores were incapable of fusing with those of colorless thalli but were capable of establishing cytoplasmic bridges (figs. 132–134) with the larger planospores from colorless thalli.

Various types of life cycles occur among the known species, and these correspond to similar cycles in *Allomyces*. For this reason the terms used for the cycles in the latter genus might be applied here, but as has been done for *Synchytrium* and *Allomyces* it seems better to distinguish the groups as subgenera: i.e. *Blastocladiella*, *Eucladiella* and *Cystocladiella* (See Karling, 1973).

Subgenus BLASTOCLADIELLA Karling

Mycopath et Mycol. Appl. 49:169, 1973.

Plates 133-137

This subgenus includes the short cycled species *B. simplex* Matthews (figs. 1–14), *B. stomatophilum* (Couch and Cox) Couch and Whiffen (figs. 15–27), *B. laevisperma* Couch and Whiffen (figs. 28–41) *B. asperosperma* Couch and Whiffen (figs. 42- 53), *B. britannica* Hornstein and Cantino (figs. 54- 65), *B. novae-zeylandiae* Karling (figs. 87–105), and *B. emersonii* Cantino and Hyatt (figs. 106–134), in which sexuality is lacking or unknown, and only zoosporangial and resistant sporangia thalli are developed, so far as is known. Thus, the subgenus *Blastocladiella* which includes the type species, *B. simplex*, corresponds to *Brachyallomyces* of *Allomyces*. In *B. emersonii*, however, Cantino and Hyatt (1953) expressed the view that "disregarding the strange lack of sexual fusions the life history of this species corresponds to a disrupted *Allomyces* cycle, but under controlled environmental conditions, it can be short circuited into a brachyallomycetous cycle which consists almost exclusively of resistant sporangial plants." Possibly some of the other species in this subgenus may be found to have a different life cycle when they are fully known. Resistant sporangia are unknown in *B. stomatophilum*, and germination of the resistant sporangia has not been seen in *B. novae-zeylandiae* and

PLATE 137

Figs. 106-134. *Blastocladiella emersonii* Cantino and Hyatt. Figs. 106, 114-118, 121-123, 126-130 drawn from living material; fig. 111 after Lovett, 1963; figs. 197, 112, 113, 119, 120, 124, 125 after Lovett, 1968, and Lessie and Lovett, 1968; figs. 108, 110 after Fuller, 1966, and Reichle and Fuller, 1967; fig. 109 after Cantino *et al.* 1963; fig. 109a after Truesdell and Cantino, 1970; figs. 131-134 after Cantino and Hornstein, 1954, 1956.

Fig. 106. Ovoid, 7 × 9 μ, planospore from living material with a nuclear cap, nucleus, side body or mitochondrion and refractive globules.

Fig. 107. Greatly enlarged semi-diagramatic sketch of planospore with the single mitochondrion, lipoidal globules, nucleus, nuclear cap, and attachment of the whiplash flagellum.

Fig. 108. Ultrastructural view of a longitudinal section of a planospore sketched from an electron photograph; NC = nuclear cap, N = nucleus, M = mitochondrion.

Fig. 109. Interpretative sketch of the structure of a planospore; N = nucleus, NU = nucleolus, L = lipoidal sac, M = mitochondrion, R = a banded rootlet, B = basal structure, G = gamma particles, F = flagellum.

Fig. 109a. Three-dimensional sketch of a gamma particle.

Fig. 110. Cross section of the flagellum with the sheath, 9 outer double fibrils with arms skewed clockwise, and a central pair of single fibrils.

Fig. 111. Nuclear caps from a purified preparation. Sketched from a photomicrograph.

Fig. 112. Semi-diagramatic sketch of an early germination stage of a planospore; nuclear cap has broken down and the ribosomes are uniformly distributed in the cytoplasm around the nucleus; mitochondrion has divided into several segments.

Fig. 113. Later germination stage with a branched rhizoid; nucleus is dividing.

Figs. 114-117. Germination of a planospore and stages in the development of sessile sporangial thalli; planospore body has enlarged to become the incipient zoosporangium.

Fig. 118. Stalked sporangial thallus growing in a liquid medium.

Figs. 119, 120. Semi-diagramatic sketches of septum development at the base of a sessile zoosporangium.

Fig. 121. Stalked sporangial thallus; protoplasm has aggregated in the apical portion, leaving bands behind in the stalk.

Fig. 122. Stalked multiseptate sporangial thallus; septa of stalk have probably developed from bands of protoplasm such as shown in fig. 121.

Fig. 123. Sessile zoosporangium with the protoplasm in the "ring" stage.

Fig. 124. Semi-diagramatic sketch of a stage in exit papilla formation.

Fig. 125. Similar sketch of a cleavage segment; ribosomes aggregating around the nucleus to form the nuclear cap.

Fig. 126. Discharge of planospores.

Fig. 127. Resting sporangial thallus with numerous refractive globules.

Fig. 128. Contraction of the protoplasm from the wall and early stage in resting sporangium development.

Fig. 129. Thick-walled, dark-brown pitted resting sporangium in upper portion of the thallus.

Fig. 130. Germination of a fairly deeply pitted resting sporangium.

Fig. 131. Free-hand sketch of a planospore with gamma particles, g, in the apex.

Figs. 132, 133. Cytoplasmic bridges between large and smaller planospores of a mutant, sketched from photographs.

Fig. 134. Enlarged sketch of 2 such planospores.

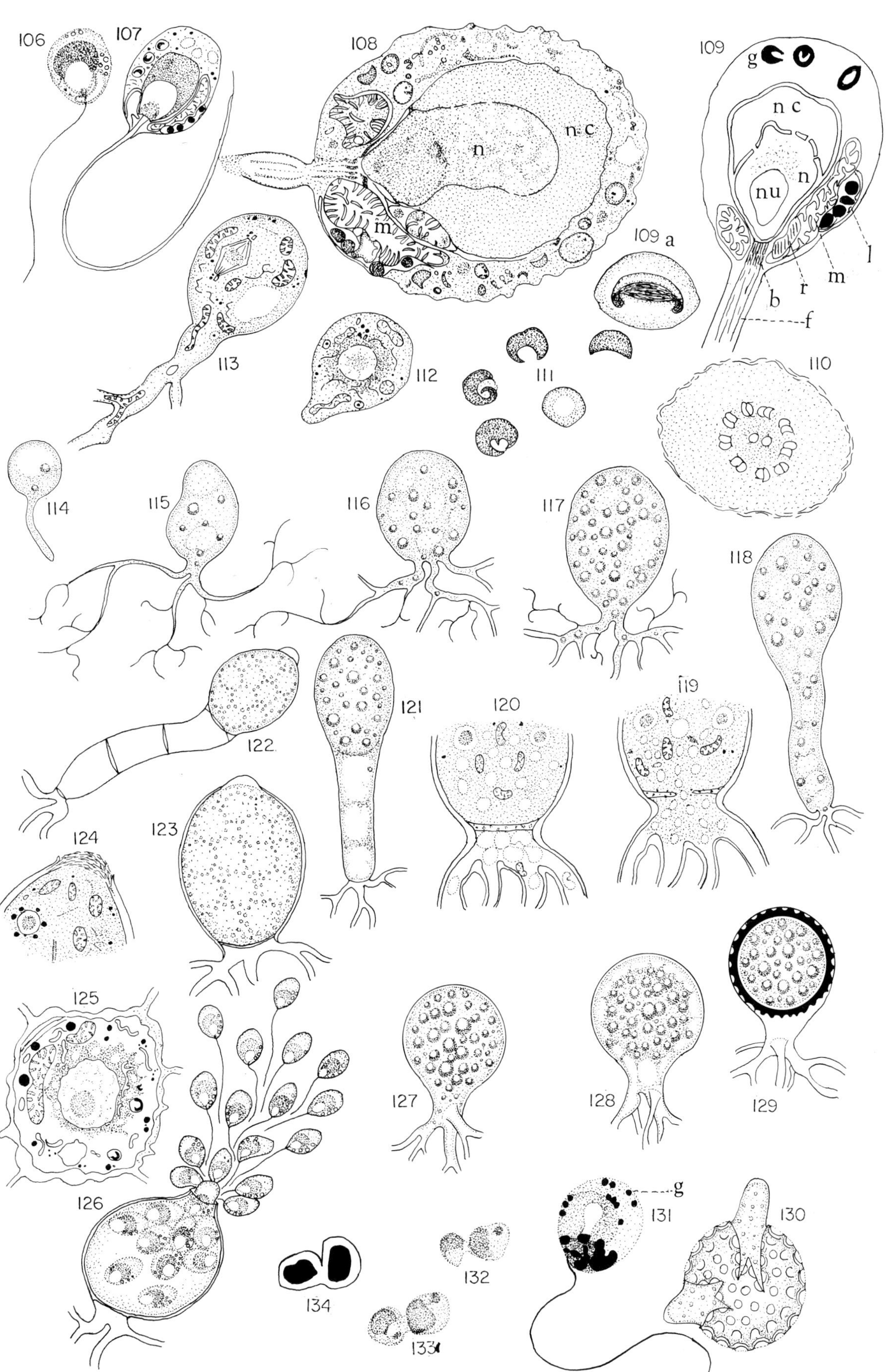

Subgenus *Blastocladiella*

B. anabaenae. In *B. brittanica* only a single germinated resistant sporangium has been observed (Cantino, 1970).

Fusion of gametes has not been observed in the species of the subgenus *Blastocladiella*, but in *B. emersonii*, as noted previously, Cantino and Hornstein (1954) found that the smaller planospores of an orange mutant strain (BEM) could establish cytoplasmic bridges or isthmuses with the larger ones of colorless thalli. They regarded this as a cytoplasmic instead of a nuclear exchange mechanism which might play a vital role in the behavioral pattern of this species, possibly in the same manner as it does in *Paramecium* and other organisms. In a footnote to the 1954 paper and in a reported later study (1956) they illustrated the presence of gamma particles in the apical portion of the planospores of the wild type of *B. emersonii* (figs. 109, 109A, 131) which are generally absent in those from the orange mutant, and they postulated that the particles "are one of the factors involved in the cytoplasmic exchange...."

Subgenus CYSTOCLADIELLA Karling

Mycopath. et Mycol. Appl. 49:171, 1973

Plate 138

This subgenus of 2 species, *B. cystogena* Couch and Whiffen and *B. microcystogena* Whiffen, is long-cycled in the sense that both gametophytic and sporophytic generations occur but thin-walled zoosporangial thalli are lacking. The encysted zygote germinates (fig. 148) to produce a stalked sporophytic thallus (figs. 149–151) which bears a terminal faintly areolate dull-yellow to dark-brown (fig. 135) or smooth and hyaline to light yellow-brown (figs. 152, 153) resistant sporangium. It germinates (figs. 136–139, 154, 155) as in *Eucladiella* species, but the uniflagellate planospores produced and discharged do not become actively motile and disperse. Instead, they may stick together or move about sluggishly for a while, and soon come to rest and encyst in clusters or tendrils (figs. 140, 154, 155). They enlarge slightly, develop an exit papilla (figs. 142, 158, 159), and are transformed holocarpically into gametangia and give rise to 4 small isoplanogametes which fuse in pairs (figs. 146, 162) to produce biflagellate zygotes (figs. 147, 163).

Accordingly, *Cystocladiella* corresponds to the subgenus *Cystogenes* of *Allomyces* in its type of isogamous sexual reproduction, but it differs by the lack of thin-walled zoosporangia in its life cycle.

Subgenus EUCLADIELLA Karling

Myopath. et Mycol. Appl. 49:170, 1973

Plates 139, 140

This subgenus includes two long-cycled species, *B. stübenii* and *B. variabilis* in which the germinating resistant sporangium produces + and − gametes which develop into isomorphic and equal gametangial thalli (figs. 177, 184). In *B. variabilis* on of these (−) is orange colored and the other (+) is colorless, and by analogy with the long cycled species of *Allomyces* the former may be considered to be male and the latter female. In *B. stübenii* both gametangia are colorless and similar to but smaller than the zoosporangia. In

PLATE 138

Figs. 135-151. *Blastocladiella cystogena* Couch and Whiffen on boiled *Paspalum* grass leaves. (Couch and Whiffen, 1942.)

Fig. 135. Mature resting sporangium with peculiar thickening in the wall, wall fairly areolate, nearly hyaline to dull-yellow or orange-brown.

Figs. 136-139. Stages in the germination of the resting sporangia and the initial discharge of planospores.

Fig. 140. Germinated resting sporangium with encysted planospores near and within the exit orifice; only a portion of the planospores are shown.

Fig. 141. Uniflagellate, sluggishly motile or non-motile planospores immediately after discharge from the resistant sporangium and before encystment.

Fig. 142. Encysted planospores or cystospores which have developed exit papillae as they became transformed into gametangia.

Fig. 143. Gametangium, 8.2-10.2 μ diam., with its content cleaved into gamete initials.

Fig. 144. Four ovoid, 4.1 × 4.9 μ, gametes emerging from a gametangium.

Figs. 145, 146. Pairing and fusion of isoplanogametes.

Fig. 147. Biflagellate, ovoid, 6 × 8 μ, zygote.

Fig. 148. Young thallus growing on agar.

Fig. 149. Young stalked resting sporangial thallus; resting sporangium developing in the apex of the thallus.

Fig. 150. Mature stalked septate thallus with the resting sporangium almost filling the apex; thallus growing out of the stoma of a *Paspalum* leaf.

Fig. 151. Mature thallus grown on hemp seed.

Figs. 152-164. *Blastocladiella microcystogena* Whiffen on grass leaf bait. Figs. 152-154, 157, 158, 164 after Whiffen, 1946; figs. 155, 156, 159-163 drawn from New Zealand material.

Fig. 152. Stalked resting sporangia thallus with a spherical resting sporangium; wall smooth, hyaline to light yellow brown; refractive globules small and evenly dispersed.

Fig. 153. Mature, sessile resting sporangial thallus; resting sporangium filling the globular portion completely.

Figs. 154, 155. Germinated resting sporangia with clusters of encysted planospores at the exit orifice.

Fig. 156, 157. Ovoid, 2.5 × 2.9 μ, isoplanogametes produced by encysted planospores or gametangia (microcystospores.)

Figs. 158, 159. Small, spherical, 4.8-6.2 μ, gametangia whose content has cleaved into 4 potential gametes.

Fig. 160. Emergence of 4 isoplanogametes.

Figs. 161, 162. Stages in the pairing and fusion of gametes.

Fig. 163. Biflagellate zygote.

Fig. 164. Germinated zygote or young thallus.

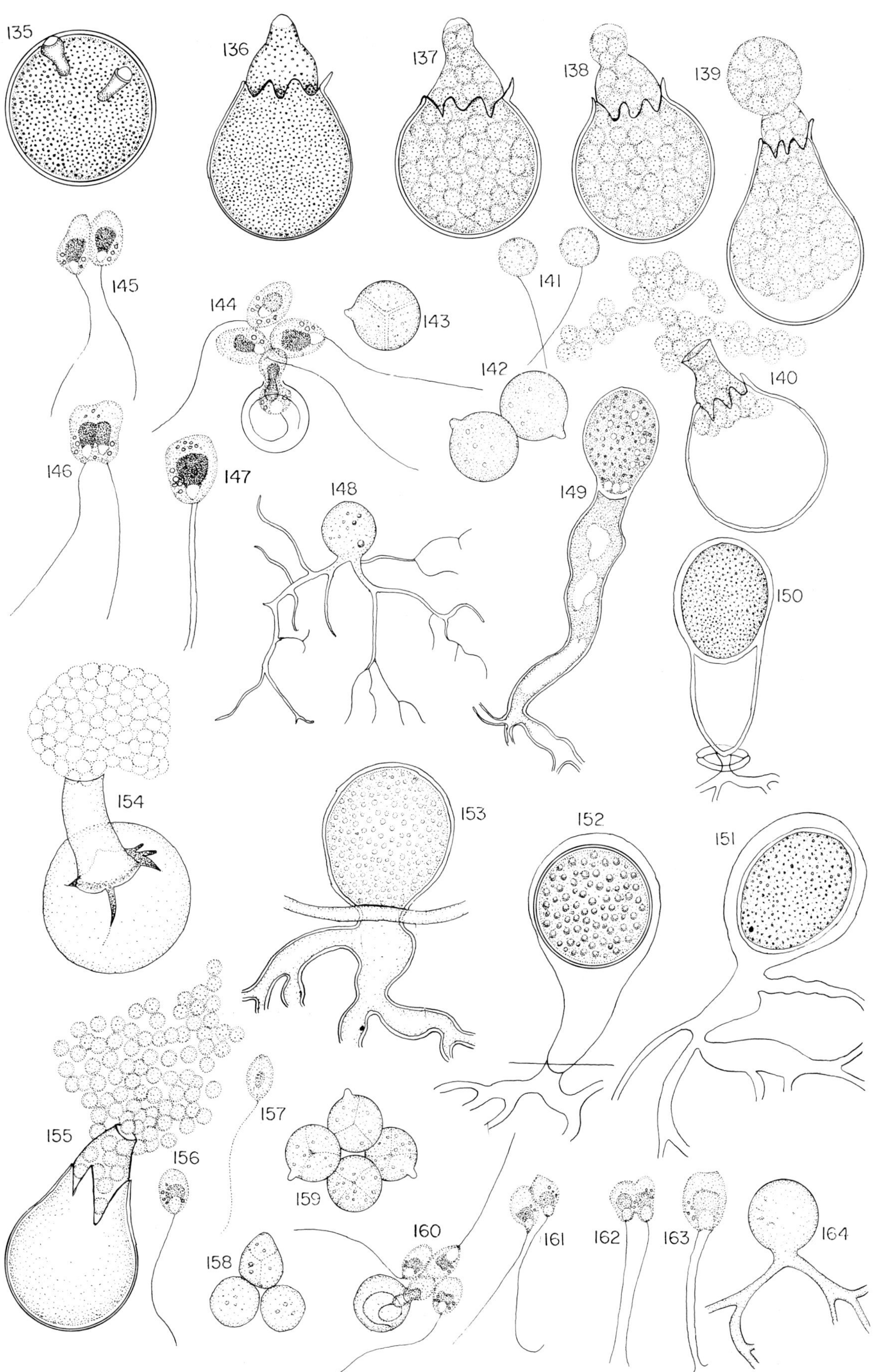

Subgenus *Cystocladiella*

both species the isoplanogametes produced in the gametangia are smaller than the planospores and fuse in pairs (figs. 174, 175) to form biflagellate zygotes (fig. 176). These germinate without delay and develop into zoosporangial or resistant sporangial thalli (sporophytes).

Accordingly, *Eucladiella* corresponds to the subgenus *Allomyces* in being long-cycled, but its sexual reproduction is isogamous instead of anisogamous as in the latter subgenus. Also, the problem of where meiosis and sex determination in the life cycle has not been solved in *Eucladiella*. Harder and Sörgel (1938) and Stüben (1939) believed that meiosis occurs during germination of the resistant sporangia and that sex determination is phenotypic, but they did not present any experimental evidence to support this belief. Emerson (1950) reported from his studies of a *B. variabilis*-like species that sex is phenotypically determined. However, it turned out that the species he studied is *B. emersonii* in which Cantino and Hyatt (1953) subsequently reported that sex is determined phenotypically.

Particularly noteworthy is that the thalli of *B. stübenii* may be monocentric and *Rhizophlyctis*-like with radiating rhizoids (fig. 168) or elongate (figs. 169, 170), or branched with several reproductive organs. In *B. variabilis*, which apparently has been seen only once, it is usually elongate and stalked with rhizoids at its base (figs. 181, 183) or sometimes branched (fig. 182).

ALLOMYCES Butler

Ann. Bot. London 25:1027, 1911

Septocladia Coker and Grant, 1922. J. Elisha Mitchell Sci. Soc. 37:180.

This genus is reported to include approximately 9 species and several varieties which are worldwide in distribution as saprophytes on animal and plant debris in soil and water. So far nearly 100 separate isolations have been made in various parts of the world and maintained in culture. Three types of life cycles occur among the known species, and these differences are the basis for grouping the species in 3 subgenera: *Allomyces* (*Euallomyces*), *Cystogenes*, and *Brachyallomyces*. The thallus structure, organization, development and variations of this genus as well as asexual and sexual reproduction and other details are adequately presented in the description and illustrations of the subgenera below and will not be summarized here.

Subgenus ALLOMYCES Karling

Mycopath. et Mycol. Appl. 49: 171, 1973

Plates 141-145, figs. 1-7

This subgenus includes the type and long-cycled species with alternating and equal haploid gametophytic (sexual) and diploid sporophytic (asexual) generations. As such it is comprised at present of *A. arbuscula* Butler, emend Barrett and Hatch, *A. macrogynus* Emerson and Wilson, *Allomyces* × *javanicus* Kniep, and possibly *A. javanicus* var. *allomorphus* Indoh and some other varieties. Other long-cycled species formerly known as *A. kniepii* Sörgel and *A. strangulata* (Barrett) Minden have been shown to be identical with *A. arbuscula*, and *A. javanicus* var. *javanicus* Emerson was found by Emerson and C. W. Wilson (1954) to be a natural hybrid, *Allomyces* × *javanicus*, between *A. arbuscula* and *A. macrogynus*. The latter two species as well as the hybrid occur in the same localities in nature, and the hybrid has been produced experimentally by crossing the above-named species. Apparently, because the

PLATE 139

Figs. 165-177. *Blastocladiella stübenii* Couch and Whiffen on sterilized house flies and agar media. (Stüben, 1939.)

Fig. 165. Ovoid, 3.5 × 4.8 μ, planospore with a side body, nuclear cap and nucleus.

Fig. 166. Germination of a planospore.

Fig. 167. Portion of a young sporangial thallus.

Fig. 168. Spherical, multirhizoidal, *Rhizophlyctis*-like sporangial thallus with 2 exit tubes; rhizoids omitted on the lower portion of the zoosporangium.

Figs. 169, 170. Abnormal sporangial thalli growing under a cover slip.

Fig. 171. Similarly grown, branched sporangial thallus; only a portion of the rhizoid are shown.

Fig. 172. Portion of resting sporangial thallus on peptone agar with a resting sporangium in the apex and surmounted by the remains of the planospore cyst.

Fig. 173. Germinated, empty resting sporangium.

Figs. 174, 175. Fusion of isoplanogametes from gametophytic thalli.

Fig. 176. Biflagellate zygote.

Fig. 177. Diagram illustrating the life cycle and alternation of gametophytic and sporophytic generations: G = gametophytes + and −, Z = zygote, SP = sporophytes.

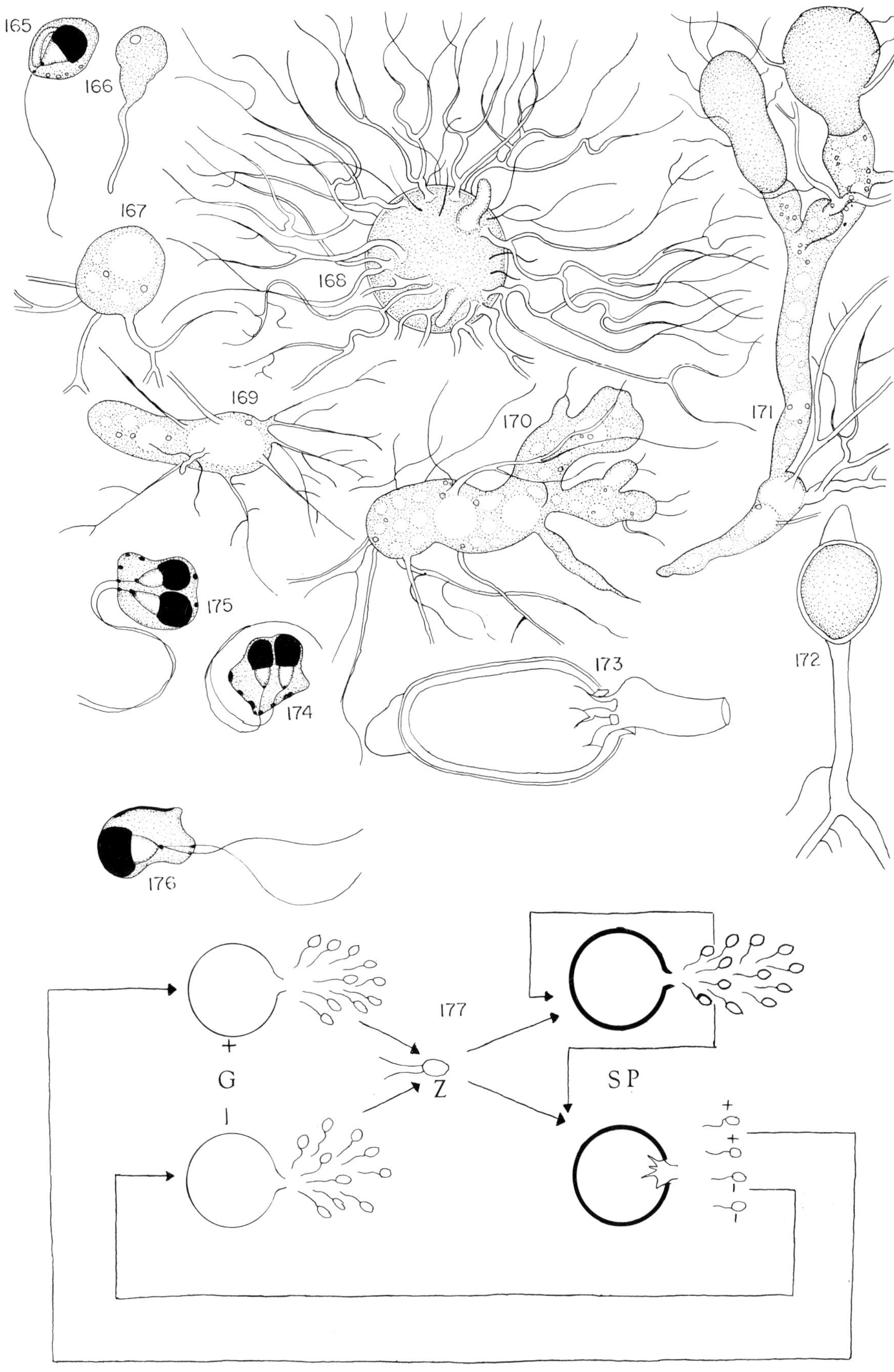

Subgenus *Eucladiella*

life cycles of these species consist of equal gametophytic and sporophytic generations which can be cultured separately, easily grown on known synthetic media, and manipulated readily under experimental conditions they have been studied more intensively than any other group of the Chytridiomycetes. Accordingly, the published papers relating to their distribution throughout the world, morphology, morphogenesis and development, cytology, ultrastructure, cytogenetics, genetics, sexuality, nutritional requirements, biochemical syntheses, and physiology are very extensive, but only a few of the pertinent references will be cited here.

In the usual and normal type of development the diploid Z.S. planospores* (figs. 1, 74) or zygotes (figs. 37, 38, 65, 71) give rise to the diploid sporophyte which bears brown, thick-walled, pitted (R.S.) resistant sporangia and thin-walled (Z.S.) zoosporangia (figs. 9, 76). The diploid Z.S. planospores germinate and usually repeat the sporophytic generation. The haploid R.S. planospores, on the other hand, from germinated resistant sporangia (figs. 16, 73) give rise to the haploid gametophyte which bears female and male gametangia, usually in alternating pairs. These two types of thalli are generally equal in size, type of branching, general organization and development, but the gametophyte has a longer lag phase of growth than the sporophyte (Machlis and Crasemann, 1956).

Several departures from the usual life cycle noted above have been reported. These include (1) apomixis or parthenogenetic development of female gametes into gametophytes or sporophytes without fusion. (2) development of sporophytes from R.S. planospores; (3) development of resistant sporangia on gametophytes; (4) transformation of male gametangia into resistant sporangia, and (5) rare fusions of Z.S. planospores.

Kniep (1929, 1930) reported that he had separated male and female gametangia in *A. javanicus* and found that the female gametes from the latter occasionally developed into mature gametophytes without fusing with male gametes. This variation was confirmed by Sörgel (1937) and Emerson (1941) in *A. arbuscula*, *A. javanicus*, and *A. macrogynus* (figs. 41, 97). Sörgel (1937) also found that parthenogenetic female gametes of *A. arbuscula* may develop into asexual sporophytic thalli (fig. 42), an observation confirmed (fig. 43) by Emerson (1941) in different strains of this species. In this connection it may be noted that Turian (1958) reported that M/20 boric acid inhibits conjugation of gametes, resulting in parthenogenetic germination of female gametes and the establishment of female colonies or sporophytes. Also, copper ion, acetate, arsenite, adenine and thymine suppressed female differentiation and led to predominantly male colonies.

Sörgel (1937) and Emerson (1941) as well as Emerson and C. W. Wilson (1954) observed that R.S. planospores frequently develop into sporophytes rather than gametophytes. This usually occurred when such planospores had an extended swimming period before settling down on the nutrient medium, whereas if the motile period was short gametophytes developed. Such behavior of the R.S. planospores as well as of the parthenogenetic female gametes noted above suggests strongly that these motile cells are facultative as in many chytrid genera and can function in several ways.

The development of resistant sporangia on the gametophyte (figs. 42, 43) occurs fairly regularly in *A. arbuscula*, according to Sörgel (1937), Emerson (1941) and Hatch (1944), and the first of these workers attributed this occurrence to a mixture of haploid gametophytic and diploid sporophytic nuclei within the same hyphae. Sörgel, also, described intergrades and direct changeover from sexual to asexual thalli, or vice-versa and presented as proof correlative measurements of nuclear volumes in such hyphae (mischhyphen). However, he did not make chromosome counts in such nuclei to support his contention, and for this reason Emerson (1941) and Stumm (1958) questioned his explanation. Pertinent to this departure from the usual method of development it may be noted that Beneke and G. B. Wilson (1950) found that treatment of zygotes of *A. macrogynus* with sodium nucleate increased the frequency with which R.S. planospores gave rise to sporophytes as well as gametophytes. Sost (1955) reported that colchicine had a similar effect, and these three workers suggested that such treatment of the zygotes induced some nuclei to become polyploid and the resulting asexual sporophytes mixiploid. Accordingly, at meiosis in the germinating resistant sporangium some of the R.S. planospores would become haploid and some diploid or polyploid. Whiffen (1951) found

*Emerson and C. W. Wilson (1954) referred to the resistant sporangia as "meiosporangia" and their haploid planospores as "meiospores." The thin-walled zoosporangia were called "mitosporangia" and their haploid planospores "mitospores". This terminology has been used by their associates and students rather generally, but in this presentation the thin-walled zoosporangia and resistant sporangia will be referred to occasionally as ZS and RS, respectively, for the sake of brevity, and their respective motile cells as Z.S. and R.S. planospores.

PLATE 140

Figs. 178-184. *Blastocladiella variabilis* Harder and Sörgel isolated from soil and grown on nutrient agar. (Harder and Sörgel, 1938.)

Fig. 178. Ovoid planospore with a conspicuous nuclear cap and globules.

Figs. 179, 180. Young subspherical and ovoid sporangial thalli with branched basal rhizoids.

Fig. 181. Elongate clavate thallus.

Fig. 182. Branched thallus.

Fig. 183. Mature clavate thallus with the apical zoosporangium delimited by a septum and bearing 2 papillae.

Fig. 184. Diagram illustrating the life cycle and alternation of gametophytic and sporophytic generations; G = gametophytes + and −, Z = zygote, SP = sporophytes, S = sporangial thallus, RS = resting sporangium thallus.

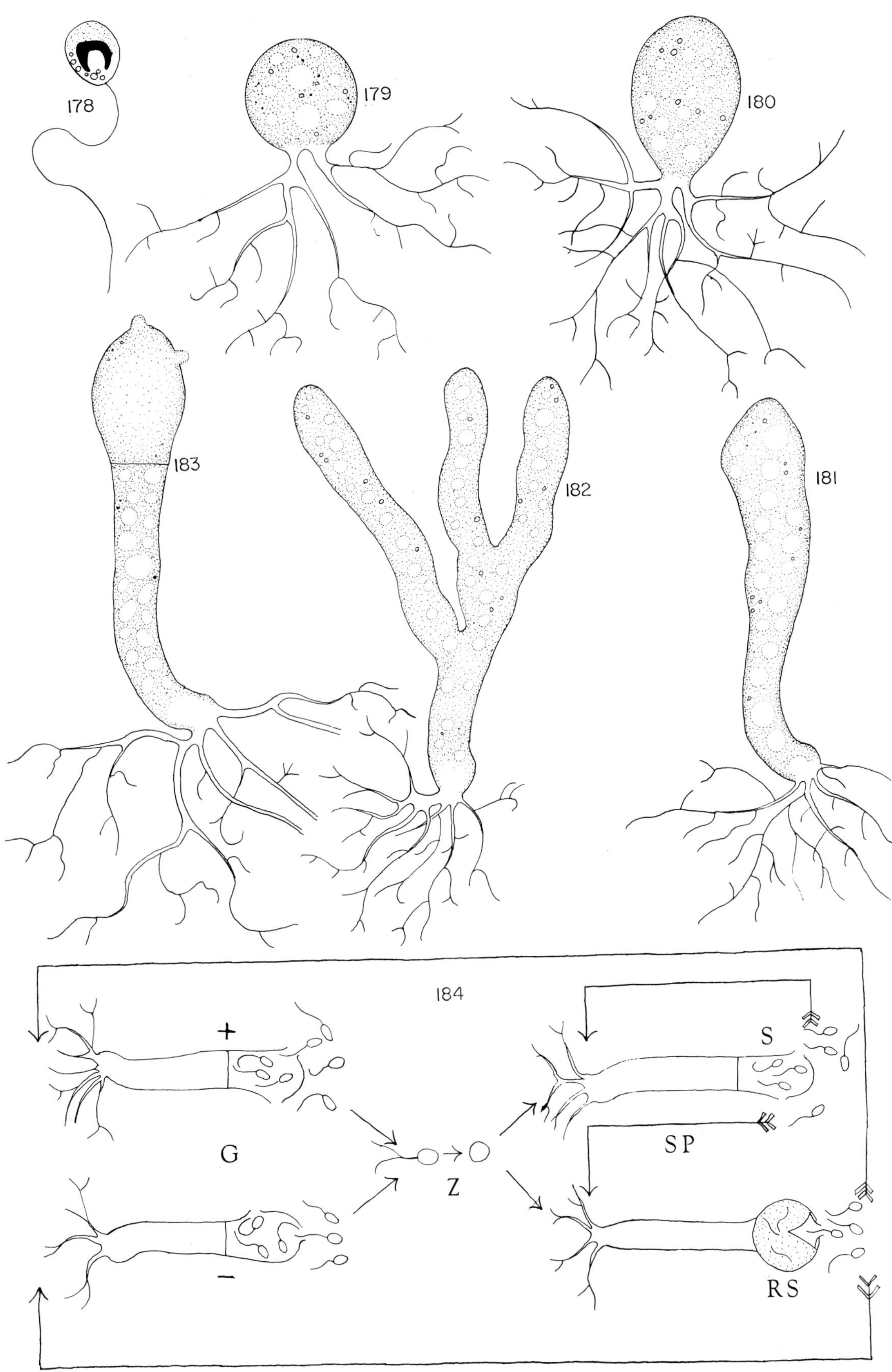

Subgenus *Eucladiella*

that when cycloheximine was added to the growth medium for *A. arbuscula*, sporophytic hyphal outgrowths occurred which bore gametangia. When transplanted such outgrowths developed into a mixture of sporophytic-gametophytic mycelia or hyphae which later became either entirely sporophytic or gametophytic. She suggested that the antibiotic may have caused separation of the chromosomes into 2 groups during meiosis without their splitting.

In *A.* (*kniepii*) Sörgel (1937) observed that in addition to male and female gametangia and resistant sporangia some thalli bore thick-walled bodies with papillae like the gametangia and had a color between that of the normal red male gametangia and the normal brown resistant sporangia (fig. 44). He concluded, therefore, that these bodies were a transition between male gametangia and resistant sporangia. Upon isolation and germination such bodies gave rise to planospores which degenerated after a short motile period. Sörgel noted, also, that single non-discharged Z.S. planospores may develop directly into a small holocarpic secondary zoosporangium within the primary one (fig. 45). In addition, he noted that Z.S. planospores may fuse in pairs or greater numbers (fig. 46), apparently without karyogamy, to form normal-sized or giant fusion products. These germinate and form multirhizoidal germlings (fig. 47, 48) which develop into sporophytes.

The Z.S. and R.S. planospores and gametes have the same general complex structure (figs. 51, 63) as has been revealed by the ultrastructural studies of Renaud and Swift (1964), Moore (1964, 1968), Turian and Kellenberger (1956), Fuller and Calhoun (1968), Hill (1969), Fuller and Olson (1971), Olson (1973), and others. However, unlike the planospores of *Blastocladiella emersonii* they contain no gamma particles and have several, instead of one, mitochondria. However, in planospores of *Allomyces* × *javanicus* Moore (1968) found a large basal mitochondrion which he believed might have a similar relation to the flagellum as the single basal one *Blastocladiella*. A similar one was reported by Fuller and Olson (1971) in *A. macrogynus* and *A. neo-monilioformis*. In addition the planospores include microtubules, a conspicuous nuclear cap composed of ribosomes, nucleus with a porous double membrane, nucleolus, kinetosome, several lipoidal globules, and a whiplash flagellum (fig. 49). Fuller and Olson (1971) reported that the side body adjacent to a portion

Plate 141

Figs. 1-29. *Allomyces arbuscula* Butler. Figs. 1, 12 after Ritchie, 1947; figs. 2, 11, 13, 16, 19 drawn from living material; figs. 3, 5-8, 14 after Barrett, 1912; fig. 4 after Butler, 1911; figs. 9, 17, 18 after Emerson, 1941; fig. 15 after Rorem and Machlis, 1959; figs. 20-23 after Hatch, 1935; figs. 10, 24-29 after Renaud and Swift, 1964.

Fig. 1. Subspherical, 10-12 μ diam., living planospore with a nucleus, nuclear cap, nucleolus, refractive lipid globules, and a whiplash flagellum.

Fig. 2. Germinated planospore.

Fig. 3. Young thallus showing dichotomous branching; most of rhizoids omitted.

Fig. 4. Portion of a mature thallus showing dichotomous branching; only rhizoidal axes shown.

Figs. 5-8. Longitudinal and cross section of hyphae showing the nature of pseudosepta.

Fig. 9. Portion of a diploid, asexual sporophytic thallus bearing thin-walled zoosporangia and a resistant sporangium.

Fig. 10. Diagram of a nucleus in a hyphal tip showing a single inpocketing centriole at the base.

Fig. 11. Thin-walled zoosporangium with its protoplasm in the "ring" or "lipid crown" stage; lipid globules outlining the peripheral nuclei.

Fig. 12. Discharge of diploid planospores from a clavate thin-walled zoosporangium.

Fig. 13. A brown resistant sporangium with finely pitted wall.

Fig. 14. Enlarged view of a portion of the wall.

Fig. 15. Diagram of a 5 μ thick fixed and stained section through an immature resistant sporangium showing the presence of chromospheres composed of 12% RNA and 60% protein.

Fig. 16. Germinating resistant sporangium discharging haploid planospores.

Fig. 17, 18. Germination of haploid planospore and young gametophyte, resp.

Fig. 19. Portion of gametophyte bearing epigynous female and hypogynous male gametangia. Note similarity of structure to the sporophyte (fig. 9).

Fig. 20. A female and male gametangium with the protoplasm in the "ring" or "lipid crown" stage; lipid globules outlining peripheral nuclei; mitochondria elongate.

Fig. 21. Cleavage in a female gametangium.

Fig. 22. Cleavage segments with the nuclei surrounded by a dark, staining reticulum composed of mitochondria, according to Harch (1935).

Fig. 23. Incipient female gametes with mitochondria (?) fused to form the nuclear cap.

Figs. 24-29. Diagrams of stages in the development of basal bodies and flagellar sheath of gametes in gametangia.

Fig. 24. Early stage, nucleus with a pair of in-pocketing centrioles; membrane of incipient gamete at the right.

Fig. 25. Elongation of one member of the pair.

Fig. 26. Later stage at the initiation of gamete formation; vesicles from the membrane fusing via pinocytosis with the basal body to form a primary vesicle.

Fig. 27. Primary vesicle stage.

Fig. 28. The start of flagellar fiber formation; a primary vesicle invaginated and enlarging by fusion with secondary vesicles.

Fig. 29. Elongation of the fibers and the primary vesicle which will form the flagellar sheath

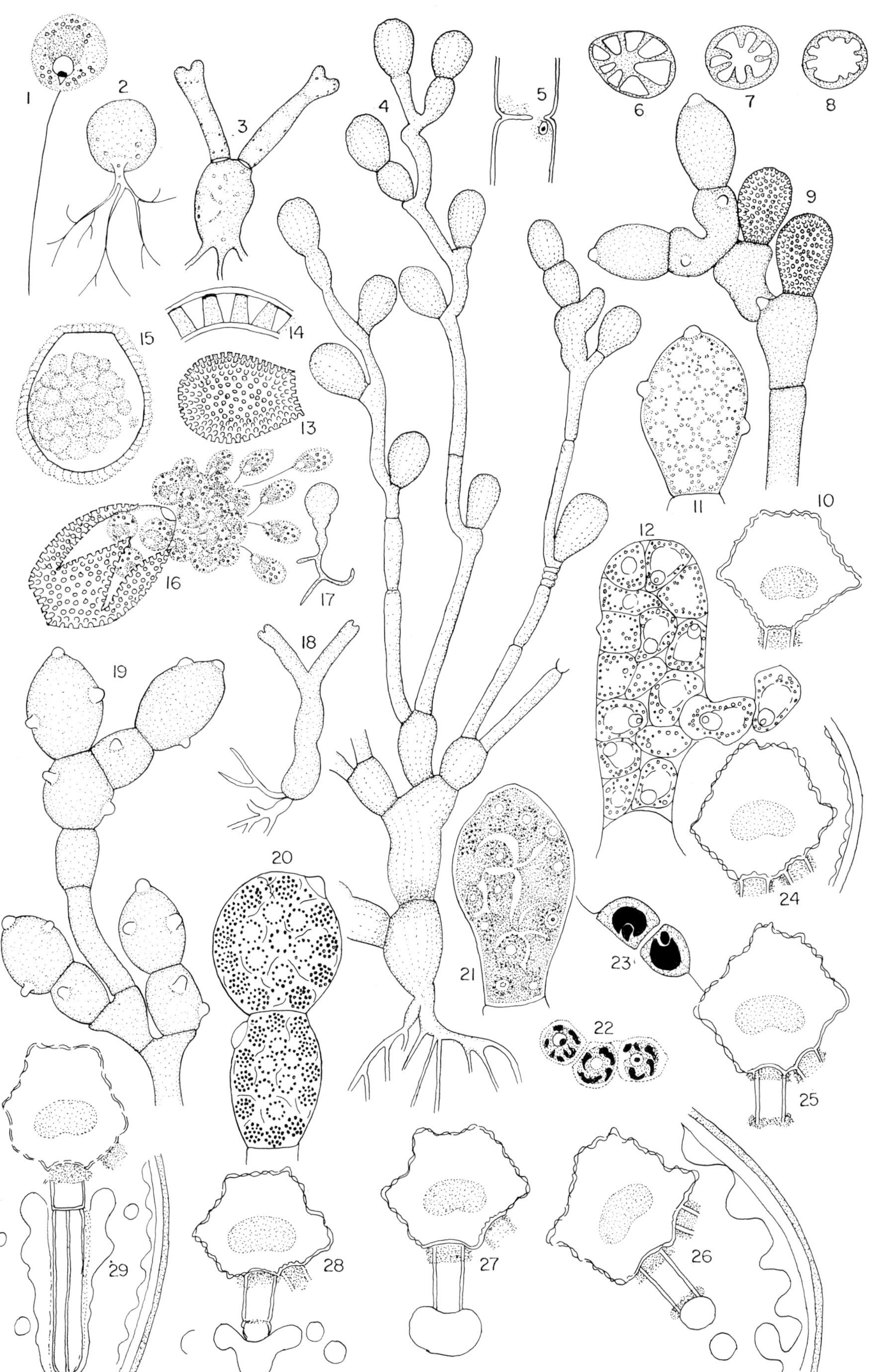

Subgenus *Allomyces*

of the basal mitochondrion and composed of several elements is the equivalent of the "lipid sac" and "side body granules" described by Cantino, et al (1968) and Lessie and Lovett (1968) in *Blastocladiella emersonii*, and renamed it the Stüben body.

In the early stages of gametangial differentiation Renaud and Swift found a deviation of the centriolar pattern from that present in the hyphal tip. In the former the centrioles occur in pairs (fig. 24) at the base of the nuclear membrane. One of them grows distally to more than 3 times its original length while the other one retains its small size (fig. 25). The larger centriole would correspond to the basal body or kinetosome of the future gamete. Gametogenesis is usually induced by transferring a ripe culture to distilled water, and within a short time vesicles are pinched off from the cell membrane (fig. 26) and come into contact with the basal body. These fuse at its base and form a large primary vesicle, after which the flagellum starts to grow and invaginates it (figs. 27, 28). As the flagellum elongates so does the vesicle by fusion with secondary vesicles, and by this process the flagellar sheath is formed (fig. 29). This discovery of two centrioles in relation to gametogenesis is particularly noteworthy in light of subsequent studies, noted in the Introduction, which show 2 centrioles in the planospores of numerous other species.

Germination of the Z.S. and R.S. planospores (figs. 2, 52–56), zygotes (figs. 17, 18, 38, 39, 67–71, 87–92) and gametes is essentially similar, and in the process the nuclear cap breaks up and dedifferentiation begins, according to Turian (1961). The ribosomes of the nuclear cap become dispersed in the cytoplasm, the mitochondria fragment and disperse, and the nucleus divides. The initial germ tube may be slender, and as it branches more and more an extensive rhizoidal system is formed. At the same time the upper part of the germinating cell elongates outward (figs. 55, 91, 92), becomes tubular or cylindrical, and thus forms the main axis or basal cell of the thallus. It usually branches dichotomously at the apex (figs. 3, 56, 71), and this type of branching may continue in the development of the mature thallus or it may become sympodial. As the thallus develops, pseudosepta are formed (figs. 5–8) which extend only partly across the lumen of the hyphae. Thus, the thallus is essentially unicellular as in *Blastocladiella*. The pseudosepta and other walls are composed of chitin, glucan and ash, contain a fraction of protein, according to Aronson and Machlis (1959), and do not react positively to any tests for cellulose.

The thin-walled zoosporangia, resistant sporangia, and gametangia are formed by inflation of the ends of the branches and are delimited by septa as they mature. Secondary zoosporangia and resistant sporangia may terminate the hyphal branches which grow up from beneath the primary reproductive organs in a sympodial manner (figs. 4, 9). Also, zoosporangia and sometimes resistant sporangia develop in basipetal succession along the length of a hypha. The former may occur singly or in chains, terminally and sympodially, with a rounded apex and truncate end, clavate, barrel-shaped, or cylindrical and occasionally Y- and T- shaped. In size they range from 27 to 50 μ in diameter by 30–118 μ in length in the various species. They as well as the gametangia develop one to several discharge papillae (figs. 9, 11, 19) at maturity, and these consist of a pulley-shaped refractive plug (Skucas, 1966) "securely placed in a hole in the wall" which can be removed by treatment with sulfuric acid. These plugs are composed of a pectic substance, probably protopectin. The protoplasm of these reproductive organs consists of numerous nuclei,

PLATE 142

Figs. 30-40, 43. *Allomyces arbuscula* Butler. Figs. 41, 42, 44-48 *Allomyces (Kniepii) arbuscula*. Figs. 30-40 after Hatch, 1938; figs. 41, 42, 44-48 after Sörgel, 1937; fig. 43 after Emerson, 1941.

Figs. 30, 31. Fixed and stained female and male planogametes, resp. showing relative sizes and structure.

Fig. 32. Initial stage of fusion at the posterior ends of the gametes.

Fig. 33. Later stage with the gametic nuclei closer together.

Fig. 34. Binucleate, biflagellate planozygote.

Fig. 35. Zygote at rest with paired nuclei; nuclear cap beginning to dissociate.

Fig. 36. Karyogamy; nuclear cap material widely dispersed.

Fig. 37. Fused, diploid nucleus with separated male and female nucleoli.

Fig. 38. Germinating zygote; male and female nucleoli fused.

Fig. 39. Intranuclear division of primary diploid nucleus of the germinating zygote, believed by Hatch (1938) to be the 1st meiotic division.

Fig. 40. Tetranucleate germinated zygote, small nuclei degenerating.

Figs. 41-48. Variations of the normal developmental cycle and some abnormalities.

Fig. 41. Parthenogenic development of a gametophyte from a female gamete which has germinated in the female gametangium.

Fig. 42. Parthenogenic development of a sporophyte from a female gamete which has germinated in the female gametangium.

Fig. 43. Portion of a gametophytic hypha or thallus bearing a resistant sporangium and gametangia.

Fig. 44. Portion of a gametophytic thallus bearing gametangia and resistant sporangia; some of the latter bear abortive exit papillae and may represent a morphological transition from gametangia to resistant sporangia.

Fig. 45. Zoosporangium in which all but one of the planospores have degenerated; one planospore has developed into a dwarf zoosporangium.

Fig. 46. Portion of a sporophyte; some of the planospores have fused to form variously sized bodies.

Figs. 47, 48. Germination of such bodies formed by fusion of planospores.

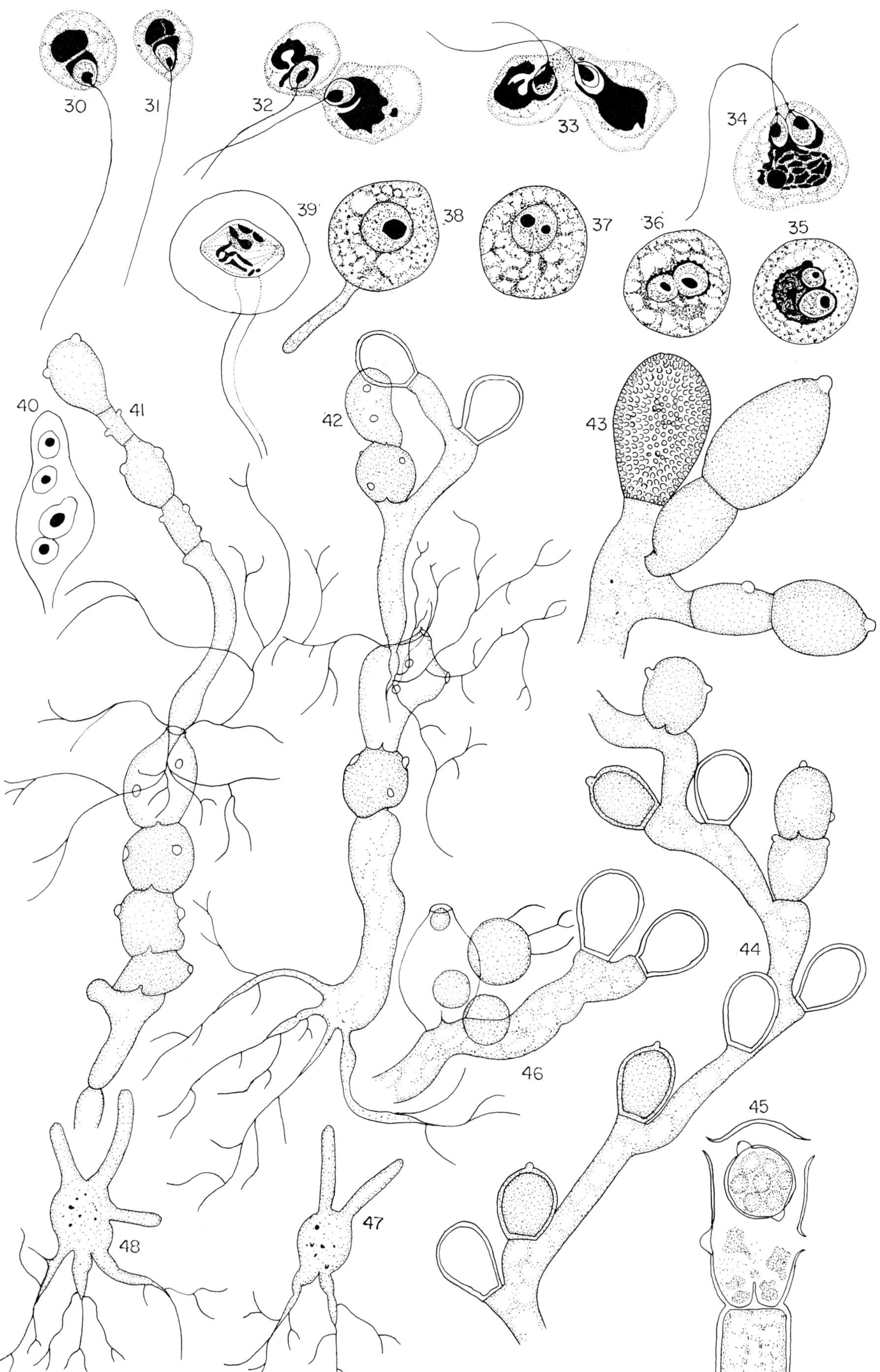

Subgenus *Allomyces*

mitochondria, endoplasmic reticulum, microtubules, vesicles, ribosomes and lipoidal globules, and at one stage the globules are arranged around the nuclei, producing the characteristic "ring" or "lipid crown" appearance under the light microscope (figs. 11, 20, 59).

In preparation for sporogenesis and gametogenesis differentiation of the protoplasm begins, and this process is essentially the reverse of dedifferentiation. The ribosomes begin to accumulate around the nucleus and become enveloped by a membrane derived possibly from the endoplasmic reticulum (Moore, 1968) to form the nuclear cap. Also, the mitochondria aggregate in the same vicinity and usually impinge on the nuclear cap (figs. 51, 63). The first evidence of cleavage into planospore or gamete initials is the aggregation of small vesicles in the cytoplasm which are filled with electron-dense material (Moore, 1968). These become oriented in a linear manner and then fuse to form cleavage planes which delimit polyhedral uninucleate segments. Following the polyhedral stage the spore initials gradually round up, separate, and mature into planospores or gametes which emerge when the pulley-shaped plug in the papilla dissolves.

The resistant sporangia originate as do the zoosporangia, but in normal wild types they are produced in much greater abundance or in a ratio of 9 to 1 (DeLong, 1956). As a result the culture soon becomes brown in color. However, DeLong found that by inducing mutants by UV irradiation the ratio between the types of sporangia may be shifted to as much as 20%. Although resistant sporangia are formed in the original hyphal wall they are not fused with it. They have a multi-layered wall of their own, the outer of which is thick and brown in color, contains melanin, and is sculptured with fine pits (figs. 9, 13, 43, 72) in a characteristic manner. In the subgenus *Allomyces* the pits are usually minute, closely spaced, and sometimes almost indistinguishable as separate entities. The resistant sporangia may remain viable for many years. They acquire their characteristic morphology within 48 to 72 hours after initiation, and at this stage they contain about a dozen nuclei and numerous spherical orcein-staining bodies (fig. 15) called chromospheres (Rorem and Machlis, 1957). However, the sporangia are not mature at this stage because they will not germinate when placed in water but require 2 weeks to months for maturation. This process involves several factors, but the disappearance of the chromospheres is a significant one. It as well as complete maturation may be hastened by as much as 70% if the fungus is grown in nutrient broth instead of on agar slants. Also, indolbutyric acid incorporated in the agar or broth reduces the time of chromosphere disappearance by 50% (Machlis and Ossia, 1953a,b).

Concommitant with the erosion of the chromospheres the nuclei enter the phosphases of meiosis, and by this time the resting sporangia become capable of germination. They do so best within a range of 20–25°C, and in this process the content swells and cracks the outer pitted wall. An inner thin-walled endosporangium protrudes and develops one to several discharge papillae (figs. 16, 73) which eventually deliquesce and release the fully formed haploid R.S. planospores. These, as noted earlier, develop into haploid gametophytes.

Meiosis is an integral part of germination in *Allomyces*, and the metaphases of meiosis I may be present within 55–65 minutes after mature resistant sporangia are placed in water. Metaphases of meiosis II appear within 80–110 minutes, and after 2 hours R.S. plano-

PLATE 143

Figs. 49-73. *Allomyces macrogynus* Emerson and Wilson. Figs. 49, 52-56, 59, 60, 71-73 drawn from living material; figs. 50, 56, 57 after Emerson, 1941; fig. 51 after Fuller and Calhoun, 1968; figs. 61, 62, 64-70 after Emerson, 1955, (sketched from photographs); fig. 63 after Turian and Kellenberger, 1956.

Fig. 49. Ovoid, 10 × 7 μ, haploid R. S. planospore (meiospore) with nucleus, nucleolus, nuclear cap, and refractive lipoid globules.

Fig. 50. Living amoeboid Z. S. planospore (mitospore) from a thin-walled zoosporangium.

Fig. 51. Diagram of a longitudinal section of a planospore showing its ultrastructure, based on an electron micrograph, L = lipoid bodies, M = mitochondria, N = nucleus, NC = nuclear cap, T = microtubules enveloping the nucleus and nuclear cap and originating near the kinetosome, K.

Figs. 52-56. Encysted diploid planospore (meiospore), its germination, and the early developmental stages of a gametophyte, resp.

Fig. 57. Small gametophyte growing in a very weak nutrient solution with an epigynous male and a hypogynous female gametangium.

Fig. 58. Portion of a gametophyte bearing paired gametangia; male gametangium small, ovoid and epigynous, female gametangium elongate, subcylindrical and hypogynous.

Fig. 59. A pair of gametangia with their protoplasm in the "ring" or "lipid crown" stage; lipid globules outlining peripheral nuclei.

Fig. 60. Simultaneous discharge of haploid, small male and larger female gametes.

Figs. 61, 62. Two amoeboid living female gametes and a male gamete, resp.

Fig. 63. Diagram of a section through the nuclear region of a gamete showing an ultrastructure similar to that of a planospore, drawn from an electron micrograph. n = nucleus with an interrupted double nuclear membrane, NU = nucleolus, NC = nuclear cap with a double convoluted membrane, M = mitochondria.

Fig. 64. Paired anisogametes.

Fig. 65. Biflagellate zygote.

Figs. 66-71. Encysted zygote, germination and early developmental stages of the diploid sporophyte, resp.

Fig. 72. Brown resistant sporangium with finely spaced pits.

Fig. 73. Germination of a resistant sporangium and discharge of haploid planospores.

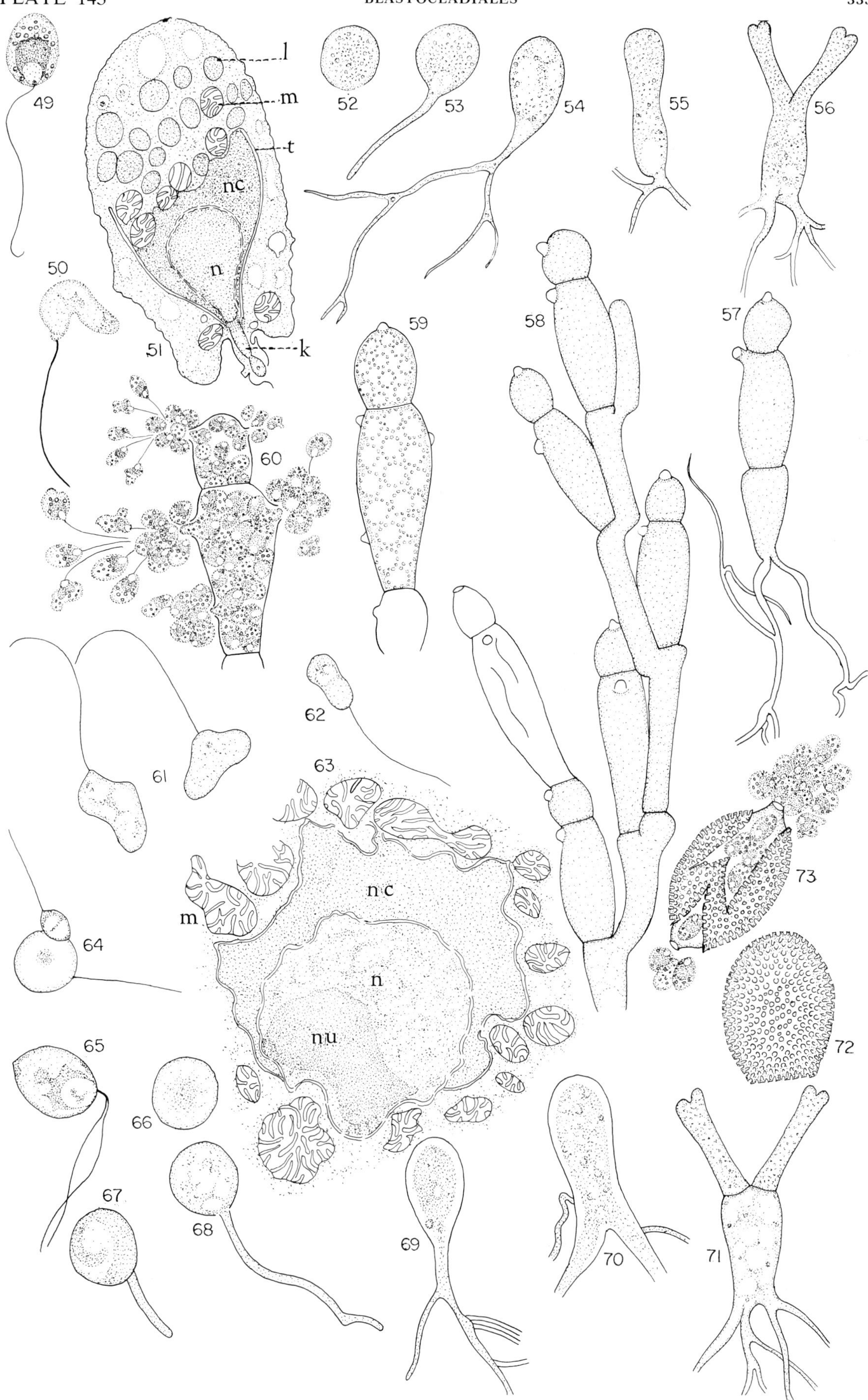

Subgenus *Allomyces*

spores are released, according to C. W. Wilson (1952). The basic haploid number of chromosomes found by C. W. Wilson (1952) and by Emerson and C. W. Wilson (1954) in *Allomyces* are as follows: natural isolates of *A. arbuscula* 8, 16, 24?, 22, 22- 26; natural isolates of *A. macrogynus* 14, 28, 50+, and the hybrid *Allomyces* × *javanicus* 20, 22, 27, 36, 42, 44 or a highly variable number. Accordingly, polyploidy occurs in natural isolates of both *A. arbuscula* and *A. macrogynus*. C. W. Wilson's (1952) cytological studies are particularly significant in that they demonstrated that meiosis occurs in the germination of the resistant sporangia and not in the germinating zygote as Hatch (1938) believed, thus confirming the earlier belief of Kniep (1938) and Sörgel (1937) which were based primarily on volumetric studies of gametophytic and sporophytic nuclei (figs. 78, 79).

As to the gametophytic generation, the gametangia are usually associated in pairs, and the relative positions of the male and female are used to a large degree in distinguishing species and as markers in genetic studies. In *A. arbuscula* the female gametangium is epigynous or terminal and the male is hypogynous (figs. 19, 20, 41, 44), while in *A. macrogynus* and *A. javanicus* var. *allomorphus* Indoh the positions are reversed (57–59). In *Allomyces* × *javanicus* (*A. javanicus* Kniep) the gametangia may be single or in pairs and irregularly arranged with the male terminal (fig. 81) or subterminal (fig. 80). The male gametangium is readily distinguishable by its salmon-pink to orange golden or reddish color. It may be barrel-shaped, ovoid, globose, clavate or cylindrical and varies from 35 μ to 56 μ long by 7.5–24 μ in diameter in the different species. The female gametangium is usually dull-greyish or colorless, ovoid, clavate, barrel-shaped or cylindrical and 3.6 to 33 μ in diameter by 7.5 to 51 μ long. Each gametangium bears one to several exit papillae which deliquesce at maturity and release male and female gametes (fig. 60). The latter are ovoid, subspherical to elongate, 3.6–8.5 × 4.8–11 μ long and colorless in the different species, and Viswanathan and Turian (1966) found that they contain twice as much RNA and slightly more DNA than the male gametes. The male gametes are smaller, 3.4–8 × 4.4–8 μ, and vary from faint orange to pink and brick red, depending on the amounts of gamma carotene, a rare isomer of carotene, present in the minute lipoid granules, according to Emerson and Fox (1940). These workers as well as Haxo (1955) regarded the presence of carotenes as circumstantial evidence that they play a role in sexual reproduction. On the other hand, Turian (1952) demonstrated that male gametes from cultures of *A. macrogynus* grown in the presence of 1:100,000 diphenylamine (DPA), an inhibitor of carotenoid production, function normally in the absence of visible amounts of the pigment, thus casting doubt on their significance in sexual reproduction. Later, however, he (1957a,b) reported that 1:600,000 DPA as well as Mn , ethanol, and butanol inhibited the development of male gametangia in *A. arbuscula* and *A. macrogynus*. Foley (1958) isolated albino strains of *A. arbuscula* that were devoid of pigment and also failed to develop male gametangia, and she concluded from these observations that carotenoids may be mediators or by-products of the morphogenesis of male gametangia. It may be noted in this connection that Fähnrich (1974a,b) found that sexual reproduction can be readily induced by starvation and that actinomycin D at 25–50 g/ml interrupts but does not block gametogenesis. Puromycin and cycloheximide, also, interrupts gametogenesis in all stages of development.

Sexual reproduction in the subgenus *Allomyces* is anisogamous and consists of fusion of a larger female (fig. 30) and a smaller male gamete (figs. 31, 82). The latter is attracted to the female by a powerful attractant, *sirenin*, which is produced by the female gamete. Its synthesis begins before the gametes are discharged from the gametangium and is released into the surrounding water (Carlile and Machlis, 1964a,b; Machlis *et al*, 1966a, b). Fusion begins at the posterior end in *A. arbuscula* (fig. 32) or side by side in *A. javanicus*, according to Kniep's (1930) figures (fig. 83). This is followed by the complete fusion of the cytoplasm and nucle-

Plate 144

Figs. 74-97. *Allomyces* × *javanicus* Kniep. Figs. 74-76, 78, 79, 92 after Kniep, 1929; fig. 77 after Wolf, 1939; figs. 93-97 after Emerson, 1941.

Fig. 74. Ovoid, 8-10 × 11-12.5 μ planospore with a conspicuous nuclear cap, nucleus, nucleolus and long flagellum.

Fig. 75. Young sporophyte (?) with a conspicuous basal cell, several rhizoidal axes, and dichotomous as well as sympodial branches.

Fig. 76. Portion of a diploid sporophyte with thin-walled zoosporangia and resistant sporangia.

Fig. 77. Ovoid, brown resistant sporangium with a rounded apex and truncate base; pits small and closely spaced.

Figs. 78, 79. Relative sizes of sporophytic and gametophytic nuclei, resp.

Fig. 80. Portion of gametophyte with an epigynous female gametangium and a male and female gametangium in tandem beneath.

Fig. 81. Epigynous male and hypogynous female gametangia discharging gametes.

Fig. 82. Fixed and stained preparations of female and male gametes showing their structure and relative dizes.

Fig. 83. Fusion of gametes side by side (?).

Fig. 84. Binucleate, biflagellate zygote with paired nuclei and nuclear caps.

Fig. 85. Zygote with fused nuclear caps; nuclei paired.

Fig. 86. Biflagellate zygote with a fusion nucleus.

Figs. 87-92. Encysted zygote, its germination, and the early developmental stages of the diploid sporophyte, resp.

Fig. 93-96. Stages in the parthenogenetic development of a female gamete into a gametophyte, resp.

Fig. 97. Apical portion of such a gametophyte showing alteration of female and male gametangia.

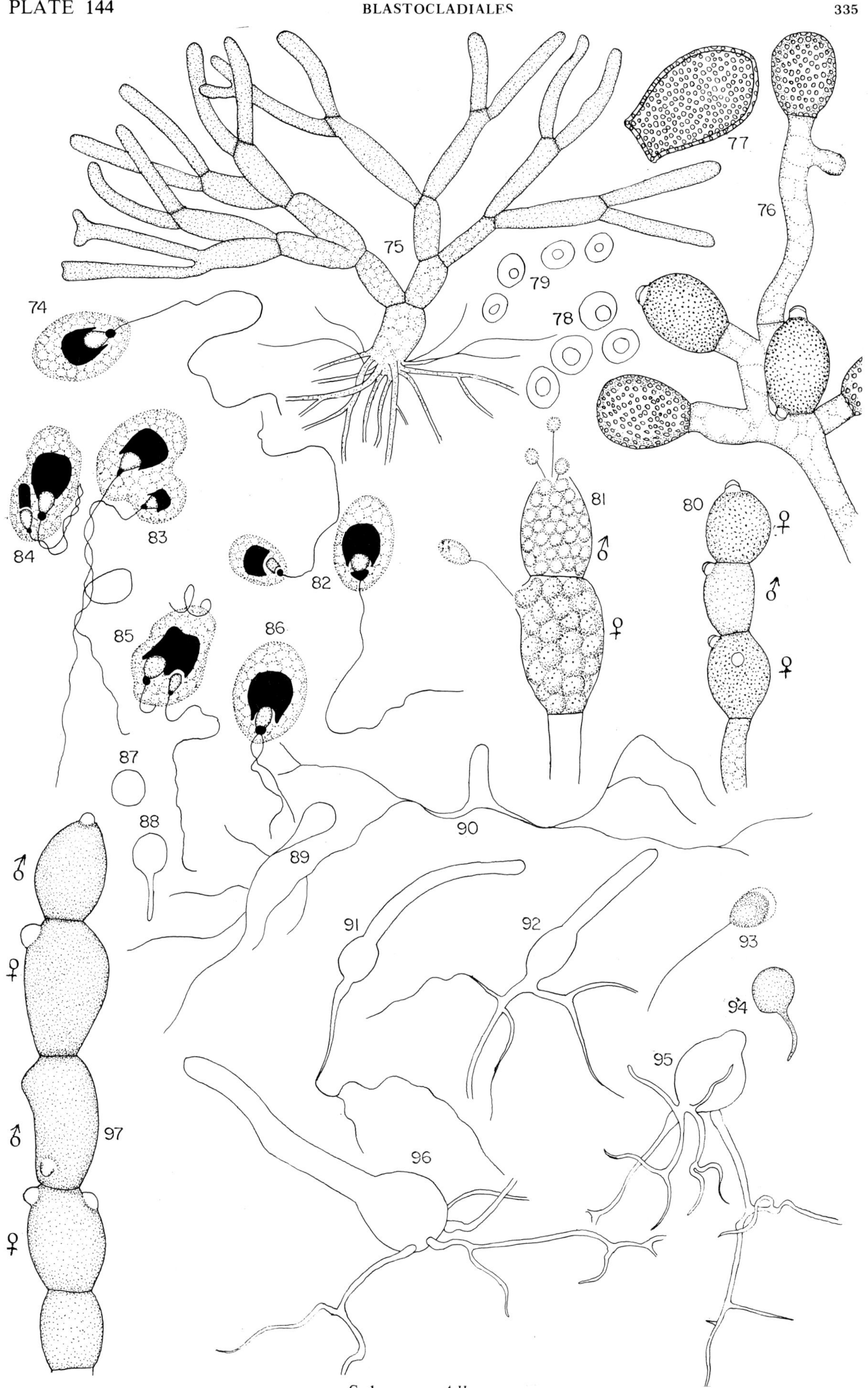

Subgenus *Allomyces*

ar caps (figs. 33, 34, 85), but the gametic nuclei remain separate for some time in the biflagellate zygote. Kniep (fig. 86) showed nuclear fusions while the zygote was still motile, but in *A. arbuscula* Hatch (1938) found that karyogamy was delayed until just before the zygote germinated (figs. 36, 37).* Thus diplophase is established, and diploid nuclei divide mitotically throughout the development of the sporophyte generation until the resistent sporangia germinate.

Species of this subgenus have been used in intensive hybridization studies by Emerson (1941) and Emerson and C. W. Wilson (1954) in which the relative position of the gametangia in *A. arbuscula* and *A macrogynus* were used primarily as genetic markers. As noted earlier the male gametangium is hypogynous and the female epigynous in *A. arbuscula*, and in *A. macrogynus* the relative positions are reversed. Inasmuch as the sporophytes of both species are so similar, inheritance of characteristics was followed in the gametophytic generation. Reciprocal matings between 4 strains of *A. arbuscula* and 3 of *A. macrogynous* in various combinations revealed that in addition to the parental types of epigyny (E) and hypogyny (H) a whole series of intermediates (I) segregate out in the progeny. From these results they concluded that the arrangement of the gametangia is controlled either by more than one pair of independently segregating alleles or by a single pair of duplicated alleles. Less than a 10th of the resistant sporangia from Fl sporophytes were viable. All of the progeny in ⅕ of the offspring exhibited pure hypogeny (H), and for several very cogent reasons Emerson and Wilson believed that hybridization failed to occur in such cases. Second and 3rd generation progeny from the true crosses perpetuated the characters E, H, or I of the gametophytes from which they were derived and generally exhibited a high level of viability of the resistant sporangia.

The results obtained from these genetic experiments were analyzed and correlated with the chromosome numbers and their behavior in the parental and hybrid strains. They found that a tetraploid series consisting of natural strains of 8, 16, 24 and 32 haploid chromosomes occur in *A. arbuscula*; while counts of 14, 28, 50+ (56?) exist in strains of *A. macrogynus*. The parental strains used in hybridization had 16 and 28 chromosomes, respectively, in the gametophytic nuclei. Hence, the sporophytes with 32 and 56 chromosomes were tetraploid. In germination of the resistant sporangia 2 patterns, A and B, of number and behavior were observed at meiosis I in Fl hybrids resulting from crosses of tetraploid parents. In type A nuclei, $16 + 28 = 44$, chromosomes were present, but only a few of these were bivalent with the remainder distributed randomly to the poles as univalents. The type B nuclei contained approximately 55 chromosomes with about half of them bivalents. The others were distributed randomly to the poles as univalents. This irregular distribution of chromosomes in types A and B resulted in the reduced viability of the resistant sporangia of FI. A direct correlation between pairing and percentage of viability was found from precise determinations within the hybrid population, i.e., a few pairs and about 1% viability in type A and many pairs, and ca. 3% viability in type B. Emerson and C. W. Wilson concluded from the observed number of chromosomes, the amount of pairing, and the fact that type B hybrids always have *A. arbuscula* as the female parent that type B nuclei arise by doubling of the *arbuscula* complement during or just prior to fusion, i.e., $2 \times 16 + 28 = 60$. Also, that the irregular distribution of univalents in the FI explains the variable haploid numbers of chromosomes of 20–44 which occur in the F_2 and F_3 progeny and suggests a mechanism through which E and H as well as I segregate out.

Chromosome pairing usually was regular and complete in the F_2 and F_3 meioses which brought resistant sporangial viability up to the high levels of the parents. Particularly significant was Emerson's and Wilson's cytological and genetic demonstration that *A. javanicus* is a natural hybrid of *A. arbuscula* and *A. macrogynus*. The natural intermediate condition of epigyny and hypogyny on the same plant was duplicated by experimental crosses which resulted in offspring with essentially the same variation or range of chromosome numbers that occur in natural crosses.

More recently genetic studies have involved the heredity of biochemically deficient irradiated mutants, the presences of resistant sporangia on gametophytes, albino gametophytes, and control of resistant sporangial development. Yaw and Cutter (1951) produced lysine-

*See Pommerville and Fuller, Arch. Mikrobiol. 109:21, 1976.

PLATE 145

Figs. 98-116. *Allomyces javanicus* var. *allomorphus* Indoh. (Indoh, 1952.)

Figs. 98-100. Portions of sporophytic thalli with zoosporangia.

Figs. 101-107. Line drawings showing some of the variations in the shape of resistant sporangia.

Figs. 108-110. Some variations in the relative positions of the female and male gametangia or gametophytes; male gametangia indicated by darkened walls.

Fig. 108. Branched hypogynous female gametangium and 2 epigynous male gametangia.

Fig. 109. Portion of a branched thallus with epigynous and hypogynous male and female gametangia.

Fig. 110. A chain of male gametangia.

Figs. 111-116. Additional relative positions of male and female gametangia.

Figs. 117-120. *Allomyces monspeliensis* Arnaud. (Arnaud, 1952.)

Fig. 117. General view of a branched thallus with clusters of resistant sporangia (?) at the apex of the branches.

Fig. 118. Zoosporangium with endospores (encysted planospores?.)

Fig. 119. Endospores of a zoosporangium (possibly encysted planospores which will give rise to gametes?.)

Fig. 120. Three resistant sporangia.

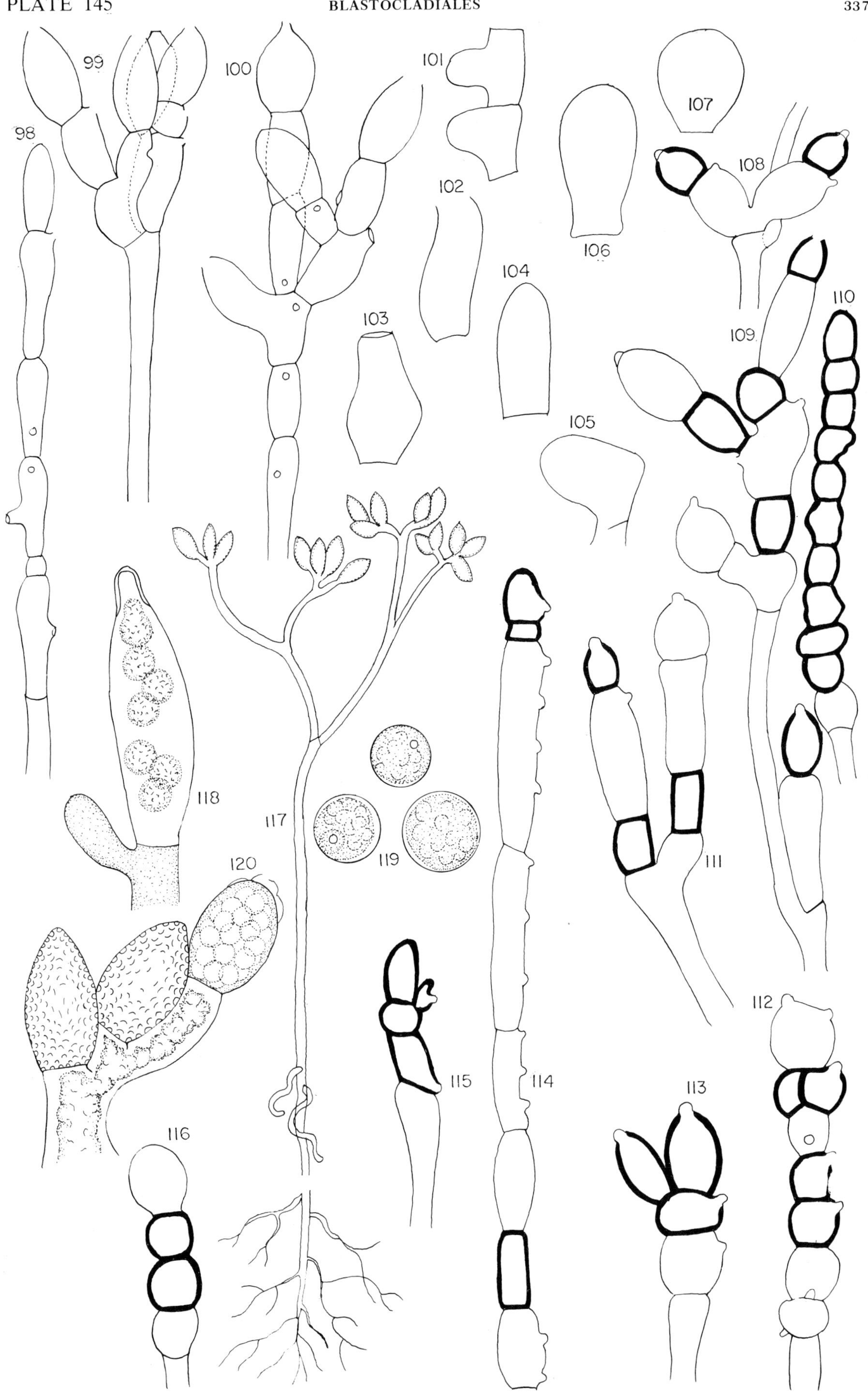

Subgenus *Allomyces*

less and argenineless strains of *A. arbuscula* and concluded from their results that lysine requirement is due to mutation of a single gene. Foley (1958) studied the genetic behavior of albino isolates of *A. arbuscula* and came to the conclusion that the inheritance of pigmentation does not appear to be controlled directly by nuclear genes but possibly by the cytoplasm. Stumm (1958) induced a mutant with X-ray which almost completely lacked expression of femaleness, and he attributed this lack to a single gene. Also, he found a spontaneous mutant in which resistant sporangia on the gametophyte was controlled by a single gene. As noted before De Long (1965) induced a mutant by UV irradiation in which the ratio of ZS to RS was shifted markedly from that of wild strains so that only 20% of the total sporangia were resistant, and she considered this shift as a result of a one-gene mutation at locus 'R'.

Subgenus CYSTOGENES Emerson

Lloydia 4:134, 1941

Plates 146-148

This subgenus is similar in many respects to species of the subgenus *Cystocladiella*, and in this respect differs from *Allomyces* by its holocarpic gametophyte which consists only of a small, spherical, 11–15 μ diam., thin-walled cyst. The sporophyte, on the other hand, is large and dominant as in the other subgenera and bears thin-walled zoosporangia and resistant sporangia. As such *Cystogenes* is reported to include *A. moniliformis* Coker and Braxton and *A. neo-monilioformis* Indoh, and a few varieties or strains which are usually listed as synonyms of the latter species. Possibly, *Allomyces monspeliensis* Arnaud on batrachian eggs belongs here because of its beaked resistant sporangia, but so little is known about its life cycle that its identification and classification are uncertain at present. Sparrow (1960) rejected it as a species of *Allomyces*. In both of the well-known species listed above, the secondary zoosporangia occur in catenulate fashion or are usually borne in basipetal succession with the successive ones decreasing in size towards the base (figs. 3, 4). *Allomyces monilioformis*, like *A. catenoides* Sparrow, is unique in that carotene pigments are distributed throughout the thallus instead of only in the male gametangia. Also, its planospore may encyst, and their contents emerge again as in members of the Saprolegniaceae.

The resistant sporangia are partly or wholly deciduous at maturity with a brown wall that is sculptured by large and widely-spaced pits (figs. 6, 7A). According to Skucas (1967) the wall of *A. neo-monilioformis* consists of 3 parts, a 5-layered outer wall, 2 layers of cementing substances, and a 1-layered inner wall (fig. 45), and is composed of glucose, glucoamine, chitin, melanin, protein, and lipoids. In germinating the resistant sporangia produce biflagellate R.S. planospores or R.S. aflagellate amoebae which are sluggish, encyst shortly after emerging, and thus become incipient holocarpic gametophytes. These function in entirety as gametangia, and their contents usually cleave into 4 uniflagellate gametes (fig. 26) or aflagellate amoebae (figs. 11-14). These emerge through a deliquesced exit papilla, fuse in pairs, and form the diploid zygote (figs. 19, 20, 30-35) and sporophyte. Apparently, the small gametophytes or cysts are homologous with the gametophyte in *Allomyces*, and C. W. Wilson (1952) suggested that they are homothallic. He inferred from the pairing of nuclei in the germinating resistant sporangia that the gametes might be physiologically differentiated as to sex.

Emerson (1938) was the first to discover the formation of biflagellate R.S. planospores, their encystment, and the development of 4 uniflagellate secondary planospores in the cysts. However, he failed to observe fusion of such planospores, and on this basis he (1941) created

PLATE 146

Figs. 1-21. *Allomyces monilioformis* Coker and Braxton. Figs. 1-7 after Coker and Braxton, 1926; fig. 7A after Wolf, 1939; figs. 8-21 after Teter, 1944 of a Trinidad strain.

Fig. 1. Three day-old portion of a thallus with zoosporangia and a resistant sporangium.

Fig. 2. Young elongate terminal primary zoosporangium.

Figs. 3, 4. Monilioid chains of empty zoosporangia in a 25 day-old culture.

Fig. 5. Encysted Z. S. planospore and empty cysts; one emerging from a cyst and exhibiting repeated emergence.

Fig. 6. Portion of a sporophyte with resistant sporangia; pits in the brown wall large and widely spaced.

Fig. 7. Partly decidious resistant sporangium.

Fig. 7A. Resistant sporangium with a pointed apex and truncate base.

Fig. 8. Germination of a resistant sporangium.

Fig. 9. Germinated resistant sporangium with encysted "amoeboid primary swarmers," some of which have developed exit papillae and become incipient gametophytes and gametangia.

Fig. 10. Enlarged view of a cyst or incipient gametangium.

Figs. 11-14. Emergence of 4 successive amoeboid aflagellate gametes, resp., from a gametangium.

Fig. 15. Pairing of aflagellate amoeboid gametes.

Figs. 16, 17. Stages of fusion.

Figs. 18-20. Changes in shape of the amoeboid zygote.

Fig. 21. Germination of an encysted zygote with an unbranched germ tube.

Figs. 22-28. *Allomyces neo-monilioformis* Indoh (*A. cystogenus*). (Emerson, 1938.)

Fig. 22. Biflagellate R. S. planospore.

Fig. 23. Encysted R. S. planospore.

Figs. 24, 25. Germination of cysts.

Fig. 26. Discharge of 4 "secondary planospores."

Figs. 27, 28. Empty cysts.

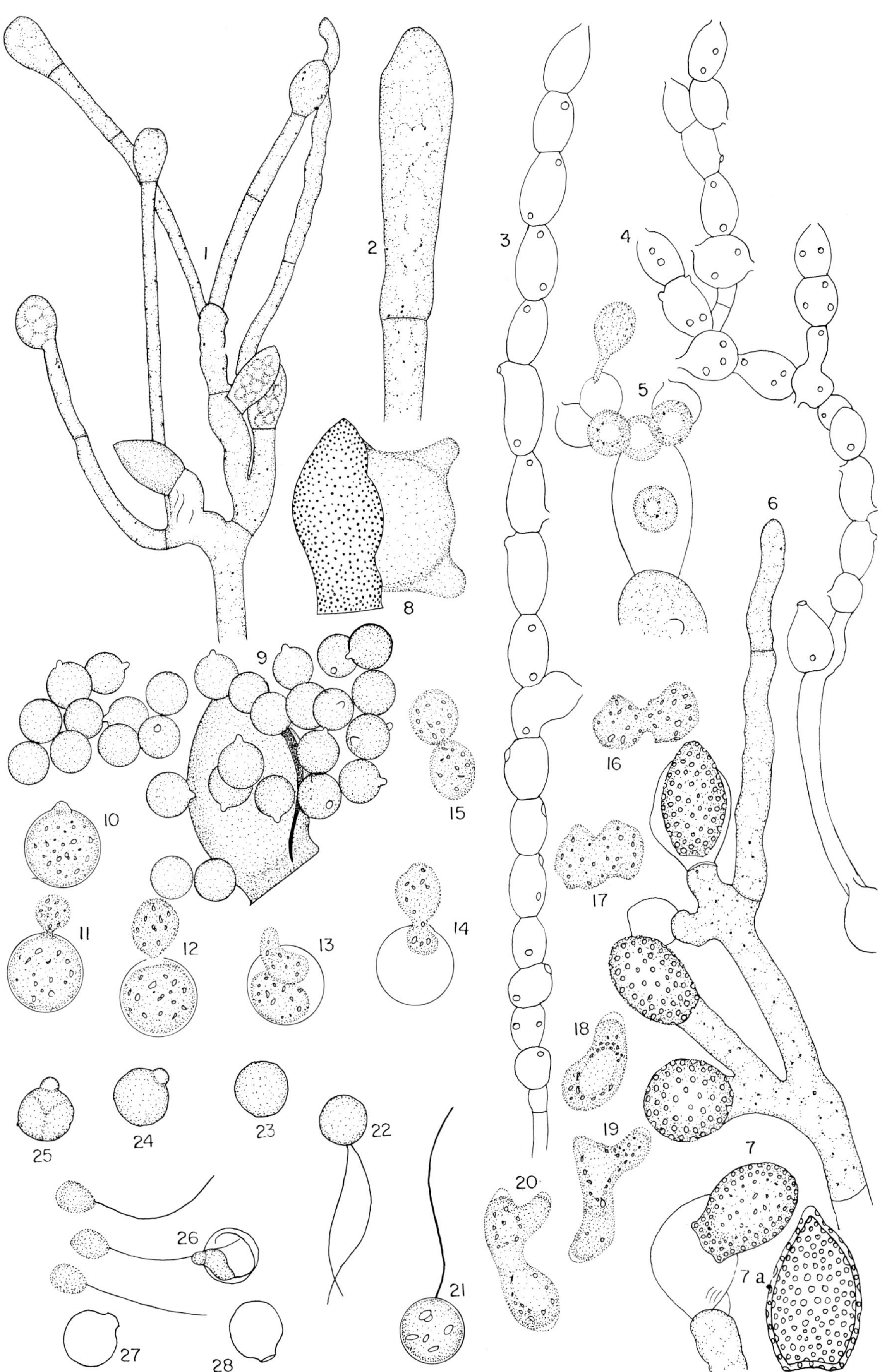

Subgenus *Cystogenes*

and diagnosed *Cystogenes* as consisting of only one generation. McCrainie (1942) and Teter (1944) soon discovered that the secondary planospores and aflagellate amoebae were gametes which fused in pairs. Thus, they demonstrated that isogamous as well as heterogamous sexual reproduction occurs in the genus *Allomyces*.

McCrainie found that in *A. neo-monilioformis* (*A. cystogenes*) the germination products of the resistant sporangia vary in size and may contain one or usually more nuclei. At discharge they are aflagellate and encyst (fig. 29) almost at once in a group at the point of discharge. They germinate within a short while and produce uniflagellate, semi-amoeboid gametes which fuse in pairs (figs. 30–33). Teter confirmed the same type of sexuality for *A. monilioformis* and found in a Trinidad strain that the R.S. germination products were aflagellate and amoeboid and gave rise to aflagellate amoeboid gametes which fused as such (figs. 11-16). However, in a North Carolina strain of *A. neo-monilioformis* (strain 3, Emerson, 1941) he found that the R.S. planospores were flagellate with 1 to 4 flagella per spore. Nevertheless, he assumed 2 flagella per spore to be normal. Such R.S. planospores encysted and subsequently produced 4 gametes which fused in pairs (figs. 34, 36).

Kato (1959) described a strain of *A. cystogenes*, presumably *A. neo-monilioformis*, from the Akita Prefecture of Japan in which several unusual or abnormal stages occur in the life cycle (Plate 148). As in other strains the Z.S. planospores repeat the sporophytic generation (figs. C, D, E, F), but the germinating resistant sporangia may form either biflagellate planospores or aflagellate amoebae (figs. I, I). Both of these may follow several pathways of development. Some of the biflagellate planospores encyst and then apparently germinate to form sporophytes (figs. W, P), while others encyst and develop into gametophytes (J, K). The latter give rise to uniflagellate gametes which fuse in pairs to form the zygote (figs. L, M, N, O), or the gametes (figs. Q, X), encyst (figs. R, Y). These give rise to secondary gametophytes (fig. S) or develop parthenogenetically (fig. Z). The secondary gametophytes form secondary gametes (figs. T, U, V) which fuse in pairs to form a biflagellate zygote (fig. O).

As for the R.S. amoebae they follow the same pathways as described for the biflagellate R.S. planospores, according to Kato. They may repeat the sporophytic generation directly (figs. W', P', E', F') or encyst (fig. J') become gametophytes, and form 4 amoeboid gametes which fuse in pairs (figs. M', N', O'). Or the gametes may encyst (figs. R', Y'), develop parthenogenetically into a germling (fig. Z) or develop into secondary gametophytes (fig. S'). The latter develop secondary amoeboid gametes (fig. T', U') which fuse in pairs (fig. V') to form an amoeboid zygote (fig. O').

Wilson's (1952) excellent cytological study confirmed the contention that 2 flagella are normally present on the R.S. planospores in *A. monilioformis* and that the number of flagella conforms to the number of nuclei in the planospore initial. He found that the resistant sporangia of the subgenera *Allomyces* and *Cystogenes* contain approximately the same number of nuclei which undergo meiosis as the resistant sporangia germinate. However, in *Cystogenes* the haploid nuclei pair just before cleavage into R.S. planospore initials, and are held together by a common nuclear cap. Thus, at cleavage binucleate initials are formed which mature, emerge as biflagellate R.S. planospores, and encyst. During encystment one mitotic division occurs, thereby increasing the number of the nuclei to 4. Individual nuclear caps are then formed, and 4 uninucleate gametes develop in each cyst.

Plate 147

Figs. 29-47. *Allomyces* (*cystogenus*) *neo-monilioformis* Indoh. Figs. 29-33 after McCrainie, 1942; figs. 34-37 after Teter, 1944; figs. 38-41, 46-49 after Emerson, 1941; figs. 42-44 after Indoh, 1940; fig. 45 after Skucas, 1967.

Fig. 29. Germinated resistant sporangium with encysted non-flagellate spores at the region of discharge, and 3 small uniflagellate gametes.

Fig. 30. Side by side pairing of gametes.

Figs. 31, 32. Zygotes with paired nuclei and nuclear caps; flagella only partly drawn in fig. 32.

Fig. 33. Biflagellate zygote with fused nuclei and nuclear caps.

Figs. 34-37. Fixed and stained gametes, zygotes and planospores of the North Carolina strain.

Fig. 34. Uniflagellate gametes.

Fig. 35. Zygote with fusing nuclear caps and nuclei.

Fig. 36. Zygote with fused nuclei and nuclear caps; flagella appearing as one.

Fig. 37. Z. S. planospores.

Fig. 38. Discharge of planospores from a zoosporangium.

Figs. 39, 40. Germinated Z. S. planospores and germling, resp.

Fig. 41. Portion of a sporophytic thallus with zoosporangia and resistant sporangia.

Fig. 42. Linear series of zoosporangia.

Fig. 43. Branch with a decidious resistant sporangium.

Fig. 44. Germinated resistant sporangium with adjacent cysts or incipient gametophytes.

Fig. 45. Section through the resistant sporangium wall. Sketch from an electron microphotograph. HS = hyphal sheath; OW = outer wall; P = pits; CS = cementing substance; CM = cell membrane; IW = inner wall; OL = oil globule.

Fig. 46. Germinating resistant sporangium.

Fig. 47. Germinated resistant sporangium with biflagellate and encysted planospores.

Fig. 48. Cluster of cysts or gametophytes at exit orifice of a germinated resistant sporangium.

Fig. 49. Gametangium with 4 gametes.

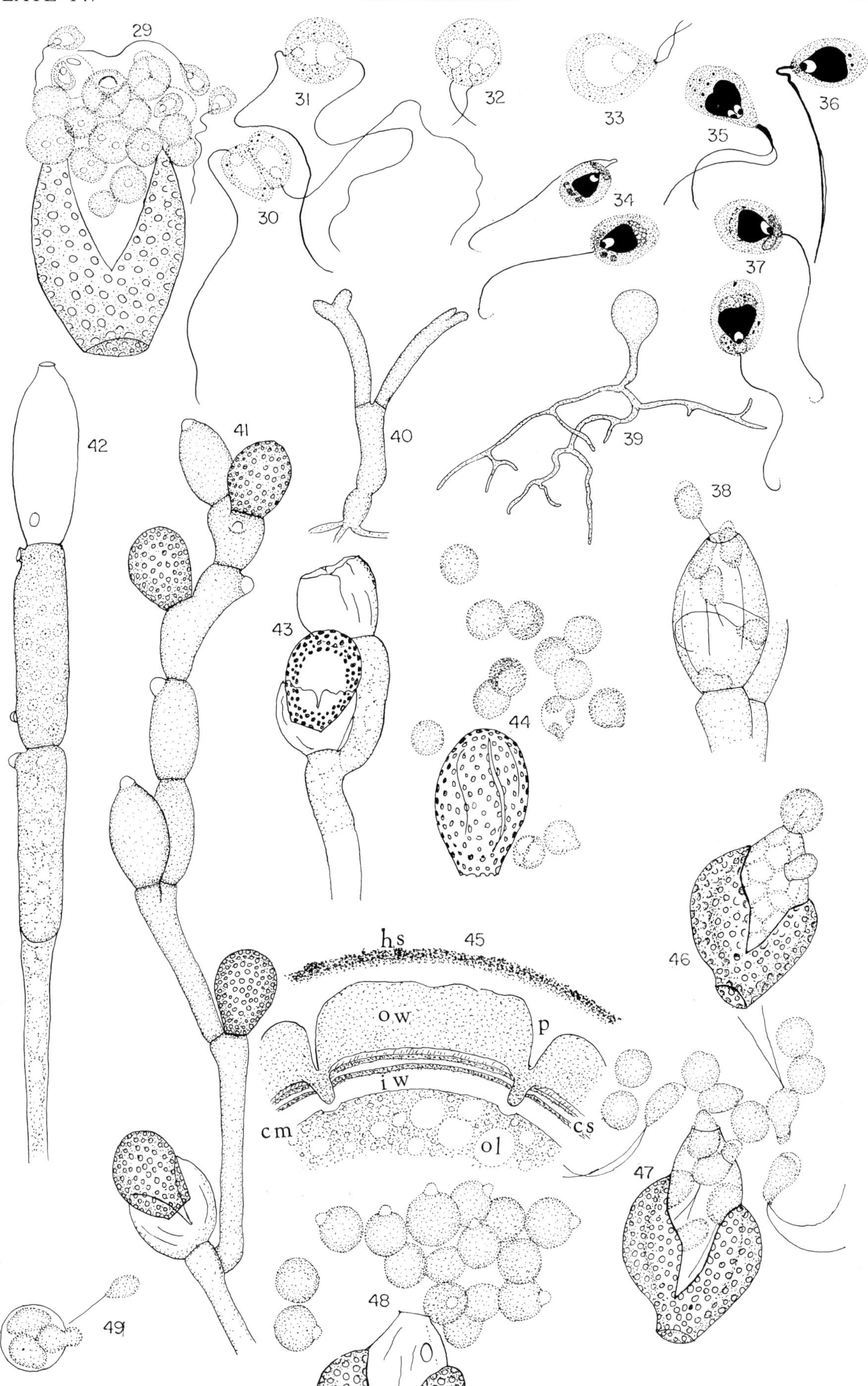

Subgenus *Cystogenes*

Subgenus BRACHYALLOMYCES Emerson

Lloydia 4:135, 1941

Plates 149, 150, Figs. 1-7

This questionable subgenus was established for several isolates in which no gametophytic or sexual thalli were known nor had developed in cultures after several years of attempts to induce their formation. Most of these isolates were usually lumped under the binomial *A. anomalus* Emerson, but subsequently some were found to develop gametophytes (Emerson, 1941) which relegated them to the subgenus *Allomyces*. Wilson and Flanagan's (1968) studies of nuclear and chromosome behavior in the resistant sporangia and hyphae of *Brachyallomyces* isolates revealed 2 ways in which the life cycle is maintained without a sexual phase. In some isolates meiosis occurs in the germinating resistant sporangia with subsequent apomixis of the R.S. planospores which thus repeats the sporophytic phase. In other isolates only mitosis occurs in the resistant sporangia, and meiosis is excluded from the life cycle.

Their studies cast additional doubts about the validity of *Brachyallomyces* and indicate what might be found in the future. Using 6 isolates from various parts of the world which had failed to develop gametophytes after weekly platings during a period of 3 years they found that if resistant sporangia of *A. anomalus* are germinated in a 0.02 m K_2 HPO_4 solution some of their planospores and resulting germlings developed into gametophytes. Also, they found that a few gametophytes developed in 2 strains or isolates, R.1 and C.T. 16, when hyphal tips were cut off and grown on fresh agar. Gametophytes of strains M.24 and ID.1, treated with the phosphate solution were like those of *A. macrogynus*, while those of R.1 and C. R. 16 were similar to those of *A. arbuscula*, both members of the subgenus *Allomyces*. Wilson and Flanagan, also, found some evidence that in resistant sporangia with meiosis the diploid number of chromosomes is maintained by their doubling during germination.

The experimental results suggest that other isolates of this subgenus may have a low and latent potential for gametophytic development which can be brought out by various means. Probably, more effective and fruitful methods and treatment for release of this potential will be discovered in the future.

Thus, at the present time *Brachyallomyces* appears to be a dumping ground for isolates which have not yet been found to develop sexual gametophytic thalli, and in this sense it might be regarded as similar to some of the form genera of the Fungi Imperfecti.

The drawings in Plate 149 which serve to illustrate *A. anomalus* and *Brachyallomyces* were made of an isolate from potting soil in Connecticut and kept in culture for 5 years. During this time it failed to develop gametophytes, and for this reason it is identified provisionally as *A. anomalus*.

PLATE 148

Allomyces (*cystogenes*) *neo-monilioformis*, diagram of the life cycle of the Mitsuseki strain. After Kato, 1959. Black continuous lines and thick black-walled structures represent the normal and diploid course of the life cycle. Fine thin-walled structures and dotted lines represent the abnormal and haploid course of the life cycle.

A = Sporophytic thallus with zoosporangia and resistant sporangia.

B = Zoosporangium with planospores which repeat the sporophytic phase, C, D, E, F, A.

G = Germinating resistant zoosporangium.

I = Germinated resistant sporangium with biflagellate planospores.

J = Encysted R.S. planospores or incipient gametophytes.

K = Gametophyte with 4 gametes.

L = Emergence of gametes.

M = Motile, flagellate gametes.

N = Pairing of gametes.

O = Biflagellate zygote.

P = Encysted zygote

W = Biflagellate R.S. planospore which may encyst.

Q, X = Uniflagellate gametes discharged from a gamete, encysting and developing into secondary cysts or incipient gametophytes at R and Y.

Y, Y′ = Secondary cysts or gametophytes developing parthenogenetically into a germling, Z.

S = Secondary cyst or gametophyte with gametes.

T = Discharge of gametes.

U = Motile uniflagellate gametes.

V = Pairing of gametes which fuse and develop into a biflagellate zygote.

I′ = Germinated resistant sporangium producing non-flagellate amoebae.

J′ = Encysted amoebae or incipient gametophytes.

K′ = Gametophyte with amoeboid gametes.

L′ = Discharge of amoeboid gametes.

M′ = Amoeboid gametes.

N′ = Pairing of amoeboid gametes.

O′ = Amoeboid zygote.

P′ = Encysted amoeboid zygote.

W′ = Amoeba from germinated resistant sporangium.

E′, F′ = Young sporophytes.

Q′, X′ = Discharge of amoebae from gametophytes.

Y′, R′ = Secondary cysts from encysted amoebae.

S′ = Secondary cyst or gametophyte discharging amoebae, U.

V′ = Paired amoebae.

O′ = Amoeboid zygote which encysts at P′.

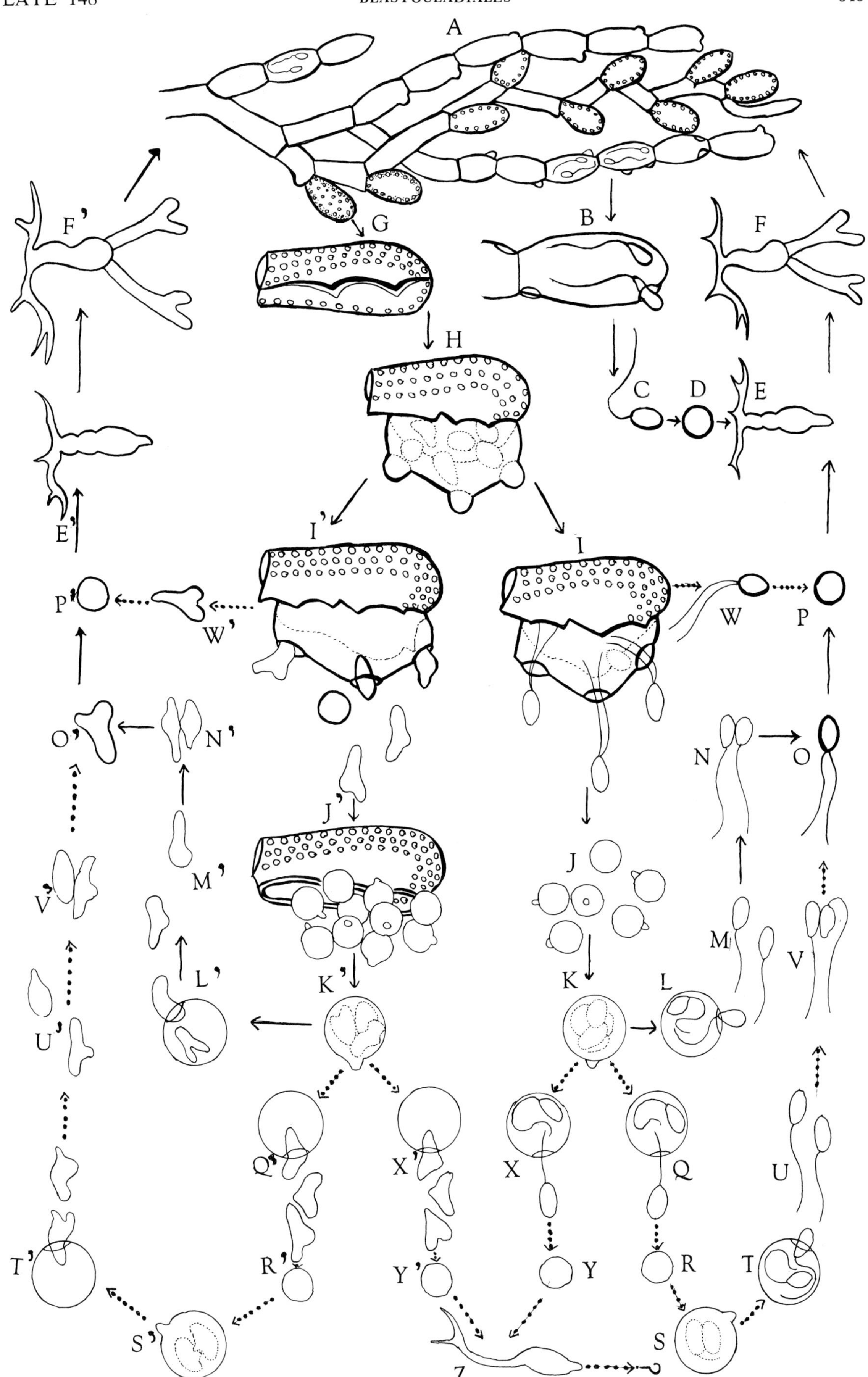

Subgenus *Cystogenes*

The species reported by Hennen (1963) and described by Emerson and Robertson (1974) as *Allomyces reticulatus* (plate 150, figs. 1–7) was assigned to this subgenus because a sexual stage is unknown. Unlike other species, the dark-brown R.S. have a reticulate or ridged instead of a pitted wall like that of *Ramocladia reticulata* Sörgel and are formed by basipetal segmentation of the hyphae.

The classification of Sparrow's (1964) *A. catenoides* is uncertain at present. So far as it is known it forms only thin-walled catenulate zoosporangia with intervening empty isthmuses in elongate *Catenaria*-like hyphae which may bear terminal or lateral rhizoids (plate 149, figs. 17–24).

MICROALLOMYCES Emerson and Robertson

Amer. J. Bot. 61:307, 1974

Plate 150, Figs. 8-16

This monotypic genus is reported to differ primarily from *Allomyces* by the total lack of pseudoseptations in the hyphae and from most *Allomyces* species by its small size, and lack of unlimited growth on laboratory media. It is furthermore distinguished by its occurrence on animal or plant substrata rich in protein. So far a sexual phase is unknown, and *M. dendroideus* Emerson and Robertson has a short brachyallomycetous type of life cycle. As in *Allomyces* the thallus consists of basal rhizoids, a main basal cylindrical trunk, and dichotomously, subdichotomously, or sympodially arranged fertile branches (fig. 8). On natural substrata such as flies or bits of hemp seed the thallus rarely becomes more than 1 mm in height and usually only about 500 μ. However, in suitable nutrient agar or in broth media, it shows the potential of continuous growth and may develop an almost mycelial structure extending 7 to 8 mm from its point of origin.

The development of the thallus from the planospore, formation of thin-walled zoosporangia at the tips of the branches, structure of the planospores, the development of finely-pitted red or yellowish-brown resistant sporangia, and their germination are essentially similar to those of *Allomyces* species (figs. 9 to 16). However, in R. S. germination broad tongue-like tubes, up to 100 μ in length, may be formed for the emission of the planospores instead of broad exit papillae (fig. 15).

As to the relationships of *Microallomyces* within the *Blastocladiaceae* Emerson and Robertson reported that in basic morphology it overlaps *Blastocladia*, particularly *B. ramosa*, and differs from *Blastocladiopsis* by its capacity for continued growth of the newly formed apices of the sympodial branches.

BLASTOCLADIOPSIS Sparrow

J. Washington Acad. Sci. 40:52, 1950

Plate 151

This genus was created for a saprophyte, *B. parva* (Whiffen) Sparrow, which was initially described as a species of *Blastocladia* and regarded by Whiffen (1943) as a transition species between the latter genus and *Blastocladiella*. It remains to be seen, however, that it is a valid genus on the basis of free-lying resistent sporangia alone. Sparrow regarded the free-lying or loose, smooth, non-punctate resistant sporangia lying in container cells or hyphal tips, the formation of several discharge papillae or tubes on the zoosporangia, the occasional occurrence of dwarf thalli, and the rapidity of development as characters which distinguish it from *Blastocladia*, but it is doubtful that these are generic distinctions. Free-lying loose or non-loose resistant sporangia with smooth or sculptured walls may occur in *Blastocladiella*, for example, showing that such characteristics may vary markedly in one genus. Be-

PLATE 149

Figs. 1-16. *Allomyces anomalus* Emerson, drawings from living material of a Connecticut isolate.

Fig. 1. Motile ovoid, 9-12 μ, Z.S. planospore.

Fig. 2. Encysted Z.S. planospore.

Figs. 3, 4. Germinated Z.S. planospores.

Figs. 5, 6. Young thalli.

Fig. 7. Older thallus showing dichotomous branching.

Fig. 8. Portion of thallus bearing resistant sporangia and zoosporangia.

Figs. 9, 10. Two variations in shapes of zoosporangia with several exit papillae.

Fig. 11. A zoosporangium with its protoplasm in the "ring" or lipid "crown" stage.

Fig. 12. Discharge of Z.S. planospores.

Fig. 13, 14. Resistant sporangia.

Fig. 15. Germinated resistant sporangium.

Fig. 16. R.S. ovoid, 8 × 10 μ, planospore.

Figs. 17-24. *Allomyces catenoides* Sparrow. (Sparrow and Morrison, 1961.)

Figs. 17-20. Various types of terminal, lateral, and intercalary pigmented zoosporangia; sterile isthmuses in fig. 19.

Fig. 21. Terminal, intercalary and lateral zoosporangia separated by sterile isthmuses.

Fig. 22. Zoosporangia discharging planospores.

Figs. 23, 24. Ovoid, 10-11 × 12-16 μ, hyaline Z.S. planospores with conspicuous nuclear caps.

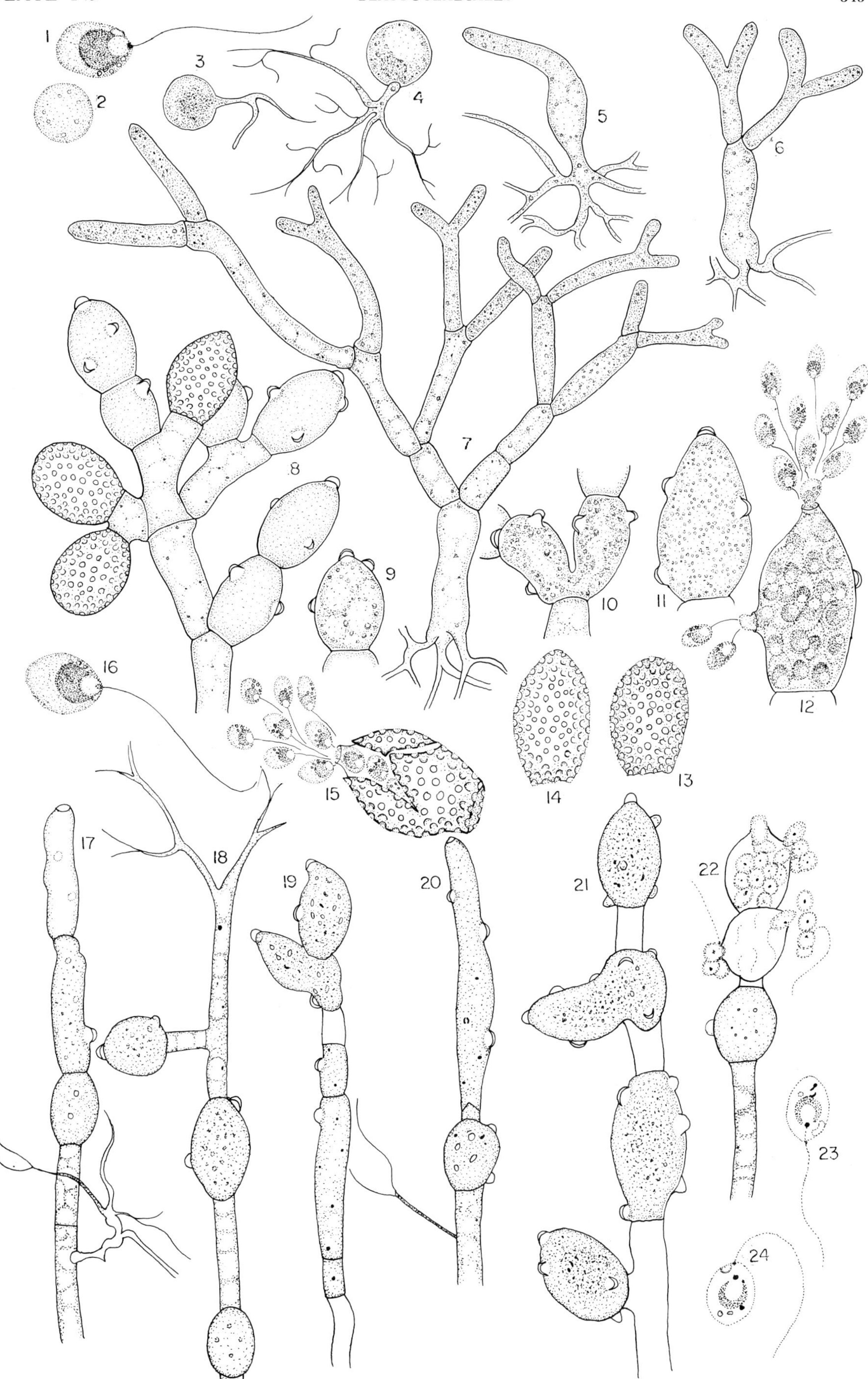

Subgenus *Brachyallomyces*

sides, Whiffen noted that some of the resistant sporangia in *B. parva* are not free-lying in the hyphal tips but fill the container cell completely. Emerson and Robertson (1974) did not regard free-lying resistant sporangia as the distinctive character of this genus, and maintained that the distinguishing and fundamental feature of *Blastcladiopsis* is the lack of capacity for renewed growth of the sympodial branches once they have formed sporangia at their apices. Emerson (personal communication) believes that another species formerly described as *Ramocladia reticulata* belongs in *Blastocladiopsis.*

Blastocladiopsis parva develops unusually fast at room temperatures and may complete its growth and maturation in 4 to 5 days. The R.S. planospores (fig. 1) germinate readily on agar media, and in this process the germ tube branches to form an extensive intramatrical rhizoidal system while the planospore body enlarges to become a globular structure (fig. 2) on the substratum. Subsequently, this structure elongates outward (figs. 3–5) to form a central axis, and branches dichotomously or subdichotomously once to several times (figs. 7, 8) at the apex. Such thalli may be up to 300 μ or more in length with the axis and branches 12–50 μ in diameter. As the thallus matures its protoplasm accumulates in the rapidly inflating tips of the branches, which are then delimited by a cross septum (figs. 4, 8). These tips usually become the so-called container cells of the resistant sporangia, or are rarely transformed into thin-walled zoosporangia. As the incipient sporangia mature the protoplasm of the tips apparently contracts from the wall as in some species of *Blastocladiella* and *Catenaria*, but this has not been observed or described. Nonetheless, this is suggested, at least, by the fact that the container cell sometimes bears a papilla or short tube as in *Catenaria*. The protoplasm then becomes invested with a smooth, pale-yellow or amber wall and matures into a resistant sporangium. As a result, apparently, of this contraction most of these sporangia lie free and loose in vesicles or container cells (figs. 9–11). These germinate readily after being dried for a while, and in the process the outer wall is cracked open by the swelling of the contents and the protrusion of one or more broad exit papillae or tubes (figs. 12, 13). The discharged R.S. planospores swim away directly without encysting at the exit orifice.

So far as is known thin-walled zoosporangia are relatively rare, and only a few of them have been seen. Whiffen found that although they are scarce they develop more abundantly at 21°C than at 24°C and 25°C which favor development of the resistant sporangia. They, nevertheless, develop on the same thallus as the latter, and their development is similar to that of the resistant sporangia with the difference that there is no contraction of the protoplasm from the wall of the hyphal tips. At maturity the zoosporangia may be irregular (figs. 14, 15) or subcylindrical (figs. 16, 17) in shape and 40 × 65–6 to 41.3 × 90.2 μ in size with 1 to 6 discharge papillae or short tubes.

Sexual reproduction is unknown in this genus, and no fusions of gametes have been observed. Whiffen mixed R.S. planospores from several different germinated resistant sporangia together but did not find any fusions in such mixtures.

BLASTOCLADIA Reinsch

Jahrb. wiss. Bot. 11:298, 1878

Plates 152-155

Although this is the earliest known genus from which the order takes its name, its species are not as well known as those of *Blastocladiella* and *Allomyces*. At present it is reported to contain approximately a dozen or more aquatic saprophytic species and a few unidentified specimens on submerged decaying twigs and fruits in freshwater where they form crisp pustules. Usually, they are characterized by a large tree-like thallus up to 2000 μ high or long in some species, which consists of short or long trunk or central axis, 6–162 μ diam. bearing short or long branches at its apex and anchored to the substratum by a system of branched rhizoids or holdfasts. However, the thalli may vary so

PLATE 150

Figs. 1-7. *Allomyces reticulatus* Emerson; figs. 8-16 *Microallomyces dendroideus*. Emerson and Robertson (1974.)

Fig. 1. Spherical, 8-9 × 10-12, planospore from a germinated R. S.

Fig. 2. "Germling with rhizoid initials and hyphal germ tube."

Fig. 3. Hyphal tip bearing zoosporangia 36-80 × 60-110 μ, with exit papillae.

Fig. 4. "Small mature thallus bearing resistant sporangia."

Fig. 5. "Chain of resistant sporangia within the parent hyphal envelope."

Fig. 6. Truncate resistant sporangium, 40-60 × 60-100 μ, showing reticulate sculpturing.

Fig. 7. Germinated resistant sporangium.

Fig. 8. "Mature thallus bearing zoosporangia."

Fig. 9. Ovoid, 9-11 × 13-19 μ, planospore from a zoosporangium.

Fig. 10. Encysted planospore.

Fig. 11. "Germling with rhizoid initials."

Fig. 12. Branch bearing zoosporangia.

Fig. 13. Resistant sporangia attached to a mature thallus.

Fig. 14. Enlarged view of truncate resistant sporangium, 16-30 × 30-60 μ, with pitted brown outer wall.

Fig. 15. Germinated resistant sporangium.

Fig. 16. Ovoid planospore from a resistant sporangium.

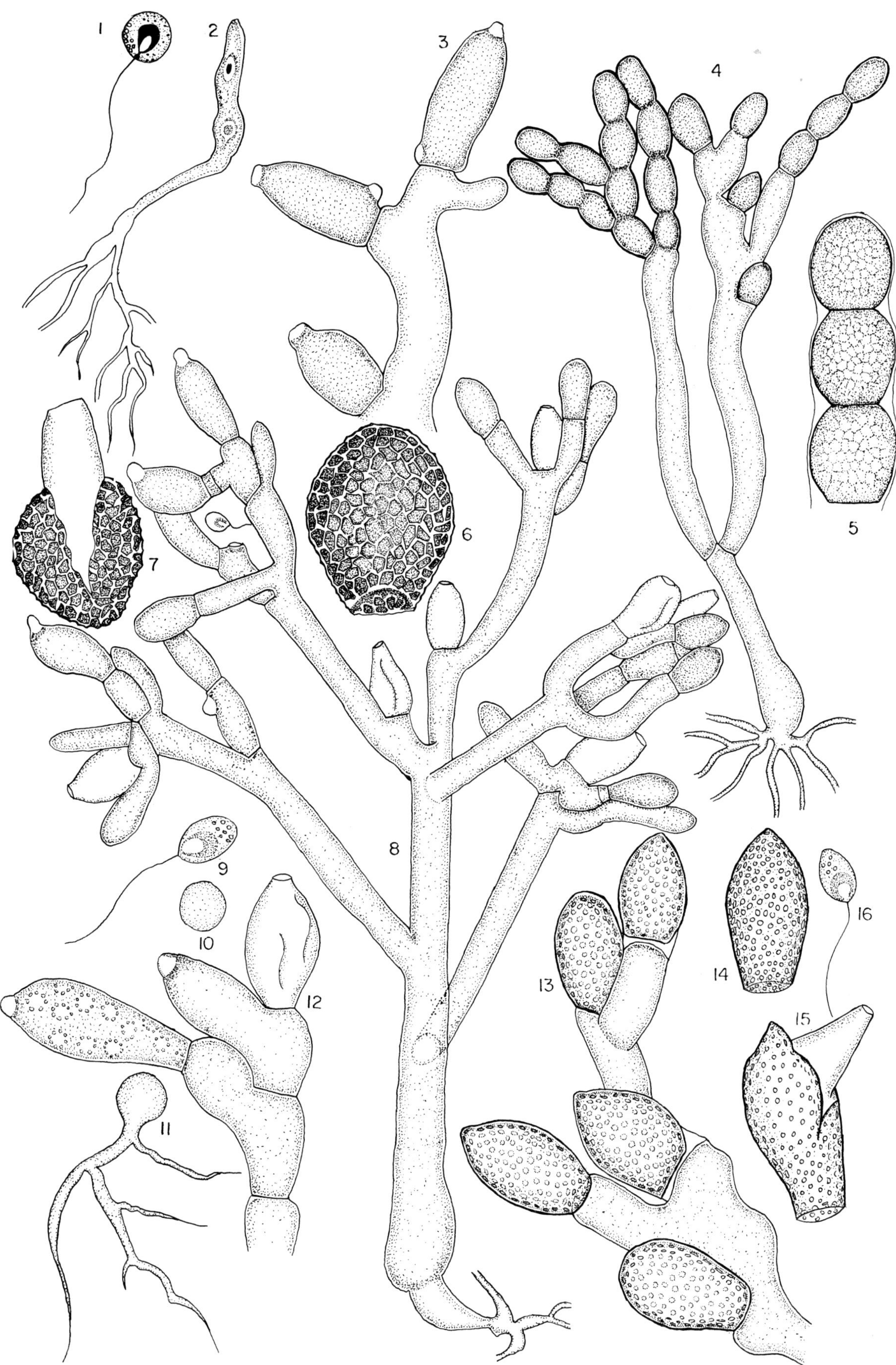

Subgenera *Brachyallomyces, Microallomyces*

markedly under certain conditions that it is difficult to identify the species with certainty. They appear to overlap to some degree and there is considerable disagreement about the number of valid species. Quite likely the number of known species will be reduced when they have been isolated and studied in axenic cultures. Nevertheless, at present species distinctions are made on the basis of shapes and sizes of the zoosporangia and resistant sporangia and to some degree on whether or not the apex of the trunk or its branches are swollen. *Blastocladia pringsheimii* Reinsch appears to be the most common species, and is the only one that has been studied fairly intensively. It and *B. ramosa* Thaxter are the only 2 that have grown in axenic cultures on synthetic media.

Among the known species the central axis, trunk or basal cell may be short and thick (fig. 53, 64) or elongate (figs. 45, 47) and inflated or lobed (figs. 54, 57, 65) or branched at the apex (figs. 1, 29, 39, 67, 69). The branches arising from it may be short and clavate to slightly tapered or elongate and cylindrical and arranged in a dichotomous, subdichotomous, sympodial, or racemose manner. These bear the zoosporangia and resistant sporangia which are delimited by a cross wall. Otherwise, the thallus is non-septate, although Waterhouse (1942) illustrated the branches of *B. racemosa* as being septate or pseudoseptate. In addition to the branches and reproductive organs simple and branched sterile setae (figs. 3A, 54, 65) may be borne on the trunk and project outward. These may be 2–6 μ in diameter, bulbous at the base, and up to 260 μ long. The rhizoids at the base of the trunk are abundant or scanty, usually richly branched, extensive, delicate or coarse and up to 5 μ in diameter.

So far as is known the developmental cycle of this genus is relatively simple, and no conclusive evidence of a gametophytic and sexual phase has been found. Bessey (1939) reported fusion of isoplanogametes in *B. pringsheimii*, but this report was refuted by Blackwell (1939, 1940) from more intensive studies. The large Z.S. planospores (figs. 2, 22, 28, 59) with a subtriangular nuclear cap, nucleus, peripheral granules, and posterior flagellum encyst and later germinate by a relatively fine germ tube (fig. 20) which branches and eventually develops a rhizoidal or holdfast system. Meanwhile, the planospore body elongates outward and expands (figs. 21, 35, 36) to become the trunk or axis, and the reproductive organs originate as papillate outgrowths from its surface or that of the branches. Thin-walled smooth zoosporangia and thick-walled pitted or punctate, brownish resistant sporangia are borne on the same (figs. 40, 50, 69) or different thalli. Thus, insofar as they are known the life cycles of species of *Blastocladia* correspond to those of the short-cycle subgenera *Blastocladiella* and *Brachyallomyces*, but it is not improbable that with more intensive study strains may be found with a sexual phase.

The zoosporangia vary from narrowly and broadly fusiform (figs. 1, 57) and ovoid broadly clavate (fig. 4) to cylindrical (figs. 61, 64) in shape with a rounded or beaked apex and truncate base (fig. 60), and may be up to 350 μ long by 6–70 μ in diameter. The apical discharge papilla often contains a plug which projects down into the protoplasm (figs. 5, 38) and is carried out by the emerging planospores (figs. 6, 7). The latter emerge individually or in a column, or in a pyriform group surrounded by an evanescent vesicle. In *B. prolifera* Minden the zoosporangia proliferate up to 5 times (figs. 37, 39), and in *B. sparrowii* Indoh (fig. 65) proliferation occurs commonly. The Z.S. planospores of different species vary in size from 3.5 to 12–15 μ in diameter and from spherical, elongate, ovoid, ellipsoidal or pyriform in shape. The reports of their size are not always consistent, and in *B. pringsheimii*, for example, they have been described as spherical, 12–14 μ diam., and ovoid, 5–6 × 6–9 μ. The smallest planospores, 3.5 μ diam., occur in *B. truncata* Sparrow.

As noted earlier the resistant sporangia may occur among zoosporangia on the same thallus or on separate thalli. So far, they are unknown in 5 species. In *B. pringsheimii* their occurrence and development depend on certain conditions. Emerson and Cantino (1948)

PLATE 151

Figs. 1-16. *Blastocladiopsis parva* (Whiffen) Sparrow isolated from soil on agar media, snake skin, and grass. Figs. 1-4, 6-9, 11-15 after Whiffen, 1943; figs. 5, 10, 16, 17 after Sparrow, 1950.

Fig. 1. Elongate, anteriorly tapered R. S. planospore, 2.8-3.2 × 5.6-6.1 μ, with a conspicuous "food body" and several oil globules.

Fig. 2. Germling 12 hrs.-old from an agar culture.

Fig. 3. Young thallus, planospore body has elongated outward from the substratum and is beginning to branch at the apex.

Fig. 4. A reduced thallus 24 hrs.-old from an agar culture.

Fig. 5. Young thallus emerging through the stoma of grass substratum.

Fig. 6. Portion of a thallus with dichotomous branching.

Fig. 7. Subdichotomously branched thallus; apices of branches enlarging to become resistant sporangia.

Fig. 8. Later stage of resistant sporangium development at the apices of the branches.

Fig. 9. Thallus with mature resistant sporangia lying free in the inflated tip of the branches.

Fig. 10. Dwarf thallus with a free-lying resistant sporangium.

Fig. 11. Enlarged view of a resistant sporangium which partly fills the container cell.

Figs. 12, 13. Stages in germination of a resistant sporangium and discharge of planospores.

Fig. 14. An empty zoosporangium with 4 exit papillae.

Fig. 15. A zoosporangium with 4 exit papillae shortly before sporogenesis.

Figs. 16, 17. Empty elongate zoosporangia with an apical and lateral exit papillae, resp.

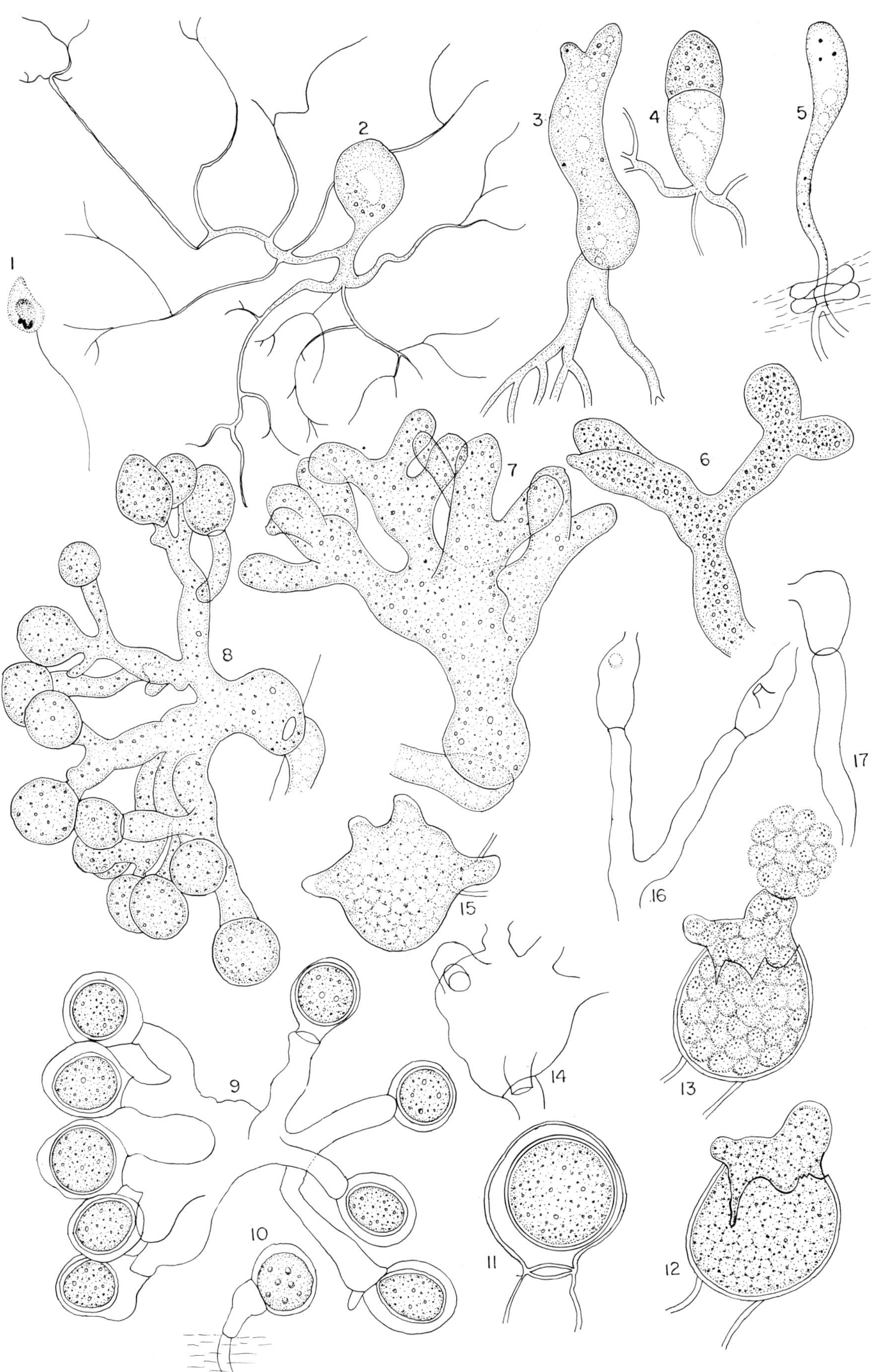

Blastocladiopsis

found that in axenic culture they were produced only when a 99.5% carbon dioxide atmosphere was maintained, or when the medium on which the species was grown contained carbonate. At maturity they are fairly large, 15–50 μ diam. by 18–99 μ long, and predominantly ovoid, ellipsoidal, fusiform, and spherical to subspherical with a rounded or beaked apex and a broadly or narrowly truncate base. The outer wall is finely pitted or punctate and brownish in color, but in *B. ramosa* it is almost hyaline or colorless.

So far germination has been observed only in *B. pringsheimii* and *B. gracilis* Kanouse, and this process is essentially similar to that of *Allomyces*. The thick outer wall cracks as the content expands, and an endosporangium protrudes (figs. 14–18). The latter bears 1 or 2 exit papillae (figs. 14, 18) and eventually discharges its planospores. These, apparently, are similar in size, shape and structure to the Z.S. planospores. On nutrient media and under cover slips they may develop into minute thalli and zoosporangia with only 4 to 8 planospores (figs. 22, 27, 28), and on tomato epidermis flat, lobed and irregular thalli develop which bear zoosporangia at the ends of the lobes (figs. 23, 24). Both of these types are strikingly unlike the very large ones (fig. 1) that develop on various kinds of submerged fruits and twigs.

INCOMPLETELY KNOWN GENERA

Plate 156

In this category are included 4 genera which Sörgel illustrated and described very briefly without Latin or other diagnoses in 1952. These were isolated from soil, fresh, and rain water, on dead flies in San Domingo, Haiti, Yugoslavia and Southern France, and grown in pure culture on synthetic media. Sörgel did not describe or give measurements of the planospores, zoosporangia, and resting sporangia and his description of the genera was not very detailed. So far as this writer knows these genera have not been reported or described subsequently*.

Ramocladia includes 2 species, *R. reticulata* Sörgel and *R. pallida* Sörgel, whose thalli consist of a relatively thick stalk or central axis which bears branches or hyphae at the apex and rhizoids at the base (figs. 1–5). According to Sörgel, this genus differs, thusly, from *Blastocladiella* by its normally branched thallus and from *Allomyces* by its limited growth and non-septate hyphae. The only septa present are those which delimit the zoosporangia and resistant sporangia. In *R. reticulata* either of these reproductive structures may branch (fig. 2), the resistant sporangia less often than the zoosporangia. Also, under unfavorable nutrient conditions unbranched thalli may develop. Zoosporangia and resting sporangia may develop on the same thallus (fig. 4), and the latter has a brown distinctly reticulate outer wall. It germinates in the same manner as the resistant sporangia of other blastocladiaceous genera by the cracking of the outer wall and protrusion of an endo-

**Ramocladia* and *Allocladia* have been rediscovered by Emerson (personal communication), and he believes *Ramocladia* is a *Blastocladiopsis* and that *Allocladia* is an *Allomyces* species.

PLATE 152

Figs. 1-28. *Blastocladia pringsheimii* Reinsch. Figs. 1, 4, 10-13 after Thaxter, 1896; fig. 2, 9 after Cotner, 1930; fig. 3 after Reinsch; 1878; fig. 3A after Kanouse, 1927; figs. 5-8 after Lloyd, 1939; figs. 14-20, 22-28 after Blackwell, 1940; fig. 21 after Emerson and Cantino, 1948.

Fig. 1. Branched thallus with a relatively short trunk, narrowly fusiform zoosporangia, and incipient resistant sporangia forming as small buds; only a part of the coarse, branched holdfast shown.

Fig. 2. Posteriorly uniflagellate ovoid, 5.5-6.5 × 6-8 μ, planospores with a conspicuous nuclear cap, nucleus, and peripheral granules.

Fig. 3. Portion of a thallus bearing 2 resistant sporangia and zoosporangia.

Fig. 3a. Hair-like filament on the thallus.

Fig. 4. Clavate zoosporangium shortly before dehiscence.

Fig. 5. Upper portion of a zoosporangium shortly before dehiscence with a plug of dense material in the apex.

Figs. 6-9. Stages in the emergence of the planospores in a vesicle; plug gradually disintegrating.

Figs. 10, 11. Ovoid and spherical brown resistant sporangia with finely pitted walls.

Fig. 12. Median section of a spherical resistant sporangium with large refractive globules.

Fig. 13. Optical section of the R. S. wall.

Fig. 14. Early stage of R. S. germination showing splitting of the wall, protrusion of the endosporangium, and globules aggregated in the center.

Figs. 15-17. Later successive stages of germination; globules gradually dispersed in the process.

Fig. 18. View of a protruding endosporangium with 2 exit papillae.

Figs. 19, 20. Germination stages of R. S. planospores.

Fig. 21. Later stage, planospore body elongating and expanding outward to become the trunk or axis of the thallus.

Fig. 22. Small thalli developing under a cover slip from planospores which encysted at orifice of the resistant sporangium and germinated; two of the sporangial thalli have formed planospores.

Fig. 23. Flat thallus, 2 days old, growing on the epidermis of a tomato and bearing 4 resistant sporangia.

Fig. 24. Earlier stage of a similar type of thallus.

Fig. 25. Elongate germling growing out of a tomato cell.

Fig. 26. A 7 day-old thallus growing on cornmeal agar; only a portion of the rhizoids shown.

Fig. 27. Small zoosporangium developed under a cover slip from R. S. planospores and bearing only 4 planospores.

Fig. 28. Similar but larger thallus; zoosporangium produced 8 planospores and a secondary zoosporangial initial or rudiment.

Blastocladia

sporangium with exit papillae. *Ramocladia pallida* (fig. 5) is distinguishable by the great size of the central axis and length of the branches or hyphae whose growth is terminated by zoosporangia or resistant sporangia. The latter is pale-golden instead of brown in color and apparently smooth, although Sörgel did not specifically report them as such. Although this species is *Allomyces*-like, its size is such that *Allomyces* looks like a dwarf in comparison. Sörgel regarded *Ramocladia* as a transitional genus between *Blastocladiella* and *Allomyces*.

Allocladia includes one species, *A. monilifera* Sörgel (fig. 6), which resembles a robust *Allomyces* species, but it differs from the latter by the lack of pseudosepta in the young developmental stages. The formation of these structures is delayed until the resistant sporangia develop. Furthermore, its branches or hyphae at the point of origin of the zoosporangia is twice the diameter of those of *Allomyces javanicus*. Sörgel regarded this genus as a connecting link between *Ramocladia* and *Allomyces*.

Leptocladia with one species, *L. laevis* Sörgel (fig. 7), has a *Ramocladia*-like thallus consisting of a short, broad central axis from which arise long narrow branches or hyphae that end in resistant sporangia. These are smooth, according to fig. 7, and apparently hyaline or colorless.

Brevicladia (fig. 8) is a primitive (?) genus which consists only of an ovoid body with a rhizoidal axis and a few branched rhizoids. Sörgel did not name any species of this genus, but he regarded it as connecting the chytrids with the Blastocladiaceae.

QUESTIONABLE GENUS

Callimastix was created by Weissenberg (1912) for a haemolympic multiflagellate, parasite, *C. cyclopis*, of *Cyclops*, which he classified among the zooflagellates. He reported it again in 1950 and expressed the belief that the flagellate stage "apparently represents the swarm spore of a protist belonging to a group of lower fungi (Chytridinea? Myxomycetes). Subsequently Vavra (1960, 1963) and Vavra and Joyon (1964) reported it in a number of genera and species of *Copepoda* and pointed out that the structure of the flagellate stage is very similar to that of the planospore of the Blastocladiales. Moore (1968), also, stated that "a recent ultrastructure study by Vavra and Joyon (1966) clearly shows it to be a member of the Blastocladiales." Martin (1971) and Whisler *et al* (1972) also, pointed out that the cone-shaped nucleus and triangular nuclear cap and the paracrystalline body shown by Vavra and Joyon (1966) is very similar to those of *Coelomomyces punctatus* and *C. psorophorae*.

According to Weissenberg, Vavra and Joyon the parasite exists for a long while in the host body cavity as a multinucleate, elongate, irregular and rarely spherical, 20 × 100 to 30 × 400 μ, plasmodium, and after a period of rapid growth fragments into spherical 10–20 μ diam., multinucleate masses. These in turn fragment into uninucleate spherical 3–6 μ diam., non-flagellate elements which fill the haemolymph of the copepod. In about 3 to 5 days they enlarge and develop 8 to 9 or less basal posteriorly directed flagella, 24 to 30 μ in length. This so-called planospore becomes quite active in the haemolymph and escapes as the thin cuticle between the segments of the cephalothorax breaks down.

Vavra and Joyon's ultrastructural study of the flagellate stage shows that its structure is strikingly similar to that of the planospores of the Blastocladiales. It lacks a Golgi apparatus, has a prominent turbinate, nuclear cap, several discrete mitochondria, a paracrystalline body, and paired centrioles as in *Allomyces arbuscula* (see Renaud and Swift, 1964), only one of which is involved in the development of the flagellum.

PLATE 153

Figs. 29-36. *Blastocladia ramosa* Thaxter. Figs. 29, 30, 33 after Thaxter, 1896; fig. 34 after Indoh, 1940; figs. 35, 36 after Emerson and Cantino, 1948; figs. 31, 32 after Sparrow, 1943.

Fig. 29. General habit of the plant.

Fig. 30. Terminal branch with 2 zoosporangia and 6 resistant sporangia.

Figs. 31-33. Resistant sporangia.

Fig. 34. Emergence of a planospore from a zoosporangium.

Figs. 35, 36. Stages in the development of the thallus on synthetic media.

Figs. 37-39. *Blastocladia prolifera* Minden. (Minden, 1916.)

Fig. 37. Dwarf thallus with proliferating zoosporangia.

Fig. 38. Zoosporangium.

Fig. 39. A larger branching thallus with proliferating zoosporangia.

Figs. 40-43. *Blastocladia rostrata* Minden. Figs. 40, 41, 43 after Minden, 1916; fig. 42 after Sparrow, 1943.

Fig. 40. Portion of a branched thallus bearing zoosporangia and resistant sporangia with prominent apical papillae and beaks.

Fig. 41. Elongate, 7.5-10 μ, planospore.

Fig. 42. Discharge of planospores.

Fig. 43. Decidious resistant sporangium falling out of the thin-walled enveloping case.

Figs. 44-46. *Blastocladia tenuis* Kanouse. (Kanouse, 1927.)

Fig. 44. Branch of a thallus with a terminal zoosporangium.

Fig. 45. Sparsely branched slender thallus with terminal zoosporangia; sculptured with rudged markings on the brittle golden-brown wall.

Fig. 46. Apical portion of a thallus.

Figs. 47-49. *Blastocladia gracilis* Kanouse. (Kanouse, 1927.)

Fig. 47. Thallus showing the typical racemose arrangement of the long subcylindrical zoosporangia.

Fig. 48. Enlarged empty zoosporangium.

Fig. 49. Resistant sporangium with a thick, brown, punctate (?) wall.

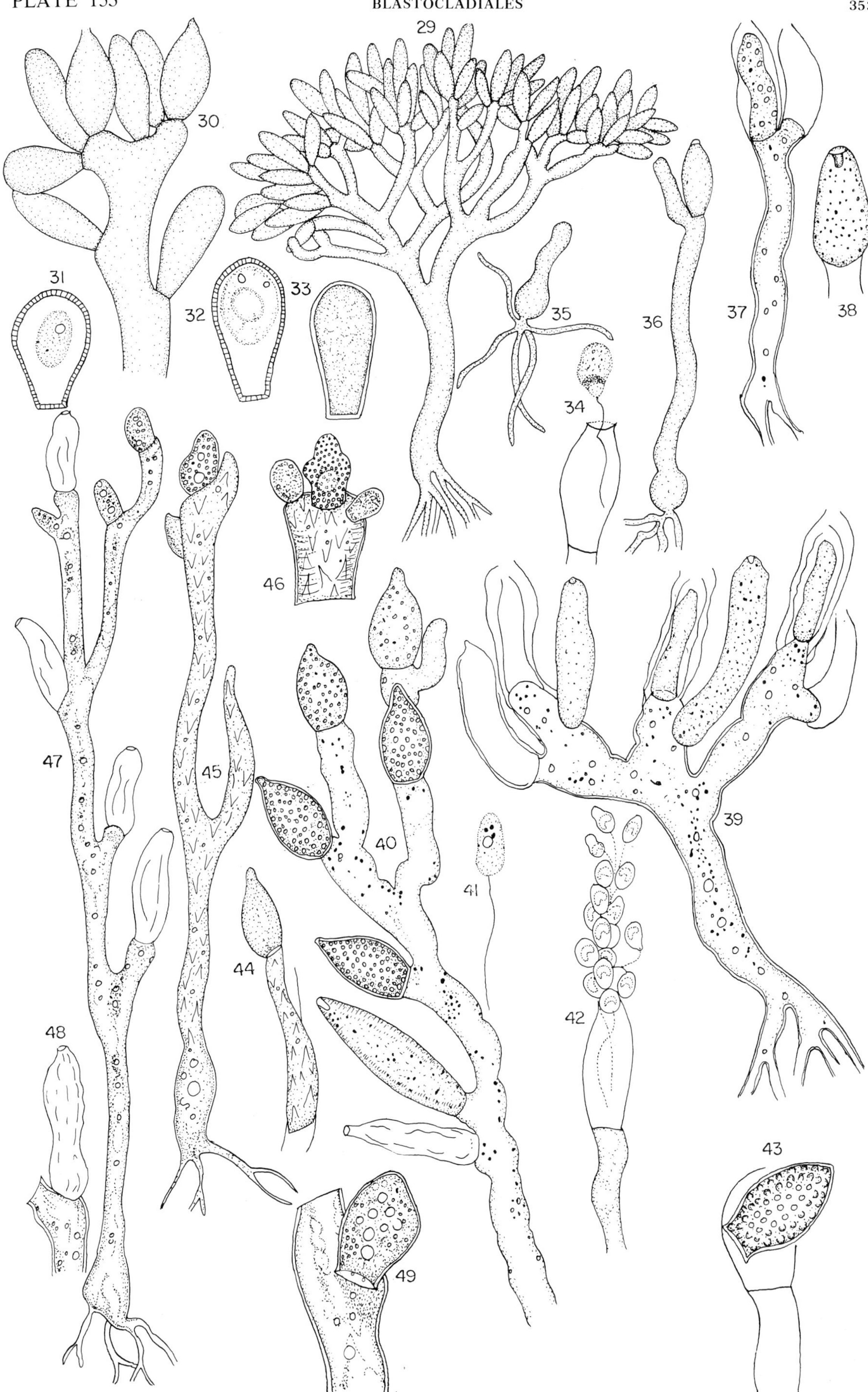

Blastocladia

Whisler *et al* (1975) noted that the copepod phase of *Coelomomyces psorophorae* is remarkably similar to *Callimastix cyclopis* in *Cyclops* and stated "the few differences that exist between the two genera may well relate to different host species and rearing conditions.

REFERENCES TO THE BLASTOCLADIACEAE

Arnaud, G. 1952. Mycologie concrète: Genera. Bull. Soc. Mycol. France 68:181-223, 8 figs.

Aronson, J. M. and L. Machlis. 1959. The chemical composition of the hyphal walls of the fungus *Allomyces*. Amer. J. Bot. 46:292-300, 5 figs.

Barrett, J. T. 1912. The development of *Blastocladia strangulata*, n. sp. Bot. Gaz. 54:353-371, pls. 18-20.

Beneke, E. S. and G. B. Wilson. 1950. The treatment of *Allomyces javanicus* var. *japonensis* Indoh with colchicine and sodium nucleate. Mycologia 42:519-522.

Bessey, E. A. 1939. Isoplanogametes in *Blastocladia*. Mycologia 31:308-309.

Blackwell, E. 1937. Germination of the resistant spores of *Blastocladia pringsheimii*. Nature 140:933.

———. 1939. The problem of gamete production in *Blastocladia*. Mycologia 31:627-628.

———. 1940. A life cycle of *Blastocladia pringsheimii* Reinsch. Trans. Brit. Mycol. Soc. 24:68-86, 9 figs.

Blondel, B. and G. Turian. 1960. Relation between basophilia and fine structure of the cytoplasm in the fungus *Allomyces macrogynus* Em. J. Biophy. Biochem. Cytology 7:127-134, pls. 48-57.

Butler, E. J. 1911. On *Allomyces* a new aquatic fungus. Ann. Bot. 25:1023-1025, 18 figs.

Canter, H. M. and L. G. Willoughby. 1964. A parasitic *Blastocladiella* from Windermere plankton. J. Roy. Micro. Soc. 83:365-372, 37 figs., pls. 159-162.

Cantino, E. C. 1950. Nutrition and phylogeny in the water molds. Quart. Rev. Biol. 25:269-277, 1 fig.

———. 1955. Physiology and phylogeny in the water molds. Ibid. 30:138-149, 1 fig.

———. 1969. The Y particle satellite ribosome package and spheroidal mitochondrion in the zoospore of *Blastocladiella emersonii*. Phytopath. 59:1071-1076.

———. 1970. Germination of a resistant sporangium of *Blastocladiella britannica* bearing on its taxonomic status. Trans. Brit. Mycol. Soc. 54:303-307, 1 fig.

———, and E. A. Hornstein. 1954. Cytoplasmic exchange without gametic copulation in the water mold, *Blastocladiella emersonii*. Amer. Natur. 88: 143-154, 3 figs.

———. 1956. Gamma and the cytoplasmic control of differentiation in *Blastocladiella*. Mycologia 48:443-446, 2 figs.

———. 1956. The relation between cellular metabolism and morphogenesis in *Blastocladiella*. Mycologia 48:225-240, 4 figs.

———, and M. O. T. Hyatt. 1953. Phenotypic "sex" determination in the life history of a new species of *Blastocladiella*, *B. emersonii*. A. V. Leeuwenhoek 19:25-70, 14 figs.

———, J. S. Lovett, L. V. Leak, and J. Lythgoe. 1963. The single mitochondrion, fine structure and germination of the spore of *Blastocladiella emersonii*. J. Gen. Microbiol. 31:393-404, 4 figs.

———, and J. P. Mack. 1969. Form and function in the zoospore of *Blastocladiella emersonii*. I. The Y particle and satellite ribosome package. Nova Hedwigia 18:115-147.

———, L. C. Truesdell, and D. S. Shaw. 1968. Life history of the motile spore of *Blastocladiella emersonii*: a study in cell differentiation. J. Elisha

PLATE 154

Fig. 50. *Blastocladia gracilis* Waterhouse. Portion of a much branched thallus bearing a few zoosporangia and numerous resistant sporangia. Some of rhizoids and resistant sporangia omitted. (Waterhouse, 1941.)

Figs. 51-53. *Blastocladia glomerata* Sparrow. Figs. 51, 52 after Sparrow, 1936; fig. 53 after Waterhouse, 1941.

Fig. 53. Thallus with a broad short basal cell and clusters of broadly ellipsoidal zoosporangia at the apex.

Figs. 51, 52. Broadly ellipsoidal zoosporangia.

Figs. 54-59. *Blastocladia globosa* Kanouse. Figs. 54-58 after Kanouse, 1927; fig. 59 after Cotner, 1930.

Fig. 54. Thallus with a globular apex bearing numerous resistant sporangia, empty zoosporangia, and setae.

Figs. 55, 56. Punctate resistant sporangia.

Fig. 57. Thallus with zoosporangia borne directly on the globular apex.

Fig. 58. Incipient resistant sporangium and zoosporangium borne on a branch from the globular apex.

Fig. 59. Broadly ovoid, 7-9 × 5-6 μ planospore with a conspicuous nuclear cap.

Fig. 60. *Blastocladia truncata* Sparrow. A tapering unbranched thallus bearing truncate zoosporangia at the apex. (Sparrow, 1932.)

Figs. 61-63. *Blastocladia angusta* Lund. Fig. 61 after Lund, 1934; figs. 62, 63 after Crooks, 1937.

Fig. 61. Portion of a thallus with a cylindrical basal cell bearing narrowly cylindrical, greatly elongate zoosporangia at the tips of short branches.

Fig. 62. Thallus bearing zoosporangia and resistant sporangia.

Fig. 63. Broadly ellipsoidal resistant sporangium with a narrowly truncate base, a beaked apex, and thin, hyaline punctate wall.

Fig. 64. *Blastocladia aspergilloides* Crooks. Unbranched brown thalli with inflated apices bearing long narrow cylindrical zoosporangia and a few sterile setae. (Crooks, 1937.)

Fig. 65. *Blastocladia sparrowii* Indoh. Thallus with a cylindrical basal cell and lobed apex bearing proliferating zoosporangia and setae; a few zoosporangia and setae omitted. (Sparrow, 1943.)

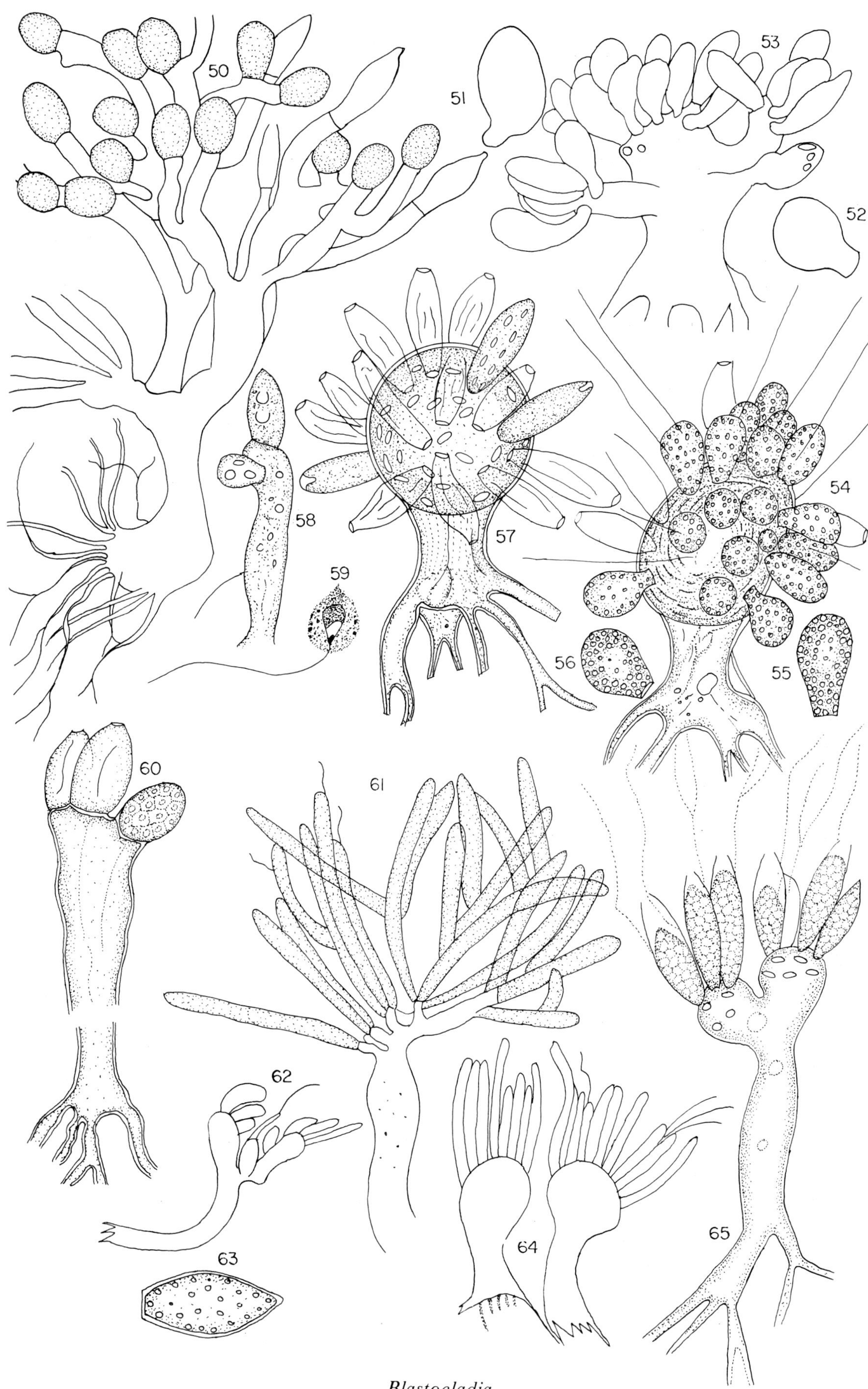

Blastocladia

Mitchell Sci. Soc. 84:125-146; 10 figs.
———, K. F. Suberkropp, and L. C. Truesdell. 1969. Form and function in the zoospore of *Blastocladiella emersonii*. II. Spheroidal mitochondria and respiration. Nova Hedwigia 18: 149-158, 3 figs.
———, and L. C. Truesdell. 1970. Organization and fine structure of the side body and its lipid sac in the zoospore of *Blastocladiella emersonii*. Mycologia 62:548-567, 9 figs.
———. 1971. Cytoplasmic Y-like particles and other ultrastructural aspects of zoospores of *Blastocladiella britannica*. Trans. Brit. Mycol. Soc, 56:169-179, pls. 15-18.
Carlile, M. J. and L. Machlis. 1965. The reponse of male gametes of *Allomyces* to the sexual hormone sirenin. Amer. J. Bot. 52:478-483, 2 figs.
Coker, W. C. and H. H. Braxton. 1926. New water molds from the soil. J. Elisha Mitchell Sci. Soc. 42:139-147, pls. 10-15.
———, and F. A. Grant, 1922. A new genus of water mold related to *Blastocladia*. Ibid. 37: 180-182, pl. 32.
Cotner, G. B. 1930. Cytological study of the zoospores of *Blastocladia*. Bot. Gaz. 89:295-309, 10 figs.
Crooks, K. M. 1937. Studies of Australian aquatic Phycomycetes. Proc. Roy. Soc. Victoria 49: a-6-232, 11 figs., pl. 10.
Couch, J. N., and A. J. Whiffen. 1942. Observations on the genus *Blastocladiella*. Amer. J. Bot. 29: 582-591, 66 figs.
Cox, H. T. 1939. A new genus of the Rhizidiaceae. J. Elisha Mitchell Sci. Soc. 55:389-397, 1 fig., pls. 45, 46.
Das-Gupta, S. N., and Rachel John. 1953. Studies on the Indian aquatic fungi. I. Some water moulds of Lucknow. Proc. Indian Acad. Sci., ser. B., 38: 165-170, 15 figs.
DeLong, S. K. 1965. Studies on the genetic control of resistant sporangium formation in *Allomyces*. Amer. J. Bot. 52:999-1005, 1 fig.
Emerson, R. 1938. A new life cycle involving cyst-formation in *Allomyces*. Mycologia 30:120-132, 11 figs.
———. 1939. Life cycles in the *Blastocladiales*. Trans. Brit. Mycol. Soc. 23:123.
———. 1941. An experimental study of the life cycles and taxonomy of *Allomyces*. Lloydia 4:77-144, 16 figs.
———. 1950. Current trends of experimental research on the aquatic Phycomycetes. Ann. Rev. Microbiol. 4:169-200.
———. 1955. The biology of water molds. *In* Aspects of Synthesis and Growth, Edited by D. Rudnick, Princeton Univ. Press, pp. 171-208, 4 figs., pls. 1-8.
Emerson, R., and E. E. Cantino. 1948. The isolation, growth and metabolism of *Blastocladia* in pure culture. Amer. J. Bot. 35:157-171, 9 figs.
———, and D. L. Fox. 1940. Y-carotene in the sexual phase of the aquatic fungus *Allomyces*. Proc. Roy. Soc. London, Ser. B. 128:275-293., 1 fig., pl. 16.
———, and J. A. Robertson. 1974. Two new members of the Blastocladiaceae I. Taxonomy, with an evaluation of genera and interrelationships in the family. Amer. J. Bot. 61:303-317, 21 figs.
———, and C. M. Wilson. 1954. Interspecific hybrids and the cytogenetics and cytotaxonomy of *Euallomyces*. Mycologia 46:393-434, 15 figs.
Fähnrich, P. 1974a. Untersuchungen zur Entwickelung der Phycomyceten *Allomyces arbuscula*. I. Einfluss von Inhibitoren der Protein - und Nucleinsäuresynthese auf die Differenzierung von Gametangien. Arch. f. Mikrobiol. 98:85-92.
———. 1974b. Untersuchungen zur Entwickelung der Phycomyceten *Allomyces arbuscula*. II. Einfluss von Inhibitoren der Protein — und Nucleinsäuresynthese auf die Gametogenese. Ibid. 99:147-153, 5 figs.
Foley, J. M. 1958. The occurrence, characteristics and genetic behavior of albino gametophytes in *Allomyces*. Amer. J. Bot. 45:639-648, 11 figs.
Fuller, M. S. 1966. Structure of the uniflagellate zoospores of aquatic Phycomycetes. Proc. 18th Symp. Colston Res. Soc. 18:67-84, 26 figs.
———, and S. A. Calhoun. 1968. Microtubule-kinetosome relationship in motile cells of the *Blastocladiales*. Zeitschr. f. Zellforsch. 87:526-533, 6 figs.
———, and L. W. Olson. 1971. The zoospore of *Allomyces*. J. Gen. Microbiol. 66:171-183, 5 pls.
Harder, R. and G. Sörgel. 1938. Uber einen neurer plano-isogamen Phycomyceten mit Generationswechsel und seine phylogenetische Bedeutung. Nachrich. Gesell. Wiss. Gottingen, Math.-Physik.

PLATE 155

Fig. 66. *Blastocladia sparrowii* Indoh. Thallus with subcylindrical, subdichotomously branched lobes; setae lacking. (Das Gupta and John, 1953.)

Figs. 67, 68. *Blastocladia incrassata* Indoh. Thalli with cylindrical basal cells, sparse rhizoids, closely branched in a dichotomous or racemose manner, bearing terminal sessile, racemose, or cymbosely arranged, cylindrical to broadly clavate zoosporangia. (Indoh, 1940.)

Figs. 69, 70. *Blastocladia* (?) sp. Arbusculate thallus with a cylindrical basal cell, dichotomous or irregular thin branches, cylindrical zoosporangia, and elliptical, to pyriform, yellowish resistant sporangia. (Indoh, 1940.)

Fig. 70. Resistant sporangium with conspicuously pitted outer wall. (Indoh, 1940.)

Fig. 71. *Blastocladia* (?) sp. sympodial thalli with zoosporangia borne singly on the ends of the branches. (Waterhouse, 1941.)

Fig. 72. *Blastocladia* sp. spherical planospore with a nuclear cap, side body, and globules. (Sparrow, 1943.)

Blastocladia

Kl., Fachgruppe VI, n.f. 5:119-127, 4 figs.

———. 1939. Same title. Veröffenl. Deutsch-Dominikanischen Tropenforschungsinstituts 1: 118-127.

Hatch, W. R. 1936. Zonation in *Allomyces arbuscula*. Mycologia 28:439-444, 5 figs.

———. 1938. Conjugation and zygote germination in *Allomyces arbuscula*. Ann. Bot. (n.s.) 2:583-614, pls. 18-22.

———. 1944. Zoosporogenesis in the resistant sporangia of *Allomyces*. Mycologia 36:650-663, 2 figs.

Haxo, F. T. 1955. Some biochemical aspects of fungal carotenoids. Forschr. Chem. Org. Naturst. 12: 169-197.

Hennen, J. F. 1963. Studies on a new species of *Allomyces*. Proc. Indiana Acad. Sci. 72:258.

Hill, E. P. 1969. The fine structure of the zoospores and cysts of *Allomyces macrogynus*. J. Gen. Microbiol. 56:125-130, pls. 1-5.

Hoerner, E. M., and K. Zetsche. 1974. Ein antagonistiche wirkung von blau-und rotlicht auf die Entrvickelung des pilzes *Blastocladiella britannica*. Pl. Sci. Letters 2:127-131.

Hornstein, E. A., and E. C. Cantino. 1961. Morphogenesis in and the effect of light on *Blastocladiella britannica* sp. nov. Trans. Brit. Mycol. Soc. 44: 185-198, 11 figs., pl. 15.

Indoh, H. 1940. Studies on Japanese aquatic fungi. II. The Blastocladiaceae. Sci. Rept. Tokyo Bunrika, Daigaku, sect. B. 4:237-284, 34 figs.

———. 1952. A new variety of *Allomyces javanicus*. Nagao 2:24-29, 4 figs.

Kanouse, B. B. 1927. A monographic study of special groups of the water molds. I. Blastocladiaceae. Amer. J. Bot. 14:287-306, pls. 32-34.

Karling, J. S. 1967. Some zoosporic fungi of New Zealand. X. Blastocladiales. Sydowia 20:144-150, pls. 29-31.

———. 1973. A note on *Blastocladiella*. Mycopath et Mycol. Appl. 23:169:172.

Kato, R. 1959. On the life history of a strain of *Allomyces cystogenus*. J. Jap. Bot. 34:100-110, 2 pls.

Knjep, H. 1929. *Allomyces javanicus* n. sp., ein anisogamer Phycomycet mit Planogameten. Ber. deut. Bot. Gesell. 47:199-212, 7 figs.

———. 1930. Uber Generationswechsel von *Allomyces*. Zeitschr. f. Bot. 22:433-441, 2 figs.

Lessie, P. E. 1967. Fine structure of zoospore differentiation in *Blastocladiella emersonii*. Masters Thesis, Purdue University.

———, and J. S. Lovett. 1968. Ultrastructural changes during sporangium formation and zoospore differentiation in *Blastocladiella emersonii*. Amer. J. Bot. 55:220-236, 56 figs.

Lloyd, D. 1938. A record of two years continuous observations on *Blastocladia pringsheimii* Reinsch. Trans. Brit. Mycol. Soc. 21:152-166, 3 figs.

Lovett, J. S. 1963. Chemical and physical characteristics of "nuclear caps"isolated from *Blastocladiella* zoospores. J. Bacteriol. 85:1235-1246, 7 figs.

———. 1968. Reactivation of ribonucleic acid and protein synthesis during germination of *Blastocladiella* zoospores. Ibid. 96:962-969. 6 figs.

Lund, A. 1934. Studies on the Danish freshwater Phycomycetes and notes on their occurrence particularly relative to the hydrogen ion concentration of the water. Kgl. Danske Vidensk. Selk. Skrift. Naturv. Math., afd. IX, 6:1-97, 39 figs.

Machlis, L., W. H. Nutting, and H. Rapport. 1966. Production, isolation, and characterization of sirenin. Biochem. 6:2147-2152.

———, W. H. Nutting, and H. Rapoport. 1968. The fine structure of sirenin. J. Amer. Chem. Soc. 90:1674-1676.

———, and E. Ossia. 1953. Maturation of the meiosporangia of *Euallomyces*. I. The effect of cultural conditions. Amer. J. Bot. 40:358-365, 2 figs.

McCrainie, J. 1942. Sexuality in *Allomyces cystogenus*. Mycologia 34:209-213, 1 fig.

Martin, W. W. 1971. The ultrastructure of *Coelomomyces punctatus* zoospores. J. Elisha Mitchell Sci. Soc. 87:209-221, 20 figs.

Matsumae. A., R. B. Myers, and E. C. Cantino. 1970. Comparative number of Y particles in the flagellate cells of various species and mutants of *Blastocladiella*. J. Gen. Appl. Microbiol. 16:443-453, 4 figs.

Matthews, V. D. 1937. A new genus of the Blastocladiaceae. J. Elisha Mitchell Sci. Soc. 53:191-195, 1 fig., pls. 20-21.

Minden, M. von. 1916. Beitrage zur Biologie und Systematik einheimischer submerser Phycomyceten. *In* Falk, Mykolog. Untersuch. Berichte 2:

PLATE 156

Figs. 1-4. *Ramnocladia reticulata* Sörgel from soil. (Sörgel, 1952.)

Figs. 1-3. Variations in thalli and branching; terminated by zoosporangia; branches arising from a central stalk; in fig. 2 the zoosporangium is branched.

Fig. 4. Thallus bearing an empty zoosporangium and a resistant sporangium with a reticulate outer wall.

Fig. 5. *Ramnocladia pallida* Sörgel. Portion of a richly branched thallus with subspherical pale-golden resistant sporangia at the end of elongate hyphe or branches. (Sörgel, 1952.)

Fig. 6. *Allocladia monilifera* Sörgel. Richly branched coarse, thallus bearing broadly fusiform zoosporangia with a truncate end. (Sörgel, 1952.)

Fig. 7. *Leptocladia laevis* Sörgel. Thallus with relatively narrow branches arising at the apex of a coarse central axis and bearing smooth, hyaline (?) resistant sporangia. (Sörgel, 1952.)

Fig. 8. *Brevicladia* sp. Thallus consisting of an ovoid zoosporangium (?), rhizoidal axis, and rhizoids. (Sörgel, 1952.)

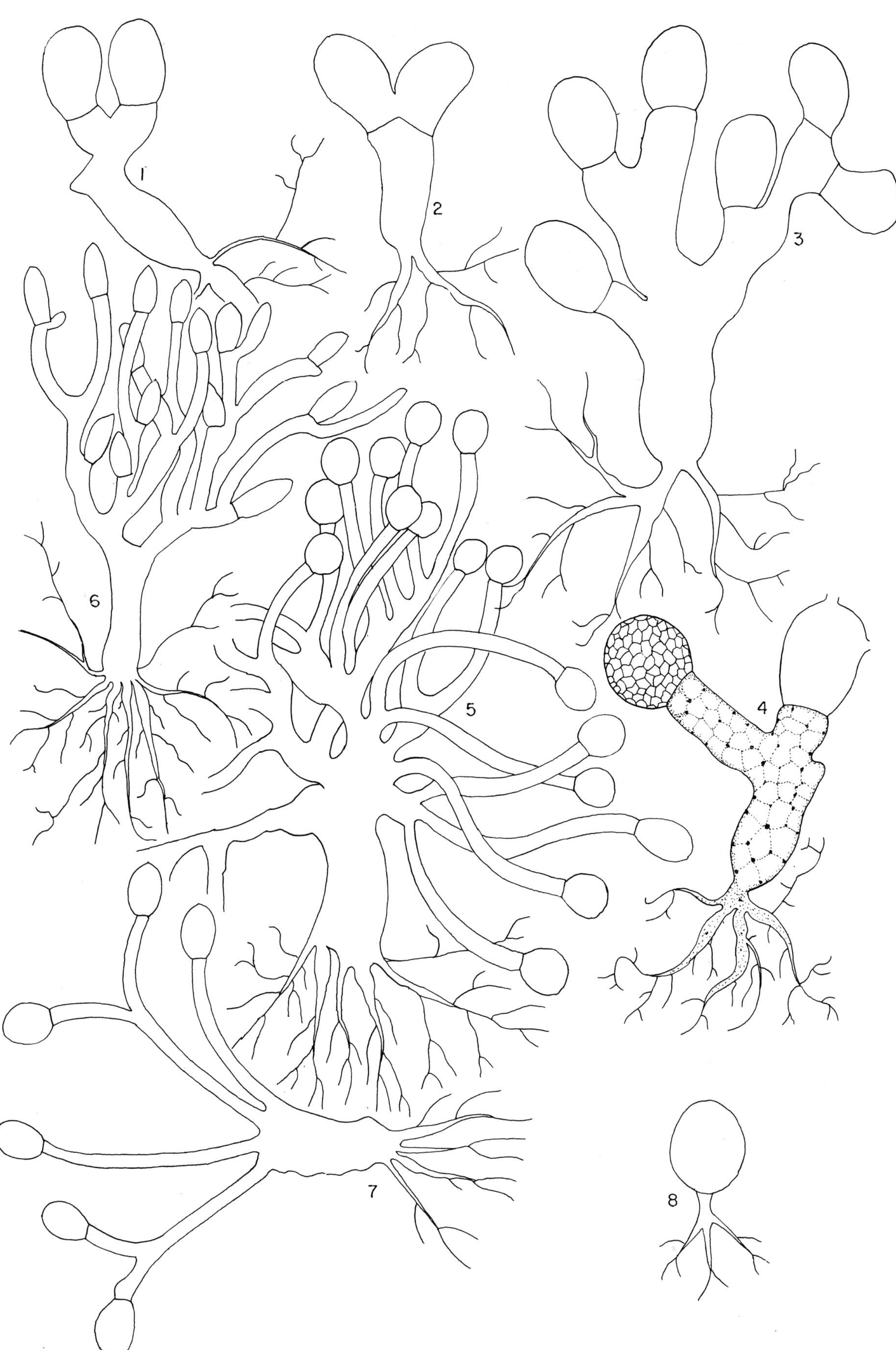

Questionable Genera

146-252, 24 figs., 8 pls.

Moore, R. T. 1964. Zoospore formation in the Phycomycete *Allomyces*. J. Cell Biol. 23:109A-110A.

———. 1968. Fine structure of mycota 13. Zoospore and nuclear cap formation in *Allomyces*. J. Elisha Mitchell Sci. Soc. 84:147-165, 22 figs.

Olson, L. W. 1973a. The meiospore of *Allomyces*. Protoplasma 78:113-127, 22 figs.

———. 1973b. Synchronized development of gametophyte germlings of the aquatic Phycomycete *Allomyces macrogynus*. Ibid. 78:129-144, 18 figs.

Quantz, L. 1943. Untersuchungen zur Ernährungsphysiologie einiger niederer Phycomyceten *Allomyces kniepii, Blastocladiella variabilis* und *Rhizophylctis rosea*. Jahrb wiss. Bot. 91:120-169, 15 figs.

Reichle, R. E. and M. S. Fuller. 1967. The fine structure of *Blastocladiella emersonii* zoospores. Amer. J. Bot. 54:81-92, 29 figs.

Reinsch, P. F. 1878. Beobachtungen über einige neue Saprolegniaeae, über die Parasiten in Desmidienzellen und über die Stachelkugeln in *Achlya*-schlauchen. Jahrb. wiss. Bot. 11:293-311, pls. 14-17.

Renaud, F. L. and H. Swift. 1964. The development of basal bodies and flagella in *Allomyces arbuscula*. J. Cell Biol. 23:339-354, 24 figs.

Ritchie, D. 1947. The formation and structure of the zoospores of *Allomyces*. J. Elisha Mitchell Sci. Soc. 63:168-205, pls. 22-26.

Rorem, E. S. and L. Machlis. 1957. The ribonucleoprotein in the meiosporangia of *Allomyces*. J. Biophys. Biochem. Cytology 3:879-888, pl. 288.

Schussnig, B. 1930. Der Generation-und phasenwechsel bei den Chlorophyceen. Oesterr. Bot Zeitschr. 79:58-77.

Shaw, D. S. and E. C. Cantino. 1969. An albino mutant of *Blastocladiella emersonii*; comparative studies of zoospore behavior and fine structure. J. Gen. Microbiol. 59:369-382, 6 pls.

Skucas, G. P. 1966. Structure and composition of zoosporangia discharge papillae in the fungus *Allomyces*. Amer. J. Bot. 53:1006-1011, 13 figs.

———. 1967. Structure and composition of the resistant sporangial wall in the fungus *Allomyces*. Ibid. 54:1152-1158, 6 figs.

———. 1968. Changes in wall and internal structure of *Allomyces*-resistant sporangia during germination. Ibid. 55:291-295, 10 figs.

Soll, D. R., Bromberg, and D. R. Sonneborn. 1969. Zoospore germination in the water mold, *Blastocladiella emersonii* 1.....Develop. Biol. 20:183-217, 13 figs.

Sörgel, G. 1937a. Uber heteroploide Mutanten by *Allomyces kniepii*. Zeitschr. f. Bot. 31:335-336.

———. 1937b. Untersuchungen über den Generationswechsel von *Allomyces*. Ibid. 31 401-446, 15 figs.

———. 1941. Uber die Verbreitung einiger niederer Phycomyceten in Erden Westindiens. Beih. Bot. Centralbl. Abt. 61:1-32, 4 figs.

———. 1952. Uber mutmassliche phylogenetische Zusammenhänge bei niederer Pilzen, insbesondere den Blastocladiales. Biol. Zentralbl. 71: 385-397, 13 figs.

Sost, H. 1955. Uber die Determination des Generationswechsels von *Allomyces arbuscula* (Butl.) (Polyploidversuche). Arch. f. Protistk. 100:541-564, 8 figs., pl. 17.

Sparrow, F. K., Jr. 1932. Observations on the aquatic fungi of Cold Spring Harbor. Myxologia 24: 268-303, 4 figs., pls. 7, 8.

———. 1936. A contribution to our knowledge of the aquatic Phycomycetes of Great Britain. J. Linn. Soc. London (Bot) 50:417-478, 7 figs., pls. 14-20.

———. 1943. The aquatic Phycomycetes. XIX-785 pp. 634 figs. Univ. of Michigan Press, Ann Arbor.

———. 1950. Some Cuban Phycomycetes. J. Wash. Acad. Sci. 40:50-55, 30 figs.

———. 1964. A new species of *Allomyces*. Mycologia 56:460-461.

———. 1965. The occurrence of *Physoderma* in Hawaii, with notes on other Hawaiian Phycomycetes. Mycopath. et Mycol. Appl. 25:119-143, pls. 1-7.

———, and B. M. Morrison. 1961. A peculiar isolate of *Allomyces*. Papers Mich. Acad. Sci., Arts and Letters, 46:175-181, pls. 1-3.

Stüben, H. 1939. Uber Entwickelungsgeschichte und Ernärungsphysiologie eines neuen niedern Phycomyceten mit Generationswechsel. "Planta," Arch. Wiss. Bot. 30:353-383, 17 figs.

Stumm, C. 1958. Die Analyse von Genmutanten mit geänderten Fortpflanzungs eigenschaften bei *Allomyces* Butl. Zeitschr. Vererbungsl. 89: 521-539.

Teter, H. E. 1944. Isogamous sexuality in a new strain of *Allomyces*. Mycologia 36:194-210, 3 figs.

Thaxter, R. 1896. New or peculiar fungi. 3. *Blastocladia*, Bot. Gaz. 21:45-52, pl. 3.

Truesdell, L. C. and E. C. Cantino. 1970. Decay of y-particles in germinating zoospores of *Blastocladiella emersonii*. Arch. f. Microbiol. 70:378-392.

Turian, G. 1952. Carotenoides et differenciation sexuale chez *Allomyces*. Experimentia 8:302.

———. 1954. L'acide borique, inhibiteur de la copulation gametique chez *Allomyces*. Ibid. 10: 498.

———. 1955. Culture de la phase gametophytique d'*Allomyces javanicus* en milieu synthetique liquide. Compt. Rendu Acad. Sci. Paris 240: 1005-1007.

———. 1956. Le corps paranucléaire des gamètes geants d'*Allomyces javanicus* traité à l'acide borique. Protoplasma 47:135-138.

———. 1957a. Recherches sur l'action anticaroténogène de la diphénylamine et ses consequences sur la morphogènese reproductive ches *Allomyces* et *Neurospora*. Physiol. Plantarum 10:667-680.

———. 1957b. Recherches sur la morphogenese sexuelle chez *Allomyces*. Bull. Soc. Bot. Suisse 67:458-486.

———. 1958. Recherches sur les bases cytochemiques et cytophysiologiques fr la morphogènese chez la champignone aquatiques *Allomyces*. Rev. Cytol. et Biol. Vég. 19:241-272.

———. 1961. Cytoplasmic differentiation and dedifferentiation in the fungus *Allomyces*. Protoplasma 54:323-327, 1 fig.

———, nd E. Kellenberger. 1956. Ultrastructure du corps paranucleaire, des mitochondries et de la membrane nucleaire des gametes d'*Allomyces macrogynus*. Exp. Cell Res. 11:417-422, 4 figs.

Vavra, J. 1960. A contribution to the knowledge of the parasitic flagellate. *Callimastix cyclopis*. J. Protozool. 7, suppl., pp. 26-27.

———. 1963. Protozoan parasites form freshwater Crustaceans in Czechoslovakia. Progress in Protozoology. Proc. Intern. Cong. on Protozoology, Prague, 1961, pp. 597-598.

———, and L. Joyon, 1966. Etude sur la morphologie le cycle evolutif et la position systématique de *Callimasix cyclopis* Weissenberg 1912. Parasitologica 2:5-16, pls. 1-6.

Villeret, G. 1952. Les Blastocladiales. Bull. Soc. Bretagne 27:113-137.

Viswanathan, M. A., and G. Turian. 1966. The nucleic acid content of the male and female gametes of *Allomyces* and the comparative saline fraction of the ribonucleic acid of the gametangia. Experimentia 22:377-378.

Waterhouse, G. M. 1941. Some water moulds of the Hogsmill River collected from 1937 to 1939. Trans. Brit. Mycol. Soc. 25:315-325, 4 figs.

Weissenberg, R. 1912. *Callimastix cyclopis*, n.g., n.s.p., ein geisseltragendes Protozoon aus dem Serum von *Cyclops*. Sitzungsb. Ges. naturf. Freunde 5:299-305, 1 fig.

———. 1950. The development and the affinities of *Callimastix cyclopsis* Wissenberg, a parasitic organism from the serum of *Cyclops*. Proc. Amer. Soc. Protozool 1:4-5.

Whiffen, A. J. 1943. New species of *Nowakowskiella* and *Blastocladia*. J. Elisha Mitchell Sci. Soc. 59: 37-43, pls. 2-4.

———. 1946. Two new terricolous Phycomyctes belonging to the genera *Lagenidium* and *Blastocladiella*. Ibid. 62:54-58, pl. 7.

———. 1951 The effect of cycloheximide on the sporophyte of *Allomyces arbuscula*. Mycologia 43:635-644, 1 fig.

Willoughby, L. G. 1959. A new species of *Blastocladiella* from Great Britain. Trans. Brit. Mycol. Soc. 42:287-291, 1 fig., pl. 18.

Wilson, C. W. 1952. Meiosis in *Allomyces*. Bull. Torrey Bot. Club 79:139-160, 23 figs.

———, and P. W. Flanagan. 1968. The life cycle of *Brachyallomyces*. Canad. J. Bot. 46:1361-1367, 38 figs.

Wolf, F. T. 1939. A study of some aquatic Phycomycetes isolated from Mexican soils. Mycologia 31: 376-387.

Yaw, K. E., and V. M. Cutter, Jr. 1951. Crosses involving biochemically deficient mutants of *Allomyces arbuscula*. Mycologia 43:156-160.

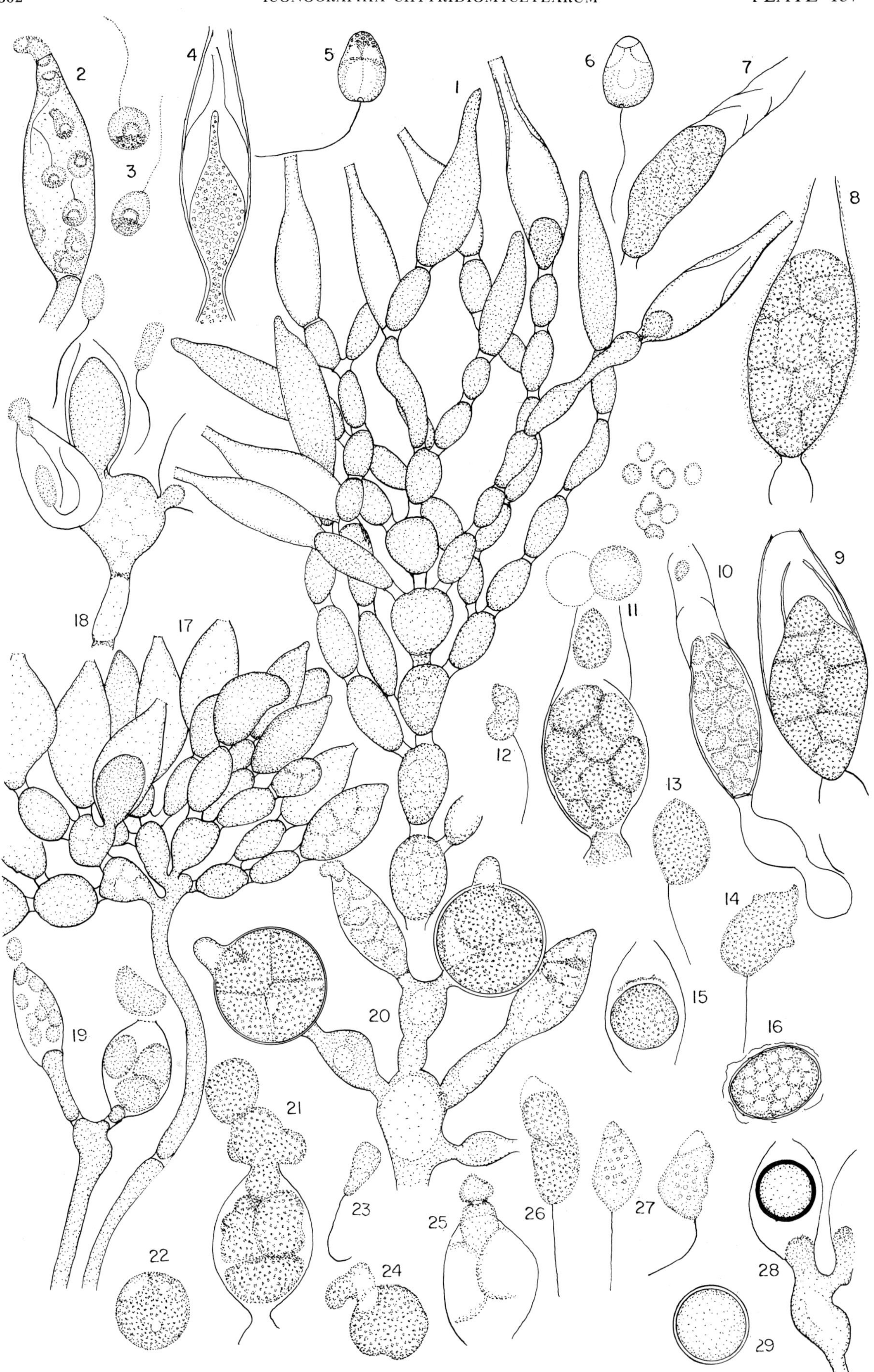

Gonopodya

Chapter XI

MONOBLEPHARIDALES

So far as is known this order appears to represent the climax of sexual reproduction in the Chytridiomycetes. Studies of this order were initiated by Cornu's (1871) significant discovery of *Monoblepharis*, an aquatic genus which unlike all other oogamous fungi possesses motile male gametes or sperms. This initial discovery of oogamous reproduction by means of sperms and eggs was the first of many subsequent discoveries of sexuality in the Blastocladiales and Monoblepharidales, and it stands as a landmark in the studies of these fungi. Apparently, some mycologists were skeptical about the existence of a group of fungi with the characteristics of *Monoblepharis*, but Thaxter's (1895) rediscovery of this genus put such doubts to rest. Since that time knowledge of the group has been extended greatly by the studies of Lagerheim (1900), Woronin (1904), Laibach (1926, 1927), Sparrow (1933, 1940), Beneke (1948), Johns and Benjamin (1954), Perrott (1955) and others.

At present the Monoblepharidales includes 2 families, Gonopodyaceae and Monoblepharidaceae, 3 or possibly 4 genera, approximately 16 species, and a few varieties. All known species are saprophytes and occur in soil and on submerged fruits, twigs, and insect cadavers in fresh water. *Monoblepharella* occurs in soil in warm latitudes and may be trapped on various baits added to soil-water cultures. *Monoblepharis* and *Gonopodya* occur more commonly on submerged twigs and apple and rose fruits in relatively cold freshwater habitats. Under favorable conditions the species produce an abundant mycelium whose extent and nature depends on the particular species. Characteristically, the protoplasm in the hyphae is vacuolate in such a manner that it has a scalariform or foamy appearance, and this makes recognition of monoblepharids easy even in the vegetative state.

The most distinctive characteristic of the order is its method of sexual reproduction in which a relatively large nonflagellate egg is fertilized by a smaller posteriorly uniflagellate motile or amoeboid antherozoid outside of a gametangium or within an oogonium. The antheridia and oogonia are usually borne at the apex of the same thallus, or some species may be normally diclinous. In *Gonopodya*, and rarely in *Monoblepharella*, several eggs are formed in the oogonium, while normally in *Monoblepharis* and *Monoblepharella* only

PLATE 157

Figs. 1-16. *Gonopodya prolifera* Fischer. Figs. 1-3 after Thaxter, 1895; fig. 4 after Reinsch, 1878; figs. 5, 6 after Sparrow, 1933; figs. 7-16 sketched from photographs after Johns and Benjamin, 1954.

Fig. 1. Apical portion of a thallus showing method of branching, ovoid segments, constrictions and pseudosepta, and obclavate proliferating zoosporangia.

Fig. 2. Broadly fusiform zoosporangium discharging planospores.

Fig. 3. Spherical and ovoid planospores.

Fig. 4. Proliferating zoosporangium.

Figs. 5, 6. Motile elliptical planospores with anterior globules which have fused in fig. 6.

Fig. 7. Proliferating male gametangium.

Fig. 8, 9. Broadly clavate proliferated female gametangia whose content has undergone cleavage into incipient female gametes.

Fig. 10. Proliferated male gametangium discharging, 3-6 × 7-10 μ, male gametes.

Fig. 11. Proliferated female gametangium discharging larger, 15-19 μ diam., non-flagellate female gametes.

Fig. 12. Posteriorly uniflagellate male gamete.

Fig. 13. Uniflagellate zygote, motile by virtue of the flagellum of the male gamete.

Fig. 14. Amoeboid uniflagellate zygote.

Fig. 15. Encysted zygote within the female gametangium.

Fig. 16. Encysted zygote with a hyaline cyst wall.

Figs. 17-29. *Gonopodya polymorpha* Thaxter. Figs. 17, 19 after Thaxter, 1895; figs. 18, 23, 27-29 after Miller, 1963; figs. 20-22, 24-26 sketched from photographs from Johns and Benjamin, 1954.

Fig. 17. Upper portion of a thallus showing method of branching, ovoid segments, constrictions and pseudosepta, and broadly ovoid zoosporangia.

Fig. 18. Proliferated zoosporangia with elongate to ovoid, 6-10 × 7 μ, planospores.

Fig. 19. Branch bearing a male and a female gametangium with male and female gametes, resp.; described by Thaxter as zoosporangia.

Fig. 20. Fasicle with 2 nearly spherical female, and 2 broadly fusiform male gametangia; content of female gametangia cleaving; male gametes in one male gametangium.

Fig. 21. Female gametangium discharging female gametes.

Fig. 22. Mature spherical, 11-18 μ diam., non-flagellate female gamete.

Fig. 23. Male, 6-10 × 7 μ, posteriorly uniflagellate gamete.

Fig. 24. Male gamete crawling over the surface of a female gamete.

Fig. 25. Male gamete entering female gametangium.

Figs. 26, 27. Amoeboid posteriorly uniflagellate zygotes.

Fig. 28. Zygote encysted within a female gametangium.

Fig. 29. Encysted zygote with a smooth brownish wall.

one is formed. In the former two genera the fertilized egg emerges and becomes motile by means of the persistent antherozoid flagellum, whereas in *Monoblepharis* the zygote is non-motile, remains in the oogonium, or it may migrate up and come to rest on the orifice of the oogonium. Fusion of the gametic nuclei is delayed in the zygote until the development of its outer wall has reached an advanced stage. Meiosis has not been observed, but it is presumed to occur during the first division of the diploid nucleus in preparation for germination, which occurs after a short or long dormant period by the formation of a hyphal germ tube.

So far as its sexual reproduction is concerned the Gonopodyaceae is the more primitive of the 2 families of this order. Reminiscent of the female gametes in species of the Blastocladiales, the oogonium of *Gonopodya* bears several eggs, and these may occasionally creep out and be fertilized at some distance from the oogonium as occurs in anisogamous planogametic reproduction. Furthermore, the zygote becomes motile for a while before encysting. In the Monoblepharidiaceae sexual reproduction is more clearly oogamous, as noted previously. A single egg is formed and fertilized in the oogonium, and the zygote does not become motile except to emerge to the orifice of the oogonium in some species.

Since the time of the discovery of *Monoblepharis* the phylogeny and relationships of this order have been the subject of conjecture. Cornu regarded *Monoblepharis* as a member of the Saprolegniaceae but suggested that it might have affinities with *Coelochaetae* and *Oedogonium*. This suggestion and the view that the Phycomycetes and other fungi might be degenerate algae received considerable attention by the early students of *Monoblepharis*. Lagerheim stressed the similarity of its sexual reproduction to that of *Oedogonium*, while Thaxter (1903) and Laibach emphasized its resemblance to that of *Vaucheria*. Atkinson (1909) on the other hand, rejected the concept that the Monoblepharidiaceae and other lower fungi originated from algal ancestors by a process of gradual degeneration. Since that time the discovery of anisogamous planogametic sexual reproduction in *Allomyces* and oogamy in *Gonopodya*, *Monoblepharella*, and *Monoblepharis* indicates a close relation of the Monoblepharidiales with the Blastocladiales. Sörgel (1952) in particular, has emphasized that the Blastocladiaceae has given rise in a direct line to the Monoblepharidales (see plate 165). His placement of *Gonopodya* at the top of the Monoblepharidaceae was made before sexual reproduction was discovered in this genus, and in light of present-day information it obviously belongs below *Monoblepharella* and *Monoblepharis* in a scheme of this nature.

GONOPODYACEAE

This is a small family of 2 genera, *Gonopodya* and *Monoblepharella*, and 7 or possibly a few more saprophytic species which occur on submerged fruits and twigs in freshwater and on organic materials in tropical soils. In the former genus the mycelium is constricted at fairly regular intervals and catenulate with pseudosepta and bears internally proliferous zoosporangia and gametangia, while in the latter it is very fine with reticulate, vacuolate scalariform protoplasm, lacks constriction and pseudosepta, and bears non-proliferating zoosporangia and gametangia. Another generic difference is the development of several eggs in the oogonium of

Plate 158

Figs. 1-32. *Monoblepharella taylori* Sparrow. Figs. 1, 6, 18-32 after Sparrow, 1940; figs. 2-5, 7-17 after Springer, 1945; figs. 24a, 29a after Sparrow, 1953.

Fig. 1. Portion of a hyphal tip with an apical incipient zoosporangium and an empty oogonium with 3 empty antherida beneath it.

Fig. 2. Portion of a mycelium showing vacuolate protoplasm and the angle of branching.

Fig. 3. Portion of a mycelium with an inflated area.

Figs. 4, 5. Two stages of the development of a zoosporangium.

Fig. 6. Discharge of planospores from a fusiform zoosporangium.

Fig. 7. Ovoid, 4.5-5 × 7-9 μ, planospore with anterior refractive globules.

Fig. 8. Stained planospore with a conspicuous nuclear cap and blepharoplast at the base of the flagellum.

Fig. 9. Stained planospore showing whiplash at the end of the flagellum.

Fig. 10. Germinating planospore.

Fig. 11. Planospores germinating within the zoosporangium.

Fig. 12. Method of the formation of a secondary zoosporangium.

Figs. 13-15. Stages in the formation of an oogonium.

Figs. 16, 17. Development and delimitation of the antheridium laterally from beneath the oogonium; refractive globules of fig. 15 have coalesced into numerous large ones.

Fig. 18. Mature oogonium with an egg, and the emergence of the last of 4 antherozoids.

Figs. 19, 20. Antherozoid entering orifice of the oogonium and its absorption by the egg.

Figs. 21, 22. Emergence of the zygote after fusion; antherozoid flagellum persistent.

Fig. 23. Emerged spherical, 8-11 μ diam., zygote with numerous large refractive globules.

Figs. 24-27. Amoeboid changes in shape of the zygote during temporary rest periods.

Fig. 24a. Fixed and stained binucleate zygote.

Fig. 28. Zygote at rest; antherozoid flagellum condensed into a small droplet.

Fig. 29. Mature encysted oospore, 8-11 μ diam., lying free in the water with a smooth light-brown and numerous large refractive globules.

Figs. 29a. Nearly mature binucleate oospore.

Fig. 30. Germination of a oospore; germ tube somewhat shortened in the copying.

Fig. 31. Oogonium bearing 6 eggs.

Fig. 32. Oogonia containing mature oospores.

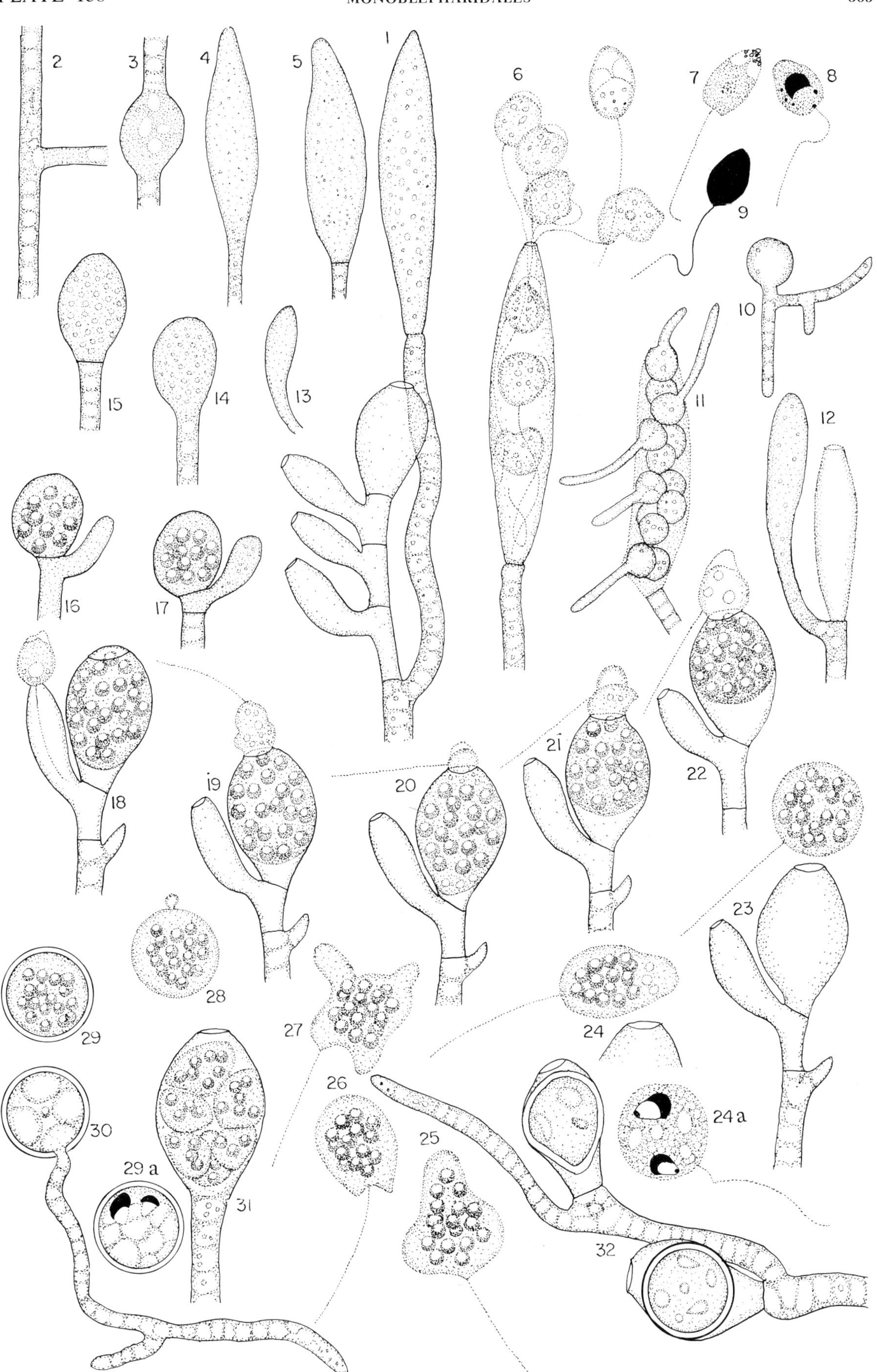

Monoblepharella

Gonopodya, and the formation typically of one egg in the oogonium of *Monoblepharella*.

Asexual reproduction occurs by the formation of posteriorly uniflagellate planospores in variously-shaped zoosporangia. As noted previously sexual reproduction is anisogamous or heterogamous and consists of the fusion of a small posteriorly uniflagellate antherozoid with a larger non-flagellate egg. In *Gonopodya* fusion usually occurs outside of the oogonium, and in *Monoblepharella* the antherozoid is partly absorbed by the egg while still in the oogonium, after which the fertilized zygote soon emerges to the outside. In both genera the zygote becomes motile outside of the oogonium by means of the persistent flagellum of the antherozoid, and after the motile period it encysts freely in the surrounding water and develops into an oospore. Germination is unknown in *Gonopodya*, but occurs by the development of a germ tube in *Monoblepharella*.

GONOPODYA Fischer

Rabenhorst. Kryptogamen-Fl. 1:382, 1892

Plate 157

This small genus includes 2 long-known species, *G. prolifera* (Cornu) Fischer and *G. polymorpha* Thaxter, and possibly a third, *G. bohemica* Cejp, which is interpreted to be a common form of *G. prolifera*. These species occur on submerged fruits of various kinds and twigs of evergreen and deciduous trees in fresh water and apparently are worldwide in distribution. *Gonopodya prolifera* appears to be the most common and variable species and forms small white pustules on submerged fruits under quite foul environmental conditions. Species distinctions are not always well-defined because the thalli, zoosporangia, and gametangia may vary markedly. The zoosporangia of *G. prolifera*, for

PLATE 159

Figs. 33-36. *Monoblepharella* spp. Fig. 33 after Koch, 1964; figs. 34-36 sketches from electron micrographs by Fuller and Reichle, 1968.

Fig. 33. Diagram of a representative monoblepharidian planospore with apical vacuoles and globules, conspicuous nuclear cap, nucleus and nucleolus.

Fig. 34. Longitudinal section of a planospore of *Monoblepharella* sp. C = annular cisternae; F = flagellum; K = kinetosome; L = lipid globules; M = mitochondrion; NU = nucleus; RR = ribosomal region traversed by endoplastic reticulum; RU = rumposome.

Figs. 35, 36. Portions of the rumposome in longitudinal and cross sections, resp.

Figs. 37-47. *Monoblepharella mexicana* Shanor. (Shanor, 1942.)

Fig. 37. Swimming ovoid, 4-5 × 6.6-10 μ, planospore with refractive globules at the anterior end.

Fig. 38. Portion of the fine mycelium showing terminal zoosporangia and a succession of oogonia and antheridia.

Fig. 39. Discharge of planospores from a zoosporangium with an apical orifice and a lateral unopened discharge tube; branching of hypha almost at right angles.

Fig. 40. Discharge of posteriorly uniflagellate antherozoids; one antherozoid at the orifice of the oogonium which bears 1 egg.

Fig. 41. Antherozoid almost completely absorbed by the egg.

Fig. 42. Uniflagellate zygote emerging from the oogonium.

Figs. 43, 44. Swimming and amoeboid posteriorly uniflagellate zygotes, resp.

Fig. 45. Mature spherical, 10-13 μ diam. oospore free in the water; wall smooth, amber to light-brown.

Fig. 46. Mature oospore formed within the oogonium in a badly contaminated culture.

Fig. 47. Oospore formed partly outside and within the oogonium in a contaminated culture.

Figs. 48-59. *Monoblepharella elongata* Springer. (Springer, 1945.)

Fig. 48. An elongate, siliquiform, 5-13 × 40-120 μ, zoosporangium with 3 lateral papillae.

Fig. 49. Oblong, 4-6 × 6.5-10.5 μ, planospore with an anterior group of refractive globules.

Fig. 50. Discharge of planospores from a short zoosporangium.

Fig. 51. Oogonium with a sublateral antheridial rudiment below.

Fig. 52. An elongate, cylindrical papillate antheridium and an oogonium with 1 egg.

Fig. 53. A large oogonium with 8 eggs.

Fig. 54. Discharge of antherozoids.

Fig. 55. Ovoid, 3-4 × 5.6 μ, antherozoid.

Fig. 56. An antherozoid in the orifice of the oogonium.

Fig. 57. Emerging posteriorly uniflagellate zygote.

Fig. 58. Mature, spherical, 9-12 μ diam., oospore with a light-brown smooth wall.

Fig. 59. Germination of an oospore.

Figs. 60-66. *Monoblepharella laruei* Springer. (Springer, 1945.)

Fig. 60. Discharge of planospores; secondary zoosporangium at the left.

Fig. 61. Elongate, 5-5.5 × 7-9 μ, planospore with an anterior group of small refractive globules.

Figs. 62, 63. Early stages in the development of a terminal or epigynous antheridium.

Fig. 64. Oogonial rudiment developing beneath the incipient antheridium as a lateral outgrowth.

Figs. 65, 66. Antheridium lateral to the oogonia.

Figs. 67-70. *Monoblepharella endogena* Sparrow. (Sparrow, 1953.)

Fig. 67. Discharge of planospores.

Fig. 68. Oogonium with 4 eggs.

Fig. 69. Portion of a hypha with a mature endogenous oospore and an immature oogonium.

Fig. 70. Apical portion of a hypha bearing a mature endogenous oospore and an oogonium.

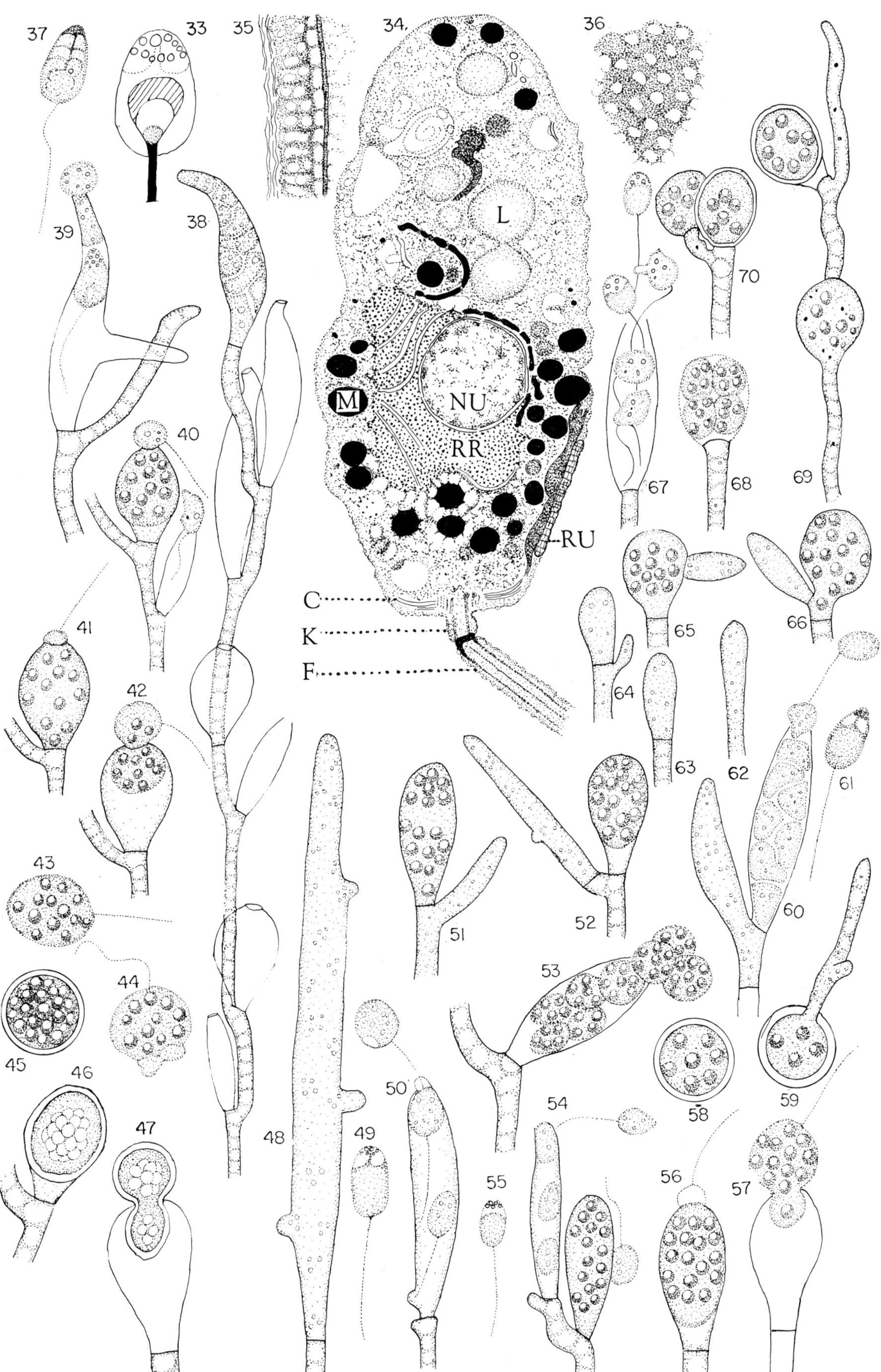

Monoblepharella

example, may be very similar at times to those of *G. polymorpha*. These two species, nonetheless, are distinguished primarily by differences in the shapes of the female gametangia, the numbers of discharge papillae on them, (usually one in *G. prolifera* and 1 to 4 in *G. polymorpha*), and the relative distinctness of the mycelial segments and constrictions.

The thallus or mycelium is usually profuse, irregularly or dichotomously branched and anchored apparently to the substratum by rhizoids. It is composed of narrowly or broadly fusiform, subcylindrical, catenulate or monilioform segments which are delimited by hyaline constricted pseudosepta (figs. 1, 17). In *G. prolifera* the segmentation is more distinct than in *G. polymorpha*, and the overall lengths of the thalli may vary from 200–500 μ in the former and 200–1000 μ in the latter species. The zoosporangia and gametangia are borne at the apex of the branches and may proliferate once to several times (figs. 4, 7, 9). In *G. prolifera* the zoosporangia are clavate, pod-shaped 22 × 130 μ, or sometimes 200 to 250 μ long (figs. 1, 2, 4) usually with an elongate tapered apex, while in *G. polymorpha* the zoosporangia are usually broadly ovoid, 12–30 × 20–60 μ, and taper abruptly with a blunt tip (figs. 17, 18). Up to 50 or more posteriorly uniflagellate, ovoid to elongate planospores (figs. 5, 6, 18) are produced in a zoosporangium, and these emerge as the apex of the latter deliquesces. They are reported to have the same internal structure and organization as those of *Monoblepharis*.

According to Petersen (1909) the planospores of *G. polymorpha* may be transformed into resting spores within the zoosporangium, but his fig. XXV may possibly relate to parthenogenetic oospores, or fertilized oospores which did not emerge. The oospores germinate to reproduce the thallus, whereas the unfused male and female gametes are said to be incapable of germination.

Sexual reproduction in *Gonopodya* is anisogamous or heterogamous and consists of the fusion of a relatively small posteriorly uniflagellate male gamete (antherozoid) and a larger non-flagellate female gamete (egg), usually outside of the gametangia. The latter structures are borne on the same thallus as the zoosporangia, according to Benjamin (1958). In *G. prolifera* he found that thalli bearing gametangia may originate from proliferated zoosporangia, which "provides circumstantial evidence that both sexual and asexual reproductive structures are, indeed, produced by the same vegetative thallus...." However, it is not known what factors operate to determine what type of reproductive organs will be formed on the thallus. Sparrow (1960) suggested that temperature may be a determining factor as in the cases of *Monoblepharella* and *Monoblepharis*.

The female gametangia in *G. prolifera* are elongately ovoid (figs. 8, 9, 11), inflated at the base, and tapering towards the apex, and vary from 25 to 35 μ in greatest diameter by 60 to 180 μ in length. The male gametangia are slenderer and vary from 10 to 15 μ in diameter by 40 to 107 μ in length. The latter bear up to 40 posteriorly uniflagellate gametes (figs. 12, 23) which emerge as the discharge papillae deliquesce. The male gametes or antherozoids are usually elliptical or slightly elongate with a conspicuous cluster of refractive granules at the anterior end. The female gametangium bears up to 20 large nonflagellate gametes or eggs (fig. 22), and these contain numerous minute refractive granules which give them a brownish color. They usually creep out of the gametangium (fig.

Plate 160

Figs. 1-9. *Monoblepharis sphaerica* Cornu. Figs. 1-5 after Cornu, 1872; figs. 6-9 after Perrott, 1955.

Fig. 1. Young oogonium developing at the tip of a hypha.

Fig. 2. Later stage; oogonium has been delimited by a cross wall; hypogynous antheridium developing beneath.

Fig. 3. Mature oogonium with the contracted egg; hypogynous antheridium with antherozoids beneath.

Fig. 4. Discharge of antherozoids, two are creeping up the side of the oogonium.

Fig. 5. Mature bullate, endogenous oospore; hypha branching beneath the empty antheridium.

Fig. 6. Sympodially branched hypha with an antheridium and a mature bullate oospore in the oogonium.

Fig. 7. Typical arrangement of the oogonia and antheridia.

Fig. 8. Two oogonia and an antheridium borne side by side at the hyphal tip.

Fig. 9. Uncommon arrangement of oogonia and antheridia.

Figs. 10-25. *Monoblepharis hypogyna* Perrott. Figs. 10-16 after Perrott, 1955; figs. 17-25 after Cornu, 1872.

Fig. 10. Oogonial rudiment with numerous refractive globules.

Fig. 11. Oogonium delimited by a cross wall; hypogynous antheridium developing below.

Fig. 12. Fusiform oogonium with an exogenous bullate oospore.

Fig. 13. Sympodially branched hypha bearing 2 sets of reproductive organs.

Fig. 14. Hypha with 3 sets of reproductive organs; oogonia bearing sparsely bullate endogenous oospores.

Fig. 15. Hypha bearing 2 superimposed oogonia, each with an endogenous oospore.

Fig. 16. Partly emerged bullate oospore.

Fig. 17. Enlarged drawing of an antherozoid.

Figs. 18-21. Discharged and creeping antherozoids, fertilization, emergence of oospore, and a mature exogenous bullate oospore, resp.; branches developing below the antheridium.

Fig. 22. Upper portion of a hypha with empty hypogenous antheridia and 2 oogonia with exogenous bullate oospores; 1 oogonium with an immature oospore.

Fig. 23. Three endogenous partly bullate oospore (chamydospores, parthenogenic oospores?.)

Fig. 24. Partly emerged smooth-walled ospore.

Fig. 25. Section through a portion of an oospore showing the outer thick bullate wall, a thin inner wall, and a portion of the protoplasm.

Monoblepharis

21) as amoeboid bodies and each fuses with the smaller male gamete. In this process the latter creeps over the surface of the egg (fig. 24) and may remain as a distinct entity for some time before fusion. Its flagellum is retained, and as a result the zygote becomes posteriorly uniflagellate. At first the zygote is amoeboid in motion (figs. 26, 27) with a hyaline anterior pseudopodium, but later it becomes actively motile or reverts intermittently to the amoeboid condition. Eventually it encysts and becomes a smooth-walled spherical resting oospore which is surrounded by a hyaline envelope in *G. prolifera* (fig. 16).

Sometimes the antherozoid may enter the female gametangium (fig. 25) and fuse with an egg, and in some such cases the fusion is incomplete. In the latter event the egg may emerge with a part of the antherozoid visible on its surface, and in the former event the zygote may encyst within the female gametangium (figs. 15, 28). So far germination of the encysted zygote or resting spore has not been observed, and nothing is known as to where karyogamy and meiosis occur in the life cycle of *Gonopodya*. The only cytological study so far reported was made by Laibach (1927), who showed that the planospore initials were uninucleate after cleavage in the zoosporangium. Presumably, the gametes are uninucleate also and formed in the same manner.

The type species, *G. prolifera*, and *Monoblepharis* were first regarded by Cornu as members of the Saprolegniaceae but Fischer (1892), Schroeter (1893), Minden (1915), and Sparrow (1943) later placed them in a separate order, the Monoblepharidales. Fischer established *Gonopodya* as a separate genus in the family Monoblepharidaceae where it remained until Peterson (1909) established the family Gonopodyaceae for it, a family recognized later by Cejp (1957) and Sparrow (1960).

MONOBLEPHARELLA Sparrow

Allen Hancock Pacific Expeditions, Publ. Univ. S. Calif. 3:103, 1940

Plates 158, 159

Monoblepharella is a small genus of 5 species which have been collected and isolated from tropical and subtropical soils in the warmer regions of the Western Hemisphere. Two other imperfectly known species,

PLATE 161

Figs. 26-42, 60. *Monoblepharis polymorpha* Cornu. Figs. 26-29, 31-35, 37-42 after Sparrow, 1933; fig. 30 after Perrott, 1955; figs. 36, 36a after Cornu, 1872.

Fig. 26. Fruiting tip of a hypha with bullate and undulate exogenous oospores on the empty oogonia and empty antheridia which originiated epigynously.

Fig. 27. Motile citriform, 7.8-10.4 × 10-13 μ planospore with anterior globules.

Fig. 28. Stained planospore showing internal structure.

Fig. 29. Insertion of the flagellum.

Fig. 30. Elongate filamentous zoosporangium with planospore initials.

Fig. 31. Discharge of planospores.

Figs. 32-34. Developmental stages of the epigynous antheridium and oogonium beneath it.

Fig. 35. Mature oogonium whose development has carried the antheridium upward and sublaterally; antherozoids in the antheridium.

Fig. 36. Mature oogonium and an almost empty antheridium.

Fig. 36A. Amoeboid antherozoid.

Fig. 37. Tapering, approx. 2.6 × 5.2 μ, antherozoid with apical granules.

Figs. 38, 39. Antherozoids entering orifice of the oogonium and fusing with the apex of the egg; flagellum still visible.

Fig. 40. Beginning of emergence of the zygote; protoplasm of the antherozoid still visible at the left.

Fig. 41. Later stage; protoplasm of antherozoid still distinguishable.

Fig. 42. Mature bullate oospore resting on a collar at the orifice of the oognoium.

Figs. 43-61. *Monoblepharis polymorpha* (*M. brachyandra*) Woronin. (Lagerheim, 1900.)

Figs. 43, 44. Gemmae.

Fig. 45. Mature oogonium with an egg and a short epigynous antheridium with antherozoid initials.

Fig. 46. Antherozoid at the orifice of the oogonium.

Fig. 47. Entrance of an antherozoid; an additional antherozoid in the axil of the antheridium and oogonium.

Fig. 48. Initial stage of the emergence of the zygote; additional antherozoid adjacent.

Fig. 48a. Enlarged view of an antherozoid.

Fig. 49. Incipient oospore at the orifice of the oogonium; the additional antherozoid has entered the oogonium and lies in its base.

Fig. 50. Mature bullate oospore.

Fig. 51. Endogenous chlamydospores (parthenogenic oospores?.)

Fig. 52. Line drawing of the fertile apex of a hypha.

Fig. 53. Pairing of gametic nuclei in the oospore.

Fig. 54-56. Fused diploid nuclei.

Fig. 57. Germination of oospore.

Fig. 58. Intercalary zoosporangium discharging planospores.

Fig. 59. Enlarged view of a planospore.

Fig. 60. Germination of a planospore.

Fig. 61. Lower portion of a hypha with rhizoids.

Figs. 62, 63. *Monoblepharis polymorpha* (*M. brachyandra* var. *longicollis*) Lagerheim. (Lagerheim, 1900.)

Fig. 62. Line drawing of the fertile portion of a hypha with a long stalk and short blunt antheridia.

Fig. 63. Mature oospore.

Fig. 64. Fruiting tip of *M. polymorpha* with a short blunt antheridium and an oospore at the orifice of the oogonium; "a combination of *polymorpha* and *brachyandra* characters."

Monoblepharis

Monoblepharis reginens Lagerheim and *M. ovigera* Lagerheim, from Sweden, may possibly be referred to this genus when their type of sexual reproduction becomes known. The growth rate of the other species is unusually slow, and after being collected a month or more is required before the species form colonies of appreciable size. At this stage they are readily recognizable as a lustrous pearl gray halo around the substratum. Species distinctions are based on the sequence of development and relative positions of the gametangia and the length and shapes of the gametangia as well as the diameter and branching of the mycelium, but these distinctions may vary so that some of the species overlap in these respects.

The thin-walled non-septate mycelium and hyphae of the various species are usually very long, fine, delicate and of a diameter of 1 to 7.5 μ, with 2 to 3 μ being most typical. At irregular intervals globose or spindle-shaped swellings (fig. 3) usually occur in the mycelium, and the number and size of these vary in the different species. In *M. taylori* Sparrow and *M. elongata* Springer they may become quite large, 17–17 × 20–35 μ. Branching occurs at right or almost at right angles to the hyphae (figs. 2, 39), and this may be relatively abundant or sparse in the different species. As is characteristic of the Monoblepharidaceae, the protoplasm in the hyphae has a distinctive regular, scalariform, or reticulate and vacuolate appearance (figs. 1, 32, 38), and in vigorously growing hyphae it appears extremely active under high magnifications.

Zoosporangia develop abundantly at temperatures of 13–36°C, and the primary ones are borne at the tips of the branches. The secondary zoosporangia usually develop sympodially. They are delimited from the hyphae by cross walls (figs. 1, 6, 38, 39, 48), and vary considerably in shape and size. Generally, they are siliquiform, subcylindrical and narrowly or broadly fusiform with thin walls and vary from 18 to 120 μ in length by 5 to 20 μ in greatest diameter among the different species. They may develop 1 to 4 papillae (fig. 48), but in planospore discharge only one is usually functional.

The planospores are fully delimited in one to several rows in the zoosporangia (figs. 38, 60) and emerge singly. They are ovoid, subcylindrical or oblong in shape and vary from 4 to 6 μ in diameter by 8 to 10.5 μ in length with a cluster of refractive granules in the anterior end (figs. 6–8, 37, 49, 61). From figures and descriptions in the literature Koch (1961) diagrammed them as having a well-defined nuclear cap (fig. 33) surrounded by a membrane. However, Fuller and Reichle's (1968) ultrastructural studies revealed that the planospores do not have a nuclear cap of the type found in the Blastocladiales. Instead, the nucleus is surrounded by an area or concentration of ribosomes (fig. 34 RR) which is partly surrounded and traversed by membranous elements (endoplasmic reticulum?). Furthermore,

PLATE 162

Figs. 65-75. *Monoblepharis insignis* var. *insignis* Thaxter. Figs. 65-72 after Thaxter, 1895; figs. 73-75 after Perrott, 1955.

Fig. 65. Tip of a fertile hypha with epigynously or subapically attached antheridia, and oogonia with smooth-walled endogenous oospores.

Fig. 66. Incipient apical antheridium delimited by a cross wall.

Fig. 67. Undehisced antheridium and an oogonium; secondary oogonium forming below.

Fig. 68. Closed oogonium over which antherozoids are creeping.

Fig. 69. Same oogonium 10 minutes later; oogonium discharging finely granular protoplasm at its apex.

Fig. 70. Empty antheridium and an oogonium with an egg and 4 antherozoids; 1 antherozoid attached to the egg.

Fig. 71. A fertilized egg; zoosporangium (?) with planospores beneath.

Fig. 72. Oogonium from which 2 spherical masses of granular protoplasm have escaped; 2 antherozoids creeping over the large mass; 1 antherozoid fusing with the egg in the oogonium.

Fig. 73. Three oogonia with mature oospores and 3 antheridia in basipetal succession.

Fig. 74. Oogonium with an epigynous antheridium borne on a short branch lateral to the oogonium.

Fig. 75. A sympodially branched hyphal tip with an oogonium and antheridium.

Figs. 76-82. *Monoblepharis insignis* var. *minor* Perrott. (Perrott, 1955.)

Fig. 76. "A chain of superimposed oogonia; oogonia bearing smooth-walled endogenous oospores and epigynous antheridia."

Fig. 77. A sympodially branched hypha bearing oogonia with oospores and antheridia.

Fig. 78. A similarly branched hypha; epigynous antheridium with 2 antherozoids.

Fig. 79. Young oogonia rich in large oil globules; 1 antheridium with antherozoids.

Fig. 80. Unusually elongate oogonium with an elongate oospore and epigynous antheridium.

Fig. 81. Three intercalary oogonia with oospores.

Fig. 82. An oogonium with a closed inflated tip.

Figs. 83-88. *Monoblepharis fasiculata* var. *fasiculata* Thaxter. Figs. 83-85 after Thaxter, 1895; figs. 86-88 after Perrott, 1955.

Fig. 83. Tip of a fertile hypha with a fasicle of antheridia and oogonia with smooth-walled endogenous oospores.

Fig. 84. Unopened oogonium and an antheridium with antherozoid initials; branch forming beneath.

Fig. 85. An oogonium with a mature oospore.

Fig. 86. A fasicle of epigynous antheridia and oogonia with oospores.

Fig. 87. Hyphal tip with an immature oogonium, an epigynous antheridium with antherozoids, and an oogonium with a mature oospore.

Fig. 88. A single oogonium from a fasicle with an open inflated tip bearing a small oospore, and a larger oospore in the base.

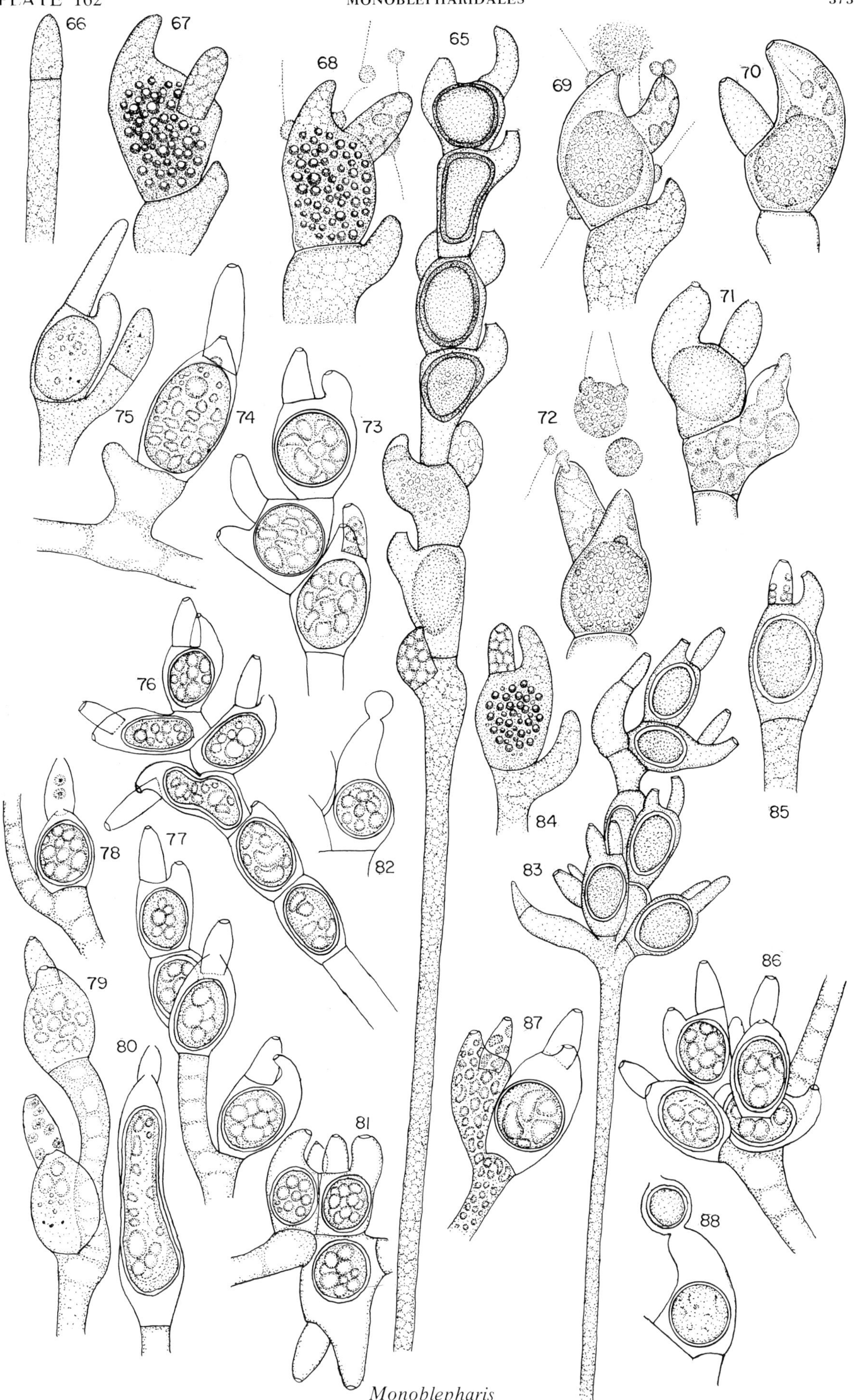

Monoblepharis

the nucleus is separated from the functional kinetosome, and a second or accessory one is present according to Reichle (1972). At one side of the planospore occurs a body composed of interconnecting tubules which Fuller (1966) called the rumposome (fig. 34, RU). Highly magnified longitudinal and cross sections of it are shown in figs. 35 and 36. Although Fuller and Reichle ascribed no particular function to it, they believed that it may be similar and comparable to the "fibrous body" found by Chambers *et al* (1967) in *Nowakowskiella profusa* (see figs. 38, 39, pl. 2).

Sexual reproduction in *Monoblepharella* is heterogamous and consists of the fusion of small posterior uniflagellate antherozoids and large non-flagellate eggs. The antheridia and oogonia may be borne on the same thallus, usually in close association (figs. 18–23, 40–42, 54, 64–66). Temperature apparently plays an important role in their formation and differentiation. Springer (1945b) found that they are formed at temperatures of 26–32°C. Also, if the environment or conditions in which a culture is growing are changed prior to the development of a delimiting wall, oogonial and antheridial rudiments may cease development and differentiation as such and become transformed into zoosporangia, indicating that such rudiments may be facultative in the early stages.

No antheridia have been found in *M. endogena* Sparrow, and the oospores are apparently formed asexually and endogenously (figs. 68–70) in the oogonium. In *M. taylori* and *M. elongata* the first gametangium formed in a pair is the oogonium, and the antheridium develops hypogenously or terminally on a short branch subtending the oogonium (figs. 16–22, 51, 52, 54). In *M. mexicana* Shanor the antheridium is formed first, and the oogonium develops terminally on a short branch subtending the antheridium. But in *M. laruei* Springer the oogonium frequently develops beneath the antheridium and later becomes epigynous (figs. 64–66). The antheridium is usually narrowly or broadly fusiform in most species, but in *M. elongata* it may be up to 10–35 μ in greatest diameter and bear lateral papillae (fig. 52) as in *M. mexicana*. The oogonium varies from clavate to broadly and narrowly obpyriform and subspherical in shape with a tapering base and rounded apex (figs. 16–22, 40–42, 51–57, 65, 66, 69, 70) which lacks a conspicuous papilla. At maturity its clear protoplasm contains numerous large refractive globules which give the oogonium a glistening appearance. Generally, a single egg is formed in the oogonium, but sometimes its content may cleave into 4, 6, and 8 eggs (figs. 31, 53, 68). The antheridium bears 2 to 8 posteriorly uniflagellate, ovoid, 2.5–4 × 3.4–6 μ, antherozoids (fig. 55) which apparently have the same structure as the planospores.

The antherozoids emerge singly as the tip of the antheridium deliquesces (figs. 50, 54) and may become actively motile or strongly amoeboid. One of these moves toward the orifice of the oogonium (figs. 19, 40), and is then incompletely engulfed by the egg during fertilization (figs. 20, 41, 56). The fertilized zygote emerges very shortly (figs. 22, 42, 57) and swims away by means of the persistent flagellum of the antherozoid (figs. 23, 24, 43). At this stage it is broadly ovoid, 8–19 × 10.5–13.6 μ, to nearly spherical, 7–10 μ diam. and binucleate (fig. 24a). Occasionally, it may encyst

PLATE 163

Figs. 89, 90. *Monoblepharis fasiculata* var. *magna* Perrott. (Perrott, 1955.)

Fig. 89. "Tip of a sympodially branched hypha bearing three oogonia arranged in a fasicle"; each oogonium with an epigynous antheridium; 2 oogonia bearing smooth-walled oospores; 3rd oogonium filled with protoplasm and globules.

Fig. 90. "A hypha with sexual reproductive organs arranged singly at the tips of the sympodially branched hypha."

Figs. 91-97. *Monoblepharis bullata* Perrott. (Perrott, 1955.)

Fig. 91. "A hyphal tip showing a very common arrangement of reproductive organs, i.e., a series of superimposed oogonia" with "one of the lateral walls attached to the base of the previously formed oogonium above; endogenous oospores bullated."

Fig. 92. Another common arrangement of oogonia with epigynous antheridia.

Fig. 93. A hypha with an antheridium and the rudiment of an oogonium forming beneath.

Fig. 94. A young oogonium filled with refractive globules, epigynous antheridium with 3 antherozoids.

Fig. 95. "A hypha with two fasicles of reporductive organs," apical oogonia filled with refractive globules.

Fig. 96. Apical oogonium with a greatly inflated tip bearing an oospore.

Fig. 97. "A series of three unusual oogonia." Inflated tip of one oogonium bearing a small oospore.

Figs. 98, 99. *Monoblepharis regignens* Lagerheim proliferating zoosporangia. (Lagerheim, 1900.)

Figs. 100-107. *Monoblepharis ovigera* Lagerheim. Figs. 100-102 after Lagerheim, 1900; figs. 103-106 after Sparrow, 1933; fig. 107 after Beneke, 1948.

Fig. 100. Terminal zoosporangium discharging planospores.

Fig. 101. Terminal and lateral intercalary zoosporangia.

Fig. 102. Ovoid planospore.

Fig. 103. Immature terminal zoosporangium.

Fig. 104. Zoosporangium with planospores.

Fig. 105. Sex organs of a *Monoblepharis* associated with the zoosporangia of *M. ovigera*.

Fig. 106. "One of many similar free-floating resting bodies found near the structures shown in" fig. 105.

Fig. 107. Portion of a thallus with a group of zoosporangia which are characteristic of *M. ovigera*.

Figs. 108-112. *Monoblepharis ovigera* (?) (*Monoblepharopsis elongata* Höhnk.) (Höhnk, 1935.)

Figs. 108, 109. Proliferating zoosporangia.

Fig. 110. Elongate zoosporangium with planospore initials.

Figs. 111, 112. Discharge of planospores. Planospores in fig. 112, 5 × 8 μ.

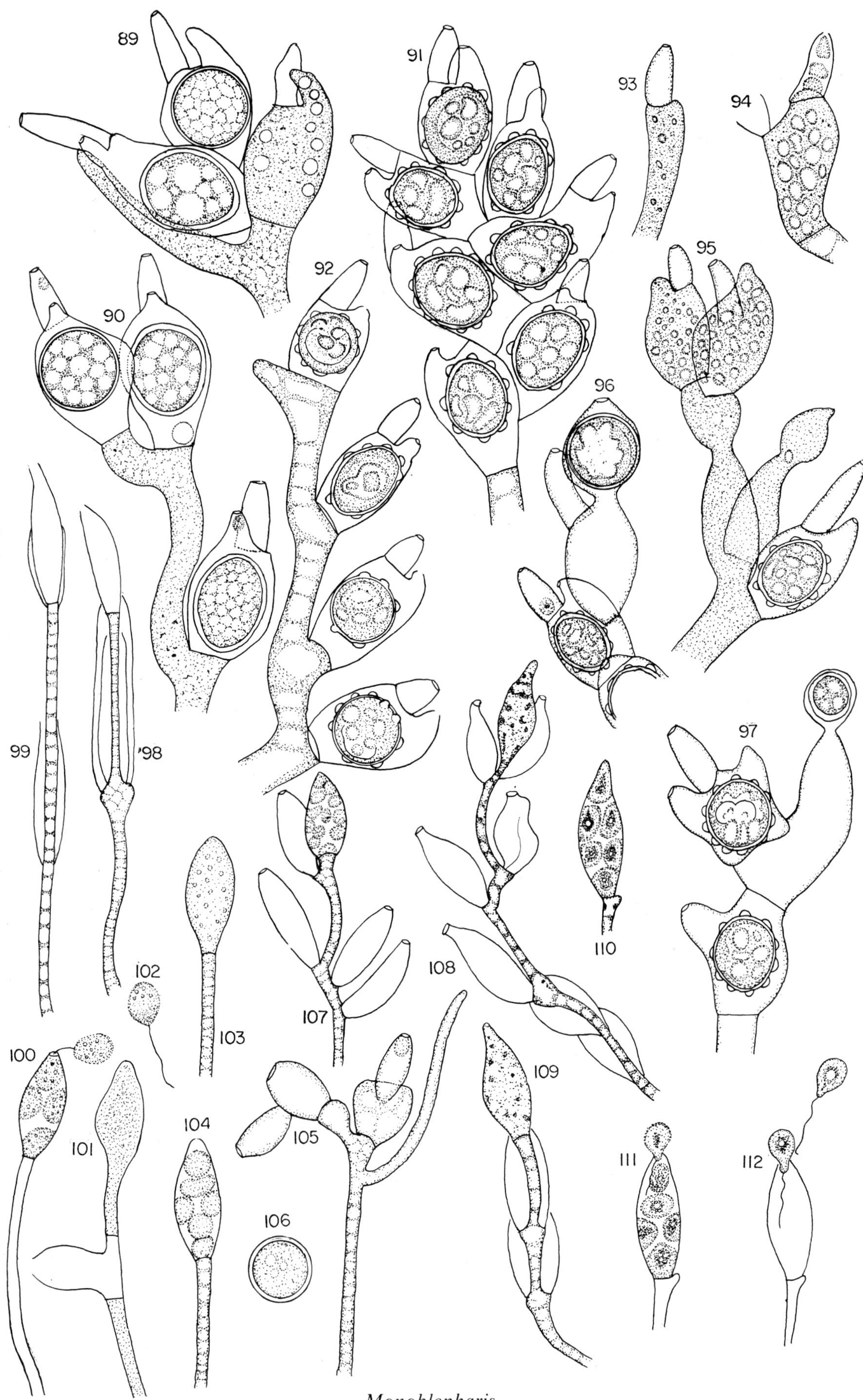

Monoblepharis

at the orifice of the oogonium, or in contaminated cultures it may encyst within or partly so in the oogonium (figs. 46, 47). During the motile period the zygote may become intermittently amoeboid (figs. 25–27, 44). Eventually, it comes to rest in the surrounding water and develops into a spherical, 7.5–13 μ diam., oospore with an amber to light brown wall (figs. 29, 45, 58) and numerous refractive globules within its protoplasm. The gametic nuclei remain separate (fig. 29a) while the wall is developing, and karyogamy is delayed until an advanced stage of wall formation as in *Monoblepharis*. Germination occurs by the formation of a tube (figs. 30, 59) from which the mycelium develops.

MONOBLEPHARIDACEAE

This is a small family of 1 or possibly 2 genera, approximately 10 species, and a few varieties. Three of the species, *M. regignens, M. ovigera,* and *Monoblepharopsis elongata* may prove to be members of *Monoblepharella*, or representatives of the genus *Monoblephariopsis* created by Laibach. Members of this family are characterized by a non-septate, sparingly or profusely and sympodially branched or unbranched mycelium which is anchored to the substratum by rhizoids and bears zoosporangia, antheridia, and oogonia at or along its apical portion, usually on the same thallus or hyphae. Characteristically, the protoplasm of the hyphae has a foamy, reticulate, scalariform appearance which makes recognition of the species relatively easy in the vegetative or non-reproductive state.

Sexual reproduction is oogamous and occurs by the fusion of a small posteriorly uniflagellate antherozoid with a large non-flagellate egg within the oogonium. Unlike species of *Gonopodya* only one egg is produced in the oogonium, and except for very rare instances the zygote does not escape from the oogonium nor does it become amoeboid or motile as in the Gonopodyaceae. However, the zygote or fertilized egg may emerge and come to rest on the orifice of the oogonium in some species, but in others it remains within the oogonium. Karyogamy is delayed until wall formation of the oospore reaches an advanced stage. The mature bullate or smooth oospore thus formed germinates by a

PLATE 164

Figs. 113-144. *Monoblepharis macandra* var. *macandra* Lagerheim. Fig. 113, 119, 120 after Sparrow, 1933; figs. 114-116, 118, 123-130, 133-139 after Laibach, 1927; figs. 117, 121, 122, 132, 140, 141 after Woronin, 1904; fig. 131 after Lagerheim, 1900; figs. 142-144 after Perrott, 1955.

Fig. 113. Fruiting tip of a hypha with 2 exogenous oospores, empty oogonia, an empty zoosporangium, 2 empty antheridia and 1 filled with antherozoids.

Fig. 114. Young zoosporangium.

Fig. 115. Later multinucleate stage.

Fig. 116. Delimited zoosporangium with content cleaving into planospore initials.

Fig. 117. Proliferating zoosporangium.

Fig. 118. Intercalary zoosporangium.

Fig. 119. Cluster of empty elongate cylindrical zoosporangia.

Fig. 120. Discharge of planospores, some of which are clustered at the exit orifice. others pulling away.

Fig. 121. Planospore.

Fig. 122. Germination of planospores within the zoosporangium.

Figs. 123-125. Developmental stages of a multinucleate antheridium and cleavage into antherozoids.

Figs. 126, 127. Developmental stages of a uninucleate oogonium.

Fig. 128. Nucleus has migrated to the apex of the oogonium.

Fig. 129. Fertilization; antherozoid has fused with the apex of the egg cell; gametic nuclei close together.

Fig. 130. Early stage of the emergence of the binucleate zygote.

Fig. 131. Rare instance of the emergence of the egg with the attached flagellum of the antherozoid.

Fig. 132. Rare instance of a emerged egg; presumably to be fertilized outside of the oogonium, according to Woronin.

Fig. 133. Gametic nuclei close together as the wall of the oospore has formed.

Fig. 134. Karyogamy; wall of the oospore in an advanced stage of development.

Fig. 135. Diploid nucleus.

Fig. 136. Later stage; diploid nucleus lying in the center of a mature oospore.

Fig. 137. Small daughter nuclei from the first division of the diploid nucleus in preparation for germination.

Fig. 138. Multinucleate oospore shortly before germination.

Fig. 139. Small germinated oospore; some nuclei have migrated out into the germ hypha.

Fig. 140. Tip of a hypha with 3 empty antheridia.

Fig. 141. Tip of a hypha with 5 empty superimposed oogonia, 2 empty zoosporangia, and apparently 1 empty antheridium.

Fig. 142. Hyphal tip branching sympodially; chlamydospore or a parthenogenetic oospore (?) in one oogonium.

Fig. 143. Cylindrical oogonium with an ovoid smooth-walled oospore and from which a normal oogonium with a bullate exogenous oospore arises.

Fig. 144. Long terminal oogonium with a smooth-walled endogenous parthenogenetic oospore.

Figs. 145-147. *Monoblepharis macandra* var. *laevis* Sparrow. Fig. 145, 146 after Sparrow, 1933; figs. 147 after Perrott, 1955.

Fig. 145. Hyphal tip with an oogonium bearing a smooth-walled exogenous oospore, an antheridium, and a branch.

Fig. 146. Hyphal tip bearing 3 antheridia and 2 oogonia with smooth-walled exogenous oospores.

Fig. 147. Hyphal tip bearing a series of 2 oogonia and 4 antheridia; 1 oogonium with a smooth-walled endogenous oospore.

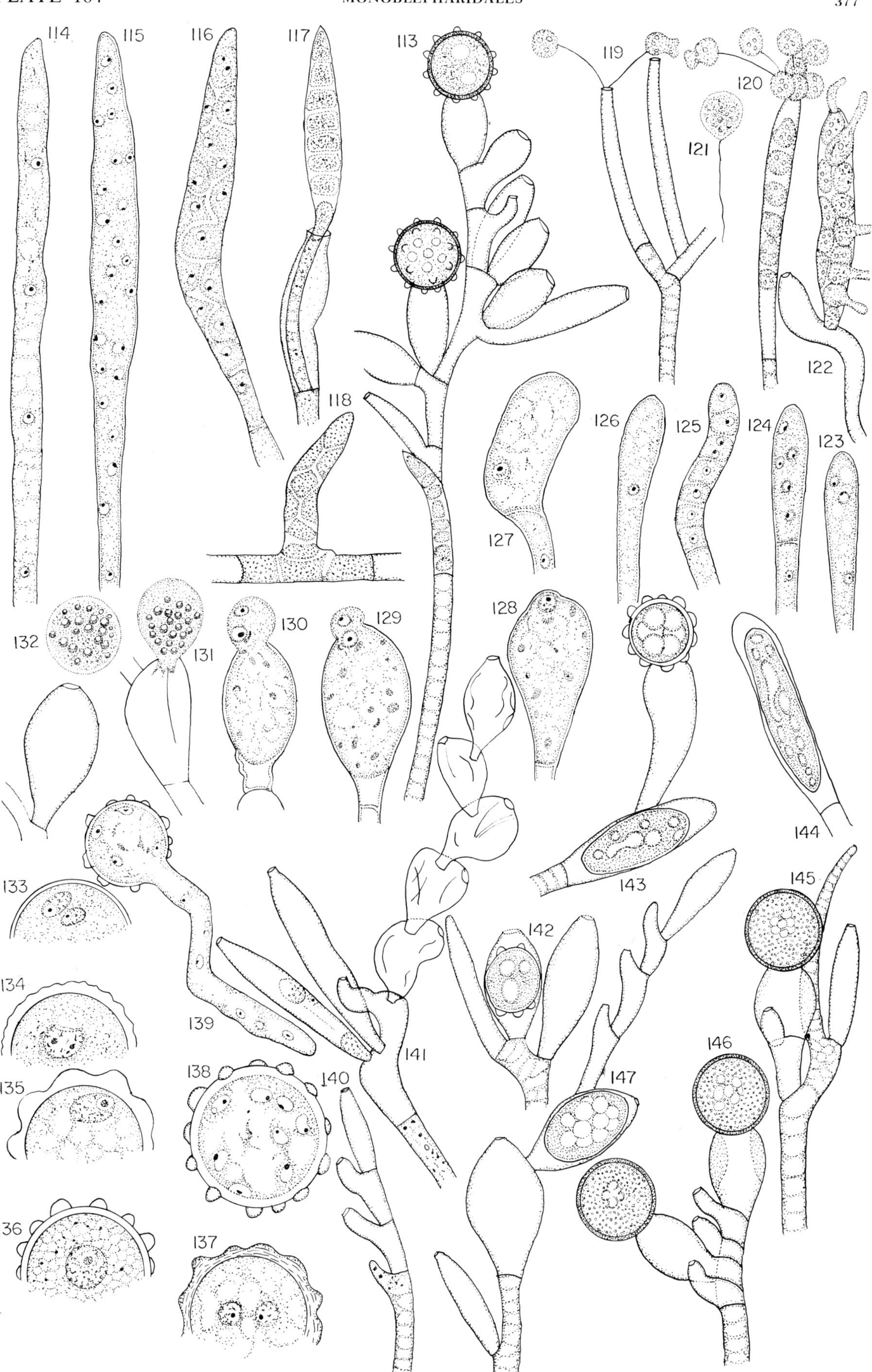

Monoblepharis

hypha which reestablishes the mycelium. Meiosis has not been observed, but it is presumed to occur during the first division of the diploid nucleus in preparation for germination of the oospore.

Myrioblepharis Thaxter is rejected as a genus of this family. Minden (1915) believed that it is a *Pythium* species parasitized by a ciliate protozoan, and this view was confirmed by Waterhouse (1945) who proved that it is an association of a ciliate with proliferating zoosporangia of *Pythium* and *Phytophthora*. Miller (1963) and Koch (1964) have subsequently confirmed Waterhouse's observations.

MONOBLEPHARIS Cornu

Bull. Soc. bot. France 18:59, 1871

Diblepharis Lagerheim, 1900. Bih. Kgl. Svensk. Vetensk-AK. Handl. 25, afd 3:39
Monoblephariopsis Laibach, 1927. Jahrb. wiss. Bot. 66: 603

Plates 160-164

This genus is reported to include approximately 10 species and a few varieties, but it is quite possible that 2 of these species, *M. regignens* Lagerheim and *M. ovigera* Lagerheim may prove to be members of *Monoblepharella* when their method of sexual reproduction has been discovered and verified. Also, *Monoblephariopsis elongata* Höhnk appears to be identical with *M. ovigera*, and *Monoblepharis lateralis* Hine is probably a saprolegniaceous species parasitized by a flagellate. These species are saprophytes on vegetable debris, particularly twigs, and animal remains in cold freshwater and may grow vigorously in temperatures as low as 3° C. Some of the species have been found by Perrott (1960) to have a distinct periodicity in nature. The oospores germinate in early spring and produce sex organs and oospores shortly after germination, and these spring-formed oospores remain dormant until late fall when they germinate. Thus, in 1 year there may be 2 periods of germination and growth, one in spring and another in autumn with dormant periods in summer and winter. Several of the species have been isolated and grown in axenic cultures on synthetic media by Perrott (1958).

Species distinctions are generally made on the basis of the mature oospore's position relative to the oogonium, i.e., exogenous or endogenous, sizes of the oospores and the ornamentation or character of their outer wall, position of the antheridia, i.e., epigynous or hypogynous and inserted or exserted, whether the sex organs are borne in fascicles or singly at the tips of hyphae, and on the diameter and relative degree of branching of the mycelium. However, some of these characteristics vary and overlap considerably. Occasionally, in species with exogenous oospores the latter may be borne endogenously, or vice versa in species with endogenous oospores, and in species with bullated oospores the wall may be undulating or smooth. Thus, it is not improbable that some of the known species and varieties may prove to be identical.

Thaxter (1895) described the planospores of *M. insignis* Thaxter and *M. fasiculata* Thaxter as being biflagellate (?) and for this reason Lagerheim (1900) established a new genus, *Diblepharis*, for these 2 species, which have subsequently been found to have uniflagellate planospores. The remaining species known at that time were retained in *Monoblepharis*, and Lagerheim created 2 subgenera for them: *Eumonoblepharis* for species with endogenous oospores, and *Exoospora* for species with exogenous oospores. He, also, regarded *M. macandra* as a variety of *M. polymorpha*, and established *M. brachyandra* with stubby and short antheridia as a new species with one variety *longicollis*. *Monoblepharis macandra* has subsequently been described as a separate species by Woronin (1904) while *M. brachyandra* and its variety have been reduced to synonyms of *M. polymorpha*. In 1927 Laibach created *Monoblephariopsis* for *M. regignens* and *M. ovigera* because of their more slender hyphae, smaller proliferating zoosporangia, and lack of sex organs. This genus was recognized by Höhnk (1935) who added another species *M. elongata* Höhnk. In her monograph of *Monoblepharis* Perrott (1955) followed Lagerheim's disposition of the species in subgenera: *Monoblepharis* for the coarse species with cylindrical oogonia (except *M. sphaerica*) and endogenous oospores, and *Exoospora* for the delicate species with more or less pyriform oogonia and exogenous oospores. These subgeneric distinctions are questionable as such in this writer's opinion. At least, they do not have the same value as the distinctions of the subgenera of *Synchytrium*, *Blastocladiella*, and *Allomyces* which are based primarily on differences in life cycles, i.e., short- or long-cycled. Perrott (1955) separated the species with exogenous oospores which Woronin had described as *M. spherica* Cornu and renamed it *M. hypogyna* Perrott. Also, she created a variety of each for *M. fasiculata* and *M. insignis*.

Species of *Monoblepharis* are characterized by a non-septate, sparingly or profusely and sympodially branched or unbranched, colorless or pale-brown mycelium which is attached to the substratum by blunt-ending rhizoids (fig. 61). It consists of somewhat rigid or flexuous, cylindrical, nearly isodiametric, slightly tapering hyphae which vary considerably in length and diameter in the different species. In *M. regignens* and *M. ovigera* (figs. 98, 108) the hyphae may be only 23 to 33 μ in length and 1.4 to 4 μ in diameter, while in *M. insignis* Thaxter (fig. 65) and other species they may be 1.5 to 2.5 mm long by 9 to 16 μ in diameter. The protoplasm in them is characteristically foamy, reticulate and scalariform in appearance (figs. 65, 83, 113) which enables one to recognize easily species of the genus in the vegetative state. According to Lagerheim, gemmae (figs. 43, 44) occur in *M. polymorpha* (var. *brachyandra*).

Asexual reproduction occurs by the formation of

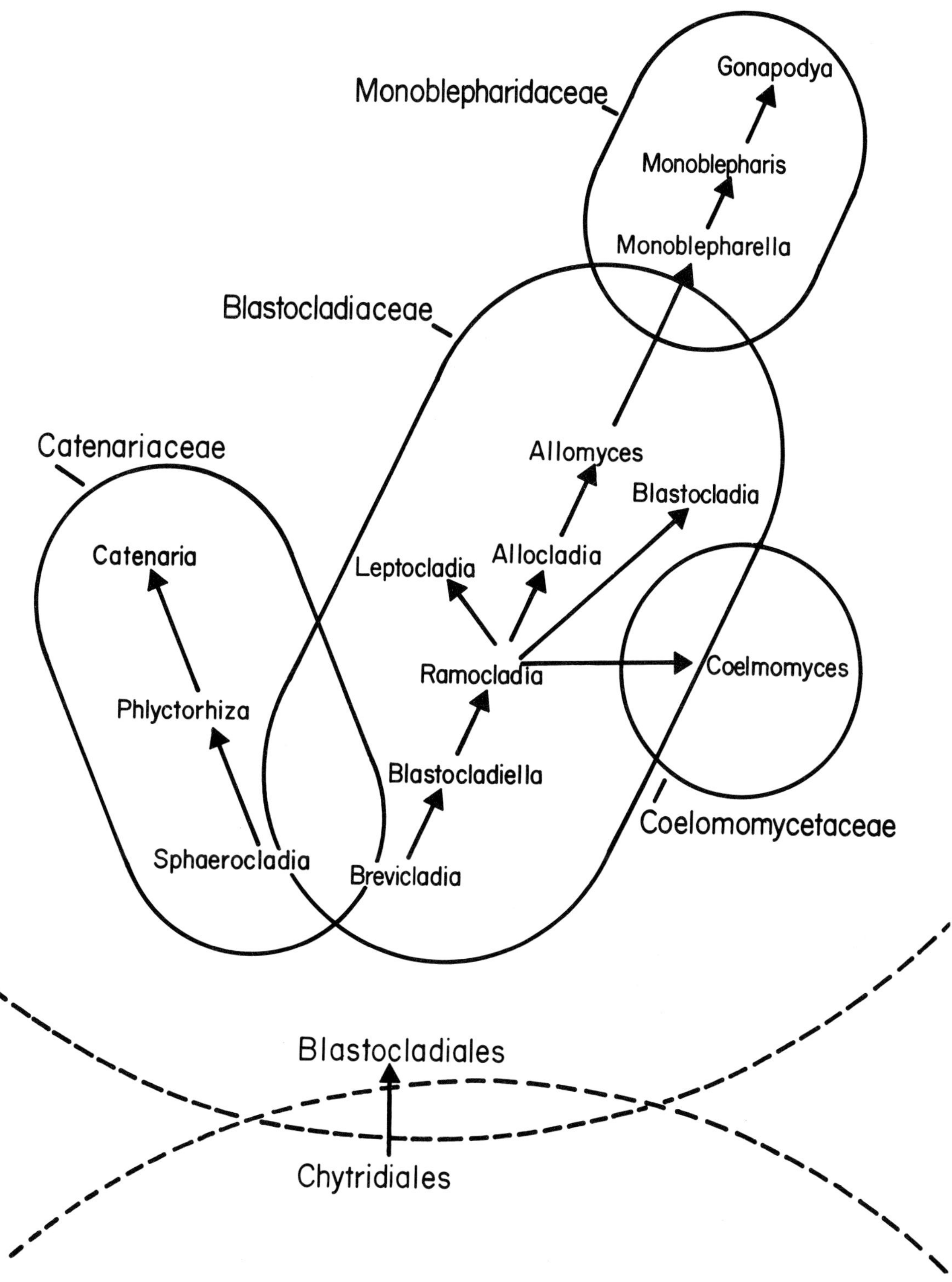

Sörgel's Concept of Interrelationships of the Blastocladiales and Monoblepharidales

posteriorly uniflagellate planospores in terminal (figs. 30, 31, 116), occasionally intercalary (figs. 58), narrowly cylindrical (figs. 119, 120), or shorter, fusiform (fig. 100, 111, 112) zoosporangia which may (fig. 98, 99, 109, 117) or may not proliferate. Sometimes the zoosporangia are formed in rows in basipetal succession. Such non-sexual organs are formed in most species at temperatures of about 8 to 11°C. These zoosporangia may be up to 234 to 346 μ long by 4 to 13 μ in diameter in *M. macandra* (Lagerheim) Woronin and *M. polymorpha* Cornu, but only 18 to 36 μ long in *M. regignens* Lagerheim (fig. 98, 99). The planospores are fully formed in single or double rows in the zoosporangia (figs. 30, 31, 110, 116, 122) and are liberated as the apex deliquesces (figs. 31, 119). They remain adherent by their flagellum to the exit orifice for varying lengths of time and may oscillate for a little while before becoming free (fig. 120).

The planospores are ovoid to oblong with a tapering anterior end and range from 7.8 to 10.4 by 10.4 to 13 μ in *M. macandra* and 5 × 8 μ in *M. regignens*. The anterior end contains several refractive globules (fig. 27) which may sometimes fuse into a single broadly cone-like body. A sharply-defined nuclear cap apparently is lacking. Quite likely ultrastructural studies will show that the dark-staining body around the nucleus as well as the other organelles are similar to those of *Monoblepharella* planospores, according to Fuller and Reichle (1968). In germinating, the planospore may produce 2 germ tubes (fig. 60), one giving rise to the rhizoids or holdfast and the other developing into the main body of the thallus.

Sexual reproduction is oogamous and consists of the fusion of a small posteriorly uniflagellate antherozoid with a large non-flagellate egg produced in antheridia and oogonia, respectively, which are borne on the same thallus. These gametangia are non-proliferous and may be arranged in fascicles (figs. 83, 86) or in superimposed series (fig. 91), or singly at the tips of hyphae (fig. 2-5) or their branches. In *M. polymorpha*, *M. fasiculata*, *M. bullata* Perrott and *M. insignis* the antheridium is epigynous and appears to be inserted on the oogonium (figs. 26, 34-36, 38-42, 65, 91, 92). This appearance results from the development of the oogonial rudiment beneath the antheridium (33, 34), and as the former organ enlarges the latter is carried upward and subapically to its definitive position on the oogonium (fig. 35). In *M. spherica* Cornu and *M. hypogyna* Perrott the antheridium is hypogynous and formed immediately beneath the oogonium (figs. 2- 4, 11- 14). In *M. macandra* vars. *macandra* and *laevis* the antheridia occur singly or in groups on separate branches from the oogonia at first, but later they develop in groups with the oogonia and are sometimes hypogynous and conspicuously exerted (figs. 113, 146). The antheridia are generally elongate or narrowly cylindrical and vary from 4 to 14 μ in diameter by 7 to 59 μ in length in the different species. They produce from 4 to 16 posteriorly uniflagellate, strongly amoeboid, antherozoids in the different species, and these apparently have the same shape and internal structure as the planospores (fig. 37) but are considerably smaller and vary from 1.5-4 by 3.5-7 μ. Protandry is a fairly common occurrence. The oogonia vary from elongate to narrowly-cylindrical, 5-21 × 21-31 μ. At maturity they contain a single large egg whose protoplasm contains numerous large and small refractive globules (figs. 1, 2, 11).

As the uninucleate oogonium (fig. 127) matures the egg contracts from the base, and the refractive globules aggregate in its center (figs. 35, 39, 67, 68, 84). An apical refractive receptive papilla develops (fig. 35), and the nucleus migrates into the apex (fig. 128). According to Thaxter the "ripe" oogonium discharges some granular protoplasms from its apex (fig. 69) which attracts the antherozoids, but this has not been confirmed by subsequent workers (Barnes and Melville, 1932; Sparrow, 1933). The discharged amoeboid antherozoid creep to the receptive papilla, but apparently fertilization is possible only after the egg has attained the proper stage of development. Frequently, antherozoids may creep over immature eggs without being able to fertilize them, or are unable to fertilize eggs which appear to be mature or "ripe" for fertilization. As the antherozoid approaches the apex of the oogonium, the peripheral collar of the oogonial wall dilates slightly and the sperm comes to rest on the receptive papilla (fig. 38). It is engulfed almost immediately by the egg (figs. 38, 39), but its flagellum may protrude stiffly for a short while from the oogonial orifice. Its protoplasm may be recognized as a darker area (figs. 40, 41) for some time after fertilization. Within a few minutes the zygote of *M. polymorpha, M. hypogyna*, and *M. macandra* vars. *macandra* and *laevis* migrates up in the oogonium (figs. 41, 49, 130, 145, 146) and comes to rest at the orifice and matures. In some species it may readily fall away later into the water. In *M. sphaerica, M. fasiculata* vars. *fasiculata* and *magna, M. insignis* and *M. bullata* it remains in the oogonium (figs. 6-9, 65, 83, 91). Rarely, according to Lagerheim, the zygote may move away from the oogonium and undergo amoeboid changes in shape. The flagellum may persist for a short time (fig. 131), but always for a briefer time than in *Monoblepharella*. Woronin, also, illustrated (fig. 132) the egg as rarely emerging from the oogonium and assumed that it is fertilized on the outside. Endogenous parthenogenetic oospores or chlamydospores (fig. 51) are formed in some species. Occasionally, the tip or neck of the oogonium becomes inflated, open or unopened, and bears a small or large oospore (figs. 88, 97).

The gametic nuclei do not fuse at once in the zygote but remain paired (fig. 133) until wall formation has reached an advanced stage and bullations, if present, have begun to develop. Karyogamy then occurs (fig. 134), and the mature oospores become uninucleate (135, 136). These mature oospores are spherical, 12-25 μ in diam., or subspherical, ellipsoidal, ovoid-oblong to subcyclindrical, 18–21 × 30–45 μ, in the different species. The wall is golden-brown to dark- and pale-brown, and may be smooth in some species (*M. fasiculata* and *M. insignis*) and bullate in others (*M. sphaerica*, *M. bullata*, *M. hypogyna*, *M. polymorpha* and *M. macandra*). In the latter species the bullae may be relatively sparse or numerous and 1 to 2 μ in height. The oospores of some bullate species, e.g., *M. polymorpha*, may lack bullations and have an undulate outer wall. Also, in some of these species the wall may be smooth. According to Woronin, the wall in *M. hypogyna* is 2-layered (fig. 25) with a thick outer bullate layer and a thin inner layer. The oospores may germinate endo- or exogenously in different species by means of a hypha (figs. 57, 139).

In preparation for germination, the diploid zygotic nucleus may divide into as many as 16 smaller ones (figs. 137, 138) and it is presumed that the first division is meiotic, although meiosis has not been observed. As the germ tube is formed the nuclei move into it (fig. 139), and the vegetative mycelium is established accordingly.

REFERENCES TO THE MONOBLEPHARIDALES

Atkinson, G. F. 1909. Some problems in the evolution of the lower fungi. Ann. Mycol. 7:441-472, figs. 1-20.

Barnes, B. and R. Melville. 1932. Notes on British aquatic fungi. Trans. Brit. Mycol. Soc. 17:82-96, 6 figs.

Benjamin, R. K. 1958. On the relation of the sexual and nonsexual phases of *Gonopodya*. Mycologia 50:789-792, 3 figs.

Beneke, E. S. 1948. The Monoblepharidaceae as represented in Illinois. Trans. Ill. Acad. Sci. 41:27-30, pl. 1.

Cejp, K. 1946. Sur les affinite's des Blastocladiaceae. Révision du genre *Gonopodya*, sa position systématique. Bull. Soc. Mycol. France 62:246-257.

————. 1947. Monografika studie Blastocladiales (Phycomycetes). Z. Vestiniku Kral. ces. Spol. Nauk. Tr. mat.-prirod. Roc. 1946:1-55, 2 pls.

————. 1957. Houby. 1. Ceskoslovenská Akad. Ved. Sekce biologicka 495 pp., 114 figs., 8 pls., Prague.

Chambers, T. C., K. Marcus, and L. G. Willoughby. 1967. The fine structure of the mature zoosporangium of *Nowakowskiella profusa*. J. Gen. Microbiol. 46:135-141.

Cornu, M. 1871. Note sur deux genres nouveaux de la famille Saprolégniées. Bull. Soc. Bot. France 18:58-59.

————. 1872. Monographie des Saprolégniées; etude physiologiques et systématique. Ann. Sci. Nat. Bot. V, 15:1-198, 7 pls.

Fischer, A. 1892. Phycomycetes. Die Pilze Deutschlands, Oesterreichs und der Schweiz. Rabenhorst. Kryptogamen-Fl. 1:490. Leipzig.

Fuller, M. S. 1966. Structure of the uniflagellate zoospores of aquatic Phycomycetes. In Colston papers. Vol. XVIII:67-84.

———— and R. E. Reichle. 1968. The fine structure of *Monoblepharella* sp. zoospores. Canad. J. Bot. 279-283, pls. 1-6.

Hine, F. B. 1878-1879. Observations on several forms of Saprolegnieae. Amer. Quart. Micro. Journ. 1:18-28, 136-146, pls. 5-7.

Höhnk, W. 1935. Saprolegniales und Monoblepharidales aus der Umgebung Bremens mit besonderer Berücksichtung der Oekologie der Saprolegniaceae. Abhandl. Naturwiss. Vereins Bremen 29: 207-237, 7 figs.

Johns, R. M., and R. K. Benjamin. 1954. Sexual reproduction in *Gonopodya*. Mycologia 46:201-208, 17 figs.

Koch, W. J. 1961. Studies on the motile cells of chytrids. III. Major types. Amer. J. Bot. 48:786-788, 8 diagrams.

————. Thaxter's *Myrioblepharis*. Mycologia 56:436-440, 4 figs.

Lagerheim, G. 1900. Mykologische Studien. II. Untersuchungen über die Monoblepharideen. Bih. Kgl. Svensk. Vetensk-Ak. Handl. 25, afd. 3, no. 8:1-42, pls. 1,2.

Laibach, F. 1926. Zur Zytologie von *Monoblepharis*. Ber deut. Bot. Gesell. 44:59-64, 3 figs.

————. 1927. Zytologische Untersuchungen über die Monoblepharideen. Jahbr. wiss. Bot. 66:596-628, 12 figs., pls. 12,13.

Miller, C. E. 1963a. Observations on sexual reproduction in *Gonopodya polymorpha* Thaxter. J. Elisha Mitchell Sci. Soc. 79:153-156, 22 figs.

————. 1963b. A fungivorous ciliate. Mycologia 55: 361-364, 10 figs.

Minden, M. von. 1915. Chytridiineae, Ancylistineae, Monoblepharidineae, Saprolegniineae. Kryptogamenfl. Mark Brandenbrug 5:476-478.

Perrott, P. E. 1955. The genus *Monoblepharis*. Trans. Brit. Mycol. Soc. 38:247-282, 17 figs., pls. 9-13.

————. 1958. Isolation and pure culture of *Monoblepharis*. Nature 182:1322-1324, 3 figs.

————. 1960. The ecology of some aquatic Phycomycetes. Trans. Brit. Mycol. Soc. 43:19-30, pl. 1.

Petersen, H. E. 1909. Studier over Ferskvands-Phycomyceter. Bot. Tidsskr. 29:345-440, 27 figs.

Reichle, R. E. 1972. Fine structure of *Oedogoniomyces* zoospores with comparative observations on *Monoblepharella* zoospores. Canad. J. Bot. 50:819-824, 25 figs.

Reinsch, P. F. 1878. Beobachtungen über einige neue Saprolegnieae, über die Parasiten in Desmidienzellen und über die Stachelkugen in *Achyla*-schlauchen. Jahrb. wiss. Bot. 11:282-311, pls. 14-17.

Schroeter, J. 1893. Phycomycetes. *In* Engler und Prantl, Die Natürl. Pflanzenf. 1:63-141.

Shanor, L. 1942. A new *Monoblepharella* from Mexico. Mycologia 34:241-247 20 figs.

Sörgel, G. 1952. Uber mutmassliche phylogenetische

Zusammenhänge bei niederer Pilzen, inbesondere den Blastocladiales. Biol. Zentralbl. 32:305-315, pl. 6.

Sparrow, F. K., Jr. 1933. The Monoblepharidales. Ann. Bot. 47:517-542, 2 figs., pl. 20.

———. 1940. Phycomycetes recovered from soil samples collected by W. R. Taylor on the Allen Hancock 1939 expedition. Publ. Univ. So. Calif. 3:101-112, 2 pls.

———. 1943. Aquatic Phycomycetes.... XIX + 785 pp. 634 figs. Univ. Michigan Press, Ann Arbor.

———. 1953a. A new species of *Monoblepharella*. Mycologia 45:592-595, 1 fig.

———. 1953b. Cytological observations on the zygote of *Monoblepharella*. Ibid. 45: 723-726, 1 fig.

———. 1960. Aquatic Phycomycetes...2nd ed. XXV-1187 pp., 91 figs. Univ. Michigan Press, Ann Arbor.

Springer, M. E. 1945a. Two new species of *Monoblepharella*. Mycologia 37:205-216, 51 figs.

———. 1945b. A morphologic study of the genus *Monoblepharella*. Amer. J. Bot. 32:259-269, 46 figs.

Thaxter, R. 1895a. New or peculiar aquatic fungi. 1. *Monoblepharis*. Bot. Gaz. 20:433-440, pl. 29.

———. 1895b. New and peculiar fungi. 2. *Gonopodya* Fischer and *Myrioblepharis* nov. gen. Ibid. 20:477-485, pl. 31.

———. 1903. Mycological notes, 1-2. Rhodora 5: 97-108, pl. 46.

Waterhouse, G. M. 1945. The true nature of *Myrioblepharis* Thaxter. Trans. Brit. Mycol. Soc. 28: 94-100, 4 figs., pl. 5.

Woronin, M. 1904. Beitrag zur Kenntniss der Monoblepharideen. Mem. Acad. Imper. Sci. St. Peters. Phys.-Math. Cl. viii, 16:1-24, pls. 1-3.

Chapter XII

APPENDIX

HYPHOCHYTRIOMYCETES

HYPHOCHYTRIALES (ANISOCHYTRIDIALES)

This order of parasitic and saprophytic aquatic fungi differs primarily from the previous orders by the formation of anteriorly uniflagellate tinsellate planospores. Nevertheless, the thallus resembles that of families of the Chytridiales in being monocentric, holocarpic and olpidioid, or monocentric and eucarpic, or polycentric and eucarpic. On such morphological and developmental grounds the species are segregated in 3 families; the Anisolpidiaceae which is strikingly similar to the Olpidiaceae, the Rhizidiomycetaceae whose species resemble many of the Rhizidiaceae, and Hyphochytriaceae with a predominantly polycentric thallus which may be regarded as comparable to some degree to that of the Cladochytriaceae. However, it remains to be seen whether or not these distinctions will be sustained as new genera and species are discovered. Even in our present state of knowledge the family distinctions are not always sharply defined because some of the polycentric species may sometimes develop monocentric holocarpic, olpidioid, and monocentric eucarpic thalli. Also, species with monocentric eucarpic thalli may occasionally become polycentric. The present classification, nonetheless, seems practical for the time being.

The development and establishment of the different types of thalli are essentially like those of many chytrids, and as in the latter fungi the wall is composed of chitin and cellulose. In most species the protoplasm of the zoosporangia emerges as a naked globular mass through a pore, papilla, or exit canal and undergoes cleavage into planospores on the outside. In a few others cleavage is endogenous, and the planospores emerge fully formed. Canter (1950), thus, pointed out that two series of sites of cleavage are present in the order, and on this basis Sparrow (1960) created the genera *Canteriomyces* in the Anisolpidiaceae and *Rhizidiomycopsis* in the Rhizidiomycetaceae and suggested creation of another genus in the Hyphochytriaceae for the species in which cleavage occurs exogenously. Inasmuch as the site of cleavage may vary considerably in some species it is questionable that this is a distinct generic characteristic. In the event it should prove to be more distinctive and characteristic the creation of subgenera perhaps might be more appropriate than the establishment of new genera.

Resting spores are known in relatively few species. In *Anisolpidium ectocarpii* Karling they are formed sexually by the fusion of pairs of isolplanogametes or uninucleate protoplasts within the host cell. Also, in the questionable genus *Reesia* and in *Chytridium mesocarpi* (Fisch) Fischer fusion of motile isogametes has been reported to precede resting spore development, but the status of this genus and species is doubtful. In *Latrostium comprimens* Zopf and *Hyphochytrium hydrodictyii* Valkanov the resting spores appear to be formed asexually and develop thick walls.

The species of the Hyphochytriales were classified among the chytrids by the earlier chytridiologists, but the confirmation of Zopf's discovery of anteriorly uniflagellate planospores eventually led to the segregation of these species in a separate order. Its origin and relationships are obscure, but several theories of its origin have been proposed. On the assumption that planospore structure is indicative of origin and relationships it has been suggested that these species arose from the anteriorly uniflagellate Monadineae and evolved in a parallel series with the posteriorly uniflagellate chytrids. On the other hand, Bessey (1942) suggested that they have evolved from anteriorly biflagellate, heterocont unicellular algae by the loss of the posteriorly directed non-tinsellate flagellum. In that event, perhaps the second centriole of "accessory blepharoplast" might be regarded as a relic of a lost flagellum.

In this connection it may be noted that Lovett and Haselby (1971) found in the Hyphochytriomycetes which they studied that this class could not be placed or classified in any of the other major classes of fungi on the basis of differences in 25 S RNA molecular weights. Lé John (1972) concluded that on the basis of enzyme control, cell wall structure, and lysine pathways the Hyphochytriomycetes are probably forerunners of the Chytridiales and Oomycetes, and not intermediates.

ANISOLPIDIACEAE

This family of 1 or possibly more genera is characterized by a monocentric endobiotic or intramatrical holocarpic thallus whose development, structure and organization are similar to those of the chytrid genus *Olpidium* and other genera of the Olpidiaceae. Some species of this family were formerly classified as species of *Olpidium, Pleotrachelus,* and *Olpidiopsis* until Karling (1943) segregated them in a distinct family. The genera *Ressia* and *Cystochytrium* were tentatively included in this family, but so far as present knowledge goes they are doubtful members. *Ressia amoeboides* Fisch in particular has been reported by Wagner (1969) to produce posteriorly uniflagellate planospores (see plate 11), and in *Cystochytrium radicale* Cook the thallus is reported to segment occasionally into several zoosporangia.

Sexual reproduction is known in only 1 species, *Anisolpidium ectocarpii* Karling. It consists of the fusion of uninucleate protoplasts (isoaplanogametes) within the host cell, and the zygote develops into a thick-walled resistant sporangium which germinates shortly by the development of an exit canal and produces anteriorly uniflagellate R.S. planospores. These are similar in size and structure to those produced in the thin-walled zoosporangia. Fusion in pairs of anteriorly uniflagellate isoplanogametes was reported by Fisch (1884) in *Ressia amoeboides*. The motile zygote infected the host and developed into a resting spore, but this observation was not confirmed by Wagner.

ANISOLPIDIUM Karling

Amer. J. Bot. 30:637, 1943

Canteriomyces Sparrow, 1960. Aquatic Phycomycetes, p. 750.

Plates 166, 167

This genus includes 5 species at present, 4 of which are parasitic in freshwater and marine algae and 1 saprophytic in dead pine pollen grains. The parasitic species cause hypertrophy of the infected cell and have been found along the coasts of the U.S.A., Denmark, Sweden, France, Italy, Greenland, and inland in England and the U.S.A. The saprophyte, *A. sapro-*

PLATE 166

Figs. 1-26. *Anisolpidium ectocarpii* Karling. Figs. 1-14 after Karling, 1943; figs. 15-26 after Johnson, 1957.

Fig. 1. Cell of *Ectocarpus mitchellae* with 3 small parasites and a germinating planospore on its surface.

Fig. 2. Naked uninucleate thallus.

Fig. 3. Young tetranucleate thallus with its nuclei in the equatorial stage of division.

Fig. 4. Profile and polar views of dividing nuclei with approximately 5 chromosomes.

Fig. 5. Larger thallus with a well-defined wall and numerous refractive globules.

Fig. 6. Later stage showing emergence of numerous vacuoles and an increase in the number of refractive globules.

Fig. 7. Small, sectioned and stained thallus showing confluence of vacuoles into a large central one; nuclei lying in the peripheral layer of cytoplasm.

Fig. 8. Zoosporangium with an irregular inflated exit tube.

Fig. 9. Vacuolate zoosporangium with 2 curved exit tubes.

Fig. 10. Centrifugal cleavage of protoplasm into planospore initials.

Fig. 11. Discharge of fully formed planospores.

Fig. 12. Various shapes undergone by the anteriorly uniflagellate planospores, 1.8-2.3 × 3-3.5 μ, from a zoosporangium.

Fig. 13. Encysted planospore.

Fig. 14. Zoosporangium occupying 2 host cells with 4 exit tubes, 1 of which has discharged planospores within the host; group of encysted planospores above 1 exit tube.

Fig. 15. Two isoaplanogametes within a host cell; empty gametic cysts on the outside.

Figs. 16, 17. Plasmogamy and pairing of gametic nuclei.

Fig. 18. Zygote with fused nuclei.

Fig. 19. Later developmental stage of zygote.

Fig. 20. First division (meiotic?) of the diploid nucleus in the zygote.

Fig. 21. Vacuolate incipinet resistant spore or sporangium; nuclei lying in the peripheral layer of protoplasm.

Fig. 22. Beginning of R. S. germination; nuclei in the peripheral layer dividing.

Fig. 23. Centrifugal cleavage into planospore initials in a germinating resistant sporangium.

Fig. 24. Discharge of R. S. planospores.

Figs. 25, 26. R. S. planospores enlarged.

Figs. 27-31. *Anisolpidium sphacelariarum* Karling in *Chaetopteris plumosa*. Figs. 27, 28 after Sparrow, 1936; figs. 29-31 after Petersen, 1905.

Fig. 27. Young thallus next to host nucleus, planospores cyst persistant.

Fig. 28. A planospore.

Fig. 29. Host cell with 9 empty zoosporangia.

Fig. 30. Incipient zoosporangium with 2 short exit tubes.

Fig. 31. Three empty zoosporangia.

Figs. 32-35. *Anisolpidium rosenvingii* Karling in *Pylaiella littoralis*. (Petersen, 1905.)

Fig. 32. Empty zoosporangium with 2 exit tubes; causing hypertrophy of the host cell and filling it completely.

Fig. 33. Young zoosporangium with alveolar protoplasm.

Fig. 34. A young and an empty zoosporangium with 1 exit tube.

Fig. 35. An empty zoosporangium with 4 exit tubes.

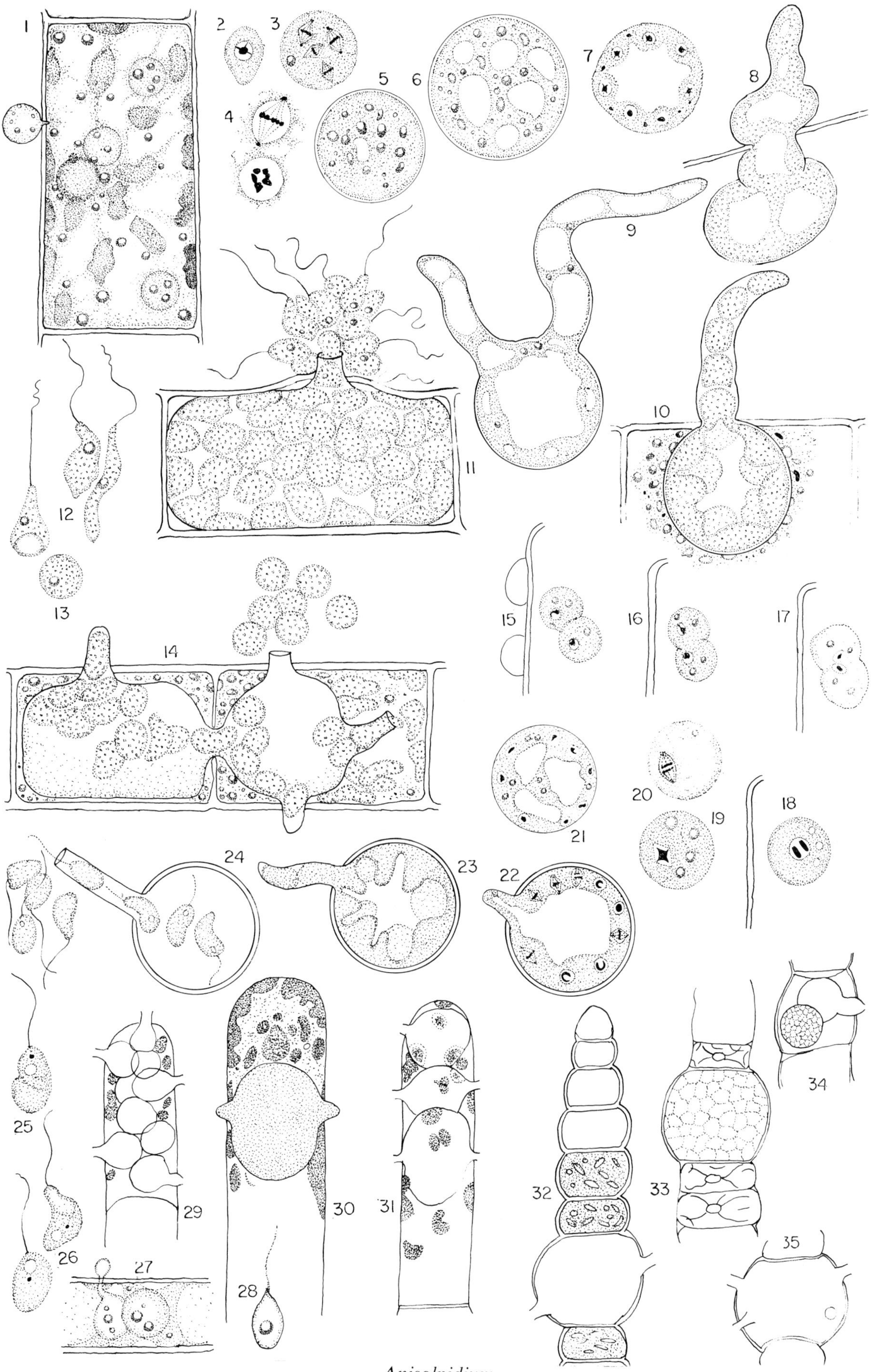

Anisolpidium

bium Karling, occurred in soil on Pitcairn Island and was trapped in pollen grains. Although a few thalli of this species retained their olpidioid character on agar and produced planospores it may possibly prove to be polycentric like *Hyphochytrium catenoides* Karling and *Hyphochytrium* sp. Persiel, polycentric species which may develop olpidioid thalli when trapped in pollen grains.

As noted previously the thallus of this genus develops in the same manner as that of *Olpidium*. The planospore comes to rest on the host cell or substratum and develops a fine germ tube (figs. 1, 40, 41, 62) through which the protoplasm enters the host cell as a naked mass, leaving the empty planospore case on the outside (fig. 63). Frequently, several planospores may infect the same cell (fig. 40, 41) which results in the development of as many as 9 zoosporangia (fig. 29) in 1 cell. The young thalli appear to be naked (figs. 1, 63), but apparently they are immiscible with the host protoplasm. In *A. ectocarpii* Karling, *A. sphacelariarum* (Kny) Karling, and *A. stigeoclonii* (de Wildemann) Canter the young thallus migrates in the host cell so that it frequently lies adjacent to the host nucleus (figs. 1, 27, 42). As it enlarges its nucleus divides mitotically with an intranuclear spindle (figs. 3, 4), and numerous refractive globules and vacuoles appear in the cytoplasm (figs. 5, 6, 53). The latter usually coalesce into a large central one (fig. 7) which is outlined by a peripheral layer of cytoplasm and nuclei.

As it matures the thallus is transformed holocarpically into an incipient zoosporangium. At this stage it is predominantly globular (figs. 9, 10, 53, 54) and varies from 8 to 52 μ in diameter in the different species. Others may be ovoid (fig. 44), oblong, elongate, or nearly cylindrical. Sometimes they have the same size and shape as the infected cell (figs. 11, 32, 33), or may bud into adjacent cells (fig. 14). At maturity they develop 1 (figs. 10, 54) to several (fig. 14) short or long tapering, straight or curved (figs. 9, 10) or irregular (fig. 8) exit tubes through which the planospores or naked protoplasm emerge. In the latter event the tip of the tube inflates (figs. 46, 52, 56) as the protoplasm begins to move upward, and its wall becomes progressively thinner and thinner with the emergence of the protoplasm until it is scarcely discernible as such.

In *A. ectocarpii* centrifugal cleavage into planospore initials occurs within the zoosporangium (fig. 10) or resistant sporangium (fig. 23) so that the planospores usually emerge fully formed (fig. 11). In *A. stigeoclonii* and *A. saprobium*, however, the sporeplasm usually emerges as naked mass (figs. 37, 47, 48, 56) and undergoes cleavage on the outside of the zoosporangium (figs. 47, 49). In the former species cleavage may be very slow and require as much as 30 minutes for completion.

Sparrow (1960) established a new genus, *Canteriomyces*, for *A. stigeoclonii* because its planospores are delimited outside of the zoosporangium. However, in this species as well as in *A. saprobium* cleavage of the sporeplasm may occur occasionally within the zoosporangium so that the planospores emerge fully formed (figs. 51, 52, 64). Accordingly, the method and site of sporogenesis are not always sharply defined and it is doubtful that such differences merit generic dis-

Plate 167

Figs. 36-52. *Anisolpidium stigeoclonii* Canter in *Stigeoclonium* Figs. 36, 37 after de Wildemann, 1931; figs. 38-51 after Canter, 1950; fig. 52 drawn from living material by Karling.

Fig. 36. Two thalli in an enlarged cell; tips of exit canals inflating.

Fig. 37. Emergence of naked protoplasm and its cleavage into planospore initials.

Fig. 38. Tinsellate planospores.

Fig. 39. Non-tinsellate planospore.

Figs. 40, 41. Germinating planospores on mucilage surrounding the algal cell.

Fig. 42. Young thallus adjacent to the host nucleus.

Fig. 43. Hypertrophied host cell with 3 parasites.

Fig. 44. Later developmental stage; protoplasm of parasites alveolar or foamy.

Fig. 45. Beginning of exit canal development.

Fig. 46. Tip of exit canal becoming inflated.

Figs. 47-49. Discharge or emergence of naked protoplasm from a zoosporangium.

Fig. 50. Mass of planospores with a halo of flagella.

Fig. 51. Zoosporangium with fully formed planospores.

Fig. 52. Zoosporangium with planospore initials.

Figs. 53-67. *Anisolpidium saprobium* Karling. (Karling, 1968.)

Fig. 53. Dead pollen grain of *Pinus sylvestris* with 3 thalli.

Fig. 54. Thallus with a long exit canal.

Fig. 55. Tip of exit canal inflated in an early stage of emerging protoplasm.

Fig. 56. Emergence of naked protoplasm.

Fig. 57. Cleavage of protoplasm into planospore initials.

Fig. 58. Group of separated planospores shortly before dispersing; flagella beating slowly.

Fig. 59. A group of enlarged pyriform, 1.8-2.3 × 3-3.5 μ, planospores with refringent globules and greyish granular protoplasm; flagella 5-6.5 μ long.

Fig. 60. Three encysted planospores.

Fig. 61. Encysted planospore in water which developed an exit canal and produced 1 flagellate planospore.

Fig. 62. Infection of a pollen grain.

Fig. 63. Later stage; empty planospore cyst on the outside of the host; the small saprophyte within.

Fig. 64. Zoosporangium discharging fully formed planospores.

Fig. 65. Encysted planospores within a zoosporangium and around its exit canal.

Fig. 66. Three monocentric holocarpic thalli developing in slush agar.

Fig. 67. Zoosporangium in slush agar which has developed a long exit canal; protoplasm beginning to emerge.

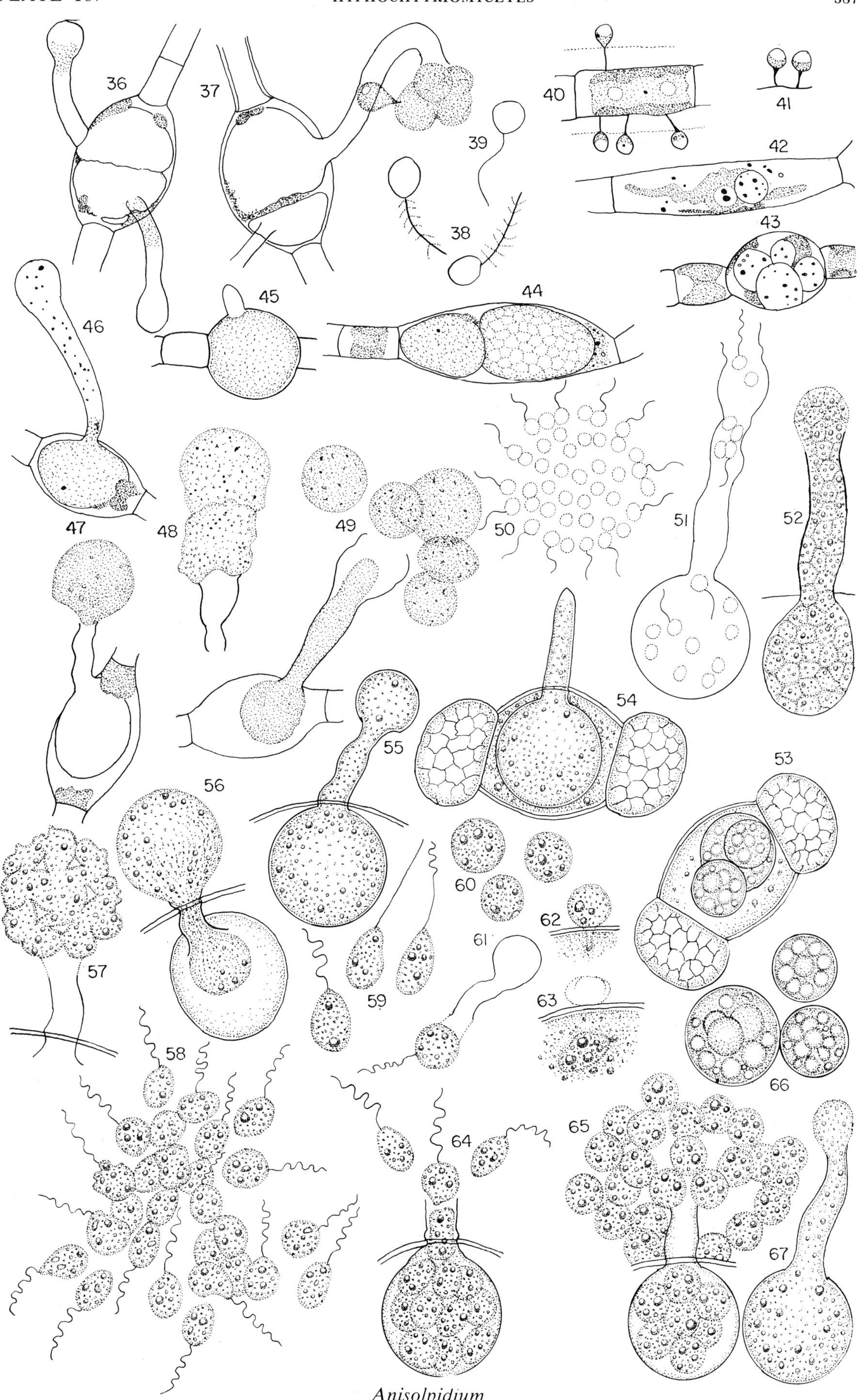

Anisolpidium

tinction. Therefore, the present writer is merging *Canteriomyces* with *Anisolpidium*.

Following cleavage in the so-called exogenous type of sporogenesis the delimited planospore initials lie quiescent in a cluster for a while as the flagella develop and beat slowly (figs. 50, 58). As this beating accelerates, the spore body begins to rock from side to side, and within a short while thereafter the planospores dart away in a characteristic manner. Generally, they are relatively small, pyriform and elliptical to clavate, 1.8–2.3 × 3.5–5 μ, with greyish granular content containing one or more glistening globules (fig. 59), and a short, 5–7 μ, tinsillate flagellum (fig. 38). However, Canter found that in *A. stigeoclonii* the flagellum may be tinsillate or non-tinsillate (fig. 39). Fairly often in *A. saprobium* no flagellar development occurs, or the flagellum is absorbed before the planospores disperse. In such instances the latter may encyst in a cluster around the exit canal and in the zoosporangium (fig. 65).

After a period of motility the planospore comes to rest, gradually absorbs its flagellum, and encysts (fig. 60). Such encysted cells may occur in abundance in water around the host of *A. stigeoclonii* but apparently degenerate unless they infect the host. In *A. saprobium*, however, they may enlarge to various sizes, "germinate" by forming an exit canal, and produce 1 (fig. 61) or more planospores, depending on their sizes. In axenic cultures on slush agar they may enlarge to mature size (fig. 66) and later develop exit canals (fig. 67) and produce planospores exogenously.

In the process of sexual reproduction in *A. ectocarpii* planospores infect the host, and their protoplasts (isoaplanogametes) become associated in pairs (fig. 15). These fuse (figs. 16, 17) and karyogamy occurs very shortly afterwards (fig. 18). The diploid nucleus divides (figs. 19, 20), and as additional mitoses occur, the zygote enlarges and becomes multi-nucleate and vacuolate (fig. 21). As it matures it develops a thick, smooth, hyaline wall and becomes a spherical, 10–14 μ diam., or ovoid resistant sporangium with coarsely granular content. Thus, its development closely parallels that of the zoosporangium except for its origin and the formation of a thick wall. It generally overwinters in a dormant state and germinates in the following spring, and in this process it functions as a sporangium. It develops 1 to 2 exit canals, the peripheral nuclei divide (fig. 22), and centrifugal cleavage of the protoplasm occurs (fig. 23). The R.S. planospores are, thus, fully

PLATE 168

Figs. 1-22. *Rhizidiomyces apophysatus* Zopf. Figs. 1, 15 after Zopf, 1884; figs. 2-14 after Karling, 1944; fig. 17 after Couch, 1941; figs. 18-21 sketched from electron micrographs, and fig. 22 after Fuller and Reichle, 1965.

Fig. 1. Oogonium of *Achlya racemosa* parasitized by 2 empty zoosporangia and another one with a long exit canal; 3 smaller thalli at upper left apparently are young stages of *R. apophysatus*, also.

Figs. 2-6. Encysted planospore, germination and developmental stages of an apophysate thallus, resp.

Fig. 7. Young ectobiotic pyriform zoosporangium with numerous refractive globules; apophysis and branched rhizoids endobiotic.

Fig. 8. Early state of a hyaline exit papilla which will elongate into an exit canal.

Fig. 9. Later stage; the more granular protoplasm is moving up into the expanding tip of the incipient canal.

Fig. 10. Zoosporangium with a long canal whose tip is expanding.

Fig. 11. Same zoosporangium in a later stage; protoplasm is emerging.

Fig. 12. Emerged naked spherical mass of protoplasm.

Fig. 13. Cleavage into planospore initials.

Fig. 14. An irregular cluster of planospores shortly before dispersing; flagella beating slowly.

Figs. 15, 16. Ovoid, 3 × 5-6 μ, swimming planospores with refractive globules and greyish granular protoplasm.

Fig. 17. Fixed and stained planospores showing lateral tinsels on the flagellum.

Fig. 18. Shadowed planospore with fine-ended tinsels; central axoneme surrounded by a membrane which is swollen at the base.

Fig. 19. Cross section of a flagellum showing the typical 9 + 2 arrangement of the fibrils.

Fig. 20. Cell wall of an encysted planospore with attached tinsels from the absorbed flagellum.

Fig. 21. Portion of a tinsel showing banding.

Fig. 22. Schematic diagram of a planospore: bb = basal body; d = dictyosome; lb = lipid body; rr = ribosome-containing region; er = endoplasmic reticulum; m = mitochondrion; fr = fiber or tubule-containing region; st = surface tubules; c = 2nd centriole associated with the basal body; v = vacuole; n = nucleus with nucleolus.

Figs. 23-36. *Rhizidiomyces ichneumon* Gobi on *Chlamydomonas globulosa*. (Gobi, 1900.)

Fig. 23. Actively motile algal cell with 4 young parasites; host flagella omitted.

Fig. 24. Incipient zoosporangium with an endobiotic apophysis.

Fig. 25. Obpyriform non-apophysate (?) incipient zoosporangium.

Figs. 26, 27. Incipient zoosporangia with epiobiotic apophyses.

Fig. 28. "Falsely" dichotomously branched rhizoidal system with blunt ends.

Fig. 29. Apophysate thallus showing the limited extent and branching of the rhizoid; zoosporangium continuous with the apophysis and rhizoids.

Figs. 30-33. Stages in the development of an exit canal and the emergence of the naked protoplasm.

Fig. 34. Cleavage into planospore initials.

Fig. 35. Ovoid, 3 μ long, planospores with granular content.

Fig. 36. Minute zoosporangium which formed 4 planospores.

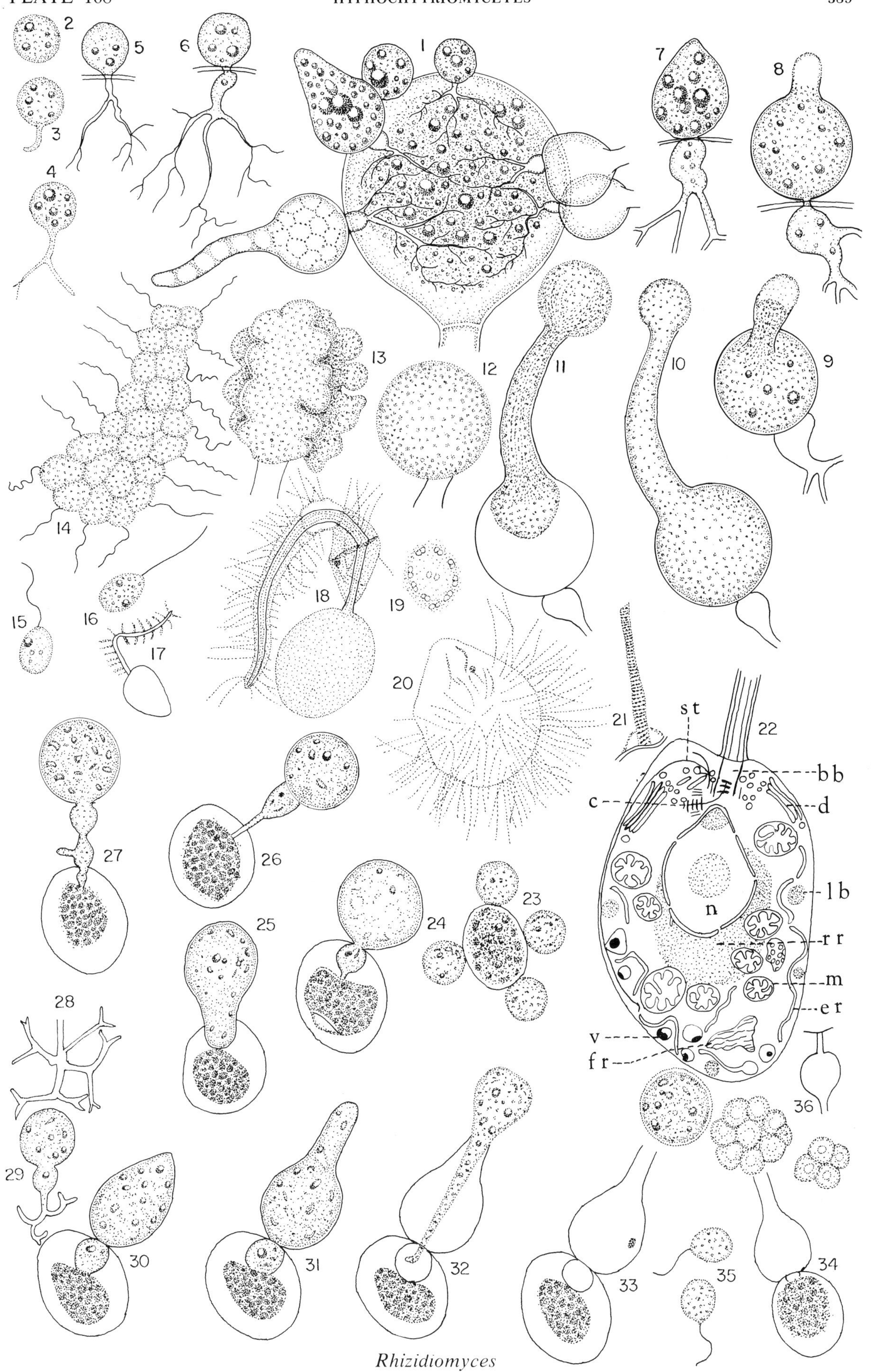

Rhizidiomyces

formed within the zoosporangium and emerge in a similar manner as those formed in the thin-walled zoosporangia (fig. 24). Also, their size, shape, and structure are similar (figs. 25, 26) according to Johnson (1957).

Accordingly, sexual reproduction in *A. ectocarpii* is similar in some respects to that of *Olpidium radicale* and possibly *O. agrostidis* as described by Schwarze and Cook (1928) and Sampson (1932) in which plasmogamy of 2 protoplasts derived from separate planospores occur within the host cell. However, in these species the gametic nuclei do not fuse at once as in *A. ectocarpii* but remain separate until just before germination of the resting zygote or sporangium when karyogamy occurs. Also, a comparison may be made in this respect with *Rozella allomycis* in which Sörgel (1952) believed that plasmogamy of sexually opposite but compatible protoplasts occur within the host cell as in *O. radicale.*

Nothing is known as to where in the life cycle meiosis and sex segregation occur, but presumably the former occurs during the first division of the zygotic nucleus (fig. 20). If so, the diploid generation in *A. ectocarpii* is very short. Possibly, the R.S. planospores are facultative and may give rise to zoosporangia within the host cell or their protoplasts fuse in pairs to form resistant sporangia.

RHIZIDIOMYCETACEAE

This is the largest family of the anisochytrids with 2 or possibly more genera and 11 or more species which are worldwide in distribution and parasitic on fungi and algae or saprophytic in soil and water. It is characterized by a monocentric eucarpic thallus that is strikingly similar to and parallels the development, structure and organization of some rhizidiaceous chytrids. The thallus consists usually of an epibiotic or extramatrical apophysate or non-apophysate zoosporangium and an endobiotic or intramatrical rhizoidal system which may be extensive and richly branched or reduced to few blunt-ending branches. In these respects the thalli may be readily mistaken for those of a chytrid unless the method of sporogenesis is observed. Occasionally, polycentric thalli occur in one species, and in another the thallus may become greatly elongate and tubular and forms several incipient zoosporangia when grown on and in agar media. Accordingly, the distinctions between this family and those of the following Hyphochytriaceae are not always clearly defined.

Resting spores *per se* are unknown except in *Latrostium comprimens* Zopf, and in this species they are apparently formed asexually. However, in some species

Plate 169

Figs. 37-49. *Rhizidiomyces bivellatus* Nabel on chitinic substrata. Figs. 37-44, 46-49 after Karling, 1944; fig. 45 after Nabel, 1939.

Fig. 37. Ovoid, 5.5 × 6 × 7-8 μ, planospores with coarsely grayish granular protoplasm and a refractive globule; flagella 12-14 μ long.

Fig. 38. Encysted planospore.

Fig. 39. Germinating planospore.

Fig. 40. Young thallus with an extramatrical sporangial rudiment and a single intramatrical rhizoidal axis.

Fig. 41. Incipient zoosporangium with 2 rhizoidal axes.

Fig. 42. Zoosporangium with 4 radically oriented rhizoidal axes.

Fig. 43. A thallus of two zoosporangia in tandem with exit canals; refractive globular material has been finely dispersed.

Fig. 44. Spherical zoosporangium with a thickened rough wall and an exit canal.

Fig. 45. Zoosporangium with an elongate irregular exit canal.

Fig. 46. Emergence of protoplasm from a rough thickened walled zoosporangium.

Fig. 47. Cleavage within and outside of zoosporangium.

Fig. 48. Cluster of planospores shortly before dispersing.

Fig. 49. Dormant thick-walled zoosporangium with numerous refractive globules.

Figs. 50-67. *Rhizidiomyces hansonii* Karling on vegetable debris. (Karling, 1944.)

Fig. 50. Large group of planospores shortly before dispersing.

Fig. 51. Ovoid, 4-4.5 × 5.5-6.5 μ, planospores with numerous orange-tinted granules.

Fig. 52. Germination of planospores.

Fig. 53. Young thallus with a single rhizoidal axis.

Fig. 54. Mature zoosporangium with numerous refractive globules.

Figs. 55, 56. Formation and elongation of the exit canal, resp.; refractive globules have been dispersed.

Fig. 57. Zoosporangium with an unusually long, 280 μ, vacuolate exit canal.

Figs. 58, 59. Emergence of the naked protoplasm.

Figs. 60-62. Invagination of the naked protoplasm and stages in centripetal cleavage.

Fig. 63. Zoosporangium with encysted planospores.

Fig. 64. "Germination" of a dormant fairly thick-walled zoosporangium by an irregular, broad exit canal.

Figs. 65, 66. Unusually thick-walled dormant zoosporangia.

Fig. 67. "Germination" of a thick-walled dormant zoosporangium.

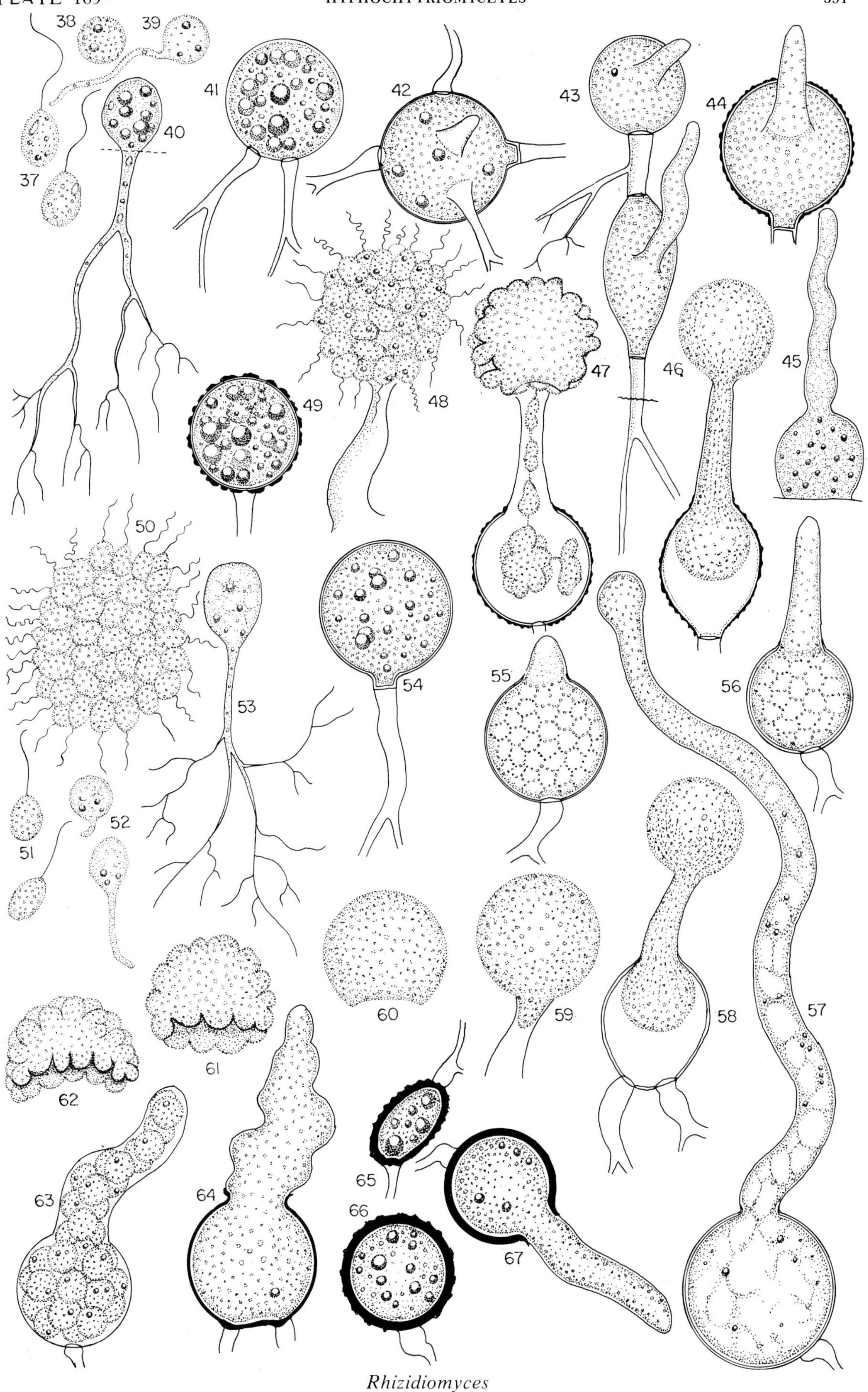

Rhizidiomyces

the zoosporangia may become thick-walled and dormant. Sexual reproduction has not been observed in this family, unless *Chytridium mesocarpi* (Fisch) Fischer proves to be a valid member. Fisch (1884) reported that the planospores of this species fuse in pairs to form a motile zygote whose content is discharged into the host and develops into a resting spore.

RHIZIDIOMYCES Zopf

Nova Acta Acad. Leop.-Carol. 47:188, 1884

Rhizidiomycopsis Sparrow 1960, Aquatic Phycomycetes p. 757

Plates 168-171

This genus is reported to include 11 species at present, but some of the saprophytic members may prove to be identical when they are studied intensively in axenic cultures. Three species, *R. apophysatus* Zopf, *R. japonicus* Kobayashi and OOkubo, and *R. parasiticus* Karling are parasites of the oogonia of water molds and sporangia of a chytrid, respectively, while *R. ichneumon* Gobi is parasitic on actively motile and encysted *Chlamydomonas globulosa*. The other species occur as saprophytes in soil and water and have been trapped on pollen grains, grass leaves, hemp seeds, and snake skin. Three of these, *R. hirsutus* Karling, *R. globosus* Karling, and *Rhizidiomyces* sp. Fuller have been isolated and grown on synthetic media.

The structure, organization, and development of the thallus in this genus are strikingly similar to that of the epi-endobiotic, eucarpic, rhizidiaceous chytrids. The planospore encysts and develops a germ tube (figs. 3, 39, 52, 73, 96) which penetrates the various hosts and substrata and develops into the rhizoidal system (figs. 4–6, 40, 53, 85, 86). At the same time the planospore body enlarges to become the incipient zoosporangium (figs. 6, 7, 97). In *R. apophysatus* (figs. 6, 7), *R. ichneumon* (fig. 24), *R. parasiticus*, and *R. coronus* Karling swelling may develop in the germ tube beneath the zoosporangium so that the latter becomes apophysate (figs. 1, 30-33), but in some such species the thalli may occasionally be non-apophysate. Also, in *R. ichneumon* the apophysis may be epibiotic (figs. 26, 27), and the zoosporangium and apophysis are continuous (fig. 29, 32).

Fairly often in liquid media the encysted planospore of *R. hirsutus* (fig. 72) and *R. bulbosus* (figs. 103, 104) may develop an exit canal and give rise to 1 or more planospores without developing rhizoids. In other thalli of *R. hirsutus* in liquid media the encysted planospore may sometimes develop several radially oriented hairs or setae (figs. 74, 75) which apparently function as absorbing organs. The planospore body then develops into a zoosporangium (fig. 76). This may occur also in *R. hansonii* (fig. 42) on vegetable debris, and such thalli have a striking resemblance to those of *Rhizophlyctis*. Occasionally in *R. hansonii* polycentric thalli may develop on the same substratum (fig. 43).

In most species the branched rhizoidal system may be fairly extensive with the branches running out to fine points (figs. 1, 40, 53, 90), but in others the system is sparse (figs. 28, 29, 105, 106) with reduced, irregularly or falsely dichotomous branching. In such cases the ends of the branches may end bluntly.

Rhizidiomyces hirsutus may vary markedly in development from that described above when grown in liquid media and on soft nutrient agar. As the germ tube grows down into the agar it may elongate to a length up to 2.4 mm, become tubular, and attain a diameter up to 24 μ. At the same time as many as 3 to 5 swellings develop along its length (fig. 79), or the tube

PLATE 170

Figs. 68-84. *Rhizidiomyces hirsutus* Karling. Figs. 68, 69, 71, 73-77 after Karling, 1945; figs. 70, 72, 78-84 after Karling, 1968.

Fig. 68. Portion of a thallus on the surface of a grass leaf; extramatrical zoosporangium with hairs; intramatrical rhizoidal axis branched with rhizoids running out to fine points.

Fig. 69. Motile ovoid, 3-4 $\times$ 6-8 μ, planospores with one fairly large and several smaller refractive granules; flagellum 14-18 μ long.

Fig. 70. Fixed and stained planospore with tinsels on the flagellum.

Fig. 71. Encysted planospore.

Fig. 72. Empty planospore cyst lying free in water with a neck which formed a single planospore.

Fig. 73. Germination of a planospore on a grass leaf.

Figs. 74, 75. Young thalli developing in a liquid medium; hairs apparently serving as absorbing organs.

Fig. 76. Similar mature thallus discharging the protoplasm through a long neck.

Fig. 77. Cleavage of protoplasm outside of the zoosporangium.

Fig. 78. Spherical zoosporangium on agar with relatively short, blunt-ended, simple and branched hairs, or rhizoids.

Fig. 79. Elongate tubular thallus growing down into agar with 3 secondary swellings or incipient zoosporangia; primary swelling or enlarged planospore body on the surface of the agar; protoplasm with numerous refractive globules.

Fig. 80. Greatly elongate thallus in agar with a terminal carrot-like swelling from which 4 secondary swellings are budding out; refractive globular material has been dispersed.

Fig. 81. Branched thallus in agar.

Fig. 82. Cluster of encysted planospore on agar above the exit canal.

Fig. 83. Cluster of encysted planospores on agar which are enlarging and developing into thalli; extent of rhizoids not shown.

Fig. 84. Spherical zoosporangium with encysted planospores.

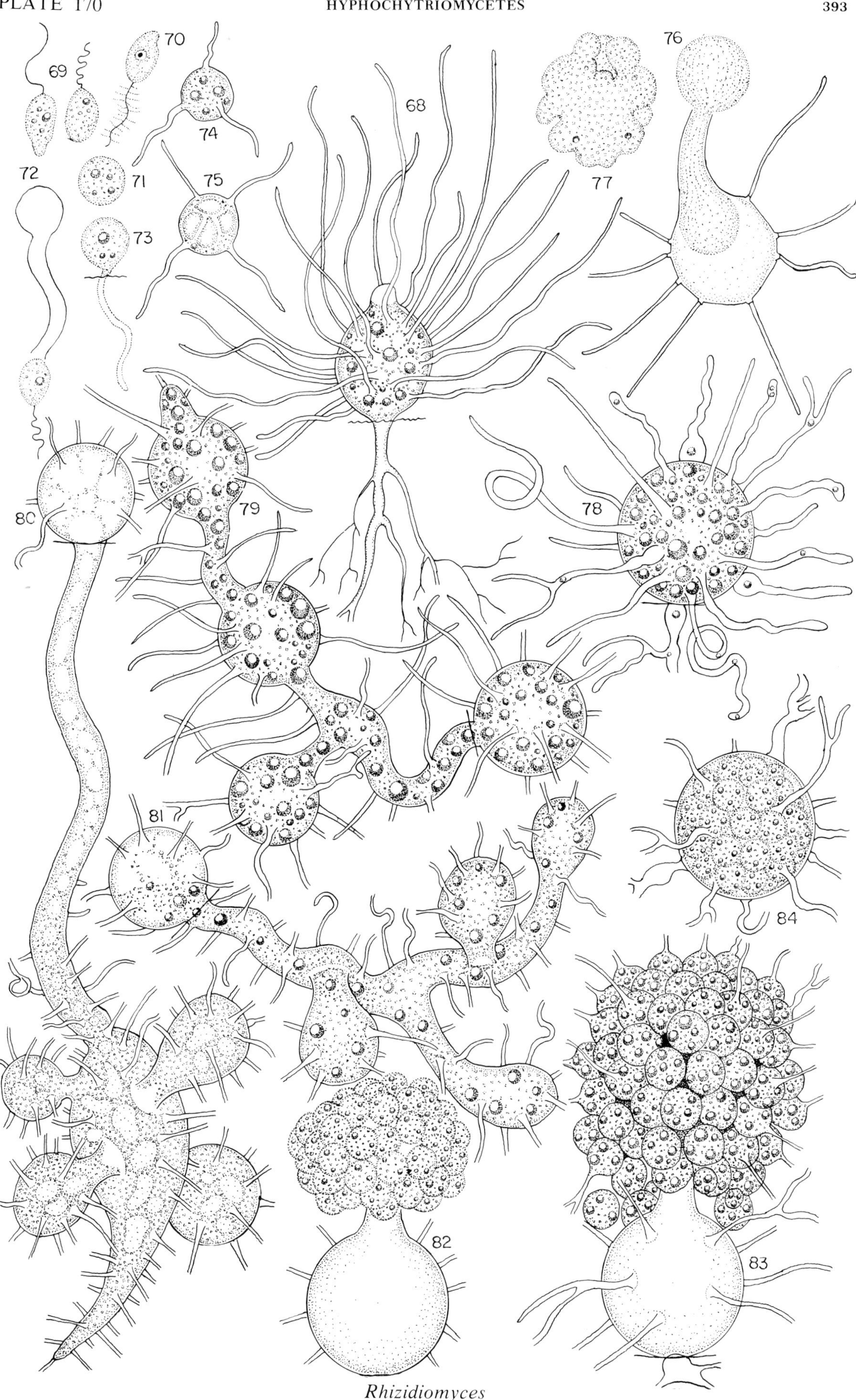

Rhizidiomyces

may branch several times with swellings (fig. 81) developing at the ends of the branches. Sometimes the end of the tube enlarges markedly and may bear several swellings (fig. 80). Simple and branched hairs, 2–3 μ diam., and up to 210 μ long, usually occur along the surface of the tube and swellings. Later, the enlarged planospore cyst or zoosporangium on the surface of the agar as well as the secondary and tertiary swellings within the agar develop exit canals and emit naked masses of protoplasm when mounted in fresh water. Thus, such thalli might be regarded as polycentric in this respect, but the swellings are not delimited by septa as are the zoosporangia of the Cladochytriaceae and Catenariaceae. So far as is known the tubular thalli are coenocytic.

The zoosporangia are predominantly spherical to subspherical, ovoid or pyriform in shape and vary markedly in size. *In R. ichneumon* they may be only 9 to 16 μ high while in *R. bivellatus*, the largest species, they may attain a diameter of 80 to 100 μ. In most species the wall is smooth and hyaline, but in *R. hirsutus* it bears numerous simple and branched hairs (figs. 68, 78–81). In *R. saprophyticus* (Karling) comb. nov. the wall is 1.5 to 2 μ thick and reddish-brown in color. In *R. bivellatus* Nabel (1939) found by microchemical tests that the wall consists of 2 layers, the inner of chitin and the outer of cellulose. His observations of the composition of the wall was confirmed by Fuller and Barshad (1960) and Fuller (1960) in *Rhizidiomyces* sp. who used X-ray analysis as well as biochemical and microchemical tests.

In the early and late developmental stages the zoosporangia contain numerous small and large refractive globules (figs. 7, 40, 41, 78, 79, 81), but as they attain maturity the refractive material becomes progressively dispersed so that the protoplasm becomes greyish-granular in appearance (figs. 8, 9, 42–44, 80, 115). Shortly after this stage 1 or more hyaline papillae develop (figs. 8, 9, 55, 68) which in most species elongate into exit canals as the granular protoplasm flows into them (figs. 9, 43, 44). However, in *R. japonicus* (fig. 87) no canal is formed, and in *R. saprophyticus* (fig. 99) and *R. bulbosus* (figs. 108, 109) the papillae do not elongate, or they become bulbous. The number, diameter and length of the canals vary. Usually, only 1 is formed, but occasionally 2 or 3 may develop. In *R. hansonii* they may rarely become up to 280 μ long, and in *R. bivellatus* Nabel reported them to be up to 300 μ in length and 10 μ in diameter. Their growth from the papillate stage on is relatively rapid and may be

PLATE 171

Figs. 85-87. *Rhizidiomyces japonicus* Kobayashi and Okubo on the oogonia of *Aplanes*. (Kobayashi and Okubo, 1954.)

Figs. 85, 86. Young and older apophysate thalli on the host.

Fig. 87. Discharge through a pore of endogenously formed ovoid planospores, 3-6 μ diam., with several refractive globules.

Figs. 88-92. *Rhizidiomyces parasiticus* Karling on the sporangium of *Rhizophlyctis* sp. (Karling, 1964.)

Fig. 88. Ovoid, 3.5-4.2 × 5.7-7 μ, planospore with refractive content.

Fig. 89. Encysted planospores.

Fig. 90. Mature epibiotic non-apophysate zoosporangium and richly branched endobiotic rhizoids.

Fig. 91. Infection.

Fig. 92. Cluster of planospores formed outside of the zoosporangium.

Figs. 93, 95. Spherical, ovoid, 4-4.2 × 5-5.5 μ, and amoeboid planospores.

Figs. 96, 97. Germinated planospore and a young non-apophysate thallus with branched rhizoids.

Fig. 98. Development of a short exit canal on a non-apophysate zoosporangium.

Fig. 99. Mature zoosporangium with 3 short exit canals and endogenously formed planospores; wall smooth thick and reddish brown.

Fig. 100. Discharge of planospores.

Figs. 101-113. *Rhizidiomyces bulbosus* Karling saprophytic on various substrata. (Karling, 1968.)

Fig. 101. Ovoid, 3-3.8 × 4-4.5 μ, planospores with several refractive globules; flagellum 12-16 μ long.

Fig. 102. Encysted planospores.

Fig. 103. Young thallus or encysted planospore with an exit canal; later it formed 2 planospores; rhizoids lacking.

Fig. 104. Non-rhizoidal thallus forming a bulbous exit canal.

Figs. 105, 106. Young and mature non-apophysate zoosporangia with irregular blunt-ended rhizoids.

Fig. 107. Underside view of a zoosporangium and rhizoids; exit papillae or canals forming.

Fig. 108. Zoosporangium with 4 bulbous exit canals; endogenous cleavage of planospores.

Fig. 109. Earlier stage of migration of protoplasm into bulbous exit canal.

Fig. 110. Planospores formed both exogenously and endogenously.

Fig. 111. Exit papilla or canal after discharge.

Fig. 112. Endo- and exogenous cleavage of protoplasm.

Fig. 113. A zoosporangium with encysted planospores within and on the outside.

Figs. 114-117. *Rhizidiomyces coronus* Karling saprophytic on various substrata. (Karling, 1968.)

Fig. 114. A young zoosporangium with a small, enveloping, striated coronum.

Fig. 115. A mature zoosporangium with a coronum and 3 exit canals; protoplasm emerging through one of them.

Fig. 116. Endo- and exogenous cleavage of the protoplasm.

Fig. 117. Ovoid, 3-3.6 × 5-5.5 μ, planospores with refractive granules; flagellum 12-15 μ long.

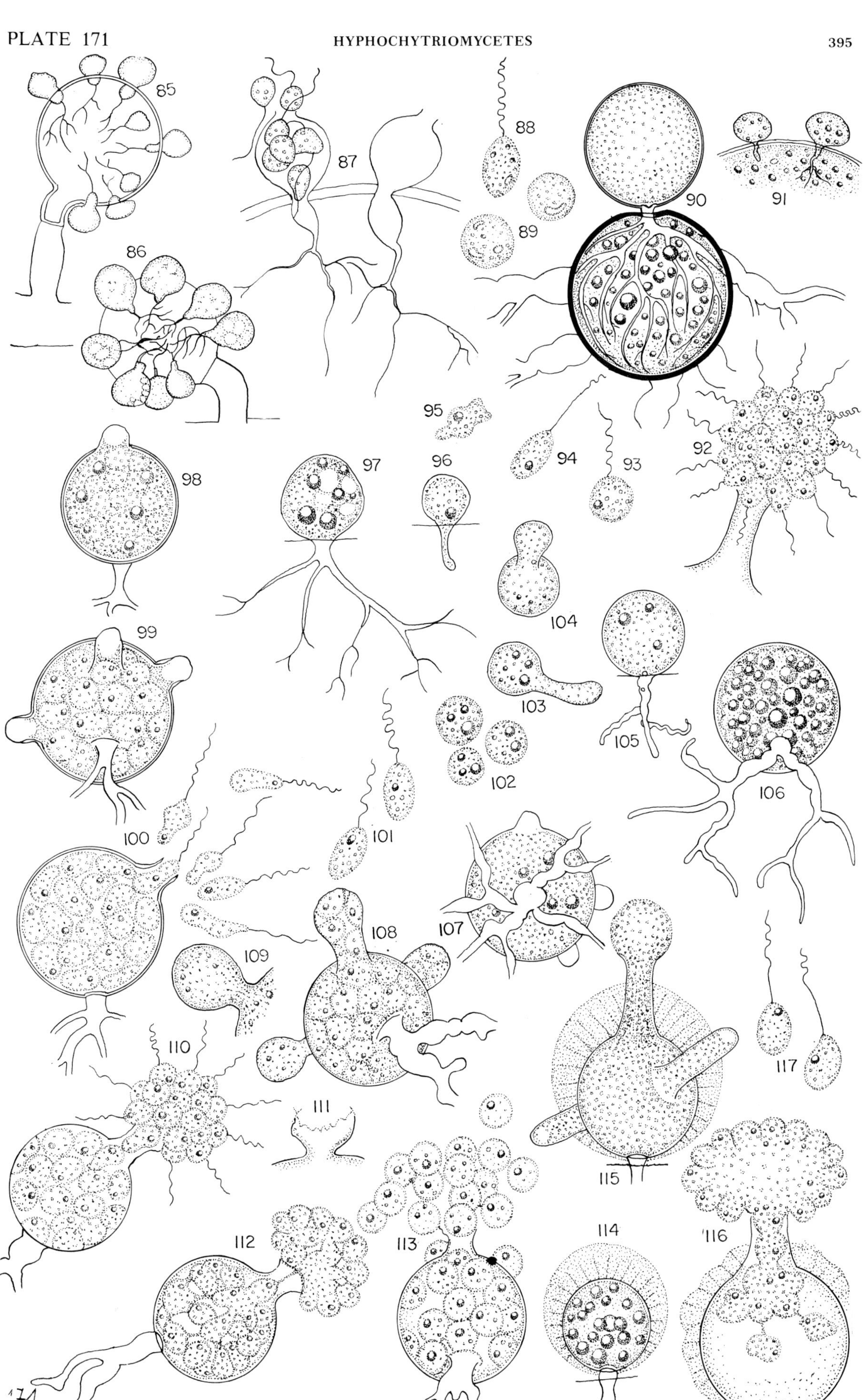

Rhizidiomyces

completed in 15 minutes. As the tube attains its full length the tip begins to expand (figs. 10, 115) as the protoplasm starts to move upward. Nevertheless, the zoosporangium is still full of protoplasm at this stage. The tip of the tube continues to expand as the upward movement of the protoplasm accelerates, and in a short while the basal portion of the zoosporangium becomes empty (figs. 11, 46, 58, 76). In the expansion of the tip its wall appears to become progressively thinner until it is no longer perceptible as such. Eventually in most species the entire protoplasmic mass emerges and lies as a globule near the exit orifice (figs. 12, 33, 59). Karling (1944) reported that its mass is naked, although Nabel (1939) and later Fuller (1962) claimed that it is enveloped by a vesicular membrane. Subsequent observations have confirmed Karling's studies in this respect.

Shortly after emergence the protoplasmic mass may undergo changes in shape which may involve invagination on the periphery nearest the exit canal (fig. 60). In the case of minute zoosporangia in which the mass is quite small the granules in the protoplasm may move about quite rapidly, and the whole mass may behave as an amoeba. Normally in large masses ceavage furrows soon appear (figs. 13, 47, 61, 62, 77) and divide the mass into planospore initials. Shortly thereafter the anterior flagellum becomes visible and begins to beat with a wave-like motion whereby the spore initials undergo a rocking movement (figs. 14, 50, 92). This beating of the flagellum and rocking movement of the spore body accelerates, and within a few minutes the planospores disperse.

Variations in the place of cleavage and sporogenesis from that described above frequently occur. If small or large portions of the protoplasm fail to emerge they usually undergo cleavage within the zoosporangium (figs. 47, 110, 112, 116), and in such cases the incipient planospores may adhere to the emerged mass by a protoplasmic filament. In other instances when the exit canal remains closed cleavage is entirely endogenous (figs. 63, 84), and the planospores encyst within. Encystment may, also, occur within and outside of the zoosporangium when the exit canal is open (figs. 83, 113). When *R. hirsutus* is grown on a soft agar medium the planospores may encyst in a mass at the exit orifice (fig. 82) without becoming motile, and these may enlarge, develop rhizoids and become secondary thalli and zoosporangia, thus forming a heaped-up mass on the agar (fig. 83). Such thalli in turn form planospores which encyst and develop into tertiary thalli and zoosporangia, whereby conspicuous mounds of thalli are formed on the surface of the agar.

In *R. japonicus* and *R. saprophyticus* cleavage is normally endogenous so that the planospores emerge fully formed (figs. 87, 100), but in the latter species it is occasionally exogenous, also. Sparrow (1960) erected the genus *Rhizidiomycopsis* for *R. japonicus* because its planospores are formed endogenously and emerge through a pore in the wall (fig. 87), and later *R. saprophyticus* was tentatively assigned to this genus (Karling, 1967). However, in view of the occasional occurrence of cleavage within the zoosporangium of other species as well as in species of *Hyphochytrium* it is doubtful that the site of cleavage and maturation of the planospores is a sharply-defined generic criterion, and for this reason *Rhizidiomycopsis* is merged with *Rhizidiomyces*.

The planospores of this genus are predominantly ovoid (figs. 15, 16, 35, 37, 69, 101, 117) but may be subspherical (fig. 35) or oblong to elongate (fig. 101) with greyish-granular protoplasm containing 1 or more small refractive granules. The tinsellate anterior flagellum is relatively short, 8–16 μ, in most species, but in *R. bivellatus* it may be up to 20 μ long. Occasionally, the planospores become amoeboid, (fig. 95), but usually they are actively motile. Their swimming movement is usually quite characteristic and different from the

PLATE 172

Figs. 1-11. *Latrostium comprimens* Zopf in oogonia of *Vaucheria* species. (Zopf, 1894.)

Fig. 1. Epibiotic ellipsoidal zoosporangium with an endobiotic tuft of rhizoids on a collapsed oospore of the host.

Fig. 2. Collapsed oospore parasitized by 4 zoosporangia.

Fig. 3. Surface view of a zoosporangium.

Fig. 4. Oospore parasitized by 2 zoosporangia and a resting spore.

Fig. 5. Half-ripe zoosporangium in profile.

Fig. 6. Same zoosporangium 18 hours later; planospore fully formed and emerging through a wide apical pore.

Fig. 7. Ovoid, 2.5-3 μ diam., planospores with a large hyaline refractive globule and a long anterior flagellum.

Fig. 8. An empty and collapsed zoosporangium.

Figs. 9, 10. Oogonia with 3 and 5 striated, thick-walled, smooth, hyaline resting spores, resp.; degenerated content of oospores of host omitted.

Fig. 11. Two broadly lenticular, 30-50 μ diam. resting spores with thick, hyaline, smooth, striated walls and containing 1 unusually large refractive globule.

Figs. 12-19. *Rhizophydium decipiens* Scherffel in the oogonia of *Oedogonium*. (Scherffel, 1926.)

Fig. 12. Empty zoosporangium on the oospore of *O. cardiacum*.

Fig. 13. A zoosporangium with 2 spherical, 3-4 μ diam., planospores.

Fig. 14. Three empty zoosporangia in the oogonium of *O. cardiacum*.

Fig. 15. A resting spore with a thick striated wall in the oogonium of *O. vaucherii*.

Fig. 16. Two ovoid resting spores in the oogonium of *O. cardiacum*.

Figs. 17-19. Enlarged views of subspherical and ovoid, 18-39 μ diam., resting spore with a thick, 2-6 μ, hyaline, smooth, occasionally striated walls and containing numerous small refractive globules.

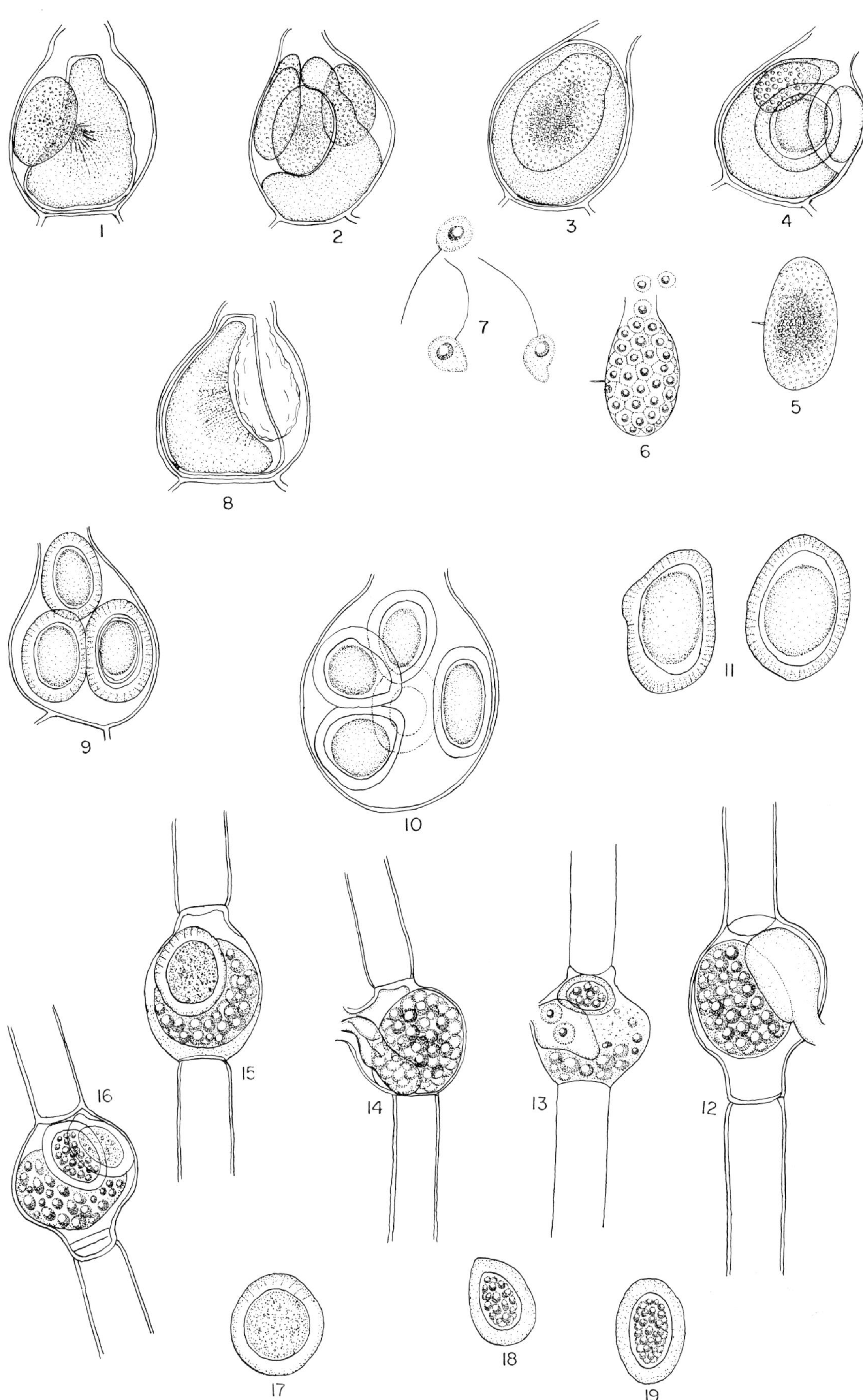

Latrostium

darting movement of most chytrid planospores. The path of movement is a spiral one. They may swim forward for some distance, stop, back up, and then go forward again.

The ultrastructural study of *R. apophysatus* by Fuller and Reichle (1965) shows that the planospore body has the usual organelles of other planospores, but like those of *Monoblepharella* lack a membrane-enclosed nuclear cap. Instead, the ribosomes lie free in the cytoplasm but are aggregated around the nucleus (fig. 22 rr). In addition the body contains numerous mitochondria (m), a collar-like dictyosome (d), a basal body (bb) with its associated centriole (c), membrane-bound groups of tubules, endoplasmic reticulum (er), lipid bodies, and numerous vacuoles. The flagellum develops in relation to the larger of the two centrioles, which becomes transformed into the basal body (bb). The blunt-ended flagellum consists of a central axoneme surrounded by a sheath which is enlarged at the base (fig. 18) and bears lateral projecting tinsels or mastigonemes. The tinsels are spirally banded for ⅔ of their length (fig. 21) and the terminal ⅓ consists of 1 or 2 fine filaments. When the flagellum is "pulled back" into the encysting planospore the tinsels become spread over about ⅔ of the cyst surface (fig. 20).

Resting spores as such have not been reported in *Rhizidiomyces,* but in old cultures of *R. bivellatus* (fig. 49) and *R. hansonii* (figs. 65, 66) thick-walled bodies occur which remain dormant for weeks and months. They "germinate" and produce planospores in the same manner as the thin-walled zoosporangia, and for this reason they are regarded as dormant zoosporangia instead of true resting spores.

LATROSTIUM Zopf

Beitr. Physiol. Morph. niederer Organismen
4:62, 1894

Plate 172

This genus includes a single species, *L. comprimens* Zopf, which is parasitic on the oospores in oogonia of *Vaucheria sessilis* and *V. terrestris*. It has been reported only 2 times as such, initially in Germany by Zopf and later in Belgium and Swtizerland by De Wildemann (1895). It is characterized by an epibiotic broadly lenticular, thin-walled, hyaline and smooth zoosporangium, which lies between the oogonial wall and the oospore, and an endobiotic bushy tuft of delicate rhizoids (fig. 1). One to 6 parasites may occur in an oogonium, and as their zoosporangia enlarge the shape of the host oospore is greatly altered (figs. 1, 2, 4, 8). The sporangial thalli occur in early spring, and they are followed in late spring and early summer by the development of resting spores with delicate rhizoids.

The method of infection is unknown, but the planospores presumably enter the oogonial opening, encyst on the oospore, and germinate. As the incipient zoosporangium or resting spore expands it collapses the oospore (figs. 1, 2, 4, 8), and the latter's wall and content eventually disintegrate. The mature zoosporangium has a broad apical papilla (fig. 5) which deliquesces and releases the fully formed planospores (fig. 6). These are irregularly ovoid (fig. 7), 2.5–3 μ diam, with a broad anterior end and a long flagellum and contain a large hyaline globule. After dehiscence is completed the empty zoosporangium collapses (fig. 8).

The resting spores are broadly lenticular and flattened with a thick, hyaline, smooth, radially striated exospore and a thin endospore and contain an unusually large fat globule that nearly fills the lumen (figs.

PLATE 173

Figs. 1-9. *Hyphochytrium infestans* Zopf on the ascocarp of a *Helotium*-like species of the Pezizaceae. (Zopf, 1884.)

Fig. 1. Ascocarp densely infected with the hyphae and zoosporangia of *R. infestans*.

Figs. 2-4. Variations in the shapes of intercalary zoosporangia and the septate hyphae.

Fig. 5. Intercalary zoosporangium showing the site of the exit pore.

Figs. 6, 7. A terminal apiculate zoosporangium in 2 stages of maturation.

Fig. 8. Discharge of planospores through a subapical pore in the same zoosporangium.

Fig. 9. Three spherical planospores with a relatively long flagellum.

Figs. 10-18. *Hyphochytrium hydrodictyii* Valkanov in young cells of *Hydrodictyon reticulatum*. (Valkanov, 1929.)

Fig. 10. Infection and some small thalli; cyst of the planospore persistent on the outside of the host.

Fig. 11. Young thallus or incipient primary zoosporangium developing 2 hyphae at its ends.

Fig. 12. Line drawing showing a zoosporangium and the branched mycelium.

Fig. 13. A late stage in the maturation of a zoosporangium.

Fig. 14. A mature of "ripe" zoosporangium with numerous refractive globules and a broad exit papilla near the base.

Figs. 15, 16. Spherical and ovoid planospores with a broad anterior end, a refractive globule, and a long flagellum.

Figs. 17, 18. Terminal apiculate and intercalary resting spores filled with refractive globules.

Figs. 19-22. *Hyphochytrium penilliae* Artemchuk and Zelezinskaya in planktonic crawfish, *Penilia avirostris.* (Artemchuk and Zelezinskaya, 1969.)

Fig. 19. Portion of the branching mycelium, 4-7 μ diameter.

Fig. 20. Germination of an encysted planospore, 4.6-6.9 μ diameter.

Figs. 21, 22. Branched hyphae bearing young and mature smooth colorless zoosporangia, 16.8-19.3 × 23.7-27.3 μ, which lack an apiculus and exit tube, and form planospores endogenously.

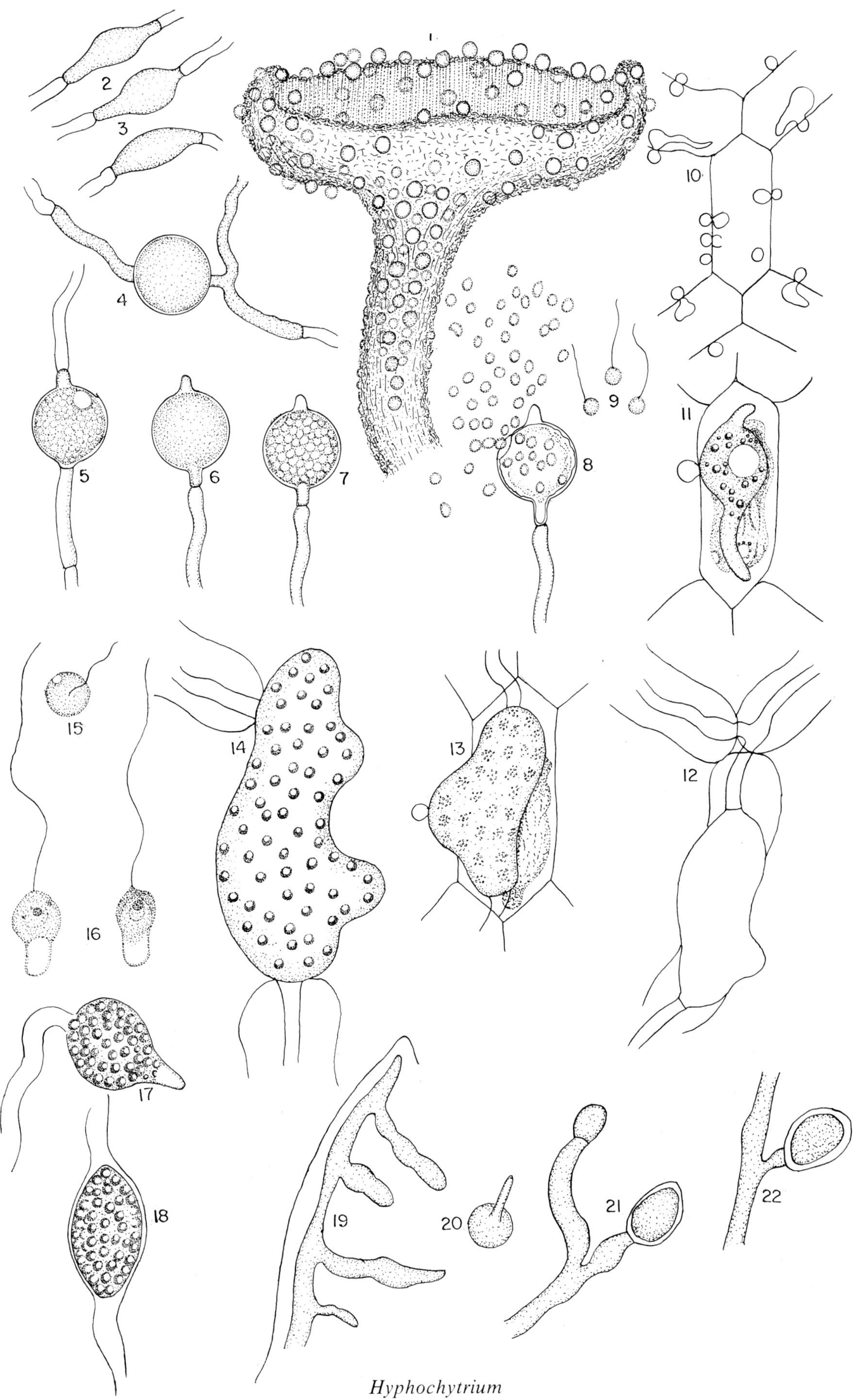

Hyphochytrium

9–11). The wall reacts *positively* for cellulose by treatment with chloro-iodide of zinc. Germination of the resting spores is unknown.

De Wildemann found no zoosporangia but only resting spores in his material, and Sparrow (1960) suggested that the former's fungus might be *Rhizophydium decipiens* which Scherffel (1926) described from the oogonia of *Oedogonium*. Minden (1916), on the other hand, suggested earlier that this species might be identical with *L. comprimens*, although the latter reported that it is a true chytrid with posteriorly uniflagellate, darting planospores. As shown in figs. 12 to 16 for comparative purposes the zoosporangia and resting spores of *R. decipiens* lie between the oogonial wall and oospore, and press it aside as they expand. However, no rhizoids were found on either the sporangial and resting spore thalli. The resting spores (figs. 15–19) have a thick hyaline, smooth and sometimes radially striated wall as in *L. comprimens* but contain numerous smaller refractive globules instead of a large one, and Scherffel regarded the latter difference together with the structure and behavior of the planospores as excluding it from *Latrostium*.

HYPHOCHYTRIACEAE

This is a small family of 1 genus and 5 identified species which are reported to differ from those of the previous family by the development of a septate mycelioid polycentric thallus with terminal and intercalary zoosporangia. However, as in the Rhizidiomycetaceae some species may additionally and occasionally develop monocentric, eucarpic thalli and even monocentric holocarpic ones as in the Anisolpidiaceae. Accordingly, marked variations in development, structure, and organization may occur, and the family distinctions are not always clearly defined. Additional details of the family characteristics, development, and variations are given below in the illustrations and description of *Hyphochytrium*.

HYPHOCHYTRIUM Zopf

Nova Acta Acad. Leop.-Carol. 47:187, 1884

Hyphophagus (Zopf) Minden, 1911. Krytogamen- fl. Mark Brandenburg 5:420

Plates 173-175

This first named genus of the order includes 5 species and an unidentified member which are parasitic or occurs as saprophytes in soil in many parts of the world. *Hyphochytrium infestans* Zopf occurs in the ascocarp of a *Helotium*-like fungus, *H. hydrodictyii* Valkanov is a destructive parasite of *Hydrodictyon reticulatum*, and *H. penilae* Artemchuk and Zelezinskaya, the only known marine species, causes a severe mycosis and death of planktonic crawfish. The other species, *H. catenoides* Karling, *H. oceanum* Karling and *Hyphochytrium* sp. (H. 28) Persiel occur in soil throughout most parts of the world and may be trapped on various substrata and grown on synthetic media. So far, *H. catenoides* appears to be the best known and widely distributed species. Siang (1949) isolated it on agar plates exposed on the roof of a building and believed that it might be airborne, a belief questioned by Barr (1970) who isolated it from several diverse habitats and made morphological and physiological studies of the various isolates.

Usually, the species are characterized by a polycentric thallus with intercalary and terminal zoosporangia borne on coarse hyphae or a mycelium. However, in *H. catenoides, H. oceanum,* and *Hyphochytrium* sp. monocentric, holocarpic, olpidioid as well as eucarpic thalli occur (figs. 40, 44, 54–60). The latter type of thallus may occur so commonly in *H.*

PLATE 174

Figs. 23-40. *Hyphochytrium catenoides* Karling. (Karling, 1939.)

Fig. 23. Portion of a large thallus with swellings and zoosporangia connected by hyphae: A = zoosporangium dishcarging fully-formed planospores; B = encysted planospores within and outside of the zoosporangium; C = spherical globule of emerged protoplasm; D = planospores shortly before dispersing; E = planospores delimited endogenously.

Fig. 24. Ovoid and elongate, 1.5-3.5 × 2.3 μ, planospores with numerous granules, and a 7-10 μ long flagellum.

Fig. 25. Spherical planospore shortly before coming to rest.

Fig. 26. Fixed and stained planospore; tinsels on flagellum omitted.

Figs. 27-31. Variations in planospore germination.

Fig. 32. Young thallus with persistent extramatrical planospore cyst.

Fig. 33-35, 37. Stages in the development of a thallus; planospore cyst disappeared early in fig. 33.

Fig. 36. Elongate thallus in which the protoplasm has accumulated in fusiform and spherical swellings.

Fig. 38. Bead-like thallus consisting of 4 inflated segments; 2 connected by short isthmuses.

Fig. 39. Emerged protoplasmic mass undergoing changes in shape before cleavage.

Fig. 40. Monocentric holocapric thallus in a pollen grain of *Pinus sylvestris* discharging its protoplasm.

Figs. 41-46. *Hyphochytrium* (H. 28) sp. (Persiel, 1960.)

Fig. 41. Ovoid, 3-4 × 5-8 μ, planospore with several refractive globules; flagellum 8-10 μ long.

Fig. 42. Infection of pollen grain of *Corylus*.

Fig. 43. Young thallus in pollen grain; planospore cyst has disappeared.

Fig. 44. Four thalli in a pollen grain; one thallus apparently holocarpic and discharging its protoplasm.

Figs. 45, 46. Developmental stages of 2 thalli.

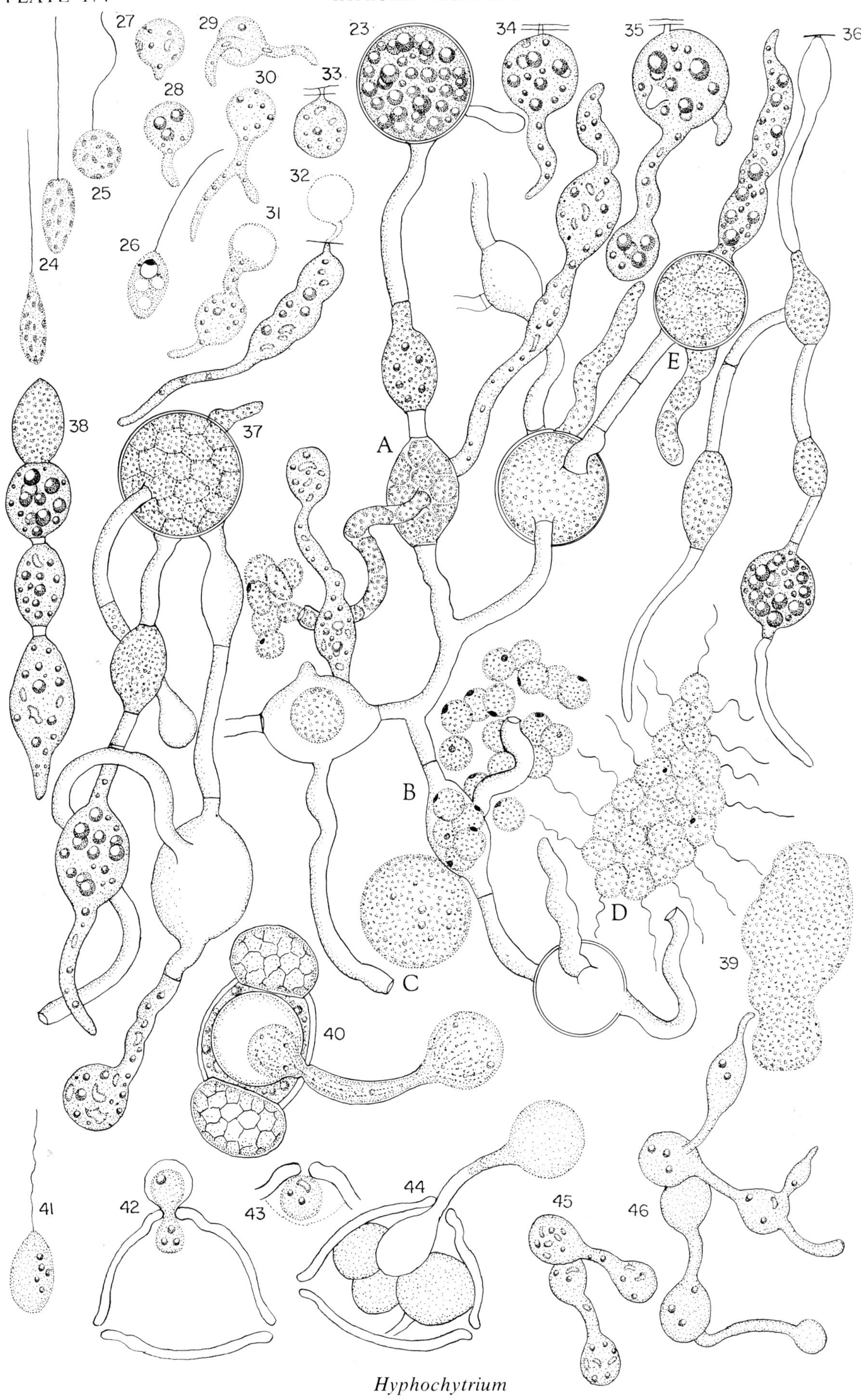

Hyphochytrium

oceanum that its inclusion in this genus and family is questionable. The mycelium may be 2.2 to 4 μ in diameter, irregular, septate, branched, (figs. 4, 12, 19, 23) and ramifies the host or substratum.

The ends of the hyphae usually end bluntly (figs. 19, 23, 46, 57–59), but in the case of some monocentric thalli they may run out to fine points (fig. 60). The zoosporangia are delimited by septa and are usually connected by short (fig. 38) or long isthmuses (fig. 23). They may be spherical, 10–75 μ, ovoid, 10–12 × 18–22 μ, spindle-shaped, elongate, irregular, 30–40 × 65–100, and deeply lobed (figs. 7, 11, 14, 37, 59, 61, 62). At maturity they develop a subapical pore (figs. 5, 8), a papilla (fig. 14), or 1 to 6 exit canals (fig. 62) which may be relatively short (fig. 59) or straight, curved, or contorted (figs. 23A, C, 37), 3 to 6 μ diam., and up to 250 μ long.

In *H. infestans* (fig. 7, 8), *H. penilae*, and apparently in *H. hydrodictyii* the planospores are delimited within the zoosporangium and emerge fully formed, but in *H. catenoides, H. oceanum* and *Hyphochytrium* sp. the protoplasm emerges as a naked mass (fig. 23C, 39, 44, 63) and cleaves into planospores. However, in the first 2 species named above cleavage may sometimes occur partly or wholly endogenously, also (figs. 23A, B, E, 37). The process of cleavage in such cases is basically similar to that reported for species of the Rhizidiomycetaceae and need not be described again.

The sizes of the planospores are known only in *H. catenoides*, 1.5–2 × 3–3.5 μ (fig. 24), *H. oceanum*, 3.3–3.8 × 5–5.5 μ (fig. 47), *H. peniliae*, 4.6–6.9 μ, and *Hyphochytrium* sp. 3–4 × 5–8 μ and have an anterior flagellum which varies from 8 to 15 μ in length in the different species or may be very long (fig. 16). The planospores are ovoid to elongate in shape with a broad or attenuated anterior end and contain one to several small refractive globules of granules. Frequently, they become amoeboid (fig. 48) and after a period of motility encyst (figs. 49, 50). After germinating the empty cyst may be persistent on the outside of the host for a long time in *H. hydrodictyii* (figs. 11, 13), but in other species it disappears very shortly (figs. 33, 43, 53, 54). In germination the germ tube may branch (fig. 52) or its tip enlarges into a globular (figs. 33–35) or elongate body along the host wall (fig. 10) from which 2 to several hyphae (figs. 11, 53, 54) arise, elongate, and branch (figs. 12, 37).

Resting spores are known only in *H. hydrodictyii*, and these are apparently formed asexually. They may be terminal or intercalary (figs. 17, 18) with a thick inner wall and filled with refractive globules. Germination has not been reported, but Valkanov found a few papillate spores which might suggest that they function as a sporangium in the process.

Hyphochytrium infestans was regarded by Vuillemin (1907) as a filamentous fungus parasitized by a chytrid, and for this reason Minden created a new genus, *Hyphophagus*, in the Cladochytriaceae for it. Nonetheless, he retained the name Hyphochytriaceae as defined by Schroeter as a family for *Macrochytrium*, *Zygochytrium* and *Tetrachytrium*. However, Valknow, Karling, and Persiel, and other investigators have confirmed Zopf's report of a group of fungi with the characteristics of *Hyphochytrium*.

IMPERFECTLY KNOWN, DOUBTFUL AND EXCLUDED GENERA

REESIA Fisch

Sitzungber Phys.-Med. Soc. Erlangen 16:41, 1884

This genus has been illustrated, described, and discussed in relation to the Olpidiaceae.

CHYTRIDIUM MESOCARPI (Fisch) Fischer

Rabenhorst Kryptogamen-Fl. 1:126, 1892

This questionable species was named *Euchytridium mesocarpii* by Fish (1884, l.c. p. 101) who described the planospores as being anteriorly uniflagellate. The

PLATE 175

Figs. 47-63. *Hyphochytrium oceanum* Karling. Figs. 47-59 after Karling, 1967; fig. 60 after Karling, 1968.

Fig. 47. Ovoid, 3.5-3.8 × 5-5.5 μ, planospores with one larger and several smaller refractive granules; flagellum 12-15 μ long.

Fig. 48. Amoeboid planospore.

Fig. 49. Encysted planospores.

Fig. 50. Same cysts after enlarging.

Fig. 51. Infection of a grass leaf through the stoma.

Fig. 52. Branching of germ tube in the substratum.

Figs. 53, 54. Developmental stages of monocentric, eucarpic thalli with several rhizoids or hyphae.

Figs. 55-57. Developmental stages of a monocentric thallus which formed only 2 blunt-ended hyphae at the base of the zoosporangium.

Fig. 58. Irregular thallus with 1 blunt-ended, curved hypha.

Fig. 59. Irregular monocentric eucarpic thallus with several blunt-ended hyphae arising from the periphery of the zoosporangium which is beginning to discharge its protoplasm.

Fig. 60. A monocentric eucarpic thallus in snake skin with blunt-ended hyphae and rhizoids running out to fine points.

Fig. 61. A polycentric thallus with a large spherical primary zoosporangium and 8 secondary ones arising from its periphery.

Fig. 62. An irregular, unusually large zoosporangium from an extensive polycentric thallus with 6 exit canals; islands of protoplasm beginning to emerge through 5 of the canals.

Fig. 63. Invagination of emerged protoplasmic mass and cleavage.

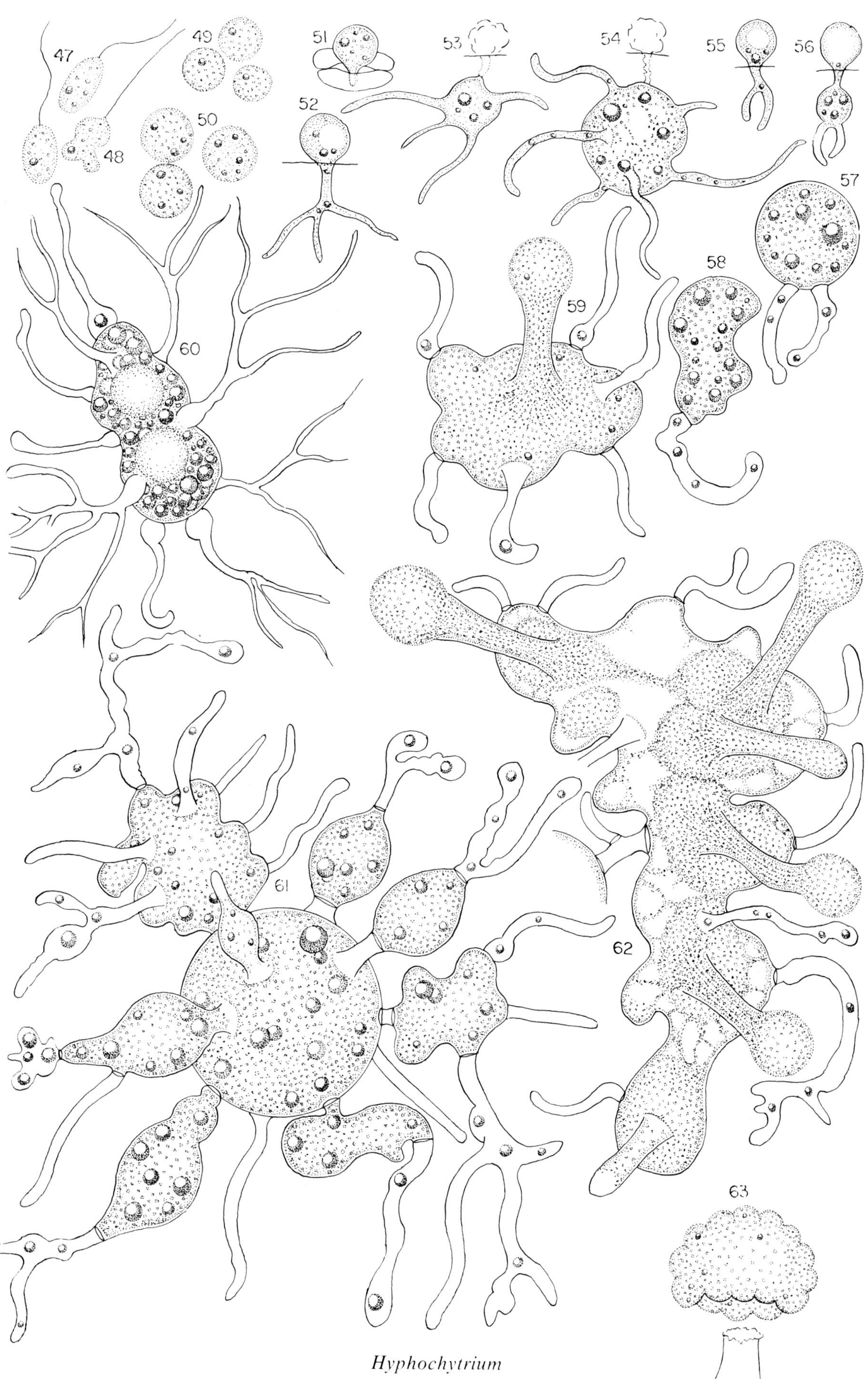

Hyphochytrium

monocentric eucarpic thallus consists of an operculate, flask-like, brownish, thick-walled, smooth zoosporangium with a broad base and tubular apex, and a delicate, rarely branched rhizoid. The comparatively large anteriorly uniflagellate planospores with greyish granular protoplasm emerge suddenly from the tip of the zoosporangium after dehiscence by a circular operculum. After a period of motility they pair and fuse at the anterior ends to form a slow swimming zygote which soon comes to rest on the host and discharges its content into the alga by a small papilla. There, the content or protoplast enlarges, develops a double wall and becomes a resting spore which germinates within a short time. In this process it produces planospores which infect other cells and form new thalli and zoosporangia. Fisch's account is not fully clear because he says that all planospores from a zoosporangium fuse, but at the same time he states that those which do not fuse never form new zoosporangia. Thus, it appears that only planospores from germinated resting spores are capable of forming new zoosporangia or gametangia.

Fish did not illustrate this species, and Fischer and Minden doubted that its planospores conjugate. In the event Fisch's observations on the presence of an operculum and anteriorly uniflagellate planospores or gametes are confirmed a new genus should be created for this species, as suggested by Sparrow (1960).

CYSTOCHYTRIUM Cook

Trans. Brit. Mycol. soc. 16:251, 1932
Arch. f. Protistenk. 66:288, 1929

Cook (1932) described the planospores of *C. radicale* Cook from roots of *Veronica beccabunga* as having an apical flagellum, and from this and his figure 1 it is assumed here that the planospores are anteriorly uniflagellate. It is on this assumption that *Cystochytrium* is discussed in relation to the anisochytrids.

The thallus is monocentric and holocarpic and resembles that of *Olpidium* in its early stages. It soon elongates until it is 10 to 25 times its diameter and may become swollen and curved with rudimentary branches or protuberances. At this stage the wall thickens, and a sheath of chitin is laid down around the thallus. The cytoplasm fills the thallus at first but later contracts from the ends. In some instances masses of protoplasm separate so that several may be present in a sheath, and these become separated by cross walls. Thus, several zoosporangia surrounded by a common membrane are formed. Each zoosporangium produces 6 to 12 large globose, 12–15 μ diam., planospores which emerge through a perforation or aperture midway in the sporangial wall. After a period of motility they may migrate into neighboring host cells. No resting spores *per se* occurred.

Cook (1932) suggested that this genus should be placed in the Hyphochytriales close to *Hyphochytrium*, but the lack of a mycelium or rhizoids militates against this suggestion. Its monocentric holocarpic thallus is more like that of *Anisolpidium* species.

Possibly, *Zygorhizidium vaucheriae* Rieth is a species of the Hyphochytriomycetes. Reith described and illustrated the planospores as being anteriorly uniflagellate, and if the flagellum turns out to be tinsellate this species should be removed from the Chytridiales and placed as a representative of a new genus with sexuality in the family Rhizidiomycetaceae. However, the walls of the thallus do not show a positive reaction when tested for cellulose.

CATENARIOPSIS non nudum Couch

Amer. J. Bot. 28 707, 1941

This genus, mentioned only by name, was regarded by Karling (1943) as a species of *Hyphochytrium*, possibly *H. catenoides*.

REFERENCES TO THE HYPHOCHYTRIOMYCETES

Artemchuk, N. Ja. and L. M. Zelezinskaya, 1969. *Hyphochytrium peniliae* n. sp. affecting planktonic crawfish *Penilia avirostris* Nana. Mikol. Fitopatol. 3:356-358, 2 figs.

Barr, D. J. S. 1970. *Hyphochytrium catenoides*: a morphological and physiological study of North American isolates. Mycologia 62:492-503, 19 figs.

Bessey, E. A. 1942. Some problems in fungus phylogeny. Mycologia 34:355-376, 5 figs.

Canter, H. M. 1950. Studies on British chytrids. IX. *Anisolpidium stigeoclonii* (de Wildemann) n. comb. Trans. Brit. Mycol. Soc. 33. 335-344, 6 figs., pls. 24-26.

Cook, W. R. J. 1929. A preliminary account of a new species of Protista. Arch. f. Protistenk. 66:285-289, pl. 7.

_______ 1932. The life-history of *Cystochytrium radicale*, occurring in the roots of *Veronica beccabunga*. Trans. Brit. Mycol. Soc. 16:246-252, 19 figs. pl. 10.

Couch, J. N. 1941. The structure and action of the cilia in some aquatic Phycomycetes, Amer. J. Bot. 28:704-713, 58 figs.

Fuller, M. S. 1960. Biochemical and microchemical study of the walls of *Rhizidiomyces* sp. Amer. J. Bot. 47:838-842, 7 figs.

_______. 1962. Growth and development of the water mold *Rhizidiomyces* in pure culture. Ibid. 49: 64-71, 24 figs.

_______, and I. Barshad. 1960. Chitin and cellulose in the walls of *Rhizidiomyces* sp. Ibid. 47:105-109, 6 figs.

_______, and R. Reichle. 1965. The zoospore and early development of *Rhizidiomyces apophysatus*. Mycologia 57:946-961, 22 figs.

Fisch, C. 1884a. Beiträge zur Kenntniss der Chytridiaceen. Sitzungsb. Phys.-Med. Soc. Erlangen 16:29-72, 39 figs., pl. 1.

———. 1884b. Ueber zwei neue Chytridiaceen. Ibid. 16:101-103.

Fischer, A. 1892. Phycomycetes. Die Pilze Deutschlands, Oesterreichs und der Schweiz. Rabenhorst Kryptogamen-Fl. 1:1-490. Leipzig.

Gobi, C. 1900. O nouom parazitnom gribkie (Ueber einen neuen parasitischen Pilze) *Rhizidiomyces ichneumon* n. sp. Scripta Bot. Horti Univ. Imp. Petro. 15:227-272, pls. 6, 7.

Johnson, T. W. 1957. Resting spore development in the marine Phycomycete *Anisolpidium ectocarpii*. Amer. J. Bot. 44:875-878, 26 figs.

Karling, J. S. 1939. A new fungus with anteriorly uniciliate zoospores: *Hyphochytrium catenoides*. Amer. J. Bot. 26:512-519, 18 figs.

———. 1943. The life history of *Anisolpidium ectocarpii* gen. nov., et. sp. nov., and a synopsis and classification of other fungi with anteriorly uniflagellate zoospores. Ibid. 30:637-648, 21 figs.

——— 1944. Brazilian anisochytrids. Ibid. 31: 391-397, 64 figs.

———. 1945. *Rhizidiomyces hirsutus* sp. nov., a hairly anisochytrid from Brazil. Bull Torrey Bot. Club. 72:47-51, 19 figs.

———. 1964. Indian anisochytrids. Sydowia 17: 193-196, 6 figs.

———. 1967. Some zoosporic fungi of New Zealand. IX. Hyphochytridiales or Anisochytridiales. Ibid. 20:137-142, pls. 24-28.

———. 1968. Zoosporic fungi of Oceania. J. Elisha Mitchell Sci. Soc. 84:166-178, 62 figs.

Kny, L. 1871. *Chytridium (Olpidium)spacellarum*. Sitzungsb. Gesell. Naturforsch. Freunde zu Berlin 1871: 93-97.

Kobayashi, Y. and M. Okubo, 1954. Studies on the aqutic fungi of the Ozegahara moor. Rept. Ozegahara Gen. Sci. Surv. Comm. 1954:561-575, 18 figs. (in Japanese).

Le' John, H. B. 1972. Enzyme regulation, lysine pathways, and cell wall structure as indicators of major lines of evolution in fungi. Nature 231:164-168.

Lovett, J. S., and J. A. Haselby. 1971. Molecular weights of the ribosomal ribonucleic acid in fungi. Arch. f. Mikrobiol. 80: 191-204.

Minden, M. von. 1911. Chytridineae, Ancylistinease, Monoblepharidineae, Saprolegniineae. Kryptogamenfl. Mark Brandenburg 5:191-630.

———.1916. Beiträge zur Biologie und Systematik einheimischer submerser Phycomyceten. *In* Falck, Mykolog. Untersuch. Ber. 2:146-255, pls. 1-8, 24 figs.

Nabel, K. 1939. Über die Membran niederer Pilze, besonders von *Rhizidiomyces bivellatus* nov. spec. Arch. f. Mikrobiol. 10:515-541, 7 figs.

Persiel, I. 1960. Beschreibung neuer Arten der Gattung *Chytriomyces* and eininger seltener niederer Phycomyceten. Ibid. 36:283-305-14 figs.

Petersen, H. E. 1905. Contributions à la connaissance des Phycomycètes Marins (Chytridinae Fischer). Oversight Kgl. Danske Vidensk. Selskabs Forhandl 1905:439-488, 11 figs.

Rieth, A. 1967. Ein Beitrag zur Kenntnis algenparasitärer Phycomycetes. Biol. Centralbl. 86:435-448, 40 figs.

Sampson, K. 1932. Observations on a new species of *Olpidium* occurring in root hairs of *Agrostis*. Trans. Brit. Mycol. Soc. 41:187-228, 73 figs.

Scherffel, A. 1926. Einiges über neue oder ungenügend bekannte Chytridineen. (Der "Beiträge zur Kenntnis der Chytridineen." Teil II). Arch. Protistenk. 54:167-260, pls. 9-11.

Schwarze, E. J. and W. R. I. Cook. 1928. The life history and cytology of a new species of *Olpidium*: *Olpidium radicale* sp. nov. Trans. Brit. Mycol. Soc. 13:205-221, pls. 13-15.

Siang, Wan-Nien. 1949. Are aquatic Phycomycetes present in the air? Nature 164:1010.

Sörgel G. 1952. Dauerorganbildung bei *Rozella allomycis* Foust, ein Beitrag zur Kenntnis der Sexualität der niederen Phycomyceten. Arch. f. Mikrobiol. 17:247, 5 figs.

Sparrow, F. K., Jr. 1960. Aquatic Phycomycetes—2nd ed. XXV-1187 pp. Univ. Michigan Press, Ann Arbor.

Valkanov, A. 1929. Protistenstudien, 5. *Hyphochytrium hydrodictii* ein neuer Algenpilz. Arch. f. Protistenk. 67:122-127, 11 figs.

Wildemann, E. de. 1895a. Notes Mycologiques IV. Ann. Soc. Belge Micro. (Mem) 19:59-80, pl. 2.

———. 1895b. Notes Mycologiques. XV. Ibid. 19: 88-117, pls. 3, 4.

———. 1900. Observations sur quelques Chytridinees nouvelles. Mem. l'Herb. Boissier 15:1-10.

———. 1931. Sur quelques Chytridinees parasites d'Algues. Bull. Acad. Roy. Belg. (Cl. Sci.) 17: 281-298, 3 figs., 2 pls.

Zopf, W. 1884. Zur Kenntniss der Phycomyceten. I. Nova Acta Acad. Leop.-Carol. 47:143-236, pls. 12-21.

———. 1894. Ueber niedere thierische und pflanzliche Organismem, welche als Krankheitserreger in Algen, Pilzen, niederer Thieren und höheren Pflanzen auftreten. Beitr. Physiol. Morph. niederer Organismem 4:43-68, pls. 2, 3.

SUBJECT INDEX

A

Achlya 22, 60
Achlyella 1, 226
flauhaultii 225
Achlygetonaceae 4, 59
Achlygeton 1, 60
entophytum 59
rostratum 59
salinum 59
solatium 60
Aerial zoosporangia 230, 232
Allochytridium 156
expandens 148
Allocladia 290, 352
monilifera 352, 358
Allomyces 290, 324
allomyces × *javanicus* 324, 334
anomalus 342, 344
arbuscula 324, 326, 328, 330, 334, 336
catenoides 344
javanicus var. *allomorphus* 324, 334
macrogynus 324, 326, 332, 334, 336
monilioformis 338, 340
monspeliensis 336
neo-monilioformis 338, 340, 342
reticulatus 346
Amoeba 24, 26, 340
Amoebidium 284
Amoebochytrium 248
rhizidioides 248
Amoebospores 1, 248
Amphicypellus 134
elegans 130, 134
Androspores 128
Anisochytridiales 9, 383
Anisolpidiaceae 9, 393
Anisolpidium 384
ectocarpii 383, 384, 386, 388
rosenvingii 384
saprobium 386, 388
sphacelariarum 384
stigeoclonii 386, 388
Antheridia 9
Antherozoids 9
Aphanistes 255
oedogoniarum 225
pellucida 225
Aphanomyces 60
Aphanomycopsis 60
Aplanochytrium 234
Aplanospores 1
Arnaudovia 172
hyponeustonica 172, 174
Asteridia 50
Asterocystis 14
Asterophlyctis 114, 118
irregularis 114
sarcoptoides 114
Astrospheres 50
Axioneme 6, 7

B

Bicricium 62
lethale 59
naso 59
transversum 59
Blastocladia 346
angusta 354
aspergilloides 354
globosa 354
glomerata 354
gracilis 350, 352, 354
incrassata 356
pringsheimii 348, 350
prolifera 348, 352
ramosa 352
rostrata 352
sparrowii 354, 356
tenuis 352
truncata 348, 354
Blastocladiaceae 9, 312
Blastocladiales 2, 8, 289, 290
Blastocladiella 314
anabaenae 314, 316, 322
asperosperma 314, 316
britannica 316, 318, 322
cystogena 322
emersonii 314, 316, 318, 320
laevisperma 314
microcystogena 322
novae-zeylandiae 316, 318
simplex 312, 314, 316
stomatophilum 312
stübenii 322, 324
variabilis 322, 326
Blastocladiopsis 344
elegans 346
parva 346, 348
Blastulidium 28, 32
paedophthorum 30, 32
Blepharoplast 2
functional 2
non-functional 2
Blyttiomyces 164
aureus 160
harderi 166
helicus 158
laevis 158
rhizophlyctidis 160
spinulosus 158, 164
Blyttiomyces sp. 166
Brachyallomyces 342
Brevicladia 290, 352, 358

C

Callimastix 312
cyclopis 352
Canteria 232
apophysata 230
Canteriomyces 384, 386, 388
Catenaria 290, 298
allomycis 292, 294, 300, 302
anguillulae 294, 296, 302
indica 294, 298
sphaerocarpa 2, 294, 298, 302
spinosa 298, 302
vermicola 294, 298
verrucosa 298
Catenariaceae 9, 290
Catenariopsis 404
Catenochytridium 166
carolinianum 6, 168
carolinianum var. *marinum* 166
kevorkiana 160
laterale 168
marinum 166
oahuensis 166
Catenomyces 290, 294
persecinus 294
Catenophlyctis 289, 292
variabilis 290, 292
Caulochytrium 1, 230
gloeosporii 228, 230
Centrioles 3
Chromosphere 332
Chrysophlyctis 50
Chytridiaceae 6, 63
Chytridiales 2, 3, 63
Chytridihaema 28, 38
cladocerarum 38
Chytrioides 36
schizophylli 36, 38
Chytridioideae 6, 63
Chytridiopsis 36
socius 36, 38
Chytridium 160
appressum 150, 154
chaetophilum 152, 162
cocconeidis 152, 162
coleochaetes 150
confervae 162
cornutum 162
curvatum 150, 162
epithemae 150
lagenula 152
lateopercumum 152
lecythii 152, 162
megastomum 152

mesocarpi 392, 402
microcystides 150, 162
neopapillatum 162
nodulosum 150
olla 150, 162
papillatum 152
perniciosum 152
pilosum 162
polysiphoniae 152
proliferum 152, 162
rhizophydii 152, 162
schenkii 162
sphaerocarpum 150
surirellae 150, 162
versatile 150
versatile var. *podochytrioides* 150
Chytriomyces
annulatus 126
appendiculatus 122, 128, 130
aureus 122
closterii 124
cosmaridis 124, 128
fructicosus 124, 128, 130
gilgaiensis 128
heliozoicola 128
hyalinus 6, 122, 130
hyalinus var. *granulata* 122
lucidus 124, 128
mammilifer 124
mortierellae 124, 130
parasiticus 122
poculatus 124
reticulatus 124
rotoruaensis 124
spinosus 124
stellatus 124, 130
tabellariae 128
vallesiacus 124
vaucheriae 126, 128
verracosus 128
willoughbyi 124, 128
Cladochytriaceae 6, 237
Cladochytrium 238
aneurae 238
aurantiacum 234, 238
aureum 238
cornutum 238
crassum 237
hyalinum 237, 238, 240
irregulare 238
polystomum 232, 238
replicatum 234, 238, 240
setigerum 232, 238
tenue 232, 240
tianum 234, 238
Clavochytrium 314
Cleavage 402
endogenous 402
exogenous 402
Coelomomyces 304
anophelisca 308
chironomi 306
dodgei 304, 308
finlayae 306
indiana 304, 308
indicus 308, 310
lativittatus 306
notonectae 308
pentangulatus 304, 306
psorophorae 306, 308, 310
punctatus 304, 306, 310
solomonis 306
stegomyiae 304
tuzetae 306
Coelomomycetaeceae 8, 303
Coelomycidium 28, 33, 38
ephemerae 33
melusinae 33
simulii 33, 34, 36
Coelosporideae 38
Coenomyces 254, 258
consuens 258
Coralliochytrium 204
scherffelii 206
Cylindrochytridium 198
endobioticum 198, 200
johnstonii 198, 200
Cyphidium 14
Cystochytrium 404
Cystocladiella 322
Cystogenes 292, 338

D

Dangeardia 96
echinulata 96
laevis 66
mammillata 96
ovata 96
sporapiculata 96
Dangeardiana 100
eudorinae 98, 102
sporapiculata 98
Dermocystidium 38
pusula 38
Diblepharis 378
Dictyomorpha 20
dioica 22
Dictyosomes 2, 398
Diplochytridium 163
aggregatum 156, 163
cejpii 156, 163
chlorobotris 163
citriforme 156, 163
deltanum 163, 164
gibbosum 156, 163
inflatum 156, 163
isthmiophilum 163
kolianum 163
lagenaria 154, 163, 164
lagenaria var. *japonense* 163
mallomonadis 163
mucronatum 156, 163
neochlamydococci 163
oedogonii 156, 163
schenkii 156, 163, 164
scherffelii 163
sexuale 154
stellatum 154, 163
Diplochytrium 164
Diplophlyctis 210
amazonensis 210
chitinophila 210, 214
complicatus 210, 212
intestina 210, 212, 214
laevis 210, 212
nephrochytrioides 210, 214
sexualis 210, 212
verrucosa 214
Diplophlyctoideae 6, 208
Dipolium 22
philosexualis 22, 24

E

Eccrina 284
Eccrinales 284
Ectochytridium 122
Endoblastidium 28, 36
caulleryi 36
legeri 36
Endochytrium 196
cystarum 196, 198
digitatum 194
multiguttulatum 196, 198
oophilum 194
operculatum 198
pseudodistomum 194, 198
ramosum 196
Endocoenobium 176
eudorinae 176
Endodesmidium 46
formosum 43
Endolpidium 14
Endoplasmic reticulum 2
Endospores 33
Entophlyctaceae 6, 189
Entophlyctis 192
apiculata 189, 192, 194, 196
aureus 192
brassicae 192
bulligera 189, 194
characearum 190, 192, 196
cienkowskiana 189, 192, 194, 196
confervae-glomeratae 1, 189, 192, 194
crenata 190, 194
helioformis 189, 194
lobata 190, 192, 194, 196
maxima 190, 192
rhizina 189

salicorniae 192
spirogyrae 192
texana 190, 192, 194, 196
tetraspora 190
vaucheriae 190, 192, 196
willoughbyii 192
woronichinii 192
Entophlyctoideae 6, 190
Euchytridium 402
Eucladiella 322, 324
Euglena 24
Euphlyctochytrium 86
Subgenus *Exosynchytrium* 56
callirrhoe 50

F

Flagella 1, 3
Fulminaria 284
mucophilum 284

G

Gametangia 63, 126
Gametes 230, 324
Gamma particles 318
Genera of doubtful affinity 28
Gonopodyaceae 9, 364
Gonopodya 366
bohemica 366
polymorpha 363, 366, 368
prolifera 363, 366, 368
Gynospores 128

H

Hapalopera 64
piriformis 65
Haplocystis 225
mirabilis 225
Harpochytriaceae 9, 283
Harpochytriales 2, 283
Harpochytrium 2, 284
apiculatum 283
botryococci 283
hedenii 2, 283
hyalothecae 283
monae 283
tenuissimum 283
vermiculare 283
Heterophrys 24
Hyphagen 303, 306
Hyphochytriaceae 400
Hyphochytriomycetes 3, 9, 383
Hyphochytrium 400
catenoides 400
hydrodictyii 398, 402
infestans 398, 400, 402
oceanum 400, 402
peniliae 398, 400, 402
Hyphophagus 400

J

Johnkarlingia 36
brassicae 36

K

Karlingia 148
asterocysta 144, 152
chitinophila 144, 150
curvispinosa 146
dubia 146
granulata 142
hyalina 144, 150, 152
lobata 146
marylandia 146, 152
rosea 1, 142, 150
spinosa 144
Karlingia sp. 146
Karlingiomyces 148
Kinetosome 2

L

Latrostium 398
comprimens 390, 396
Leptocladia 290, 352
laevis 352, 358
Lipoidal sac 2, 318
Loborhiza 78
metzneri 78, 80

M

Macrochytrium 228
botrydioides 226, 228
Mastigochytrium 226
saccardiae 226
Megachytriaceae 6
Megachytrium 252
westonii 25, 254
Meiosis 332
Subgenus *Mesochytrium* 54
desmodiae 48, 54
Microallomyces 344
dendroideus 344, 346
Micromyces 48
furcata 44
grandis 44
longispinosus 44
ovalis 44
petersenii 44
zygogonii 44
Micromycopsis 46
cristata var. *cristata* 46
fischerii 4, 46
intermedia 46
Subgenus *Microsynchytrium* 52
australe 46
endobioticum 46
fulgens 46
Microtubules 2
Mitochondria 2, 3
Mitochytridium 208
ramosum 208, 210
Miyabella 50
Monoblepharella 370
elongata 366
endogena 366
lareui 366
mexicana 366
taylori 364
Monoblepharidaceae 9, 376
Monoblepharidales 2, 290, 363
Monoblepharidopsis 378
elongata 378
lateralis 378
Monoblepharis 378
bullata 374, 380, 381
fasiculata 372, 374, 381
hypogena 378, 381
insignis 378
macandra 376, 380, 381
ovigera 374, 378
polymorpha 370, 380, 381
reginens 374, 378, 380
sphaerica 378
Morella 24, 33
endamoebae 33, 34
dinobryoni 34
hoari 34
hypertrophica 33
nucleophaga 33
Myceliochytrium 280
Myiophagus 62, 304
ucrainica 60, 62
Myzocytium 60

N

Nephrochytrium 214
amazonense 216, 218
appendiculatum 214
aurantium 214, 216, 218
buttermerense 216, 218
stellatum 216
Nephromyces 280
Nowakowskia 110
hormothecae 108
Nowakowskiella 242
atkinsii 244, 246
crassa 244
delica 244, 246
elegans 240, 244
elongata 242, 244
granulata 242, 244, 246
hemisphaerospora 242, 246
macrospora 244, 246

multispora 244, 246
pitcairnensis 244, 246
profusa 1, 6, 244
ramosa 6, 240, 244, 246
sculptura 242, 244, 246
Nuclear caps 2, 316, 318, 332
Nuclearia 24
Nucleophaga 1, 32
intestinalis 30
peranemae 30, 32

O

Obelidium 104
hamatum 104
megarrhizum 104
mucronatum 104
Oedogoniomyces 2, 284, 285
lymnaeae 285, 286
Olpidiaceae 4, 13
Olpidiaster 14
Olpidiella 14
Olpidiomorpha 26
pseudosporae 26
Olpidium 14
agrostidis 13
allomycetos 13
appendiculatum 14
askaulos 14
bornovanus 18
bothriospermi 13
brassicae 1, 6, 18
cucurbitarum 18
endogenum 14
euglenae 16
fulgens 18
granulatum 14
gregarium 14
hantzschiae 16
hyalothecae 14
indum 14
leptophrydis 16
longicollum 14
luxurians 16
nematodeae 14
pendulum 16
protonemae 14
pseudosporearum 16
radicale 13
rhizophlyctidis 16
rotiferum 14
saccatum 14
synchytrii 16
trifolii 13
uredinis 14
utriculiforme 14
vampyrellae 16
viciae 13
virulentus 18
wildemanni 18
zopfianus 18

P

Planospores 1, 2, 3, 4, 5, 6, 7
Paracrystalline body 310
Parasexual 237, 246
Parthenospores 6
Perirhiza 292
Phlyctidiaceae 6, 63
Phlyctidium 64
anatropum 65
brevipes 65
bumilleriae 65
globosum 64
keratinophilum 64
marinum 64
mycetophagum 64
olla 65
obpiriformis 64
piriformis 65
spinulosum 65
tenue 65
Phlyctochytrium 84
africanum 88
articum 1
aureliae 86
biporosum 88
bryopsidis 90
bullatum 86
chaetiferum 88
circulidentatum 92
cladophorae 90
dentiferum 86
desmidiacearum 90
dichotomum 1, 92
furcatum 92
halli 88
hirsutum 88
hydrodictyii 86
indicum 90
irregulare 1, 6, 80
kniepii 1, 88
megastomum 90
mucronatum 86
multidentatum 92
palustre 88
planicorne 86, 92
punctatum 1, 4, 6, 90
quadricorne 86
rheinboltae 88
semiglobiferum 90
spectabile 90
synchytrii 88
urceolare 86
variabile 90
vaucheriae 90
zygnematis 86
Phlyctorhiza 200
endogena 200, 202
peltata 200
variabilis 200
Physocladia 240
obscura 238, 240
Physoderma 262
alfalfae 262, 268, 273
australasica 262, 264
australasica var. *sparrowii* 262, 264, 270
butomi 264
calami 266
calthae 268
commelinae 264
corchori 264
dulichii 268
flammulae 264, 273
hydrocotylidis 272
lycopi 266, 270
maculare 4, 261, 270
majus 266
maydis 264, 270, 273
menthae 268
menyanthes 264, 273, 274
pluriannulata 262, 268, 270, 272, 273
potteri 264
pulposum 261, 270, 272, 273
Physoderma sp. 266, 268, 270
Physodermataceae 6, 261
Physorhizophydium 80
pachydermum 82
Phytophthora 3
Planospores 1
Plasmophagus 22
oedogoniorum 22, 24
Pleolpidium 18
Pleotrachelus 14
Podochytrium 82
chitinophilum 84
clavatum 84
cornutum 84
ellerbeckense 84
emmanuelense 84
lanceolatum 84
Polycarum 38
Polychytrium 248
aggregatum 248, 250
Polychytrium sp. 250
Polyphagoideae 168
Polyphagus 170
asymmetricus 168, 172
elegans 168
euglenae 4, 168
forminii 168, 172
laevis 168
parasiticus 170
ramosus 168, 172
serpentinus 168
starri 168, 172
Polyphlyctis 92
unispina 94

Polyploidy 334
Pringsheimiella 20
dioica 20
Prosorus 46, 48, 50, 52, 54, 56
Pseudolpidiella 14
Pseudopileum 132, 134
unum 126
Pseudosepta 330
Pseudosphaerita 24, 26
Pseudospora 26
volvocis 26, 28, 30
Subgenus *Pycnochytrium* 56
aureum 56
macrosporum 52
Pyroctonum 238

R

Ramocladia 290, 352
pallida 350, 358
reticulata 350, 358
Reesia 28, 402
amoeboides 28, 30, 32
cladophorae 28
lemnae 28
Relative sexuality 44
Rhizidiaceae 6, 63
Rhizidiocystis 280
ananasi 280
Rhizidiomyces 392
apophysatus 388, 392, 398
bivellatus 390, 394, 398
bulbosus 394
coronus 392, 394
globosus 392
hansonii 390, 392, 398
hirsutus 392
ichneumon 388, 392, 394
japonicus 392, 394, 396
parasiticus 392, 394
saprophyticus 394, 396
Rhizidiomycetaceae 9, 390
Rhizidiomycopsis 392, 396
Rhizidiopsis 82
Rhizidium 104
algaecolum 104
braunii 102
braziliensis 100, 106
chitinophilum 100
elongatum 102
endosporangiatum 102
equitanus 104
laevis 100
leptorhizum 104
lignicola 104
mycophilum 106
nowakowskii 100
ramosum 100
reniformis 102
richmondense 102, 106
spirogyrae 104
variabile 102
varians 104
verrucosum 102
windermerense 102, 106, 108
Rhizoclosmatium 112
aurantiacum 110, 112
globosum 110, 112
hyalinum 110, 112
marinum 110, 112
Rhizophlyctis 140
aurantiaca 146
boneseyi 136, 144
boninense 140
borneensis 140
fusca 138, 140
harderi 134, 140, 142
hirsutus 138
ingoldii 138, 142, 144
lovetti 138, 140
mastigotrichis 134, 140
oceanis 136, 142
palmellacearum 140
petersenii 134, 144
petersenii var. *appendiculata* 134
polythricis 140
reynoldsii 146
serpentinus 146
tolypothricis 134
varians 136, 140, 142
Rhizophlyctis spp. 136, 140, 144
Rhizophydium 64
achnanthis 66
anatropum 66, 74
androdioctes 74
angulosum 65, 74
anomalum 70
apiculatum 65, 72
beauchampii 74
brebissonii 66
brevipes 66
bumilleriae 65
bullatum 70
carpophilum 4
chaetiferum 68
collapsum 70
columnaris 76
condylosum 4, 74
conicum 72
coronum 72
contractophilum 70, 76
difficile 70, 76
elyensis 72
ephippium 70, 76
fulgens 76
gibbosum 68
globosum 68
goniosporum 68
graminis 74
granulosporum 68, 74
keratinophilum 74
laterale 66
macroporosum 65, 72
macrosporum 72
melosirae 68
mischocci 68
mougeotiae 76
mycetophagum 72
nobile 68, 70
nodulosum 74
novae-zeylandiensis 65
oblongum 70, 72
obpyriformis 66
olla 65
ovatum 68, 74
patellarium 2, 70
piriformis 66
pollenis-pini 68
polystomum 65, 72
rarotoruaensis 4, 72, 74
racemosum 74, 76
scenedesmi 65
simplex 76
skujai 64
sphaerocarpum 68
sphaerocystidis 72
sphaerotheca 1, 68
spinulosum 66
tenue 66, 76
tetragenum 76
uniguttulum 70, 76
vaucheriae 68
venustum 70
Rhizophyton 64
Rhizoplast 2
Rhizosiphon 218
akinetum 218, 220, 222
anabaenae 218, 220, 222
crassum 218, 220
Rhopalophlyctis 136, 314
sarcoptoides 132
Ribosomes 2
Rozea 18
Rozella 18
allomycis 1, 20
chytriomycii 20
cladochytrii 20
marina 20
Rozia 18
Rumposome 2, 374

S

Saccomyces 176
dangeardii 178
endogena 178
Saccomyces sp. 178
Saccopodium 278
gracile 278
Sagittospora 38

Scherffelia 204
Scherffeliomyces 204
appendiculatus 204, 206
leptorhiza 204, 206
parasitans 204, 206
Scherffeliomycopsis 206
coleochaetis 4, 206, 208
Septocarpus 82
Septochytrium 250
macrosporum 254
marylandicum 254
plurilobulum 252
variabile 252
Septocladia 324
Septolpidiaceae 60
Septolpidium 60
lineare 59
Septosperma 78
anomala 78
irregularis 78
multiforme 78
rhizophydii 78
spinosa 78
Serumsporidium 33
Sexuality 72, 74, 76
Siphonaria 114
petersenii 112
sparrowii 112
variabilis 112
Siphonochytrium 198
Sirenin 334
Sirolpidium 59
Solutoparies
pythii 106, 110
Solutoparies sp. 106, 110
Sorus 46, 48, 50, 52, 54, 56
Sparrowia 130
parasitica 128
subcruciformis 130, 132
Sphaerita 1
dangeardii 24, 26
endogena 24, 26
simplex 24, 26
trachelomonadis 26
Sphaerocladia 290, 314
Sphaerostilidium 64
Sporophlyctidium
africanum 82
Sporophlyctis 174
chinensis 174
rostrata 4, 174, 176
Synchytriaceae 4, 43
Synchytrium 50
Subgenus *Synchytrium* (*Eusynchytrium*) 54
desmodiae 48
taraxaci 48
Systematics 3

T

Tetrachytrium 275, 278
triceps 276, 278
Traustotheca
clavata 22
Truittella 202
setifera 202, 204
Tylochytrium 64

U

Urophlyctis 262, 273, 274

W

Subgenus *Woroninella* 50, 56
decipiens 54
minutum 54

Z

Zographia 304
Zygochytrium 275, 276
aurantiacum 276
Zygorhizidium 122
affluens 118
chlorophycidis 116
cystogenum 116
melosirae 116
parallelasede 116
parvum 116
planktonicum 116
vaucheriae 120
venustum 118
verrucosum 116
willei 116

AUTHOR INDEX

A

Ajello 86, 248
Alexopoulos 237, 246
Anthony 304
Antikajian 106, 120, 140, 158
Aronson 330
Artemchuk 400
Atkinson 66, 86, 92, 283, 285, 364

B

Bahnweg 234
Ball 33
Balsac 30, 32
Barksdale 22
Barnes 285, 380
Barr 1, 2, 18, 68, 96, 163, 166, 192, 194, 300, 400
Barrett 328
Barshad 394
Bartlett 264
Bartsch 164, 172
Batko 98, 100, 192, 204
Becherelle 18
Becker 26, 33, 34
Beckett 308
Beneke 326, 363
Benjamin 363, 368
Berdan 6, 166, 168, 238, 240, 250, 252
Bessey 3, 276, 348, 383
Birchfield 298
Blackwell 348
Bland 304
Blytt 164
Bogoyavlensky 308
Booth 166, 192, 194
Borzi 110
Bostick 128
Bradley 192
Braun 14, 60, 64, 65, 68, 86, 100, 104, 160, 192
Broadsky 4, 24, 26, 34
Brug 30, 32
Brumpt 33
Buckley 296
Busgen 264
Butler 20, 244, 246, 296, 324

C

Calhoun 328
Campbell 1, 2
Canter 1, 4, 32, 34, 43, 44, 46, 48, 64, 65, 66, 68, 70, 72, 76, 78, 84, 86, 96, 98, 102, 106, 108, 118, 122, 126, 132, 162, 163, 164, 170, 220, 222, 232, 383, 388
Cantino 316, 318, 320, 322, 330, 348
Carlile 334
Cejp 14, 24, 254, 370
Chambers 1, 148, 244, 246, 374
Chasman 304
Chatton 4, 6, 24, 26, 32, 33, 34
Chen 238
Chong 1, 2, 300
Cienkowski 192
Clinton 4, 261, 274
Codreanu 36
Cohn 54, 68
Coker 324, 328
Cook 13, 18, 192, 208, 390, 404
Cooke 6
Corda 18
Cornu 18, 20, 364, 370, 378
Couch 4, 20, 22, 44, 50, 65, 66, 68, 74, 76, 88, 163, 198, 208, 214, 216, 264, 280, 292, 296, 300, 302, 303, 304, 306, 308, 320, 404
Cox 314, 320
Crasemann 228, 326
Crouch 34
Curtis 46
Cutter 336

D

Dangeard 4, 16, 24, 26, 30, 32, 44, 48, 50, 59, 66, 76, 90, 192, 208, 283, 285, 296
Debaisieux 33, 36
DeBalsac 32
DeBary 3, 48, 50, 54, 56, 86, 92, 148, 264
DeBeauchamp 38
Deckenbach 254, 256
Delong 332, 338
DeMeillon 308
DeWildemann 386, 398, 400
Dobell 34
Dodge 306
Doflein 30, 32
Dogma 8, 78, 90, 110, 118, 120, 132, 140, 146, 148, 156, 158, 166, 196, 198, 210, 212, 214, 216, 250
Domjan 1, 163, 204
Duboscq 36
DuPlessis 56

E

Emerson 283, 284, 285, 324, 326, 334, 336, 338, 342, 344, 346, 348, 356
Engler 54

F

Fähnrich 334
Fisch 28, 30, 32, 104, 192, 384, 392, 402
Fischer, E. C. 16, 18, 28, 62, 64, 66, 68, 106, 140, 192, 194, 225, 248, 285, 366, 370, 402
Fischer, H. 62
Fitzpatrick 54, 112, 276, 285
Flanagan 342
Foley 334, 338
Fott 64, 65, 66, 163
Foust 20
Fox 334
Friedman 65, 66, 68, 82, 84
Fritsch 285
Fuchs 1
Fuller 1, 2, 3, 4, 6, 246, 300, 316, 328, 336, 374, 394, 396, 398

G

Gaertner 1, 4, 66, 74, 76, 88, 92, 206, 296
Gäumann 54, 56
Geitler 1, 4, 98, 100, 194
Giard 280, 281
Gimesi 74
Gobi 284, 302
Golberg 33
Gojdics 26
Goldi-Smith 162
Gomez 24, 26
Graff 174
Grant 324
Griffins 262, 268
Gwynne-Vaughan 285

H

Hanson 72, 78, 80, 168, 200, 294, 298
Harant 281
Harder 314, 324
Harper 54
Haselby 383
Haskins 148, 150, 192, 210, 212, 214
Hatch 326, 334
Hawksworth 8
Haxo 334
Hazard 304
Heim 44
Held 1
Hennen 344
Hill 328
Hillegas 250
Hoerner 316
Hohnk 226, 378

Holland 92
Hollande 30, 32
Homma 50
Hornstein 316, 320, 322
Hovasse 74, 80
Humphries 296
Hyatt 318, 320

I

Indoh 334
Ingold 1, 4, 18, 118, 134, 162, 176
Ito 50
Ivanic 24
Iyengar 308

J

Jaag 170
Jacobsen 18
Jaczewaski 192
Jahn 26
Jane 283, 285
Jirovec 30, 33
Johanson 148, 250, 252, 281, 390
Johns 100, 170, 172, 204, 206, 268, 263
Johnson 8, 16, 20, 64, 228, 296
Joyon 352

K

Kanouse 350
Karling 4, 6, 8, 13, 14, 16, 20, 22, 24, 26, 32, 33, 43, 44, 46, 48, 50, 52, 54, 56, 59, 60, 62, 64, 65, 66, 70, 72, 74, 78, 86, 88, 90, 92, 94, 100, 102, 104, 106, 118, 120, 128, 130, 132, 136, 142, 144, 148, 150, 152, 154, 158, 160, 162, 163, 166, 194, 196, 198, 200, 202, 210, 212, 214, 216, 232, 237, 238, 242, 244, 248, 250, 254, 262, 264, 274, 280, 283, 284, 290, 292, 298, 300, 302, 320, 321, 324, 384, 396, 402, 404
Kato 340
Kazma 1, 2
Keilin 303, 308
Kellenberger 318, 328
Kirby 32, 33
Kniep 172, 326, 334
Knox 64, 84, 86, 160, 298
Kny 386
Kobayashi 64, 65, 90, 140, 162, 163, 166, 192, 285, 286, 392
Koch 1, 2, 3, 4, 6, 8, 20, 22, 24, 88, 90, 92, 130, 136, 162, 163, 164, 372
Köhler 43, 92
Kole 46, 50, 52
Komarek 254
Konno 140, 162
Korschikoff 283
Krieger 266
Kusano 13, 18, 43, 44, 46, 52, 54, 56

L

Lagerheim 14, 226, 283, 284, 363, 364, 378
Laibach 363, 364, 370, 376, 378
Laird 304
Lavier 33
Ledingham 74
Leger 36
Le'John 383
Lesemann 1
Lessie 316, 318, 330
Lindau 104
Lingappa, B. T. 44, 46, 264, 270
Lingappa, Y. 261, 262, 270, 273
Loubes 34
Lovett 148, 316, 318, 330, 383
Lowenthal 98, 122, 124
Lubinsky 34, 38

M

Machlis 326, 330, 332, 334
Madelin 304, 308
Magnus 14, 273
Mannier 34, 36
Manton 1
Markus 8, 244
Martin 3, 59, 298, 300, 306, 308, 352
Masters 163, 164
Mattes 34
Matthews 314, 320
Maurand 36
McAlpine 262
McCrainie 340
McNitt 1
Melville 380
Metz 3
Milanez 92
Miller 8, 44, 50, 64, 136
Minden 28, 62, 64, 130, 140, 178, 194, 226, 228, 276, 285, 370, 400, 402
Miyabe 264
Moniez 38
Moore, E. D. 136
Moore, R. T. 328, 332, 352
Morini 18
Mullins 20, 22
Munasinghe 48
Murray 148
Muspratt 304, 306

N

Nabel 394, 396
Nemec 192
Nipkov 170
Nolan 290
Noller 33
Nowakowski 4, 108, 170, 178, 238, 240, 242

O

O'Conners 34
Olive 230
Olson 1, 2, 3, 4, 5, 88, 92, 328
Ookubo 64, 65, 90, 163, 166, 192, 212, 285, 286, 392
Ossia 332

P

Parmlee 264
Pascher 14, 283, 285
Paterson 64, 92, 94, 96, 98, 134, 176
Pavgi 36, 300
Penard 34
Percival 46
Perez 30, 32, 33, 38
Perez Reyes 24, 26, 33
Perrott 363, 378
Persiel 72, 76, 88, 92, 130, 166, 402
Petersen 18, 112, 114, 118, 120, 174, 368, 370
Pfitzer 82, 84
Pommerville 336
Pongratz 64, 65, 66, 76
Powell 2
Prantl 54
Prunet 238
Pumaly 24

Q

Quantz 148

R

Rabenhorst 14, 18, 64, 66, 68
Raciborski 50, 172
Ramsbottum 194
Rasin 308
Reichle 2, 3, 246, 316, 374, 380, 398
Reinsch 346
Renaud 3, 326, 352
Rhathe 218
Réyes 33
Richards 212
Rieth 4, 44, 48, 50, 90, 92, 126, 166, 172, 404
Roane 64, 160
Robertson, J. A. 312, 344, 346
Robertson, M. 4
Rorem 332
Rosen 86
Rothwell 280

Rubcov 36
Rytz 48

S

Sahtiyanic 13, 18
Salas Gomez 26
Salkin 156
Sampson 13, 18, 390
Saville 264
Sawada 13, 18
Schcherban 33
Schenk 59, 60, 64, 162, 194, 210
Scherffel 1, 14, 16, 26, 43, 46, 48, 64, 68, 82, 86, 90, 122, 162, 163, 218, 222, 285, 400
Schnepf 1
Schilberszki 50, 52
Schneider 38
Scholz 65, 70
Schröder 96, 98, 140
Schroeter 14, 28, 50, 54, 64, 86, 92, 112, 140, 210, 242, 248, 266, 276, 285, 370
Schulz 162
Schussnig 290
Schwartz 13, 18
Serbinov 1, 4, 26, 64, 68, 174, 176, 194
Seymour 78
Shanor 246, 374
Sharma 273
Shaw 354
Shen 174, 238
Siang 238, 400
Sideris 280
Singh 36, 300
Skucas 330, 338
Skuja 218, 283, 284
Skvortzow 14, 16, 26
Soll 318
Sörgel 20, 290, 312, 314, 326, 328, 334, 350, 352, 364, 379, 390
Sorokin 14, 59, 62, 104, 192, 225, 276, 278, 296, 298
Sost 326
Sparrow 1, 16, 24, 28, 43, 48, 59, 60, 64, 65, 66, 68, 70, 76, 82, 84, 86, 92, 96, 100, 104, 112, 114, 120, 140, 142, 148, 162, 163, 166, 168, 176, 178, 192, 194, 196, 200, 204, 212, 216, 225, 226, 234, 240, 244, 248, 252, 254, 256, 266, 268, 272, 273, 274, 292, 298, 338, 344, 348, 363, 368, 370, 372, 380, 383, 384, 386, 392, 396
Springer 372
Stanier 148
Stempel 36
Stüben 314
Stumm 326, 338
Swift 3, 328, 352

T

Teixeira 272
Teter 340
Thaxter 60, 62, 363, 364, 378
Thind 274
Thirumalacher 238, 268
Timmink 1, 2
Tisdale 264
Tomaschek 16, 163
Torrey 62
Townley 94, 96
Travland 2, 283, 284, 285
Tregouboff 36, 38
Truesdell 318
Turian 318, 328, 330, 334

U

Uebelmesser 14, 16, 90, 92, 140, 142, 144
Ulken 84
Umphlett 1, 8, 88, 92, 304, 306, 308

V

Valkanov 76, 98, 100, 170, 172, 174, 402
Vavra 352
Viegas 272
Viswanathan 334
Voos 230
Vuillemin 402

W

Wagner 28, 32, 383
Walker 264, 304
Wallroth 261, 262
Waterhouse 348
Waterston 62
Weiser 33, 34
Weissenberg 352
Weston 148
Whiffen 78, 108, 114, 118, 120, 160, 178, 214, 237, 244, 248, 320, 326, 344, 346
Whisler 2, 283, 284, 285, 304, 308, 310, 326, 352, 354
Wildemann 14, 24, 68, 196, 238, 264
Wille 162, 283
Willoughby 1, 8, 78, 82, 84, 94, 96, 108, 126, 130, 148, 156, 162, 166, 192, 194, 198, 200, 210, 242, 244, 250, 283, 284, 298
Wilson, C. W. 290, 324, 334, 336, 338, 342
Wilson, G. B. 326
Wilson, O. T. 273
Wize 60, 62
Woronin 48, 54, 148, 378, 381

Y

Yaw 336

Z

Zelezinskaya 400
Zeller 163
Zetsche 316
Zizka 34
Zopf 1, 4, 16, 18, 64, 68, 82, 84, 100, 102, 104, 192, 210, 238, 248, 392, 398, 400, 402
Zukal 140